Animal nutrition

We work with leading authors to develop the strongest
educational materials in biology, bringing cutting-edge
thinking and best learning practice to a global market.

Under a range of well-known imprints, including
Prentice Hall, we craft high-quality print and
electronic publications which help readers
to understand and apply their content,
whether studying or at work.

To find out more about the complete range of our
publishing, please visit us on the World Wide Web at:
www.pearsoned.co.uk

ANIMAL NUTRITION

SIXTH EDITION

P McDonald
Formerly Reader in Agricultural Biochemistry, University of Edinburgh, and Head of the Department of Agricultural Biochemistry, Edinburgh School of Agriculture

R A Edwards
Formerly Head of the Department of Animal Nutrition, Edinburgh School of Agriculture

J F D Greenhalgh
Emeritus Professor of Animal Production and Health, University of Aberdeen

C A Morgan
Animal Nutritionist, Scottish Agricultural College, Edinburgh

An imprint of **Pearson Education**

Harlow, England · London · New York · Reading, Massachusetts · San Francisco
Toronto · Don Mills, Ontario · Sydney · Tokyo · Singapore · Hong Kong · Seoul
Taipei · Cape Town · Madrid · Mexico City · Amsterdam · Munich · Paris · Milan

Pearson Education Limited
Edinburgh Gate
Harlow
Essex CM20 2JE

and Associated Companies throughout the world

Visit us on the World Wide Web at:
www.pearsoned.co.uk

First edition published by Oliver & Boyd 1966
Second edition 1973
Third edition 1981
Fourth edition 1988
Fifth edition 1995
Sixth edition published 2002

ISBN: 978-0-582-41906-3

British Library Cataloguing-in-Publication Data

A catalogue record for this book is available from the British Library

Library of Congress Cataloging-in-Publication Data

Animal nutrition / P. McDonald... [*et al.*].–6th ed.
    p. cm.
  Includes bibliographical references.
  ISBN 0-582-41906-9
  1. Animal nutrition. 2. Feeds. I. McDonald, Peter, 1926–

SF95 .A634 2001
636.08′52–dc21

10   9   8
10   09   2001036295

Typeset by 68
Printed and bound in Malaysia (CTP-VVP)

Contents

Preface to the sixth edition   xi
Acknowledgements   xii

1	**The animal and its food**	**1**

1.1	Water	2
1.2	Dry matter and its components	2
1.3	Analysis and characterisation of foods	4
	Summary	12
	Further reading	13

2	**Carbohydrates**	**14**

2.1	Classification of carbohydrates	14
2.2	Monosaccharides	16
2.3	Monosaccharide derivatives	19
2.4	Oligosaccharides	22
2.5	Polysaccharides	25
2.6	Lignin	29
	Summary	30
	Further reading	31

3	**Lipids**	**32**

3.1	Classification of lipids	32
3.2	Fats	33
3.3	Glycolipids	44
3.4	Phospholipids	45
3.5	Waxes	48
3.6	Steroids	49
3.7	Terpenes	53
	Summary	53
	Further reading	54

4	**Proteins, nucleic acids and other nitrogenous compounds**	**55**

4.1	Proteins	55
4.2	Amino acids	55
4.3	Peptides	61
4.4	Structure of proteins	62

4.5	Properties of proteins	63
4.6	Classification of proteins	64
4.7	Nucleic acids	65
4.8	Other nitrogenuous compounds	68
4.9	Nitrates	70
4.10	Alkaloids	70
	Summary	71
	Further reading	72

5 Vitamins 73

5.1	Introduction	73
5.2	Fat-soluble vitamins	77
5.3	Water-soluble vitamins: the vitamin B complex	90
5.4	Vitamin C	104
5.5	Hypervitaminosis	105
	Summary	106
	Further reading	107

6 Minerals 108

6.1	Functions of minerals	108
6.2	Natural and supplementary sources of minerals	112
6.3	Acid-base balance	115
6.4	Major elements	117
6.5	Trace elements	127
6.6	Other elements	143
	Summary	144
	Further reading	145

7 Enzymes 146

7.1	Classification of enzymes	147
7.2	Nature of enzymes	149
7.3	Mechanism of enzyme action	152
7.4	Specific nature of enzymes	155
7.5	Factors affecting enzyme activity	156
7.6	Nomenclature of enzymes	160
	Summary	161
	Further reading	162

8 Digestion 163

8.1	Digestion in monogastric mammals	163
8.2	Microbial digestion in ruminants and other herbivores	179
8.3	Alternative sites of microbial digestion	195

Summary   196
Further reading   197
Historical reference   198

9 Metabolism   199

9.1 Energy metabolism   201
9.2 Protein synthesis   223
9.3 Fat synthesis   229
9.4 Carbohydrate synthesis   235
9.5 Control of metabolism   241
    Summary   243
    Further reading   244

10 Evaluation of foods (A) Digestibility   245

10.1 Measurement of digestibility   246
10.2 Special methods for measuring digestibility   248
10.3 Validity of digestibility coefficients   252
10.4 Digestion and digestibility in the various sections of the tract 253
10.5 Factors affecting digestibility   255
10.6 Alternative measures of the digestibility of foods   258
10.7 Availability of mineral elements   259
     Summary   260
     Problems   261
     Further reading   261

11 Evaluation of foods (B) Energy content of foods and the partition of food energy within the animal   263

11.1 Demand for energy   263
11.2 Supply of energy   264
11.3 Animal calorimetry: methods for measuring heat production and energy retention   271
11.4 Utilisation of metabolisable energy   281
     Summary   290
     Problems   291
     Further reading   291

12 Evaluation of foods (C) Systems for expressing the energy value of foods   292

12.1 Energy systems and energy models   293
12.2 Energy systems for ruminants   294
12.3 Energy systems for pigs and poultry   305
12.4 Energy systems for horses   307
12.5 Predicting the energy value of foods   308

	Summary	310
	Problems	310
	Further reading	311
	Historical references	312

13 Evaluation of foods (D) Protein — 313

13.1	Crude protein (CP)	313
13.2	Digestible crude protein (DCP)	315
13.3	Determination of endogenous nitrogen	316
13.4	Measures of protein quality for monogastric animals	318
13.5	Measures of food protein used in practice in the feeding of pigs and poultry	327
13.6	Measures of protein quality for ruminant animals	328
13.7	The UK metabolisable protein system	343
	Summary	345
	Further reading	346

14 Feeding standards for maintenance and growth — 348

14.1	Feeding standards for maintenance	350
14.2	Animal growth and nutrition: feeding standards for growth and wool production	365
14.3	Mineral and vitamin requirements for maintenance and growth	379
14.4	Nutritional control of growth	383
	Summary	386
	Further reading	387
	Historical reference	388

15 Feeding standards for reproduction — 389

15.1	Nutrition and the initiation of reproductive ability	390
15.2	Plane of nutrition, fertility and fecundity	392
15.3	Egg production in poultry	396
15.4	Nutrition and the growth of the foetus	400
	Summary	408
	Further reading	409

16 Lactation — 410

16.1	Sources of milk constituents	411
16.2	Nutrient requirements of the lactating dairy cow	416
16.3	Nutrient requirements of the lactating dairy goat	442
16.4	Nutrient requirements of the lactating ewe	447
16.5	Nutrient requirements of the lactating sow	452
16.6	Nutrient requirements of the lactating mare	457
	Summary	461
	Further reading	462

17	**Voluntary intake of food**	**464**
17.1	Food intake in monogastric animals	465
17.2	Food intake in ruminants	471
17.3	Prediction of food intake	477
	Summary	479
	Further reading	479

18	**Animal nutrition and the consumers of animal products**	**481**
18.1	Comparative nutrition	481
18.2	The contribution of animal products to human requirements	483
18.3	Objections to the use of animal products	487
18.4	Future trends in the consumption of animal products	492
	Summary	493
	Further reading	494

19	**Grass and forage crops**	**495**
19.1	Pastures and grazing animals	495
19.2	Grasses	496
19.3	Legumes	506
19.4	Other forages	509
	Summary	513
	Further reading	513

20	**Silage**	**515**
20.1	Silage, ensilage and silos	515
20.2	Role of plant enzymes in ensilage	516
20.3	Role of microorganisms in ensilage	517
20.4	Nutrient losses in ensilage	521
20.5	Classification of silages	523
20.6	Nutritive value of silages	527
	Summary	534
	Further reading	535

21	**Hay, artifically dried forages, straws and chaff**	**536**
21.1	Hay	536
21.2	Artificially dried forages	542
21.3	Straws and related by-products	544
	Summary	549
	Further reading	550

22	**Roots, tubers and related by-products**	**551**
22.1	Roots	552
22.2	Tubers	556
	Summary	558
	Further reading	559

23	**Cereal grains and cereal by-products**	**560**
23.1	The nutrient composition of grains	560
23.2	Barley	562
23.3	Maize	570
23.4	Oats	572
23.5	Wheat	574
23.6	Other cereals	575
23.7	Cereal processing	578
	Summary	582
	Further reading	582

24	**Protein concentrates**	**583**
24.1	Oilseed cakes and meals	583
24.2	Oilseed residues of minor importance	596
24.3	Leguminous seeds	597
24.4	Animal protein concentrates	600
24.5	Milk products	606
24.6	Single-cell protein	607
24.7	Synthetic amino acids	609
24.8	Non-protein nitrogen compounds as protein sources	609
	Summary	614
	Further reading	614

25	**Food additives**	**616**
25.1	Antibiotics	616
25.2	Probiotics	618
25.3	Oligosaccharides	621
25.4	Enzymes	622
25.5	Organic acids	625
25.6	Modifiers of rumen fermentation	627
	Summary	629
	Further reading	629
	Appendix I Solutions to problems	630
	Appendix II Notes on tables	632
	Index	669

Preface to the sixth edition

Since the first edition of this book was published, in 1966, the authors – driven or encouraged by the publishers – have revised the text every seven years. Revision for even-numbered editions has been intended to comprise only minor changes, and this rule has been followed for many of the chapters of this, the sixth edition. Nevertheless, two new chapters have been added. The first of these (Chapter 18, Animal nutrition and the consumers of animal products) has been introduced principally to discuss contemporary concerns for the need for, and safety of, meat, milk and eggs in human diets. In addition, this chapter provides a comparison of the nutrient requirements of man with those of some of his domestic animals. The second new chapter (Chapter 25, Food additives) revives part of a chapter on growth-stimulating substances that was included in the first two editions of the book. The reason for re-introducing this topic is that the additives available have changed markedly in the past 40 years, from substances unacceptable today (such as oestrogenic hormones) to substances chosen with more regard for the anticipated views of the human consumer.

During the lives of the authors, and the life of this book, the science of nutrition has changed its emphasis from the effects of food on the whole animal to the impact of nutrients on selected tissues and organs. Thus the emphasis in research today is less on outcomes and more on mechanisms, which may be investigated down to the level of the cell and its components. The authors would like to be able to extend the text in the direction of this 'cellular nutrition' (and may do so in future editions), but for the sixth edition they are mindful of the need to restrict the book to a reasonable size. Thus applied nutrition – meaning the feeding of the whole animal – continues to be the theme of this edition.

But applied nutrition itself is subject to evolution, most notably in the feeding systems that link the energy and protein contents of foods to the requirements and responses of animals. The authors have attempted to update the central chapters of the book to reflect the increasing complexity of feeding systems. The inclusion of computer programs for systems was considered but was rejected for the reason that such programs are outdated too rapidly to be relevant to the intended seven-year life of the sixth edition. Other changes to this edition include the description of new methods of food analysis and the addition of more material on the nutrition of the horse, including feeding standards for this species. The format of the text has been modified to include boxes of specialised material, lists of chapter contents, and chapter summaries. For chapters dealing with the calculation of rations, etc., problems have been added (with solutions at the end of the book).

The sixth edition is the first produced without the involvement of Dr Peter McDonald, the instigator of 'Animal Nutrition'. He is alive and well, but decided that the time had come for him to leave the revision to others. His co-authors miss him, not least for his superior ability to meet deadlines, but are glad that much of his contribution to earlier editions is still present in the sixth edition.

As with previous editions, the authors are grateful for advice and support from their colleagues and friends. Among these, Dr Mitch Lewis of the Scottish Agricultural College, Edinburgh, deserves special thanks.

J F D Greenhalgh, R A Edwards and C A Morgan
April 2001

Acknowledgements

We are grateful to the following for permission to reproduce copyright material:

Figure 1.1 after 'Response in the yield of milk constituents to the intake of nutrients by dairy cows', Report No. 11 from Agricultural and Food Research Council Technical Committee, CABI Publishing International, Wallingford, Oxford; Table 1.3 adapted from *Principles of Pig Science* (Table 10.4) reprinted by permission of Dierick N A, University of Ghent; Table 3.6 adapted from *Textbook of Biochemistry with Chemical Correlations*, 4th edn. (Table 2.12), p. 56 by permission of John Wiley & Sons Inc. (ed. Delvin T M, 1997); Figure 5.1 Roche Vitec, *Animal Nutrition and Vitamin News*, Vol. 1, A1-10/2, Nov. 1984, reprinted by permission of Roche Vitamins (UK) Ltd; Table 5.2 Some typical values for liver reserves of vitamin A in different species (Moore T, 1969) *Fat Soluble Vitamins,* (Morten R A ed.) published by Pergamon Press: 233; Figure 6.1 'Mineral Metabolism' (Symonds H W and Forbes J M) in *Quantitative Aspects of Ruminant Digestion and Metabolism* (Forbes J M and France J, eds), CAB International, Wallingford; Figure 12.2 from *New Zealand Journal of Agricultural Research*, 18: 295 (Joyce J P et. al., 1975); Table 13.1 Factors for converting nitrogen to crude protein (Jones D B, 1931); Table 13.2 Novel methods for determining protein and amino acid digestibilities in feedstuffs, (Sauer W C and de Lange K, 1992) in *Modern Methods in Protein Nutrition and Metabolism,* London, Academic Press (Nissen S ed.); Table 13.3 Nitrogen balance for a Large White/Landrace pig on a diet of wheat, Soya bean meal and herring meal from (Morgan C A and Whittemore C T, unpublished); Table 13.4 Calculations of biological value for maintenance and growth of the rat from (Mitchell H H, 1924); Figure 13.4 (Figure 1) and Table 13.8 (Table 13.1) after *Recent Advances in Animal Nutrition,* by Chapula W and Sniffen C J in (eds Garnsworthy P C and Cole D J A, 1994), Nottingham University Press; Table 13.5 from *Journal of Animal Science*, 14: 53 (Armstong D G and Mitchell H H, 1955); Table 13.7 from *The Nutrient Requirements of Pigs,* Agricultural Research Council, 1981, Commonwealth Agricultural Bureaux, Farnham Royal; Table 13.11 after *Journal of Dairy Science*, 75: 2304–23 p. 2306 (Table 2), American Dairy Science Association (Clark J H, Klusmeyer T H and Cameron M R, 1992); Table 13.12 after *Journal of Dairy Science*, 66: 2198–207 p. 2200 (Table 2), American Dairy Science

Association (Muscato T V, Sniffen C J, Krisnamoorthy U and Van Soest P J, 1983); Table 14.4 Composition of gains in empty body weight from *The Nutrient Requirements of Ruminant Livestock, Commonwealth Agricultural Bureaux*; Table 14.6 Percentage composition and energy content of the gains from (Mitchell H H, 1962), *Comparative Nutrition of Man and Domestic Animals*, published by Elsevier Science; Table 14.7 Differences between breeds and sexes in the body composition from the data calculated, PhD Thesis, (Ayala H J, 1974); Table 15.1 Age and size at puberty of Holstein cattle from Sorenson A M 1959 No. 936 (Bratton R W, 1959) and No. 940, published by Cornell University; Table 15.4 from *The Nutrient Requirements of Ruminant Livestock* Agricultural Research Council, 1980, Commonwealth Agricultural Bureaux, Farnham Royal; Table 15.5 from *The Nutrient Requirements of Pigs*, Agricultural Research Council, 1981, Commonwealth Agricultural Bureaux, Farnham Royal; Table 15.6 from *Journal of Agricultural Sciences*, Cambridge 45: 229 (Gill J C and Thomson W, 1954); Table 16.2 Fatty acid composition from (Bickerstaff R, 1970); Table 16.3 Total fatty acid composition of bovine milk lipids (Jensen R G, Ferrier A M and Lammi-Keefe C J, 1991) *Journal of Dairy Science* 74: 3228; Table 16.4 and Table 16.5 after *Dairy Facts and Figures* (Table 16) p. 30, 1999, National Milk Records, Wiltshire; Figure 16.5 *Recent Advances in Animal Nutrition*, pp. 49–69, (Broster W and Thomas C, 1981) London, Butterworths in Haresign W ed; Tables 16.6 and 16.7 from *Dairy Science Abstracts*, 23: 251 (Rook J A F, 1961); Figure 16.8 from *Journal of Agriculture and Science*, Cambridge, 38: 93 (Wallace L R, 1948); Figure 16.9 from *Journal of Agriculture and Science*, Cambridge, 79: 303 (Peart J N *et. al.*, 1972); Tables 16.13 and 16.14 after Agricultural and Food Research Council Technical Committee, No. 10, 1998, *The Nutrition of Goats* (Table 2.9) p. 17, CABI Publishing International, Wallingford, Oxford; Figure 16.6 Cumulative balances of calcium and phosphorus during lactation (47 weeks) and dry period (Ellenberger H B, Newlander J A and Jones C H, 1931), published by Cornell University; Table 16.18 Variation in yield and composition of sows' milk; Table 16.19 Effect of litter size on mild yield in the sow (Elsley F W H, 1970) *Nutrition and Lactation in the sow*, published by Butterworth Heinemann, Oxford; Table 16.22 after *Equine Nutrition and Feeding*, 2nd edn. (Frape D, 1999), Blackwell Science Ltd.; Table 17.2 Diet selection of young pigs from data of Kyriazakis I, Emmans G C. Commonwealth Agricultural Bureaux, published by Elsevier Science; Table 17.3 The effects on intake and digestibility of grinding (Greenhalgh J F D and Reid G W, 1973) *Animal Production* published by Prentice Hall; Table 18.1 adapted from *Human Nutrition and Dietectics* (Garrow S and James W P T, eds, 1993), Harcourt Publishers Ltd; Tables 18.3, 18.4 and 18.5 from Food and Agricultural Organisation of the United Nations, Viale delle Terme di Caracalla, Rome, Italy; Table 18.6 after *National Survey of Domestic Food Consumption and Expenditure*, Ministry of Agricultural Food and Fisheries © crown copyright, reproduced by permission of the controller of HMSO; Table 18.7 after *Proceedings of the Nutrition Society*, 58: 219–34 (Rosegrant M W, Leach N and Gerpacio R V, 1999), The Nutrition Society; Table 19.6 Composition (g/kg DM) and ME values from (Gohl B, 1981) *Tropical Feeds*, published by FAO, Italy; Table 19.7 Composition of the dry matter of Lucerne, from MAFF 1975 Energy Allowances and Feeding Systems for Ruminants, p. 70 HMSO, London; Table 19.8 Composition of whole barley at different stages of growth (Edwards R A, Donaldson E and MacGregor A W, 1986) pp. 650–66, published by John Wiley & Sons, USA; Table 20.4 Typical composition of well-preserved silages made from perennial (Donaldson E and Edwards R, 1976) pp. 536–44, published by Cambridge University Press; Table 20.5 The composition of two badly-preserved silages made from either cocksfoot or Lucerne (McDonald P, Hender A R and Heron S J E, 1991), *The Biochemistry of Silage*, Marlow, Chalcombe Publications: 271; Table 20.6 Composition and nutritive value of grass silages inoculated with lactic acid bacteria compared with an untreated control (Henderson A R *et al.*, 1990), Proceeding Eurobank Conference, Uppsalar, August 1986 pp. 93–9; Table 20.7 from *Animal Feed, Science and Technology,* 7: 305–14 (Henderson A R, McDonald P and Anderson D H, 1982); Table 20.9 after *Animal Feed Science and Technology,* 28: 115–28 p. 126 (Table 6), (Barber G D, Givens D I, Kridis M S, Offer N W and Murray I, 1990) Elsevier Science reprinted with permission from Excerpta Medica Inc.; Table 20.11 after *Animal Science,* 66: 357–67 p. 363 (Table 4) British Society of Animal Science (Offer N W, Percival D S, Dewhurst R J and Thomas C, 1998); Table 21.1 Changes in nitrogenous components of ryegrass/clover from Grass and Forage Science by (Carpintero M C, Henderson A R and McDonald P, 1979) pop. 311, published by Blackwell Science, Oxford; Table 21.5 Some mineral contents of hay and barley straw p. 23 (Mackenzie E J and Purves D, 1967) published by The Edinburgh School of Agriculture; Table 21.6 from *Agricultural Progress,* 58: 11 (Greenhalgh J F D, 1983); Table 21.7 from *Animal Feed Science and Techonology,* 27: 17 (Djajanegara A and Doyle P T, 1989); Table 23.3 from 'The nutritional value of fresh brewers' grains from Occasional, publications No. 3 (Barber W P and Lonsdale C R, 1980) published by British Society of Animal Science, Midlothian; Table 23.4 The nutritive value of distiller's grain *from Distillery By-Products as Feeds for Livestock* (Black H *et al.*, 1991) published by Scottish Agricultural College; Table 23.5 adapted from *British Journal of Nutrition* 32: 59 (Ørskov E R, Fraser C and Gordon J G, 1974) and *Animal Production* 18: 85 (Ørskov E R, Fraser C and McHattie I, 1974); Table 24.1 after *Feed Facts Quarterly,* No. 1 (Table 21, 1999 and (Table 23, 2000), HGM Publications see http//:www.hgmpubs.com; Figure 25.1 and 25.3 after *The Living Gut* (Figure 7.8, p. 105 and Figure 8.9, p. 142), (Ewing W N and Cole D J A, 1994), Context Publications, 53 Mill Street, Packington, Leics. LE65 1WN, Tel: 01530 415228, context@totalise.co.uk; Table 25.1 adapted from *Alltech European Lecture Tour,* Feb.–Mar. (Rotter B A, Marquardt R R and Guenter W, 1989); Figure 25.2 after *Feed Compounder,* Jan. 1991, pp. 16–19, (Offer N W, 1991), HGM Publications, see http//:www.hgmpubs.com; Appendix Tables 11 and 12 adapted from *Nutrition Requirements of Horses,* 5th rev. edn. (Tables 5.1 and 5.3), 1989, reprinted by permission of the National Academy Sciences, courtesy of the National Academy Press, Washington DC.

Whilst every effort has been made to trace the owners of copyright material, in a few cases this has proved impossible and we take this opportunity to offer our apologies to any copyright holders whose rights we may have unwittingly infringed.

1 The animal and its food

1.1   Water

1.2   Dry matter and its components

1.3   Analysis and characterisation of foods

Food is material which, after ingestion by animals, is capable of being digested, absorbed and utilised. In a more general sense we use the term 'food' to describe edible material. Grass and hay, for example, are described as foods, but not all their components are digestible. Where the term 'food' is used in the general sense, as in this book, then those components capable of being utilised by animals are described as *nutrients*.

The animals associated with man cover the spectrum from herbivores, the plant eaters (ruminants, horses and small animals such as rabbits and guinea pigs); omnivores, which eat all types of foods (pigs and poultry); to carnivores, which eat chiefly meat (dogs and cats). Under the control of man these major classes of animal still pertain, but the range of foods that animals are now offered is far greater than they might normally consume in the wild (for example, ruminants are given plant by-products of various human food industries and some dog foods contain appreciable amounts of cereals). Nevertheless, plant and plant products form the major source of nutrients in animal nutrition.

The diet of farm animals in particular consists of plants and plant products, although some foods of animal origin such as fishmeal and milk are used in limited amounts. Animals depend upon plants for their existence and consequently a study of animal nutrition must necessarily begin with the plant itself.

Plants are able to synthesise complex materials from simple substances such as carbon dioxide from the air, and water and inorganic elements from the soil. By means of photosynthesis, energy from sunlight is trapped and used in these synthetic processes. The greater part of the energy, however, is stored as chemical energy within the plant itself and it is this energy that is used by the animal for the maintenance of life and synthesis of its own body tissues. Plants and animals contain similar types of chemical substances, and we can group these into classes according to constitution, properties and function. The main components of foods, plants and animals are:

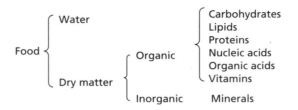

1.1 Water

The water content of the animal body varies with age. The newborn animal contains from 750 to 800 g/kg water but this falls to about 500 g/kg in the mature fat animal. It is vital to the life of the organism that the water content of the body be maintained: an animal will die more rapidly if deprived of water than if deprived of food. Water functions in the body as a solvent in which nutrients are transported about the body and in which waste products are excreted. Many of the chemical reactions brought about by enzymes take place in solution and involve hydrolysis. Because of the high specific heat of water, large changes in heat production can take place within the animal with very little alteration in body temperature. Water also has a high latent heat of evaporation, and its evaporation from the lungs and skin gives it a further role in the regulation of body temperature.

The animal obtains its water from three sources: drinking water, water present in its food and metabolic water, this last being formed during metabolism by the oxidation of hydrogen-containing organic nutrients. The water content of foods is very variable and can range from as little as 60 g/kg in concentrates to over 900 g/kg in some root crops. Because of this great variation in water content, the composition of foods is often expressed on a dry matter basis, which allows a more valid comparison of nutrient content. This is illustrated in Table 1.1, which lists a few examples of plant and animal products.

The water content of growing plants is related to the stage of growth, being greater in younger plants than in older plants. In temperate climates the acquisition of drinking water is not usually a problem and animals are provided with a continuous supply. There is no evidence that under normal conditions an excess of drinking water is harmful, and animals normally drink what they require.

1.2 Dry matter and its components

The dry matter (DM) of foods is conveniently divided into organic and inorganic material, although in living organisms there is no such sharp distinction. Many organic compounds contain mineral elements as structural components. Proteins, for example, contain sulphur, and many lipids and carbohydrates contain phosphorus.

Table 1.1 Composition of some plant and animal products expressed on a fresh basis and a dry matter basis

	Water	Carbohydrate	Lipid	Protein	Ash
Fresh basis (g/kg)					
Turnips	910	70	2	11	7
Grass (young)	800	137	8	35	20
Barley grain	140	730	15	93	22
Groundnuts	60	201	449	268	22
Dairy cow	570	2	206	172	50
Milk	876	47	36	33	8
Muscle	720	6	44	215	15
Egg	667	8	100	118	107
Dry matter basis (g/kg)					
Turnips	0	778	22	122	78
Grass (young)	0	685	40	175	100
Barley grain	0	849	17	108	26
Groundnuts	0	214	478	285	23
Dairy cow	0	5	479	400	116
Milk	0	379	290	266	65
Muscle	0	21	157	768	54
Egg	0	24	300	355	321

It can be seen from Table 1.1 that the main component of the DM of pasture grass is carbohydrate, and this is true of all plants and many seeds. The oilseeds, such as groundnuts, are exceptional in containing large amounts of protein and lipid material. In contrast the carbohydrate content of the animal body is very low. One of the main reasons for the difference between plants and animals is that, whereas the cell walls of plants consist of carbohydrate material, mainly cellulose, the walls of animal cells are composed almost entirely of lipid and protein. Furthermore, plants store energy largely in the form of carbohydrates such as starch and fructans, whereas an animal's main energy store is in the form of lipid.

The lipid content of the animal body is variable and is related to age, the older animal containing a much greater proportion than the young animal. The lipid content of living plants is relatively low, that of pasture grass, for example, being 40–50 g/kg DM.

In both plants and animals, proteins are the major nitrogen-containing compounds. In plants, in which most of the protein is present as enzymes, the concentration is high in the young growing plant and falls as the plant matures. In animals, muscle, skin, hair, feathers, wool and nails consist mainly of protein.

Like proteins, nucleic acids are also nitrogen-containing compounds and they play a basic role in the synthesis of proteins in all living organisms. They also carry the genetic information of the living cell.

The organic acids that occur in plants and animals include citric, malic, fumaric, succinic and pyruvic acids. Although these are normally present in small quantities, they nevertheless play an important role as intermediates

in the general metabolism of the cell. Other organic acids occur as fermentation products in the rumen, or in silage, and these include acetic, propionic,
butyric and lactic acids.

Vitamins are present in plants and animals in minute amounts, and many
of them are important as components of enzyme systems. An important
difference between plants and animals is that, whereas the former can synthesise all the vitamins they require for metabolism, animals cannot, or have
very limited powers of synthesis, and are dependent upon an external supply.

The inorganic matter contains all those elements present in plants and
animals other than carbon, hydrogen, oxygen and nitrogen. Calcium and
phosphorus are the major inorganic components of animals, whereas
potassium and silicon are the main inorganic elements in plants.

1.3 Analysis and characterisation of foods

Originally the most extensive information about the composition of foods
was based on a system of analysis described as the *proximate analysis of
foods*, which was devised over 100 years ago by two German scientists,
Henneberg and Stohmann. Recently, new analytical techniques have been
introduced and the information about food composition is rapidly expand-
ing (see below). However, the system of proximate analysis still forms the
basis for the statutory declaration of the composition of foods in Europe.

Proximate analysis of foods

This system of analysis divides the food into six fractions: moisture, ash,
crude protein, ether extract, crude fibre and nitrogen-free extractives.

The moisture content is determined as the loss in weight that results from
drying a known weight of food to constant weight at 100 °C. This method
is satisfactory for most foods, but with a few, such as silage, significant losses of volatile material may take place.

The ash content is determined by ignition of a known weight of the food
at 550 °C until all carbon has been removed. The residue is the ash and is
taken to represent the inorganic constituents of the food. The ash may, however, contain material of organic origin such as sulphur and phosphorus
from proteins, and some loss of volatile material in the form of sodium chloride, potassium, phosphorus and sulphur will take place during ignition.
The ash content is thus not truly representative of the inorganic material in
the food either qualitatively or quantitatively.

The crude protein (CP) content is calculated from the nitrogen content
of the food, determined by a modification of a technique originally devised
by Kjeldahl over 100 years ago. In this method the food is digested with sulphuric acid, which converts to ammonia all nitrogen present except that in
the form of nitrate and nitrite. This ammonia is liberated by adding sodium
hydroxide to the digest, distilled off and collected in standard acid, the

quantity so collected being determined by titration or by an automated colorimetric method. It is assumed that the nitrogen is derived from protein containing 16 per cent nitrogen, and by multiplying the nitrogen figure by 6.25 (i.e. 100/16) an approximate protein value is obtained. This is not 'true protein' since the method determines nitrogen from sources other than protein, such as free amino acids, amines and nucleic acids, and the fraction is therefore designated crude protein.

The ether extract (EE) fraction is determined by subjecting the food to a continuous extraction with petroleum ether for a defined period. The residue, after evaporation of the solvent, is the ether extract. As well as lipids it contains organic acids, alcohol and pigments. In the current official method, the extraction with ether is preceded by hydrolysis of the sample with sulphuric acid and the resultant residue is the acid ether extract.

The carbohydrate of the food is contained in two fractions, the crude fibre (CF) and the nitrogen-free extractives (NFE). The former is determined by subjecting the residual food from ether extraction to successive treatments with boiling acid and alkali of defined concentration; the organic residue is the crude fibre.

When the sum of the amounts of moisture, ash, crude protein, ether extract and crude fibre (expressed in g/kg) is subtracted from 1000, the difference is designated the nitrogen-free extractives. The crude fibre fraction contains cellulose, lignin and hemicelluloses, but not necessarily the whole amounts of these that are present in the food: a variable proportion, depending upon the species and stage of growth of the plant material, is contained in the nitrogen-free extractives. The nitrogen-free extractives fraction is a heterogeneous mixture of all those components not determined in the other fractions. It includes sugars, fructans, starch, pectins, organic acids and pigments, in addition to those components mentioned above.

Modern analytical methods

In recent years the proximate analysis procedure has been severely criticised by many nutritionists as being archaic and imprecise, and in the majority of laboratories it has been partially replaced by other analytical procedures. Most criticism has been focused on the crude fibre, ash and nitrogen-free extractives fractions. The newer methods have been developed to characterise foods in terms of the methods used to express nutrient requirements. In this way, an attempt is made to use the analytical techniques to quantify the potential supply of nutrients from the food. For example, for ruminants, analytical methods are being developed that describe the supply of nutrients for the rumen microbes and the host digestive enzyme system (Fig. 1.1).

Starch and sugars

Inadequacies in the nitrogen-free extractives fraction have been addressed by the development of methods to quantify the non-structural carbohydrates,

Food measurements   **Rumen parameters**   **Absorbed nutrients**

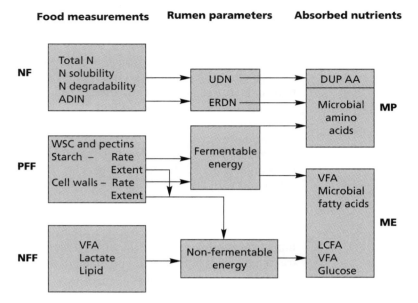

Fig. 1.1 Proposed model for characterisation of foods for ruminants: AA = amino acids, ADIN = acid detergent insoluble nitrogen, DUP = digestible undegradable protein, ERDN = effective rumen degradable nitrogen, LCFA = long chain fatty acids, ME = metabolisable energy, MP = metabolisable protein, N = nitrogen, NF = nitrogen fraction, NFF = non-fermentable fraction, PFF = potentially fermentable fraction, UDN = undegradable nitrogen, VFA = volatile fatty acids, WSC = water-soluble carbohydrates. From Agricultural and Food Research Council 1998. Technical Committee on Responses to Nutrients, Report No. 11, Wallingford, UK, CAB International.

which are mainly starches and sugars. Sugars can be determined colorimetrically after combination with a reagent such as anthrone. Starch is determined by dilute acid hydrolysis of the sample followed by polarimetric determination of the released sugars. This gives a figure for total sugars (i.e. those originating from the hydrolysed starch plus the simple sugars in the food). Sugars *per se* are determined by extracting the sample with ethanol, acidifying the filtrate and taking a second polarimeter reading. The starch content is calculated from the difference between the two readings multiplied by a known factor for the starch source.

Fibre

Alternative procedures for fibre have been developed by Van Soest (Table 1.2). The *neutral-detergent fibre* (NDF), which is the residue after extraction with boiling neutral solutions of sodium lauryl sulphate and ethylenediaminetetraacetic acid (EDTA), consists mainly of lignin, cellulose and hemicellulose and can be regarded as a measure of the plant cell wall material. The analytical method for determining NDF was originally devised for forages, but can also be used for starch-containing foods provided an amylase treatment is

Table 1.2 Classification of forage fractions using the detergent methods of Van Soest (After Van Soest P J 1967 *Journal of Animal Science* **26**: 119)

Fraction	Components
Cell contents (soluble in neutral detergent)	Lipids Sugars, organic acids and water-soluble matter Pectin, starch Non-protein N Soluble protein
Cell wall constituents (fibre insoluble in neutral detergent)	
1. Soluble in acid detergent	Hemicelluloses Fibre-bound protein
2. Acid-detergent fibre	Cellulose Lignin Lignified N Silica

included in the procedure. By analogy with the nitrogen-free extractives fraction discussed above, the term *non-structural carbohydrate* (NSC) is sometimes used for the fraction obtained by subtracting the sum of the amounts (g/kg) of CP, EE, ash and NDF from 1000.

The *acid-detergent fibre* (ADF) is the residue after refluxing with 0.5 M sulphuric acid and cetyltrimethyl-ammonium bromide, and represents the crude lignin and cellulose fractions of plant material but also includes silica.

The determination of ADF is particularly useful for forages as there is a good statistical correlation between it and the extent to which the food is digested (digestibility). In the UK the ADF method has been modified slightly, the duration of boiling and acid strength being increased. The term *modified acid-detergent fibre* (MADF) is used to describe this determination.

The *acid-detergent lignin* determination involves the preparation of acid-detergent fibre as the preparatory step. The ADF is treated with 72 per cent sulphuric acid, which dissolves cellulose. Ashing the residue determines crude lignin, including cutin.

The Van Soest methods of fibre analysis are used in the system of food analysis for ruminants developed at Cornell University (see Box 1.1).

In monogastric, and particularly human, nutrition the term *dietary fibre* is often used. This is defined as lignin plus those polysaccharides that cannot be digested by monogastric endogenous enzymes. By virtue of its definition, dietary fibre is difficult to determine in the laboratory and the alternative term *non-starch polysaccharides* (NSP) has been adopted. The NSP in most foods, along with lignin, are considered to represent the major components of cell walls. Methods for measurement of NSP fall into two categories:

1. Enzymic–gravimetric methods, which measure a variety of components and give no details of polysaccharide type. In the method of the Association of Official Analytical Chemists for total dietary fibre,

---

Box 1.1
The Cornell net carbohydrate and protein system

The fractionation of the carbohydrates by analysis is currently most fully developed in the Cornell net carbohydrate and protein system for ruminant diets. This is based on the Van Soest analytical system, with the addition of other standard techniques, to derive the following fractions in foods:

1. Total carbohydrate = 100 − (crude protein + fat + ash)
2. Non-structural carbohydrate (NSC) = 100 − (crude protein + fat + (NDF − NDF protein) + ash)
3. Sugar as a proportion of NSC
4. Starch, pectin, glucans, volatile fatty acids = NSC − sugar
5. Lignin

The carbohydrates are then classified according to their degradation rate by rumen microbes: fraction A – fast (comprising the sugars), fraction B1 – intermediate (starch, pectin, β-glucans), fraction B2 – slow (available cell wall material represented by lignin-free NDF) and fraction C – indigestible (unavailable cell wall in the form of lignin).

---

samples are gelatinised by heating and treated with enzymes to remove starch and proteins. The total dietary fibre is precipitated with ethanol and the residue is dried and weighed.

2. Enzymic–chromatographic methods, which identify the individual carbohydrates in the diet. The Englyst method can be used to determine total, soluble and insoluble dietary fibre. Measurement of NSP by this method involves removal of starch with the enzymes pullulanase and α-amylase. After precipitation with ethanol, the NSP residue is then hydrolysed with 12 M sulphuric acid. The individual monomeric neutral sugar constituents are determined by gas–liquid chromatography (see below) with separate determination of uronic acids. Alternatively, the total sugars are determined colorimetrically after reaction with dinitrosalicylate solution. Total NSP and insoluble NSP are determined directly by analysis of separate subsamples and the soluble NSP are calculated by difference. The major constituents of NSP are rhamnose, arabinose, xylose, glucose, galactose, mannose and glucuronic and galacturonic acids. Cellulose is the major source of glucose, and hemicellulose provides xylose, mannans and galactose. The degradation of pectins releases arabinose, galactose and uronic acids. A comparison of the dietary fibre contents for a range of food types is given in Table 1.3.

In recent years attention has focused on the importance of both these forms of fibrous material in the human diet. Water-soluble NSP is known to lower serum cholesterol, and insoluble NSP increases faecal bulk and speeds up the rate of colonic transit. This last effect is thought to be beneficial in preventing a number of diseases including cancer of the bowel.

The NSP of foods may be degraded in the gut of pigs by microbial fermentation, yielding volatile fatty acids, which are absorbed and contribute to the energy supply. A further benefit relates to the volatile fatty

Table 1.3 The fibre components (g/kg DM) of some common foods (Adapted from Dierick N A and Decuypere J A 1994 Enzymes and growth in pigs. In: Cole D J A Wiseman J and Varley M A (eds) *Principles of Pig Science*, Loughborough, Nottingham University Press, pp 169–95 and Appendix Table 1.1 for CF)

Food	NSP	Rhamnose	Fucose	Arabinose	Xylose	Mannose	Galactose	Glucose	Uronic acids	Lignin	NDF	ADF	CF
Wheat	102	2	1	23	37	5	4	27	7	11	105	35	26
Barley	158	1	1	25	50	4	3	75	12	33	210	89	53
Maize gluten feed	348	2	0	66	96	4	17	102	29	31	400	114	39
Peas	154	3	0	32	10	2	8	80	23	8	194	110	63
Soya bean meal	196	4	2	25	17	10	49	59	36	30	115	83	58
Rapeseed meal	221	3	2	43	18	4	16	64	48	100	256	206	152
Sugar beet pulp	602	15	1	163	20	10	40	193	161	63	490	276	203
Grass meal	485	1	6	28	128	4	12	253	29	50	723	389	210
Wheat straw	512	1	0	21	169	5	7	315	18	171	752	465	417

NSP   non-starch polysaccharides
NDF   neutral-detergent fibre
ADF   acid-detergent fibre
CF   crude fibre

acid, butyric acid, which is reported to be an important source of energy for the growth of cells in the epithelium of the colon. Thus the presence of this acid will promote development of the cells and enhance absorption. The extent of degradation depends on the conformation of the polymers and their structural association with non-carbohydrate components, such as lignin. In addition the physical properties of the NSP, such as water-holding capacity and ion exchange properties, can influence the extent of fermentation. The gel-forming NSPs, such as β-glucan, reduce the absorption of other nutrients from the small intestine and depress digestibility and adversely affect faecal consistency in pigs and poultry. On a positive note, the water-holding properties lead to beneficial effects on the behaviour of pregnant sows by increasing time spent eating and resting due to increased gut fill and reducing inappropriate behaviour, such as bar chewing.

Minerals

A simple ash determination provides very little information about the exact mineral make-up of the food and, when this is required, analytical techniques involving spectroscopy are generally used. In *atomic absorption spectroscopy*, an acid solution of the sample is heated in a flame and the vaporised atoms absorb energy, which brings about transitions from the ground state to higher energy levels. The source of energy for this transition is a cathode lamp, containing the element to be determined, which emits radiation at a characteristic wavelength. The radiation absorbed by the atoms in the flame is proportional to the concentration of the element in the food sample.

Flame emission spectroscopy measures the radiation from solutions of the sample heated in air/acetylene or oxygen/acetylene flames. Each element emits radiation at specific wavelengths and there are published tables of flame emission spectra. Atomic absorption and flame emission spectrometry are being replaced by *inductively coupled plasma emission spectroscopy*, as this has a greater sensitivity for the relatively inert elements and can be used to determine several elements simultaneously or sequentially. Energy from the inductively coupled plasma source is absorbed by argon ions and elements to form a conducting gaseous mixture at temperatures up to 10 000 °C. The electromagnetic radiation emitted from atoms and ions within the plasma is then measured.

Just as with other nutrients, a measure of the concentration of the element alone is not sufficient to describe its usefulness to the animal. Attempts have been made to assess the availability of minerals using chemical methods, such as solubility in water or dilute acids, but these have had little success. At present animal experiments are the only reliable way to measure mineral availability (see Chapter 10).

Amino acids, fatty acids and sugars

Knowledge of the crude protein content of a food is not a sufficient measure of its usefulness for non-ruminants. The amino acid composition of the protein is required in order to assess how a food can meet the essential amino acid requirements (see Chapter 4). Similarly, the total ether extract content does not give sufficient information on this fraction since it is important to know its fatty acid composition. In non-ruminants, this has large effects on the composition of body fat and, if soft fat is to be avoided, the level of unsaturated fatty acids in the diet must be controlled. In ruminants, a high proportion of unsaturates will depress fibre digestion in the rumen. When detailed information on the amino acid composition of protein, the fatty acid composition of fat or the individual sugars in NSP is required, then techniques involving chromatographic separation can be used. In *gas–liquid chromatography*, the stationary phase is a liquid held in a porous solid, usually a resin, and the mobile phase is a gas. Volatile substances partition between the liquid and the vapour and can be effectively isolated. This form of chromatography is, however, usually a slow process and in order to speed up the separation procedure, *high-performance liquid chromatography* has been developed. In this technique, pressure is used to force a solution, containing the compounds to be separated, rapidly through the resin held in a strong metal column. In addition to speeding up the process, high resolution is also obtained. Gas–liquid chromatography and high-performance liquid chromatography can also be used for the determination of certain vitamins (e.g. A, E, B_6, K), but the measurement of available vitamins requires biological methods.

An example of the application of high-performance liquid chromatography is seen with food proteins, which are hydrolysed with acid and the released amino acids are then determined using either:

1. Ion exchange chromatography – by which the amino acids are separated on the column, then mixed with a derivatisation agent, which reacts to give a complex that is detected by a spectrophotometer or fluorimeter.
2. Reverse phase chromatography – in which the amino acids react with the reagent to form fluorescent or ultraviolet-absorbing derivatives, which are then separated using a more polar mobile phase (e.g. acetate buffer with a gradient of acetonitrile) and a less polar stationary phase (e.g. octadecyl-bonded silica). The availability of amino acids to the animal can be estimated by chemical methods. For example, for lysine there are colorimetric methods which depend on the formation of compounds between lysine and dyes (see Chapter 13).

Measurement of protein in foods for ruminants

The new methods of expressing the protein requirements of ruminants (see Chapter 13) require more information than just the crude protein (nitrogen) content of the food. The unavailable nitrogen is measured as acid detergent insoluble nitrogen. Information on the rate of degradation in the rumen of the available nitrogen is also required and this can be estimated by biological methods. In the Cornell net carbohydrate and protein system, the neutral and acid detergent extractions of Van Soest, described above, are used in combination with extraction with a borate–phosphate buffer and trichloracetic acid solution to derive several protein fractions. These fractions describe the components that are degraded in the rumen or digested in the small intestine (see Chapter 13).

Spectroscopy

More recently, in some laboratories, procedures combining statistical regression techniques with *near-infrared reflectance spectroscopy* (NIRS) have been introduced. Energy in the wavelength range from 1100 to 2500 nm is directed on to a cell containing the dried milled sample, and the diffuse reflected energy is measured. This is related to the chemical nature of the sample, since the bonds $-CH$, $-OH$, $-NH$ and $-SH$ absorb energy at specific wavelengths. A spectrum of reflected energy is recorded and, by use of a computer, the spectral data obtained from a calibration set of samples are related by multiple linear regression to their known composition determined by traditional methods. These relationships are then used to relate the reflectance readings obtained for individual foods to their composition. This technique is now used routinely to determine a range of food characteristics, including those that are the resultant of a number of nutrient concentrations such as digestibility, metabolisable energy and nitrogen degradability in the rumen (see Chapters 12 and 13).

Nuclear magnetic resonance spectroscopy is a complex technique, which is also used to determine the constituents of foods. This method makes use of the fact that some compounds contain certain atomic nuclei, which can be identified from a nuclear magnetic resonance spectrum, which measures variations in frequency of electromagnetic radiation absorbed. It provides more specific and detailed information of the conformational structure of compounds than, for example, NIRS but is more costly and requires more time and skill on the part of the operator. For these reasons, it is more suited to research work and for cases in which the results from simpler spectroscopy techniques require further investigation. Nuclear magnetic resonance spectroscopy has been useful in the investigation of the soluble and structural components of forages.

Summary

1. Water is an important component of animal foods. It contributes to the water requirements of animals and dilutes the nutrient content of foods. Water content varies widely between foods.

2. The constituents of dry matter comprise carbohydrates (sugars, starches, fibres), nitrogen containing compounds (proteins, amino acids, non-protein nitrogen compounds), lipids (fatty acids, glycerides), minerals and vitamins.

3. Analytical techniques have been developed from simple chemical/gravimetric determinations.

4. Modern analytical techniques attempt to measure nutrients in foods in terms of the nutrient requirements of the animal.

5. Starch is determined by polarimetry.

6. Fibrous constituents can be determined by application of detergent solutions and weighing the residue or by the use of enzymes followed by weighing or gas–liquid chromatography.

7. Individual mineral elements are measured by atomic absorption spectroscopy, flame photometry or inductively coupled plasma emission spectroscopy.

8. Gas–liquid chromatography is used to determine individual amino acids, fatty acids and certain vitamins.

9. Near-infrared reflectance spectroscopy is used routinely to determine food characteristics and to predict nutritive value. Nuclear magnetic resonance spectroscopy is a research technique for determining the chemical structure of food components.

Further reading

Agricultural and Food Research Council 1987 Technical Committee on Responses to Nutrients, Report No. 2. Characterisation of feedstuffs: nitrogen. *Nutrition Abstracts and Reviews, Series B: Livestock Feeds and Feeding* **57**: 713–36.

Agricultural and Food Research Council 1988 Technical Committee on Responses to Nutrients, Report No. 3. Characterisation of feedstuffs: other nutrients. *Nutrition Abstracts and Reviews, Series B: Livestock Feeds and Feeding* **58**: 549–71.

Asp N-G and Johansson C-G 1984 Dietary fibre analysis. *Nutrition Abstracts and Reviews* **54**: 735–51.

Association of Official Analytical Chemists 1990 *Official Methods of Analysis*, 15th edn, Washington, DC.

Chalupa W and Sniffen C J 1994 Carbohydrate, protein and amino acid nutrition of lactating dairy cattle. In: Garnsworthy P C and Cole D J A (eds) *Recent Advances in Animal Nutrition*, Loughborough, Nottingham University Press, pp 265–75.

Coultate T P 1989 *Food – The Chemistry of its Components*, 2nd edn, London, Royal Society of Chemistry.

Kritchevsky D, Bonfield C and Anderson J W 1988 *Dietary Fiber*, New York, Plenum Press.

Ministry of Agriculture, Fisheries and Food 1985 *The Analysis of Agricultural Materials*, Ref. Book 427, London, HMSO.

The Feeding Stuffs (Sampling and Analysis) Regulations 1999, London, HMSO.

Van Soest P J 1994 *Nutritional Ecology of the Ruminant*, 2nd edn, Ithaca, NY, Comstock.

2 Carbohydrates

2.1   Classification of carbohydrates

2.2   Monosaccharides

2.3   Monosaccharide derivatives

2.4   Oligosaccharides

2.5   Polysaccharides

2.6   Lignin

In general, carbohydrates are neutral chemical compounds containing the elements carbon, hydrogen and oxygen and have the empirical formula $(CH_2O)_n$, where n is three or more. However, some compounds with general properties of the carbohydrates also contain phosphorus, nitrogen or sulphur; and others, e.g. deoxyribose ($C_5H_{10}O_4$), do not have hydrogen and oxygen in the same ratio as that in water. The carbohydrate group contains polyhydroxy aldehydes, ketones, alcohols and acids, their simple derivatives, and any compound that may be hydrolysed to these.

2.1   Classification of carbohydrates

The carbohydrates may be classified as shown in Table 2.1. The simplest sugars are the monosaccharides, which are divided into sub-groups: trioses ($C_3H_6O_3$), tetroses ($C_4H_8O_4$), pentoses ($C_5H_{10}O_5$), hexoses ($C_6H_{12}O_6$) and heptoses ($C_7H_{14}O_7$), depending upon the number of carbon atoms present in the molecule. The trioses and tetroses occur as intermediates in the metabolism of other carbohydrates and their importance will be considered in Chapter 9. Monosaccharides may be linked together, with the elimination of one molecule of water at each linkage, to produce di-, tri-, tetra- or polysaccharides or containing, respectively, two, three, four or larger numbers of monosaccharide units.

The term 'sugar' is generally restricted to those carbohydrates containing less than ten monosaccharide residues, while the name oligosaccharides (from the Greek *oligos*, a few) is frequently used to include all sugars other than the monosaccharides.

Polysaccharides, also called glycans, are polymers of monosaccharide units. They are classified into two groups, the homoglycans, which contain only a single type of monosaccharide unit, and the heteroglycans, which on

Table 2.1 Classification of carbohydrates

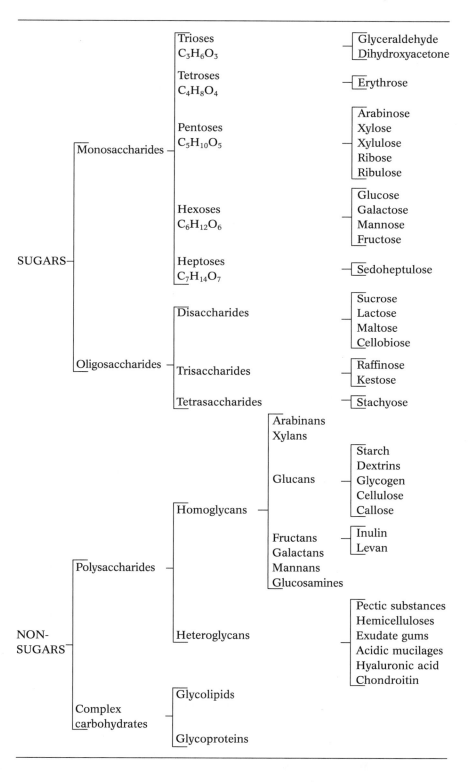

SUGARS

Monosaccharides

Trioses
$C_3H_6O_3$ — Glyceraldehyde
Dihydroxyacetone

Tetroses
$C_4H_8O_4$ — Erythrose

Pentoses
$C_5H_{10}O_5$ — Arabinose
Xylose
Xylulose
Ribose
Ribulose

Hexoses
$C_6H_{12}O_6$ — Glucose
Galactose
Mannose
Fructose

Heptoses
$C_7H_{14}O_7$ — Sedoheptulose

Oligosaccharides

Disaccharides — Sucrose
Lactose
Maltose
Cellobiose

Trisaccharides — Raffinose
Kestose

Tetrasaccharides — Stachyose

NON-SUGARS

Polysaccharides

Homoglycans — Arabinans
Xylans

Glucans — Starch
Dextrins
Glycogen
Cellulose
Callose

Fructans — Inulin
Levan
Galactans
Mannans
Glucosamines

Heteroglycans — Pectic substances
Hemicelluloses
Exudate gums
Acidic mucilages
Hyaluronic acid
Chondroitin

Complex carbohydrates

Glycolipids

Glycoproteins

hydrolysis yield mixtures of monosaccharides and derived products. The molecular weight of polysaccharides varies from as little as about 8000 in some plant fructans to as high as 100 million in the amylopectin component of starch. Hydrolysis of these polymers to their constituent sugars can be effected by the action of either specific enzymes or acids.

The complex carbohydrates are an ill-defined group of compounds which contain carbohydrates in combination with non-carbohydrate molecules. They include the glycolipids and glycoproteins. The structure and biological importance of these two groups of compounds are discussed in Chapters 3 and 4, respectively.

2.2 Monosaccharides

Structure

The monosaccharide sugars occur in a number of isomeric forms. Thus glucose and fructose (both hexoses) are structural isomers, glucose having an aldehyde group and fructose, a ketone group. Both of these sugars occur in two mirror image, stereoisomeric forms, dextro and laevo (D- and L-), according to the orientation of the OH group at carbon atom 5. Biologically the D- forms are the more important.

$$
\begin{array}{cccc}
^1CHO & CHO & ^1CH_2OH & CH_2OH \\
| & | & | & | \\
H^2COH & HOCH & ^2C{=}O & C{=}O \\
| & | & | & | \\
HO^3CH & HCOH & HO^3CH & HCOH \\
| & | & | & | \\
H^4COH & HOCH & H^4COH & HOCH \\
| & | & | & | \\
H^5COH & HOCH & H^5COH & HOCH \\
| & | & | & | \\
^6CH_2OH & CH_2OH & ^6CH_2OH & CH_2OH \\
\text{D-Glucose} & \text{L-Glucose} & \text{D-Fructose} & \text{L-Fructose}
\end{array}
$$

Under physiological conditions, sugars exist mainly in another isomeric form, as ring or cyclic structures, rather than straight chains. Glucose forms a pyranose ring and fructose, most commonly, a furanose ring. Each ring structure can occur in two isomeric forms, designated α and β. Starch and glycogen are polymers of the α- form, while cellulose is a polymer of the β- form.

Properties of the monosaccharides

Because of the presence of an active aldehyde or ketone grouping, the monosaccharides act as reducing substances. The reducing properties of

α-D-Glucose β-D-Glucose α-D-Fructose

these sugars are usually demonstrated by their ability to reduce certain metal ions, notably copper or silver, in alkaline solution. The aldehyde and ketone groups may also be reduced chemically, or enzymatically, to yield the corresponding sugar alcohols. Examples of oxidation and reduction products are given in the section dealing with monosaccharide derivatives (p. 19).

Pentoses

The most important members of this group of simple sugars are the aldoses L-arabinose, D-xylose and D-ribose, and the ketoses D-xylulose and D-ribulose.

α-L-Arabinose α-D-Xylose α-D-Ribose

L-Arabinose occurs as pentosans in arabinans. It is a component of hemicelluloses and it is found in silage as a result of their hydrolysis. It is also a component of gum arabic and other gums. *D-Xylose* also occurs as pentosans in xylans. These compounds form the main chain in grass hemicelluloses. Xylose, along with arabinose, is produced in considerable quantities when herbage is hydrolysed with normal sulphuric acid. *D-Ribose* is present in all living cells as a constituent of ribonucleic acid (RNA), and it is also a component of several vitamins and coenzymes.

```
      CH2OH            CH2OH
       |                |
       C=O              C=O
       |                |
      HOCH             HCOH
       |                |
      HCOH             HCOH
       |                |
      CH2OH            CH2OH

    D-Xylulose       D-Ribulose
```

The phosphate derivatives of D-xylulose and D-ribulose occur as intermediates in the pentose phosphate metabolic pathway (see p. 209).

Hexoses

Glucose and fructose are the most important naturally occurring hexose
sugars, while mannose and galactose occur in plants in a polymerised form
as mannans and galactans.

D-*Glucose*, grape sugar or dextrose, exists in the free state as well as in combined form. The sugar occurs free in plants, fruits, honey, blood, lymph and
cerebrospinal fluid, and is the sole or major component of many oligosaccharides, polysaccharides and glucosides. In the pure state, glucose is a white
crystalline solid and like all sugars is soluble in water.

D-*Fructose*, fruit sugar or laevulose, occurs free in green leaves, fruits and
honey. It also occurs in the disaccharide sucrose and in fructans. Green
leafy crops usually contain appreciable amounts of this sugar both free and
in polymerised form. The free sugar is a white crystalline solid and has a
sweeter taste than sucrose. The exceptionally sweet taste of honey is due to
this sugar.

D-*Mannose* does not occur free in nature but exists in polymerised form
as mannan and also as a component of glycoproteins. Mannans are found
widely distributed in yeasts, moulds and bacteria.

D-*Galactose* does not occur free in nature except as a breakdown product during fermentation. It is present as a constituent of the disaccharide
lactose, which occurs in milk. Galactose also occurs as a component of the
anthocyanin pigments, galactolipids, gums and mucilages.

Heptoses

D-*Sedoheptulose* is an important example of a monosaccharide containing
seven carbon atoms and occurs, as the phosphate, as an intermediate in the
pentose phosphate metabolic pathway (see p. 209).

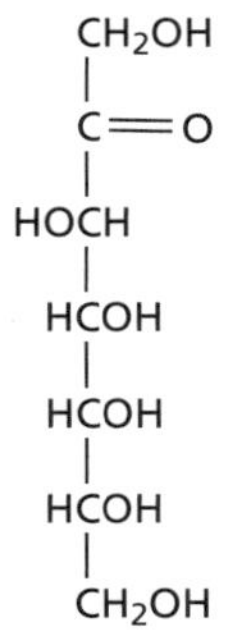

$$CH_2OH$$
$$|$$
$$C=O$$
$$|$$
$$HOCH$$
$$|$$
$$HCOH$$
$$|$$
$$HCOH$$
$$|$$
$$HCOH$$
$$|$$
$$CH_2OH$$

D-Sedoheptulose

2.3 Monosaccharide derivatives

Phosphoric acid esters

The phosphoric acid esters of sugars play an important role in a wide variety of metabolic reactions in living organisms (see Chapter 9). The most commonly occurring derivatives are those formed from glucose, the esterification occurring at either carbon atoms 1 or 6 or both.

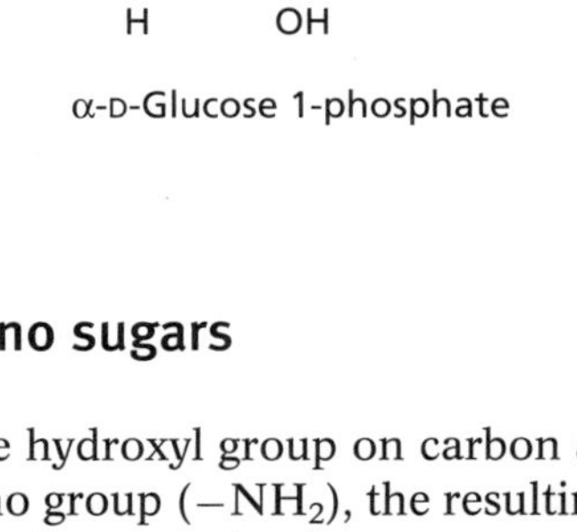

α-D-Glucose 1-phosphate 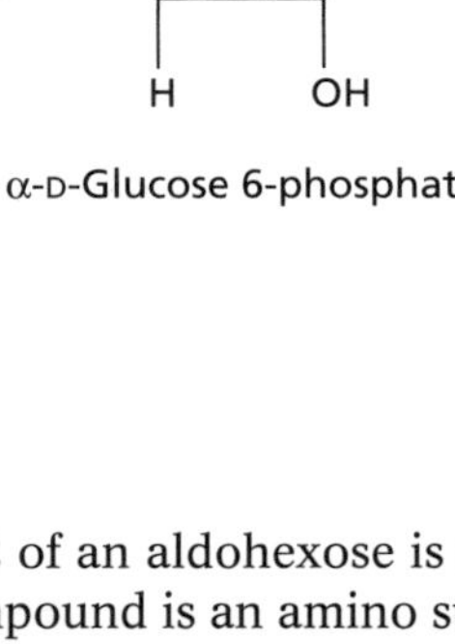   α-D-Glucose 6-phosphate

Amino sugars

If the hydroxyl group on carbon atom 2 of an aldohexose is replaced by an amino group ($-NH_2$), the resulting compound is an amino sugar. Two such naturally occurring important compounds are D-glucosamine, a major component of chitin (see p. 28), and D-galactosamine, a component of the polysaccharide of cartilage.

β-D-Glucosamine   β-D-Galactosamine

Deoxy sugars

Replacement of a hydroxyl group by hydrogen yields a deoxy sugar. The derivative of ribose, deoxyribose, is a component of deoxyribonucleic acid (DNA). Similarly, deoxy derivatives of the two hexoses, galactose and mannose, occur as fucose and rhamnose, respectively, these being components of certain heteropolysaccharides.

α-D-Deoxyribose   α-L-Rhamnose

Sugar acids

The aldoses can be oxidised to produce a number of acids, of which the most important are:

Aldonic acids   Aldaric acids   Uronic acids

In the case of glucose, the derivatives corresponding to these formulae are gluconic, glucaric and glucuronic acids, respectively. Of these compounds, the uronic acids, particularly those derived from glucose and galactose, are important components of a number of heteropolysaccharides.

Sugar alcohols

Simple sugars can be reduced to polyhydric alcohols; for example, glucose yields sorbitol, galactose yields dulcitol, while both mannose and fructose yield mannitol. This last-named alcohol occurs in grass silage and is formed by the action of certain anaerobic bacteria on the fructose present in the grass.

$$
\begin{array}{ccccc}
\text{CH}_2\text{OH} & & \text{CH}_2\text{OH} & & \text{CHO} \\
| & & | & & | \\
\text{C}=\text{O} & & \text{HOCH} & & \text{HOCH} \\
| & & | & & | \\
\text{HOCH} & \xrightarrow{+2\text{H}} & \text{HOCH} & \xleftarrow{+2\text{H}} & \text{HOCH} \\
| & & | & & | \\
\text{HCOH} & & \text{HCOH} & & \text{HCOH} \\
| & & | & & | \\
\text{HCOH} & & \text{HCOH} & & \text{HCOH} \\
| & & | & & | \\
\text{CH}_2\text{OH} & & \text{CH}_2\text{OH} & & \text{CH}_2\text{OH} \\
\text{D-Fructose} & & \text{D-Mannitol} & & \text{D-Mannose}
\end{array}
$$

Glycosides

If the hydrogen of the hydroxyl group attached to the carbon 1 atom of glucose is replaced by esterification, or by condensation, with an alcohol (including a sugar molecule) or a phenol, the derivative so produced is termed a glucoside. Similarly galactose forms galactosides and fructose forms fructosides. The general term glycoside is used collectively to describe these derivatives and the linkage is described as a glycosidic bond.

Oligosaccharides and polysaccharides are classed as glycosides, and these compounds yield sugars or sugar derivatives on hydrolysis. Certain naturally occurring glycosides contain non-sugar residues. For example, the nucleosides contain a sugar combined with a heterocyclic nitrogenous base (see Chapter 4).

The cyanogenetic glycosides liberate hydrogen cyanide (HCN) on hydrolysis, and because of the toxic nature of this compound, plants containing this type of glycoside are potentially dangerous to animals. The glycoside itself is not toxic and must be hydrolysed before poisoning occurs. However, the glycoside is easily broken down to its components by means of an enzyme which is usually present in the plant. An example of a cyanogenetic glycoside is linamarin (also called phaseolunatin), which occurs in linseed, Java beans and cassava. If wet mashes or gruels containing these

Table 2.2 Some important naturally occurring cyanogenetic glycosides

Name	Source	Hydrolytic products in addition to glucose and hydrogen cyanide
Linamarin (phaseolunatin)	Linseed (*Linum usitatissimum*) Java beans (*Phaseolus lunatus*) Cassava (*Manihot esculenta*)	Acetone
Vicianin	Seeds of wild vetch (*Vicia angustifolia*)	Arabinose, benzaldehyde
Amygdalin	Bitter almonds, kernels of peach, cherries, plums, apples and fruits of Rosaceae	Benzaldehyde
Dhurrin	Leaves of the great millet (*Sorghum vulgare*)	p-Hydroxy-benzaldehyde
Lotaustralin	Trefoil (*Lotus australis*) White clover (*Trifolium repens*)	Methylethyl ketone

foods are given to animals, it is advisable to boil them when mixing in order to inactivate any enzyme present. On hydrolysis linamarin yields glucose, acetone and hydrogen cyanide.

Examples of other cyanogenetic glycosides and their sources are shown in Table 2.2.

2.4 Oligosaccharides

Disaccharides

A large number of disaccharide compounds are theoretically possible, depending upon the monosaccharides present and the manner in which they are linked. The most nutritionally important disaccharides are sucrose, maltose, lactose and cellobiose, which on hydrolysis yield two molecules of hexoses:

$$C_{12}H_{22}O_{11} + H_2O \rightarrow 2C_6H_{12}O_6$$

Sucrose is formed from one molecule of α-D-glucose and one molecule of β-D-fructose joined together through an oxygen bridge between their respective carbon atoms 1 and 2. As a consequence, sucrose has no active reducing group.

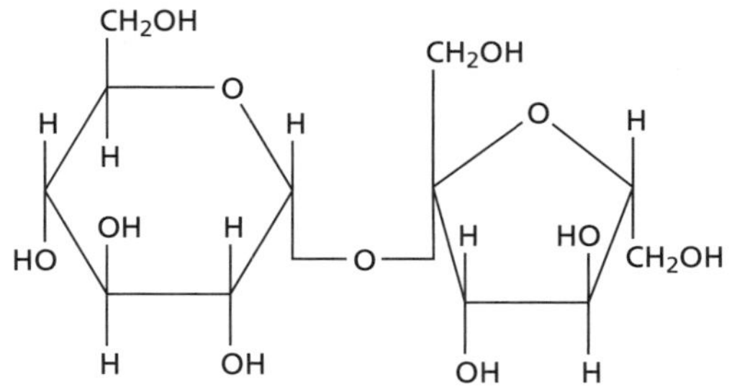

Sucrose

Sucrose is the most ubiquitous and abundantly occurring disaccharide in plants, where it is the main transport form of carbon. This disaccharide is found in high concentration in sugar cane (200 g/kg) and in sugar beet (150–200 g/kg); it is also present in other roots such as mangels and carrots, and occurs in many fruits. Sucrose is easily hydrolysed by the enzyme sucrase or by dilute acids. When heated to a temperature of 160 °C it forms barley sugar and at a temperature of 200 °C it forms caramel.

Lactose, or milk sugar, is a product of the mammary gland. Cow's milk contains 43–48 g/kg. It is not as soluble as sucrose and is less sweet, imparting only a faint sweet taste to milk. Lactose is formed from one molecule of β-D-glucose joined to one of β-D-galactose in a β-(1→4)-linkage and has one active reducing group.

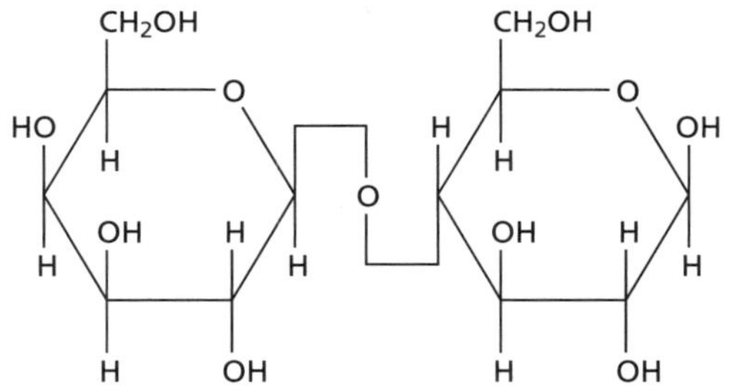

Lactose

Lactose readily undergoes fermentation by a number of organisms, including *Streptococcus lactis*. This organism is responsible for souring milk by converting the lactose into lactic acid ($CH_3.CHOH.COOH$). If lactose is heated to 150 °C it turns yellow, and at a temperature of 175 °C the sugar is changed into a brown compound, lactocaramel. On hydrolysis lactose produces one molecule of glucose and one molecule of galactose.

Maltose, or malt sugar, is produced during the hydrolysis of starch and glycogen by dilute acids or enzymes. It is produced from starch during the germination of barley by the action of the enzyme amylase. The barley, after controlled germination and drying, is known as malt and is used in the manufacture of beer and Scotch malt whisky. Maltose is water-soluble, but it is not as sweet as sucrose. Structurally it consists of two α-D-glucose residues linked in the α-1,4 positions and it has one active reducing group.

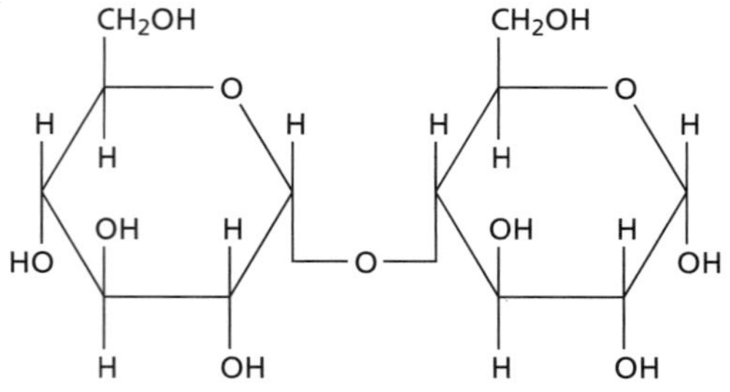

Maltose

Cellobiose does not exist naturally as a free sugar, but is the basic repeating unit of cellulose. It is composed of two β-D-glucose residues linked through a β-(1→4)-bond. This linkage cannot be split by mammalian digestive enzymes. It can, however, be split by microbial enzymes. Like maltose, cellobiose has one active reducing group.

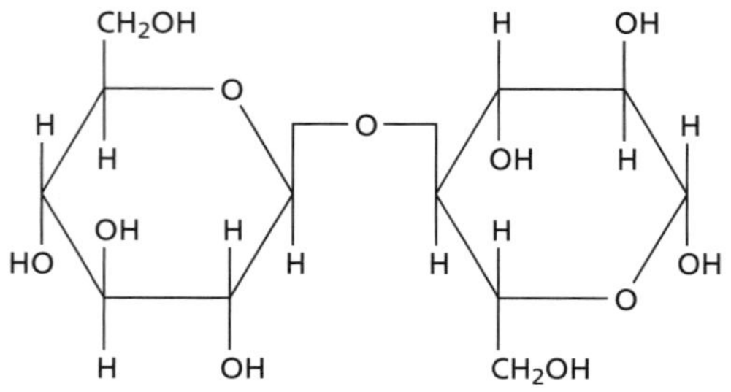

Cellobiose

Trisaccharides

Raffinose and *kestose* are two important naturally occurring trisaccharides. They are both non-reducing and on hydrolysis produce three molecules of hexose sugars:

$$C_{18}H_{32}O_{16} + 2H_2O \rightarrow 3C_6H_{12}O_6$$

Raffinose is the commonest member of the group, occurring almost as widely in plants as sucrose. It exists in small amounts in sugar beet and accumulates in molasses during the commercial preparation of sucrose. Cotton seed contains about 80 g/kg of raffinose. On hydrolysis, this sugar produces glucose, fructose and galactose.

Kestose, and its isomer isokestose, occurs in the vegetative parts and seeds of grasses. These two trisaccharides consist of a fructose residue attached to a sucrose molecule.

Tetrasaccharides

Tetrasaccharides are made up of four monosaccharide residues. *Stachyose*, a member of this group, is almost as ubiquitous in higher plants as raffinose and has been isolated from about 165 species. It is a non-reducing sugar and on hydrolysis produces two molecules of galactose, one molecule of glucose and one of fructose:

$$C_{24}H_{42}O_{21} + 3H_2O \rightarrow 4C_6H_{12}O_6$$

2.5 Polysaccharides

Homoglycans

These carbohydrates are very different from the sugars. The majority are of high molecular weight, being composed of large numbers of pentose or hexose residues. Homoglycans do not give the various sugar reactions characteristic of the aldoses and ketoses. Many of them occur in plants either as reserve food materials such as starch or as structural materials such as cellulose.

Arabinans and xylans

These are polymers of arabinose and xylose, respectively. Although homoglycans based on these two pentoses are known, they are more commonly found in combination with other sugars as constituents of heteroglycans.

Glucans

Starch is a glucan and is present in many plants as a reserve carbohydrate. It is most abundant in seeds, fruits, tubers and roots. Starch occurs naturally in the form of granules, whose size and shape vary in different plants. The granules are built up in concentric layers, and although glucan is the main component of the granules they also contain minor constituents such as protein, fatty acids and phosphorus compounds, which may influence their properties.

Starches differ in their chemical composition and, except in rare instances, are mixtures of two structurally different polysaccharides, amylose and amylopectin. The proportions of these present in natural starches depend upon the source, although in most starches, amylopectin is the main component, amounting to about 70–80 per cent of the total. An important qualitative test for starch is its reaction with iodine: amylose produces a deep blue colour and amylopectin solutions produce a blue-violet or purple colour.

Amylose is mainly linear in structure, the α-D-glucose residues being linked between carbon atom 1 of one molecule and carbon atom 4 of the adjacent molecule. A small proportion of α-(1→6)-linkages may also be present. Amylopectin has a bush-like structure containing primarily α-(1→4)-linkages, but also an appreciable number of α-(1→6)-linkages.

Part of amylose molecule showing 1,4-linkages

Starch granules are insoluble in cold water, but when a suspension in water is heated the granules swell and eventually gelatinise. On gelatinisation, potato starch granules swell greatly and then burst open; cereal starches swell but tend not to burst.

Animals consume large quantities of starch in cereal grains, cereal by-products and tubers.

Glycogen is a term used to describe a group of highly branched polysaccharides isolated from animals or microorganisms. Glycogens occur in liver, muscle and other animal tissues. They are glucans, analogous to amylopectin in structure, and have been referred to as 'animal starches'. Glycogen is the main carbohydrate storage product in the animal body and plays an essential role in energy metabolism.

The molecular weights of glycogen molecules vary considerably according to the animal species, the type of tissue and the physiological state of the animal. The glycogen of rat liver, for example, has molecular weights in the range 1 to 5×10^8, whereas that from rat muscle has a rather lower molecular weight of about 5×10^6.

Dextrins are intermediate products of the hydrolysis of starch and glycogen:

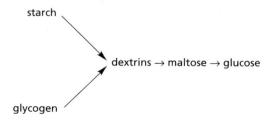

Dextrins are soluble in water and produce gum-like solutions. The higher members of these transitional products produce a red colour with iodine, while the lower members do not give a colour. The presence of dextrins gives a characteristic flavour to bread crust, toast and partly charred cereal foods.

Cellulose is the most abundant single polymer in the plant kingdom, forming the fundamental structure of plant cell walls. It is also found in a nearly pure form in cotton. Pure cellulose is a homoglycan of high molecular weight in which the repeating unit is cellobiose. Here the β-glucose residues are 1,4-linked.

Cellulose

In the plant, cellulose chains are formed in an ordered manner to produce compact aggregates (microfibrils) which are held together by both inter- and intra-molecular hydrogen bonding. In the plant cell wall, cellulose is closely associated, physically and chemically, with other components, especially hemicelluloses and lignin.

Callose is a collective term for a group of polysaccharides consisting of β-(1,3)- and frequently β-(1,4)-linked glucose residues. These β-glucans occur in higher plants as components of special walls appearing at particular stages of development. A large part of the endosperm cell wall of cereal grains is composed of β-glucans of this type. They are also deposited by higher plants in response to wounding and infection.

Fructans

These occur as reserve material in roots, stems, leaves and seeds of a variety of plants, but particularly in the Compositae and Gramineae. In the Gramineae, fructans are found only in temperate species. These polysaccharides are soluble in cold water and are of a relatively low molecular weight. All known fructans contain β-D-fructose residues joined by 2,6 or 2,1 linkages. They can be divided into three groups: (1) the levan group, characterised by 2,6 linkages; (2) the inulin group, containing 2,1 linkages; and (3) a group of highly branched fructans found, for example, in couch grass (*Agropyron repens*) and in wheat endosperm. This group contains both types of linkage.

Most fructans on hydrolysis yield, in addition to D-fructose, a small amount of D-glucose, which is derived from the terminal sucrose unit in the fructan molecule. The structure of a typical grass fructan is depicted below.

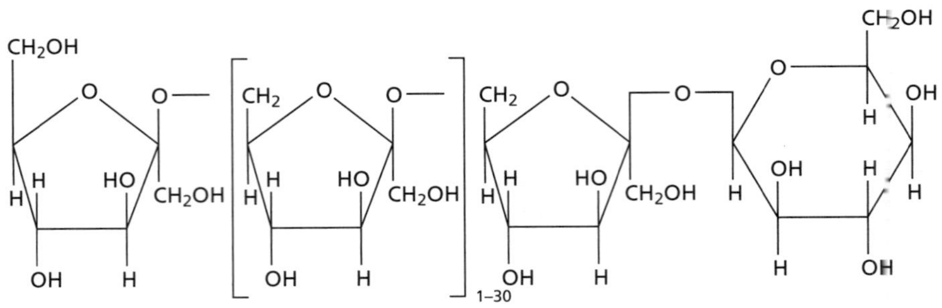

Grass fructan   (Sucrose residue)

Galactans and mannans

These are polymers of galactose and mannose, respectively, and occur in the cell walls of plants. A mannan is the main component of the cell walls of palm seeds, where it occurs as a food reserve and disappears during germination. A rich source of mannan is the endosperm of nuts from the South American tagua palm tree (*Phytelephas macrocarpa*); the hard endosperm of this nut is known as 'vegetable ivory'. The seeds of many legumes, including clovers, trefoil and lucerne, contain galactans.

Glucosaminans

Chitin is the only known example of a homoglycan containing glucosamine, being a linear polymer of acetyl-D-glucosamine. Chitin is of widespread occurrence in lower animals and is particularly abundant in Crustacea, in fungi and in some green algae. Next to cellulose, it is probably the most abundant polysaccharide of nature.

Heteroglycans

Pectic substances

Pectic substances are a group of closely associated polysaccharides, which are soluble in hot water and occur as constituents of primary cell walls and intercellular regions of higher plants. They are particularly abundant in soft tissues such as the peel of citrus fruits and sugar beet pulp. Pectin, the main member of this group, consists of a linear chain of D-galacturonic acid units in which varying proportions of the acid groups are present as methyl esters. The chains are interrupted at intervals by the insertion of L-rhamnose residues. Other constituent sugars, e.g. D-galactose, L-arabinose, D-xylose, are attached as side chains. Pectic acid is another member of this class of compounds; it is similar in structure to pectin but is devoid of ester groups. Pectic substances possess considerable gelling properties and are used commercially in jam making.

Hemicelluloses

Hemicelluloses are defined as alkali-soluble cell wall polysaccharides that are closely associated with cellulose. The name hemicellulose is misleading and implies erroneously that the material is destined for conversion to cellulose. Structurally, hemicelluloses are composed mainly of D-glucose, D-galactose, D-mannose, D-xylose and L-arabinose units joined together in different combinations and by various glycosidic linkages. They may also contain uronic acids.

Hemicelluloses from grasses contain a main chain of xylan made up of β-(1→4)-linked D-xylose units with side chains containing methylglucuronic acid and frequently glucose, galactose and arabinose.

Exudate gums and acid mucilages

Exudate gums are often produced from wounds in plants, although they may arise as natural exudations from bark and leaves. The gums occur naturally as salts, especially of calcium and magnesium, and in some cases a proportion of the hydroxyl groups are esterified, usually as acetates. Gum arabic (acacia gum) has long been a familiar substance; on hydrolysis it yields arabinose, galactose, rhamnose and glucuronic acid. Acidic mucilages are obtained from the bark, roots, leaves and seeds of a variety of plants. Linseed mucilage is a well-known example which produces arabinose, galactose, rhamnose and galacturonic acid on hydrolysis.

Hyaluronic acid and chondroitin

These two polysaccharides have a repeating unit consisting of an amino sugar and D-glucuronic acid. Hyaluronic acid, which contains acetyl-D-glucosamine, is present in the skin, the synovial fluid and the umbilical cord. Solutions of this acid are viscous and play an important part in the lubrication of joints. Chondroitin is chemically similar to hyaluronic acid but contains galactosamine in place of glucosamine. Sulphate esters of chondroitin are major structural components of cartilage, tendons and bones.

2.6 Lignin

Lignin, which is not a carbohydrate but is closely associated with this group of compounds, confers chemical and biological resistance to the cell wall, and mechanical strength to the plant. Strictly speaking the term 'lignin' does not refer to a single, well-defined, compound but is a collective term which embraces a whole series of closely related compounds.

Lignin is a polymer that originates from three derivatives of phenylpropane: coumaryl alcohol, coniferyl alcohol and sinapyl alcohol. The lignin molecule is made up of many phenylpropanoid units associated in a complex cross-linked structure.

$$CH = CH.CH_2OH$$

R R_1

OH

(1) Coumaryl alcohol, where $R = R_1 = H$.
(2) Coniferyl alcohol, where $R = H$, $R_1 = OCH_3$.
(3) Sinapyl alcohol, where $R = R_1 = OCH_3$.

Lignin is of particular interest in animal nutrition because of its high resistance to chemical degradation. Physical incrustation of plant fibres by lignin renders them inaccessible to enzymes that would normally digest them. There is evidence that strong chemical bonds exist between lignin and many plant polysaccharides and cell wall proteins which render these compounds unavailable during digestion. Wood products, mature hays and straws are rich in lignin and consequently are poorly digested unless treated chemically to break the bonds between lignin and other carbohydrates (see p. 257).

Summary

1. Carbohydrates are compounds containing carbon, hydrogen and oxygen, with the last two elements present in the same proportions as in water, and are found especially in plant foods.

2. They range from the simple sugar molecules (monosaccharides) with between three and seven carbon atoms to combinations of two, three or four molecules (di-, tri- and tetrasaccharides) and finally to complex polymers of the sugar molecules (polysaccharides).

3. The monosaccharides have an active aldehyde or ketone group and can take the mirror image D- or L-formation.

4. Under physiological conditions the monosaccharides exist mainly in cyclic forms, which can be α- or β-isomers.

5. The carbohydrate group also contains molecules derived from sugars which also contain phosphorus, nitrogen and sulphur and other derivatives that feature in intermediary metabolism, nucleic acid structure and substances such as cyanogenetic glucosides.

6. The main carbohydrates occurring in foods include the monosaccharides glucose, fructose, arabinose, xylose and ribose, the disaccharides sucrose, maltose and lactose and the polysaccharides starch, hemicellulose and cellulose.

7. Lignin, although not itself a carbohydrate, is associated with carbohydrates in plant cell walls.

Further reading

Aspinall G O (ed.) 1982–85 *The Polysaccharides*, Vols 1–3, New York, Academic Press.

Binkley R W 1988 *Modern Carbohydrate Chemistry*, New York, Marcel Dekker.

Dey P M and Dixon R A (eds) 1985 *Biochemistry of Storage Carbohydrates in Green Plants*, London, Academic Press.

Duffus C M and Duffus J H 1984 *Carbohydrate Metabolism in Plants*, London, Longman.

Stumpf P K, Conn E E and Preiss J (eds) 1988 *The Biochemistry of Plants, Vol. 14, Carbohydrates*, New York, Academic Press.

Tipson R S and Horton D (eds) *Advances in Carbohydrate Chemistry and Biochemistry*, (annual volumes since 1945), New York, Academic Press.

3 Lipids

3.1   Classification of lipids

3.2   Fats

3.3   Glycolipids

3.4   Phospholipids

3.5   Waxes

3.6   Steroids

3.7   Terpenes

3.1   Classification of lipids

The lipids are a group of substances found in plant and animal tissues. They are insoluble in water but soluble in common organic solvents such as benzene, ether and chloroform. They act as electron carriers, as substrate carriers in enzymic reactions, as components of biological membranes and as sources and stores of energy. In the proximate analysis of foods they are included in the ether extract fraction. They may be classified as shown in Table 3.1.

Plant lipids are of two main types: structural and storage. The structural lipids are present as constituents of various membranes and protective surface layers, and make up about 7 per cent of the leaves of higher plants. The surface lipids are mainly waxes, with relatively minor contributions from long chain hydrocarbons, fatty acids and cutin. The membrane lipids, present in mitochondria, the endoplasmic reticulum and the plasma membranes, are mainly glycolipids (40–50 per cent) and phosphoglycerides. Plant storage lipids occur in fruits and seeds and are, predominantly, triacylglycerols. Over 300 different fatty acids have been isolated from plant tissues but only about seven are of common occurrence. The most abundant is α-linolenic acid; the most common saturated acid is palmitic and the most common monounsaturated acid is oleic.

In animals, lipids are the major form of energy storage, mainly as fat, which may constitute up to 97 per cent of the adipose tissue of obese animals. The yield of energy from the complete oxidation of fat is about 39 MJ/kg DM compared with about 17 MJ/kg DM from glycogen, the major carbohydrate form of stored energy. In addition, stored fat is almost anhydrous, whereas stored glycogen is highly hydrated. Weight for weight,

Table 3.1 Classification of the lipids

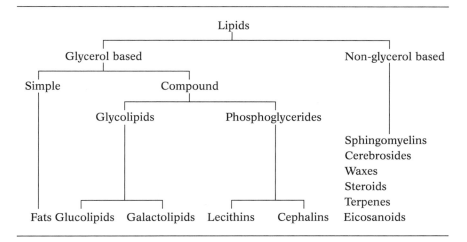

fat is, therefore, about six times as effective as glycogen as a stored energy source.

The structural lipids of animal tissues, mainly phosphoglycerides, constitute between 0.5 and 1 per cent of muscle and adipose tissue; the concentration in the liver is usually between 2 and 3 per cent. The most important non-glyceride, neutral lipid fraction of animal tissue is made up of cholesterol and its esters, which together make up 0.06 to 0.09 per cent of muscle and adipose tissue.

3.2 Fats

Fats and oils are constituents of both plants and animals, and are important sources of stored energy. Both have the same general structure but have different physical and chemical properties. The melting points of the oils are such that at ordinary room temperatures they are liquid and they tend to be more chemically reactive than the more solid fats. The term 'fat' is frequently used in a general sense to include both groups. As well as its major function of supplying energy, stored fat is important as a thermal insulator and, in some warm-blooded animals, as a source of heat for maintaining body temperature. This is especially important in animals that are born hairless, those that hibernate and those that are cold-adapted. Such animals have special deposits of 'brown fat' in which oxidation is uncoupled from ATP production (Chapter 9) and all the energy is liberated as heat. Palmitate oxidised to produce ATP would yield about 13 MJ/kg as heat, compared with the uncoupled yield of 39 MJ/kg. In these tissues, the mitochondria are liberally supplied with respiratory electron carriers, particularly cytochromes, which accounts for their brown colour.

Structure of fats

Fats are esters of fatty acids with the trihydric alcohol glycerol, and are also referred to as glycerides or acylglycerols. When all three alcohol groups are esterified by fatty acids, the compound is a triacylglycerol (triglyceride):

$$^{sn\text{-}1}CH_2OH$$
$$|$$
$$^{sn\text{-}2}CH\,OH \quad + \quad 3R.COOH \quad \longrightarrow$$
$$|$$
$$^{sn\text{-}3}CH_2OH$$

$$^{sn\text{-}1}CH_2.O.CO.R$$
$$|$$
$$^{sn\text{-}2}CH.O.CO.R \quad + \quad 3H_2O$$
$$|$$
$$^{sn\text{-}3}CH_2.O.CO.R$$

| Gycerol | Fatty acid | Triacylglycerol |

It is important to appreciate that, in stereochemical terms, the positions occupied by the acid chains are not identical. Under the stereospecific numbering system the positions are designated sn-1, sn-2 and sn-3 as shown. They are readily distinguished by enzymes and this may lead to preferential reactivity at one or more of the positions. Phosphorylation, for example, always takes place at carbon atom sn-3 rather than carbon atom sn-1. Although triacylglycerols are predominant, mono- and diacylglycerols do occur naturally but in much smaller amounts.

Triacylglycerols differ in type according to the nature and position of the fatty acid residues. Those with three residues of the same fatty acid are termed simple triacylglycerols as illustrated above. When more than one fatty acid is concerned in the esterification then a mixed triacylglycerol results:

$$CH_2.O.CO.R_1$$
$$|$$
$$CH.O.CO.R_2$$
$$|$$
$$CH_2.O.CO.R_3$$

Mixed triacylglycerol

R_1, R_2 and R_3 represent the chains of different fatty acids. Naturally occurring fats and oils are mixtures of such mixed triacylglycerols. Soyabean oil has been estimated to contain about 79 per cent of mixed triacylglycerols compared with about 21 per cent of the simple type. Comparable figures for linseed oil are 75 and 25 per cent, respectively. Triacylglycerols with residues of one fatty acid only do occur naturally; laurel oil, for example, contains about 31 per cent of the triacylglycerol of lauric acid.

Most of the naturally occurring fatty acids have an even number of carbon atoms, which is to be expected in view of their mode of formation (Chapter 9). The majority contain a single carboxyl group and an unbranched carbon, chain which may be saturated or unsaturated. The unsaturated acids contain either one (monoenoic), two (dienoic), three (trienoic) or many (polyenoic) double bonds. Fatty acids with more than one double bond are frequently referred to as polyunsaturated fatty acids (PUFA). The unsaturated acids possess different physical and chemical properties from the saturated; they have lower melting points and are more chemically reactive.

The presence of a double bond in a fatty acid molecule means that the acid can exist in two forms depending upon the spatial arrangement of the hydrogen atoms attached to the carbon atoms of the double bond. When the hydrogen atoms lie on the same side of the double bond, the acid is said to be in the *cis* form, whereas it is said to be in the *trans* form when they lie on opposite sides, as shown below:

Most naturally occurring fatty acids have the *cis* configuration.

The fatty acids are named by replacing the final -e of the name of the parent hydrocarbon by the suffix -oic. Thus a saturated 18-carbon acid would be named octadecanoic after the parent octadecane. An 18-carbon acid with one double bond would be octadecenoic after octadecene. The position of the double bond is indicated by reference to the carboxyl carbon atom (carbon atom 1). Thus 9,octadecenoic acid would have 18 carbon atoms and a double bond between carbon atoms 9 and 10. Similarily, 9,12,15-octadecatrienoic acid would have 18 carbon atoms and double bonds between carbon atoms 9 and 10, 12 and 13, and 15 and 16. The names may be abbreviated by stating the number of carbon atoms followed by a colon, followed by the number of double bonds (Δ), the positions of which are stated as a superscript. Thus octadecatrienoic acid would be designated $18:3^{\Delta 9,12,15}$. Alternatively it may be written 9,12,15-18:3. Carbon atoms 2 and 3 are designated alpha (α) and beta (β), respectively, and the methyl carbon at the distal end of the chain as the omega (ω) carbon atom. In nutritional work, the unsaturated acids are frequently named in relation to the terminal methyl as carbon atom 1. Under this system 9,12,15-octadecatrienoic acid would become ω-3,6,9-octadecatrienoic acid since carbon atoms 3, 6 and 9 correspond to carbon atoms 16, 13 and 10 under the former system. The abbreviated designation would be ω-3,6,9-18:3. It has become common practice to use n instead of ω and we then have n-3,6,9-18:3 and frequently 18:3(n-3). In addition the configuration of the double bonds is indicated by the use of the prefixes *cis* and *trans*. Thus, α-linolenic acid would be, all *cis*-9,12,15-octadecatrienoic, or more simply, all *cis* 9,12,15-18:3.

For certain purposes the PUFA are grouped into families, based on oleic (n-9–18:1), linoleic (n-6,9-18:2) and α-linolenic (n-3,6,9-18:3) as precursors. The families are called omega-9 (ω-9), omega-6 (ω-6) and omega-3 (ω-3), referring to the positions of the double bonds nearest to the omega carbon atom in these acids. Again, n is frequently substituted for ω. Some of the nutritionally important fatty acids are shown in Table 3.2.

Table 3.2 Common fatty acids of natural fats and oils

Acid	Formula	Melting point (°C)
Saturated		
Caprylic (octanoic)	$C_7H_{15}\cdot COOH$	16.3
Capric (decanoic)	$C_9H_{19}\cdot COOH$	31.2
Lauric (dodecanoic)	$C_{11}H_{23}\cdot COOH$	43.9
Myristic (tetradecanoic)	$C_{13}H_{27}\cdot COOH$	54.1
Palmitic (hexadecanoic)	$C_{15}H_{31}\cdot COOH$	62.7
Stearic (octadecanoic)	$C_{17}H_{35}\cdot COOH$	69.6
Unsaturated		
Palmitoleic (9-hexadecenoic) (9-16:1 or 16:3n-7)	$C_{15}H_{29}\cdot COOH$	0
Oleic (octadecenoic) (9-18:1 or 18:1n-9)	$C_{17}H_{33}\cdot COOH$	13
Linoleic (octadecadienoic) (9,12-18:2 or 18:2n-6)	$C_{17}H_{31}\cdot COOH$	−5
α-Linolenic (9,12,15-octadecatrienoic) (9,12,15-18:3 or 18:3n-3)	$C_{17}H_{29}\cdot COOH$	−14.5
Arachidonic (eicosatetraenoic) (5,8,11,14-20:4 or 20:4n-6)	$C_{19}H_{31}\cdot COOH$	−49.5
Timnodonic (eicosapentaenoic) (5,8,11,14,17-20:5 or 22:5n-3)	$C_{19}H_{29}\cdot COOH$	
Docosahexaenoic (5,8,11,14,17,20-22:5 or 22:5n-3)	$C_{21}H_{36}\cdot COOH$	

Two low molecular weight, saturated fatty acids, namely butyric ($C_3H_7.COOH$) and caproic ($C_4H_{10}.COOH$), are found in significant amounts in the milk fats of ruminants, and caproic along with caprylic acid is present in a few oils such as palm kernel and coconut. Other fatty acids containing two carboxyl groups, odd numbers of carbon atoms and branched chains have been isolated from natural fats but they are not considered to be of great importance.

Triacylglycerols are named according to the fatty acids they contain, e.g:

$$CH_2.O.CO.C_{17}H_{33}$$
$$|$$
$$CH.O.CO.C_{17}H_{33}$$
$$|$$
$$CH_2.O.CO.C_{17}H_{33}$$

$$CH_2.O.CO.C_{15}H_{31}$$
$$|$$
$$CH.O.CO.C_{17}H_{33}$$
$$|$$
$$CH_2.O.CO.C_{17}H_{35}$$

Trioleoylglycerol (Triolein)   1-palmitoyl 2-oleoyl 3-stearoylglycerol (Palmito-oleostearin)

The fatty acid residues are not randomly distributed between the alcohol groups of the parent glycerol. Thus, in cows' milk fat, for example, the short chain acids are concentrated at position 3. In human milk fat, the unsaturated acids are predominantly at position 1 and the saturated at position 2.

Animal depot fats tend to have saturated acids at position 1, and unsaturated and short chain acids at position 2; PUFA tend to accumulate at position 3.

There is evidence that the configuration of the constituent triacylglycerols of fats can influence the extent to which they are digested. Thus palmitate (hexadecanoate) randomly distributed throughout the 1, 2 and 3 positions was found to be less digestible than that which occupied position 2, the favoured position for attack by pancreatic lipase.

The fatty acid composition of the triacylglycerols determines their physical nature. Those with a high proportion of low molecular weight (short chain) and unsaturated acids have low melting points. Thus tristearin is solid at body temperature whereas triolein is liquid.

Composition of fats

It is frequently important in nutritional investigations to assess the quality of the fat being produced under a certain treatment. When the effect of the diet is considerable, the results may be obvious in a softening or hardening of the fat. Less obvious changes may occur and for these a more objective assessment is necessary. Differences between fats are a function of their fatty acid composition since glycerol is common to all fats. The logical method of following changes in fats is, therefore, to measure their fatty acid constitution. Analysis of fats for individual fatty acids has presented great problems in the past but the introduction of techniques such as gas chromatography has allowed determinations to be made more easily and accurately. As well as providing a most valuable research tool, gas chromatographic analysis has given detailed quantitative information on the fatty acid constitution of many different fats. This provides a more certain identification and characterisation of fats, and a more accurate method of detecting and quantitatively estimating the adulteration of a given fat or oil than was previously available. As well as its major role as an energy source, fat has a vital role in providing individual fatty acids which have specific nutritional roles within the animal body. Information on fatty acid composition is, therefore, a prerequisite in the evaluation of fats in this context.

Some typical values for a number of important fats and oils are given in Table 3.3.

In general, plant and marine oils, especially those of fish, are more highly unsaturated than those of mammalian origin. This is because of the presence of varying amounts of linoleic and linolenic acids in addition to the monounsaturated oleic (cis-9-octadecenoic), which is quantitatively the major fatty acid in most natural fats. In addition, the fish oils have significant concentrations of highly unsaturated C_{20} and C_{22} acids. In mammalian depot fat, the proportion of the more unsaturated acids is lower and there is a higher proportion of high molecular weight saturated acids such as palmitic and stearic, with smaller but significant contributions from lauric (dodecanoic) and myristic (tetradecanoic) acids. For this reason, fats like lard, and beef and mutton tallow are firm and hard, whereas fish and plant oils are softer and frequently are oils in the true sense.

Table 3.3 Fatty acid composition (g/100 g) of some common fats and oils

	Rape-seed	Soya bean	Rye-grass	Cocks-foot	Butter fat	Lard	Beef tallow	Men-haden	Cod liver	Her-ring
4:0	–	–	–	–	3	–	–	–	–	–
6:0	–	–	–	–	2	–	–	–	–	–
8:0	–	–	–	–	1	–	–	–	–	–
10:0	–	–	–	–	3	–	–	–	–	–
12:0	–	–	–	–	4	–	–	–	–	–
14:0	Tr	Tr	Tr	Tr	12	Tr	3	8	1	9
16:0	4	10	12	11	31	32	26	22	19	15
18:0	1	4	2	3	10	8	19	3	5	1
20:0	1	Tr	–	–	–	–	–	–	–	–
22:0	Tr	Tr	–	–	–	–	–	–	4	–
16:1	2	Tr	2	2	2		6	11	4	7
18:1n-9	54	25	15	–	23	48	40	21	15	16
20:1n-9	–	–	–	–	–	–	–	2	10	16
22:1n-9	–	–	–	–	–	–	–	2	2	23
18:2n-6	23	52	68	79	2	11	5	2	2	1
18:3n-3	10	7	–	–	Tr	Tr	–	–	–	–
20:4n-6	–	–	–	–	–	–	–	2	1	Tr
20:5n-3	–	–	–	–	–	–	–	14	6	3
22:6n-3	–	–	–	–	–	–	–	10	27	3
Others	5	2	7	0	1	1	4	1	2	6

Within individual animals, subcutaneous fats contain a higher proportion of unsaturated acids and are thus softer than deep-body fat. The physical nature of fat varies between animals; marine mammals having softer body fat than land mammals. The reason in both cases is that animal fat has to maintain a degree of malleability at the temperature of the tissue, which is influenced by ambient temperatures. Thus the fats of the feet and ears, which are inclined to be colder than the interior of the body, tend to be unsaturated.

Ruminant milk fats are characterised by their high content of low molecular weight fatty acids, these sometimes forming as much as 20 per cent of the total acids present. As a result they are softer than the depot fats of the respective animals but not as soft as fats of vegetable and marine origin, being semi-solid at ordinary temperatures. Milk fats of non-ruminants resemble the depot fat of the particular animal.

In most commercially important edible plant oils, the dominant fatty acids are oleic, linoleic and linolenic. Coconut oil is an exception in having the saturated 12:0 lauric acid as its major acid. Families of plants tend to produce characteristic oils which frequently contain unusual fatty acids. Examples are the erucic acid of rapeseed; ricinoleic, the 18-carbon, monoenoic, hydroxy acid of the castor bean; and vernolic, the 18-carbon, trienoic, epoxy acid of the Compositae.

Essential fatty acids

In 1930, linoleic (*cis, cis*-9,12-octadecadienoic) acid was shown to be effective in preventing the development of certain conditions in rats given diets almost devoid of fat. These animals showed a scaly appearance of the skin and suboptimal performance in growth, reproduction and lactation; eventually they died. More recent work has demonstrated a wide range of symptoms in a variety of animals, including some in human beings under certain circumstances (Table 3.4).

Arachidonic (all *cis*5,8,11,14-eicosatetraenoic acid) acid has been shown to have equivalent or even greater activity than linoleic, and linolenic (all *cis*9,12,15-octadecadienoic) is about 1.5 times as effective as linoleic. Mammals cannot synthesise fatty acids with double bonds closer than carbon atom 9 from the terminal methyl group. Such acids have to be supplied in the diet. Linoleic acid (18:2 n-6) and linolenic acid (18:3 n-3) are thus dietary essentials. Arachidonic acid is synthesised in the body from linoleic acid. However, one of the steps in the synthesis, a Δ-6 desaturation, is rate-limiting and production may be slow and an exogenous supply advantageous. Linoleic and α-linolenic acids are referred to as the essential fatty acids (EFA). Like other polyunsaturated acids they form part of various membranes and play a part in lipid transport and certain lipoprotein enzymes. In addition, they are the source materials for the synthesis of the eicosanoids. These include the prostaglandins, thromboxanes and leukotrienes, hormone-like substances that regulate many functions including blood clotting, blood pressure, smooth muscle contraction and the immune response. They are, also, the source of other important C_{20} acids in the form of eicosapentaenoic (EPA), hydroxy-eicosatrienoic (HETrR) and docosahexaenoic (DHA) acids. All are involved in maintaining the fluidity of mammalian cell membranes. EPA is the precursor of the 3-series of prostaglandins and thromboxanes, and the 5-series of leukotrienes. DHA is thought to play an important role in brain and retinal function, and EPA and HETrR have a modulating effect on the production of eicosanoids from arachidonic acid.

Table 3.4 Symptoms associated with essential fatty acid deficiencies

1. Growth retardation
2. Increased permeability to water and increased water consumption
3. Increased susceptibility to bacterial infections
4. Sterility
5. Less stable biomembranes
6. Capillary fragility
7. Kidney damage, haematuria and hypertension
8. Decreased visual acuity
9. Decreased myocardial contractility
10. Decreased ATP synthesis in liver and heart
11. Decreased nitrogen retention

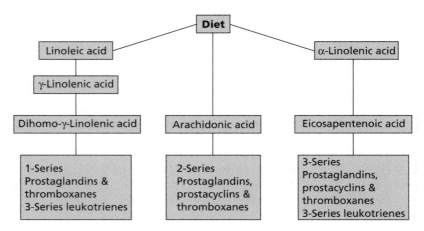

Fig. 3.1 Relationship between the essential fatty acids and the eicosanoids.

The relationship between the essential fatty acids and the eicosanoids is illustrated in Fig. 3.1.

The 1- and 3-series prostaglandins are anti-inflammatory and inhibit platelet aggregation, whereas the 2-series are pro-inflammatory and pro-aggregatory. The 1- and 3-series thromboxanes mildly stimulate platelet aggregation and stimulate the contraction of respiratory, intestinal and vascular smooth muscle, as do the leukotrienes. The 2-series thromboxanes have a much more powerful action in this respect.

As a general rule, mammals are considered to have an EFA requirement of 3 per cent of the energy requirement (3en%) as linoleic acid, although estimates have ranged from 1 to 15 per cent. Estimates for individual species have been more specific. Thus, *The Nutrient Requirements of Pigs* (see Further Reading, Chapter 12) gives the requirements of pigs under 30-kg liveweight as 3en% as linoleic acid or 2en% as arachidonic acid. For pigs of 30 to 90 kg, the figures are 1.5en% as linoleic and 1en% as arachidonic.

The oilseeds are generally rich sources of linoleic acid, and linseed is a particularly good source of α-linolenic acid. Pigs and poultry, which normally have considerable quantities of oilseed residues in their diets, will, therefore, receive an adequate supply of the essential fatty acids.

Ruminant animals are largely dependent on grasses and forages for their nutritional needs and are thereby supplied with liberal quantities of linoleic and α-linolenic acids. Although considerable hydrogenation of unsaturated acids takes place in the rumen, with consequent overall reduction of EFA supply, the possibility of ruminants suffering from a deficiency is remote. A certain proportion of dietary EFA escapes hydrogenation and this, allied to very efficient utilisation and conservation of EFA by ruminants, is enough to ensure adequacy under normal conditions. Frank EFA deficiency is rare in human beings although, under certain conditions, it does occur in infants and the elderly.

Properties of fats

Hydrolysis

Fats may be hydrolysed by boiling with alkalis to give glycerol and soaps:

$$
\begin{array}{c}
\text{CH}_2.\text{O}.\text{CO}.\text{R} \\
| \\
\text{CH}.\text{O}.\text{CO}.\text{R} \quad + \quad 3\text{KOH} \quad \longrightarrow \\
| \\
\text{CH}_2.\text{O}.\text{CO}.\text{R} \\
\text{Fat}
\end{array}
\qquad
\begin{array}{c}
\text{CH}_2\text{OH} \\
| \\
\text{CHOH} \quad + \quad 3\text{R}.\text{COOK} \\
| \\
\text{CH}_2\text{OH} \\
\text{Glycerol} \qquad \text{Soap}
\end{array}
$$

Such a hydrolysis is termed saponification since it produces soaps, which are sodium and potassium salts of the fatty acids. The process may take place naturally under the influence of enzymes, collectively known as lipases, when it is termed lipolysis. The enzymes may have a certain specificity and preferentially catalyse hydrolysis at particular positions in the molecule. Removal of the fatty acid residue attached to carbon atom 2 of an acylglycerol is more difficult than those at positions 1 and 3. Under natural conditions the products of lipolysis are usually mixtures of mono- and diacylglycerols with free fatty acids. Most of these acids are odourless and tasteless but some of the lower ones, particularly butyric and caproic, have extremely powerful tastes and smells; when such a breakdown takes place in an edible fat it may frequently be rendered completely unacceptable to the consumer. The lipases are mostly derived from bacteria and moulds, which are chiefly responsible for this type of spoilage, commonly referred to as rancidity. Extensive lipolysis of dietary fats takes place in the duodenum and in their absorption from the small intestine. Lipolysis also precedes the hydrogenation of fats in the rumen, and the oxidation of fats in the body.

Oxidation

The unsaturated fatty acids readily undergo oxidation at the carbon atom adjacent to the double bond to form hydroperoxides:

$$
\begin{array}{c}
—\text{CH}_2.\text{CH}:\text{CH}.\text{CH}_2.\text{CH}_2. \\
| \longleftarrow \text{O}_2 \\
\downarrow \\
—\text{CH}_2.\text{CH}:\text{CH}.\text{CH}_2.\text{CH}_2. \\
| \\
\text{OOH}
\end{array}
$$

These break down to give shorter chain products, including free radicals, which then attack other fatty acids much more readily than does the original oxygen. More free radicals are produced with the result that the speed of the oxidation increases exponentially. Eventually the concentration of free radicals becomes such that they react with each other and the reaction is terminated. Such a reaction, in which the products catalyse the reaction, is described as autocatalytic. This particular reaction is an autoxidation. The formation of the free radicals is catalysed by ultraviolet light and certain metal ions, particularly copper, and the presence of either increases the rate of oxidation dramatically.

The products of oxidation include shorter chain fatty acids, fatty acid polymers, aldehydes (alkanals), ketones (alkanones), epoxides and hydrocarbons. The acids and alkanals are major contributors to the smells and flavours associated with oxidised fat, and significantly reduce its palatability. The potency of these compounds is typified by deca-2,4 dienal, which is detectable in water at concentrations of as little as 1 in 10 000 million.

Oxidation of saturated fatty acids results in the development of a sweet, heavy taste and smell commonly known as ketonic rancidity. This is due to the presence of the methyl ketones resulting from the oxidation, which may be represented as follows:

$$
\begin{array}{ccccc}
CH_3 & & CH_3 & & CH3 \\
| & & | & & | \\
CH_2 & & CH_2 & & CH_2 \\
| & & | & & | \\
CH_2 + O \longrightarrow & & CH_2 \longrightarrow & & CH_2 + CO_2 \\
| & & | & & | \\
CH_2 & & C{:}O & & C{:}O \\
| & & | & & | \\
CH_2 & & CH_2 & & CH3 \\
| & & | & & \\
COOH & & COOH & &
\end{array}
$$

Caproic acid   Pentanone

Similar reactions following mould-induced lipolysis are responsible for the characteristic flavours of various soft and blue cheeses.

Antioxidants

Natural fats possess a certain degree of resistance to oxidation, owing to the presence of compounds termed antioxidants. These prevent the oxidation of unsaturated fats until they themselves have been transformed into inert products. A number of compounds have this antioxidant property, including phenols, quinones, tocopherols, gallic acid and gallates. In the United Kingdom, propyl, octyl or dodecyl-gallate, butylated hydroxyanisole, butylated hydroxytoluene and ethoxyquinone may be added to edible oils in amounts specified in *The Feedingstuffs Regulations* 2000. Another eight substances such as synthetic α-, γ- and δ-tocopherols and various derivatives of ascorbic acid may be used without limit.

The most important naturally occurring antioxidant is vitamin E, which protects fat by preferential acceptance of free radicals. The possible effects of fat oxidation in diets in which vitamin E levels are marginal are of considerable importance.

Hydrogenation

This is the process whereby hydrogen is added to the double bonds of the unsaturated acids of a fat, thereby converting them to their saturated analogues. Oleic acid, for example, yields stearic acid as shown:

$$
\begin{array}{c}
CH_2 \\
| \\
(CH_2)_7 \\
| \\
CH \\
\| \\
CH \\
| \\
(CH_2)_7 \\
| \\
COOH
\end{array}
\quad + \quad
\begin{array}{c}
H \\
| \\
H
\end{array}
\quad \longrightarrow \quad
\begin{array}{c}
CH_3 \\
| \\
(CH_2)_{16} \\
| \\
COOH
\end{array}
$$

Oleic acid   Stearic acid

The process (hardening) is important commercially for producing firm hard fats from vegetable and fish oils in the manufacture of margarine. The hardening results from the higher melting points of the saturated acids. For the rate of reaction to be practicable a catalyst has to be used, usually finely divided nickel. Hardening has the added advantage of improving the keeping quality of the fat since removal of the double bonds eliminates the chief centres of reactivity in the material.

Dietary fats consumed by ruminants first undergo hydrolysis in the rumen and this is followed by progressive hydrogenation of the unsaturated free fatty acids (mainly 18:2 and 18:3 acids) to stearic acid. This helps to explain the apparent anomaly that, whereas their dietary fats are highly unsaturated, the body fats of ruminants are highly saturated.

Hydrogenation results in the production not only of saturated acids but also of *trans* acids. In addition a redistribution of double bonds within the fatty acid chain takes place, accounting for the presence in ruminant fats of vaccenic (*trans*-11,18:1) and elaidic (*trans*-9,18:1) acids. A similar transformation occurs in the industrial hydrogenation of plant and fish oils. Partially hydrogenated vegetable oils, for example, commonly contain from 3 to 5 g *trans* acids/100 g of the total fatty acids, and partially hydrogenated fish oils about 20 g.

Digestion, absorption and metabolism of the *trans* acids is comparable with that of their counterparts. They have higher melting points than their *cis* analogues and their incorporation into ruminant body fats contributes to the hardness of the latter. *Trans* acids do not possess essential fatty acid activity but there is evidence that some may enter pathways leading to eicosanoid formation and give rise to substances of unknown physiological effects. There is evidence, too, that they decrease the activity of the desaturases involved in EFA metabolism. However, it would appear that, as long as EFA intake is adequate and *trans* acids intake is not excessive, they do not have any significant effect on EFA status. *Trans* acids may be incorporated into biological membranes without detrimental effect.

3.3 Glycolipids

In these compounds two of the alcohol groups of the glycerol are esterified by fatty acids and the other is linked to a sugar residue. The lipids of grasses and clovers, which form the major part of the dietary fat of ruminants, are predominantly (about 60 per cent) galactolipids. Here the sugar is galactose and we have:

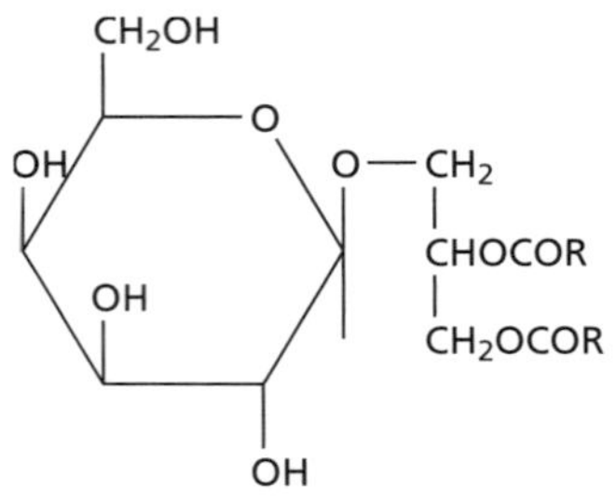

Galactolipid

Galactolipids

The galactolipids of grasses are mainly of the monogalactosyl type illustrated above, but smaller quantities of the digalactosyl compounds are also present. These have two galactose residues at the first carbon atom. The fatty acids of the galactosides of grasses and clovers consist largely of linoleic and α-linolenic acids as shown in Table 3.5.

Table 3.5 Fatty acid composition of some forage lipids (g/100 g)

	Perennial ryegrass	Cocksfoot	Red clover	White clover	Pasture
14:0	Tr	1.4	1.5	1.1	1.1
16:0	11.9	11.2	14.2	6.5	16
16:1	1.7	6.4	–	2.5	2.5
18:0	2.2	2.6	3.7	0.5	2
18:1	14.6	–	–	6.6	3.4
18:2	68.2	76.5	5.6	18.5	13.2
18:3	–	–	72.3	60.7	61.3

Rumen microorganisms are able to break down the galactolipids to give galactose, fatty acids and glycerol. Preliminary lipolysis appears to be a prerequisite for the galactosyl glycerides to be hydrolysed by the microbial galactosidases.

In animal tissues, glycolipids are present mainly in the brain and nerve fibres. The glycerol of the plant glycolipids is here replaced as the basic unit by the nitrogenous base sphingosine:

$$\underset{\text{Sphingosine}}{CH_3.(CH_2)_{12}.CH{:}CH.\overset{\overset{\displaystyle OH}{|}}{CH}.\overset{\overset{\displaystyle NH_2}{|}}{CH}.CH_2OH}$$

In their simplest form, the cerebrosides, the glycolipids have the amino group of the sphingosine linked to the carboxyl group of a long chain fatty acid and the terminal alcohol group to a sugar residue, usually galactose. The typical structure is:

Acid

$$CH_3.(CH_2)_{12}.CH{:}CH.\overset{\overset{\displaystyle OH}{|}}{CH}.\overset{\overset{\displaystyle NH{-}CO.R}{|}}{CH}.CH_2{-}O{-}\text{(Galactose ring)}{-}O$$

Sphingosine   Galactose

More complex substances, the gangliosides, are found in the brain. They have the terminal alcohol group linked to a branched chain of sugars with sialic acid as the terminal residue of at least one of the chains.

3.4 Phospholipids

The role of the phospholipids is primarily as constituents of the lipoprotein complexes of biological membranes. They are widely distributed, being particularly abundant in the heart, kidneys and nervous tissues. Myelin of the nerve axons, for example, contains up to 55 per cent of phospholipid. Eggs are one of the best animal sources and, among the plants, soya beans contain relatively large amounts. The phospholipids contain phosphorus in addition to carbon, hydrogen and oxygen.

Phosphoglycerides

These are esters of glycerol in which only two of the alcohol groups are esterified by fatty acids, with the third esterified by phosphoric acid. The parent compound of the phosphoglycerides is, thus, phosphatidic acid, which may be regarded as the simplest phosphoglyceride.

$CH_2.O.CO.R_1$
|
$CH.O.CO.R_2$
| O
$CH_2.O.\overset{..}{P}.OH$
     |
    OH

Phosphatidic acid

Phosphoglycerides are commonly referred to as phosphatides. In the major biologically important compounds, the phosphate group is esterified by one of several alcohols, the commonest of which are serine, choline, glycerol, inositol and ethanolamine. The chief fatty acids present are the 16-carbon saturated and the 18-carbon saturated and monoenoic, although others with 14–24 carbon atoms do occur. The most commonly occurring phosphoglycerides in higher plants and animals are the lecithins and the cephalins.

Lecithins

Lecithins have the phosphoric acid esterified by the nitrogenous base choline and are more correctly termed phosphatidylcholines. A typical example would have the formula:

$CH_2.O.CO.C_{15}H_{31}$
|
$CH.O.CO.C_{17}H_{33}$
|
$CH_2.O. PO_3.CH_2.CH_2.N^+(CH_3)_3$

Lecithin

The fatty acid residues at sn-1 are mostly palmitic (16:0) or stearic (18:0). At sn-2 they are primarily oleic (18:1), linoleic (18:2) or α-linolenic (18:3) acid.

Cephalins

Cephalins differ from the lecithins in having ethanolamine instead of choline and are correctly termed phosphatidylethanolamines. Ethanolamine has the following formula:

NH_2
|
$CH_2.CH_2OH$

The fatty acids at sn-1 are the same as in lecithin but those at sn-2 are unsaturated, mainly linoleic, eicosatetraenoic and docosahexaenoic.

Phosphoglycerides are white waxy solids which turn brown when exposed to the air owing to oxidation followed by polymerisation. When placed in water the phosphoglycerides appear to dissolve. However, the true solubility is very low, the apparent solubility being due to the formation of micelles.

They are hydrolysed by naturally occurring enzymes, the phospholipases, which specifically cleave certain bonds within the molecule to release fatty acids, the phosphate ester, the alcohol and glycerol. The release of choline, when followed by a further oxidative breakdown, has been considered to be responsible for the development of fishy taints by the release of the trimethylamine group or its oxide; currently these taints are considered to be the result of fat oxidation and not of lecithin breakdown.

The phosphoglycerides combine within the same molecule both the hydrophilic (water-loving) phosphate ester groups and the hydrophobic fatty acid chains. They are therefore surface-active and play a role as emulsifying agents in biological systems, for example in the duodenum. Their surface-active nature also explains their function as constituents of various biological membranes.

Sphingomyelins

Sphingomyelins belong to a large group, the sphingolipids, which have sphingosine instead of glycerol as the parent material. They differ from the cerebrosides in having the terminal hydroxyl group linked to phosphoric acid instead of a sugar residue. The phosphoric acid is esterified by either choline or ethanolamine. The sphingomyelins also have the amino group linked to the carboxyl group of a long chain fatty acid by means of a peptide linkage:

$$\text{Sphingosine} \qquad\qquad \text{Phosphoryl choline}$$

$$CH_3.(CH_2)_{12}.CH{:}CH.CHOH.CH.CH_2.O.PO_3^-.CH_2.CH_2.N^+(CH_3)_3$$
$$|$$
$$NH.CO.R$$

$$\text{Sphingomyelin}$$

Like the lecithins and cephalins, the sphingomyelins are surface-active and are important as components of membranes, particularly in nervous tissue. They may constitute up to 25 per cent of the total lipid in the myelin sheath which protects the nerve cells, but are absent from, or present in very low concentrations only in, energy-generating tissue.

Ether phospholipids

Ether phospholipids are glycerol based but have an alkyl rather than an acyl group at carbon atom 1, as is the case in the glycerides. Typical are the plasmologens, which have a vinyl ether grouping as shown:

$$CH_2.O.CH:CH.R$$
$$|$$
$$CH.O.CO.R$$
$$|$$
$$CH_2.O.PO_3^-.CH_2.CH_2.NH_3^+$$

Such compounds may form up to 50 per cent of the phospholipids of heart tissue but their function is unclear. An ether phospholipid called platelet activating factor is a highly potent aggregator of blood platelets.

3.5 Waxes

Waxes are simple, relatively non-polar lipids consisting of a long chain fatty acid combined with a monohydric alcohol of high molecular weight. They are usually solid at ordinary temperatures. The fatty acids present in waxes are those found in fats, although acids lower than lauric are very rare; higher acids such as carnaubic ($C_{23}H_{47}.COOH$) and mellissic ($C_{30}H_{61}.COOH$) may also be present. The most common alcohols found in waxes are carnaubyl ($C_{24}H_{49}.OH$) and cetyl ($C_{16}H_{33}OH$).

Natural waxes are usually mixtures of a number of esters. Beeswax is known to consist of at least five esters, the main one being myricyl palmitate:

$$C_{15}H_{31}.COOH + C_{31}H_{63}OH \rightarrow C_{15}H_{31}.COOC_{31}H_{63} + H_2O$$

Waxes are widely distributed in plants and animals, where they often have a protective function. The hydrophobic nature of the wax coating reduces water losses caused by transpiration in plants, and provides wool and feathers with waterproofing in animals. Among better-known animal waxes are lanolin, obtained from wool, and spermaceti, a product of marine animals. In plants, waxes are usually included in the cuticular fraction, where they form a matrix in which cutin and suberin are embedded. The term 'wax' is used here in the collective sense and, although true waxes are always present, the major part is made up of a complex mixture of substances. Alkanes (from C_{21} to C_{37}) make up a large proportion of the whole with odd-chain compounds predominating. Branched chain hydrocarbons, aldehydes, free fatty acids (from C_{12} to C_{36}) and various ketols are commonly occurring though minor constituents. Free alcohols are usually of minor importance but may form up to half of some waxes.

Cutin is a mixture of polymers of C_{16} and C_{18} monomers, commonly 16-hydroxypalmitic- and 10,16-dihydroxypalmitic acids. Phenolic constituents such as para-coumaric and ferulic acids are usually present but in small amounts only. Suberin is found in the surfaces of the underground parts of plants and on healed wound surfaces. The major aliphatic constituents are ω-hydroxy acids, the corresponding dicarboxylic acids and very long chain acids and alcohols. There are, also, substantial amounts of phenolic substances, mainly p-coumaric acid, which form a phenolic core to which the acids are attached. Both cutin and suberin are highly resistant to break-down and are not of any significant nutritional value. The waxes, too, are resistant to breakdown and are poorly utilised by animals. Their presence in foods in large amounts leads to high ether extract figures and may result in the nutritive value being overestimated.

3.6 Steroids

The steroids include such biologically important compounds as the sterols, the bile acids, the adrenal hormones and the sex hormones. They have a common structural unit of a phenanthrene nucleus linked to a cyclopentane ring (Fig. 3.2).

The individual compounds differ in the number and positions of their double bonds and in the nature of the side chain at carbon atom 17.

Sterols

These have eight to ten carbon atoms in the side chain, an alcohol group at carbon atom 3, but no carbonyl or carboxyl groups. They may be classified into:

- the phytosterols of plant origin,
- the mycosterols of fungal origin,
- the zoosterols of animal origin.

The phytosterols and the mycosterols are not absorbed from the gut and are not found in animal tissues.

Cholesterol

Cholesterol is a zoosterol which is present in all animal cells. It has a low solubility in water, about 0.2 mg/100 ml. It is the major sterol in human

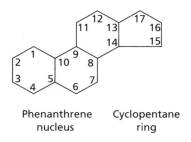

Phenanthrene   Cyclopentane
   nucleus   ring

Fig. 3.2 Basic steroid structural unit.

beings and is important as a constituent of various biological membranes. It is particularly important in the myelinated structures of the brain and central nervous system and may constitute up to 170 g/kg. It is the precursor of the steroid hormones. It is also the precursor of the bile acids.

Normal concentrations in the blood plasma range from 1200 to 2200 mg/litre. Some 30 per cent of this is in the free state, the remainder being bound to lipoproteins. These are complexes of proteins and lipids held together by non-covalent bonds. Each has a characteristic size, molecular weight, chemical composition and density. They are classified on the basis of their density. The five classes, of which one, the chylomicrons, occur only in the post-absorptive state, are shown in Table 3.6.

In the plasma, the lipoproteins exist as spherical structures with a core of triacyl glycerols and cholesterol esters. This is surrounded by a shell, about 20 Å thick, containing proteins, unesterified cholesterol and phosphatidyl cholines. Since they have a greater surface to volume ratio the smaller particles have a higher protein to lipid ratio and are more dense. Thus the HDLP fraction has about 45 per cent protein and 55 per cent lipid, whereas the VLDLP fraction has about 10 per cent protein and 90 per cent lipid. Cholesterol is very insoluble and prolonged high levels in blood result in its deposition on the walls of the blood vessels. These deposits eventually harden to atherosclerotic plaque. This narrows the blood vessel and serves as a site for clot formation and may precipitate myocardial infarction or heart attack.

There is strong evidence that the risk of coronary heart disease is directly related to the plasma concentration of LDL-cholesterol and inversely related to that of HDL-cholesterol, and that the risk is reduced significantly by lowering elevated serum cholesterol levels. It has been known for many years that one of the most important dietary factors regulating serum cholesterol levels is the ratio of polyunsaturated (PUFA) to saturated fatty acids (SFA). The SFA increase and the PUFA decrease cholesterol levels, except for the *trans* PUFA, which have a similar effect to the SFA. A ratio of 0.5–0.9 SFA : PUFA is considered to be satisfactory. It is important to

Table 3.6 Density-based classes of lipoproteins

Class	Density (g/ml)	Molecular weight (daltons)	Diameter (Å)
High-density lipoproteins (HDLP)	1.063–1.210	$4–2 \times 10^5$	50–130
Low-density lipoproteins (LDLP)	1.019–1.063	2×10^6	200–280
Intermediate-density lipoproteins (IDLP)	1.006–1.019	4.5×10^6	250
Very low-density lipoproteins (VLDLP)	0.95–1.006	$5 \times 10^6–10^7$	250–750
Chylomicrons	<0.95	$10^9–10^{10}$	$10^3–10^4$

appreciate that the different families of PUFA affect lipid metabolism in different ways. Thus the ω-6 acids significantly decrease serum cholesterol levels and have a minor effect only on triacyl glycerol levels, whereas the ω-3 acids have a minor effect on serum cholesterol but significantly lower triacyl glycerol levels. This is important in the light of recent evidence that high serum triacylglycerol level *per se* is an important risk factor in coronary heart disease. The ω-3 acids are the precursors of the 1-series of prostaglandins and 3-series of thromboxanes. The former strongly inhibit platelet aggregation and the latter are weakly pro-aggregating. The ω-6 acids are precursors of the series-2 prostaglandins and thromboxanes, the former being strongly pro-aggregating and the latter weakly anti-aggregating. On balance, from this point of view, the ω-3 acids may be regarded as having a more beneficial effect than the ω-6 acids. They have a further beneficial effect in that they inhibit the transformation of the ω-6 acids to their eicosanoid products.

7-Dehydrocholesterol

This substance, which is derived from cholesterol, is important as the precursor of vitamin D_3, which is produced when the sterol is exposed to ultraviolet light (Fig. 3.3).

This is a good illustration of how relatively small changes in chemical structure may bring about radical changes in physiological activity.

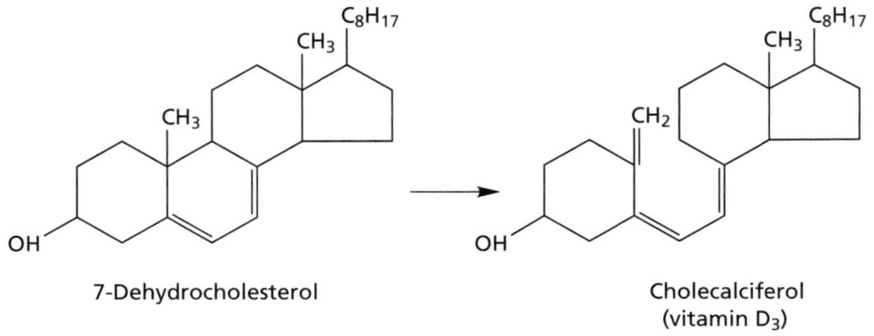

Fig. 3.3 Formation of vitamin D_3.

Ergosterol

This phytosterol is widely distributed in brown algae, bacteria and higher plants. It is important as the precursor of ergocalciferol or vitamin D_2, into which it is converted by ultraviolet irradiation. The change is the same as that which takes place in the formation of vitamin D_3 from 7-dehydrocholesterol and involves opening of the second phenanthrene ring.

Bile acids

The bile acids have a five-carbon side chain at carbon atom 17 which terminates in a carboxyl group bound by an amide linkage to glycine or taurine (Fig. 3.4).

The bile acids are synthesised from cholesterol and this constitutes the major end point of cholesterol metabolism. Under physiological conditions the acids exist as salts. They are produced in the liver, stored in the gall bladder and secreted into the upper small intestine. They are important in several ways:

- They provide the major excretory pathway for cholesterol, which cannot be catabolised to carbon dioxide and water by mammals. Bile contains high concentrations of free cholesterol, about 390 mg/100 ml.
- The bile salts assist, along with the detergent action of phospholipids, in preventing the cholesterol in the bile fluid from crystallising out of solution.
- They act as emulsifying agents in preparing dietary triacylglycerols for hydrolysis, by pancreatic lipase, in the process of digestion.
- They may have a role in activating pancreatic lipase.
- They facilitate the absorption, from the digestive tract, of the fat-soluble vitamins.

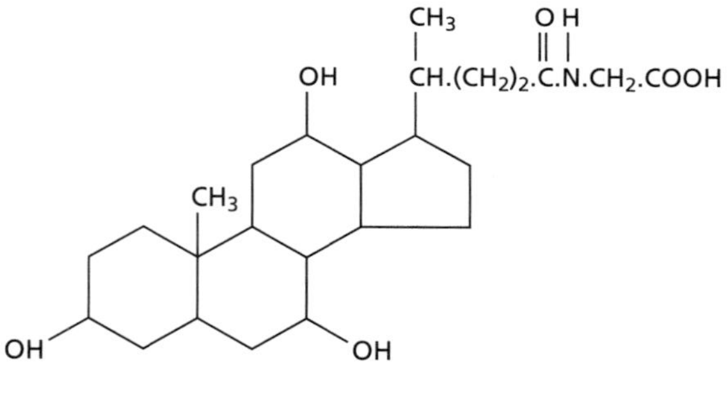

Fig. 3.4 Glycocholic acid.

Steroid hormones

These include the female sex hormones (oestrogens), the male sex hormones (androgens) and progesterone, as well as cortisol, aldosterone and corticosterone, which are produced in the adrenal cortex. The adrenal hormones have an important role in the control of glucose and fat metabolism.

3.7 Terpenes

Terpenes are made up of a number of isoprene units linked together to form chains or cyclic structures. Isoprene is a five-carbon compound with the following structure:

CH$_2$:C.CH:CH$_2$
|
CH$_3$

Isoprene

Many terpenes found in plants have strong, characteristic odours and flavours, and are components of essential oils such as lemon or camphor oil. The word 'essential' is used to indicate the occurrence of the oils in essences and not to imply that they are required by animals. Among the more important plant terpenes are the phytol moiety of chlorophyll, the carotenoid pigments, plant hormones such as giberellic acid and vitamins A, E and K. In animals, some of the coenzymes, including those of the coenzyme Q group, are terpenes.

Summary

1. Lipids are a group of substances, insoluble in water but soluble in common organic solvents. They include the fats and oils, the glycolipids, the phospholipids, the lipoproteins, the steroids and the terpenes.

2. Fats and oils are major sources of stored energy in both plants and animals. They are esters of fatty acids with glycerol. Their physical and chemical nature is determined by their fatty acid composition; high molecular weight saturated acids confer chemical stability and physical hardness, whereas unsaturated acids confer chemical reactivity and physical softness.

3. Linoleic and linolenic acids are termed the essential fatty acids, although only linoleic is truly a dietary essential. They are the source materials of the eicosanoids, which include the prostaglandins, the thromboxanes and the leukotrienes.

4. Phospholipids are phosphorus-containing compounds based on fatty acids esterified with glycerol or a nitrogeous base. They are important as constituents of the lipoprotein complexes of biological membranes.

5. Waxes are mixtures of esters of high molecular weight fatty acids with high molecular weight alcohols. They are chemically inert, have little or no nutritive value and are mainly protective in function.

6. The steroids have a basic structural unit of a phenanthrene nucleus linked to a cyclopentane ring. They include the sterols, the bile acids and the adrenal and sex hormones.

7. Cholesterol is the precursor of many sterols. It is present in all animal cells and is particularly important in the myelinated structures of the

brain and nervous tissue. There is strong evidence that the risk of coronary heart disease is directly related to the concentration of low-density lipoprotein cholesterol in the blood.

8. 7-Dehydrocholesterol and ergosterol are important as the precursors of vitamins D_3 and D_2, respectively.

9. The lipids in the form of triacylglycerols, cholesterol esters, low-density lipoproteins and the lipid-derived eicosanoids are intimately involved in the aetiology of heart disease.

Further reading

Garton G A 1969 Lipid metabolism of farm animals. In: Cuthbertson D P (ed.) *Nutrition of Animals of Agricultural Importance*, Oxford, Pergamon Press.

Harwood J L 1997 Plant lipid metabolism. In: Dey P M and Harborne J (eds) *Plant Biochemistry*, London, Academic Press.

Palmquist D L 1988 The feeding value of fats. In: Ørskov E R (ed.) *World Animal Science*, B 4 Amsterdam, Elsevier.

Devlin T M (ed.) 1997 *Textbook of Biochemistry with Clinical Correlations*, 4th edn, New York, John Wiley & Sons.

4 Proteins, nucleic acids and other nitrogenous compounds

4.1   Proteins

4.2   Amino acids

4.3   Peptides

4.4   Structure of proteins

4.5   Properties of proteins

4.6   Classification of proteins

4.7   Nucleic acids

4.8   Other nitrogenous compounds

4.9   Nitrates

4.10   Alkaloids

4.1   Proteins

Proteins are complex organic compounds of high molecular weight. In common with carbohydrates and fats they contain carbon, hydrogen and oxygen, but in addition they all contain nitrogen and generally sulphur.

Proteins are found in all living cells, where they are intimately connected with all phases of activity that constitute the life of the cell. Each species has its own specific proteins, and a single organism has many different proteins in its cells and tissues. It follows therefore that a large number of proteins occur in nature.

4.2   Amino acids

Amino acids are produced when proteins are hydrolysed by enzymes, acids or alkalis. Although over 200 amino acids have been isolated from biological materials, only 20 of these are commonly found as components of proteins.

Amino acids are characterised by having a basic nitrogenous group, generally an amino group ($-NH_2$), and an acidic carboxyl unit ($-COOH$).

Most amino acids occurring naturally in proteins are of the α- type, having the amino group attached to the carbon atom adjacent to the carboxyl group, and can be represented by the general formula:

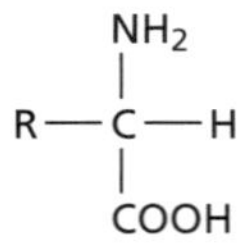

$$R - \underset{\underset{COOH}{|}}{\overset{\overset{NH_2}{|}}{C}} - H$$

The exception is proline, which has an imino (−NH) instead of an amino group. The nature of the 'R' group, which is referred to as the side chain, varies in different amino acids. It may simply be a hydrogen atom, as in glycine, or it may be a more complex radical containing, for example, a phenyl group.

The chemical structures of the 20 amino acids commonly found in natural proteins are shown in Table 4.1.

Special amino acids

Some proteins contain special amino acids which are derivatives of common amino acids. For example collagen, the fibrous protein of connective tissue, contains hydroxyproline and hydroxylysine, which are the hydroxylated derivatives of proline and lysine, respectively.

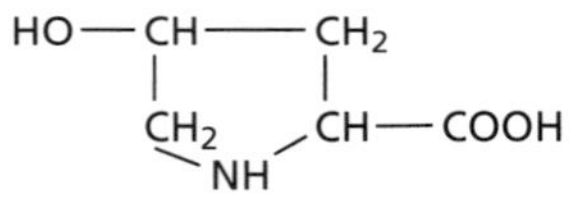

Hydroxyproline

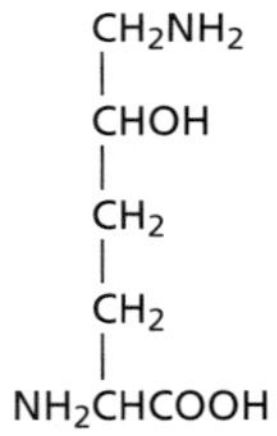

Hydroxylysine

Two iodine derivatives of tyrosine, triiodothyronine and tetraiodothyronine (thyroxine), act as important hormones in the body and are also amino acid components of the protein thyroglobulin (see p. 134).

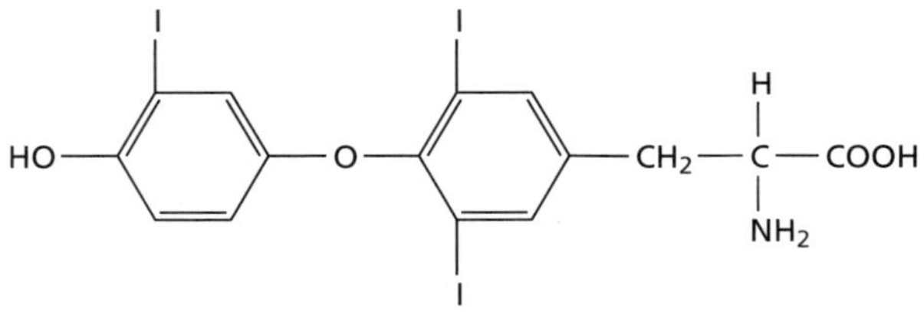

Triiodothyronine

Table 4.1 Amino acids commonly found in proteins

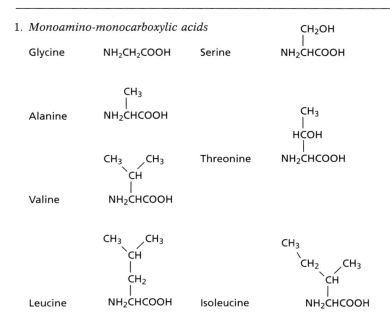

1. *Monoamino-monocarboxylic acids*

   Glycine NH_2CH_2COOH

   Serine

$$CH_2OH$$
$$|$$
$$NH_2CHCOOH$$

   Alanine

$$CH_3$$
$$|$$
$$NH_2CHCOOH$$

   Threonine

$$CH_3$$
$$|$$
$$HCOH$$
$$|$$
$$NH_2CHCOOH$$

   Valine

$$CH_3 \diagdown \diagup CH_3$$
$$CH$$
$$|$$
$$NH_2CHCOOH$$

   Leucine

$$CH_3 \diagdown \diagup CH_3$$
$$CH$$
$$|$$
$$CH_2$$
$$|$$
$$NH_2CHCOOH$$

   Isoleucine

$$CH_3$$
$$\diagdown CH_2 \diagup CH_3$$
$$CH$$
$$|$$
$$NH_2CHCOOH$$

2. *Sulpur-containing amino acids*

   Cysteine

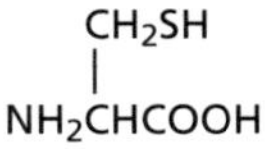

$$CH_2SH$$
$$|$$
$$NH_2CHCOOH$$

   Methionine

$$CH_3$$
$$|$$
$$S$$
$$|$$
$$CH_2$$
$$|$$
$$CH_2$$
$$|$$
$$NH_2CHCOOH$$

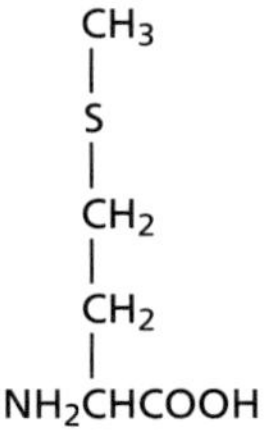

3. *Monoamino-dicarboxylic acids and their amine derivatives*

   Aspartic acid

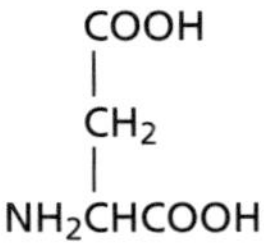

$$COOH$$
$$|$$
$$CH_2$$
$$|$$
$$NH_2CHCOOH$$

   Glutamic acid

$$COOH$$
$$|$$
$$CH_2$$
$$|$$
$$CH_2$$
$$|$$
$$NH_2CHCOOH$$

   Asparagine

$$CO-NH_2$$
$$|$$
$$CH_2$$
$$|$$
$$NH_2CHCOOH$$

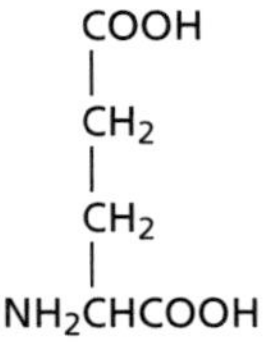

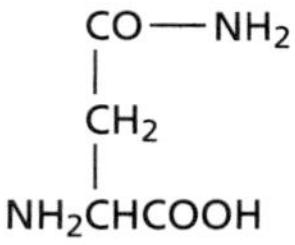

   Glutamine

$$CO-NH_2$$
$$|$$
$$CH_2$$
$$|$$
$$CH_2$$
$$|$$
$$NH_2CHCOOH$$

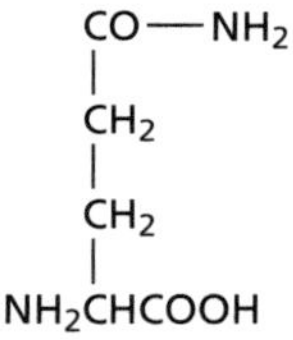

Table 4.1 (cont.)

4. *Basic amino acids*

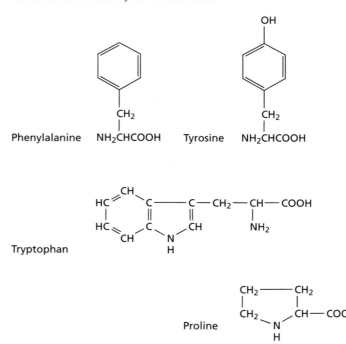

Lysine

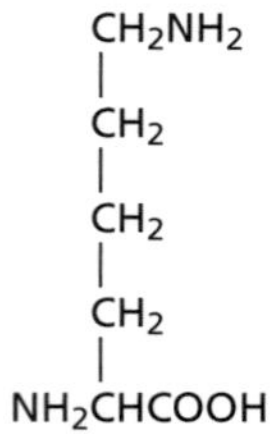

Histidine

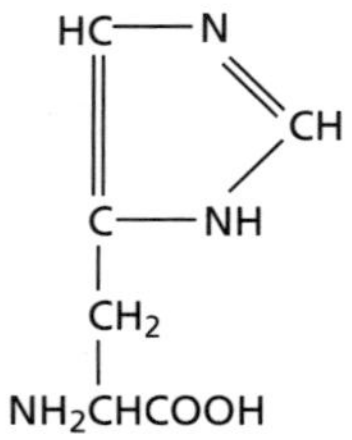

Arginine

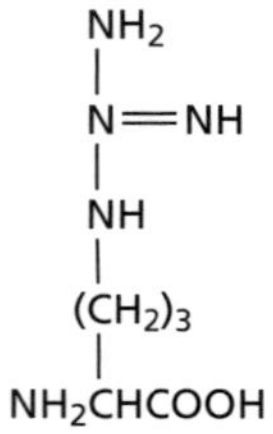

5. *Aromatic and heterocyclic amino acids*

Phenylalanine

Tyrosine

Tryptophan

Proline

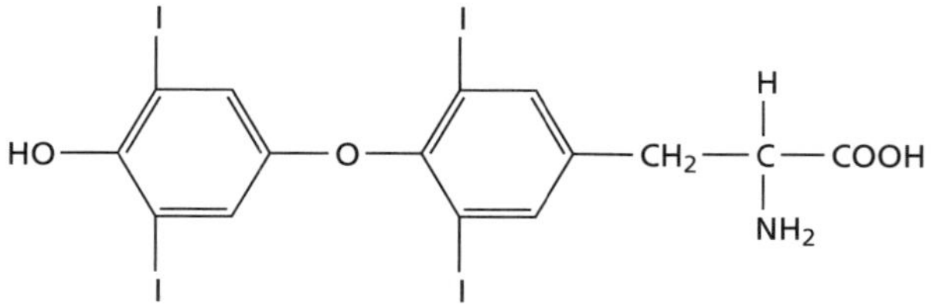

Tetraiodothyronine (thyroxine)

A derivative of glutamic acid, γ-carboxyglutamic acid, is an amino acid present in the protein thrombin. This amino acid is capable of binding calcium ions and plays an important role in blood clotting (see p. 89).

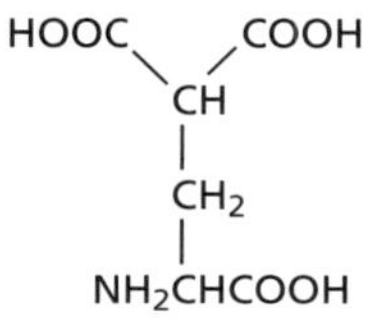

γ-Carboxyglutamic acid

The amino acid γ-aminobutyric acid functions in the body as a neuro-transmitter. It is also found in silage as a fermentation product of glutamic acid (see p. 518).

$$NH_2CH_2CH_2CH_2COOH$$

γ-Aminobutyric acid

The sulphur-containing amino acid cysteine also requires special mention. It may occur in protein in two forms, either as itself or as cystine, in which two cysteine molecules are joined together by a disulphide bridge:

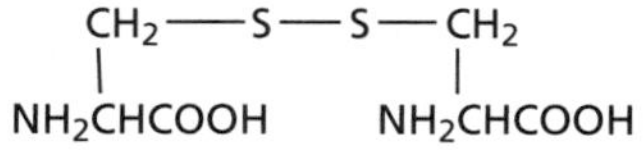

Cystine

Properties of amino acids

Because of the presence of an amino group and a carboxyl group, amino acids are amphoteric, i.e. have both basic and acidic properties. Molecules such as these, with basic and acidic groups, may exist as uncharged molecules, or as dipolar ions with opposite ionic charges, or as a mixture of these. Amino acids in aqueous solution exist as dipolar ions or 'zwitter ions' (from the German *Zwitter*, a hermaphrodite):

$$
\begin{array}{c}
NH_3^+ \\
| \\
R - C - H \\
| \\
COO^-
\end{array}
$$

In a strongly acid solution an amino acid exists largely as a cation, while in alkaline solution it occurs mainly as an anion. There is a pH value for a given amino acid at which it is electrically neutral; this value is known as the *isoelectric point*. Because of their amphoteric nature, amino acids act as buffers, resisting changes in pH.

All the α-amino acids except glycine are optically active. As with the carbohydrates (Chapter 2), amino acids can take two mirror image forms, D- and L-.

$$
\begin{array}{cc}
\begin{array}{c}
COOH \\
| \\
H - C - NH_2 \\
| \\
R
\end{array}
&
\begin{array}{c}
COOH \\
| \\
NH_2 - C - H \\
| \\
R
\end{array} \\
\text{D-Amino acid} & \text{L-Amino acid}
\end{array}
$$

All the amino acids involved in protein structure have an L- configuration of the carbon atom. If supplied in the D- form, some amino acids can be converted to the L- form by deamination of the amino acid to the keto acid and reamination to the L- form (Chapter 9).

Indispensable amino acids

Essential AA

for rat

Plants and many microorganisms are able to synthesise proteins from simple nitrogenous compounds such as nitrates. Animals cannot synthesise the amino group, and in order to build up body proteins they must have a dietary source of amino acids. Certain amino acids can be produced from others by a process known as transamination (see Chapter 9), but the carbon skeletons of a number of amino acids cannot be synthesised in the animal body and these are referred to as indispensable or essential amino acids.

Most of the early work in determining which amino acids could be classed as indispensable was carried out with rats fed on purified diets. The following ten indispensable amino acids are required for growth in the rat:

Arginine	Methionine
Histidine	Phenylalanine
Isoleucine	Threonine
Leucine	Tryptophan
Lysine	Valine

The chick requires a dietary supply of the ten amino acids listed above but in addition needs a dietary source of glycine. Birds require arginine because their metabolism does not include the urea cycle (Chapter 9), which would normally supply this amino acid. The list of indispensable amino acids required by the pig is similar to that for the rat, with the exception of arginine, which can be synthesised by the pig. It has been reported that rapidly growing animals may respond to arginine because the very active metabolism of the liver results in little of the amino acid being available to the general circulation. Cats require a dietary supply of arginine owing to their limited ability to synthesise ornithine from glutamate. Poultry have a limited capacity to synthesise proline. The actual dietary requirement of certain indispensable amino acids is dependent upon the presence of other amino acids. For example, the requirement for methionine is partially dependent on the cysteine content of the diet (see p. 441).

In the case of the ruminant, all the indispensable amino acids can be synthesised by the rumen microorganisms, which theoretically makes this class of animal independent of a dietary source once the rumen microorganisms have become established. However, maximum rates of growth or milk production cannot be achieved in the absence of a supply of dietary amino acids in a suitable form.

4.3 Peptides

Peptides are built up from amino acids by means of a linkage between the α-carboxyl of one amino acid and the α-amino group of another acid as shown below:

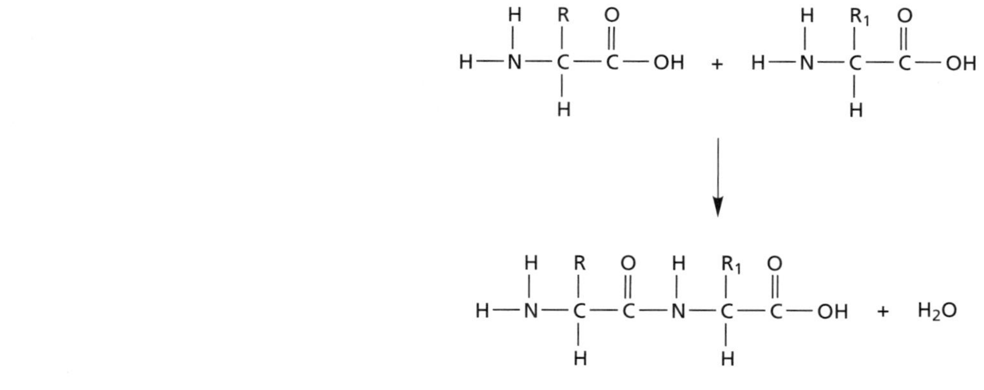

This type of linkage is known as the *peptide linkage*; in the example shown, a dipeptide has been produced from two amino acids. Large numbers of amino acids can be joined together by this means, with the elimination of one molecule of water at each linkage, to produce polypeptides.

Besides being important building blocks in the construction of proteins, some peptides possess their own biological activity. Milk, in particular, is a source of many biologically active peptides. The enzymatic hydroysis of the milk protein, casein, releases opioid peptides, which have pharmacological activities such as analgesia and sleep-inducing effects. Other peptides derived from casein are involved in calcium flow in tissues and modification of the immune system response, and other milk peptides stimulate growth of desirable bacteria and suppress harmful bacteria, and some act as growth factors for intestinal cells.

Other peptides are important in the control of food intake and these include bombesin, enterostatin, glucagon and leptin.

4.4   Structure of proteins

For convenience the structure of proteins can be considered under four basic headings.

Primary structure

The sequence of amino acids along the polypeptide chain of a protein, as described above, is called the 'primary structure' of the protein.

Secondary structure

The secondary structure of proteins refers to the conformation of the chain of amino acids resulting from the formation of hydrogen bonds between the imino and carbonyl groups of adjacent amino acids, as shown in Fig. 4.1.

The secondary structure may be regular, in which case the polypeptide chains exist in the form of an α-helix or a β-pleated sheet, or it may be irregular and exist, for example, as a random coil.

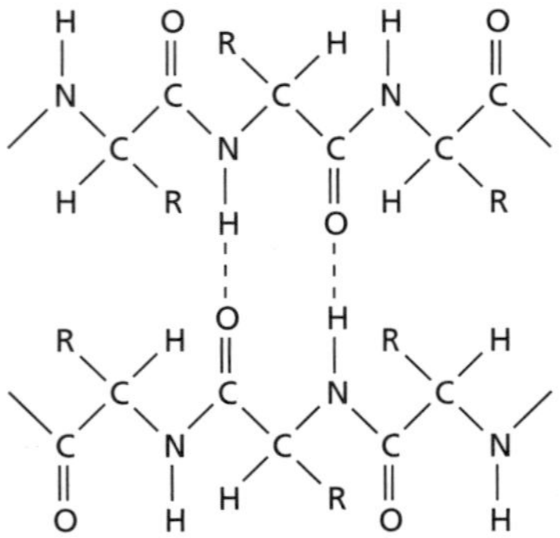

Fig. 4.1 Configuration of polypeptide chain. Dotted lines represent possible hydrogen bonds.

Tertiary structure

The tertiary structure describes how the chains of the secondary structure further interact through the R groups of the amino acid residues. This interaction causes folding and bending of the polypeptide chain, the specific manner of the folding giving each protein its characteristic biological activity.

Quaternary structure

Proteins possess quaternary structure if they contain more than one polypeptide chain. The forces that stabilise these aggregates are hydrogen bonds and electrostatic or salt bonds formed between residues on the surfaces of the polypeptide chains.

4.5   Properties of proteins

All proteins have colloidal properties; they differ in their solubility in water, ranging from insoluble keratin to albumins, which are highly soluble. Soluble proteins can be precipitated from solution by the addition of certain salts such as sodium chloride or ammonium sulphate. This is a physical effect and the properties of the proteins are not altered. On dilution the proteins can easily be redissolved.

Although the amino and carboxyl groups in the peptide linkage are nonfunctional in acid–base reactions, all proteins contain a number of free amino and carboxyl groups, either as terminal units or in the side chain of amino acid residues. Like amino acids, proteins are therefore amphoteric. They exhibit characteristic isoelectric points, and have buffering properties.

All proteins can be *denatured* or changed from their natural state. Denaturation has been defined by Neurath and co-workers as 'any nonproteolytic modification of the unique structure of a native protein, giving rise to definite changes in chemical, physical or biological properties'. Products of protein hydrolysis are not included under this term. Several agents can bring about denaturation of proteins; these include heat, acids, alkalis, alcohols, urea and salts of heavy metals. The effect of heat on proteins is of special interest in nutrition as this results in new linkages within and between peptide chains. Some of these new linkages resist hydrolysis by proteases produced in the digestive tract and impede their access to adjacent peptide bonds.

Susceptibility of proteins to heat damage is increased in the presence of various carbohydrates, owing to the occurrence of Maillard-type reactions, which initially involve a condensation between the carbonyl group of a reducing sugar with the free amino group of an amino acid or protein. Lysine is particularly susceptible. With increasing severity of heat treatment, further reactions, involving protein side chains, can occur and result in the browning of foods. The dark coloration of overheated hays and silages is symptomatic of these types of reactions.

4.6 Classification of proteins

Proteins may be classified into two main groups: simple proteins and conjugated proteins.

Simple proteins

These proteins produce only amino acids on hydrolysis. They are subdivided into two groups, fibrous and globular proteins, according to shape, solubility and chemical composition.

Fibrous proteins

These proteins, which in most cases have structural roles in animal cells and tissues, are insoluble and are very resistant to animal digestive enzymes. They are composed of elongated, filamentous chains joined together by cross-linkages. The group includes collagens, elastin and keratins.

Collagens are the main proteins of connective tissues and constitute about 30 per cent of the total proteins in the mammalian body. As mentioned earlier (p. 56), the amino acid hydroxyproline is an important component of collagen. Hydroxylation of proline to hydroxyproline involves vitamin C and if the vitamin is deficient, collagen fibres are weakened and may give rise to gum and skin lesions (see p. 105). The indispensable amino acid tryptophan is not found in these proteins.

Elastin is the protein found in elastic tissues such as tendons and arteries. The polypeptide chain of elastin is rich in alanine and glycine and is very flexible. It contains cross-links involving lysine side chains, which prevent the protein from extending excessively under tension and allow it to return to its normal length when tension is removed.

Keratins are classified into two types. The α-keratins are the main proteins of wool and hair. The β-keratins occur in feathers, skin, beaks and scales of most birds and reptiles. These proteins are very rich in the sulphur-containing amino acid cysteine; wool protein, for example, contains about 4 per cent of sulphur (see p. 377).

Globular proteins

Globular proteins are so called because their polypeptide chains are folded into compact structures. The group includes all the enzymes, antigens and those hormones which are proteins. Its first sub-group, *albumins,* are water-soluble and heat-coagulable and occur in milk, the blood, eggs and in many plants. *Histones* are basic proteins which occur in cell nuclei, where they are associated with deoxyribonucleic acid (see p. 67). They are soluble in salt solutions, are not heat-coagulable, and on hydrolysis yield large quantities of arginine and lysine. *Protamines* are basic proteins of relatively low molecular weight, which are associated with nucleic acids and are found in large quantities in the mature, male germ cells of verte-brates. Protamines are rich in arginine, but contain no tyrosine, tryptophan or sulphur-containing amino acids. *Globulins* occur in milk, eggs and blood, and are the main reserve protein in many seeds.

Conjugated proteins

Conjugated proteins contain, in addition to amino acids, a non-protein moi-ety termed a prosthetic group. Some important examples of conjugated pro-teins are glycoproteins, lipoproteins, phosphoproteins and chromoproteins.

Glycoproteins are proteins with one or more heteroglycans as prosthetic groups. In most glycoproteins the heteroglycans contain a hexosamine, either glucosamine or galactosamine or both; in addition galactose and mannose may also be present. Glycoproteins are components of mucous secretions, which act as lubricants in many parts of the body. The storage protein in egg white, ovalbumin, is a glycoprotein.

Lipoproteins, which are proteins conjugated with lipids such as triacyl-glycerols and cholesterol, are the main components of cell membranes and are also the form in which lipids are transported in the bloodstream to tis-sues, either for oxidation or for energy storage. They can be classified into five main categories in increasing order of density: chylomicrons, very low-density lipoproteins (VLDL), low-density lipoproteins (LDL), intermediate-density lipoproteins and high-density lipoproteins (HDL) (see Chapter 3).

Phosphoproteins, which contain phosphoric acid as the prosthetic group, include the caseins of milk (see p. 411) and phosvitin in egg yolk.

Chromoproteins contain a pigment as the prosthetic group. Examples are haemoglobin and cytochromes, in which the prosthetic group is the iron-containing compound haem, and flavoproteins, which contain flavins (see p. 93).

4.7 Nucleic acids

Nucleic acids are high molecular weight compounds which play a funda-mental role in living organisms as a store of genetic information, and they are the means by which this information is utilised in the synthesis of pro-teins. On hydrolysis, nucleic acids yield a mixture of basic nitrogenous com-pounds (purines and pyrimidines), a pentose (ribose or deoxyribose) and phosphoric acid.

The main pyrimidines found in nucleic acids are cytosine, thymine and uracil. The relationships between these compounds and the parent material, pyrimidine, are given below:

Pyrimidine

Cytosine   Thymine   Uracil

Adenine and guanine are the principal purine bases present in nucleic acids.

Purine   Adenine   Guanine

The compound formed by linking one of the above nitrogenous compounds to a pentose is termed a *nucleoside*. For example:

Adenine   D-Ribose   Adenosine

If nucleosides such as adenosine are esterified with phosphoric acid they form *nucleotides*, e.g. adenosine monophosphate (AMP):

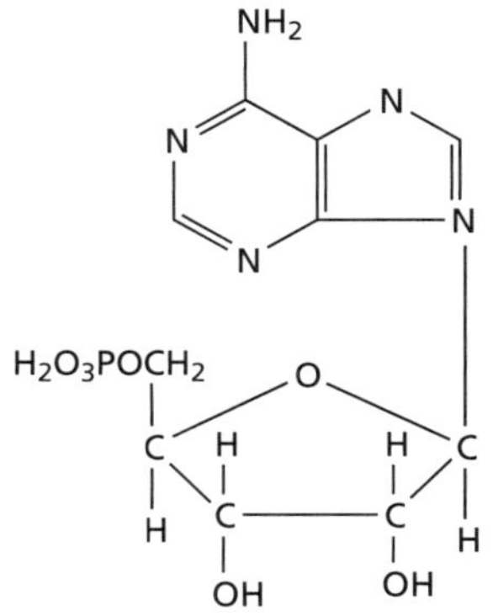

Nucleic acids are polynucleotides of very high molecular weight, generally measured in several millions. A nucleotide containing ribose is termed ribonucleic acid (RNA), while one containing deoxyribose is referred to as deoxyribonucleic acid (DNA).

The nucleotides are arranged in a certain pattern; DNA normally consists of a double-strand spiral or helix (Fig. 4.2). Each strand consists of alternate units of the deoxyribose and phosphate groups. Attached to each sugar group is one of the four bases, cytosine, thymine, adenine or guanine. The bases on the two strands of the spiral are joined in pairs by hydrogen bonds, the thymine on one strand always being paired with the adenine on the other and the cytosine with the guanine. The sequence of bases along these strands carries the genetic information of the living cell (see p. 226). DNA is found in the nuclei of cells as part of the chromosome structure.

There are several distinct types of ribonucleic acid, which are defined in terms of molecular size, base composition and functional properties. They differ from DNA in the nature of their sugar moiety and also in the types of nitrogenous base present. RNA contains the pyrimidine uracil in place of thymine. There is evidence to indicate that unlike DNA, most RNA molecules exist in the form of single, folded chains arranged spirally. There are three main forms of RNA, termed messenger RNA, ribosomal RNA and transfer RNA. The functions of these three forms of RNA are dealt with in the protein synthesis section of Chapter 9.

Apart from their importance in the structure of nucleic acids, nucleotides exist free as monomers and many play an important role in cellular metabolism.

Reference has been made previously to the phosphorylation of adenosine to form adenosine monophosphate (AMP). Successive additions of phosphate residues give adenosine diphosphate (ADP) and then the

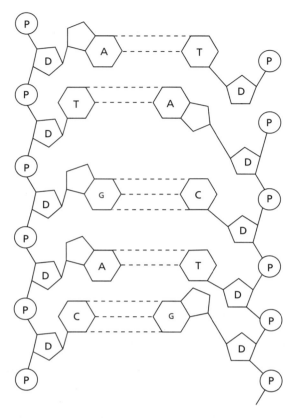

Fig. 4.2 Diagrammatic representation of part of the ladder-like DNA molecule, showing the two strands of alternate phosphate (P) and deoxyribose (D) molecules. The horizontal rods represent the pairs of bases held by hydrogen bonds (represented by dotted lines). A = adenine, T = thymine, C = cytosine, G = guanine.

triphosphate (ATP). The importance of ATP in energy transformations is described in Chapter 9.

4.8 Other nitrogenous compounds

A considerable variety of nitrogen-containing compounds, other than proteins and nucleic acids, occur in plants and animals. In plants, free amino acids are usually present; those in greatest amount include glutamic acid, aspartic acid, alanine, serine, glycine and proline. Other compounds are nitrogenous lipids, amines, amides, purines, pyrimidines, nitrates and alkaloids. In addition, most members of the vitamin B complex contain nitrogen in their structure.

It is clearly impossible to deal with these compounds in any detail and only some of the important ones not previously mentioned will be discussed.

Amines

Amines are basic compounds present in small amounts in most plant and animal tissues. Many occur as decomposition products in decaying organic matter and have toxic properties.

A number of microorganisms are capable of producing amines by decarboxylation of amino acids (Table 4.2). These may be produced in the rumen under certain conditions and can occur in fermented foods such as cheese, wine, sauerkraut and sausage. They are termed biogenic amines and may give rise to physiological symptoms; histamine, for example, is an amine formed from the amino acid, histidine, and in cases of anaphylactic shock is found in the blood in relatively large amounts. Histamine has also been implicated in dietary-induced migraine. Silages in which clostridia have dominated the fermentation usually contain appreciable amounts of amines (see Chapter 20).

Several metabolic pathways (e.g. lipid metabolism, creatine and carnitine synthesis) require methyl groups and these can be supplied by choline or methionine. During the process of transmethylation betaine, a tertiary amine, is formed by the oxidation of choline. Betaine can be added to the diet to act as a more direct supply of methyl groups, thus sparing choline for its other functions of lecithin and acetylcholine formation, and methionine for protein synthesis. Betaine occurs in sugar beet, and the young leaves may contain about 25 g/kg; it is this amine which is responsible for the 'fishy' aroma frequently associated with the commercial extraction of sugar from beet. In the animal body, betaine may be transformed into trimethylamine, and it is this which gives the fishy taint to milk produced by cows that have been given excessive amounts of sugar beet by-products.

Table 4.2 Some important amines and their parent amino acids

Amino acid	Amine
Arginine	Putrescine
Histidine	Histamine
Lysine	Cadaverine
Phenylalanine	Phenylethylamine
Tyrosine	Tyramine
Tryptophan	Tryptamine

Amides

Asparagine and glutamine are important amide derivatives of the amino acids aspartic acid and glutamic acid. These two amides are also classed as amino acids (Table 4.1) and occur as components of proteins. They also occur as free amides and play an important role in transamination reactions.

Urea is an amide which is the main end-product of nitrogen metabolism in mammals, but it also occurs in many plants and has been detected in wheat, soya bean, potato and cabbage.

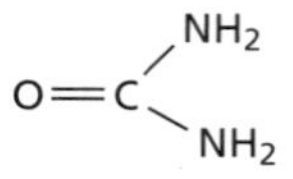

Urea

In man and other primates, uric acid is the end-product of purine metabolism and is found in the urine. In subprimate mammals the uric acid is oxidised to *allantoin* before being excreted.

In birds, uric acid is the principal end-product of nitrogen metabolism and thus corresponds, in its function, to urea in mammals.

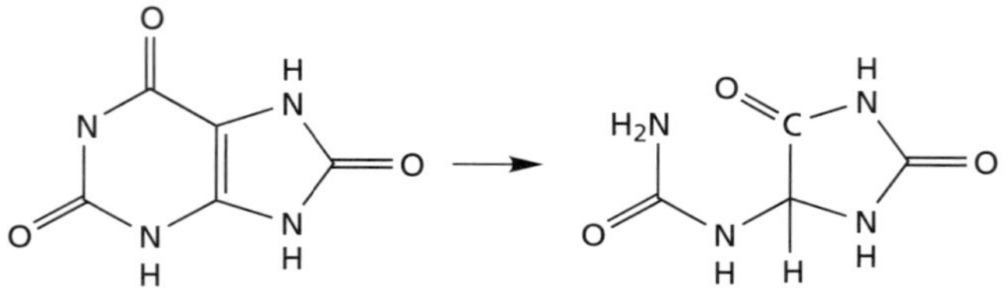

Uric acid   Allantoin

4.9 Nitrates

Nitrates may be present in plant materials and, whereas nitrate itself may not be toxic to animals, it is reduced readily under favourable conditions, as in the rumen, to nitrite, which is toxic. 'Oat hay poisoning' is attributed to the relatively large amounts of nitrate present in green oats.

Quite high levels of nitrate have been reported in herbage given heavy dressings of nitrogenous fertilisers (see Chapter 19).

4.10 Alkaloids

These compounds are of particular interest since many of them have poisonous properties. In plants, their presence is restricted to a few orders of the dicotyledons. A number of the more important alkaloids, with their sources, are listed in Table 4.3. The alkaloid in ragwort, for example, attacks the liver and much of this organ can be destroyed before symptoms appear. Another nutritionally significant source of alkaloids is the fungus ergot, which grows on cereal grains (see Chapter 23).

Table 4.3 Some important alkaloids occurring in plants

Name	Source
Coniine	Hemlock
Nicotine	Tobacco
Ricinine	Castor plant seeds
Atropine	Deadly nightshade
Cocaine	Leaves of coca plant
Jacobine	Ragwort
Quinine	Cinchona bark
Strychnine	Seeds of *Nux vomica*
Morphine	Dried latex of opium poppy
Solanine	Unripe potatoes and potato sprouts

Summary

1. Proteins are complex organic compounds of high molecular weight containing carbon, hydrogen, oxygen, nitrogen and generally sulphur.

2. Proteins are made up from a pool of 20 amino acids, about ten of which are essential (indispensable) for non-ruminants.

3. Amino acids join together by a peptide linkage between the α-carboxyl group of one acid and the α-amino group of another.

4. The primary structure of a protein refers to the sequence of amino acids in the polypeptide chain; the conformation of the chain as a result of hydrogen bonding is the secondary structure; folding of the chain gives the tertiary structure; the quaternary structure refers to the configuration of those proteins with more than one polypeptide chain.

5. Proteins have colloidal properties and can be denatured by heat, creating new linkages between the chains, some of which are resistant to hydrolysis.

6. Proteins may be simple, either fibrous or globular, or conjugated to a non-protein molecule.

7. Nucleic acids act as a store of genetic information when arranged to form DNA and RNA.

8. Amines are derivatives of amino acids and some possess physiological activity.

9. Urea is an important member of the amide group of compounds.

10. Nitrate is converted to the toxic nitrite in the rumen.

11. Alkaloids are poisonous nitrogen-containing compounds formed by some plants and fungi.

Further reading

Creighton T E 1992 *Proteins: Structures and Molecular Properties,* 2nd edn, Oxford, W H Freeman.

D'Mello J P F 1995 *Amino Acids in Animal Nutrition*, Oxford, CABI Publishing.

Horton H R, Moran L A, Ochs R S, Rawn J D and Scrimgeour K G 1993 *Principles of Biochemistry*, Englewood Cliffs, NJ, Prentice Hall.

Lehninger A L, Nelson D L and Cox M M 1993 *Principles of Biochemistry*, 2nd edn, New York, Worth.

Mathews C K and van Holde K E 1990 *Biochemistry*, Redwood City, CA, Benjamin Cummings Publishing Co.

Neurath H and Hill R L (eds) 1982 *The Proteins*, New York, Academic Press.

5 Vitamins

5.1 Introduction

5.2 Fat-soluble vitamins

5.3 Water-soluble vitamins: the vitamin B complex

5.4 Vitamin C

5.5 Hypervitaminosis

5.1 Introduction

Discovery of vitamins

The discovery and isolation of many of the vitamins were originally achieved through work on rats which had been given diets of purified proteins, fats, carbohydrates and inorganic salts. Using this technique, Hopkins in 1912 showed that a synthetic diet of this type was inadequate for the normal growth of rats, but that when a small quantity of milk was added to the diet the animals developed normally. This proved that there was some essential factor, or factors, lacking in the pure diet.

About this time the term 'vitamines', derived from 'vital amines', was coined by Funk to describe these accessory food factors, which he thought contained amino-nitrogen. It is now known that only a few of these substances contain amino-nitrogen and the word has been shortened to vitamins, a term which has been generally accepted as a group name.

Although the discovery of the vitamins dates from the beginning of the twentieth century, the association of certain diseases with dietary deficiencies had been recognised much earlier. In 1753 Lind, a British naval physician, published a treatise on scurvy proving that this disease could be prevented in human beings by including salads and summer fruits in their diet. The action of lemon juice in curing and preventing scurvy had been known, however, since the beginning of the seventeenth century. The use of cod-liver oil in preventing rickets has long been appreciated, and Eijkmann knew at the end of the nineteenth century that beri-beri, a disease common in the Far East, could be cured by giving the patients brown rice grain rather than polished rice.

Vitamins and biochemistry

Vitamins are usually defined as organic compounds that are required in small amounts for normal growth and maintenance of animal life. But this definition ignores the important part that these substances play in plants, and their importance generally in the metabolism of all living organisms. Unlike the nutrients covered in Chapters 2–4, vitamins are not merely building blocks or energy-yielding compounds but are involved in, or are mediators of, the biochemical pathways (Fig. 5.1). For example, many of the B vitamins act as co-factors in enzyme systems but it is not always clear how the symptoms of deficiency are related to the failure of the metabolic pathway.

In addition to avoiding explicit vitamin deficiency symptoms (see below) or a general depression in production due to a subclinical deficiency, some vitamins are added to the diet at higher levels in order to enhance the quality of the animal product, e.g. vitamin D for egg shell strength and vitamin E for prolonging the shelf life of carcasses.

Vitamins are required by animals in very small amounts compared with other nutrients; for example, the vitamin B_1 (thiamin) requirement of

Box 5.1
Vitamin supplementation of diets

Most food mixes prepared as supplements for ruminants or as the sole food for pigs and poultry are supplemented with vitamins. With other nutrients, such as energy and protein, it is possible to demonstrate a response to increments in intake, which can be evaluated against the cost of the increment. This is not possible with vitamins, for which the cost is relatively small in relation to the consequences of deficiency. Therefore, vitamins are usually supplied at levels greater than those shown to be required under experimental conditions. This oversupply allows for uncertainties met under practical conditions (e.g. variable vitamin content and availability in foods, loss of vitamin potency in storage, range of management practices, quality of the environment, health status, extra requirements due to stress). This is not to say that such safety margins should be excessive, since this would be wasteful: in addition, an excess of one vitamin may increase the requirement for another. For example, the fat-soluble vitamins share absorption mechanisms and compete with each other. Thus an excess of vitamin A will increase the dietary requirements of vitamins E, D and K.

Originally, vitamins for supplements were isolated from plant products. However, yields from such sources are low and the vitamins can be expensive. Yields can be increased when vitamins are produced from microorganisms by fermentation. Nowadays many vitamins are produced in multi-stage chemical processes which are controllable and the yield is predictable.

For ease of handling in the feed mill the vitamin supplement needs to (a) be free-flowing, (b) not be dusty and (c) mix homogeneously with other diet ingredients; the vitamin must remain stable and yet be biologically available when consumed by the animal. Some of these criteria are incompatible and a compromise has to be reached. Oily vitamins are absorbed on to silica, others are coated or micro-encapsulated and antioxidants are added to prevent breakdown of those vitamins that are susceptible to oxidation. The manufacturers also make use of stable derivatives of vitamins (e.g. the acetate form of α-tocopherol as opposed to the alcohol form).

Maintenance of vitamin activity in the supplement is affected by temperature, humidity, acidity/alkalinity, oxygen, ultraviolet light, the presence of some trace minerals (dietary supplements are usually combinations of vitamins, minerals and trace elements), physical factors such as hammer milling and the length of time the supplement is stored. For example, choline chloride can destroy other vitamins during storage.

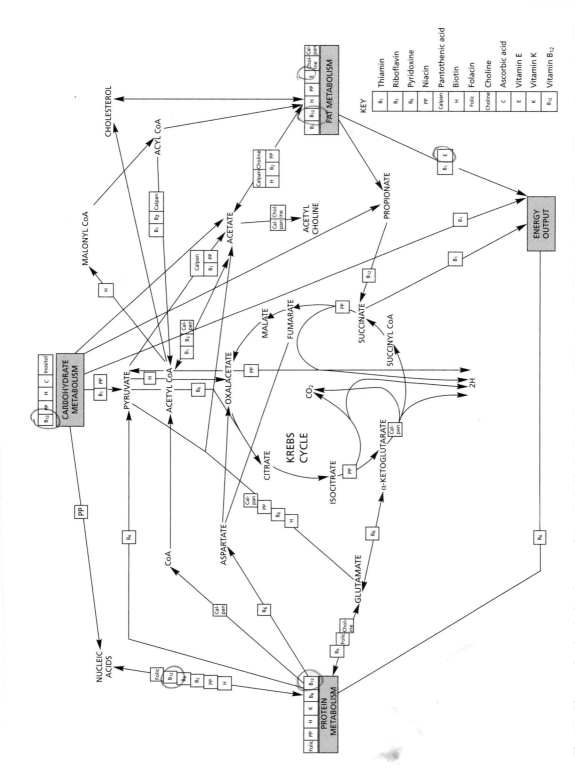

Fig. 5.1 Diagram showing the involvement of vitamins in biochemical pathways. (Adapted from *Roche Vitec Animal Nutrition and Vitamin News* **1**: A1–10/2 Nov. 1984.)

a 50 kg pig is only about 3 mg/day. Nevertheless a continuous deficiency in the diet results in disordered metabolism and eventually disease.

Some compounds function as vitamins only after undergoing a chemical change; such compounds, which include β-carotene and certain sterols, are described as provitamins or vitamin precursors.

Many vitamins are destroyed by oxidation, a process speeded up by the action of heat, light and certain metals such as iron. This fact is important since the conditions under which a food is stored will affect the final vitamin potency. Some commercial vitamin preparations are dispersed in wax or gelatin, which act as a protective layer against oxidation (for further details of vitamin supplementation of diets, see Box 5.1).

The system of naming the vitamins by letters of the alphabet was most convenient and was generally accepted before the discovery of their chemical nature. Although this system of nomenclature is still widely used with some vitamins, the modern tendency is to use the chemical name, particularly in describing members of the B complex.

There are at least 14 vitamins which have been accepted as essential food factors and a few others have been proposed. Only those that are of nutritional importance are dealt with in this chapter.

It is convenient to divide the vitamins into two main groups: fat-soluble and water-soluble. Table 5.1 lists the important members of these two groups.

Table 5.1 Vitamins important in animal nutrition

Vitamin	Chemical name
Fat-soluble vitamins	
A	Retinol
D_2	Ergocalciferol
D_3	Cholecalciferol
E	Tocopherol[a]
K	Phylloquinone[b]
Water-soluble vitamins	
B complex	
B_1	Thiamin
B_2	Riboflavin
	Nicotinamide
B_6	Pyridoxine
	Pantothenic acid
	Biotin
	Folic acid
	Choline
B_{12}	Cyanocobalamin
C	Ascorbic acid

[a] A number of tocopherols have vitamin E activity.
[b] Several naphthoquinone derivatives possessing vitamin K activity are known.

5.2 Fat-soluble vitamins

Vitamin A

Chemical nature

Vitamin A ($C_{20}H_{29}OH$), known chemically as retinol, is an unsaturated monohydric alcohol with the following structural formula:

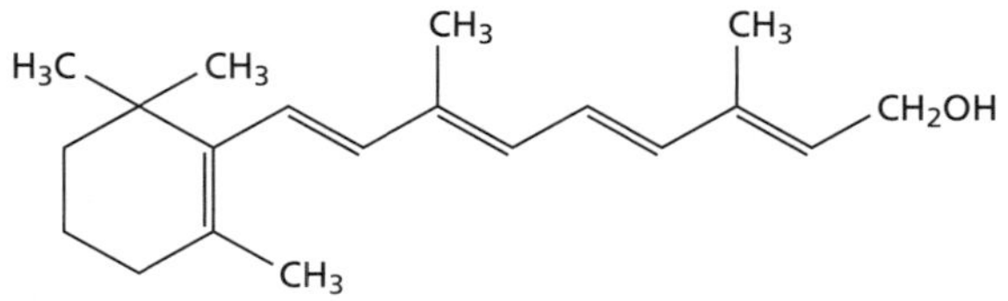

Vitamin A (all-*trans* form)

The vitamin is a pale yellow crystalline solid, insoluble in water but soluble in fat and various fat solvents. It is readily destroyed by oxidation on exposure to air and light. A related compound with the formula $C_{20}H_{27}OH$, originally found in fish, has been designated dehydroretinol or vitamin A_2.

Sources

Vitamin A accumulates in the liver and this organ is likely to be a good source; the amount present varies with species of animal and diet. Table 5.2 shows some typical liver reserves of vitamin A in different species, although these values vary widely within each species.

The oils from livers of certain fish, especially cod and halibut, have long been used as an important dietary source of the vitamin. Egg yolk and milk fat also are usually rich sources, although the vitamin content of these depends, to a large extent, upon the diets of the animals by which they have been produced.

Vitamin A is manufactured synthetically and can be obtained in a pure form.

Table 5.2 Some typical values for liver reserves of vitamin A in different species[a] (Adapted from Moore T 1969 In: Morton R A (ed.) *Fat Soluble Vitamins*, Oxford, Pergamon Press, p 233)

Species	Vitamin A (μg/g liver)	Species	Vitamin A (μg/g liver)
Pig	30	Horse	180
Cow	45	Hen	270
Rat	75	Codfish	600
Man	90	Halibut	3000
Sheep	180	Polar bear	6000
		Soup-fin shark	15000

[a] In every species wide individual variations are to be expected.

Provitamins

Vitamin A does not exist as such in plants, but is present as precursors or provitamins in the form of certain carotenoids which can be converted into the vitamin. At least 600 naturally occurring carotenoids are known, but only a few of these are precursors of the vitamin.

In plants, carotenoids have yellow, orange or red colours but their colours are frequently masked by the green colour of chlorophyll. When ingested, they are responsible for many of the varied and natural colours which occur in crustaceans, insects, birds and fish. They are also found in egg yolk, butterfat and the body fat of cattle and horses, but not in sheep or pigs. Carotenoids may be divided into two main categories: carotenes and xanthophylls. The latter include a wide range of compounds, for example lutein, cryptoxanthin and zeaxanthin, most of which cannot be converted into vitamin A. Of the carotenes, β-carotene is the most important member and this compound forms the main source of vitamin A in the diets of farm animals. Its structure is shown below:

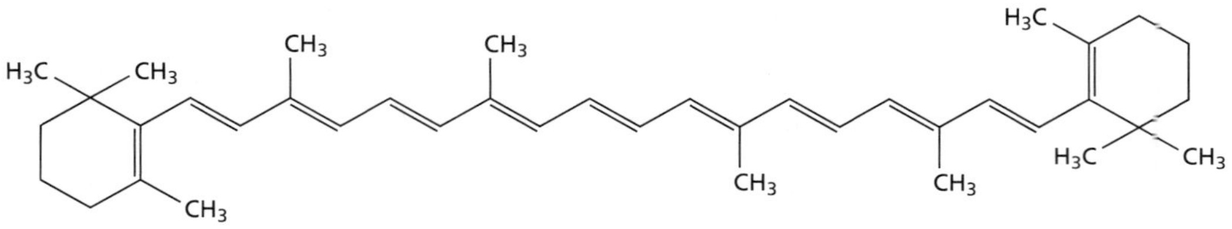

β-carotene

The long, unsaturated hydrocarbon chains in carotenes (and vitamin A) are easily oxidised to by-products that have no vitamin potency. Oxidation is increased by heat, light, moisture and the presence of heavy metals. Consequently, foods exposed to air and sunlight rapidly lose their vitamin A potency, so that large losses can occur during the sun-drying of crops. For example, lucerne hay has around 15 mg β-carotene/kg, but artificially dried lucerne and grass meals have 95 and 155 mg/kg. Fresh grass is an excellent source (250 mg/kg), but this is halved during ensilage.

Conversion of carotene into vitamin A can occur in the liver, but usually takes place in the intestinal mucosa. Theoretically, hydrolysis of one molecule of the C_{40} compound β-carotene should yield two molecules of the C_{20} compound retinol, but although central cleavage of this type is thought to occur, it is considered likely that the carotene is degraded from one end of the chain by step-wise oxidation until only one molecule of the C_{20} compound retinol remains. Although the maximum conversion measured in the rat is 2 mg β-carotene into 1 mg retinol, the conversion efficiency in other animals ranges from $3:1$ to $12:1$. Ruminants convert about 6 mg of β-carotene into 1 mg of retinol. The corresponding conversion efficiency for pigs and poultry is usually taken as $11:1$ and $3:1$, respectively. Recent work with pig intestinal tissue indicated a more efficient conversion, closer to $2:1$, suggesting that the efficiency of absorption of β-carotene is poorer than previously thought. The vitamin A values of foods are often stated in terms of international units (iu), one iu of vitamin A being defined as the activity of 0.3 μg of crystalline retinol.

Metabolism

Vitamin A appears to play two different roles in the body according to whether it is acting in the eye or in the general system.

In the retinal cells of the eye, vitamin A (all-*trans*-retinol) is oxidised to the aldehyde (all-*trans*-retinaldehyde), which is converted into the 11-*cis* isomer. The latter then combines with the protein opsin to form rhodopsin (visual purple), which is the photo-receptor for vision at low light intensities. When light falls on the retina, the *cis*-retinaldehyde molecule is converted back into the all-*trans* form and is released from the opsin. This conversion results in the transmission of an impulse up the optic nerve. The all-*trans*-retinaldehyde is isomerised in the dark to 11-*cis*-retinaldehyde, which is then recaptured by opsin to reconstitute rhodopsin, thus continually renewing the light sensitivity of the retina (Fig. 5.2).

In its second role, in the regulation of cellular differentiation, vitamin A is involved in the formation and protection of epithelial tissues and mucous membranes. In this way it has particular importance in growth, reproduction and immune response. Vitamin A is important in the resistance to disease and promotion of healing through its effect on the immune system and epithelial integrity. In addition, it acts, along with vitamins E and C and β-carotene, as a scavenger of free radicals (see Box 5.2, p. 87).

Deficiency symptoms

Ability to see in dim light depends upon the rate of resynthesis of rhodopsin, and when vitamin A is deficient rhodopsin formation is impaired. One of the earliest symptoms of a deficiency of vitamin A in all animals is a lessened ability to see in dim light, commonly known as 'night blindness'.

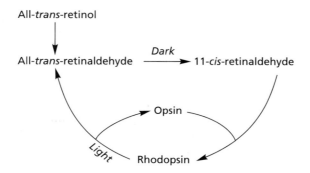

Fig. 5.2   The role of vitamin A (retinol) in the visual cycle.

It has long been realised that vitamin A plays an important role in combating infection, and it has been termed the 'anti-infective vitamin'. In several species, vitamin A deficiency has been shown to be accompanied by low levels of immunoglobulins, although the exact function of the vitamin in the formation of these important proteins is uncertain.

In adult cattle, a mild deficiency of vitamin A is associated with roughened hair and scaly skin. If it is prolonged the eyes are affected, leading to excessive watering, softening and cloudiness of the cornea and development of xerophthalmia, which is characterised by a drying of the conjunctiva. Constriction of the optic nerve canal may cause blindness in calves. In breeding animals a deficiency may lead to infertility, and in pregnant animals to abortion, short gestation, retained placenta or to the production of dead, weak or blind calves. Less severe deficiencies may result in metritis and dermatitis and calves born with low reserves of the vitamin, and it is imperative that colostrum, rich in antibodies and vitamin A, should be given at birth; otherwise the susceptibility of such animals to infection leads to scours, and if the deficiency is not rectified they frequently die of pneumonia.

In practice, severe deficiency symptoms are unlikely to occur in adult animals except after prolonged deprivation. Grazing animals generally obtain more than adequate amounts of provitamin from pasture grass and normally build up liver reserves. If cattle are fed on silage or well-preserved hay during the winter months, deficiencies are unlikely to occur. Cases of vitamin A deficiency have been reported among cattle fed indoors on high cereal rations, and under these conditions a high vitamin supplement is recommended.

In ewes, in addition to night blindness, severe cases of deficiency may result in lambs being born weak or dead. A deficiency is not common in sheep, however, because of adequate dietary intakes on pasture.

In pigs, eye disorders such as xerophthalmia and blindness may occur. A deficiency in pregnant animals may result in the production of weak, blind, dead or deformed litters. In view of the apparent importance of vitamin A in preventing reproductive disorders in pigs, it has been suggested that the retinoids may have a role in embryo development (cell differentation, gene transcription). Alternatively, they may regulate ovarian steroid production and influence the establishment and maintenance of pregnancy. In less severe cases of deficiency, appetite is impaired and growth retarded. Where pigs are reared out of doors and have access to green food, deficiencies are unlikely to occur except possibly during the winter. Pigs kept indoors on concentrates may not receive adequate amounts of vitamin A in the diet and supplements may be required.

In poultry consuming a diet deficient in vitamin A, the mortality rate is usually high. Early symptoms include retarded growth, weakness, ruffled plumage and a staggering gait. In mature birds, egg production and hatchability are reduced. Since most concentrated foods present in the diets of poultry are low or lacking in vitamin A or its precursors, vitamin A deficiency may be a problem unless precautions are taken. Yellow maize, dried grass or other green food, or alternatively cod- or other fish-liver oils or vitamin A concentrate, can be added to the diet.

In horses, the signs of deficiency include the catalogue of symptoms seen in other farm animals: night blindness, keratinisation of the skin and cornea, susceptibility to infection and infertility.

It has been suggested that, in addition to vitamin A, some species may have a dietary requirement for β-carotene *per se*. The ovaries of bovine species are known to contain high concentrations of β-carotene during the luteal phase – indeed it is an integral component of the mucosal membrane of luteal cells – and it has been postulated that certain fertility disorders in dairy cattle, such as retarded ovulation and early embryonic mortality, may be caused by a deficiency of the provitamin in the diet. In sows, injections of β-carotene have reduced embryonic mortality and increased litter sizes. It is suggested that it influences steroidogenesis and, through its antioxidant properties, it may protect the highly active ovarian cells from damage by free radicals.

Vitamin D

Chemical nature

A number of forms of vitamin D are known, although not all of these are naturally occurring compounds. The two most important forms are ergocalciferol (D_2) and cholecalciferol (D_3). The term D_1 was originally suggested by earlier workers for an activated sterol which was found later to be impure and to consist mainly of ergocalciferol, which had already been designated D_2. The result of this confusion is that in the group of D vitamins the term vitamin D_1 has been abolished. The structures of vitamins D_2 and D_3 are:

Vitamin D_2 (ergocalciferol)

Vitamin D_3 (cholecalciferol)

The D vitamins are insoluble in water but soluble in fats and fat solvents. The sulphate derivative of vitamin D present in milk is a water-soluble form of the vitamin. Both D_2 and D_3 are more resistant to oxidation than vitamin A, D_3 being more stable than D_2.

Sources

The D vitamins are limited in distribution. They rarely occur in plants except in sun-dried roughages and the dead leaves of growing plants. In the animal kingdom vitamin D_3 occurs in small amounts in certain tissues and is abundant only in some fishes. Halibut-liver and cod-liver oils are rich sources of vitamin D_3. Egg yolk is also a good source, but cows' milk is normally a poor source, although summer milk tends to be richer than winter milk. Colostrum usually contains from six to ten times the amount present in ordinary milk.

Clinical manifestations of avitaminosis D, and other vitamin deficiencies, are frequently treated by injection of the vitamin into the animal.

Provitamins

Reference has been made (p. 51) to two sterols, ergosterol and 7-dehydro-cholesterol, as being precursors of vitamins D_2 and D_3, respectively. The provitamins, as such, have no vitamin value and must be converted into calciferols before they are of any use to the animal. For this conversion it is necessary to impart a definite quantity of energy to the sterol molecule, and this can be brought about by the ultraviolet light present in sunlight, by artificially produced radiant energy or by certain kinds of physical treatment. Under natural conditions activation is brought about by irradiation from the sun. The activation occurs most efficiently with light of wavelength between 290 and 315 nm, so that the range capable of vitamin formation is small. The amount of ultraviolet radiation which reaches the earth's surface depends upon latitude and atmospheric conditions: the presence of clouds, smoke and dust reduces the radiation. Ultraviolet radiation is greater in the tropics than in the temperate regions, and the amount reaching the more northern areas in winter may be slight. Since ultraviolet light cannot pass through ordinary window glass, animals housed indoors receive little, if any, suitable radiation for the production of the vitamin. Irradiation is apparently more effective in animals with light-coloured skins. If irradiation is continued for a prolonged period, then the vitamin may itself be altered to compounds which can be toxic.

The chemical transformation occurs in the skin and also in the skin secretions, which are known to contain the precursor. Absorption of the vitamin can take place from the skin, since deficiency can be treated successfully by rubbing cod-liver oil into the skin.

Vitamin D requirements are often expressed in terms of international units (iu). One iu of vitamin D is defined as the vitamin D activity of 0.025 μg of crystalline vitamin D_3.

Metabolism

Dietary vitamins D_2 and D_3 are absorbed from the small intestine and are transported in the blood to the liver, where they are converted into 25-hydroxycholecalciferol. The latter is then transported to the kidney, where it is converted into 1,25-dihydroxycholecalciferol, the most biologically active form of the vitamin. This compound is then transported in the blood to the various target tissues, the intestine, bones and the eggshell gland in birds. The compound 1,25-dihydroxycholecalciferol acts in a similar way to a steroid hormone, regulating DNA transcription in the intestinal micro-villi, inducing the synthesis of specific messenger RNA (see Chapter 9) which is responsible for the production of calcium-binding protein. This protein is involved in the absorption of calcium from the intestinal lumen. The various pathways involved in these transformations are summarised in Fig. 5.3.

The amount of 1,25-dihydroxycholecalciferol produced by the kidney is controlled by the parathyroid hormone. When the level of calcium in the blood is low (hypocalcaemia), the parathyroid gland is stimulated to secrete more parathyroid hormone, which induces the kidney to produce more 1,25-dihydroxycholecalciferol, which in turn enhances the intestinal absorption of calcium.

 In addition to increasing intestinal absorption of calcium, 1,25-dihydroxy-cholecalciferol increases the absorption of phosphorus from the intestine and also enhances calcium and phosphorus reabsorption from the kidney and bone.

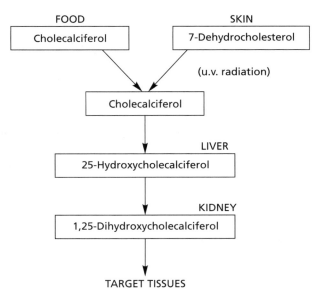

Fig. 5.3 Metabolic pathway showing production of the hormonally active form of vitamin D.

Deficiency symptoms

A deficiency of vitamin D in the young animal results in rickets, a disease of growing bone in which the deposition of calcium and phosphorus is disturbed; as a result the bones are weak and easily broken and the legs may be bowed. In young cattle the symptoms include swollen knees and hocks and arching of the back. In pigs the symptoms are usually enlarged joints, broken bones, stiffness of the joints and occasionally paralysis. The growth rate is generally adversely affected. The term 'rickets' is confined to young growing animals; in older animals vitamin D deficiency causes osteomalacia, in which there is reabsorption of bone already laid down. Osteomalacia due to vitamin D deficiency is not common in farm animals, although a similar condition can occur in pregnant and lactating animals, which require increased amounts of calcium and phosphorus. Rickets and osteomalacia are not specific diseases necessarily caused by vitamin D deficiency; they can also be caused by lack of calcium or phosphorus or an imbalance between these two elements.

In poultry, a deficiency of vitamin D causes the bones and beak to become soft and rubbery; growth is usually retarded and the legs become weak. Egg production is reduced and eggshell quality deteriorates. Most foods of pigs and poultry, with the possible exception of fishmeal, contain little or no vitamin D, and the vitamin is generally supplied to these animals, if reared indoors, in the form of fish-liver oils or synthetic preparations.

The need for supplementing the diets of cattle and sheep with vitamin D is generally not so great as that for pigs and poultry. Adult ruminants can receive adequate amounts of the vitamin from hay in the winter months, and from irradiation while grazing. However, since the vitamin D content of hays is extremely variable, it is possible that vitamin D supplementation may be desirable, especially with young growing animals or pregnant animals, on winter diets. There is a considerable lack of information about the vitamin D needs of farm animals under practical conditions.

For cattle, sheep and pigs, vitamins D_2 and D_3 have the same potency, but for poultry vitamin D_2 has only about 10 per cent of the potency of D_3.

Certain foods, such as fresh green cereals and yeast, have been shown to have rachitogenic (rickets-causing) properties for mammals, and raw liver and isolated soya bean protein have a similar effect on poultry. In one study it was shown that in order to overcome the rachitogenic activity of whole raw soya bean meal, a tenfold increase in vitamin D supplement was necessary. Heating destroys the rachitogenic activity.

Vitamin E

Chemical nature

Vitamin E is a group name which includes a number of closely related active compounds. Eight naturally occurring forms of the vitamin are known and these can be divided into two groups according to whether the side chain of the molecule, as shown below, is saturated or unsaturated.

The four saturated vitamins are designated α-, β-, γ- and δ-tocopherol and of these the α form is the most biologically active and most widely distributed.

α-Tocopherol

The β, γ and δ forms have only about 45, 13 and 0.4 per cent of the activity of the α form, respectively. The unsaturated forms of the vitamin have been designated α-, β-, γ- and δ-tocotrienols and of these only the α form appears to have any significant vitamin E activity, and then only about 13 per cent of its saturated counterpart.

Sources

Vitamin E, unlike vitamin A, is not stored in the animal body in large amounts for any length of time and consequently a regular dietary source is important. Fortunately, the vitamin is widely distributed in foods. Green fodders are good sources of α-tocopherol, young grass being a better source than mature herbage. The leaves contain 20–30 times as much vitamin E as the stems. Losses during haymaking can be as high as 90 per cent, but losses during ensilage or artificial drying are low.

Cereal grains are also good sources of the vitamin but the tocopherol composition varies with species. Wheat and barley grain resemble grass in containing mainly α-tocopherol, but maize contains, in addition to α-tocopherol, appreciable quantities of γ-tocopherol. During the storage of moist grain in silos, the vitamin E activity can decline markedly. Reduction in the concentration of the vitamin from 9 to 1 mg/kg DM has been reported in moist barley stored for 12 weeks.

Animal products are relatively poor sources of the vitamin, although the amount present is related to the level of vitamin E in the diet. Synthetic α-tocopherol and the acetate are available as commercial preparations.

The vitamin E values of foods are often stated in terms of international units, one iu of vitamin E being defined as the specific activity of 1 mg of synthetic racemic α-tocopherol acetate.

Metabolism

Vitamin E functions in the animal mainly as a biological antioxidant; in association with the selenium-containing enzyme glutathione peroxidase and other vitamins and trace-element-containing enzymes, it protects cells against oxidative damage caused by free radicals. Free radicals are formed during cellular metabolism and, as they are capable of damaging cell membranes, enzymes and cell nuclear material, they must be converted into less reactive substances if the animal is to survive. This protection is particularly important in preventing oxidation of polyunsaturated fatty acids, which function as primary constituents of subcellular membranes and precursors of prostaglandins. Oxidation of unsaturated fatty acids produces hydroperoxides, which also damage cell tissues, and more lipid free radicals (see p. 87), so that prevention of such oxidation is of vital importance in maintaining the health of the living animal. The animal has complementary methods of protecting itself against oxidative damage: scavenging of radicals by vitamin E and destruction of any peroxides formed by glutathione peroxidase (see Box 5.2).

Vitamin E also plays an important role in the development and function of the immune system. In studies with several species, supplementation of diets with the vitamin provided some protection against infection with pathogenic organisms.

Deficiency symptoms

The most frequent and, from a diagnostic point of view, the most important manifestation of vitamin E deficiency in farm animals is muscle degeneration (myopathy). Nutritional myopathy, also known as muscular dystrophy, frequently occurs in cattle, particularly calves, when they are turned out on to spring pasture. It is associated with low vitamin E and selenium intakes during the in-wintering period and possibly the relatively high concentration of polyunsaturated fatty acids in the young grass lipids. The requirement for the vitamin increases with increasing concentrations of polyunsaturated fatty acids in the diet. The myopathy primarily affects the skeletal muscles and the affected animals have weak leg muscles, a condition manifested by difficulty in standing and, after standing, a trembling and staggering gait. Eventually, the animals are unable to rise and weakness of the neck muscles prevents them from raising their heads. A popular descriptive name for this condition is 'white muscle disease', owing to the presence of pale patches or white streaks in the muscles. The heart muscle may also be affected and death may result. Serum creatine phosphokinase and glutamic oxaloacetic transaminase levels are elevated in vitamin E deficient animals.

Nutritional myopathy also occurs in lambs, with similar symptoms to those of calves. The condition is frequently referred to as 'stiff lamb disease'.

In pigs, the two main diseases associated with vitamin E and selenium deficiency are myopathy and cardiac disease. Nutritional myopathy affects, in particular, young fast-growing pigs but may occur at any age. The pigs demonstrate an uncoordinated staggering gait, or are unable to rise. In contrast to other animals, it is the pig's heart muscle that is more often affected. Sudden cardiac failure occurs and on post-mortem examination large

Box 5.2
Free radicals and antioxidants

Antioxidants are required to protect the animal's cells from damage due to the presence of free radicals. These are highly reactive molecules containing one or more unpaired electrons and can exist independently (e.g. superoxide, O_2^{*-}, and hydroxyl, OH^*). Their high reactivity is a result of their trying to lose or gain an electron to achieve stability. Within cells hydrogen peroxide (H_2O_2) can easily break down, especially in the presence of transition ions (e.g. Fe^{2+}), to produce the hydroxyl radical, which is the most reactive and damaging of the free radicals:

$$H_2O_2 + Fe^{2+} \rightarrow OH^* + OH^- + Fe^{3+}$$

Free radicals are generated during normal cellular metabolism owing to leakage from the electron transport chain in mitochondria and leakage from peroxidation of polyunsaturated fatty acids in the pathway of conversion of arachidonic acid to prostaglandins and related compounds. Also O_2^{*-} plays an essential role in the extracellular killing of microorganisms by activated phagocytes, and activation of this system can lead to further leakage.

All classes of biological molecules are vulnerable to free radical damage but especially lipids, proteins and DNA. Cell membranes are an important target because of the enzyme systems contained within them. Lipids are the most susceptible and oxidative destruction of polyunsaturated fatty acids can be extremely damaging since it proceeds as a self-perpetuating chain reaction. The more active cells, such as muscle cells, are at greatest risk of damage because they depend on the utilisation of lipids as energy sources.

To maintain cell integrity the animal's cells require protection mechanisms and these are provided by the antioxidant system, which involves a group of vitamins and enzymes containing trace elements working in series. The initial line of defence is by the enzymes superoxide dismutase (containing copper), glutathione peroxidases (containing selenium) and catalase. Superoxide dismutase eliminates superoxide radicals formed in the cell and prevents the reaction of the radical with biological membranes or their participation in the production of more powerful radicals. Glutathione peroxidase detoxifies lipid hydroperoxides that are formed in the membrane during lipid peroxidation. Catalase can also break down hydrogen peroxide.

If large amounts of radicals are produced the enzyme systems will be insufficient to prevent damage and the second antioxidant system is brought into action. Antioxidants break the chain reaction by scavenging peroxyl radicals and thus interfere with the propagation steps in the lipid peroxidation process. Vitamin E is the main antioxidant but the carotenoids, vitamin A and vitamin C are also involved. In mammalian cells vitamin E is located in the mitochondria and endothelial reticulum. It donates a hydrogen atom to the free radical to form a stable molecule, thereby breaking the chain. The amount of vitamin E in the cell membranes is low and it must be regenerated so that there is sufficient to act against other radicals. The regeneration is probably carried out by reaction with vitamin C and the ascorbate radical in turn is reduced by NADH-dependent enzymes. It has been reported that vitamin C also acts as an antioxidant in extracellular fluid, where it operates as a scavenger preventing the initiation of lipid peroxidation. It contributes up to one-quarter of the total antioxidant activity in plasma.

amounts of fluid are found around the heart and lungs and the lesions of the cardiac muscles are seen as haemorrhagic and pale areas. This condition is commonly known as 'mulberry heart disease'. Sometimes the liver is also affected and it becomes enlarged and mottled. Supplemental vitamin E has improved litter size in pigs, probably through its antioxidant properties protecting arachidonic acid and maintaining the fuctional integrity of the reproductive organs.

Vitamin E deficiency in chicks may lead to a number of distinct diseases: myopathy, encephalomalacia and exudative diathesis. In nutritional myopathy the main muscles affected are the pectorals, although the leg muscles also may be involved. Nutritional encephalomalacia or 'crazy chick disease' is a condition in which the chick is unable to walk or stand, and is accompanied by haemorrhages and necrosis of brain cells. Exudative diathesis is a vascular disease of chicks characterised by a generalised oedema of the subcutaneous fatty tissues, associated with an abnormal permeability of the capillary walls. Both selenium and vitamin E appear to be involved in nutritional myopathy and in exudative diathesis but the element does not seem to be important in nutritional encephalomacia. It should be stressed that selenium itself is a very toxic element and care is required in its use as a dietary additive. The toxic nature of selenium is discussed in the next chapter.

In horses, vitamin E deficiency results in the previously mentioned problems, i.e. lameness and muscle rigidity ('tying up') associated with skeletal and heart muscles. The red blood cells become fragile and the release of myoglobin from damaged muscle cells gives rise to coffee-coloured urine.

Vitamin K

Vitamin K was discovered in 1935 to be an essential factor in the prevention of haemorrhagic symptoms in chicks. The discovery was made by a group of Danish scientists, who gave the name 'koagulation factor' to the vitamin, which became shortened to the K factor and eventually to vitamin K.

Chemical nature

A number of forms of vitamin K are known to exist. All compounds exhibiting vitamin K activity possess a 2-methyl-1,4-naphthoquinone ring (menadione), which animals are unable to synthesise but plants and bacteria can.

Menadione (2-methyl-1,4-naphthoquinone)

The form of the vitamin present in plants is 2-methyl-3-phytyl-1,4-naphthoquinone, generally referred to as phylloquinone or vitamin K_1.

The compound originally isolated from putrified fishmeal and designated vitamin K_2 is now known to be only one of a series of K vitamins with unsaturated side chains synthesised by bacteria and referred to as mena-quinones. The predominant vitamins of the menaquinone series contain six to ten isoprenoid (CH_2:CCH_3:CH:CH_2) side-chain units.

Vitamins K are relatively stable at ordinary temperatures but are rapidly destroyed on exposure to sunlight.

Sources

Phylloquinone is present in most green leafy materials, with lucerne, cabbage and kale being good sources. The amounts present in foods of animal origin are usually related to the diet, but egg yolk, liver and fishmeal are generally good sources. Menaquinones are synthesised by bacteria in the digestive tract of animals.

Metabolism

Vitamin K is necessary for the synthesis of prothrombin in the liver. In the blood-clotting process, prothrombin is the inactive precursor of thrombin, an enzyme that converts the protein fibrinogen in blood plasma into fibrin, the insoluble fibrous protein that holds blood clots together. Prothrombin normally must bind to calcium ions before it can be activated. If the supply of vitamin K is inadequate, the prothrombin molecule is deficient in γ-carboxyglutamic acid, a specific amino acid responsible for calcium binding. Proteins containing γ-carboxyglutamic acid, dependent on vitamin K for their formation, are also present in bone, kidney and other tissues.

Deficiency symptoms

Symptoms of vitamin K deficiency have not been reported in ruminants, horses and pigs under normal conditions, and it is generally considered that bacterial synthesis in the digestive tract supplies sufficient vitamin for the animal's needs. A number of microorganisms are known to synthesise vitamin K, including *Escherichia coli*. Medicines that affect the bacteria in the gut may depress the production of vitamin K. A disease of cattle called 'sweet clover disease' is associated with vitamin K. Sweet clover (*Melilotus albus*) naturally contains compounds called coumarins which, when the crop is preserved as hay or silage, may be converted by a variety of fungi, such as the *Aspergillus* species, to dicoumarol. This compound lowers the prothrombin content of the blood and thereby impairs the blood-clotting process. The disease can be overcome by administering vitamin K to the animals. For this reason dicoumarol is sometimes referred to as an 'anti-vitamin'.

The symptoms of vitamin K deficiency in chicks are anaemia and a delayed clotting time of the blood; birds are easily injured and may bleed to death. It is doubtful whether, in birds, microbially synthesised vitamin K is available by direct absorption from the digestive tract, because the site of its formation is too distal to permit absorption of adequate amounts except by ingestion of faecal material (coprophagy).

5.3 Water-soluble vitamins: the vitamin B complex

The vitamins included under this heading are all soluble in water and most of them are components of coenzymes (see Table 5.3). Although the mechanism of action in this role is known, the connection between the observed deficiency symptoms and the failure of the metabolic pathways is not always clear.

Unlike the fat-soluble vitamins, members of the vitamin B complex, with the exception of cyanocobalamin, are not stored in the tissues in appreciable amounts and a regular exogenous supply is essential. In ruminants, all the vitamins in this group can be synthesised by microbial action in the rumen and generally this will provide satisfactory amounts for normal metabolism in the host and secretion of adequate quantities into milk. For example, it has been estimated that the amount of thiamin synthesised in the rumen is equal to the thiamin intake. However, under certain conditions, deficiencies of thiamin and cyanocobalamin can occur in ruminants. In horses too, the B vitamins synthesised by the microbial population of the gut plus those vitamins occurring in the food can meet the requirements of most adult animals.

Table 5.3 Some coenzymes and enzyme prosthetic groups involving the B vitamins

Vitamin	Coenzyme or prosthetic group	Enzyme or other function
Thiamin	Thiamin pyrophosphate (TPP)	Oxidative decarboxylation
Riboflavin	Flavin mononucleotide (FMN)	Hydrogen carrier
Riboflavin	Flavin adenine dinucleotide (FAD)	Hydrogen carrier
Nicotinamide	Nicotinamide adenine dinucleotide (NAD)	Hydrogen carrier
Nicotinamide	Nicotinamide adenine dinucleotide phosphate (NADP)	Hydrogen carrier
Pyridoxine	Pyridoxal phosphate	Transaminases, decarboxylases
Pantothenic acid	Coenzyme A (CoA)	Acyl transfer
Folic acid	Tetrahydrofolic acid	One carbon transfer
Biotin	Biotin	Carbon dioxide transfer
Cyanocobalamin	Methylcobalamin	Isomerases, dehydrases

Thiamin

Chemical nature

Thiamin (vitamin B_1) is a complex nitrogenous base containing a pyrimidine ring joined to a thiazole ring. Because of the presence of a hydroxyl group at the end of the side chain, thiamin can form esters. The main form of thiamin in animal tissues is the diphosphate ester, commonly known as thiamin pyrophosphate (TPP). The vitamin is very soluble in water and is fairly stable in mildly acidic solution but readily decomposes in neutral solutions.

Thiamin chloride

Sources

Thiamin is widely distributed in foods. It is concentrated in the outer layers of seeds, the germ, and in the growing areas of roots, leaves and shoots. Fermentation products, such as brewer's yeast, are rich sources. Animal products rich in thiamin include egg yolk, liver, kidney and pork muscle. The synthetic vitamin is available, usually marketed as the hydrochloride.

Metabolism

Thiamin pyrophosphate (or thiamin diphosphate) is a coenzyme involved in (a) the oxidative decarboxylation of pyruvate to acetyl coenzyme A (enzyme : pyruvate dehydrogenase), (b) the oxidative decarboxylation of α-ketoglutarate to succinyl coenzyme A (α-ketoglutarate dehydrogenase) in the tricarboxylic acid cycle, (c) the pentose phosphate pathway (transketolase) and (d) the synthesis of branched chain amino acids such as valine (branch chain keto acid dehydrogenase) in bacteria, yeasts and plants.

Thiamin triphosphate is involved in the activation of the chloride ion channel in the membranes of nerves, possibly by phosphorylation of the channel protein.

Deficiency symptoms

Early signs of thiamin deficiency in most species include loss of appetite, emaciation, muscular weakness and a progressive dysfunction of the nervous system. In pigs, appetite and growth are adversely affected and the animals may vomit and have respiratory troubles.

Chicks reared on thiamin-deficient diets have poor appetites and are consequently emaciated. After about ten days they develop polyneuritis, which is characterised by head retraction, nerve degeneration and paralysis.

Many of these deficiency conditions in animals can be explained in terms of the role of TPP in the oxidative decarboxylation of pyruvic acid. On a thiamin-deficient diet animals accumulate pyruvic acid and its reduction product lactic acid in their tissues, which leads to muscular weakness. Nerve cells are particularly dependent on the utilisation of carbohydrate and for this reason a deficiency of the vitamin has a particularly serious effect on nervous tissue. Since acetyl coenzyme A is an important metabolite in the synthesis of fatty acids (see p. 230), lipogenesis is reduced. The pentose phosphate pathway is also impaired by a deficiency of thiamin but there is little effect on the activity of the citric acid cycle.

Because thiamin is fairly widely distributed in foods and, in particular, because cereal grains are rich sources of the vitamin, pigs and poultry are in practice unlikely to suffer from thiamin deficiency.

In ruminants, microbial synthesis of the vitamin in the digestive tract, together with that present in the diet, will normally provide adequate amounts of thiamin to satisfy the animal's requirements. However, under certain conditions, bacterial thiaminases can be produced in the rumen which destroy the vitamin thereby causing the deficiency condition known as cerebrocortical necrosis (CCN). This condition is characterised by circling movements, head pressing, blindness and muscular tremors. There are two types of thiaminase, one of which splits the molecule in two whereas the other substitutes a N-containing ring for the thiazole ring. The resulting compound is absorbed and blocks the reactions involving thiamin. It has been suggested that lactic acidosis caused by feeding with rapidly fermentable foods may be an important factor in the production of thiaminases. Young animals appear to be the most susceptible.

Thiaminase is present in bracken (*Pteridium aquilinum*) and thiamin deficiency symptoms have been reported in horses consuming this material. Raw fish also contains the enzyme, which destroys the thiamin in foods with which the fish is mixed. The activity of the thiaminase is, however, destroyed by cooking.

Riboflavin

Chemical nature

Riboflavin (vitamin B_2) consists of a dimethyl-isoalloxazine nucleus combined with ribitol. Its structure is shown below.

It is a yellow, crystalline compound, which has a yellowish-green fluorescence in aqueous solution. Riboflavin is only sparingly soluble in water; it is heat-stable in acid or neutral solutions, but is destroyed by alkali. It is unstable to light, particularly ultraviolet light.

$$CH_2 - \overset{\overset{\displaystyle HO}{|}}{\underset{\underset{\displaystyle H}{|}}{C}} - \overset{\overset{\displaystyle OH}{|}}{\underset{\underset{\displaystyle H}{|}}{C}} - \overset{\overset{\displaystyle OH}{|}}{\underset{\underset{\displaystyle H}{|}}{C}} - CH_2OH$$

Riboflavin

Sources

Riboflavin occurs in all biological materials. The vitamin can be synthesised by all green plants, yeasts, fungi and most bacteria, although the lactobacilli are a notable exception and require an exogenous source. Rich sources are yeast, liver, milk (especially whey) and green leafy crops. Cereal grains are poor sources.

Metabolism

Riboflavin is an important constituent of the flavoproteins. The prosthetic group of these compound proteins contains riboflavin in the form of the phosphate (flavin mononucleotide or FMN) or in a more complex form as flavin adenine dinucleotide (FAD). There are several flavoproteins which function in the animal body; they are all concerned with chemical reactions involving the transport of hydrogen. Further details of the importance of flavoproteins in carbohydrate and amino acid metabolism are discussed in Chapter 9. Flavin adenine dinucleotide plays a role in the oxidative phosphorylation system (see Fig. 9.2) and forms the prosthetic group of the enzyme succinic dehydrogenase, which converts succinic acid to fumaric acid in the citric acid cycle. It is also the coenzyme for acyl CoA dehydrogenase.

Deficiency symptoms

In pigs, deficiency symptoms include poor appetite with consequent retardation in growth, vomiting, skin eruptions and eye abnormalities. Riboflavin is essential in the diet of sows to maintain normal oestrus activity and prevent premature parturition. Chicks reared on a riboflavin-deficient diet grow slowly and develop 'curled toe paralysis', a specific symptom caused by peripheral nerve degeneration, in which the chicks walk on their hocks with the toes curled inwards. In breeding hens, a deficiency reduces hatchability. Embryonic abnormalities occur, including the characteristic 'clubbed down' condition in which the down feather continues to grow inside the follicle, resulting in a coiled feather.

The vitamin is synthesised in the rumen and deficiencies in animals with functional rumens are unlikely to occur. However, riboflavin deficiencies have been demonstrated in young calves and lambs. Symptoms include loss of appetite, diarrhoea and lesions in the corners of the mouth.

Nicotinamide

Chemical nature

Another member of the B vitamin complex, nicotinamide is the amide derivative of nicotinic acid (pyridine 3-carboxylic acid) and is the form in which it functions in the body. The relationship between nicotinic acid, nicotinamide and the amino acid tryptophan, which can act as a precursor, is shown below:

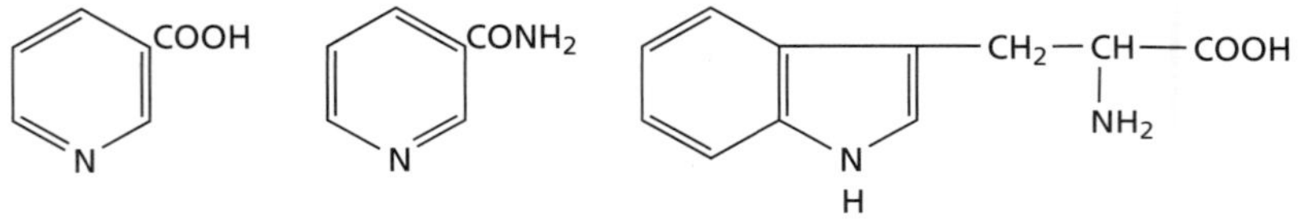

Nicotinic acid   Nicotinamide   Tryptophan

Nicotinamide is a stable vitamin and is not easily destroyed by heat, acids, alkalis or oxidation.

Sources

Nicotinic acid can be synthesised from tryptophan in the body tissues, and since animals can convert the acid to the amide-containing coenzyme (see below), it follows that if the diet is adequately supplied with proteins rich in tryptophan, then the dietary requirement for the vitamin itself should be low. However, the efficiency of conversion of tryptophan into nicotinamide is poor. Studies with chicks have shown that the amino acid was converted into the vitamin at a ratio of only 45 : 1 on a weight basis and with some foods, such as soya bean meal, the conversion ratio may be even greater. Because of this it is generally considered that an exogenous source of the vitamin is also necessary. Rich sources of the vitamin are liver, yeast, groundnut and sunflower meals. Although cereal grains contain the vitamin, much of it is present in a bound form which is not readily available to pigs and poultry. Milk and eggs are almost devoid of the vitamin although they contain the precursor tryptophan.

Metabolism

Nicotinamide functions in the animal body as the active group of two important coenzymes: nicotinamide adenine dinucleotide (NAD) and nicotinamide adenine dinucleotide phosphate (NADP). These coenzymes are involved in the mechanism of hydrogen transfer in living cells (see Chapter 9): NAD is involved in the oxidative phosphorylation system, the TCA cycle and the metabolism of many molecules including pyruvate, acetate, β-OH butyrate, glycerol, fatty acids and glutamate: NADPH is the hydrogen acceptor in the pentose phosphate pathway.

Deficiency symptoms

In pigs, deficiency symptoms include poor growth, anorexia, enteritis, vomiting and dermatitis. In fowls, a deficiency of the vitamin causes bone disorders, feathering abnormalities and inflammation of the mouth and upper part of the oesophagus.

Deficiency symptoms are particularly likely in pigs and poultry if diets with a high maize content are used, since maize contains very little of the vitamin or of tryptophan.

In dairy cows, supplements of nicotinic acid have been shown to reduce the incidence of ketosis in early lactation. This supplementation affects rumen function by increasing microbial protein synthesis in the rumen, the proportion of propionate in ruminal volatile fatty acids and nutrient turnover, with the result that the cow has an increased blood glucose concentration and reduced β-OH butyrate and free fatty acids.

Vitamin B$_6$

Chemical nature

The vitamin exists in three forms, which are interconvertible in the body tissues. The parent substance is known as pyridoxine, the corresponding aldehyde derivative as pyridoxal and the amine as pyridoxamine. The term vitamin B$_6$ is generally used to describe all three forms.

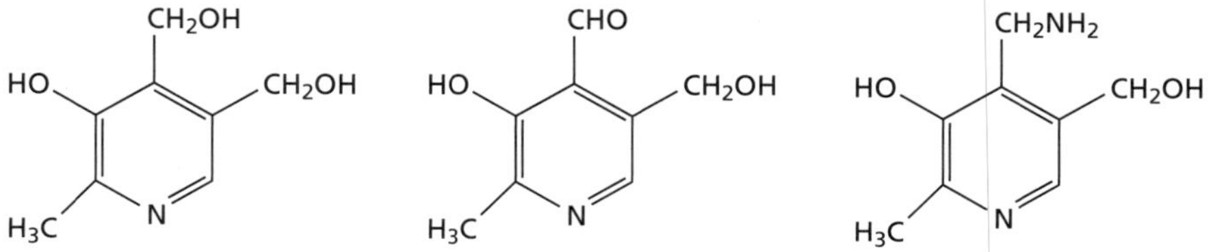

Pyridoxine   Pyridoxal   Pyridoxamine

The amine and aldehyde derivatives are less stable than pyridoxine and are destroyed by heat.

Sources

The vitamin is present in plants as pyridoxine, whereas animal products may also contain pyridoxal and pyridoxamine. Pyridoxine and its derivatives are widely distributed: yeast, pulses, cereal grains, liver and milk are rich sources.

Metabolism

Of the three related compounds, the most actively functioning one is pyridoxal in the form of the phosphate. Pyridoxal phosphate plays a central role as a coenzyme in the reactions by which a cell transforms nutrient amino acids into mixtures of amino acids and other nitrogenous compounds required for its own metabolism. These reactions involve the activities of transaminases and decarboxylases (see p. 218), and over 50 pyridoxal phosphate-dependent enzymes have been identified. In transamination, pyridoxal phosphate accepts the α-amino group of the amino acid to form pyridoxamine phosphate and a keto acid. The amino group of pyridoxamine phosphate can be transferred to another keto acid, regenerating pyridoxal phosphate. The vitamin is believed to play a role in the absorption of amino acids from the intestine.

Deficiency symptoms

Because of the numerous enzymes requiring pyridoxal phosphate, a large variety of biochemical lesions are associated with vitamin B_6 deficiency. These lesions are concerned primarily with amino acid metabolism and a deficiency affects the animal's growth rate. Convulsions may also occur, possibly because a reduction in the activity of glutamic acid decarboxylase results in an accumulation of glutamic acid. In addition, pigs reduce their food intake and may develop anaemia. Chicks on a deficient diet show jerky movements; while in adult birds, hatchability and egg production are adversely affected. In practice, vitamin B_6 deficiency is unlikely to occur in farm animals because of the vitamin's wide distribution.

Pantothenic acid

Chemical nature

Pantothenic acid, another member of the vitamin B complex, is an amide of pantoic acid and β-alanine and has the formula:

$$HOCH_2-\underset{\underset{CH_3}{|}}{\overset{\overset{CH_3}{|}}{C}}-\underset{\overset{OH}{|}}{CH}-CONHCH_2CH_2COOH$$

Pantothenic acid

Sources

The vitamin is widely distributed; indeed the name is derived from the Greek *pantothen*, 'from everywhere', indicating its ubiquitous distribution. Rich sources are liver, egg yolk, groundnuts, peas, yeast and molasses. Cereal grains and potatoes are also good sources of the vitamin. The free acid is unstable and the synthetically prepared calcium pantothenate is the commonest product used commercially.

Metabolism

Pantothenic acid is a constituent of coenzyme A, which is the important coenzyme in fatty acid oxidation, acetate metabolism, and cholesterol and steroid synthesis. It forms the prosthetic group of acyl carrier protein in fatty acid synthesis. Chemically, coenzyme A is 3-phospho-adenosine-5-diphospho-pantotheine. The importance of this coenzyme in metabolism is discussed in Chapter 9.

Coenzyme A

Deficiency symptoms

Deficiency of pantothenic acid in pigs causes slow growth, diarrhoea, loss of hair, scaliness of the skin and a characteristic 'goose-stepping' gait; in severe cases animals are unable to stand. In the chick, growth is retarded and dermatitis occurs. In mature birds, hatchability is reduced. Pantothenic acid, like all the B complex vitamins, can be synthesised by rumen microorganisms; *Escherichia coli*, for example, is known to produce this vitamin. Pantothenic acid deficiencies are considered to be rare in practice because of the wide distribution of the vitamin, although deficiency symptoms have been reported in commercial herds of Landrace pigs.

Folic acid

Chemical nature

This B complex vitamin was first discovered in the 1930s when it was found that a certain type of anaemia in human beings could be cured by treatment with yeast or liver extracts. The active component in the extracts, which was also shown to be essential for the growth of chicks, was found to be present in large quantities in green leaves and was named folic acid (Latin *folium*, a leaf).

The chemical name for folic acid is pteroylmonoglutamic acid and it is made up of three moieties: *p*-aminobenzoic acid, glutamic acid and a pteridine nucleus.

Pteroylmonoglutamic acid

Several active derivatives of the vitamin are known to occur, these containing up to 11 glutamate residues in the molecule. The monoglutamate form is readily absorbed from the digestive tract but the polyglutamates must be degraded by enzymes to the monoglutamate form before they can be absorbed.

Sources

Folic acid is widely distributed in nature; green leafy materials, cereals and extracted oilseed meals are good sources of the vitamin. Folic acid is reasonably stable in foods stored under dry conditions but it is readily degraded by moisture, particularly at high temperatures. It is also destroyed by ultraviolet light.

Metabolism

After absorption into the cell, folic acid is converted into tetrahydrofolic acid, which functions as a coenzyme in the mobilisation and utilisation of single-carbon groups (e.g. formyl, methyl) that are added to, or removed from, such metabolites as histidine, serine, glycine, methionine and purines. It is involved in the synthesis of RNA, DNA and neurotransmitters.

Deficiency symptoms

A variety of deficiency symptoms in chicks and young turkeys have been reported, including poor growth, anaemia, poor bone development and poor egg hatchability. Folic acid deficiency symptoms rarely occur in other farm animals because of synthesis by intestinal bacteria. Injections of folic acid in sows has increased litter size. In one experiment, dietary supplements resulted in higher foetal survival, thought to be related to prostaglandin activity, but the response has not been substantiated in other experiments, possibly because of the variable content of folic acid in foods. Its role in nucleic acid metabolism concurs with the view that supplements might be beneficial at times of growth and differentiation of embryonic tissue.

Biotin

Chemical nature

A part of the vitamin B complex, biotin is chemically 2-keto-3,4-imidazolido-2-tetrahydrothiophene-n-valeric acid. Its structure is:

Biotin

Sources

Biotin is widely distributed in foods; liver, milk, yeast, oilseeds and vegetables are rich sources. However, in some foods much of the bound vitamin may not be released during digestion and hence may be unavailable. Studies with chicks and pigs have shown that the availability of biotin in barley and wheat is very low, whereas the biotin in maize and certain oilseed meals, such as soya bean meal, is completely available.

Metabolism

Biotin serves as the prosthetic group of several enzymes which catalyse the transfer of carbon dioxide from one substrate to another. In animals there are three biotin-dependent enzymes of particular importance: pyruvate carboxylase (carbohydrate synthesis from lactate), acetyl coenzyme A carboxylase (fatty acid synthesis) and propionyl coenzyme A carboxylase (the pathway of conversion of propionate to succinyl CoA). The specific role of these enzymes in metabolism is discussed in Chapter 9.

Deficiency symptoms

In pigs, biotin deficiency causes foot lesions, alopecia (hair loss) and a dry scaly skin. In growing pigs, both growth rate and food utilisation are adversely affected. In breeding sows, a deficiency of the vitamin can adversely influence reproductive performance.

In poultry, biotin deficiency causes reduced growth, dermatitis, leg bone abnormalities, cracked feet, poor feathering and fatty liver and kidney syndrome (FLKS). This last condition, which mainly affects two- to five-week-old chicks, is characterised by a lethargic state with death frequently following within a few hours. On autopsy, the liver and kidneys, which are pale and swollen, contain abnormal depositions of lipid.

Biotin deficiency can be induced by giving animals avidin, a protein present in the raw white of eggs, which combines with the vitamin and prevents its absorption from the intestine. Certain bacteria of the *Streptomyces* spp. that are present in soil and manure produce streptavidin and stravidin, which have a similar action to the egg white protein. Heating inactivates these antagonist proteins.

Choline

Chemical nature

The chemical structure of choline is:

$$CH_3 - \overset{\overset{\displaystyle CH_3}{\overset{+}{|}}}{\underset{\underset{\displaystyle CH_3}{|}}{N}} - CH_2CH_2OH$$

Choline

Sources

Green leafy materials, yeast, egg yolk and cereals are rich sources of choline.

Metabolism

Unlike the other B vitamins, choline is not a metabolic catalyst but forms an essential structural component of body tissues. It is a component of lecithins, which play a vital role in cellular structure and activity. It also plays an important part in lipid metabolism in the liver, where it converts excess fat into lecithin or increases the utilisation of fatty acids, thereby preventing the accumulation of fat in this organ. Choline is a component of acetylcholine, which is responsible for the transmission of nerve impulses. Finally, it serves as a donor of methyl groups in transmethylation reactions which involve folic acid or vitamin B_{12}. Although other compounds, such as methionine or betaine, can also act as methyl donors, they cannot replace choline in its other functions.

Choline can be synthesised in the liver from methionine and the exogenous requirement for this vitamin is therefore influenced by the level of methionine in the diet.

Deficiency symptoms

Deficiency symptoms, including slow growth and fatty infiltration of the liver, have been produced in chicks and pigs. Choline is also concerned with the prevention of perosis or slipped tendon in chicks. The choline requirement of animals is unusually large for a vitamin, but in spite of this, deficiency symptoms are not common in farm animals because of its wide distribution and its high concentrations in foods, and because it can be readily derived from methionine.

Vitamin B_{12}

Chemical nature

Vitamin B_{12} has the most complex structure of all the vitamins. The basic unit is a corrin nucleus, which consists of a ring structure comprising four five-membered rings containing nitrogen. In the active centre of the nucleus is a cobalt atom. A cyano group is usually attached to the cobalt as an artefact of isolation and, as this is the most stable form of the vitamin, it is the form in which the vitamin is commercially produced.

Vitamin B_{12} (cyanocobalamin)

In the animal, the cyanide ion is replaced by a variety of ions, e.g hydroxyl (hydroxocobalamin), methyl (methylcobalamin) and 5-deoxyadenosyl (5-deoxyadenosylcobalamin), the last two forms acting as coenzymes in animal metabolism.

Sources

Vitamin B_{12} is considered to be synthesised exclusively by microorganisms and its presence in foods is thought to be ultimately of microbial origin. The main natural sources of the vitamin are foods of animal origin, liver being a particularly rich source. Its limited occurrence in higher plants is still controversial, since many think that its presence in trace amounts may result from contamination with bacteria or insect remains.

Metabolism

Before vitamin B_{12} can be absorbed from the intestine it must be bound to a highly specific glycoprotein, termed the intrinsic factor, which is secreted by the gastric mucosa. In man, the intrinsic factor may be lacking, which leads to poor absorption of the vitamin and results in a condition known as pernicious anaemia.

The coenzymic forms of vitamin B_{12} function in several important enzyme systems. These include isomerases, dehydrases and enzymes involved in the biosynthesis of methionine from homocysteine. Of special interest in ruminant nutrition is the role of vitamin B_{12} in the metabolism of propionic acid into succinic acid. In this pathway, the vitamin is necessary for the conversion of methylmalonyl coenzyme A into succinyl coenzyme A (see p. 211).

Deficiency symptoms

Adult animals are generally less affected by a vitamin B_{12} deficiency than are young growing animals, in which growth is severely retarded and mortality high.

In poultry, in addition to the effect on growth, feathering is poor and kidney damage may occur. Hens deprived of the vitamin remain healthy, but hatchability is adversely affected.

On vitamin B_{12} deficient diets, baby pigs grow poorly and show lack of coordination of the hind legs. In older pigs, dermatitis, a rough coat and suboptimal growth result. Intestinal synthesis of the vitamin occurs in pigs and poultry. Organisms which synthesise vitamin B_{12} have been isolated from poultry excreta and this fact has an important practical bearing on poultry housed with access to litter, where a majority, if not all, of the vitamin requirements can be obtained from the litter.

Vitamin B_{12} and a number of biologically inactive vitamin B_{12} analogues are synthesised by microorganisms in the rumen and, in spite of poor absorption of the vitamin from the intestine, the ruminant normally obtains an adequate amount of the vitamin from this source. However, if levels of cobalt in the diet are low, a deficiency of the vitamin can arise and cause reduced appetite, emaciation and anaemia (see p. 133). If cobalt levels are adequate, then except with very young ruminant animals, a dietary source of the vitamin is not essential. Horses also are supplied with sufficient B_{12} from microbial fermentation when sufficient cobalt is supplied. Parasitised horses have responded to vitamin B_{12} supplementation, presumably as a result of impaired digestive activity.

Other growth factors included in the vitamin B complex

A number of other chemical substances of an organic nature have been included in the vitamin B complex. These include inositol, orotic acid, lipoic acid, rutin, carnitine and pangamic acid, but it is doubtful if these compounds have much practical significance in the nutrition of farm animals.

5.4   Vitamin C

Chemical nature

Vitamin C is chemically known as L-ascorbic acid and has the following formula:

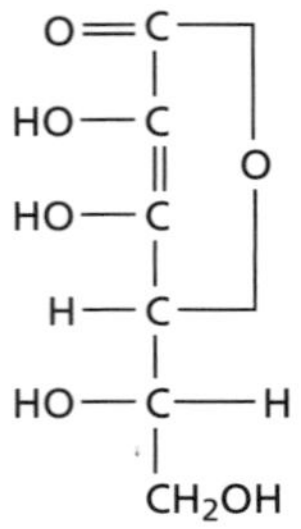

L-Ascorbic acid

The vitamin is a colourless, crystalline, water-soluble compound having acidic and strong reducing properties. It is heat-stable in acid solution but is readily decomposed in the presence of alkali. The destruction of the vitamin is accelerated by exposure to light.

Sources

Well-known sources of this vitamin are citrus fruits and green leafy vegetables. Synthetic ascorbic acid is available commercially.

Metabolism

Ascorbic acid plays an important part in various oxidation–reduction mechanisms in living cells. The vitamin is necessary for the maintenance of normal collagen metabolism. It also plays an important role in the transport of iron ions from transferrin, found in the plasma, to ferritin, which acts as a store of iron in the bone marrow, liver and spleen. As an antioxidant, ascorbic acid works in conjunction with vitamin E in protecting cells against oxidative damage caused by free radicals (see Box 5.2). The vitamin is required in the diet of only a few vertebrates – man, other primates, the guinea pig, the red-vented bulbul bird and the fruit-eating bat (both native to India) and certain fishes. Some insects and other invertebrates also require a dietary source of vitamin C. Other species synthesise the vitamin from glucose, via glucuronic acid and gulonic acid lactone; the enzyme L-gulonolactone oxidase is required for the synthesis, and species requiring ascorbic acid are genetically deficient in this enzyme.

Deficiency symptoms

The classic condition in man arising from a deficiency of vitamin C is scurvy. This is characterised by oedema, emaciation and diarrhoea. Failure in collagen formation results in structural defects in bone, teeth, cartilage, connective tissues and muscles. Resistance to infection is reduced.

Since farm animals can synthesise this vitamin, deficiency symptoms will normally not arise. However, it has been suggested that under certain conditions, e.g. climatic stress in poultry, the demand for ascorbic acid becomes greater than can be provided for by normal tissue synthesis and a dietary supplement may be beneficial.

5.5 Hypervitaminosis

Hypervitaminosis is the name given to pathological conditions resulting from an overdose of vitamins. Under 'natural' conditions it is unlikely that farm animals will receive excessive doses of vitamins, although when synthetic vitamins are added to diets there is always the risk that abnormally large amounts may be ingested if errors are made during mixing. There is experimental evidence that toxic symptoms can occur if animals are given excessive quantities of vitamin A or D.

Clinical signs of hypervitaminosis A in young chicks kept under experimental conditions and given very high doses of vitamin A include loss of appetite, poor growth, diarrhoea, encrustation around the mouth and reddening of the eyelids. In pigs, toxic symptoms include rough coat, scaly skin, hyperirritability, haemorrhages over the limbs and abdomen, periodic tremors and death.

Excessive intakes of vitamin D cause abnormally high levels of calcium and phosphorus in the blood, which result in the deposition of calcium salts in the arteries and organs. Symptoms of hypervitaminosis D have been noted in cattle and calves. In the UK, the maximum amount of vitamin D supplement added to diets for farm animals is controlled by legislation.

Depression in growth and anaemia caused by excessive doses of menadione (vitamin K) have been reported.

Summary

Vitamins are involved in metabolic pathways as coenzymes, and some act as protectors in antioxidant and immune systems. The sources and functions of individual vitamins, and the disorders caused by their deficiencies, are summarised below.

Vitamin	Source	Actions	Deficiency symptoms
A, retinol	Fish-liver oil	Sight, epithelial tissues	Blindness, epithelial infection
D_3, cholecalciferol	Fish-liver oil, sun-dried roughage	Calcium absorption	Rickets
E, α-tocopherol	Green foods, cereals	Antioxidant	Muscle degeneration, liver damage
K, menadione	Green foods, egg yolk	Prothrombin synthesis	Anaemia, delayed clotting
B_1, thiamin	Seeds	Carbohydrate and fat metabolism	Poor growth, polyneuritis
B_2, riboflavin	Green foods, milk	Carbohydrate and amino acid metabolism	Poor growth, curled toe paralysis
Nicotinamide	Yeast, liver, tryptophan	Hydrogen transfer (NAD and NADH)	Poor growth, dermatitis
B_6, pyridoxine	Cereals, yeast	Amino acid metabolism	Poor growth, convulsions
Pantothenic acid	Liver, yeast, cereals	Acetate and fatty acid metabolism (Coenzyme A)	Poor growth, scaly skin, goose-stepping in pigs
Folic acid	Green foods, cereals, oilseed meals	Metabolism of single carbon compounds	Poor growth, anaemia, poor hatchability
Biotin	Liver, vegetables	Carbon dioxide transfer	Foot lesions, hair loss, FLKS
Choline	Green foods, cereals, methionine	Component of lecithin	Poor growth, fatty liver, perosis
B_{12}, cyanocobalamin	Microorganisms, liver	Propionate metabolism	Poor growth, anaemia, poor coat/feathering
C, ascorbic acid	Citrus fruits, leafy vegetables	Oxidation/reduction reactions	Reduced resistance to infection

Further reading

Aurbach G D *et al.* (eds) *Vitamins and Hormones: Advances in Research and Applications* (annual or biennial volumes since 1942), New York, Academic Press.

Bieber-Wlaschny M 1988 Vitamin requirements of the dairy cow. In: Garnsworthy P C (ed.) *Nutrition and Lactation in the Dairy Cow*, London, Butterworth, pp 135–56.

Chew B P 1995 The influence of vitamins on reproduction in pigs. In: Garnsworthy P C and Cole D J A (eds) *Recent Advances in Animal Nutrition 1995*, Nottingham, Nottingham University Press, pp 223–39.

Diplock A T (ed.) 1985 *Fat Soluble Vitamins: Their Biochemistry and Applications*, London, Heinemann.

Latscha T 1990 *Carotenoids – Their Nature and Significance in Animal Feeds*, Basel, Hoffmann–La Roche.

Machlin L J (ed.) 1984 *Handbook of Vitamins*, New York, Marcel Dekker.

Whitehead C C 1986 Requirements for vitamins. In: Fisher C and Boorman K N (eds) *Nutrient Requirements of Poultry and Nutrition Research*, London, Butterworth, pp 173–89.

6 Minerals

6.1　Functions of minerals

6.2　Natural and supplementary sources of minerals

6.3　Acid–base balance

6.4　Major elements

6.5　Trace elements

6.6　Other elements

6.1　Functions of minerals

Although most of the naturally occurring mineral elements are found in animal tissues, many are thought to be present merely because they are constituents of the animal's food and may not have an essential function in the animal's metabolism. The term 'essential mineral element' is restricted to a mineral element which has been proved to have a metabolic role in the body. Before an element can be classed as essential it is generally considered necessary to prove that purified diets lacking the element cause deficiency symptoms in animals and that those symptoms can be eradicated or prevented by adding the element to the experimental diet. Most research on mineral nutrition has been carried out in this way. However, some of the mineral elements required by animals for normal health and growth are needed in such minute amounts that the construction of deficient diets is often difficult to achieve and deficiency has been demonstrated only in laboratory animals under special conditions. In such studies it is necessary not only to monitor food and water supplies but also to ensure that animals do not obtain the element under investigation from cages, troughs, attendants or dust in the atmosphere.

Requirements or allowances for minerals may be derived factorially from the amounts required to meet endogenous losses and the mineral retained or excreted in products, or, more usually for the trace elements, empirically from dose responses to levels in the diet. The main problem for the minerals with more than one function is deciding on the criterion for adequacy. Dietary levels of mineral that are sufficient for one function may be inadequate for another. Adequacy for the short period of growth of animals slaughtered for meat may not cover the long-term needs of breeding and longevity of adult stock. Also, for the minerals involved in bone formation, there are many criteria, e.g. bone dimensions, strength, histology and composition. Depletion of body reserves in times of undernutrition will supply the mineral for its metabolic functions and this needs to be considered if requirements are to be measured accurately.

Until 1950, 13 mineral elements were classified as essential; these comprised the major elements (calcium, phosphorus, potassium, sodium, chlorine, sulphur, magnesium) and the micro or trace elements (iron, iodine, copper, manganese, zinc and cobalt). By 1970, molybdenum, selenium, chromium and fluorine had been added to the list and subsequently arsenic, boron, lead, lithium, nickel, silicon, tin, vanadium, rubidium and aluminium have been included, the list varying slightly according to the different authorities. Plant and animal tissues contain a further 30 mineral elements, in small quantities, for which no essential function has been found. They may be acquired from the environment, but it has been suggested that as many as 40 or more elements may have metabolic roles in mammalian tissues. Fortunately, many of these trace elements, especially those of more recent discovery, are required in such minute quantities, or are so widely distributed in foods for animals, that deficiencies are likely to be extremely rare under normal practical conditions.

The classification of the essential minerals into major elements and trace elements depends upon their concentration in the animal or amounts required in the diet. Normally trace elements are present in the animal body in a concentration not greater than 50 mg/kg and are required at less than 100 mg/kg diet. Those essential mineral elements that are of particular nutritional importance together with their approximate concentrations in the animal body are shown in Table 6.1.

The minerals are held in different forms in the body, which can be considered as compartments. There is a central reserve or interchange compartment, which is usually blood plasma, and one or more compartments which interchange the mineral with the central compartment at various rates, e.g. compartments easy or difficult to mobilise. Metabolic processes take place via the central reserve (plasma), which receives minerals from other compartments, the digestive tract and the difficult to mobilise compartment. The central reserve secretes mineral into the readily mobilised

Table 6.1 Nutritionally important essential mineral elements and their approximate concentration in the animal

Major elements	g/kg	Trace elements	mg/kg
Calcium	15	Iron	20–80
Phosphorus	10	Zinc	10–50
Potassium	2	Copper	1–5
Sodium	1.6	Molybdenum	1–4
Chlorine	1.1	Selenium	1–2
Sulphur	1.5	Iodine	0.3–0.6
Magnesium	0.4	Manganese	0.2–0.5
		Cobalt	0.02–0.1

compartments, the difficult to mobilise compartment, the gastrointestinal tract, the kidneys and milk. The flux between the compartments can be measured by a combination of balance trials and injection of radioactive marker followed by sampling the tissues over time. An example of the body compartments of copper is shown in Fig. 6.1.

Nearly all the essential mineral elements, both major and trace, are believed to have one or more *catalytic* functions in the cell. Some mineral elements are firmly bound to the proteins of enzymes (see Box 6.1), while others are present in prosthetic groups in chelated form. A chelate is a cyclic compound which is formed between an organic molecule and a metallic ion, the latter being held within the organic molecule as if by a claw ('chelate' is derived from the Greek work meaning 'claw'). Examples of naturally occurring chelates are the chlorophylls, cytochromes, haemoglobin and vitamin B_{12}.

Elements such as sodium, potassium and chlorine have primarily an electrochemical or *physiological* function and are concerned with the maintenance of acid–base balance, membrane permeability and the osmotic control of water distribution within the body. Some elements have a *structural* role, for example calcium and phosphorus are essential components of the skeleton and sulphur is necessary for the synthesis of structural proteins. Finally, certain elements have a *regulatory* function in controlling cell replication and differentiation. Zinc acts in this way by influencing the transcription process, in which genetic information in the nucleotide sequence of DNA is transferred to that of an RNA molecule. It is not uncommon for

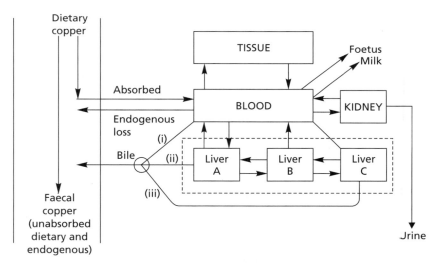

Fig. 6.1 Diagram of the possible routes of movement of copper in the ruminant body. A is a temporary storage compartment for copper in the liver destined for exchange with blood and excretion into bile, B represents a temporary storage for incorporation into caeruloplasmin and C represents a long-term storage compartment. (Adapted from Symonds H W and Forbes J M 1993 Mineral metabolism. In: Forbes J M and France J (eds) *Quantitative Aspects of Ruminant Digestion and Metabolism*, Wallingford, UK, CAB International.)

Box 6.1
Mineral elements and enzymes

Some examples of the involvement of minerals in enzymes (summarised from Underwood and Suttle – see Further Reading).

Mineral	Enzyme	Function
Iron	Succinate dehydrogenase	Aerobic oxidation of carbohydrates
	Cytochromes a, b and c	Electron transfer
Copper	Cytochrome oxidase	Terminal oxidase
	Ceruloplasmin (ferroxidase)	Iron utilisation: copper transport
	Superoxide dismutase	Dismutation of superoxide radical O_2^-
Zinc	Carbonic anhydrase	CO_2 formation
	Alcohol dehydrogenase	Alcohol metabolism
	Carboxypeptidase A	Protein digestion
	Alkaline phosphatase	Hydrolysis of phosphate esters
Manganese	Pyruvate carboxylase	Pyruvate metabolism
	Superoxide dismutase	Antioxidant by removing O_2^-
Molybdenum	Xanthine dehydrogenase	Purine metabolism
	Sulphite oxidase	Sulphite oxidation
Selenium	Glutathione peroxidases	Removal of H_2O_2 and hydroperoxides
	Type I and III deiodinases	Conversion of thyroxine to active form

an element to have a number of different roles; for example, magnesium functions catalytically, electrochemically and structurally.

A number of elements have unique functions. Iron is important as a constituent of haem, which is an essential part of a number of cytochromes important in respiration. Cobalt is a component of vitamin B_{12} and, as mentioned above, iodine forms part of the hormone thyroxine.

Some elements, for example calcium and molybdenum, may interfere with the absorption, transport, function, storage or excretion of other elements. There are many ways in which minerals may interact but the three major ones involve the formation of unabsorbable compounds, competition for metabolic pathways and the induction of metal-binding proteins. The interaction of minerals with each other is an important factor in animal nutrition, and an imbalance of mineral elements – as distinct from a simple deficiency – is important in the aetiology of certain nutritional disorders of farm animals. The use of radioactive isotopes in recent years has advanced our knowledge of mineral nutrition, although there are many nutritional diseases associated with minerals whose exact causes are still unknown.

Although we have been considering the essential role of minerals in animal nutrition, it is important to realise that many are toxic – causing illness or death – if given to the animal in excessive quantities. This is particularly true of copper, selenium, molybdenum, fluorine, vanadium and arsenic. Copper is a cumulative poison, the animal body being unable to excrete it efficiently; small amounts of the element given in excess of the animal's daily needs will, in time, produce toxic symptoms. This also applies to the element fluorine. Supplementation of any diet with minerals should always be carried out with great care, and the indiscriminate use of trace elements in particular must be avoided. Ideally, the supplement should be tailored to the target animal and a blanket oversupply should be avoided as being wasteful and potentially dangerous.

6.2 Natural and supplementary sources of minerals

Plants and plant products form the main supply of nutrients to animals, and the composition of plants will influence the animal's mineral intake. Thus the species and stage of maturity of the plant, the type of soil and climate and seasonal conditions are important factors. Legumes tend to be richer in the major minerals and certain trace elements than grasses and this is also the case with the seeds of legumes compared with the seeds of grasses and cereals. Soil conditions (e.g. pH) and mineral content affect the uptake of minerals by plants and this can be further influenced by fertiliser application and liming. The main animal products used in animal feeding, fishmeal, whey and skimmed milk, are good sources of the major minerals.

Usually, diets for farm animals contain a mineral/trace element/vitamin supplement and, on occasions, it is necessary to include extra supplies of some minerals, e.g. calcium for laying hens. Common sources of minerals used in mineral supplements are limestone for calcium, dicalcium phosphate for phosphorus, common salt for sodium and calcined magnesite for magnesium. Trace elements are usually supplied in a salt form, e.g. selenium as sodium selenite. When considering sources of mineral the cost, chemical and physical form and freedom from impurities are taken into account. It is also necessary to take account of the availability of the element in question (Chapter 10). Calcium tends to have a high availability from most sources and the phosphorus in *ortho-* and *meta*-phosphates has an availability of between 80 and 100 per cent. The availability of phosphorus from rock phosphates can be very low. Magnesium from calcined magnesite has an availability of 50–60 per cent whereas that from magnesium sulphate is up to 70 per cent. Sulphur from sulphates is 50–90 per cent available. The availability of trace elements in the form of sulphate, chloride or nitrate salts is high because they are water-soluble. Table 6.2 shows the availability of minerals in a number of sources relative to a standard source. In these examples the criteria used to assess the relative availability of the different minerals vary, e.g. absorption, accumulation in tissues, production of metabolically active compounds; hence some sources in Table 6.2 have values greater than 100 per cent when compared with the standard source of the mineral.

Table 6.2 Examples of the relative availability (%) of mineral elements from mineral compounds. (Summarised from Ammerman C B, Henry P R and Miles R D 1998 Supplemental organically-bound mineral compounds in livestock nutrition. In: Garnsworthy P C and Wiseman J (eds) *Recent Advances in Animal Nutrition*, Nottingham, Nottingham University Press, pp 67–91)

Mineral compound	Poultry	Pigs	Cattle	Sheep
Cobaltous sulphate	–	–	–	100
Cobalt glucoheptate	–	–	–	85
Cobalt carbonate	–	–	–	100
Cupric sulphate	100	100	100	100
Copper-lysine	105	–	100	–
Copper-methionine	90	110	–	–
Cupric carbonate	65	85	–	–
Cupric chloride	110	–	115	115
Ferrous sulphate ($7H_2O$)	100	100	100	100
Ferric citrate	75	150	110	–
Ferric oxide	10	10	–	5
Ferrous chloride	100	–	–	–
Iron-methionine	–	185	–	
Iron-proteinate	–	125	–	
Manganese sulphate	100	–	–	100
Manganese carbonate	55	–	–	30
Manganese-methionine	120	–	–	125
Manganous chloride	100	–	–	–
Sodium selenite	100	100	100	100
Seleno-cystine	110	–	–	–
Seleno-methionine	$80^a/115^b$	$120^a/150^b$	–	–
Seleno-yeast	–	–	290	100
Zinc chloride	100	100	–	–
Zinc sulphate	100	–	100	100
Zinc-lysine	–	100	–	–
Zinc-methionine	125	100	–	100

[a] Assessed from glutathione peroxidase production or incidence of exudative diathesis.
[b] Assessed from whole body or tissue selenium retention or incidence of pancreatic fibrosis.

Mineral supplementation of animals at pasture can be a problem. Minerals can be incorporated into free-access feed blocks, which also provide a source of energy and nitrogen. However, individual animal intake can be variable. Intake depends on season, weather conditions, the siting and number of blocks (to minimise competition), frequency of renewal and availability of water. The inclusion of oil or molasses improves palatability.

Spraying the pasture with soluble salts of trace elements (e.g. cobalt sulphate) can increase the element content of the pasture. Alternatively, trace elements can be included in fertilisers in order to increase the herbage content via the soil. However, if the deficiency is due to poor availability of the element, then this type of application will not be successful.

For some of the trace elements, e.g. cobalt and copper, a solid bolus can be deposited into the rumen using a dosing gun. This bolus then slowly dissolves over a period of months, giving a steady release of the element. There can be problems of regurgitation of the bolus and they may become coated, thereby reducing the effective release of the mineral. Recently, boluses formed from soluble glass have been produced which are not susceptible to this coating action. The glass boluses may contain more than one element (e.g. cobalt and selenium) and anthelmintics, vitamins A, D or E. Needles of copper oxide, which have a high specific gravity and are retained in the abomasum, have also been used in this way. For certain minerals (e.g. copper and selenium), oral doses, drenches or injections with solutions or pastes can be given at appropriate intervals, but the labour requirement is high.

Free ions from inorganic sources can form complexes with other dietary constituents, resulting in low absorbability and availability to the animal. Minerals in 'chelated' or 'organic' form (where the element is in combination with an organic molecule such as an amino acid) are protected from reaction with other constituents and theoretically have a greater absorbability than inorganic sources. Chelates were mentioned above and the addition of chelated minerals as supplements to diets is currently an active area of research. One of the most potent chelating agents is the synthetic compound ethylene diamine tetraacetic acid (EDTA), which has the property of forming stable chelates with heavy metals. *In vitro*, however, chelates of cobalt were not more effective than cobalt chloride in stimulating microbial synthesis of vitamin B_{12}, and oral supplements of cobalt EDTA and cobalt sulphate gave similar liver and serum vitamin B_{12} contents. The addition of chelating agents such as EDTA to poultry diets may in some cases improve the availability of the mineral element. Some experiments have shown this to happen with zinc and manganese in diets rich in phytic acid, which binds minerals in an unabsorbable form (see p. 120), but this response is not consistent across experiments. Similar variable results have been observed with pigs. However, the bonding between the metal and EDTA is strong, and work with copper chelates in sheep has shown that the copper is no more available than in inorganic salts. Amino acid and peptide chelates are absorbed efficiently, possibly because they are taken up by the peptide absorption mechanism rather than the active transport mechanism for minerals (Chapter 8). Levels of iron in the tissues of piglets and piglet growth have been improved by giving an iron proteinate to the sow. However, the piglets also had access to the sow's faeces and may have obtained iron from these. Again, although zinc methionine has improved tissue levels of zinc in pigs, results have not been consistent. The replacement of sodium selenite by selenium-enriched yeast has improved the selenium content of serum, milk and tissues of gilts and the selenium status of their offspring. Selenium yeast and selenium in brewer's grains in diets for lactating cows have increased the transfer of selenium to milk when compared with selenite. However, inorganic selenium supplements are also effective and at a lower cost. Owing to the different properties of the chelating agents, their use as mineral supplements is controversial. There are many reports in the popular press claiming increased responses. A recent critical review by Underwood and Suttle (1999) suggests caution at this stage until evidence of consistent responses is published in scientific journals. In view of their cost, it is unlikely that they will replace inorganic sources of mineral, except for special applications.

6.3 Acid–base balance

Normally in nutrition, minerals, including those which have electrolytic properties, are considered functionally as separate entities. In physiological terms, however, the electrolytes need to be taken together since cells require a specific balance of anions and cations to function efficiently. Physiological processes operate within a narrow range of conditions, especially with respect to pH. In addition enzyme systems, and hence cell metabolism, are sensitive to pH. Thus changes in acid–base status have a wide influence on cell function and the animal must regulate the input and output of ions to maintain acid–base homeostasis. Failure to maintain the correct electrolyte balance within the cell means that metabolic pathways are unable to function efficiently and resources are diverted to achieving homeostasis at the expense of growth. The diet is important in the maintenance of the correct intracellular electrolyte balance owing to the metabolisable anions and cations that it contains and which consume or generate acid during metabolism. Thus an excess of anions will result in the production of hydrogen ions to counterbalance them, giving metabolic acidosis, whereas an excess of cations requires ions such as acetate and bicarbonate and causes alkalosis. These effects are independent of the specific metabolic or physiological roles of the particular element. The balance of acids and bases influences many functions such as growth rate, appetite, amino acid and energy metabolism, calcium utilisation, vitamin metabolism, intestinal absorption and kidney function. Changes in cellular pH are often accompanied by changes in blood and urine pH. Dietary influence in this respect may be assessed by measuring the dietary electrolyte balance, defined as:

$$Na^+ + K^+ - Cl^-$$

where Na^+, K^+ and Cl^- are the concentrations of the elements in mequiv per unit weight. The dietary electrolyte balance is commonly used to assess the diets of pigs and poultry. Pigs are more susceptible to excess anions than cations and the optimal dietary electrolyte balance is around 250 mequiv/kg. In poultry, eggshell formation has an effect on the acid–base balance as hydrogen ions are generated when calcium carbonate is being synthesised. A dietary electrolyte balance of 200–300 mequiv/kg is recommended for laying hens. In conditions of heat stress, elevated respiration rate (panting) leads to respiratory alkalosis and in poultry and dairy cows the acid–base balance of the diet has been adjusted to alleviate this. Ideally, other elements making a contribution to electrolyte balance should be considered and a more sophisticated assessment may be achieved by calculating $(Na^+ + K^+ + Ca^{++} + Mg^{++}) - (Cl^- + H_2PO_4^- + HPO_4^{--} + SO_4^{--})$.

This is termed the dietary undetermined anion. The latter requires a substantial analytical facility and in practice the less comprehensive dietary electrolyte balance is generally considered to be sufficient. Recently, in ruminant nutrition the cation–anion balance or cation–anion difference (CAD) has been used to assess the electrolyte status of diets. This is calculated as $(Na^+ + K^+) - (Cl^- + SO_4^{--})$ or alternatively as $(Na^+ + K^+) - (Cl^- + S^{--})$. Manipulation of the cation–anion difference is now recommended as part of the dietary management of dairy cows in order to avoid hypocalcaemia or milk fever (see Box 6.2).

In the above it can be seen that there can be confusion with terminology and the method of calculation (i.e. which ions are included) of the acid–base status of the diet. Some authorities suggest the inclusion of NH_4^+, HCO_3^- and CO_3^{--}, while others recommend that the absorption coefficient should also be accounted for. However, such refinement will require much more research and information before this type of model can be applied.

Box 6.2
Cation–anion difference and milk fever

The metabolic acid–base balance affects the sensitivity of bone to parathyroid hormone (PTH) and the synthesis of 1,25-dihydroxycholecalciferol. Therefore, acknowledging this when balancing the mineral content of the diet can have an effect on the incidence of milk fever. Conditions which promote an alkalotic state (high dietary cations, Na^+ and K^+) reduce the sensitivity of bone to PTH and can limit the release of calcium. Conversely, an acidotic state (high dietary anions, Cl^- and S^{--}) increases the sensitivity to PTH, increases 1,25-dihydroxycholecalciferol production and hence increases the calcium supply. Through these metabolic responses, manipulation of the acid–base balance in the diet of the pre-calving cow has been successful in reducing the incidence of milk fever.

Early work used the dietary electrolyte balance calculation ($Na^+ + K^+ - Cl^-$), but subsequently it was found that the inclusion of SO_4^{--} or S^{--} was beneficial. Thus diet cation–anion difference (CAD) as $(Na^+ + K^+) - (Cl^- + SO_4^{--})$ or $(Na^+ + K^+) - (Cl^- + S^{--})$ is used when calculating and manipulating the acid–base balance of the diet. The recommended target for the latter calculation is around –100 mequiv/kg.

In practice, manipulation of the diet involves minimising foods high in potassium and sodium. Grass silage is the major problem, often containing 30–40 g potassium/kg DM, and alkali-treated grain and molasses should also be avoided. The cereal by-product, brewer's grains, is a food which is low in sodium and potassium. Unlike the low-calcium diet strategy, the CAD strategy requires a moderate to high calcium intake to be maintained. Supplementary anionic salts (e.g. ammonium chloride, ammonium sulphate, magnesium chloride, calcium chloride) can be used to adjust the CAD but they tend to be unpalatable. Practical application of this stategy requires knowledge of the mineral contents of the foods. The magnesium content of the diet is adjusted to 3.5 g/kg DM using magnesium sulphate or magnesium chloride and the sulphur content is set at 4 g/kg DM using calcium or ammonium sulphate. The CAD can then be manipulated with ammonium or calcium chloride. Calcium intake is maintained at 120–150 g/day and phosphorus at 50 g/day. This approach to the problem of milk fever requires careful management, since the high quantities of anionic salts involved may reduce food intake and precipitate other metabolic problems. The effectiveness of the strategy can be checked by measuring urine pH, which should be slightly acidic at around 6.5. It is recommended that the diets are given for at least two but no more than four weeks before calving.

Certain pathological conditions may cause disturbances in electrolyte balance, e.g. vomiting (loss of chloride), diarrhoea (loss of bicarbonate) and excessive amino acid oxidation (excess acid production). These, however, are largely outside the control of the nutritionist.

6.4 Major elements

Calcium

Calcium is the most abundant mineral element in the animal body. It is an important constituent of the skeleton and teeth, in which about 99 per cent of the total body calcium is found; in addition, it is an essential constituent of living cells and tissue fluids. Calcium is essential for the activity of a number of enzyme systems including those necessary for the transmission of nerve impulses and for the contractile properties of muscle. It is also concerned in the coagulation of blood. In blood, the element occurs in the plasma; the plasma of mammals usually contains 80–120 mg calcium/litre, but that of laying hens contains more (between 300 and 400 mg/litre).

Composition of bone

Bone is highly complex in structure, the dry matter consisting of approximately 460 g mineral matter/kg, 360 g protein/kg and 180 g fat/kg. The composition varies, however, according to the age and nutritional status of the animal. Calcium and phosphorus are the two most abundant mineral elements in bone; they are combined in a form similar to that found in the mineral hydroxyapatite, $3Ca_3(PO_4)_2 \cdot Ca(OH)_2$. Bone ash contains approximately 360 g calcium/kg, 170 g phosphorus/kg and 10 g magnesium/kg.

The skeleton is not a stable unit in the chemical sense, since large amounts of the calcium and phosphorus in bone can be liberated by reabsorption. This takes place particularly during lactation and egg production, although the exchange of calcium and phosphorus between bones and soft tissue is always a continuous process. Resorption of calcium is controlled by the action of the parathyroid gland. If animals are fed on a low-calcium diet, the ionic calcium concentration in the extracellular fluid falls, the parathyroid gland is stimulated and the hormone produced causes resorption of bone, liberating calcium to meet the requirements of the animal. Since calcium is combined with phosphorus in bone, the phosphorus is also liberated and excreted by the animal.

The parathyroid hormone also plays an important role in regulating the amount of the calcium absorbed from the intestine by influencing the production of 1,25-dihydroxycholecalciferol, a derivative of vitamin D, which is concerned with the formation of calcium-binding protein (see p. 83). Finally, the hormone stimulates the kidney to resorb urinary calcium.

Deficiency symptoms

If calcium is deficient in the diet of young growing animals, then satisfactory bone formation cannot occur and the condition known as rickets is produced. The symptoms of rickets are misshapen bones, enlargement of the joints, lameness and stiffness. In adult animals, calcium deficiency produces osteomalacia, in which the calcium in the bone is withdrawn and not replaced. In osteomalacia, the bones become weak and are easily broken. In hens, deficiency symptoms are soft beak and bones, retarded growth and bowed legs; the eggs have thin shells and egg production may be reduced. The symptoms described above for rickets and osteomalacia are not specific for calcium and can also be produced by a deficiency of phosphorus, an abnormal calcium : phosphorus ratio or a deficiency of vitamin D (p. 84). It is obvious that a number of factors can be responsible for subnormal calcification.

Milk fever (parturient paresis) is a condition which most commonly occurs in dairy cows shortly after calving. It is characterised by a lowering of the serum calcium level, muscular spasms and, in extreme cases, paralysis and unconsciousness. The exact cause of hypocalcaemia associated with milk fever is obscure, but it is generally considered that, with the onset of lactation, the parathyroid gland is unable to respond rapidly enough to increase calcium absorption from the intestine to meet the extra demand. Normal levels of blood calcium can be restored by intravenous injections of calcium gluconate, but this may not always have a permanent effect. It has been shown that avoiding excessive intakes of calcium while maintaining adequate dietary levels of phosphorus during the dry period reduces the incidence of milk fever. Deliberate use of low-calcium diets to increase calcium absorption in the practical prevention of milk fever requires a good estimate of calving date, or calcium deficiency may occur. Also low-calcium diets are hard to achieve with forages unless straw is used. Recently, success in controlling milk fever has been achieved by manipulating the acid–base balance of the diet (see Box 6.2). Administration of large doses of vitamin D_3 for a short period prior to parturition has also proved beneficial, but the timing is critical. Hypocalcaemia in ewes bearing twins often occurs before lambing.

Sources of calcium

Milk, green leafy crops, especially legumes, and sugar beet pulp are good sources of calcium; cereals and roots are poor sources. In some lucerne crops, calcium associated with oxalates is unavailable. Animal by-products containing bone, such as fishmeal, are excellent sources. Calcium-containing mineral supplements that are frequently given to farm animals, especially lactating animals and laying hens, include ground limestone, steamed bone flour and dicalcium phosphate. If rock calcium phosphate is given to animals it is important to ensure that fluorine is absent, as otherwise this supplement may be toxic. High levels of fat in the diet of monogastric animals result in the formation of calcium soaps of fatty acids, which reduce the absorbability of calcium.

Calcium : phosphorus ratio

When giving calcium supplements to animals it is important to consider the calcium : phosphorus ratio of the diet, since an abnormal ratio may be as harmful as a deficiency of either element in the diet. The calcium : phosphorus ratio considered most suitable for farm animals other than poultry is generally within the range 1 : 1 to 2 : 1, although there is evidence which suggests that ruminants can tolerate rather higher ratios providing the phosphorus requirements are met. In the case of the calcium and phosphorus requirements for ruminants recently published by the Agriculture and Food Research Council's Technical Committee on Responses to Nutrients, the requirement for phosphorus can exceed that of calcium in some circumstances. The proportion of calcium for laying hens is much larger, since they require great amounts of the element for eggshell production. The calcium is usually given to laying hens as ground limestone mixed with the diet or, alternatively, calcareous grit may be given *ad libitum*. Granular limestone is more effective since the large particles are retained in the gizzard for a longer time.

Phosphorus

Phosphorus has more known functions in the animal body than any other mineral element. The close association of phosphorus with calcium in bone has already been mentioned. In addition it occurs in phosphoproteins, nucleic acids and phospholipids. The element plays a vital role in energy metabolism in the formation of sugar-phosphates and adenosine di- and triphosphates (see Chapter 9). The importance of vitamin D in calcium and phosphorus metabolism has already been discussed in the previous chapter. The phosphorus content of the animal body is rather smaller than the calcium content. Whereas 99 per cent of the calcium found in the body occurs in the bones and teeth, the proportion of the phosphorus in these structures is about 80–85 per cent of the total. The remainder is in the soft tissues and fluids, where it serves the essential functions given above. The control of phosphorus metabolism is different from that of calcium. If in an available form, phosphorus is absorbed well even when there is an excess over requirement. The excess is excreted via the kidney or the gut (via saliva). In monogastric animals, the kidney is the primary route of excretion. Plasma phosphorus diffuses into saliva and in ruminants the large amount of chewing during rumination results in saliva being the major input of phosphorus into the rumen rather than the food.

Deficiency symptoms

Extensive areas of phosphorus-deficient soils occur throughout the world, especially in tropical and subtropical areas, and a deficiency of this element can be regarded as the most widespread and economically important of all the mineral disabilities affecting grazing livestock.

Like calcium, phosphorus is required for bone formation and a deficiency can also cause rickets or osteomalacia. 'Pica' or depraved appetite has been noted in cattle when there is a deficiency of phosphorus in their diet; the affected animals have abnormal appetites and chew wood, bones, rags and other foreign materials. Pica is not specifically a sign of phosphorus deficiency, since it may be caused by other factors. Evidence of phosphorus deficiency may be obtained from an analysis of blood serum, which would show a phosphorus content lower than normal. In chronic phosphorus deficiency, animals may have stiff joints and muscular weakness.

Low dietary intakes of phosphorus have also been associated with poor fertility, apparent dysfunction of the ovaries causing inhibition, depression or irregularity of oestrus. There are many examples, throughout the world, of phosphorus supplementation increasing fertility in grazing cattle. In cows, a deficiency of this element may also reduce milk yield. In hens, there is a reduced egg yield, hatchability and shell thickness.

Subnormal growth in young animals and low liveweight gains in mature animals are characteristic symptoms of phosphorus deficiency in all species. Phosphorus deficiency is usually more common in cattle than in sheep, as the latter tend to have more selective grazing habits and choose the growing parts of plants, which happen to be richer in phosphorus.

Sources of phosphorus

Milk, cereal grains and fishmeal products containing bone are good sources of phosphorus; the content in hays and straws is generally very low. Considerable attention has been paid to the availability of phosphorus. Much of the element present in cereal grains is in the form of phytates, which are salts of phytic acid, a phosphoric acid derivative:

Phytic acid

Insoluble calcium and magnesium phytates occur in cereals and other plant products. Experiments with chicks have shown that the phosphorus of calcium phytate is utilised only 10 per cent as effectively as disodium phosphate. In studies with laying hens, phytate phosphorus was utilised about half as well as dicalcium phosphate. Certain plant foods, such as wheat, contain phytase and in the pig stomach some of the phytate phosphorus is made available by the action of this enzyme. However, it is likely that the phytase is destroyed in the acid conditions once the acid secreted penetrates the food mass in the stomach. Intestinal phytase activity from the microflora has been observed, but it appears to be of little importance in the pig. It has been shown with sheep that hydrolysis of phytates by bacterial phytases occurs in the rumen. Bacteria in the hind gut also have phytase activity, but its significance in phosphorus supply for the monogastric is not clear. Phytate phosphorus appears therefore to be utilised by ruminants as readily as other forms of phosphorus, although studies using radioactive isotopes indicate that the availability of phosphorus may range from 0.33 to 0.90. Recent studies with a fungal source of phytase added to the diet of pigs have shown significant increases in ileal and total tract digestibility of phytate phosphorus (see Chapter 25).

Feeding with high levels of phosphorus should be avoided as the excess is excreted and contributes to pollution by encouraging the growth of algae in water courses. In addition, high phosphorus intakes in association with magnesium can lead to the formation of mineral deposits in the bladder and urethra (urolithiasis) and blockage of the flow of urine in male sheep and cattle.

Potassium

Potassium plays a very important part, along with sodium, chlorine and bicarbonate ions, in osmotic regulation of the body fluids and in the acid–base balance in the animal. Whereas sodium is the main inorganic cation of extracellular tissue fluids, potassium functions principally as the cation of cells. Potassium plays an important part in nerve and muscle excitability, and is also concerned in carbohydrate metabolism.

Deficiency symptoms

The potassium content of plants is generally very high, that of grass, for example, being frequently above 25 g/kg DM, so that it is normally ingested by animals in larger amounts than any other element. Consequently, potassium deficiency is rare in farm animals kept under natural conditions. One exception to this is provided by distiller's grains ('draff'; p. 566) which, as a result of the removal of the liquid after fermentation, is deficient in several soluble elements including potassium. Appropriate supplementation is necessary where draff forms a large proportion of the diet.

There are certain areas in the world where soil potassium levels are naturally low. Such areas occur in Brazil, Panama and Uganda, and it is suggested that in these tropical regions, potassium deficiencies may arise in grazing animals especially at the end of the long dry season, when potassium levels in the mature herbage are low.

Deficiency symptoms have been produced in chicks given experimental diets low in potassium. They include retarded growth, weakness and tetany, followed by death. Deficiency symptoms, including severe paralysis, have also been recorded for calves given synthetic milk diets low in potassium.

A dietary excess of potassium is, normally, rapidly excreted from the body, chiefly in the urine. Some research workers believe that high intakes of the element may interfere with the absorption and metabolism of magnesium in the animal, which may be an important factor in the aetiology of hypomagnesaemic tetany (see p. 126).

Sodium

Most of the sodium of the animal body is present in the soft tissues and body fluids. Like potassium, sodium is concerned with the acid–base balance and osmotic regulation of the body fluids. Sodium is the chief cation of blood plasma and other extracellular fluids of the body. The sodium concentration within the cells is relatively low, the element being replaced largely by potassium and magnesium. Sodium also plays a role in the transmission of nerve impulses and in the absorption of sugars and amino acids from the digestive tract (see p. 175).

Much of the sodium is ingested in the form of sodium chloride (common salt) and it is also mainly in this form that the element is excreted from the body. There is evidence that sodium rather than chlorine is the chief limiting factor in salt-deficient diets of sheep and cows.

Deficiency symptoms

Sodium deficiency in animals occurs in many parts of the world, but especially in the tropical areas of Africa and the arid inland areas of Australia, where pastures contain very low concentrations of the element. A deficiency of sodium in the diet leads to a lowering of the osmotic pressure, which results in dehydration of the body. Symptoms of sodium deficiency include poor growth and reduced utilisation of digested proteins and energy. In hens, egg production and growth are adversely affected. Rats given experimental diets low in sodium had eye lesions and reproductive disturbances, and eventually died.

Sources of sodium

Most foods of vegetable origin have comparatively low sodium contents; animal products, especially foods of marine origin, are richer sources. The commonest mineral supplement given to farm animals is common salt.

Chlorine

Chlorine is associated with sodium and potassium in acid–base relationships and osmotic regulation. Chlorine also plays an important part in the gastric secretion, where it occurs as hydrochloric acid as well as chloride salts. Chlorine is excreted from the body in the urine and is also lost from the body, along with sodium and potassium, in perspiration.

A dietary deficiency of chlorine may lead to an abnormal increase of the alkali reserve of the blood (alkalosis) caused by an excess of bicarbonate, since inadequate levels of chlorine in the body are partly compensated for by increases in bicarbonate. Experiments with rats on chlorine-deficient diets have shown that growth was retarded, but no other symptoms developed.

Sources of chlorine

With the exception of fishmeals, the chlorine content of most foods is comparatively low. The chlorine content of pasture grass varies widely and figures ranging from 3 to 25 g/kg DM have been reported. The main source of this element for most animals is common salt.

Salt

Since plants tend to be low in both sodium and chlorine, it is the usual practice to give common salt to herbivores. Unless salt is available, deficiencies are likely to occur in both cattle and sheep. Experiments carried out in the USA with dairy cows on salt-deficient diets showed that animals did not exhibit immediate ill effects, but eventually appetite declined, with subsequent loss in weight and lowered milk production. The addition of salt to the diet produced an immediate cure.

Salt is also important in the diet of hens, and it is known to counteract feather picking and cannibalism. Salt is generally given to pigs on vegetable diets, but if fishmeal is given the need for added salt is reduced. Swill can also be a rich source of salt, although the product is very variable and can contain excessive amounts. Too much salt in the diet is definitely harmful and causes excessive thirst, muscular weakness and oedema. Salt poisoning is quite common in pigs and poultry, especially where fresh drinking water is limited. When the concentration of salt in the diet of hens exceeds 40 g/kg DM and the supply of drinking water is limited, then death may occur. Hens can tolerate larger amounts of salt if plenty of water is available. Chicks cannot tolerate salt as well as adults can, and 20 g/kg DM in the diet should be regarded as the absolute maximum. This value should also not be exceeded in the diets of pigs. Turkey poults are even less tolerant, and 10 g/kg of salt in the diet should not be exceeded.

Sulphur

Most of the sulphur in the animal body occurs in proteins containing the amino acids cystine, cysteine and methionine. The two vitamins biotin and thiamin, the hormone insulin and the important metabolite coenzyme A also contain sulphur. The structural compound chondroitin sulphate is a component of cartilage, bone, tendons and the walls of blood vessels. Sulphur-containing compounds are also important in elements of the respiratory process from haemoglobin through to cytochromes. Only a small amount of sulphur is present in the body in inorganic form, though sulphates are known to occur in the blood in small quantities. Wool is rich in cystine and contains about 4 per cent of sulphur.

Traditionally, little attention has been paid to the importance of sulphur in animal nutrition since the intake of this element is mainly in the form of protein, and a deficiency of sulphur would indicate a protein deficiency. In recent years, however, with the increasing use of urea as a partial nitrogen replacement for protein nitrogen and as a method for treating cereal grains (see Chapter 23), it has been realised that the amount of sulphur present in the diet may be the limiting factor for the synthesis in the rumen of cystine, cysteine and methionine. Under these conditions the addition of sulphur to urea-containing rations is beneficial. There is evidence that sulphate (as sodium sulphate) can be used by ruminal microorganisms more efficiently than elemental sulphur. The mean of the estimates of the ratio of sulphur to nitrogen in microbial protein is around 0.07, which, incidentally, is approximately the ratio found in animal tissue and milk protein. The UK Agricultural Research Council recommended that the requirement for rumen-degradable sulphur should be calculated by multiplying the rumen-degradable nitrogen requirement by 0.07 (i.e. equivalent to a N : S ratio of 14 : 1). The ratio of sulphur to nitrogen in wool protein is higher at 0.2 : 1 and the supply of sulphur-containing amino acids may be limiting for sheep with a high rate of wool production. This limitation cannot be overcome by increasing the ratio of rumen-degradable sulphur to degradable nitrogen, since this will not alter the rate of microbial protein production. However, the limitation can be alleviated by supplying sulphur containing amino acids in forms which bypass the rumen or in proteins which have low rumen degradability.

Consideration of sulphur in animal nutrition is important in areas of intensive livestock production where sulphur in soils is not replaced regularly by fertiliser application.

Inorganic sulphur seems to be of less practical importance for monogastric animals, although studies with pigs, chicks and poults have indicated that inorganic sulphate can have a sparing effect on the requirement for sulphur-containing amino acids in the diet.

Toxicity can result from excess dietary sulphur, which is converted to hydrogen sulphide, a toxic agent, by the gastrointestinal flora. This reduces rumen motility and causes nervous and respiratory distress.

Magnesium

Magnesium is closely associated with calcium and phosphorus. About 70 per cent of the total magnesium is found in the skeleton but the remainder, which is distributed in the soft tissues and fluids, is of crucial importance to the well-being of the animal. Magnesium is the commonest enzyme activator, for example in systems with thiamin pyrophosphate as a cofactor, and oxidative phosphorylation is reduced in magnesium deficiency. Magnesium is an essential activator of phosphate transferases (e.g. creatine kinase) and it activates pyruvate carboxylase, pyruvate oxidase and reactions of the tricarboxylic acid cycle. Therefore, it is essential for the efficient metabolism of carbohydrates and lipids. In addition, it is involved in cellular respiration and many other cellular reactions, forming complexes with adenosine tri-, di- and monophosphates. The formation of cyclic AMP and other secondary messengers requires magnesium. Magnesium ions moderate neuromuscular activity and, through binding to phospholipid, are involved in cell membrane integrity. Thus it can be seen that magnesium is a key element in cellular biochemistry and function. Magnesium is absorbed from the small and large intestine of monogastric animals and requirements are usually met with cereal and soya bean meal diets. In ruminants, absorbability can be low and potassium reduces the efficiency of absorption by inhibiting the two active transport systems in the rumen wall which carry magnesium against the electrochemical gradient. Potassium does not affect absorption beyond the rumen.

Deficiency symptoms

Symptoms due to a simple deficiency of magnesium in the diet have been reported for a number of animals. In rats fed on purified diets, the symptoms include increased nervous irritability and convulsions. Experimental low-magnesium milk diets for calves caused low serum magnesium levels, depleted bone magnesium, tetany and death. The condition is not uncommon in milk-fed calves about 50–70 days old. Colostrum is high in magnesium content but milk is low and this is compounded by a reduction in the efficiency of absorption of magnesium as the calf ages.

In adult ruminants, a condition known as hypomagnesaemic tetany, associated with low blood levels of magnesium (hypomagnesaemia), has been recognised since the early 1930s. A great deal of attention has been given to this condition in recent years, since it is widespread and the death rate is high.

Hypomagnesaemic tetany has been known under a variety of names, including magnesium tetany, lactation tetany and grass staggers, but most of these terms have been discarded because the disease is not always associated with lactation or with grazing animals. The condition can affect stall-fed dairy cattle, hill cattle and cattle at grass as well as sheep. There is some evidence of a breed susceptibility in the UK, where the condition appears to be more common in Ayrshire and least common in Jersey animals. Most cases occur in grazing animals and, in Europe and North America, the trouble is particularly common in the spring when the animals are turned out on to young, succulent pasture. Because the tetany can develop within a day or two of animals being turned out to graze, the condition has been referred to as the acute form. In this acute type, blood magnesium levels fall so rapidly that the reserve of magnesium within the body cannot be mobilised rapidly enough. In the chronic form of the disease, plasma magnesium levels fall over a period of time to low concentrations. This type is not uncommon in suckler herds. Clinical signs of the disease are often brought on by 'stress' factors such as cold, wet and windy weather. In adult animals, bone magnesium is not as readily available as it is in the young calf.

In New Zealand, where cows are pastured throughout the year, hypomagnesaemic tetany occurs most frequently in late winter and early spring. In Australia, a high incidence of the disease has been associated with periods of rapid winter growth of pastures.

The normal magnesium content of blood serum in cattle is within the range of 17–40 mg magnesium/litre, but levels below 17 frequently occur without clinical symptoms of disease. Tetany is usually preceded by a fall in blood serum magnesium to about 5 mg/litre. Urine magnesium is a better indicator of deficiency than serum magnesium because levels of the latter do not fall until there is a severe deficiency. However, a lack of magnesium is immediately reflected in urine levels, for which 10 mg/100 ml is satisfactory, 2–10 mg/100 ml is inadequate and <2 mg/100 ml indicates a severe deficiency. Subcutaneous injection of magnesium sulphate, or preferably magnesium lactate, can generally be expected to cure the animal if given early, but in practice this is sometimes difficult. Treatment of this kind is not a permanent cure and oral treatment with magnesium oxide, as described below, should be started immediately. Typical symptoms of tetany are nervousness, tremors, twitching of the facial muscles, staggering gait and convulsions.

The exact cause of hypomagnesaemic tetany in ruminant animals is unknown, although a dietary deficiency of magnesium may be a contributory factor. Some research workers consider the condition to be caused by a cation–anion imbalance in the diet and there is evidence of a positive relationship between tetany and heavy dressings of pasture with nitrogenous and potassic fertilisers. It has been suggested that the concentration of potassium in spring pasture should not exceed 25 g/kg DM and the application of potassium fertilisers should be carefully managed.

The use of radioactive magnesium in tracer studies indicates that the magnesium present in food is poorly absorbed from the alimentary canal; in some cases only 50 g/kg of the herbage magnesium can be utilised by the ruminant. Why this is so in ruminants is not known. Since adult animals have only very small readily available reserves of body magnesium they are dependent upon a regular dietary supply.

Although the exact cause of hypomagnesaemia is still uncertain, the primary factor would appear to be inadequate absorption of magnesium from the digestive tract. A high degree of success in preventing hypomagnesaemia may be obtained by increasing the magnesium intake. This can be effected by feeding with magnesium-rich mineral mixtures or, alternatively, by increasing the magnesium content of pasture by the application of magnesium fertilisers.

Sources of magnesium

Wheat bran, dried yeast and most vegetable protein concentrates, especially cottonseed cake and linseed cake, are good sources of magnesium. Clovers are usually richer in magnesium than grasses, although the magnesium content of forage crops varies widely. As mentioned previously, draff is deficient in soluble minerals and high levels of this food in the diet require appropriate supplementation. The mineral supplement most frequently used is magnesium oxide, which is sold commercially as calcined magnesite. When hypomagnesaemic tetany is likely to occur it is generally considered that about 50 g of magnesium oxide per head per day should be given to cows as a prophylactic measure. The daily prophylactic dose for calves is 7–15 g of the oxide, while that for lactating ewes is about 7 g. The mineral supplement can be given mixed with the concentrate ration. Alternatively, a mixture of magnesium acetate solution and molasses can be used, which is frequently made available on a free-choice basis from ball feeders placed in the field.

6.5 Trace elements

Iron

More than 90 per cent of the iron in the body is combined with proteins, the most important being haemoglobin, which contains about 3.4 g/kg of the element. Iron also occurs in blood serum in a protein called transferrin, which is concerned with the transport of iron from one part of the body to another. Ferritin, a protein containing up to 200 g/kg of the element, is present in the spleen, liver, kidney and bone marrow and provides a form of storage for iron. Haemosiderin is a similar storage compound which may contain up to 350 g/kg of the element. Iron has a major role in a host of biochemical reactions, particularly in connection with enzymes of the electron transport chain (cytochromes). Electrons are transported by the oxidation and reduction activity of bound iron. Among the enzymes containing or activated by iron are catalase, peroxidases, phenylalanine hydroxylase and many others, including all the tricarboxylic acid cycle enzymes.

Deficiency symptoms

Since more than half the iron present in the body occurs as haemoglobin, a dietary deficiency of iron would clearly be expected to affect the formation of this compound. The red blood corpuscles contain haemoglobin, and these cells are continually being produced in the bone marrow to replace those red cells destroyed in the animal body as a result of catabolism. Although the haemoglobin molecule is destroyed in the catabolism of these red blood corpuscles, the iron liberated is made use of in the resynthesis of haemoglobin, and because of this the daily requirement of iron by a healthy animal is usually small. If the need for iron increases, as it does after prolonged haemorrhage or during pregnancy, then haemoglobin synthesis may be inadequate and anaemia will result. Anaemia due to iron deficiency occurs most commonly in rapidly growing sucklings, since the iron content of milk is usually very low. This can occur in piglets housed in pens without access to soil. The piglet is born with very limited iron reserves and sow's milk provides only about 1 mg per day. The rapidly growing piglet's requirement is 15 mg per day, which, in extensive systems, could be obtained by ingestion of soil. Providing the sow with supplementary iron in gestation does not increase the foetal piglet's liver iron or the amount in the milk. Therefore, it is routinely supplied by intramuscular injection as a dextran complex or gleptoferron, by 3 days of age. Usually 200 mg of iron is injected. Alternatively, oral iron supplements are available in the form of a paste of the citrate or fumarate or granules of iron dextran but these may not be eaten or the iron may be lost if diarrhoea occurs. Attempts have been made to increase the supply of iron to the piglet by supplementing the sow's diet and relying on the piglet's consumption of the sow's faeces during exploration activity. However, this has produced uneven uptake, and injection of iron compounds is more effective. Anaemia in piglets is characterised by poor appetite and growth. Breathing becomes laboured and spasmodic – hence the descriptive term 'thumps' for this condition. Although iron deficiency is not common in older animals, increased supplementation is required when high levels of copper are used for growth promotion.

Iron deficiency anaemia is not common in lambs and calves because in practice it is unusual to restrict them to a milk diet without supplementary feeding. It does, however, sometimes occur in laying hens, since egg production represents a considerable drain on the body reserves.

Sources of iron

Iron is widely distributed in foods. Good sources of the element are green leafy materials, most leguminous plants and seed coats. Feeds of animal origin, such as blood and fish meals, are excellent sources of iron. As mentioned previously, milk is a poor source of the element.

Iron is absorbed throughout the gastro-intestinal tract, but mainly in the duodenum and jejunum. Absorption is poor and is, to a large extent, independent of the dietary source. The efficiency of absorption is increased during periods of iron need and decreased during periods of iron overload. The mechanisms whereby the body carries this out are not fully understood. A number of theories have been advanced and one of these, the 'mucosal block' theory, propounded in 1943, is still widely held by many to explain the mechanism. According to this theory, the mucosal cells of the gastro-intestinal tract absorb iron and convert it into ferritin, and when the cells become physiologically saturated with ferritin, further absorption is impeded until the iron is released and transferred to plasma. Another more recent theory proposes that the main regulator of iron uptake is the iron concentration in the epithelial cells of the duodenal mucosa.

The adult's need for iron is normally low, as the iron produced from the destruction of haemoglobin is made available for haemoglobin regeneration, only about 10 per cent of the element escaping from this cycle.

Iron toxicity

Iron toxicity is not a common problem in farm animals, but it can result from prolonged oral administration of the element. Ferrous ions are reported to generate oxygen-based free radicals, contributing to oxidative stress in the cell (see Chapter 5). Normally iron is protein-bound or in chemical form which prevent it causing oxidation. Chronic iron toxicity results in alimentary disturbances, reduced growth and phosphorus deficiency.

Copper

Evidence that copper is a dietary essential was obtained in 1924, when experiments with rats showed that copper was necessary for haemoglobin formation. Although copper is not actually a constituent of haemoglobin it is present in certain other plasma proteins such as ceruloplasmin which are concerned with the release of iron from the cells into the plasma. A deficiency of copper impairs the animal's ability to absorb iron, mobilise it from the tissues and utilise it in haemoglobin synthesis. Copper is also a component of other proteins in blood. One of these, erythrocuprein, occurs in erythrocytes, where it plays a role in oxygen metabolism. The element is also known to play a vital role in many enzyme systems; for example, it is a component of cytochrome oxidase, which is important in oxidative phosphorylation. It is also a component of superoxide dismutase, which forms part of the cell's antioxidant system. The element also occurs in certain pigments, notably turacin, a pigment of feathers. Copper is necessary for the normal pigmentation of hair, fur and wool. It is thought to be present in all body cells, being particularly concentrated in the liver, which acts as the main copper storage organ of the body. Copper has been shown to reduce the susceptibility to infection in lambs directly.

Deficiency symptoms

Since copper performs many functions in the animal body there are a variety of deficiency symptoms. These include anaemia, poor growth, bone disorders, scouring, infertility, depigmentation of hair and wool, gastro-intestinal disturbances and lesions in the brain stem and spinal cord. The lesions are associated with muscular incoordination and occur, especially, in young lambs. A copper deficiency condition known as 'enzootic ataxia' has been known for some time in Australia. The disorder there is associated with pastures low in copper content (2–4 mg/kg DM) and can be prevented by feeding with a copper salt. A similar condition which affects lambs occurs in the

UK and is known as 'swayback'. The signs of swayback range from complete paralysis of the newborn lamb to a swaying staggering gait which affects, in particular, the hind limbs. The condition can occur in two forms, one congenital, in which the signs are apparent at birth and are due to the failure of the myelin sheath of nerves to develop, and the other in which the onset of the clinical disease is delayed for several weeks. The congenital form of the condition is irreversible and can only be prevented by ensuring that the ewe receives an adequate level of copper in her diet. Delayed swayback can be prevented or retarded in copper-deficient lambs by parenteral injection of small doses of copper complexes.

Although the dietary level of copper is an important factor in the aetiology of swayback, the condition does not appear to be invariably caused by a simple dietary deficiency of the element. Swayback has been reported to occur on pastures apparently normal or even high (7–15 mg/kg DM) in copper content. One important factor is that the efficiency of absorption of dietary copper is very variable. For example, there is about a tenfold variation in the efficiency with which Scottish Blackface ewes absorb copper from autumn pasture (1.2 per cent) and from leafy brassicas (13.2 per cent). It is also known that genetic factors influence the concentration of copper in the blood, liver and brain of the sheep, and hence the incidence of swayback can be affected by genotype. Blackface lambs given a copper-supplemented barley and fishmeal diet retained 6 per cent of the dietary copper in the liver, whereas Texel lambs retained 13 per cent. Finnish Landrace and Suffolk lambs were intermediate at 8–9 per cent retention.

Copper plays an important role in the production of 'crimp' in wool. The element is present in an enzyme which is responsible for the disulphide bridge in two adjacent cysteine molecules. In the absence of the enzyme the protein molecules of the wool do not form their bridge and the wool, which lacks crimp, is referred to as 'stringy' or 'steely' (see p. 379).

Nutritional anaemia resulting from copper deficiency has been produced experimentally in young pigs by diets very low in the element and this type of anaemia could easily arise in such animals fed solely on milk. In older animals, copper deficiency is unlikely to occur and copper supplementation of practical rations is generally considered unnecessary. There are, however, certain areas in the world where copper deficiency in cattle occurs. A condition in Australia known locally as 'falling disease' was found to be related to a progressive degeneration of the myocardium of animals grazing on copper-deficient pastures.

Copper–molybdenum–sulphur interrelations

Certain pastures on calcareous soils in parts of England and Wales have been known for over 100 years to be associated with a condition in cattle known as 'teart', which is characterised by unthriftiness and scouring. A similar disorder occurs on reclaimed peat lands in New Zealand, where it is known as 'peat scours'. Molybdenum levels in teart pasture are of the order of 20–100 mg/kg DM compared with 0.5–3.0 mg/kg DM in normal pastures, and teart was originally regarded as being a straightforward molybdenosis. In the late 1930s, however, it was demonstrated that feeding with copper sulphate controlled the scouring and hence a molybdenum–copper relationship was established.

It is now known that the effect of molybdenum is complex, and it is considered that the element exerts its limiting effect on copper retention in the animal only in the presence of sulphur. Sulphide is formed by ruminal microorganisms from dietary sulphate or organic sulphur compounds; the sulphide then reacts with molybdate to form thiomolybdate, which in turn combines with copper to form an insoluble copper thiomolybdate ($CuMoS_4$) thereby limiting the absorption of dietary copper. In addition, it is considered likely that if thiomolybdate is formed in excess, it may be absorbed from the digestive tract and exert a systemic effect on copper metabolism in the animal.

Sources of copper

Copper is widely distributed in foods and under normal conditions the diet of farm animals is likely to contain adequate amounts. The copper content of crops is related to some extent to the soil copper level, but is also affected by other factors such as drainage conditions and the herbage species. Seeds and seed by-products are usually rich in copper, but straws contain little. The normal copper content of pasture ranges from about 4 to 8 mg/kg DM. The copper content of milk is low, and hence it is customary when dosing young animals, especially piglets, with an iron salt to include a trace of copper sulphate.

Copper toxicity

It has long been known that copper salts given in excess to animals are toxic. Continuous ingestion of copper in excess of nutritional requirements leads to an accumulation of the element in the body tissues, especially in the liver. Copper can be regarded as a cumulative poison, so that considerable care is required in administering copper salts to animals. The tolerance to copper varies considerably between species. Pigs are highly tolerant (see Box 6.3) and cattle relatively so. On the other hand, sheep are particularly susceptible and chronic copper poisoning has been encountered in housed sheep on concentrate diets. There is a gradual accumulation of copper in the livers of sheep until the danger level of about 1000 mg/kg fat-free DM is reached. Poisoning has been known to occur in areas where the herbage contains copper of the order of 10–20 mg/kg DM and low levels of molybdenum. Chronic copper poisoning results in necrosis of the liver cells, jaundice, loss of appetite and death from hepatic coma. The slow accumulation of copper in the liver causes damage to the organ without overt symptoms. There is leakage of enzymes from the damaged cells into the blood. Eventually there is a sudden release of copper and haemolysis, which can

Box 6.3
Copper as growth promoter

In the late 1950s and early 1960s Barber, Braude, Mitchell and their colleagues at the National Institute for Research in Dairying at Reading demonstrated that pigs given high levels of copper (up to 250 mg/kg) in their diet had faster growth rates and better food conversion efficiency than unsupplemented pigs. Most of this copper is not absorbed but passes through the digestive tract, achieving its effect by altering the microbial population in much the same way as antibiotic growth promoters, although its effect is independent of and in addition to antibiotics.

Concern about pollution of the environment has resulted in restrictions on the use of copper as a growth promoter in Europe. The maximum permitted dietary levels are 175 mg/kg for pigs up to 16 weeks of age and then 100 mg/kg up to six months of age. Furthermore, it is essential that sheep do not have access to pasture that has recently been fertilised with pig slurry.

occur spontaneously or as a result of stressors such as parturition or infection. There is a genetic variation in animals' susceptibility to copper poisoning related to the efficiency of retention, with the Scottish Blackface being the least susceptible and continental breeds, such as the Texel, being highly susceptible. The EU maximum permitted level for copper in sheep diets is 15 mg/kg and this should not be exceeded in sheep diets if toxicity is to be avoided. For susceptible breeds a dietary level of 10 mg/kg can be excessive. It is unwise to administer copper supplements to sheep unless deficiency conditions are liable to occur – many cases of death due to copper poisoning caused by the indiscriminate use of copper-fortified diets have been reported. Chronic copper poisoning in sheep has occurred under natural conditions in parts of Australia where the copper content of the pasture is high. Care should be taken when sheep are given antiprotozoal compounds such as monensin, which may eliminate the protozoa that produce the sulphide that normally reduces copper availability.

Cobalt

A number of disorders of cattle and sheep, characterised by emaciation, anaemia and listlessness, have been recognised for many years and have been described as 'pining', 'salt sick', 'bush sickness' and 'wasting disease'. These disorders occur in Europe, Australia, New Zealand and the USA. In the UK, 'pining pastures' occur in many counties and are particularly common in the border counties of England and Scotland.

As early as 1807 Hogg, the Ettrick shepherd, recognised 'pining' or 'vinquish' as being a dietary upset. Pining is associated with a dietary deficiency of cobalt caused by low concentrations of the element in the soil and herbage. Pining can be prevented in these areas by feeding with small amounts of cobalt.

The physiological function of cobalt was discovered only when vitamin B_{12} was isolated and was shown to contain the element. Cobalt is required by microorganisms in the rumen for the synthesis of vitamin B_{12}, and if the element is deficient in the diet then the vitamin cannot be produced in the rumen in amounts sufficient to satisfy the animal's requirements and symptoms of pining occur. Pining is therefore regarded as being due to a deficiency of vitamin B_{12}. There is evidence for this, since injections of vitamin B_{12} into the blood alleviate the condition, whereas cobalt injections have little beneficial effect. Although vitamin B_{12} therapy will prevent pining occurring in ruminant animals, it is more convenient and cheaper in cobalt-deficient areas to supplement the diet with the element, allowing the microorganisms in the rumen to synthesise the vitamin for subsequent absorption by the host.

When ruminants are confined to cobalt-deficient pastures it may be several months before any manifestations of pine occur because of body reserves of vitamin B_{12} in the liver and kidneys. When these are depleted there is a gradual decrease in appetite with consequent loss of weight followed by muscular wasting, pica, severe anaemia and eventually death. If the deficiency is less severe then a vague unthriftiness, difficult to diagnose, may be the only sign. Deficiency symptoms are likely to occur where levels of cobalt in the herbage are below 0.1 mg/kg DM. Under grazing conditions, lambs are the most sensitive to cobalt deficiency, followed by mature sheep, calves and mature cattle in that order.

Ruminants have a higher requirement for the element than non-ruminants because some of the element is wasted in microbial synthesis of organic compounds with no physiological activity in the host's tissues. Furthermore, vitamin B_{12} is poorly absorbed from the digestive tract of ruminants, the availability in some cases being as low as 0.03. The ruminant has an additional requirement for the vitamin because of its involvement in the metabolism of propionic acid (see p. 210), an important acid absorbed from the rumen.

There is evidence that the intestinal microorganisms in non-ruminants also can synthesise vitamin B_{12}, although in pigs and poultry this synthesis may be insufficient to meet their requirements. It is common practice to include in pig and poultry diets some animal protein food rich in vitamin B_{12} in preference to including a cobalt salt.

Apart from the importance of cobalt as a component of vitamin B_{12}, the element is believed to have other functions in the animal body as an activating ion in certain enzyme reactions.

Sources of cobalt

Most foods contain traces of cobalt. Normal pasture herbages have a cobalt content within the range 100–250 µg/kg DM.

Cobalt deficiency in ruminants can be prevented by dosing the animals with cobalt sulphate, although this form of treatment has to be repeated at short intervals, or by allowing the animals access to cobalt-containing salt licks.

A continuous supply from a single dose can be obtained by giving a cobalt bullet containing 900 g cobaltic oxide/kg; the bullet remains in the reticulum and slowly releases the element over a long period. Some of this cobalt is not utilised by the animal and is excreted, and this of course has the effect of improving the cobalt status of the pasture. Alternatively, deficient pastures can be top-dressed with small amounts of cobalt sulphate (about 2 kg/ha).

Cobalt toxicity

Although an excess of cobalt can be toxic to animals, there is a wide margin of safety between the nutritional requirement and the toxic level. Cobalt toxicosis is extremely unlikely to occur under practical farming conditions. Unlike copper, cobalt is poorly retained by the body tissues and an excess of the element is soon excreted. The toxic level of cobalt for cattle is 1 mg cobalt/kg body weight daily. Sheep are less susceptible to cobalt toxicosis than cattle and have been shown to tolerate levels up to 3.5 mg/kg. Excessive cobalt supplementation of ruminant diets can lead to the production of analogues of vitamin B_{12} and a reduction in the quantity of the true vitamin.

Iodine

The concentration of iodine present in the animal body is very small and in the adult is usually less than 600 µg/kg. Although the element is distributed throughout the tissues and secretions its only known role is in the synthesis of the two hormones, triiodothyronine (T_3) and tetraiodothyronine (T_4, thyroxine) produced in the thyroid gland (see p. 56).

Iodine is removed from iodides in the blood and combined with the amino acid tyrosine to form monoiodotyrosine (T_1) and diiodotyrosine (T_2). Two molecules of T_2 are condensed to produce T_4, the physiologically inactive transport form of the hormone, which is stored in the thyroid gland. T_4 is released into the blood capillaries as required and is activated by deiodinase enzymes to produce the physiologically active T_3. The enzymes are selenium-dependent (see p. 139) and occur in the periphery where the hormome is needed, mainly in the liver and kidneys, but also in the skin.

The thyroid hormones accelerate reactions in most organs and tissues in the body, thus increasing the basal metabolic rate, accelerating growth and increasing the oxygen consumption of the whole organism. They also control the development of the foetus and are involved in immune defence, digestion, muscle function and seasonality of reproduction.

Deficiency symptoms

When the diet contains insufficient iodine the production of thyroxine is decreased. The main indication of such a deficiency is an enlargement of the thyroid gland, termed endemic goitre, and is caused by compensatory hypertrophy of the gland. The thyroid being situated in the neck, the deficiency condition in farm animals manifests itself as a swelling of the neck, 'big neck'. Reproductive abnormalities are one of the most outstanding consequences of reduced thyroid function; breeding animals deficient in iodine give birth to hairless, weak or dead young; brain development is impaired; oestrus is suppressed or irregular and male fertility is reduced.

A dietary deficiency of iodine is not the sole cause of goitre: it is known that certain foods contain goitrogenic compounds and cause goitre in animals if given in large amounts. These foods include most members of the *Brassica* genus, especially kale, cabbage and rape, and also soya beans, linseed, peas and groundnuts. Goitrogens have been reported in milk of cows fed on goitrogenic plants. A goitrogen present in brassicas has been identified as L-5-vinyl-2-oxazolidine-2-thione (goitrin), which inhibits the iodination of tyrosine and thus interferes with thyroxine synthesis. Therefore, it cannot be overcome by adding more iodine to the diet. Thiocyanate, which may also be present in members of the *Brassica* genus, is known to be goitrogenic and may be produced in the tissues from a cyanogenetic glycoside present in some foods. Goitrogenic activity of the thiocyanate type is prevented by supplying adequate iodine in the diet. It has been reported that high dietary nitrate levels inhibit the uptake of iodine.

Sources of iodine

Iodine occurs in traces in most foods and is present mainly as inorganic iodide, in which form it is absorbed from the digestive tract. The richest sources of this element are foods of marine origin and values as high as 6 g/kg DM have been reported for some seaweeds; fishmeal is also a rich source of the element. The iodine content of land plants is related to the amount of iodine present in the soil and consequently wide variations occur in similar crops grown in different areas.

In areas where goitre is endemic, precautions are generally taken by supplementing the diet with the element, usually in the form of iodised salt. This contains the element either as sodium or potassium iodide or as sodium iodate.

Iodine toxicity

The minimum toxic dietary level of iodine for calves of 80–112 kg body weight has been shown to be about 50 mg/kg, although some experimental animals have been adversely affected at lower levels. Symptoms of toxicity include depressions in weight gain and feed intake. In studies with laying hens, diets with iodine contents of 312–5000 mg/kg DM stopped egg production within the first week at the higher level and reduced egg production at the lower level. The fertility of the eggs produced was not affected, but early embryonic death, reduced hatchability and delayed hatching resulted.

Pigs seem to be more tolerant of excess iodine and the minimum toxic level is considered to lie between 400 and 800 mg/kg.

Manganese

The amount of manganese present in the animal body is extremely small. Most tissues contain traces of the element, the highest concentrations occurring in the bones, liver, kidney, pancreas and pituitary gland. Manganese is important in the animal body as an activator of many enzymes such as hydrolases and kinases and as a constituent of arginase, pyruvate carboxylase and manganese superoxide dismutase.

Deficiency symptoms

Manganese deficiency has been found in ruminants, pigs and poultry. The effects of acute deficiency are similar in all species and include retarded growth, skeletal abnormalities, ataxia of the newborn and reproductive failure. Manganese, through its activation of glycosyl transferases, is required for the formation of the mucopolysaccharide which forms the organic matrix of bone.

Deficiencies of this element in grazing ruminants are likely to be rare, although the reproductive performance of grazing Dorset Horn ewes in Australia was improved by giving manganese over two consecutive years. Low-manganese diets for cows and goats have been reported to depress or delay oestrus and conception, and to increase abortion. Manganese is an important element in the diet of young chicks, a deficiency leading to perosis or 'slipped tendon', a malformation of the leg bones. It is not, however, the only factor involved in the aetiology of this condition, as perosis in young birds may be aggravated by high dietary intakes of calcium and phosphorus or a deficiency of choline. Another link between manganese and choline deficiencies is shown in fatty infiltration of the liver and changes in the ultrastructure of the liver.

Manganese deficiency in breeding birds reduces hatchability and shell thickness, and causes head retraction in chicks. In pigs, lameness is a symptom. Other abnormalities associated with deficiency include impaired glucose utilisation and a reduced vitamin K-induced blood-clotting response.

Sources of manganese

The element is widely distributed in foods and most forages contain 40–200 mg/kg DM. The manganese content of pasture herbages, however, can vary over a much wider range and in acid conditions may be as high as 500–600 mg/kg DM. Seeds and seed products contain moderate amounts, except for maize, which is low in the element. Yeast and most foods of animal origin are also poor sources of manganese. Rich sources are rice bran and wheat offals. Most green foods contain adequate amounts.

Manganese toxicity

There is a wide margin of safety between the toxic dose of manganese and the normal level in foods. Levels as high as 1 g/kg DM in the diet have been given to hens without evidence of toxicity. Growing pigs are less tolerant, levels of 0.5 g/kg DM having been shown to depress appetite and retard growth.

Zinc

Zinc has been found in every tissue in the animal body. The element tends to accumulate in the bones rather than the liver, which is the main storage organ of many of the other trace elements. High concentrations have been found in the skin, hair and wool of animals. Several enzymes in the animal body are known to contain zinc; these include carbonic anhydrase, pancreatic carboxypeptidase, lactate dehydrogenase, alcohol dehydrogenase, alkaline phosphatase and thymidine kinase. In addition, zinc is an activator of several enzyme systems. It is involved in cell replication and differentiation, particularly in nucleic acid metabolism. Among the other physiological functions of zinc are the production, storage and secretion of hormones, involvement in the immune system and electrolyte balance.

Deficiency symptoms

Zinc deficiency in pigs is characterised by subnormal growth, depressed appetite, poor food conversion and parakeratosis. The latter is a reddening of the skin followed by eruptions which develop into scabs. A deficiency of this element is particularly liable to occur in young, intensively housed pigs offered a dry diet *ad libitum*, though a similar diet given wet may not cause the condition. It is aggravated by high calcium levels in the diet and reduced by decreased calcium and increased phosphorus levels. Pigs given diets supplemented with high levels of copper, for growth promotion, have an increased requirement for zinc. Gross signs of zinc deficiency in chicks are retarded growth, foot abnormalities, 'frizzled' feathers, parakeratosis and a bone abnormality referred to as the 'swollen hock syndrome'.

Symptoms of zinc deficiency in calves include inflammation of the nose and mouth, stiffness of the joints, swollen feet and parakeratosis. The response of severely zinc-deficient calves to supplemental zinc is rapid and dramatic. Improvements in skin condition are usually noted within two–three days.

Manifestations of zinc deficiency, responsive to zinc therapy, have been observed in growing and mature cattle in parts of Guyana, Greece, Australia and Scandinavia. As levels in the pasture herbage are apparently comparable with those of other areas, the deficiency is believed to be conditioned by some factor in the herbage or general environment.

Sources of zinc

The element is fairly widely distributed. Yeast is a rich source, and zinc is concentrated in the bran and germ of cereal grains. Animal protein by-products, such as fishmeal, are usually richer sources of the element than plant protein supplements.

Zinc toxicity

Although cases of zinc poisoning have been reported, most animals have a high tolerance for this element. Excessive amounts of zinc in the diet are known to depress food consumption and may induce copper deficiency.

Molybdenum

The first indication of an essential metabolic role for molybdenum was obtained in 1953 when it was discovered that xanthine oxidase, important in purine metabolism, was a metalloenzyme containing molybdenum. Subsequently the element was shown to be a component of two other enzymes, aldehyde oxidase and sulphite oxidase. The biological functions of molybdenum, apart from its reactions with copper previously mentioned (p. 130), are concerned with the formation and activities of these three enzymes. In addition to being a component of xanthine oxidase, molybdenum participates in the reaction of the enzyme with cytochrome C and also facilitates the reduction of cytochrome C by aldehyde oxidase.

Deficiency symptoms

In early studies with rats, low-molybdenum diets resulted in reduced levels of xanthine oxidase but did not affect growth or purine metabolism Similar molybdenum-deficient diets have been given to chicks without adverse effect but when tungstate (a molybdenum antagonist) was added, growth was reduced and the chick's ability to oxidise xanthine to uric acid was impaired. These effects were prevented by the addition of molybdenum to the diet. A significant growth response has been obtained in young lambs by the addition of molybdate to a semi-purified diet low in the element. It has been suggested that this growth effect could have arisen indirectly by stimulation of cellulose breakdown by ruminal microorganisms. Molybdenum deficiency has not been observed under natural conditions in any species.

Molybdenum toxicity

The toxic role of molybdenum in the condition known as 'teart' is described under the element copper (p. 130). All cattle are susceptible to molybdenosis, with milking cows and young animals suffering most. Sheep are less affected and horses are not affected on teart pastures. Scouring and weight loss are the dominant manifestations of the toxicity.

Selenium

The nutritional importance of selenium became evident in the 1950s when it was shown that most myopathies in sheep and cattle, and exudative diathesis in chicks, could be prevented by supplementing the diet with the element or vitamin E (p. 85). A biochemical role of selenium in the animal body was demonstrated in 1973 when it was discovered that the element was a component of glutathione peroxidase, an enzyme which catalyses the removal of hydrogen peroxide, thereby protecting cell membranes from oxidative damage. Glutathione peroxidase contains four selenium atoms and forms a second line of defence after vitamin E, since some peroxidases remain even if vitamin E levels are adequate. Selenium has a sparing effect on vitamin E by ensuring normal absorption of the vitamin. This is due to its role in preserving the integrity of the pancreas and thereby ensuring satisfactory fat digestion. Selenium also reduces the amount of vitamin E required to maintain the integrity of lipid membranes and aids the retention of vitamin E in plasma. Conversely, vitamin E spares selenium by maintaining the element in its active form and preventing its loss. It reduces the production of hydroperoxides and thus the amount of glutathione peroxidase needed to protect cells. However, there are limits to the mutual substitution of selenium and vitamin E.

Vitamin E and selenium have roles in the immune system and protect against heavy metal toxicity. Other mutual functions and effects of deficiency in farm animals are discussed in the section on vitamin E (pp. 84–8).

The other major role of selenium is in the production of the thyroid hormones (p. 134), for which it is a component of the enzyme type I iodothyronine deiodinase (ID_1) which converts T_4 to the physiologically active T_3. When there is a deficiency of selenium the ratio of $T_4 : T_3$ increases. The enzyme is primarily found in the liver and kidney and not in the thyroid of farm animals. Type II iodothyronine deiodinase (ID_2) does not contain selenium and also converts T_4 to T_3 but as it is under feedback control from T_4, an increase in the latter, when selenium is deficient, compounds the problem. The major enzyme in ruminants is ID_1 and in non-ruminants ID_2. A third selenium-containing enzyme, ID_3, has been found in the placenta. ID_1 is particularly important in the brown adipose tissue of newborn ruminants and releases T_3 for use in other tissues.

In parts of Australia and New Zealand a condition known as 'ill thrift' occurs in lambs, beef cattle and dairy cows at pasture. The clinical signs include loss of weight and sometimes death. Ill thrift can be prevented by selenium treatment with, in some instances, dramatic increases in growth and wool yield. Similar responses with sheep have also been noted in experiments carried out in Scotland, Canada and the USA. In hens, selenium deficiency reduces hatchability and egg production.

The main form of selenium in most foods is protein-bound selenomethionine. Supplements of selenium are provided by mineral salts containing sodium selenite, slow-release capsules and selenium-enriched yeast.

Selenium toxicity

The level of selenium in foods of plant origin is extremely variable and depends mainly on the soil conditions under which they are grown. Normal levels of the element in pasture herbage are usually between 100 and 300 μg/kg DM. Some species of plants which grow in seleniferous areas contain very high levels of selenium. One such plant, *Astragalus racemosus*, grown in Wyoming, was reported to contain 14 g selenium/kg DM, while the legume *Neptunia amplexicaulis* grown on a selenised soil in Queensland contained over 4 g/kg DM of the element. Localised seleniferous areas have also been identified in Ireland, Israel and South Africa. Selenium is a highly toxic element and a concentration in a dry diet of 5 mg/kg or 500 μg/kg in milk or water may be potentially dangerous to farm animals. Alkali disease and blind staggers are localised names for chronic diseases of animals grazing certain seleniferous areas in the USA. Symptoms include dullness, stiffness of the joints, loss of hair from mane or tail and hoof deformities. Acute poisoning, which results in death from respiratory failure, can arise from sudden exposure to high selenium intakes.

Fluorine

The importance of fluorine in the prevention of dental caries in man is well established. Fluorine was added to the list of essential elements in 1972 when it was shown that the growth rate of rats was improved when small amounts were added to a low-fluorine diet. Under normal conditions a straightforward deficiency syndrome in farm animals has not been observed and even the results of studies with rodents are equivocal.

Most plants have a limited capacity to absorb fluorine from the soil and normal levels in pasture herbage range from about 2 to 20 mg/kg DM. Cereals and other grains usually contain about 1–3 mg/kg DM only.

Fluorine is a very toxic element, with ruminants being more susceptible than non-ruminants. Levels of more than 20 mg/kg DM in the diet of cattle have resulted in dental pitting and wear, leading to exposed pulp cavities. Further increases in fluorine cause depression of appetite, lameness and reduced production. Bone and joint abnormalities also occur, probably owing to ingested fluorine being deposited in the bone crystal lattice as calcium fluoride. The commonest sources of danger from this element are fluoride-containing water, herbage contaminated by dust from industrial pollution and the use of soft or raw rock phosphate supplements. Processed phosphates are generally safe.

Silicon

Rats previously fed on specially purified foods showed increased growth rates from the addition to the diet of 500 mg/kg of silicon (as sodium metasilicate).

Similar results have been obtained with chicks. Silicon is essential for growth and skeletal development in these two species. The element is believed to function as a biological cross-linking agent, possibly as an ether derivative of silicic acid of the type $R_1-O-Si-O-R_2$. Such bridges are important in the strength, structure and resilience of connective tissue. In silicon-deficient rats and chicks, bone abnormalities occur because of a reduction in mucopolysaccharide synthesis in the formation of cartilage. Silicon is required for maximal activity of the enzyme prolyl hydroxylase, which is involved in collagen synthesis. It is also thought to be involved in other processes involving mucopolysaccharides such as the growth and maintenance of arterial walls and the skin.

Silicon is so widely distributed in the environment and in foods that it is difficult to foresee any deficiencies of this element arising under practical conditions. Whole grasses and cereals may contain as much as 14–19 g Si/kg DM, with levels of up to 28 g/kg DM in some range grasses.

Silicon toxicity (silicosis) has long been known as an illness of miners caused by the inhalation of silical particles into the lungs. Under some conditions, part of the silicon present in urine is deposited in the kidney, bladder or urethra to form calculi or uroliths. Silica urolithiasis occurs in grazing wethers in Western Australia and in grazing steers in western Canada and northwestern parts of the USA. Excessive silica in feeds, for example in rice straw, is known to depress organic matter digestibility. In mature forage, silicon is in the form of solid particles which are harder than dental tissue and lead to teeth wear in sheep.

Chromium

Chromium was first shown to be essential for normal glucose utilisation in rats. Mice and rats fed on diets composed of cereals and skimmed milk and containing 100 μg chromium/kg wet weight were subsequently shown to grow faster if given a supplement of chromium acetate. Chromium appears to have a role in glucose tolerance, possibly forming a complex between insulin and its receptors. It has restored glucose tolerance in malnourished children. Chromium may also play a role in lipid synthesis and experiments have shown decreased serum cholesterol and increased high-density lipoprotein cholesterol in cases of deficiency. There is also thought to be an involvement in protein and nucleic acid metabolism. There have been few experiments with farm animals, but investigations with pigs have supported the above putative roles, with increased lean and decreased fat deposition as a result of more efficient glucose metabolism and sparing of protein catabolism. Improved reproduction in sows has been reported with supplements of organic sources of chromium. Whether the element is likely to have any practical significance in the nutrition of farm animals is still being investigated. Chromium is not a particularly toxic element and a wide margin of safety exists between the normal amounts ingested and those likely to produce deleterious effects. Levels of 50 mg chromium/kg DM in the diet have caused growth depression and liver and kidney damage in rats. Chromium in the form of its insoluble oxide is often used as a marker in digestibility trials (p. 249).

Vanadium

No specific biochemical function has been identified for vanadium but it may have a role in the regulation of the activity of sodium–potassium ATPase, phosphoryl transferase enzymes, adenyl cyclase and protein kinase. It may also act as a cofactor for certain enzymes. Vanadium deficiency has been demonstrated in rats, goats and chicks. Deficiency symptoms included impaired growth and reproduction, and disturbed lipid metabolism. In chicks consuming diets containing less than 10 μg vanadium/kg DM, growth of wing and tail feathers was significantly reduced. Subsequent studies with chicks demonstrated a significant growth response when dietary vanadium concentrations were increased from 30 μg to 3 mg/kg DM. In goats fed on diets containing less than 10 μg vanadium/kg there was no effect on growth, but there was increased incidence of abortion, reduced milk fat production and a high death rate in the kids.

There is little information about the vanadium content of foods, but levels ranging from 30 to 110 μg/kg DM have been reported for ryegrass. Herring meal appears to be a relatively rich source containing about 2.7 mg/kg DM. Vanadium is a relatively toxic element. When diets containing 30 mg/kg DM of the element were given to chicks, growth rate was depressed; at levels of 200 mg/kg DM the mortality rate was high.

Nickel

A discrete biochemical function has not been firmly established for nickel but it is thought it may be a cofactor or structural component in metalloenzymes. It may also play a role in nucleic acid metabolism.

Physiological symptoms of nickel deficiency have been produced in chicks, rats and pigs kept under laboratory conditions. Chicks given a diet containing the element in a concentration of less than 400 μg/kg DM developed skin pigmentation changes, dermatitis and swollen hocks. Diets with a low nickel content have produced scaly and crusty skin in pigs, similar to the parakeratosis seen in zinc deficiency, which suggests an involvement in zinc metabolism. Supplements of nickel have increased rumen bacterial urease activity.

Normal levels of the element in pasture herbage are 0.5–3.5 mg/kg DM while wheat grain contains 0.3–0.6 mg/kg DM. Nickel is a relatively non-toxic element, is poorly absorbed from the digestive tract and does not normally present a serious health hazard.

Tin

In 1971 it was reported that a significant growth effect, in rats maintained on purified amino acid diets in a trace element-free environment, was obtained when the diets were supplemented with tin. These studies suggested that tin was an essential trace element for mammals. The element is normally present in foods in amounts less than 1 mg/kg DM, the values in pasture herbage grown in Scotland, for example, being of the order of 300–400 μg/kg DM. The nutritional importance of this element has yet to be determined but it is suggested that tin contributes to the tertiary structure of protein or other macromolecules. Tin is poorly absorbed from the digestive tract, especially in the inorganic form, and ingested tin has a low toxicity.

Arsenic

Arsenic is widely distributed throughout the tissues and fluids of the body but is concentrated particularly in the skin, nails and hair. It has been shown that the element is essential for the rat, chick, pig and goat. It is needed to form metabolites of methionine, including cystine. Animals given an arsenic-deficient diet had rough coats and slower growth rates than control animals given a supplement of arsenic. A long-term study with goats showed interference with reproduction (abortion, low birth weights) and milk production, and sudden death. The toxicity of the element is well known; symptoms include nausea, vomiting, diarrhoea and severe abdominal pain. The toxicity of its compounds differs widely; trivalent arsenicals, which block lipoate-dependent enzymes, are more toxic than the pentavalent compounds.

6.6 Other elements

The essentiality of other elements, listed by some authorities, is the subject of debate. Often the levels required to produce a deficiency are so low as to be of no practical significance in normal animal nutrition. For example, boron, which is essential to plants, has been shown to increase growth rate and tibia weight and strength in broilers but the response was obtained at levels found only in boron-deficient plants. Lithium has been shown to be essential for goats, where it prevented growth retardation, impaired fertility and low birth weights, but again deficiency is unlikely under practical conditions.

Summary

Minerals fulfil physiological, structural and regulatory functions. Mineral supplements take various forms: mineral salts, rumen boluses, 'organic' compounds, pasture applications. The roles of individual mineral elements, and the effects of their deficiencies, are summarised below:

Mineral element	Role	Effects of deficiency
Calcium	Bone and teeth, transmission of nerve impulses	Rickets, osteomalacia, thin eggshells, milk fever
Phosphorus	Bone and teeth, energy metabolism	Rickets, osteomalacia, depraved appetite, poor fertility
Potassium	Osmoregulation, acid–base balance, nerve and muscle excitation	Retarded growth, weakness
Sodium	Acid–base balance, osmoregulation	Dehydration, poor growth, poor egg production
Chlorine	Acid–base balance, osmoregulation, gastric secretion	Alkalosis
Sulphur	Structure of amino acids, vitamins and hormones, chondroitin	Equivalent to protein deficiency (urea-supplemented diets)
Magnesium	Bone, activator of enzymes for carbohydrate and lipid metabolism	Nervous irritability and convulsions, hypomagnesaemia
Iron	Haemoglobin, enzymes of electron transport chain	Anaemia
Copper	Haemoglobin synthesis, enzyme systems, pigments	Anaemia, poor growth, depigmentation of hair and wool, swayback
Cobalt	Component of vitamin B_{12}	Pining (emaciation, anaemia, listlessness)
Iodine	Thyroid hormones	Goitre; hairless, weak or dead young
Manganese	Enzyme activation	Retarded growth, skeletal abnormality, ataxia
Zinc	Enzyme component and activator	Parakeratosis, poor growth, depressed appetite
Selenium	Component of glutathione peroxidase, iodine metabolism, immune function	Myopathy, exudative diathesis

Further reading

Agricultural and Food Research Council 1991 Technical Committee on Responses to Nutrients, Report No. 6. A Reappraisal of the Calcium and Phosphorus Requirements of Sheep and Cattle. *Nutrition Abstracts and Reviews, Series B* **61**: 573–612.

Ammerman C B, Henry P R and Miles R D 1998 Supplemental organically-bound mineral compounds in livestock nutrition. In: Garnsworthy P C and Wiseman J (eds) *Recent Advances in Animal Nutrition*, Nottingham, Nottingham University Press, pp 67–91.

Gawthorne J M, Howell J McC and White C L (eds) 1982 *Trace Element Metabolism in Man and Animals*, Berlin, Springer-Verlag.

Georgievskii V I, Annenkov B N and Samokhim V I 1982 *Mineral Nutrition of Animals*, London, Butterworth.

McDowell L R 1992 *Minerals in Animal and Human Nutrition*, New York, Academic Press.

National Research Council 1980 *Mineral Tolerances of Domestic Animals*, Washington, DC, National Academy of Sciences.

Thompson J K and Fowler V R 1990 The evaluation of minerals in the diets of farm animals. In: Wiseman J and Cole D J A (eds) *Feedstuff Evaluation*, London, Butterworth, pp 235–59.

Underwood E J 1977 *Trace Elements in Human and Animal Nutrition*, New York, Academic Press.

Underwood E J and Suttle N F 1999 *The Mineral Nutrition of Livestock*, 3rd edn, Wallingford, CABI Publishing.

7 Enzymes

7.1   Classification of enzymes

7.2   Nature of enzymes

7.3   Mechanism of enzyme action

7.4   Specific nature of enzymes

7.5   Factors affecting enzyme activity

7.6   Nomenclature of enzymes

The existence of living things involves a continuous series of chemical changes. Thus green plants elaborate chemical compounds such as sugars, starch and proteins and in doing so fix and store energy. Subsequently these compounds are broken down by the plants themselves, or the animals which consume them, and the stored energy is utilised. The complex reactions involved in these processes are reversible and when not associated with living organisms are often very slow. Extremes of temperature and/or pressure are required to increase their velocity to practicable levels. In living organisms such conditions do not exist. Yet the storage and release of energy in such organisms must take place quickly when required, necessitating a high velocity of the reactions involved. The required velocity is achieved through the activity of numerous catalysts present in the organisms.

A catalyst in the classical chemical sense is a substance which affects the velocity of a chemical reaction without appearing in the final products; characteristically the catalyst remains unchanged in mass upon completion of the reaction. The catalysts elaborated by living organisms are organic in nature and are known as enzymes. They are capable of increasing the rate of reaction by a factor of as much as 10^9-10^{12} times that of the non-catalysed reaction. The reactions catalysed by enzymes are theoretically reversible and should reach equilibrium. In living cells, reaction products are removed and the reactions are largely unidirectional and do not reach equilibrium. Rather, they reach a steady state in which the concentrations of the reactants remain relatively constant. Reactions will speed up under demand, or slow down when the products are not removed quickly enough to maintain the steady state. Enzymes affect both the forward and the reverse reactions equally so that the steady state is not changed. It is, however, attained more quickly. When one or more of the products is decomposed, the reaction may become virtually irreversible.

Each living cell contains hundreds of enzymes and can function efficiently only if their action is suitably coordinated. It is important to appreciate that within the cell the enzymes exist in different compartments; the cell is not a bag of randomly distributed enzymes. Thus the enzymes of the first stage in the oxidation of glucose (the glycolytic pathway) are present in the cytoplasm, whereas those involved in the formation of acetyl-CoA from pyruvate and its subsequent oxidation via the Krebs cycle are limited to the mitochondria (Chapter 9).

7.1 Classification of enzymes

Enzymes are classified into six major groups according to their mode of action.

1. The *oxidoreductases* catalyse the transfer of hydrogen, oxygen or electrons from one molecule to another. For example, lactate is oxidised to pyruvate in the presence of lactate dehydrogenase. In the process two electrons and two hydrogen atoms are removed from the alcohol group, leaving a ketone, and are transferred to form $NADH(+H^+)$.

$$
\begin{array}{l}
CH_3 \\
| \\
CHOH \\
| \\
COO^-
\end{array}
\; + \; NAD^+ \; \longrightarrow \;
\begin{array}{l}
CH_3 \\
| \\
C{:}O \\
| \\
COO^-
\end{array}
\; + \; NADH(+H^+)
$$

Lactate   Nicotinamide   Pyruvate   Reduced NAD^+
          adenine
          dinucleotide

The group includes:

- dehydrogenases
- oxidases
- peroxidases
- catalases
- oxygenases
- hydroxylases

2. The *transferases* are a large group of enzymes that catalyse the transfer of groups such as acetyl, amino and phosphate from one molecule to another. For example, in the formation of citrate from oxalacetate during the release of energy in the body, addition of an acetyl group takes place in the presence of citrate synthetase:

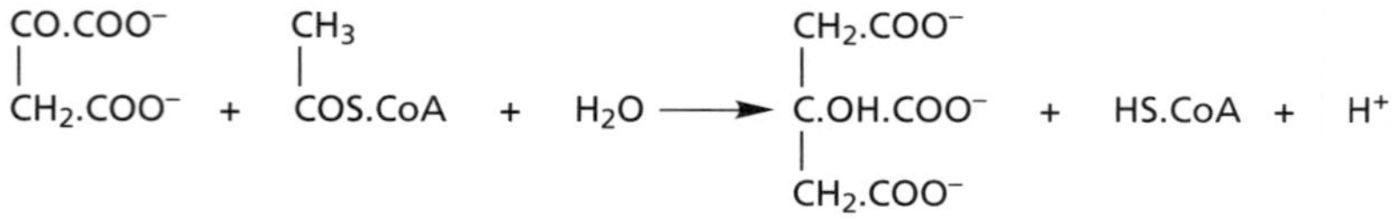

$$
\begin{array}{ccccccccc}
\text{CO.COO}^- & & \text{CH}_3 & & & & \text{CH}_2\text{.COO}^- & & \\
| & & | & & & & | & & \\
\text{CH}_2\text{.COO}^- & + & \text{COS.CoA} & + & \text{H}_2\text{O} & \longrightarrow & \text{C.OH.COO}^- & + \quad \text{HS.CoA} \; + \; \text{H}^+ \\
& & & & & & | & & \\
& & & & & & \text{CH}_2\text{.COO}^- & &
\end{array}
$$

| Oxalacetate | Acetyl-CoA | Citrate | Coenzyme A |

The group includes:

- transaldolases and transketolases
- acyl, glucosyl and phosphoryl transferases
- kinases
- phosphomutases

3. The *hydrolases* catalyse hydrolytic cleavage. Typical are the hydrolyses of fats and proteins, which are essential for the normal functioning of the organism. A fat may be broken down to glycerides (acylglycerols) and fatty acids under the influence of a lipase:

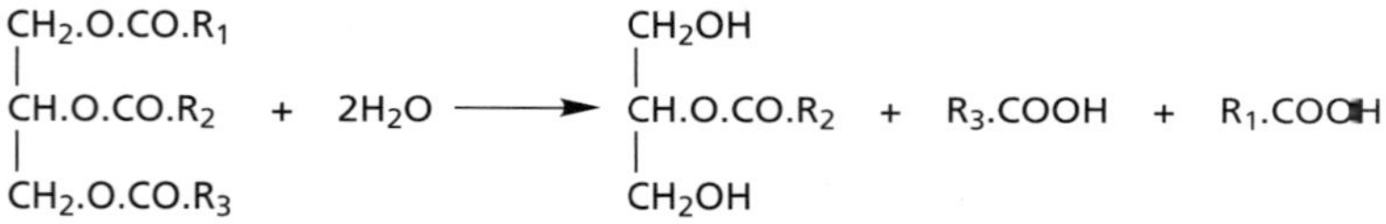

$$
\begin{array}{ccccccccc}
\text{CH}_2\text{.O.CO.R}_1 & & & & \text{CH}_2\text{OH} & & & & \\
| & & & & | & & & & \\
\text{CH.O.CO.R}_2 & + & 2\text{H}_2\text{O} & \longrightarrow & \text{CH.O.CO.R}_2 & + & \text{R}_3\text{.COOH} & + & \text{R}_1\text{.COOH} \\
| & & & & | & & & & \\
\text{CH}_2\text{.O.CO.R}_3 & & & & \text{CH}_2\text{OH} & & & &
\end{array}
$$

| Triglyceride (Triacylglycerol) | Monoglyceride (Monoacylglycerol) | Fatty acids |

Similarly peptidases split proteins by hydrolysis of the peptide linkages between the constituent amino acids.

This is a large group including:

- esterases
- glycosidases
- peptidases
- phosphatases
- thiolases
- phospholipases
- amidases
- deaminases
- ribonucleases

4. The *lyases* are enzymes that catalyse non-hydrolytic decompositions involving the removal of certain groups as in decarboxylation and deamination. Pyruvate decarboxylase, for example, catalyses the conversion of a 2-oxo-acid to an aldehyde, and carbon dioxide is removed:

$$
\begin{array}{ccc}
R & & R \\
| & & | \\
C{:}O & \longrightarrow & CHO \quad + \quad CO_2 \\
| & & \\
COOH & &
\end{array}
$$

In addition to the above, the group includes:

- aldolases
- hydratases
- dehydratases
- synthases
- lyases

5. The *isomerases* catalyse intramolecular rearrangement in optical and positional isomers. Typical of this class are the epimerases such as uridine diphosphate glucose 4-epimerase. This catalyses the change of configuration at the fourth carbon atom of glucose, and galactose is produced (Chapter 9). The group includes:

- racemases
- isomerases
- some of the mutases

6. The *ligases*, as the name implies, catalyse reactions where two molecules are bound together with the breakdown of high-energy phosphate bonds as in ATP, which provides the energy for the reaction to take place. The production of acetyl-CoA from acetate by acetyl coenzyme A synthetase is typical:

$$
\begin{array}{ccccccccccc}
CH_3 & & H & & & & CH_3 & & & & \\
| & + & | & + & ATP & \longrightarrow & | & + & H_2O & + & AMP \quad + \quad PP \\
COOH & & S.COA & & & & COS.CoA & & & &
\end{array}
$$

Acetic acid   Coenzyme A   Acetyl-CoA   Pyrophosphate

7.2 Nature of enzymes

The majority of enzymes are based on complex, high molecular weight proteins, but there are exceptions, such as the ribozymes, which are RNA-based. Some proteins may themselves act as efficient catalysts but many require the assistance of smaller entities to achieve this. Such substances, bound to the protein, are termed cofactors.

Coenzymes

The organic cofactors are known as coenzymes. They are relatively few in number but each may be associated with a number of different enzymes and so function in a large number of reactions. Among the more important coenzymes are nicotinamide adenine dinucleotide, thiamine pyrophosphate, pyridoxal phosphate and flavin mononucleotide. In such cases, the protein, referred to as the apoprotein, is catalytically inactive. When bound to the coenzyme it forms the active holoenzyme.

Flavin mononucleotide (Fig. 7.1) is a typical example of a coenzyme.

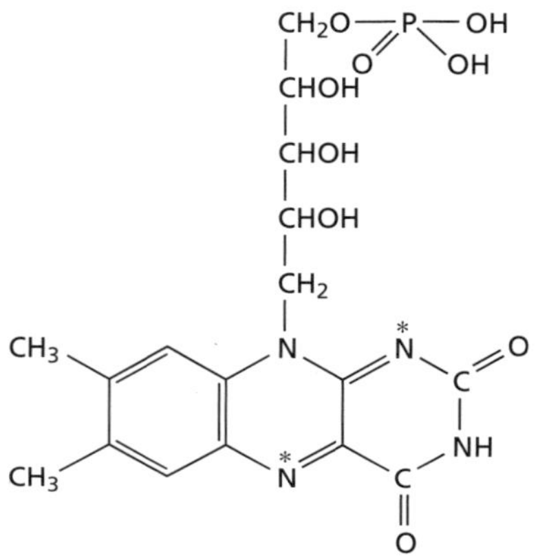

Fig. 7.1 Flavin mononucleotide.

Exchange of hydrogen atoms takes place at the positions marked *.

In some cases the coenzyme does not remain permanently attached to the apoenzyme but is released after the reaction is completed. Thus the oxidised form of adenine dinucleotide is strongly bound to the apoenzyme in dehydrogenase systems, but when the oxidation is complete the reduced form is released from the enzyme and the oxidised form is regenerated by reaction with other electron acceptors. In such cases the coenzyme is acting more as a second substrate than a true coenzyme. Many enzymes act out either or both of these roles, depending upon the reaction with which they are concerned.

Metal cofactors

Some two-thirds of enzymes require metal cofactors if they are to carry out their proper function. In some cases the metals are attached by coordinate covalent linkages to the enzyme protein or they may form part of prosthetic groups within the enzyme, as in the case of iron in the haem proteins (Fig. 7.2). Good examples of protein-bound metal cofactors are zinc and iron. Carboxypeptidase is a zinc protease which catalyses the hydrolytic cleavage of peptide bonds. The zinc generates hydrogen and hydroxyl ions, from bound water, which attack the peptide bond as shown:

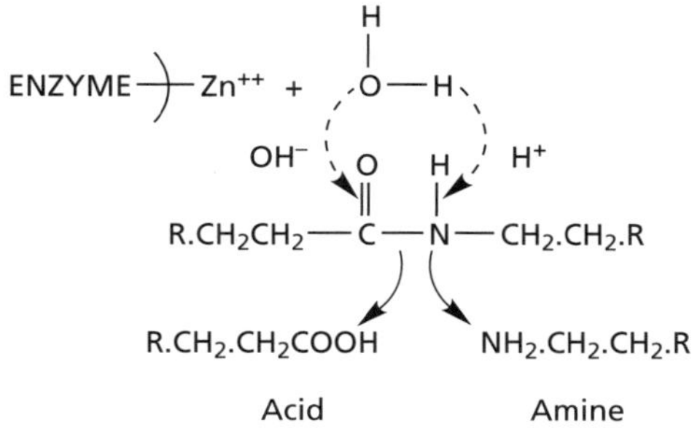

ENZYME $\rightarrow$ Zn^{++} +

$$R.CH_2CH_2 - C - N - CH_2.CH_2.R$$

$$R.CH_2.CH_2COOH \qquad NH_2.CH_2.CH_2.R$$

Acid   Amine

The cytochromes are important in certain oxidation reactions in the body in which they accept electrons from reduced substances, which are consequently oxidised. The iron forms part of the haem grouping within the cytochrome molecule as shown in Fig. 7.2.

The exchange of electrons takes place at the iron atom.

Metal cofactors do not always bind to the enzyme but rather to the primary substrate. The resulting substrate–metal complex binds to the enzyme and facilitates its activity. Creatine kinase catalyses the transfer of phosphoryl groups from adenosine triphosphate, which is broken down to adenosine diphosphate. The reaction requires the presence of magnesium ions. These, however, do not bind to the enzyme but to ATP, forming an ATP–Mg complex. It is this complex which binds to the enzyme and allows transfer of the phosphoryl group:

$$ATP + Mg^{++} \longrightarrow ATP\text{–}Mg \text{ complex}$$

$$ATP\text{–}Mg \text{ complex} + Enzyme \longrightarrow ATP\text{–}Mg\text{–}Enzyme \text{ complex}$$

$$Creatine \longrightarrow Phosphocreatine$$

$$ATP\text{–}Mg\text{–}Enzyme \qquad ADP + Enzyme + Mg^{++}$$

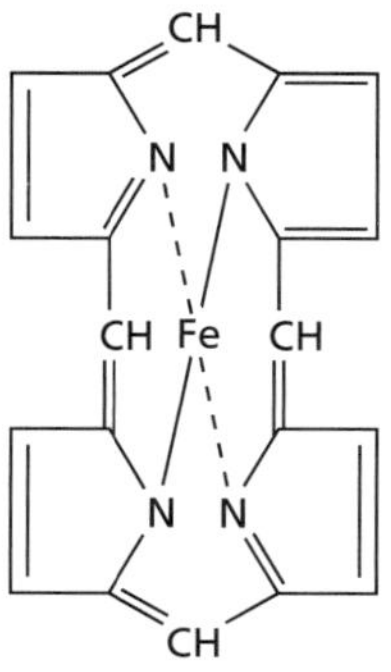

Fig. 7.2   Haem grouping.

Some enzymes are present in an inactive form which is changed to an active state at the place and time when its action is required. Thus trypsinogen is synthesised in the pancreas, transported to the small intestine and there changed to the active digestive enzyme trypsin. This kind of mechanism confers considerable control over the siting and timing of enzyme activity. The inactive precursors are known as zymogens.

7.3 Mechanism of enzyme action

If a chemical reaction is to take place, the reacting molecules must pass through a high-energy transition state. This may be envisaged as an intermediate stage in the reaction in which the molecules are distorted or strained, or have an unfavourable electronic arrangement. At any particular moment the molecules in a sample have differing energies and a few only may be able to pass the energy barrier represented by the transition state. In this situation the reaction may not proceed or may only do so very slowly. In the gaseous state, or in solution, the energies of the molecules may be raised by the provision of an outside source of energy in the form of heat; more molecules will pass the energy barrier and the reaction is speeded up. In the body this is not an option owing to the strict control exerted on body temperature. In such a situation, the same end may be achieved by lowering the barrier instead of increasing the energy of the molecules. This is the function performed by a catalyst.

In Fig. 7.3 the unbroken line illustrates the effect, on the rate of an uncatalysed reaction, of increasing inputs of energy. The apex of the curve

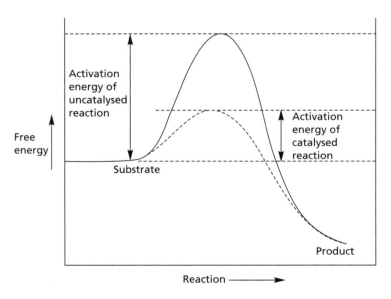

Fig. 7.3 Mechanism of enzyme catalysis.

represents the transition state and the energy which a molecule must possess in order that the reaction may take place. The free energy which must be provided to achieve this is the activation energy. The broken line shows a similar plot for the catalysed reaction. In this case the energy of the transition state is lower, more molecules are able to attain it and the reaction rate is increased. It is important to appreciate that the catalyst does not change the equilibrium but merely the speed with which it is attained.

Enzyme action involves the formation of a complex between the enzyme and the substrate or substrates. The complex then undergoes breakdown yielding the products and the unchanged enzyme:

$$E + S \rightleftharpoons ES \rightleftharpoons E + P$$

The complexes are formed between the substrate and relatively few active sites on the enzyme surface. The bonding may be ionic:

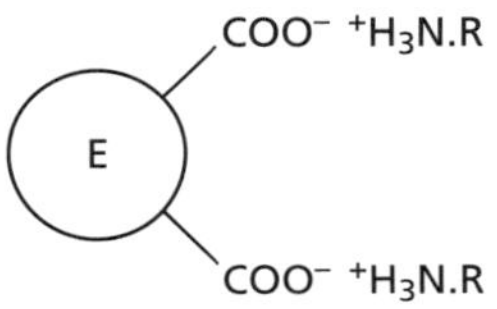

or by hydrogen bonds:

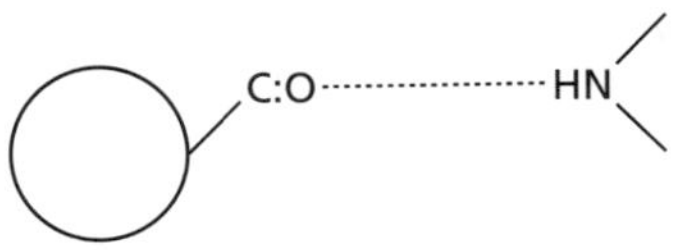

or by covalent bonds typified by sulphydryl groups on the enzyme:

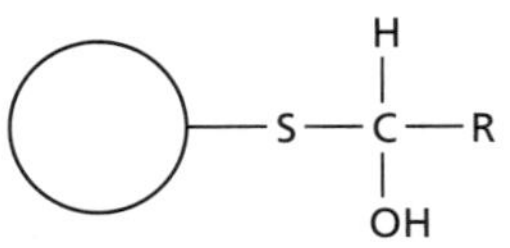

or by van der Waals bonds. These arise as a result of non-specific attractive forces between atoms when the distance between them is from 3 to 4 Å. They are weaker than the ionic or hydrogen bonds but may be important because of the large number which may be formed when conditions are favourable.

The type of reaction envisaged for enzyme catalysis is illustrated by that suggested for the chymotrypsin-catalysed hydrolysis of an ester illustrated in Fig. 7.4, where the groups involved in the catalysis are the alcohol group of serine and the imidazole of histidine.

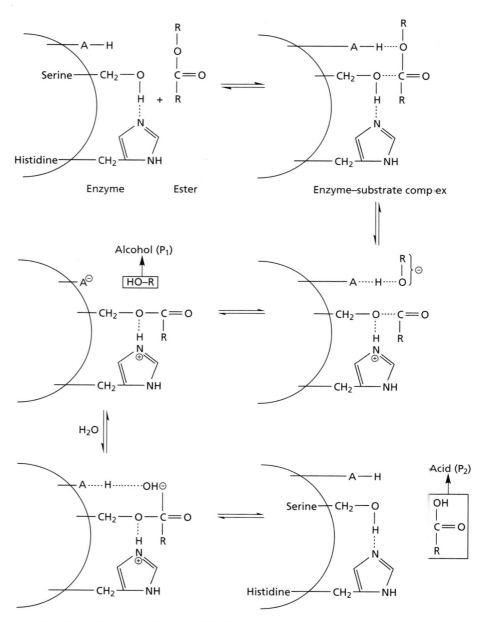

Fig. 7.4 Chymotrypsin-catalysed hydrolysis of an ester. (After Westheimer F H 1962 *Advances in Enzymology* **24**: 464.)

The active sites are always small compared with the enzyme molecule and are three-dimensional. They are usually clefts, crevices or pockets and have specific shapes. They are usually surrounded by amino acid chains, some of which help to bind the substrate and others which play a part in the actual catalytic process.

7.4 Specific nature of enzymes

Enzyme specificity is said to be absolute if the action of the enzyme is limited to one substrate only. An example is urease, which catalyses only the breakdown of urea to carbon dioxide and ammonia.

$$\begin{array}{c} NH_2 \\ \diagdown \\ C{:}O \ + \ H_2O \longrightarrow 2NH_3 \ + \ CO_2 \\ \diagup \\ NH_2 \end{array}$$

In most cases the enzyme is able to catalyse more than one of a group of substrates. In such cases the specificity is said to be relative. Such group specificity may be of a low order, as in the case of the digestive enzymes trypsin and pepsin, which catalyse the rupture of peptide bonds. In other cases it may be much higher. Chymotrypsin, for example, catalyses the hydrolytic cleavage only of peptide bonds in which the carboxyl residue is derived from an aromatic amino acid.

Enzyme specificity arises from the need for spatial conjunction of the active groups of the substrate with the active site of the enzyme. This requires an exact fit of the substrate into the active site as a key fits into a lock. In the lock and key model of enzyme action (Fig. 7.5), the necessary structures are considered to be preformed.

Whereas the model accounts for enzyme specificity it does not explain certain other aspects and the induced-fit model is currently pre-eminent. This predicates that the reacting sites need not be fully preformed but only so far as to allow the substrate to position itself close to an active site on the enzyme. Interaction then causes distortion of both enzyme and substrate, as a result of which structures are produced which allow of complete conjunction of the two and successful complex formation (Fig. 7.6).

As a result of the distortion involved in complex formation, strain is induced in the enzyme and this helps to pull the substrate into the transition state conformation.

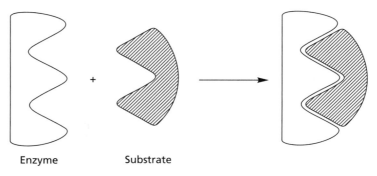

Enzyme   Substrate

Fig. 7.5 Lock and key model of enzyme–substrate complex formation.

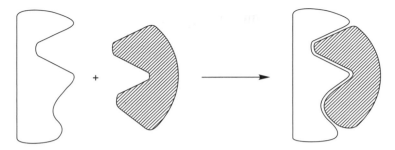

Fig. 7.6 Induced-fit model of enzyme–substrate complex formation.

In addition to distorting their substrates, enzymes frequently position groups in the right position for the catalytic action to take place, especially in the case of acid–base catalysis. In other instances, the enzyme sets a metal ion in just the right position to allow metal ion catalysis.

7.5 Factors affecting enzyme activity

Substrate concentration

In a system where the enzyme is in excess and the concentration remains constant, an increase in substrate concentration increases the velocity of a reaction. This is due to increased utilisation of the available active centres of the enzyme. If the substrate concentration continues to increase, utilisation of the available active centres becomes maximal and there is no further increase in the rate of reaction. In fact, continued increases in substrate concentration may lead to a reduced rate owing to an incomplete linkage of enzyme and substrate resulting from competition for the active centres by the excess substrate molecules.

The effect of substrate concentration on the velocity of an enzyme-catalysed reaction is frequently represented in terms of the Michaelis–Menten constant, K_m. This is the molar concentration of substrate at which half the centres of the enzyme are occupied by substrate and the rate of reaction is half the maximum. Above K_m, the effect of an increase in concentration of the substrate on the reaction decreases as the maximum is approached. At substrate concentrations below K_m, increases in concentration will give large responses in terms of reaction rate. When physiological substrate concentrations are much above an enzyme's K_m, substrate is unlikely to be the controlling factor of a metabolic pathway. Many enzymes have K_m values which approximate to their physiological concentrations. Changes in the latter will result in significant changes in reaction rate and are important in the control of metabolism.

Enzyme concentration

In a system where the substrate is present in excess, an increase in enzyme concentration gives a straight-line response in reaction velocity owing to the provision of additional active centres for the formation of enzyme–substrate complexes. Further increase in the enzyme concentration may result in some limiting factor, such as availability of coenzyme, becoming operative. Enzymes are rarely saturated with substrate under physiological conditions.

Inhibitors

A wide variety of substances can act as inhibitors of enzyme activity. They fall into two main groups: reversible and non-reversible.

Reversible inhibition

Such inhibition involves non-covalent bonding of the inhibitor to the enzyme. The group divides into three subgroups: competitive, non-competitive and uncompetitive.

A *competitive inhibitor* resembles the substrate in its chemical structure and is able to combine with the enzyme to form an enzyme–inhibitor complex. In so doing it competes with the substrate for the active sites of the enzyme, and formation of the enzyme–substrate complex is inhibited. This type of inhibition may be reversed by the addition of excess substrate, which displaces the inhibitor, forming normal enzyme–substrate complexes. One of the best-known examples is provided by the sulphonamide drugs. The synthesis of folic acid from *p*-aminobenzoic acid (PABA) is a vital metabolic process in the bacteria controlled by these drugs. The similarity between PABA and sulphanilamide, released by the sulphonamides, is obvious:

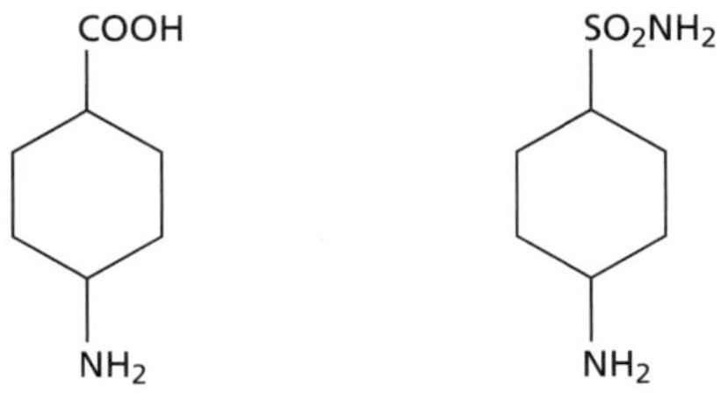

COOH SO_2NH_2

NH₂   NH₂

Para-aminobenzoic acid   Sulphanilamide

The sulphanilamide forms a complex with the relevant enzyme, thus preventing normal enzyme–substrate combination and the formation of folic acid. Addition of excess PABA overcomes the inhibition, since the formation of the enzyme–sulphanilamide complex is reversible. The situation may be visualised as shown in Fig. 7.7.

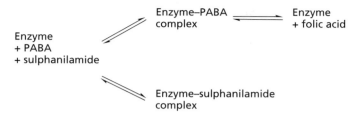

Fig. 7.7 Competitive inhibition of PABA.

The extent of the inhibition will depend upon the relative concentrations of the true substrate and the inhibitor.

A *non-competitive inhibitor* does not bind to the active site but at some other site on the enzyme surface and may cause distortion of the enzyme and reduce catalytic activity. In the simplest case the inhibitor binds readily with both enzyme and substrate and three types of complex are formed:

- enzyme–substrate (ES),
- enzyme–inhibitor (EI),
- enzyme–inhibitor–substrate (EIS).

The reaction rate will be slower owing to the removal of enzyme from the system. The EI complex will be catalytically inert. The EIS complex may, however, be susceptible of reconversion to ES and make some contribution to catalytic activity. Non-competitive inhibition cannot be reversed by excess substrate but may be reversed by exhaustive dialysis.

In *uncompetitive inhibition*, the inhibitor binds to the enzyme–substrate complex and renders it inactive.

Irreversible inhibition

Such inhibition involves covalent bonding at the active site and cannot be reversed by excess substrate or by dialysis. The site is therefore blocked and made catalytically inactive. Most inhibitors in this group are highly toxic, e.g. the organophosphorus nerve poisons. Thus, diisopropyl fluorophosphate (DFP) reacts irreversibly with the hydroxyl group of serine:

$$
\begin{array}{ccc}
\text{CH.(CH}_3)_2 & & \text{CH.(CH}_3)_2 \\
| & | & | \quad | \\
\text{C:O} \quad\quad \text{O} & & \text{C:O} \quad \text{O} \\
| \quad\quad | & & | \quad\quad | \\
\text{CH.CH}_2\text{OH} \;+\; \text{F.P:O} & \longrightarrow & \text{CH.CH}_2.\text{O.P:O} \;+\; \text{HF} \\
| \quad\quad | & & | \quad\quad | \\
\text{NH} \quad\quad \text{O} & & \text{NH} \quad \text{O} \\
| \quad\quad | & & | \quad\quad | \\
& \text{CH.(CH}_3)_2 & & \text{CH.(CH}_3)_2
\end{array}
$$

Enzyme   Diisopropyl phosphofluoridate   Inactive enzyme–inhibitor complex

Any enzyme having an essential serine at its active site will be irreversibly inactivated by it. The serine proteases and acetylcholinesterases are typical examples. The latter is essential for nerve conduction and its inactivation results in rapid paralysis of vital functions. It is this action which makes the organophosphates such potent toxins.

Allosteric effects

The activity of some enzymes can be affected by molecules which bind to the enzymes but which do not work in the same way as either competitive or non-competitive enzyme inhibitors. Such molecules which bind to macromolecules are known as ligands and may be small and simple or large and complex like proteins. They may be activators or inhibitors and may even be substrates. They are commonly referred to as effectors, modifiers or modulators.

Typically modulators bind to unique *allosteric* sites, which are quite distinct from the sites binding the substrate. The binding brings about a conformational change in the enzyme, resulting in a change in the affinity of the enzyme for the substrate. The modulators may increase or decrease the affinity and bind to *activator* and *inhibitory* sites, respectively.

The vast majority of allosteric enzymes are oligomers, i.e. they consist of several subunits or protomers. Binding of a modulator to one protomer may affect binding at the other protomers so that the effect of an activator on the affinity of an enzyme for the substrate on one protomer has the same effect on the others.

Almost all metabolic pathways employ allosteric enzymes. The complex feedback control characterising these pathways is, in large part, the result of the action of allosteric inhibitors and activators.

Environmental factors

A number of factors including temperature, acidity and salt concentration may influence enzyme activity, but in the living animal these are not likely to be of much importance.

Temperature

Over a limited temperature range, the efficiency of enzyme-catalysed reactions is increased by raising temperature. Very approximately, the speed of reaction is doubled for each increase of 10 °C. As the temperature rises, a complicating factor comes into play because denaturation of the enzyme protein begins. This is a molecular rearrangement which causes a loss of the active centres on the enzyme surface and a decrease in efficiency. Above 50 °C destruction of the enzyme becomes more rapid, and most enzymes are destroyed when heated to 100 °C. The time for which the enzyme is subjected to a given temperature affects the magnitude of the loss of activity. Each has an optimum temperature, which approximates to that of the cells in which it occurs and at which it is most effective. Thus enzymes of microorganisms adapted to cold conditions are able to function efficiently at temperatures close to zero, and others adapted to life in hot springs have optima in the region of 100 °C.

Acidity

Hydrogen ion concentration has an important effect on the efficiency of enzyme action. Many enzymes are most effective in the region of pH 6–7, which is that of the cell. Extracellular enzymes may show maximum activity in the acid or alkaline pH range, but the actual range in which an individual enzyme works is only about 2.5–3.0 units; outside this range the activity declines very rapidly. The reduction in efficiency brought about by a change in pH is due to changes in the degree of ionisation of the substrate and the enzyme. Where the linkage between active centres is electrostatic, the facility with which the intermediate complex is formed and consequently the efficiency of enzyme action is affected. In addition, highly acidic or alkaline conditions bring about denaturation and subsequent loss of activity.

7.6 Nomenclature of enzymes

Some enzymes were named in the early days of the study of enzymes and have non-systematic names such as those of the digestive enzymes pepsin, trypsin and ptyalin.

In 1972, the International Union of Pure and Applied Chemistry and the International Union of Biochemistry recommended the use of two systems of nomenclature for enzymes, one systematic and one working or recommended. The recommended name was not required to be very systematic but had to be short enough for convenient use. The systematic name formed in accordance with definite rules, was to show the action of the enzyme as exactly as possible, thus identifying it precisely. It was to consist of two parts, the first naming the substrate and the second, ending -ase, indicating the type of reaction catalysed. In addition, code numbers were allocated to enzymes according to the following scheme:

- the first number shows to which of the six main classes the enzyme belongs,
- the second number shows the subclass,
- the third number shows the sub-subclass,
- the fourth number identifies the enzyme.

The following are examples of the system:

Reaction	Recommended name	Systematic name	Code number
L-Lactate + NAD$^+$ to pyruvate + NADH + H$^+$	Lactate dehydrogenase	L-Lactate: NAD$^+$ oxidoreductase	1.1.1.27
Hydrolysis of terminal non-reducing 1,4-linked α-D-glucose residues with release of α-glucose, e.g. maltose to glucose	α-Glucosidase	α-D-Glucoside glucohydrolase	3.2.1.20
L-Glutamate to 4-aminobutyrate + CO$_2$	Glutamate decarboxylase	L-Glutamate 1-carboxylyase	4.1.1.15

Summary

1. Enzymes are organic catalysts. They are classified into six major groups according to the functions they perform, which are:

   (i) oxidoreductases, which catalyse the transfer of hydrogen, oxygen or electrons from one molecule to another;
   (ii) tranferases, which catalyse transfer of groups from one molecule to another;
   (iii) hydrolases, which catalyse hydrolytic cleavage;
   (iv) lyases, which catalyse non-hydrolytic decompositions such as decarboxylation and deamination;
   (v) isomerases, which catalyse intramolecular rearrangements;
   (vi) ligases, which catalyse bond formation, the energy for which is derived from the breakdown of high-energy compounds such as ATP.

2. The majority of enzymes are based on complex high molecular weight proteins, many of which need organic cofactors (coenzymes) if they are to work efficiently. Others need metallic cofactors, either bound to the enzyme by covalent bonds or forming an integral part of the molecule. Others do not bind to the molecule but to the primary substrate.

3. An enzyme may be present in an inactive form (zymogen), which is changed to the active form as and when required. A number of the digestive enzymes, trypsin for example, are of this type.

4. Enzymes function by lowering the activation energy of reactions.

5. Most enzymes are able to catalyse the reactions of more than one group of substances and are described as being relatively specific. Others may catalyse one reaction for one substance, in which case the specificity is described as absolute.

6. The rate of enzyme action is influenced by:

   (i) substrate concentration,
   (ii) enzyme concentration,
   (iii) temperature,
   (iv) acidity,
   (v) environment.

7. The first part of the name of an enzyme indicates the substrate attacked and the second the type of reaction carried out. In addition numbers are used to delineate class, subclass and sub-subclass.

Further reading

Devlin T M (ed.) 1997 *Textbook of Biochemistry with Clinical Correlations*, 4th edn, New York, John Wiley & Sons.

Enzyme Nomenclature 1973 *Recommendations (1972) of the International Union of Pure and Applied Chemistry and the International Union of Biochemistry*, New York, American Elsevier.

Mathews C K and van Holde K E 1999 *Biochemistry*, 3rd edn, Redwood City, CA, Benjamin Cummings Publishing Co.

Stryer L 2000 *Biochemistry*, 4th edn, San Francisco, W H Freeman.

8 Digestion

8.1 Digestion in monogastric mammals

8.2 Microbial digestion in ruminants and other herbivores

8.3 Alternative sites of microbial digestion

Many of the organic components of food are in the form of large insoluble molecules which have to be broken down into simpler compounds before they can pass through the mucous membrane of the alimentary canal into the blood and lymph. The breaking-down process is termed 'digestion', the passage of the digested nutrients through the mucous membrane 'absorption'.

The processes important in digestion may be grouped into mechanical, chemical and microbial. The mechanical activities are mastication and the muscular contractions of the alimentary canal. The main chemical action is brought about by enzymes secreted by the animal in the various digestive juices, though it is possible that plant enzymes present in unprocessed foods may in some instances play a minor role in food digestion. Microbial digestion of food, also enzymic, is brought about by the action of bacteria, protozoa and fungi, microorganisms which are of special significance in ruminant digestion. In monogastric animals, microbial activity occurs mainly in the large intestine, although there is a low level of activity in the crop of birds and the stomach and small intestine of pigs.

8.1 Digestion in monogastric mammals

The alimentary canal

The various parts of the alimentary canal of the pig, which will be used as the reference animal, are shown in Fig. 8.1. The digestive tract can be considered as a tube extending from mouth to anus, lined with mucous membrane, whose function is the prehension, ingestion, comminution, digestion and absorption of food, and the elimination of solid waste material. The various parts are mouth, pharynx, oesophagus, stomach, and small and large intestine. The movement of the intestinal contents along the tract is produced by peristaltic waves, which are contractions of the circular muscle of the intestinal wall. The contractions are involuntary and are under overall autonomic nervous control. The nervous plexus within the tissue

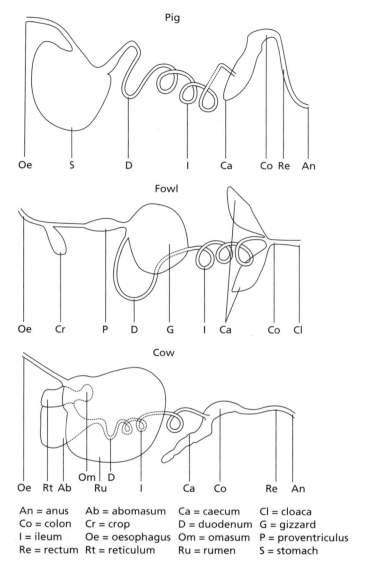

Fig. 8.1 Diagrammatic representation of the digestive tracts of different farm animals.

layers of the gut wall integrates the activity of the muscles. Several different kinds of movement of the intestinal wall are recognised, the functions of these being the transport of materials along the tract, the mixing of the digestive juices with the food, and the bringing of the digested nutrients into contact with the intestinal mucous membrane for subsequent absorption.

The small intestine, which comprises the duodenum, the jejunum and the ileum, is the main absorption site and contains a series of finger-like projections, the villi, which greatly increase the surface area available for absorption of nutrients. Each villus contains an arteriole and venule, together with a drainage tube of the lymphatic system, a lacteal. The venules ultimately drain into the hepatic portal system, and the lacteals into the thoracic lymphatic duct. The luminal side of each villus is covered with projections, the microvilli, which are often referred to as the brush border.

There are a number of secretions which flow into the alimentary canal and many of these contain enzymes which bring about the hydrolysis of the various food components (see Table 8.1). Some of the proteolytic enzymes present in the secretions are initially in the form of inactive precursors termed zymogens. These are activated after secretion into the tract.

Digestion in the mouth

mechanical

This is mainly mechanical, mastication helping to break up large particles of food and to mix it with saliva, which acts as a lubricant and is a medium for taste perception. The lower incisor teeth are used for rooting and the inward-curved upper incisors grasp and shear food items. The premolars and molars are employed to crush the food. The pig has taste buds throughout the oral cavity and they are concentrated on the tongue. The saliva is secreted into the mouth by three pairs of salivary glands: the parotids, which are sited in front of each ear; the submandibular (submaxillary) glands, which lie on each side of the lower jaw; and the sublingual glands, which are underneath the tongue. Saliva is about 99 per cent water, the remaining 1 per cent consisting of

---

Box 8.1
Studies of digestion

To study digestion in farm animals, especially ruminants, it is often necessary to obtain samples of digesta from various sections of the alimentary tract. Stomach contents can be sampled by means of a tube inserted via the oesophagus. However, the sampling of this and other sections of the tract generally requires surgical modification of the animal by the formation of fistulas. A *fistula* is an opening created between a digestive organ and the exterior of the animal, and it is maintained by means of a rubber or plastic insert known as a *cannula*. For example, in ruminants a fistula may be formed between the posterior dorsal sac of the rumen and the animal's flank. Such a fistula is normally fitted with a cannula 25–125 mm in diameter, with a screw cap. Removal of the cap allows samples of rumen contents to be taken. In both ruminants and pigs, similar, but smaller, cannulae may be inserted into the true stomach and into selected points of the small or large intestines. The intestines may also be fitted with a device known as a re-entrant cannula. For this the intestine is severed and both ends are brought close to the skin surface and joined by a tube running outside the animal. With this tube in position the digesta flow normally from the proximal to the distal portions of the intestine. However, if the tube is opened, digesta may be collected from the proximal part, measured, sampled and returned to the distal part. The re-entrant cannula therefore allows the flow of digesta to be measured directly. In pigs, for example, a cannula at the terminal ileum is used to determine how much protein digestion has occurred before the food residues are attacked by microorganisms in the caecum and colon.

Cannulated animals live a normal life and with care can be kept free of discomfort or pain for long periods. Sheep with a rumen cannula are known to have survived for more than ten years, well beyond their normal life span.

Table 8.1 Main digestive enzymes

Recommended name	Trivial name	Systematic name	Number	Source	Substrate
A. Enzymes hydrolysing peptide links					
Pepsin	–	–	3.4.23	Gastric mucosa	Proteins and peptides
Chymosin	Rennin	–	3.4.23.4	Gastric mucosa (young calves)	Proteins and peptides
Trypsin	–	–	3.4.21.4	Pancreas	Proteins and peptides
Chymotrypsin	–	–	3.4.21.1	Pancreas	Proteins and peptides
Carboxypeptidase A	Carboxypeptidase	Peptidyl-L-amino-acid hydrolase	3.4.12.2	Small intestine	Peptides
Carboxypeptidase B	Protaminase	Peptidyl-L-lysine (-L-arginine) hydrolase	3.4.12.3	Small intestine	Peptides
Aminopeptidases	–	α-Aminoacyl-peptide hydrolases	3.4.11	Small intestine	Peptides
Dipeptidases	–	Dipeptide hydrolases	3.4.13	Small intestine	Dipeptides
B. Enzymes hydrolysing glycoside links					
α-Amylase	Diastase	1,4-α-D-Glucan glucanohydrolase	3.2.1.1	Saliva, pancreas	Starch, glycogen, dextrin
α-Glucosidase	Maltase	α-D-Glucoside glucohydrolase	3.2.1.20	Small intestine	Maltose
Oligo-1,6-glucosidase	Isomaltase	Dextrin 6-α-glucanohydrolase	3.2.1.10	Small intestine	Dextrins
β-Galactosidase	Lactase	β-D-Galactoside galactohydrolase	3.2.1.23	Small intestine	Lactose
β-Frutofuranosidase	Sucrase	β-D-Fructofuranoside fructohydrolase	3.2.1.26	Small intestine	Sucrose
C. Enzymes acting on ester links					
Triacylglycerol lipase	Lipase	Triacylglycerol acylhydrolase	3.1.1.3	Pancreas	Triacylglycerols
Cholesterol esterase	–	Sterol-ester hydrolase	3.1.1.13	Pancreas and small intestine	Cholesterol esters
Phospholipase A₂	Lecithinase A	Phosphatide 2-acyl-hydrolase	3.1.1.4	Pancreas and small intestine	Lecithins and cephalins
Lysophospholipase	Lysolecithinase	Lysolecithin acyl-hydrolase	3.1.1.5	Small intestine	Lysolecithin
Deoxyribonuclease	DNase	Deoxyribonucleate 5′-oligonucleotidohydrolase	3.1.4.5	Pancreas and small intestine	DNA
Ribonuclease 1	RNase	Ribonucleate 3′-pyrimidino-oligonucleotidohydrolase	3.1.4.22	Pancreas and small intestine	RNA
Nucleosidase	–	N-Ribosyl-purine ribohydrolase	3.2.2.1	Small intestine	Nucleosides
Phosphatases	–	–	3.1.3	Small intestine	Orthophosphoric acid esters

mucin, inorganic salts and the enzymes α-amylase and the complex lysozyme. Some animals such as the horse, cat and dog lack salivary α-amylase, whereas the saliva of other species, man included, has strong α-amylase activity. The enzyme is present in the saliva of the pig, but the activity is low. It is doubtful whether much digestion occurs in the mouth, since the food is quickly swallowed and passed along the oesophagus to the stomach, where the pH is unfavourable for α-amylase activity. It is possible, however, that some digestion of starch by the enzyme can occur in the stomach, since the food mass is not immediately mixed intimately with the gastric juice. The pH of pig's saliva is about 7.3, which is only slightly above the value regarded as optimal for α-amylase activity. This enzyme hydrolyses the α-(1→4)-glucan links in polysaccharides containing three or more α-(1→4)-linked D-glucose units. The enzyme therefore acts on starch, glycogen and related polysaccharides and oligosaccharides. When amylose, which contains exclusively α-(1→4)-glucosidic bonds (p. 26), is attacked by α-amylase, random cleavages of these bonds give rise to a mixture of glucose and maltose. Amylopectin, on the other hand, contains in addition to α-(1→4)-glucosidic bonds a number of branched α-(1→6)-glucosidic bonds, which are not attacked by α-amylase, and the products include a mixture of branched and unbranched oligosaccharides (termed 'limit dextrins') in which α-(1→6)-bonds are abundant.

The enzyme lysozyme has been detected in many tissues and body fluids. It is capable of hydrolysing the β-(1→4)-N-acetyl-glucosaminidic linkage of the repeating disaccharide unit in the polysaccharides of the cell walls of many species of bacteria, thereby killing and dissolving them.

Digestion in the stomach — Storage

The stomach of the adult pig has a capacity of about 8 litres and consists of a simple compartment, which functions not only as an organ for the digestion of food but also for storage. Viewed from the exterior, the stomach can be seen to be divided into the cardia (entrance), fundus and pylorus (terminus), the cardia and pylorus being sphincters controlling the passage of food through the stomach. The inner surface of the stomach is increased in area by an infolding of the epithelium and has four distinct areas. The oesophageal area is an extension of the oesophagus into the stomach and the surface has no glands. Here α-amylase activity may continue and there is an active microbial population, mainly lactobacilli and streptococci. The cardia area covers about one-third of the surface and secretes an alkaline, enzyme-free, viscous mucus formed of a gel-forming glycoprotein which protects the epithelium from acid attack. The gastric gland region covers a further third of the surface and secretes a glycoprotein and fucolipid mucus and contains the oxyntic cells which produce hydrochloric acid. In addition, this region also produces pepsinogen. The fourth area is the pyloric region, which is prior to the entry to the small intestine. This area has glands, like those in the cardia region, which secrete a protective mucus. Thus the gastric juice consists of water, pepsinogens, inorganic salts, mucus, hydrochloric acid and the intrinsic factor important for the efficient absorption of vitamin B_{12}.

A number of factors are concerned in the stimulation of the glands to secrete gastric juice. Initially, in the cephalic phase, stimuli such as the sight and smell of food act via the vagus nerve. Then, in the gastric phase, secretion is maintained by chemical sensors and distension of the stomach. Finally, the presence of digesta in the duodenum elicits secretion by neural and hormonal messages.

The pig's stomach is rarely completely empty between meals and the slow mixing conditions are conducive to microbial fermentation at the oesophageal end and gastric digestion at the pyloric end. Pepsinogens are the inactive forms of pepsins which hydrolyse proteins. The acid concentration of the gastric juice varies with the diet but is generally about 0.1 M, which is sufficient to lower the pH to 2.0. The acid activates the pepsinogens, converting them into pepsins by removing low molecular weight peptides from each precursor molecule. Four pepsins have been found in the pig which have optimum activity at two different pH levels, 2.0 and 3.5. Pepsins preferentially attack those peptide bonds adjacent to aromatic amino acids, e.g. phenylalanine, tryptophan and tyrosine, but also have a significant action on linkages involving glutamic acid and cysteine. Pepsins also have a strong clotting action on milk. Rennin or chymosin, an enzyme which occurs in the gastric juice of the calf and the young piglet, resembles pepsins in its activity. The products of protein digestion in the stomach are mainly polypeptides of variable chain length and a few amino acids.

The emptying of the stomach contents into the duodenum is controlled by osmotic sensors in the duodenum. In addition the presence of excess lipid reduces the emptying rate.

The epithelial surface of the pig's stomach is susceptible to ulceration related to the degree of processing of cereals in the diet (see p. 579).

Digestion in the small intestine

The partially digested food leaving the stomach enters the small intestine, where it is mixed with secretions from the duodenum, liver and pancreas. The majority of digestion and absorption occurs in the small intestine, the duodenal area being the site for mixing digesta and secretions and the jejunal area being the site of absorption. The duodenal (Brunner's) glands produce an alkaline secretion, which enters the duodenum through ducts situated between the villi. This secretion acts as a lubricant and also protects the duodenal wall from the hydrochloric acid entering from the stomach.

Bile is secreted by the liver and passes to the duodenum through the bile duct. It contains the sodium and potassium salts of bile acids, chiefly glycocholic and taurocholic (see p. 52), phospholipids, the bile pigments biliverdin and bilirubin, which are the end-products of haem catabolism, cholesterol and mucin. In all farm animals except the horse, bile is stored in the gall bladder until required. The bile salts play an important part in digestion by activating pancreatic lipase and emulsifying fats. The daily requirement for bile acids is greater than the synthetic capacity of the liver and an enterohepatic circulation exists to maintain the supply.

The pancreas is a gland which lies in the duodenal loop and has two secretory functions: the endocrine process for the production of insulin and the exocrine process for the production of digestive enzymes (from the acinar cells), water and electrolytes (from the duct cells), which together form the pancreatic juice, which is secreted into the duodenum through the pancreatic duct. The proportions of the different enzymes change in response to the nature of the diet.

A number of factors induce the pancreas to secrete its juice into the duodenum. When acid enters the duodenum, the hormone secretin is liberated from the epithelium of the small intestine into the blood. When it reaches the pancreatic circulation it stimulates the pancreatic cells to secrete a watery fluid containing a high concentration of bicarbonate ions, but very little enzyme. Another hormone, cholecystokinin (pancreozymin), is also liberated from the mucosa when peptides and other digestive products reach the duodenum. This hormone stimulates the secretion into the pancreatic juice of proenzymes and enzymes such as trypsinogen, chymotrypsinogen, procarboxypeptidases A and B, proelastase, α-amylase, lipase, lecithinases and nucleases. Unlike pepsin, these enzymes have pH optima around 7–9. The inactive zymogen trypsinogen is converted to the active trypsin by enterokinase, an enzyme liberated from the duodenal mucosa. This activation is also catalysed by trypsin itself, thus constituting an autocatalytic reaction. The activation process results in the liberation of a hexapeptide from the amino terminal end of trypsinogen. Trypsin is very specific and only acts upon peptide linkages involving the carboxyl groups of lysine and arginine. Trypsin also converts chymotrypsinogen into the active enzyme chymotrypsin, which has a specificity towards peptide bonds involving the carboxyl groups of tyrosine, tryptophan, phenylalanine and leucine. Procarboxypeptidases are converted by trypsin into the proteolytic enzymes carboxypeptidases, which attack the peptides from the end of the chain, splitting off the terminal amino acid, which has a free α-carboxyl group. Such an enzyme is classified as an exopeptidase, as distinct from trypsin and chymotrypsin, which attack peptide bonds in the interior of the molecule and which are known as endopeptidases.

Pancreatic α-amylase is similar in function to the salivary amylase and attacks the α-$(1\rightarrow4)$-glucan links in starch and glycogen.

The breakdown of fats is achieved by pancreatic lipase. This enzyme does not completely hydrolyse triacylglycerols and the action stops at the monoacylglycerol stage.

$$\begin{array}{ccccc}
\alpha'\ CH_2OOC.R_1 & & CH_2OOC.R_1 & & CH_2OH \\
| & R_3.COOH & | & R_1.COOH & | \\
\beta\ CHOOC.R_2 \longrightarrow & & CHOOC.R_2 \longrightarrow & & CHOOC.R_2 \\
| & & | & & | \\
\alpha\ CH_2OOC.R_3 & & CH_2OH & & CH_2OH \\
\text{Triacylglycerol} & & \text{Diacylglycerol} & & \text{Monoacylglycerol}
\end{array}$$

Dietary fat leaves the stomach in the form of relatively large globules which are difficult to hydrolyse rapidly. Fat hydrolysis is helped by emulsification, which is brought about by the action of bile salts. These bile salts are detergents or amphipaths, the sterol nucleus being lipid-soluble and the hydroxyl groups and ionised conjugate of glycine or taurine being water-soluble (see p. 52). These amphipaths, in addition to being emulsifying agents, also have the property of being able to aggregate together to form micelles. Although triacylglycerols are insoluble in these micelles, monoacylglycerols and most fatty acids dissolve, forming mixed micelles. Some fatty acids, e.g. stearic, are not readily soluble in pure bile salt micelles but dissolve in mixed micelles.

Lecithinase A is an enzyme which hydrolyses the bond linking the fatty acid to the β-hydroxyl group of lecithin (see p. 46). The product formed from this hydrolysis, lysolecithin, is further hydrolysed by lysolecithinase (lecithinase B) to form glycerolphosphocholine and a fatty acid. Cholesterol esterase catalyses the splitting of cholesterol esters.

The nucleic acids DNA and RNA (see p. 65) are hydrolysed by the polynucleotidases deoxyribonuclease (DNase) and ribonuclease (RNase), respectively. These enzymes catalyse the cleavage of the ester bonds between the sugar and phosphoric acid in the nucleic acids. The end-products are the component nucleotides. Nucleosidases attack the linkage between the sugar and the nitrogenous bases, liberating the free purines and pyrimidines. Phosphatases complete the hydrolysis by separating the orthophosphoric acid from the ribose or deoxyribose.

The hydrolysis of oligosaccharides to monosaccharides and of small peptides to amino acids is brought about by enzymes associated with the intestinal villi. Only a small proportion of hydrolysis occurs intraluminally and arises from enzymes present in aged cells discarded from the intestinal mucosa. Most of the enzymatic hydrolysis occurs at the luminal surface of the epithelial cells, although some peptides are absorbed by the cells before being broken down by enzymes present in the cytoplasm.

Enzymes produced by the villi are sucrase, which converts sucrose to glucose and fructose; maltase, which breaks down maltose to two molecules of glucose; lactase, which hydrolyses lactose to one molecule of glucose and one of galactose; and oligo-1,6-glucosidase, which attacks the α-$(1 \rightarrow 6)$-links in limit dextrins. Aminopeptidases act on the peptide bond adjacent to the free amino group of simple peptides, whereas dipeptidases complete the breakdown of dipeptides to amino acids.

Although the large intestine is recognised as the site of major microbial fermentation (see below), there is a microbial population in the small intestine. Recent work with sugar beet pulp given to pigs fitted with ileal cannulae showed that a large proportion (47 per cent) of the neutral detergent fibre fraction was digested prior to the terminal ileum. This breakdown is the result of microbial activity in the stomach and small intestine and acid hydrolysis of some of the fibre fractions.

Digestion in the large intestine

The main site of absorption of digested nutrients is the small intestine; by the time the food material has reached the entrance to the colon most of the hydrolysed nutrients have been absorbed. With normal diets there is always a certain amount of material which is resistant to the action of the enzymes secreted into the alimentary canal. The large intestine plays an important role in the retrieval of nutrients, electrolytes and water in the digesta. Pigs have a short caecum and long colon compared with other monogastric omnivores. The mucosal surface does not have villi, as in the small intestine, but there are small projections which increase the surface area. As the ileal contents enter the large intestine the fluid and fine particles are preferentially retained by the ascending colon, whereas the coarser particles move distally at a faster rate. Cellulose and many of the hemicelluloses are not attacked by any of the enzymes present in the digestive secretions of the pig. In addition certain starches, such as raw potato starch, are resistant to hydrolysis by amylase. Lignin is known to be completely unaffected and is thus indigestible. It is also conceivable that lignified tissues may trap proteins and carbohydrates and protect them from the action of digestive enzymes. The glands of the large intestine are mainly mucus glands, which do not produce enzymes, and digestion in the large intestine is therefore brought about by enzymes which have been carried down in the food from the upper part of the tract, or occurs as a result of microbial activity.

Extensive microbial activity occurs in the large intestine, especially the caecum. Here the slow rate of passage and abundant nutrient sources encourage the prolific growth of bacteria. There is a complex population of aerobic and obligate anaerobic bacteria including lactobacilli, streptococci, coliforms, bacteroides, clostridia and yeasts. These metabolise a wide range of nitrogen and carbohydrate sources from both dietary and endogenous residues, resulting in the formation of a number of products including indole, skatole, phenol, hydrogen sulphide, amines, ammonia and the volatile fatty acids, acetic, propionic and butyric. As with the rumen bacteria, the relative numbers of the species change in response to the material available for fermentation. More volatile fatty acids are produced from the finer particles as they have a larger surface area for attack by the bacteria. The digestion of cellulose and other higher polysaccharides is nevertheless small compared with that taking place in the horse and ruminants, which have digestive systems adapted to deal with fibrous foods. With conventional pig diets, microbial fermentation accounts for 8–16 per cent of the organic matter disappearing from the gastro-intestinal tract. The products of microbial breakdown of polysaccharides are not sugars but mainly the volatile fatty acids listed above. Lactic acid can be produced under some circumstances. The volatile fatty acids are absorbed and contribute to the energy supply of the pig.

Bacterial action in the large intestine may have a beneficial effect owing to the synthesis of some of the B vitamins, which may be absorbed and utilised by the host. Synthesis of most of the vitamins in the digestive tract of the pig is, however, insufficient to meet the daily requirements and a dietary source is needed.

The waste material, or faeces, voided from the large intestine via the anus consists of water, undigested food residues, digestive secretions, epithelial cells from the tract, inorganic salts, bacteria and products of microbial decomposition.

Digestion in the young pig

From birth until about the age of five weeks the concentration and activity of many digestive secretions in the young pig are different from those in the adult animal. During the first few days after birth the intestine is permeable to native proteins. In the young pig, as in other farm animals, this is essential for the transfer of γ-globulins (antibodies) via the mother's milk to the newborn animal. The ability of the young pig to absorb these proteins declines rapidly and is low by 24 hours *post partum*.

The piglet stomach initially produces only a limited amount of hydrochloric acid and pepsinogen, but it does secrete chymosin. This operates at pH 3.5 to break the peptide bonds between phenylalanine and methionine in casein. It clots milk, thereby avoiding flooding the small intestine with nutrients. As the piglet develops, pepsinogen and hydrochloric acid secretion increases. Table 8.2 shows the activity of some of the important carbohydrases in the young pig.

The activity of lactase is high at birth and reaches a maximum in the first week of life and then slowly declines over the third or fourth week. Maltase activity increases from the fourth week, while sucrase reaches a constant level between weeks 4 and 8. The activity of α-amylase is present at birth but remains low until about four weeks of age.

These differences in enzyme activities are of special significance where piglets are reared on early weaning diets. If young pigs are weaned at 14 days of age, their diet, especially regarding the types of carbohydrates, should be different from that for animals weaned later. Early weaning mixtures usually include a high proportion of dried milk products containing lactose. For later weaning at three to four weeks cooked cereals are included in the diet since raw starch is incompletely digested in the small intestine and passes to the large intestine, where it is fermented by bacteria, causing diarrhoea.

Table 8.2 Weight of disaccharide hydrolysed per kg bodyweight per hour by small intestine enzymes in young pigs

	Lactose (g)	Sucrose (g)	Maltose (g)
Newborn	5.9	0.06	0.3
Five weeks	0.8	1.3	2.5

Digestion in the fowl

The enzymes present in the digestive secretions of the fowl are similar to those of mammals, although lactase has not been detected. However, the digestive tract of the fowl differs in a number of respects from that of the pig (see Fig. 8.1). In the fowl the lips and cheeks are replaced by the beak, the teeth being absent. Taste is limited, the taste buds being located on the back half of the tongue and the adjacent pharynx. The crop is a diverticulum of the oesophagus, situated about two-thirds down its length and just before its entry into the thorax. It is a pear-shaped sac formed as a single lobe whose main function is to act as a reservoir for holding food. It is filled and emptied by peristalsis. The crop wall does not have mucus-secreting glands. It is not essential to the bird, but its presence gives more flexibility in feeding activity. Salivary amylase is known to occur in the fowl and the action of this enzyme on starch continues in the crop. In addition, microbial activity occurs there during the storage of food. Lactobacilli predominate, adhering to the crop wall. The major products of fermentation are lactic and acetic acids.

The oesophagus terminates at the proventriculus or glandular stomach. This produces hydrochloric acid and pepsinogen. The proventriculus has minimal inherent motility and food passes through as a result of oesophageal contractions. It leads to the gizzard, a muscular organ with internal ridges which undergoes rhythmic contractions and grinds the food, with moisture, into a smooth paste. The gizzard wall produces koilin, a protein–polysaccharide complex similar in its amino acid composition to keratin, which hardens in the presence of hydrochloric acid. Digesta particles pass to the small intestine when ground sufficiently: reflux of intestinal digesta into the gizzard can also occur. The presence of grit in the gizzard, although not essential, has been shown to increase the breakdown of whole grains by about 10 per cent. Proteolysis occurs in the lumen of the gizzard. Thus the proventriculus and gizzard are equivalent in function to the mammalian stomach.

The duodenum encloses the pancreas as in mammals. In the fowl, the three pancreatic and two bile ducts (one from the gall bladder and one from the right lobe of the liver) open into the intestine at the termination of the duodenum. The arrangement and number of ducts differ among fowls, geese and turkeys.

The pancreatic juice of fowls contains the same enzymes as the mammalian secretion, and the digestion of proteins, fats and carbohydrates in the small intestine is believed to be similar to that occurring in the pig. The intestinal mucosa produces mucin, α-amylase, maltase, sucrase and proteolytic enzymes.

Unlike young pigs, chicks have maltase and sucrase activities in their small intestine and, since they perform well on diets containing uncooked cereals, it can be assumed that they possess satisfactory amylase activity.

Where the small intestine joins the large intestine there are two long blind sacs known as the caeca. These function as absorptive organs but are not essential to the fowl, since surgical removal causes no harmful effects. There are bacteria associated with the mucosal surface of the caeca and peristaltic activity mixes these with the digesta and leads to its fermentation, with the production of volatile fatty acids. Experiments with adult fowls indicate that the cellulose present in cereal grains is not broken down by microbial activity to any great extent during its passage through the digestive tract, although some hemicellulose breakdown occurs. Similarly with geese, ligation of the caeca does not alter crude fibre digestibility so it is unlikely that the volatile fatty acids make a large contribution to satisfying the energy requirement of poultry.

The caeca empty by peristaltic contraction into the relatively short colon, whose main function is the transport of digesta to its termination at the cloaca. The cloaca, from which faeces and urine are excreted together, combines the function of the rectum and bladder.

Pre-caecal digestion in the horse

The horse is a non-ruminant herbivore having a simple monogastric digestive tract similar to that of the pig but with a much enlarged hind gut (especially the caecum) which contains a microbial population. The small intestine is the main site for the digestion of non-fibrous carbohydrate, protein and fat, and the microbes in the large intestine ferment fibrous materials, as in the ruminant (see p. 179). Thus, unlike the ruminant, the horse has enzymic digestion before microbial fermentation and it falls between the pig and the ruminant in its ability to digest fibrous foods. The horse spends a long time chewing its food, during which copious amounts of mucus-containing saliva are added to ease swallowing. The saliva has no digestive enzyme activity but does contain bicarbonate, which buffers the swallowed digesta in the proximal region of the stomach. The stomach is relatively small, being only about 10 per cent of the total gastro-intestinal tract, and is suited to frequent consumption of small quantities of food. Its major role is to regulate the flow of digesta to the small intestine, ensuring efficient digestion there. The stomach is rarely empty and the retention time of digesta in the stomach is short (2–6 hours), particularly in the grazing horse. There is limited microbial fermentation in the oesophageal and fundic regions of the stomach, with the production of lactic acid. Towards the pyloric region hydrochloric acid is added and the pH of the digesta falls. The small intestine represents about 30 per cent of the volume of the gastro-intestinal tract and 75 per cent of its length, and movement of digesta through it is rapid. It is the main site of digestion of protein to amino acids and their absorption. Soluble carbohydrates, such as starches and sugars, are exposed to pancreatic amylase and α-glucosidase secreted by the cells in the wall of the small intestine. Amylase activity is low compared with that of the pig and may limit the rate of digestion of starches, allowing some to reach the large intestine if given in large quantities. The disaccharidase activity is similar to that in the pig. The horse has no gall bladder and thus cannot store bile: the presence of hydrochloric acid in the duodenum causes stimulation of bile secretion from the liver. The lack of a gall bladder does not seem to affect the digestion of fat. It is not certain whether the pancreatic secretions have lipase activity. As in the pig, the fatty acid composition of body lipids resembles that of the diet. The remaining undigested material then passes to the large intestine and is exposed to microbial fermentation (see p. 195). Despite the rapid transit of digesta through the small intestine, digestion and absorption are usually efficient and the composition of the material entering the large intestine is fairly uniform. This is important for the correct functioning of the large intestine and instances of disturbance often result from inefficient digestion in the small intestine.

Absorption of digested nutrients

The main organ for the absorption of dietary nutrients by the monogastric mammal is the small intestine. This part of the tract is specially adapted for absorption because its inner surface area is increased by folding and the presence of villi. Although the duodenum has villi, this is primarily a mixing and neutralising site and the jejunum is the major absorptive site. Absorption of a nutrient from the lumen of the intestine can take place by passive transport, involving simple diffusion, provided there is a high concentration of the nutrient outside the cell and a low concentration inside. The vascular system in the villi is arranged so that the concentration gradient is maximised. Nutrients such as monosaccharides, amino acids and small peptides are absorbed at a faster rate than can be accounted for by passive diffusion. The absorption of such molecules is aided by specific carrier transport systems in which carrier proteins reversibly bind to and transfer the nutrient across the brush border and basolateral membranes of the epithelial cells. The transfer may be by *facilitative transport,* in which the carrier transports the molecule down its concentration gradient. Alternatively, absorption may be by a process of *active transport* (or co-transport). Here the carrier has two specific binding sites and the organic nutrient is attached to one of these while the other site picks up a sodium ion in the case of monosaccharides and amino acids, or a hydrogen ion in the case of dipeptides. The sodium (or hydrogen) ion travels down the chemical gradient and the loaded carrier thus moves across the intestinal membrane and deposits the organic nutrient and the sodium inside the cell. The empty carrier then returns across the membrane free to pick up more nutrients. The sodium ion is then actively pumped back to the lumen by Na^+/K^+ transporting ATPase. In the case of the dipeptide carrier the hydrogen ion gradient is maintained by a system involving Na^+ and H^+ which generates an acidic microclimate on the small intestine surface. A number of different carriers are thought to exist although some may carry more than one nutrient; for example, xylose can be bound by the same carrier as glucose. The carrier proteins are large complex molecules and have binding sites for specific nutrients or related groups of nutrients. The molecular weight can exceed 50 000 and the sodium-linked glucose transporter is made up of 664 amino acids.

Since there are costs to the animal in synthesising and maintaining the carrier proteins, it would be inefficient to maintain a specific level of carrier when the substrate is not present. On the other hand, if insufficient carrier is available the transport pathway will limit the assimilation of nutrients.

Therefore, the carrier proteins are regulated and show adaptation to the level of nutrient present in the gut.

A third method of absorption is by *pinocytosis* (cell drinking), in which cells have the capacity to engulf large molecules in solution or suspension. Such a process is particularly important in many newborn suckled mammals in which immunoglobulins present in colostrum are absorbed intact.

Carbohydrates

The digestion of carbohydrates by enzymes secreted by the pig and other monogastric animals results in the production of monosaccharides. The formation of these simple sugars from disaccharides takes place on the surface of the microvillus membrane. The aldoses, such as glucose, are actively transported across the cell after attachment to the specific carrier and carried by the portal blood systems to the liver. The mechanism for the absorption of ketoses is unclear, although the existence of a facilitative carrier for fructose has been established. The rates of absorption of various sugars differ. At equal concentrations, galactose, glucose, fructose, mannose, xylose and arabinose are absorbed in decreasing order of magnitude.

Fats

After digestion fats are present in the small intestine in the solubilised form of mixed micelles. Efficient absorption requires a rapid movement of the highly hydrophobic molecule through the unstirred water layer adjacent to the mucosa. This is the rate-limiting stage of absorption. The mixed bile salt micelles, with their hydrophilic groups, aid this process. Absorption across the brush border membrane of the intestinal cells is by passive diffusion and is at its maximum in the jejunum. The bile salts are absorbed by an active process in the distal ileum. Following absorption there is a resynthesis of triacylglycerols, a process which requires energy, and they are formed into chylomicrons (minute fat droplets), which then pass into the lacteals of the villi, enter the thoracic duct and join the general circulation. Medium and short chain fatty acids, such as those occurring in butterfat, require neither bile salts nor micelle formation as they can be absorbed very rapidly from the lumen of the intestine directly into the portal blood stream. The entry of these fatty acids is sodium-dependent and takes place against a concentration gradient by active transport. In fowls, the lymphatic system is negligible and most of the fat is transported in the portal blood as low-density lipoproteins.

Proteins

The products of protein digestion in the lumen of the intestine are free amino acids and small (oligo-)peptides. The latter enter the epithelial cells of the small intestine where, in the main, they are hydrolysed by specific di- and tri-peptidases. However, some small peptides are absorbed intact and subsequently appear in the portal blood. The amino acids, which subsequently pass into the portal blood and thence to the liver, are absorbed from the small intestine by an active transport mechanism which in most cases is sodium-dependent. In the case of glycine, proline and lysine the sodium molecule is unnecessary. Several systems have been described for amino acid transfer and these can be classified into four main groups. One is concerned with the transfer of neutral amino acids and separate carriers transfer dicarboxylic and basic amino acids. In addition, there is a fourth system for the movement of the imino acids and glycine. These mechanisms, however, are not completely rigid and some amino acids can be transferred by more than one system. The rates of absorption of the amino acids differ; for example, the rate is higher for methionine than valine, which in turn is higher than that for threonine.

Reference has already been made to the absorption of intact proteins, such as immunoglobulins, in the newborn animal by pinocytosis.

Minerals

Absorption of mineral elements is either by simple diffusion or by carrier-mediated transport. The exact mechanisms for all minerals have not been established, but the absorption of calcium, for example, is regulated by 1,25-dehydroxycholecalciferol (see p. 83). Low alimentary pH favours calcium absorption but it is inhibited by a number of dietary factors such as the presence of oxalates and phytates. An excess of either calcium or phosphorus interferes with the absorption of the other. The absorption of calcium is also influenced by the requirements of the animal. For example, the absorption of calcium from the digestive tract of laying hens is much greater when shell formation is in progress than when the shell gland is inactive.

The absorption of iron is to a large extent independent of the dietary source. The animal has difficulty in excreting iron from the body in any quantity and a method exists for regulating iron absorption to prevent excessive amounts entering the body (see p. 128). In adults the absorption of the element is generally low, but after severe bleeding and during pregnancy the requirement for iron is increased so that absorption of the element is also increased. Anaemia due to iron deficiency may, however, develop on low-iron diets. Experiments carried out with dogs have shown that the absorption of iron by anaemic animals may be 20 times as great as that by normal healthy dogs.

Another example of the mechanism of mineral absorption is shown by zinc. This mineral is absorbed through the small intestine by a carrier-mediated process, uptake at the brush border membrane being the rate-limiting step. In rats and man, the carrier becomes saturated at levels of zinc below that normally seen in the diet and zinc uptake from aqueous solutions above this saturation level is by passive diffusion. Calcium is believed to inhibit the absorption of zinc.

It is thought that iodine in organic combination is less well absorbed than the inorganic form. Plants contain a higher proportion of inorganic iodide than foods of animal origin.

Vitamins

The fat-soluble vitamins A, D, E and K pass through the intestinal mucosa mainly by the same passive diffusion mechanism as for fats. Within the cells they may combine with proteins and enter the general circulation as lipoproteins.

Vitamin A is more readily absorbed from the digestive tract than its precursor carotene, although it is thought that vitamin A esters must first be hydrolysed by an esterase to the alcohol form before being absorbed. Phytosterols are poorly absorbed, and it is generally considered that unless ergosterol has been irradiated to vitamin D_2 before ingestion it cannot be absorbed from the tract in any quantity.

Water-soluble vitamins are believed to be absorbed both by simple diffusion and by carrier-mediated transport, which is sodium-dependent. Vitamin B_6 is absorbed by passive diffusion, mainly in the small intestine, and the amount absorbed is linearly related to the amount in the digesta. The importance of a carrier glycoprotein (intrinsic factor) for the absorption of vitamin B_{12} has already been stressed in an earlier chapter (p. 103).

Detoxification in the alimentary tract

Many food constituents are potentially toxic to the animal consuming them. Microbial contaminants are an obvious example; the digestive enzymes kill many bacteria, but some organisms may damage the gut, which allows them or the toxins they produce to invade the animal's tissues. Foreign proteins, especially those with endocrine activity, could harm the animal if absorbed, but the gut provides an effective barrier to prevent their absorption before they are hydrolysed. The same is true of nucleic acids (whose breakdown is a matter of concern today, when some animal foods may be derived from genetically modified plants). Some of the toxic constituents of pasture plants are broken down in the rumen of cattle, sheep and goats (see p. 509). As mentioned above, the animal body avoids excessive intakes of the mineral elements calcium and iron by selective absorption.

Some larger molecules are able to bypass the barrier presented by the gut. This may be desirable in some instances (examples: absorption of protein antibodies in the newborn animal and absorption of antibiotics administered orally), but undesirable in others (apparent absorption of the proteins known as prions, which are responsible for bovine spongiform encephalopathy: see p. 600).

8.2 Microbial digestion in ruminants and other herbivores

The foods of ruminants, forages and fibrous roughages, consist mainly of β-linked polysaccharides such as cellulose, which cannot be broken down by mammalian digestive enzymes. Ruminants have therefore evolved a special system of digestion that involves microbial fermentation of food prior to its exposure to their own digestive enzymes. This part of Chapter 8 will describe the anatomical and physiological adaptations of the ruminant that facilitate microbial digestion, and will outline the biochemistry of this form of digestion and its nutritional consequences. Herbivores other than ruminants, such as the horse, have adopted systems of microbial digestion that differ from those of ruminants, and these will be briefly described.

Anatomy and physiology of ruminant digestion

The stomach of the ruminant is divided into four compartments (see Fig. 8.1). In the young suckling the first two compartments, the rumen and its continuation the reticulum, are relatively undeveloped, and milk, on reaching the stomach, is channelled by a tube-like fold of tissue, known as the oesophageal or reticular groove, directly to the third and fourth compartments, the omasum and abomasum. As the calf or lamb begins to eat solid food the first two compartments (often considered together as the reticulo-rumen) enlarge greatly, until in the adult they comprise 85 per cent of the total capacity of the stomach. In the adult, the oesophageal groove does not function under normal feeding conditions, and both food and water pass into the reticulo-rumen. However, the reflex closure of the groove to form a channel can be stimulated even in adults, particularly if they are allowed to drink from a teat.

The food is diluted with copious amounts of saliva, first during eating and again during rumination: typical quantities of saliva produced per day are 150 litres in cattle and 10 litres in sheep. Rumen contents contain 850–930 g water/kg on average, but they often exist in two phases: a lower liquid phase, in which the finer food particles are suspended, and a drier upper layer of coarser solid material. The breakdown of food is accomplished partly by physical and partly by chemical means. The contents of the rumen are continually mixed by the rhythmic contractions of its walls, and during rumination material at the anterior end is drawn back into the oesophagus and returned by a wave of contraction to the mouth. Any liquid is rapidly swallowed again, but coarser material is thoroughly chewed before being returned to the rumen. The major factor inducing the animal to ruminate is probably the tactile stimulation of the epithelium of the anterior rumen; some diets, notably those containing little or no coarse roughage, may fail to provide sufficient stimulation for rumination. The time spent by the animal in rumination depends on the fibre content of the food. In grazing cattle it is commonly about eight hours per day, or about equal to the time spent in grazing. Each bolus of food regurgitated is chewed 40–50 times and thus receives a much more thorough mastication than during eating.

The reticulo-rumen provides a continuous culture system for anaerobic bacteria, protozoa and fungi. Food and water enter the rumen and the food is partially fermented to yield principally volatile fatty acids, microbial cells and the gases methane and carbon dioxide. The gases are lost by eructation (belching) and the volatile fatty acids are mainly absorbed through the rumen wall. The microbial cells, together with undegraded food components, pass to the abomasum and small intestine; there they are digested by enzymes secreted by the host animal, and the products of digestion are absorbed. In the large intestine there is a second phase of microbial digestion. The volatile fatty acids produced in the large intestine are absorbed, but microbial cells are excreted – with undigested food components – in the faeces.

Like other continuous culture systems, the rumen requires a number of homeostatic mechanisms. The acids produced by fermentation are theoretically capable of reducing the pH of rumen liquor to 2.5–3.0, but under normal conditions the pH is maintained at 5.5–6.5. Phosphate and bicarbonate contained in the saliva act as buffers; in addition, the rapid absorption of the acids (and also of ammonia: see below) helps to stabilise the pH. The osmotic pressure of rumen contents is kept near that of blood by the flux of ions between them. Oxygen entering with the food is quickly used up and anaerobiosis is maintained. In the absence of oxygen, carbon is the ultimate acceptor of hydrogen ions, hence the formation of methane. The temperature of rumen liquor remains close to that of the animal (38–42 °C). Finally, the undigested components of the food, together with soluble nutrients and bacteria, are eventually removed from the rumen by the passage of digesta through the reticulo-omasal orifice.

Rumen microorganisms

The bacteria number 10^9–10^{10} per ml of rumen contents. Over 200 species have been identified, and for descriptions of them the reader is referred to the works listed at the end of this chapter. Most are non-spore-forming anaerobes. Table 8.3 lists a number of the more important species and indicates the substrate they utilise and the products of the fermentation. This information is based on studies of isolated species *in vitro* and is not completely applicable *in vivo*. For example, it appears from Table 8.3 that succinic acid is an important end-product, but in practice this is converted into propionic acid by other bacteria such as *Selenomonas ruminantium* (Fig. 8.3, see p. 184); such interactions between microorganisms are an important feature of rumen fermentation. A further point is that the activities of a given species of bacteria may vary from one strain of that species to another. The total number of bacteria and the relative population of individual species vary with the animal's diet; for example, diets rich in concentrate foods promote high total counts and encourage the proliferation of lactobacilli.

Table 8.3 Typical rumen bacteria, their energy sources and fermentation products *in vitro*

Species	Description	Typical energy source	Typical fermentation products (excluding gases)						Alternative energy sources
			Acetic	Propionic	Butyric	Lactic	Succinic	Formic	
Fibrobacter succinogenes	Gram-negative rods	Cellulose	+				+	+	Glucose (starch)
Ruminococcus flavefaciens	Catalase-negative streptococci with yellow colonies	Cellulose	+			+	+	+	Xylan
Ruminococcus albus	Single or paired cocci	Cellobiose	+					+	Xylan
Streptococcus bovis	Gram-positive, short chains of cocci, capsulated	Starch				+			Glucose
Prevotella ruminicola	Gram-negative, oval or rod	Glucose	+				+	+	Xylan, starch
Megasphaera elsdenii	Large cocci, paired or in chains	Lactate	+	+	+				Glucose, glycerol
Lachnospira multipara	Gram-positive curved rods	Pectins	+			+			Glucose, fructose

Protozoa are present in much smaller numbers (10^6 per ml) than bacteria but, being larger, may equal the latter in total mass. Over 100 species have been identified in the rumen. In adult animals, most of the protozoa are ciliates belonging to two families. The Isotrichidae, commonly called the holotrichs, are ovoid organisms covered with cilia; they include the genera *Isotricha* and *Dasytricha*. The Ophryoscolecidae, or oligotrichs, include many species that vary considerably in size, shape and appearance; they include the genera *Entodinium*, *Diplodinium*, *Epidinium* and *Ophryoscolex*. The oligotrichs can ingest food particles and cannot utilise cellulose. A normal rumen flora (bacteria) and fauna (protozoa) is established quite early in life, as early as six weeks of age in calves.

The fungi of the rumen have been studied for less than 30 years, and their place in the rumen ecosystem has yet to be fully characterised. They are strictly anaerobic, and their life cycle includes a motile phase (as a zoospore) and a vegetative phase (sporangium). During the latter phase they become attached to food particles by rhizoids, which can penetrate cell walls. At least 12 species or strains have been identified, typically those belonging to the genus *Neocallimastix*. The rumen fungi are capable of utilising most polysaccharides and many soluble sugars; some carbohydrates not used by these fungi are pectin, polygalacturonic acid, arabinose, fucose, mannose and galactose. The contribution of the rumen fungi to the fermentation of food has not yet been quantified, but it is known that they are most numerous (constituting 10 per cent of the microbial biomass) when diets are rich in fibre (i.e. not cereal diets or young pasture herbage).

The rumen microorganisms can be envisaged as operating together, as so-called consortia, to attack and break down foods. Some, like the fungi, are capable of invading and colonising plant tissues; others follow up to ferment the spoils of the invasion. Detailed studies, including some involving electron micrography, have shown that 75 per cent of rumen bacteria are attached to food particles.

As the microbial mass synthesised in the rumen provides about 20 per cent of the nutrients absorbed by the host animal, the composition of microorganisms is important. The bacterial dry matter contains about 100 g N/kg, but only 80 per cent of this is in the form of amino acids, the remaining 20 per cent being present as nucleic acid N. Moreover, some of the amino acids are contained in the peptidoglycan of the cell wall membrane and are not digested by the host animal.

Digestion of carbohydrates

The diet of the ruminant contains considerable quantities of cellulose, hemicelluloses, starch and water-soluble carbohydrates that are mainly in the form of fructans. Thus in young pasture herbage, which is frequently the sole food of the ruminant, each kilogram of dry matter may contain about 400 g cellulose and hemicelluloses, and 200 g of water-soluble carbohydrates. In mature herbage, and in hay and straw, the proportion of cellulose and hemicelluloses is much higher, and that of water-soluble carbohydrates is much lower. The β-linked carbohydrates are associated with lignin, which may comprise 20–120 g/kg DM. All the carbohydrates, but not lignin, are attacked by the rumen microorganisms; the principal bacterial species involved (see Table 8.3) are *Fibrobacter succinogenes* and the *Ruminococci*, and the fungi are also thought to play an important role.

The breakdown of carbohydrates in the rumen may be divided into two stages, the first of which is the digestion of complex carbohydrates to simple sugars. This is brought about by extracellular microbial enzymes and is thus analogous to the digestion of carbohydrates in non-ruminants. Cellulose is decomposed by one or more β-1,4-glucosidases to cellobiose, which is then converted either to glucose or, through the action of a phosphorylase, to glucose-1-phosphate. Starch and dextrins are first converted by amylases to maltose and isomaltose and then by maltases, maltose phosphorylases or 1,6-glucosidases to glucose or glucose-1-phosphate. Fructans are hydrolysed by enzymes attacking 2,1 and 2,6 linkages to give fructose, which may also be produced – together with glucose – by the digestion of sucrose (see Fig. 8.2).

Pentoses are the major product of hemicellulose breakdown, which is brought about by enzymes attacking the β-1,4 linkages to give xylose and uronic acids. The latter are then converted to xylose. Uronic acids are also produced from pectins, which are first hydrolysed to pectic acid and methanol by pectinesterase. The pectic acid is then attacked by polygalacturonidases to

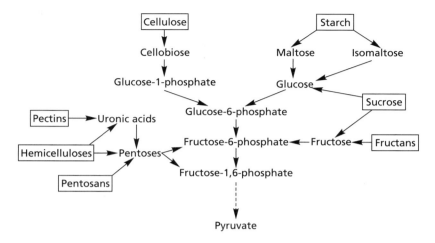

Fig. 8.2 Conversion of carbohydrates to pyruvate in the rumen.

give galacturonic acids, which in turn yield xylose. Xylose may also be produced from hydrolysis of the xylans which may form a significant part of the dry matter of grasses.

The simple sugars produced in the first stage of carbohydrate digestion in the rumen are rarely detectable in the rumen liquor because they are immediately taken up and metabolised intracellularly by the microorganisms. For this second stage, the pathways involved are in many respects similar to those involved in the metabolism of carbohydrates by the animal itself, and are thus discussed in Chapter 10. However, the main pathways are outlined in Fig. 8.3. The key intermediate (i.e. linking the pathways of Fig. 8.2 with those of Fig. 8.3) is pyruvate, and Fig. 8.3 shows the pathways that link pyruvate with the major end-products of rumen carbohydrate digestion, which are acetic, propionic and butyric acids, carbon dioxide and methane. Additional fatty acids are also formed in the rumen, generally in small quantities, by deamination of amino acids; these are isobutyric acid from valine, valeric acid from proline, 2-methyl butyric acid from isoleucine and 3-methyl butyric acid from leucine. Figure 8.3 shows that propionate can be produced from pyruvate by several alternative pathways. The pathway through lactate and acrylate predominates when the ruminant's diet includes a high proportion of concentrates, and pathways through succinate are employed when the diet consists mainly of fibrous roughages. With concentrate diets, lactate produced by the first pathway may accumulate in the rumen and threaten the animal with acidosis.

The overall equation for the rumen fermentation of hexoses is:

$$57 \text{ Hexose} \longrightarrow 114 \text{ Pyruvate} \longrightarrow \begin{array}{l} 65 \text{ Acetate} \\ 21 \text{ Propionate} \quad + \quad 20\,CO_2 \quad + \quad 73\,CH_4 \\ 14 \text{ Butyrate} \end{array}$$

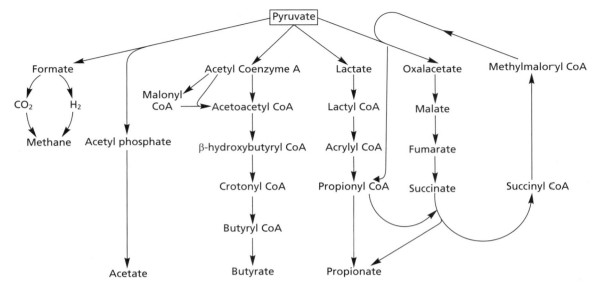

Fig. 8.3 Conversion of pyruvate to volatile fatty acids in the rumen.

Thus the molar proportions of the three volatile fatty acids (VFAs) derived from hexose are acetate 0.65, propionate 0.21 and butyrate 0.14. Some examples of the actual concentrations and proportions of VFA in the rumen are given in Table 8.4. They differ from the proportions given above because carbohydrates other than hexoses (and also amino acids) are fermented in the rumen. The relative concentrations of VFAs are often assumed to represent their relative rates of production (which are more difficult to measure: see Box 8.2), but this may be misleading if individual VFAs are absorbed at different rates. Their total concentration varies widely according to the animal's diet and the time that has elapsed since the previous meal, but is normally in the range of 70–150 mmol/litre (equivalent to approximately 5–10 g/litre). The relative proportions of the acids also vary. Mature fibrous forages give rise to VFA mixtures containing a high proportion (about 70 per cent) of acetic acid. Less mature forages tend to give a

Box 8.2
Measurement of volatile fatty acid (VFA) production

The production rate of VFAs in the rumen can be measured by infusion of isotopically labelled forms of the acids into the rumen via a cannula and by recording their dilution by newly formed VFA. If only one VFA is infused, production of the others can be estimated from their relative proportions in rumen liquor. However, infusion of all three major acids provides more reliable estimates because it allows for differences between them in rates of production and/or absorption.

Table 8.4 Volatile fatty acids (VFAs) in the rumen liquor of cattle or sheep fed on various diets

Animal	Diet	Total VFA (m moles/ litre)	Individual VFA (molar proportions)			
			Acetic	Propionic	Butyric	Others
Sheep	Young ryegrass herbage	107	0.60	0.24	0.12	0.04
Cattle	Mature ryegrass herbage	137	0.64	0.22	0.11	0.03
Cattle	Grass silage	108	0.74	0.17	0.07	0.03
Sheep	Chopped lucerne hay	113	0.63	0.23	0.10	0.04
	Ground lucerne hay	105	0.65	0.19	0.11	0.05
Cattle	Long hay (0.4), concentrates (0.6)	96	0.61	0.18	0.13	0.08
	Pelleted hay (0.4), concentrates (0.6)	140	0.50	0.30	0.11	0.09
Sheep	Hay : concentrate					
	1 : 0	97	0.66	0.22	0.09	0.03
	0.8 : 0.2	80	0.61	0.25	0.11	0.03
	0.6 : 0.4	87	0.61	0.23	0.13	0.02
	0.4 : 0.6	76	0.52	0.34	0.12	0.03
	0.2 : 0.8	70	0.40	0.40	0.15	0.05
Cattle	Barley (no ciliate protozoa in rumen)	146	0.48	0.28	0.14	0.10
Cattle	Barley (ciliates present in rumen)	105	0.62	0.14	0.18	0.06

rather lower proportion of acetic and a higher proportion of propionic acid. The addition of concentrates to a forage also increases the proportion of propionic at the expense of acetic, this effect being particularly strong with diets containing a high proportion (0.6) of concentrates. With all-concentrate diets, the proportion of propionic may even exceed that of acetic. However, even with this type of diet, acetic acid predominates if the rumen ciliate protozoa survive. Grinding and pelleting of forage has little effect on the VFA proportions when the diet consists of forage alone, but causes a switch from acetic to propionic if the diet also contains concentrates. Table 8.4 also shows that the proportion of butyric acid is less affected by diet than are the shorter chain acids.

The total weight of acids produced may be as high as 4 kg per day in cows. Much of the acid produced is absorbed directly from the rumen, reticulum and omasum, although 10–20 per cent may pass through the abomasum and be absorbed in the small intestine. In addition, some of the products of carbohydrate digestion in the rumen are used by the microorganisms to form their own cellular polysaccharides, but the amounts of these passing to the small intestine are probably small and hardly significant.

The rate of gas production in the rumen is most rapid immediately after a meal and in the cow may exceed 30 litre/hour. The typical composition of rumen gas is carbon dioxide, 40 per cent; methane, 30–40 per cent; hydrogen, 5 per cent; together with small and varying proportions of oxygen and nitrogen (from ingested air). Carbon dioxide is produced partly as a by-product of fermentation and partly by the reaction of organic acids with the bicarbonate present in the saliva. The basic reaction by which methane is formed is the reduction of carbon dioxide by hydrogen, some of which may be derived from formate. Methanogenesis, however, is a complicated process that involves folic acid and vitamin B_{12}. About 4.5 g of methane is formed for every 100 g of carbohydrate digested, and the ruminant loses about 7 per cent of its food energy as methane (see Chapter 11).

Most of the gas produced is lost by eructation; if gas accumulates it causes the condition known as bloat, in which the distension of the rumen may be so great as to result in the collapse and death of the animal. Bloat occurs most commonly in dairy cows grazing on young, clover-rich herbage, and is due not so much to excessive gas production as to the failure of the animal to eructate. Frequently the gas is trapped in the rumen in a foam, whose formation may be promoted by substances present in the clover. It is also possible that the reflex controlling eructation is inhibited by a physiologically active substance that is present in the food or formed during fermentation. Bloat is a particularly serious problem on the clover-rich pastures of New Zealand, where it is prevented by dosing the cows or spraying the pasture with anti-foaming agents, such as vegetable oils. Another form of bloat, termed 'feed lot bloat', occurs in cattle fattened intensively on diets containing much concentrate and little roughage. Bloat-preventing agents are discussed in Chapter 25.

The extent to which cellulose is digested in the rumen depends particularly on the degree of lignification of the plant material. Lignin, and also the related substance cutin, is resistant to attack by anaerobic bacteria, probably because of its low oxygen content and its condensed structure (which inhibits hydrolysis). Lignin appears to hinder the breakdown of the cellulose with which it is associated. Thus, in young pasture grass containing only 50 g lignin/kg DM, 80 per cent of the cellulose may be digested, but in older herbage with 100 g lignin/kg the proportion of cellulose digested may be less than 60 per cent. Ruminant diets based on cereals may contain as much as 500 g/kg of starch (and sugars), of which over 90 per cent may be fermented in the rumen and the rest digested in the small intestine. This fermentation is rapid, and the resulting fall in the pH of rumen liquor inhibits cellulose-fermenting organisms and thus depresses the breakdown of cellulose.

The breakdown of cellulose and other resistant polysaccharides is undoubtedly the most important digestive process taking place in the rumen. Besides contributing to the energy supply of the ruminant, it ensures that other nutrients which might escape digestion are exposed to enzyme action. Although the main factor in the process is the presence of microorganisms in the rumen, there are other factors of importance. The great size of the rumen (its contents normally contribute 10–20 per cent of the liveweight of ruminants) allows food to accumulate and ensures that sufficient time is available for the rather slow breakdown of cellulose. In addition, the movements of the reticulo-rumen and the act of rumination play a part by breaking up the food and exposing it to attack by microorganisms.

Digestion of protein

The digestion of protein in the rumen is illustrated in Fig. 8.4. Food proteins are hydrolysed to peptides and amino acids by rumen microorganisms, but some amino acids are degraded further, to organic acids, ammonia and carbon dioxide. An example of the deamination of amino acids is provided by valine, which, as mentioned above, is converted to isobutyric acid. Thus the branched chain acids found in rumen liquor are derived from amino acids.

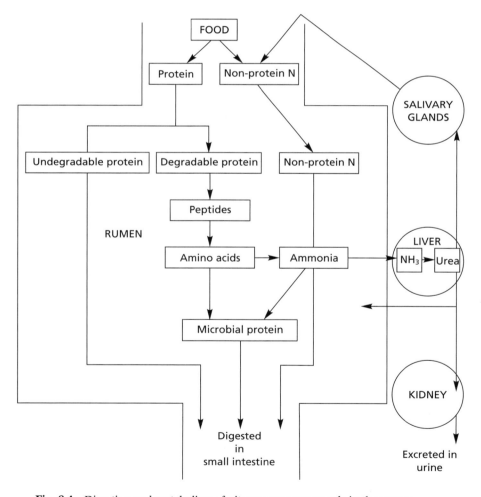

Fig. 8.4 Digestion and metabolism of nitrogenous compounds in the rumen.

The main proteolytic organisms are *Prevotella ruminicola*, *Peptostreptococci* species and the protozoa. The ammonia produced, together with some small peptides and free amino acids, is utilised by the rumen organisms to synthesise microbial proteins. Some of the microbial protein is broken down in the rumen and its nitrogen is recycled (i.e. taken up by microorganisms). When the organisms are carried through to the abomasum and small intestine their cell proteins are digested and absorbed. An important feature of the formation of microbial protein is that bacteria are capable of synthesising indispensable as well as dispensable amino acids, thus rendering their host independent of dietary supplies of the former.

The extent to which dietary protein is degraded to ammonia in the rumen, and conversely the extent to which it escapes rumen degradation and is subsequently digested in the small intestine, will be discussed in later chapters (10 and 13). At this point it is sufficient to emphasise that with most diets, the greater part (and sometimes all) of the protein reaching the ruminant's small intestine will be microbial protein of reasonably constant composition. The lesser part will be undegraded food protein, which will vary in amino acid composition according to the nature of the diet.

The ammonia in rumen liquor is the key intermediate in microbial degradation and synthesis of protein. If the diet is deficient in protein, or if the protein resists degradation, the concentration of rumen ammonia will be low (about 50 mg/litre) and the growth of rumen organisms will be slow; in consequence, the breakdown of carbohydrates will be retarded. On the other hand if protein degradation proceeds more rapidly than synthesis, ammonia will accumulate in rumen liquor and the optimum concentration will be exceeded. When this happens, ammonia is absorbed into the blood, carried to the liver and converted to urea (see Fig. 8.4). Some of this urea may be returned to the rumen via the saliva and also directly through the rumen wall, but the greater part is excreted in the urine and thus wasted.

Estimates of the optimum concentration of ammonia in rumen liquor vary widely, from 85 to over 300 mg/litre. Rather than expressing the optimum as the concentration in rumen liquor, it would probably be more realistic to relate ammonia to fermentable organic matter, since it is known that for each kilogram of organic matter fermented, an approximately constant quantity of N is taken up by rumen bacteria as protein and nucleic acid (see below).

If the food is poorly supplied with protein and the concentration of ammonia in rumen liquor is low, the quantity of nitrogen returned to the rumen as urea from the blood (see Fig. 8.4) may exceed that absorbed from the rumen as ammonia. This net gain in 'recycled' nitrogen is converted to microbial protein, which means that the quantity of protein reaching the intestine may be greater than that in the food. In this way the ruminant is able to conserve nitrogen by returning to the rumen urea that would otherwise be excreted in urine.

Although digestion is primarily the breakdown of complex molecules into simpler substances, a key feature of the digestive processes in ruminants is the production of microbial cells and hence the synthesis of microbial protein. If this synthesis is for any reason inefficient, food protein will be wasted and the host animal will subsequently be presented with a mixture of digestible nutrients that is unbalanced with respect to protein. In practice the rumen microorganisms synthesise protein in proportion to the quantities of nutrients that they ferment. With most feeds each kilogram of organic matter digested in the rumen yields approximately 200 g of microbial protein. Some rapidly fermented foods, such as immature forages rich in soluble carbohydrates, yield more microbial protein (up to 260 g/kg organic matter digested). Conversely, foods that contain relatively high proportions of digestible nutrients that are not fermented in the rumen give a lower yield of microbial protein (about 130 g/kg organic matter digested). Foods rich in fats are in this category, but are generally not given to ruminants. Silages, however, contain nutrients that have already been fermented or partly fermented; one of the major products of silage fermentation is lactic acid (see Chapter 20), and although this may be further metabolised in the rumen, the yield of rumen microbial protein per unit of organic matter digested is lower with silages than with other foods.

The rumen microbes thus have a 'levelling' effect on the protein supply of the ruminant; they supplement, in both quantity and quality, the protein of such foods as low-quality roughages but have a deleterious effect on protein-rich concentrates. It is possible to take additional advantage of the synthesising abilities of the rumen bacteria by adding urea to the diet of ruminants (see below). A more recent development, discussed on p. 194, is the protection of good-quality proteins from degradation in the rumen.

Utilisation of non-protein nitrogen compounds by the ruminant

Dietary protein is not the only contributor to the ammonia pool in the rumen. As much as 30 per cent of the nitrogen in ruminant foods may be in the form of simple organic compounds such as amino acids, amides and amines (see Chapter 4) or of inorganic compounds such as nitrates. Most of these are readily degraded in the rumen, their nitrogen entering the ammonia pool. In practice it is possible to capitalise on the ability of rumen microorganisms to convert non-protein nitrogenous compounds to protein, by adding such compounds to the diet. The substance most commonly employed is urea, but various derivatives of urea, and even ammonium salts, may also be used.

Urea entering the rumen is rapidly hydrolysed to ammonia by bacterial urease, and the rumen ammonia concentration is therefore liable to rise considerably. For this ammonia to be efficiently incorporated in microbial protein, two conditions must be met. First, the initial ammonia concentration must be below the optimum (otherwise the ammonia 'peak' produced will simply be absorbed and lost from the animal as described above); and, second, the microorganisms must have a readily available source of energy for protein synthesis. Feeding practices intended to meet these conditions include mixing urea with other foods (to prolong the period over which it is ingested and deaminated). Such foods should be low in rumen-degradable protein and high in readily fermentable carbohydrate.

It is important to avoid accidental overconsumption of urea, since the subsequent rapid absorption of ammonia from the rumen can overtax the ability of the liver to re-convert it to urea, hence causing the ammonia concentration of peripheral blood to reach toxic levels.

Derivatives of urea have been used for animal feeding with the intention of retarding the release of ammonia. Biuret is less rapidly hydrolysed than urea but requires a period of several weeks for rumen microorganisms to adapt to it. However, neither biuret nor isobutylidene diurea (IBDU) nor urea–starch compounds have consistently proved superior to urea itself.

An additional non-protein nitrogenous compound which can be utilised by rumen bacteria, and hence by the ruminant, is uric acid. This is present in high concentration in poultry excreta, and these are sometimes dried for inclusion in diets for ruminants, although in some countries the use of excreta as a food is restricted or prohibited.

The practical significance of these non-protein nitrogenous substances as potential protein sources is discussed in Chapter 24.

Digestion of lipids

The triacylglycerols present in the foods consumed by ruminants contain a high proportion of residues of the C_{18} polyunsaturated acids, linoleic and linolenic. These triacylglycerols are to a large extent hydrolysed in the rumen by bacterial lipases, as are phospholipids. Once they are released from ester combination, the unsaturated fatty acids are hydrogenated by bacteria, yielding first a monoenoic acid and, ultimately, stearic acid. Both linoleic and linolenic acid have all-*cis* double bonds, but before they are hydrogenated one double bond in each is converted to the *trans* configuration; thus, *trans* acids can be detected in rumen contents. The rumen microorganisms also synthesise considerable quantities of lipids, which contain some unusual fatty acids (such as those containing branched chains); these acids are eventually incorporated in the milk and body fats of ruminants.

The capacity of rumen microorganisms to digest lipids is strictly limited. The lipid content of ruminant diets is normally low (i.e. <50 g/kg) and if it is increased above 100 g/kg the activities of the rumen microbes are reduced. The fermentation of fibre is retarded and food intake falls. Saturated fatty acids affect rumen fermentation less than do unsaturated. Calcium salts of fatty acids have little effect on rumen fermentation and are used as fat supplements for ruminants.

Unlike their short chain counterparts, long chain fatty acids are not absorbed directly from the rumen. When they reach the small intestine they are mainly saturated and unesterified, but some – in the bacterial lipids – are esterified. Monoacylglycerols, which play an important role in the formation of mixed micelles in non-ruminants, are replaced in ruminants by lysophosphatidyl choline.

As mentioned earlier (p. 37), it is possible in non-ruminants to vary the fatty acid composition of body fats by altering the composition of dietary fats. In ruminants, this is normally not the case, and the predominating fatty acid of ruminant depot fats is the stearic acid resulting from hydrogenation in the rumen. However, it is possible to treat dietary lipids in such a way that they are protected from attack in the rumen but remain susceptible to enzymic hydrolysis and absorption in the small intestine. If such lipids contain unsaturated acids, they will modify the composition of body (and milk) fats.

Synthesis of vitamins

The synthesis by rumen microorganisms of all members of the vitamin B complex and of vitamin K has already been mentioned (see Chapter 5). In ruminants receiving foods well supplied with B vitamins the amounts synthesised are relatively small, but they increase if the vitamin intake in the diet decreases. The adult ruminant is therefore independent of a dietary source of these vitamins, but it should be remembered that adequate synthesis of vitamin B_{12} will take place only if there is sufficient cobalt in the diet.

Dynamics of digestion in the ruminant

Food enters the rumen in the form of particles of various shapes and sizes, which are at first suspended in the liquid phase. The soluble constituents of these particles (such as sugars) are quickly dissolved and can thus be rapidly degraded by the rumen microbes. The insoluble constituents are colonised by microbes and slowly broken down. In addition to these two fractions, a third component will consist of cell walls so heavily encrusted with lignin or silica as to be undegradable in the rumen. The dynamics of the rumen must be such that potentially degradable material is retained long enough to be digested, and that the products of digestion (together with undegradable material) are passed out of the stomach, either by transit to the lower gut or by absorption through the rumen wall. The partition required is aided by some of the physical characteristics of the rumen contents. The liquid phase of rumen contents may be envisaged as a tank of fixed volume, so that liquid or food entering it causes a corresponding volume to flow out through the reticulo-omasal orifice. This liquid carries with it some of the soluble constituents of the food, some bacteria, the volatile fatty acids that have not been absorbed through the rumen wall, and also some fine particles of food.

Food particles that are large, irregular in shape (e.g. long and thin plant fragments) and of low specific gravity tend to move to the top of the rumen (in some cases floating on the liquid phase) and are retained; this is desirable as large particles will not have been subjected to mechanical and microbial breakdown. When these particles are reduced to smaller and denser fragments, they descend in the rumen liquor and can be washed out, as described above.

A critical size for particles passing out of the rumen of sheep has been estimated by sieving digesta as about 1 mm (i.e. particles held on a sieve with holes of 1 mm are regarded as being too large to leave the rumen). For cattle, the critical size is considered likely to be 3–4 mm. However, the passage of particles from the rumen is too complicated to be explained solely in terms of sieve dimensions. The reticulo-omasal orifice is not a sieve and is large enough to allow the passage of particles much greater in diameter than 1 mm. What seems to happen is that the mass of food particles itself acts as a sieve, by what is termed a filter-bed effect, with large particles trapping smaller ones.

The rate of passage of liquids through the rumen is faster with roughage diets than with those containing concentrates like cereal grains, because greater rumination of roughages adds more saliva to the digesta. Adding salts to the diet increases water intake and thus increases fluid flow through the rumen. An increased rate of passage of liquid may 'wash out' bacteria, thus reducing cellulolysis and increasing the proportion of propionate in the mixture of volatile fatty acids produced. More generally, anything that increases the rate of passage of digesta from the rumen is liable to reduce the extent of digestion in that organ; for the fibre constituents of foods this is likely to be disadvantageous, but may be an advantage with constituents such as protein and starch, which can be more efficiently digested in the lower gut (see p. 193). As food intake rises, the ruminant responds by increasing either or both the quantity held in the rumen (rumen 'fill') and rate of passage, but eventually limits are reached that impose a restriction on intake. The feeding of ruminants for greater productivity often depends on the raising of these limits, for example by replacing slowly digested forages with rapidly digested concentrates, or by grinding forages to produce smaller, faster-moving particles.

Box 8.3
Measures of rumen dynamics

The movement of digesta from the rumen can be expressed either as a *rate of passage* or as its reciprocal, a *retention time*. The rate of passage of liquids is also known as the *dilution rate*. These can be measured by the use of *marker* substances, which are either natural constituents of foods or indigestible additives (see p. 248). An example of the latter is the soluble but unabsorbable high molecular weight polymer, polyethylene glycol. If a dose of this marker is injected into the rumen and allowed to mix uniformly, the subsequent decline in its concentration over time can be used to calculate the fractional rate of passage of liquid from the rumen. Typically this might be in the range 0.05–0.20, indicating that 5–20 per cent of the rumen liquor flows out of the organ each hour. The passage of particulate matter from the rumen can be followed by marking or labelling some of the food with the rare earth element ytterbium and then taking sequential samples of duodenal digesta. An older method employed dyes to mark food particles, which allowed undigested particles in digesta or faeces to be identified and counted under a microscope. Typical rates of passage of particulate matter are 0.012–0.030 per hour, and retention times, 30–80 hours. Studies of rumen dynamics have led to the development of complex mathematical models (see Further reading at the end of this chapter).

Control and modification of rumen fermentation

As investigations have revealed the mechanisms of rumen digestion, attempts have been made to alter the patterns of digestion in ways that should improve the nutrition of ruminants. The primary approach has been to modify the microbial population in order to suppress undesirable processes (e.g. methane production; see Chapter 11) or stimulate desirable processes (e.g. microbial protein synthesis). A secondary approach has been to protect nutrients from rumen fermentation in order that they should be digested in the small intestine. Changing the bacterial population through the introduction of specific organisms has generally proved difficult to achieve, or if achieved has failed to yield nutritional benefits. Changing an existing population by adding antibiotics to feeds has proved more effective, although the use of many antibiotics has been prohibited because of their value for treating diseases in animals and man, and the possibility that their wider application would lead to the evolution of resistant strains of disease-causing organisms. The antibiotics used today are mostly of the ionophore type, examples being monensin and salinomycin. These are active against gram-negative bacteria; by stimulating the production of propionate and reducing the production of acetate and butyrate, they improve the efficiency of utilisation of feed by growing ruminants. Another recent development has been the use of substances known as probiotics, such as live yeast cultures, to stimulate bacterial activity in the rumen. In some circumstances these can stabilise rumen pH, increase propionate and reduce acetate production, and reduce methane and ammonia production. The use of antibiotics and probiotics as additives to the diet is discussed in Chapter 25.

The advent of genetic engineering has raised hopes of providing, for example, bacteria with improved cellulolytic capabilities, or organisms able to manufacture specific nutrients, such as the sulphur-containing amino acids needed for wool growth (see Chapter 14). Establishing such modified organisms in the rumen ecosystem is still a problem, although one successful application of this technology has been to modify bacteria to produce the enzyme fluoracetate dehalogenase, which then protects ruminants against fluoracetate poisoning.

The rumen protozoal population has proved to be more susceptible to modification than the bacterial population, principally because protozoa can be totally eliminated from the rumen. Ruminants reared from birth in isolation from other ruminants do not develop a protozoal population. Existing populations of protozoa can be eradicated by the use of high-starch diets, or by the administration of defaunating agents such as copper sulphate. The ionophore monensin, which was originally used to kill coccidia in poultry, also kills protozoa in ruminants (although there is some evidence that rumen protozoa can adapt to and tolerate monensin).

The contribution of protozoa to rumen digestion, and hence to the nutrition and productivity of ruminants, has long been a matter of controversy. Although protozoa make a significant contribution to the digestion of polysaccharides, they are retained in the rumen and thus have the undesirable effect of 'locking up' microbial protein in the rumen and preventing its passage to the small intestine. So although defaunation of the rumen reduces the digestion of polysaccharides (especially the hemicelluloses), it increases the quantity of microbial protein reaching the duodenum by about 25 per cent. If protozoa are absent from the rumen their fibrolytic activity may be taken over by fungi. There is, however, some evidence that protozoa have another valuable role in aiding the absorption of the minerals calcium, magnesium and phosphorus from the gut.

The current view of rumen protozoa is that with low-protein, forage-based diets their presence is detrimental to the host, and defaunation can therefore increase animal productivity. With concentrate-based diets, better supplied with protein, the presence of protozoa is beneficial. Ironically, it is difficult to keep ruminants free of protozoa when they are on a forage diet and difficult to maintain protozoa on a concentrate diet!

A major objective in controlling rumen fermentation is to restrict the activity of microbes to the components of the diet that the host animal cannot digest with its own enzymes (principally the β-linked polysaccharides) or cannot utilise without microbial intervention (non-protein nitrogenous compounds). The strategy for achieving this objective is based on the protection of other nutrients (soluble carbohydrates, such as starch and sugars, and high-quality proteins) from attack in the rumen. We shall see later in this book that rumen fermentation of sugars is energetically inefficient for the host animal and may also cause it to be short of glucose. Tactics for protecting nutrients from attack in the rumen are often based on heat treatment or chemical treatment of foods. Treatment with tannins or formaldehyde modifies the structure of proteins in such a way as to impede microbial attack but still permit digestion by mammalian digestive enzymes. There are difficulties in achieving precisely the right degree of protection, however, and a more practical way of getting proteins past the rumen is to use foods to which rumen microorganisms are unaccustomed and therefore unadapted; these are principally foods of animal origin, such as fishmeals. Individual amino acids can be protected with agents based on polymers or fats. Protecting soluble carbohydrates such as starch from rumen fermentation is much more difficult, although the starch in some foods (rice by-products and, to a much lesser degree, maize) partially escapes rumen fermentation. If intensively fed ruminants are given an energy-rich supplement this is often in the form of the one class of nutrients that naturally escapes rumen fermentation, namely the triglyceride fats.

Although man has had some success in arranging for nutrients to 'bypass' the rumen, he is less successful than nature. In the young suckled ruminant, the oesophageal groove mechanism (see p. 179) allows the high-quality nutrients contained in milk to avoid rumen fermentation, even though the lamb or calf is consuming pasture herbage and digesting it in a fully functional rumen. Attempts by man to prolong this efficient partition into the adult life of the ruminant have been experimentally successful but fail commercially because of the high cost of milk or milk substitutes.

Another aspect of the control of rumen fermentation arises from the need to synchronise the supplies of energy and protein to the microorganisms. As mentioned earlier, the bacteria need a supply of energy to synthesise their cell proteins from degraded food proteins, and their energy sources, the carbohydrates, are made available at varying rates. It is therefore possible, for example, to supplement a rapidly degraded source of protein with a rapidly degradable source of carbohydrate. An alternative approach to synchronisation would be to give rapidly and slowly degraded foods at different times of the day. However, studies of rumen synchrony have yet to be fully translated into feeding practice.

8.3   Alternative sites of microbial digestion

Its great size and its location at the anterior end of the gut make the rumen pre-eminent as a digestive organ for fibrous foods. Large quantities of food can be stored rapidly for later mastication and fermentation; cell contents are released at an early stage; the main products of fermentation have ample opportunity to be absorbed in the remainder of the tract. These advantages of rumen digestion are, however, diminished by the disadvantage of all food constituents being exposed to fermentation. If fermentation is delayed until food reaches the large intestine, this disadvantage is overcome, but some of the benefits of the rumen are lost.

The parts of the large intestine that are capable of sustaining a significant microbial population are the colon and the caecum (Fig. 8.1). The caecum is blind-ended, and is duplicated in the fowl; in some animals the walls of both caecum and colon are sacculated. The digestive capacity of these organs depends on their volume relative to the rest of the tract, as that determines the time for which food residues may be delayed for fermentation. The substrate for the large intestine differs from that entering the rumen, because most of the more readily digestible nutrients will have been removed, and also some endogenous materials (such as mucopolysaccharides and enzymes) will have been added. However, as described earlier (p. 171), microbial digestion in the large intestine is similar to that occurring in the rumen. Volatile fatty acids are produced and absorbed; methane and other gases are present. Proteins and non-protein sources of nitrogen (such as urea from the bloodstream) are re-formed into microbial proteins; in some cases, but not invariably, these undergo proteolysis to amino acids, which may be absorbed. Water-soluble vitamins are synthesised and inorganic elements and water are reabsorbed. But in general, hind gut fermentation is less effective than rumen digestion, because digesta are not held for sufficient time and because many of the products of digestion (particularly proteins and vitamins) are not absorbed. Some species of animal overcome the last problem by practising coprophagy (the consumption of faeces). The rabbit has perfected this practice by producing two types of faeces, the normal hard pellets (which are not eaten) and the soft faeces or caecotrophes, which contain well-fermented material from the caecum, and which are consumed. Ruminants have a substantial capacity for hind gut fermentation, and this is used to good effect when the diet or the level of feeding causes much fibrous material to reach the caecum (see Table 10.2).

The horse is capable of living solely on fibrous forages. Unlike the ruminant, however, it has only one opportunity to chew its food and must therefore chew it thoroughly as it ingests it. At this stage large amounts of saliva are added, and the food is sufficiently buffered to permit a limited amount of fermentation in the stomach. However, most microbial digestion takes place in the enlarged colon, which has a capacity exceeding 60 litres, and the caecum with a capacity of 25–35 litres. These organs contain bacteria and protozoa which digest the food constituents in the same way as rumen microorganisms. It has been estimated that in the horse hind gut fermentation accounts for 30 per cent of the digestion of dietary protein, 15–30 per cent of that of soluble carbohydrate and 75–85 per cent of that of cell wall carbohydrate. With diets of hay and concentrates, horses digest about 85 per cent of the organic matter that would be digested by ruminants. In the pig, the hind gut is less enlarged than in the horse, and forages are poorly digested. Nevertheless, the pig can digest as much as 50 per cent of the cellulose and hemicellulose of cereal grains and their by-products. If starch grains escape digestion in the small intestine, as happens when pigs are fed on uncooked potatoes, these will also be fermented.

Although poultry have two caeca and a colon in which to ferment food residues, they gain little or nothing from hind gut fermentation when fed on their usual concentrate diets. Indeed, it has been suggested that in poultry the intestinal flora are more of a handicap than an advantage, as 'germ-free' birds (i.e. reared in isolation and with no microorganisms) tend to grow larger than normal birds.

Summary

1. In a simple-stomached animal, such as the pig, digestion begins when food reaches the stomach. Proteins are hydrolysed by *pepsins* at pH 2.0–3.5 to polypeptides and a few amino acids.

2. In the small intestine, enzymes secreted by the mucosa and by the pancreas continue the breakdown of protein to amino acids through the action of trypsin and carboxypeptidases. Fats are made water-soluble through emulsification by the bile salts (secreted by the liver) and by partial hydrolysis by lipases. Starch (and related α-linked polysaccharides) are hydrolysed by amylase to disaccharides, which are further hydrolysed to monosaccharides by specific enzymes (e.g. maltose by maltase, sucrose by sucrase, etc.).

3. In the large intestine, microorganisms produce enzymes that hydrolyse and ferment β-linked polysaccharides, mainly to volatile fatty acids.

4. In the young pig, enzyme activities are restricted because the sole food is milk. In the stomach, chymosin causes partial breakdown and clotting of casein. Lactase activity is high, but the activity of starch-digesting enzymes, amylase and maltase, is low until the fourth week of life.

5. In the fowl, digestion is aided by a storage organ anterior to the proventriculus (the crop) and a grinding organ posterior to the stomach (the gizzard). As part of the hind gut, the two blind-ended caeca facilitate microbial digestion, but this is not quantitatively important.

6. Absorption of nutrients occurs mainly in the small intestine. Carbohydrates are absorbed as monosaccharides by active transport, a process involving carrier proteins. Amino acids and fatty acids are also absorbed by active transport, but emulsified triglycerides are absorbed by passive diffusion. Large molecules, especially the immunoglobulins present in colostrum, are absorbed by a process known as pinocytosis. Many minerals and vitamins require special processes of absorption.

7. The alimentary tract has a detoxifying role, as it can inactivate some of the potentially damaging constituents of foods.

8. In ruminants, the large stomach compartment known as the reticulo-rumen allows regurgitation of food for mechanical breakdown by rumination, and acts as a continuous fermentation system for anaerobic bacteria, protozoa and fungi.

9. In the rumen, *carbohydrates* (both α- and β-linked) are hydrolysed by various routes to pyruvic acid, which is then fermented to acetic, propionic and butyric acids. The extent of breakdown and the proportions of these acids are determined by the nature of the food. Lignin is not digested and tends to interfere with the digestion of other nutrients. The volatile fatty acids are absorbed through the rumen wall. Methane and carbon dioxide are by-products of rumen fermentation.

10. Proteins in the rumen are hydrolysed to peptides and amino acids, and the latter may be deaminated to yield ammonia. Microorganisms use these products, and also non-protein sources of nitrogen, such as urea and uric acid, to synthesise microbial proteins. Microbial cells pass from the rumen to the abomasum (true stomach) and small intestine, where they are digested by the host animal's enzymes.

11. Lipids in the rumen are hydrolysed, and unsaturated fatty acids are then converted to saturated fatty acids by hydrogenation.

12. For optimal rumen fermentation, foods must be held in the organ to allow time for the slow breakdown of plant cell walls, and the microorganisms must receive balanced supplies of nitrogen (as ammonia) and energy (as carbohydrate).

13. Rumen breakdown of food protein is not always desirable, because microbial protein may be inferior to it in both quantity and quality. Thus food proteins are sometimes protected from attack by rumen microorganisms. In suckled ruminants, there is a device known as the oesophageal groove, which channels milk directly to the abomasum.

14. Microbial digestion continues in the large intestine of ruminants; volatile fatty acids and microbial protein are produced, but the protein cannot be subsequently digested and absorbed by the host animal.

15. In the horse, the large intestine is the principal site of microbial digestion.

Further reading

Annison E F and Brydon W L 1998 Perspectives of ruminant nutrition and metabolism. 1. Metabolism in the rumen. *Nutrition Research Reviews* **11**: 173–98.

Church D C 1976 *Digestive Physiology and Nutrition of Ruminants, Vol. 1. Digestive Physiology*, 2nd edn, Corvallis, OR, O and B Books.

Cronje P 2000 *Ruminant Physiology: Digestion, Metabolism, Growth and Reproduction*, Proceedings of the Ninth International Symposium on

Ruminant Physiology, Wallingford, UK, CAB International. (See also other volumes in this series.)

Czerkawski J W 1986 *An Introduction to Rumen Studies*, Oxford, Pergamon Press.

Forbes J M and France J (eds) 1993 *Quantitative Aspects of Ruminant Digestion and Metabolism*, Wallingford, UK, CAB International.

Frape D L 1998 *Equine Nutrition and Feeding*, 2nd edn, London, Blackwell Science.

Freeman B M (ed.) 1983 *Physiology and Biochemistry of the Domestic Fowl*, London, Academic Press.

Hobson P N and Stewart C S 1997 *The Rumen Microbial Ecosystem*, 2nd edn, London, Blackie Academic and Professional.

Kidder D E and Manners M J 1978 *Digestion in the Pig*, Bristol, Scientechnica.

Murray R K, Granner D K, Mayes P A and Rodwell V W 1993 *Harper's Biochemistry*, 23rd edn, USA, Appleton and Large.

Moran E T Jr 1982 *The Comparative Nutrition of Fowl and Swine. The Gastrointestinal Systems*, University of Guelph, Ontario.

Sandford P A 1982 *Digestive System Physiology*, London, Edward Arnold.

Swenson M J (ed.) 1984 *Dukes's Physiology of Domestic Animals*, Ithaca, NY, Comstock.

Van Soest P J 1994 *Nutritional Ecology of the Ruminant*, 2nd edn, Ithaca, NY, Comstock.

Historical reference

Hungate R E 1966 *The Rumen and its Microbes*, New York, Academic Press.

9 Metabolism

9.1   Energy metabolism
9.2   Protein synthesis
9.3   Fat synthesis
9.4   Carbohydrate synthesis
9.5   Control of metabolism

Metabolism is the name given to the sequence, or succession, of chemical processes that take place in the living organism. Some of the processes involve the degradation of complex compounds to simpler materials and are designated by the general term *catabolism*. *Anabolism* describes those metabolic processes in which complex compounds are synthesised from simpler substances. Waste products arise as a result of metabolism and these have to be chemically transformed and ultimately excreted; the reactions necessary for such transformations form part of the general metabolism. As a result of the various metabolic processes energy is made available for mechanical work and for chemical work such as the synthesis of carbohydrates, proteins and lipids. Figure 9.1 summarises the sources of the major metabolites available to the body and their subsequent metabolism.

The starting points of metabolism are the substances produced by the digestion of food. For all practical purposes we may regard the end-products of carbohydrate digestion in the simple-stomached animal as glucose, together with very small amounts of galactose and fructose. These are absorbed into the portal blood and carried to the liver. In ruminant animals, the major part of the carbohydrate is broken down in the rumen to acetic, propionic and butyric acids, together with small amounts of branched chain and higher volatile acids. Butyric acid is changed in its passage across the rumen wall and passes into the portal blood as (D-)β-hydroxybutyric acid (BHBA). Acetic acid and BHBA pass from the liver via the systemic blood to various organs and tissues, where they are used as sources of energy and fatty acids. Propionic acid is converted to glucose in the liver and joins the liver glucose pool. This may be converted partly into glycogen and stored, or to fatty acids, reduced coenzymes and L-glycerol-3-phosphate and used for triacylglycerol synthesis. The remainder of the glucose enters the systemic blood supply and is carried to various body tissues, where it may be used as an energy source, as a source of reduced coenzymes for fatty acid synthesis and for glycogen synthesis.

Digestion of proteins results in the production of amino acids and small peptides, which are absorbed via the intestinal villi into the portal blood and

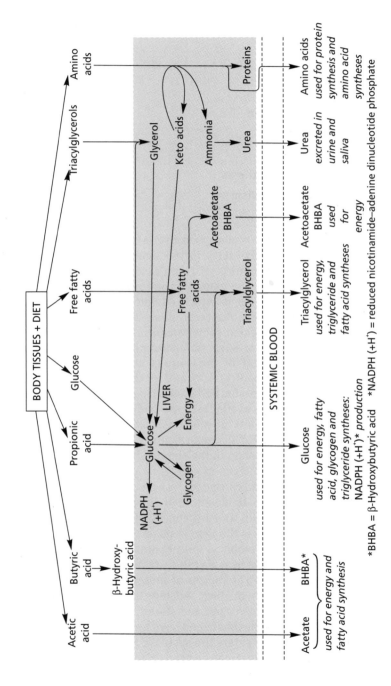

Fig. 9.1 Sources and fates of major body metabolites.

are carried to the liver, where they join the amino acid pool. They may then be used for protein synthesis *in situ* or may pass into the systemic blood, where they join the amino acids produced as a result of tissue catabolism in providing the raw material for the synthesis of proteins and other biologically important nitrogenous compounds. Amino acids in excess of this requirement are carried to the liver and broken down to ammonia and keto acids. The latter may be used for amino acid synthesis or the production of energy. Some of the ammonia may be used in amination but practically all of it is converted into urea and either excreted in the urine or recycled in the saliva. In the ruminant animal a considerable amount of ammonia may be absorbed from the rumen into the portal blood, transformed into urea by the liver and then excreted or recycled via the saliva or the rumen wall.

Most dietary lipids enter the lacteals as chylomicrons, which enter the venous blood vessels via the thoracic duct. The chylomicrons are about 500 nm in diameter with a thin lipoprotein envelope. A very small proportion of dietary triacylglycerols may be hydrolysed to glycerol and low molecular weight acids in the digestive tract and these are absorbed directly into the portal blood supply. Circulating chylomicrons are absorbed by the liver and the triacylglycerols are hydrolysed. The fatty acids so produced, along with free fatty acids absorbed from the blood by the liver, may be catabolised for energy production or used for the synthesis of triacylglycerols. These then re-enter the blood supply in the form of lipoprotein and are carried to various organs and tissues, where they may be used for lipid synthesis, for energy production and for fatty acid synthesis. Except in the case of the liver, hydrolysis is a prerequisite for absorption. Fatty acids catabolised in excess of the liver's requirement for energy are changed to (L-)β-hydroxybutyrate and acetoacetate, which are transported to various tissues and used as sources of energy.

9.1 Energy metabolism

Energy may be defined as the capacity to do work. There are various forms of energy such as chemical, thermal, electrical and radiant, all of which are interconvertible by suitable means. Thus the radiant energy of the sun is utilised by green plants to produce complex plant constituents and is stored as such. The plants are consumed by animals and the constituents broken down, releasing energy, which is used by the animals for doing mechanical work, for transport, for maintaining the integrity of cell membranes, for synthesis and for providing heat under cold conditions.

Traditionally, heat units have been used to represent the various forms of energy involved in metabolism, since all the forms are convertible into heat. The basic unit used has been the thermochemical calorie (cal), based on the calorific value of benzoic acid as the reference standard. Usually the kilocalorie (1000 cal) or the megacalorie (1 000 000 cal) were used in practice since the calorie was inconveniently small. The International Union of Nutritional Sciences and the National Committee of the International

Union of Physiological Sciences have suggested the joule (J) as the unit of energy for use in nutritional, metabolic and physiological studies. This suggestion has been almost universally adopted and is followed in this book. The joule is defined as 1 newton per metre, and 4.184 J = 1 cal.

The chemical reactions taking place in the animal body are accompanied by changes in the energy of the system. That portion of the energy change which is available to do work is termed the free energy change, designated ΔG. When ΔG is negative the reaction is said to be *exergonic* and takes place spontaneously, and when ΔG is large and negative the reaction proceeds almost to completion. When ΔG is positive, the reaction is termed *endergonic* and free energy has to be fed into the system in order for the reaction to take place. When ΔG is large and positive there is little tendency for the reaction to take place. Most of the synthetic reactions of the body are endergonic and the energy needed to drive them is obtained from exergonic catabolic reactions. Before the energy released by these changes can be utilised for syntheses and other vital body processes, a link between the two must be established. This is provided by mediating compounds which take part in both processes, picking up energy from one and transferring it to the other. Typical are adenosine triphosphate (ATP), guanosine triphosphate (GTP), cytidine triphosphate (CTP) and uridine triphosphate (UTP). By far the most important of these nucleotides is ATP. Adenosine is formed from the purine base adenine and the sugar D-ribose. Phosphorylation of the hydroxyl group at carbon atom 5 of the sugar gives adenosine monophosphate (AMP) (see Chapter 4); successive additions of phosphate residues give adenosine diphosphate (ADP) and then the triphosphate. In its reactions within the cell, ATP functions as a complex with magnesium. The addition of the last two phosphate bonds requires a considerable amount of energy, which may be obtained directly by reaction of AMP or ADP with an energy-rich material. For example, in carbohydrate breakdown one of the steps is the change of phosphoenolpyruvate to pyruvate, which results in one molecule of ATP being produced from ADP.

Phosphoenolpyruvate   Pyruvate

When production of ATP from ADP takes place directly during a reaction, as in this case, the process is known as *substrate level phosphorylation*.

Alternatively ATP may be produced indirectly. Most biological oxidations involve the removal of hydrogen from a substrate but the final combination with oxygen to form water occurs only at the end of a series of reactions. A typical example is the removal of hydrogen coupled to nicotinamide

---

*Throughout this chapter, energy inputs are shown by an open box (□) and energy outputs by a black box (■).

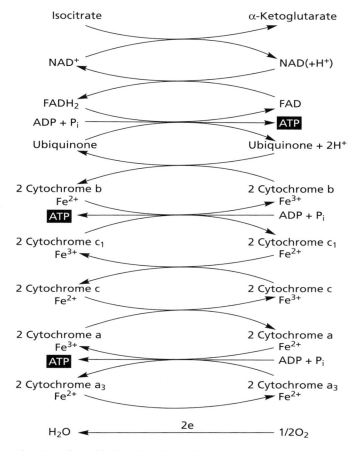

Fig. 9.2 The oxidative phosphorylation system.

adenine dinucleotide (NAD^+), as illustrated in Fig. 9.2 for the oxidation of isocitrate to α-ketoglutarate.

In this example, hydrogen removed from isocitrate is accepted by NAD^+ and is then passed to the flavin coenzyme. This donates two electrons to ubiquinone and two protons ($2H^+$) are formed. The electrons are then transferred via the sequential cytochromes to cytochrome a_3, which is capable of transferring the electrons to oxygen. The negatively charged oxygen finally unites with the two protons, yielding water. During the operation of this pathway ATP is produced from ADP and inorganic phosphate, the process being called *oxidative phosphorylation*. It is confined to the mitochondria and to the reduced NAD^+ produced within them. Considerations of energy release indicate that production of ATP takes place at the transfer of hydrogen from reduced FAD to ubiquinone, at the transfer of electrons from cytochrome b to c_1 and from cytochrome a to cytochrome a_3. The series of reactions may be represented as:

$$NADH(+H^+) + O + 3ADP + 3Pi \longrightarrow NAD^+ + 3\ ATP + H_2O$$

Oxidation of each mole of reduced NAD⁺ yields three moles of ATP from ADP and inorganic phosphate, and each mole of reduced FAD yields two moles of ATP. Since oxidative phosphorylation is confined to the mitosol, reduced NAD⁺ produced in the cytosol must cross the mitochondrial wall to be processed in this way. Reduced NAD⁺ itself is not capable of doing so. Passage into the mitosol, of the reducing equivalents which it represents, is achieved by the operation of the malate–aspartate shuttle or the glycerophosphate shuttle. In the former case, no energy cost is involved. In the case of the glycerophosphate shuttle the reducing equivalents enter the oxidative phosphorylation pathway at FAD⁺ stage and only two moles of ATP are produced. The malate–aspartate shuttle is predominant in the liver, whereas muscle makes greater use of the glycerophosphate pathway.

The energy fixed as ATP may be used for doing mechanical work during the performance of essential life processes in maintaining the animal. Both contraction and relaxation of muscle involve reactions which require a supply of energy which is provided by breakdown of ATP to ADP and inorganic phosphate. The energy fixed in ATP may also be used to drive reactions in which the terminal phosphate group is donated to a large variety of acceptor molecules. Among these is D-glucose:

$$\boxed{\text{ATP}} + \text{D-glucose} \rightarrow \text{ADP} + \text{D-glucose phosphate}$$

In this way the glucose is energised for subsequent biosynthetic reactions. In others, such as the first stage of fatty acid synthesis, ATP provides the energy and is broken down to AMP and inorganic phosphate:

$$\text{Acetate} + \text{Coenzyme A} + \boxed{\text{ATP}} \leftrightarrow \text{Acetyl-coenzyme A} + \text{pyrophosphate} + \text{AMP}$$

The role of ATP in trapping and utilising energy may be illustrated diagrammatically as shown in Fig. 9.3.

The quantity of energy that is made available by the rupture of each of the two terminal phosphate bonds of ATP varies according to the conditions under which the hydrolyses take place. Most authorities agree that, under the conditions pertaining in intact cells, it is about 50 kJ/mole but will vary with pH, magnesium ion concentration and the concentrations of

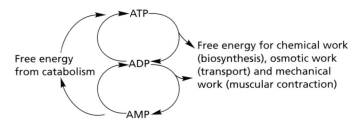

Fig. 9.3 The role of ATP in the utilisation of energy.

ATP, ADP and phosphate. The phosphate bonds are commonly referred to as high-energy bonds, represented by ~[P]. The term is not thermodynamically accurate and many workers prefer to use the term high group transfer potential.

Fixation of energy in the form of ATP is a transitory phenomenon and any energy produced in excess of immediate requirements is stored in more permanent form in such compounds as the phosphocreatine of muscle, which is formed from creatine when ATP is in excess:

$$
\begin{array}{cccc}
\text{NH}_2 & & \text{NH} \sim \textcircled{P} & \\
| & & | & \\
\text{C:NH}_2^+ & & \text{C:NH}_2^+ & \\
| & \xrightarrow{\textit{Creatine kinase}} & | & \\
\text{N.CH}_3 & + \boxed{\text{ATP}} \; \rightleftharpoons & \text{N.CH}_3 & + \text{ADP} \\
| & & | & \\
\text{CH}_2 & & \text{CH}_2 & \\
| & & | & \\
\text{COO}^- & & \text{COO}^- & \\
\text{Creatine} & & \text{Phosphocreatine} &
\end{array}
$$

When the supply of ATP is insufficient to meet the demands for energy, more ATP is produced from phosphocreatine by the reverse reaction.

Even materials like phosphocreatine are minor, temporary stores of energy only. Most energy is stored in the body as depot fat together with small quantities of carbohydrate in the form of glycogen. In addition protein may be used to provide energy under certain circumstances.

In addition to using this stored energy, the body derives energy directly from nutrients absorbed from the digestive tract. Chief of these in simple-stomached animals is glucose; in ruminant animals, the volatile fatty acids occupy this position.

Glucose as an energy source

The major pathway whereby glucose is metabolised to give energy has two stages. The first, known as glycolysis, can occur under anaerobic conditions and results in the production of pyruvate. The sequence of reactions (Fig. 9.4), often referred to as the Embden–Meyerhof pathway, takes place in the cytosol.

All the reactions in the pathway are reversible, but reactions 1, 3, 8 and 11 have large negative ΔG values under physiological conditions and are essentially irreversible. Two moles of ATP are used in the initial phosphorylations of steps 1 and 3, and the fructose diphosphate so formed breaks down to yield two moles of glyceraldehyde-3-phosphate. Subsequently, one mole of ATP is produced directly at each of steps 8 and 11. Four moles of ATP are thus produced from one mole of glucose. Since two moles of ATP are used up, the net production of ATP from ADP is two moles per mole of glucose. Under aerobic conditions the reduced NAD^+, produced at step 7,

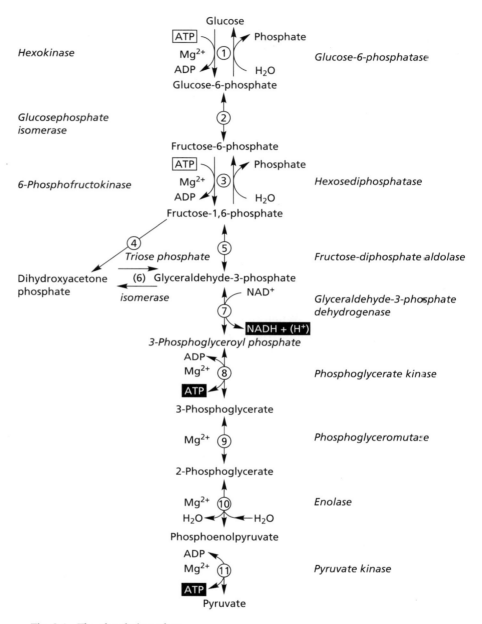

Fig. 9.4 The glycolytic pathway.

may be oxidised via the oxidative phosphorylation pathway and, assuming the operation of the malate–aspartate shuttle, three moles of ATP are produced per mole. The pyruvate produced by glycolysis is transported into the mitosol, without energy cost, and is oxidised to carbon dioxide and water with further production of energy. The first step in this process is the oxidative decarboxylation of pyruvate in the presence of thiamin diphosphate:

$$CH_3$$
$$|$$
$$C:O \quad + \quad HS.CoA \quad \xrightarrow{\textit{Pyruvate dehydrogenase}} \quad$$
$$|$$
$$COO^-$$

$$CH_3 \quad + \quad CO_2$$
$$|$$
$$COS.CoA$$

NAD$^+$ → NADH (+H$^+$)

Pyruvate   Coenzyme A   Acetyl-coenzyme A

The hydrogen is removed via the normal NAD$^+$ pathway. The acetyl-coenzyme A is then oxidised to carbon dioxide and water via the tricarboxylic acid cycle as shown in Fig. 9.5.

This involves four dehydrogenations, three of which are NAD$^+$ linked and one FAD linked, and results in 11 moles of ATP being formed from ADP. In addition, one mole of ATP arises directly with the change of succinyl-coenzyme A to succinate. The oxidation of each mole of pyruvate thus yields 15 moles of ATP. The total ATP production from the oxidation of one mole of glucose is then:

	Moles +	ATP −
1 mole of glucose to 2 moles of pyruvate	10	2
2 moles of pyruvate to 2 moles acetyl-CoA	6	0
2 moles acetyl-CoA to $CO_2 + H_2O$	24	0
Total per mole of glucose (net)		38

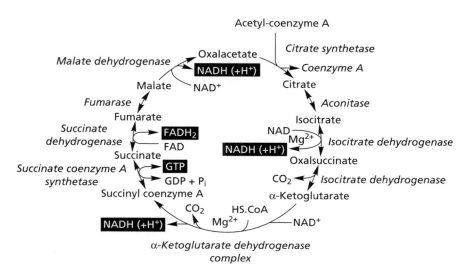

Fig. 9.5 The tricarboxylic acid cycle.

The capture of energy represented by the formation of 38 high-energy phosphate bonds may be calculated as $38 \times 50 = 1900$ kJ/mole of glucose. The total free energy content of glucose is 2870 kJ/mole. The efficiency of free energy capture by the body is thus $1900/2870 = 0.66$. Where transfer of the reduced NAD^+ produced during glycolysis involves the glycerophosphate shuttle, the yield of ATP is two moles per mole only and the oxidation of each mole of glucose yields 36 moles of ATP. The efficiency of free energy capture is then $36 \times 50/2870 = 0.63$. Such calculations assume perfect coupling of reactions and normal environmental cell conditions.

Glycolysis takes place in the cytosol, whereas the decarboxylation of pyruvate and the subsequent oxidation of acetyl-coenzyme A via the tricarboxylic acid cycle take place in the mitochondrial matrix.

Under anaerobic conditions, oxygen is not available for the oxidation of the reduced NAD^+ by oxidative phosphorylation. In order to allow the release of a small amount of energy by a continuing breakdown of glucose to pyruvate, reduced NAD^+ must be converted to the oxidised form. If not, step 7 of Fig. 9.4 will not take place and energy production will be blocked. Oxidation of reduced NAD^+ may be achieved under such conditions by the formation of lactate from pyruvate in the presence of lactate dehydrogenase:

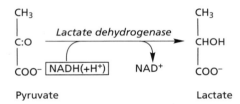

When glucose is used as an energy source under anaerobic conditions, lactate accumulates, eventually diffuses into the bloodstream and is carried to highly aerobic tissues such as the heart and the liver. Here it may undergo oxidative breakdown to carbon dioxide and water with further release of energy or it may be reconverted into glucose. Recent evidence suggests that even in highly aerobic muscle tissue much of the glucose used for energy is converted to lactate.

Another pathway by which glucose is metabolised within the body is that known variously as the pentose phosphate pathway, the phosphogluconate oxidative pathway and the hexose phosphate shunt. Although the system encompassing glycolysis and the tricarboxylic acid cycle is the major pathway of glucose metabolism in the body, the pentose phosphate pathway is of considerable importance in the cytosol of the cells of the liver, adipose tissue and the lactating mammary glands. The steps of this pathway are shown in Fig. 9.6.

The net result of this series of reactions is the removal of one carbon atom of glucose as carbon dioxide and the production of two moles of reduced $NADP^+$. Oxidation of one mole of glucose may be represented as:

$$\text{Glucose-6-phosphate} + 12NADP^+ \longrightarrow 6CO_2 + PO_4^{3-} + \boxed{12NADPH(-H^+)}$$

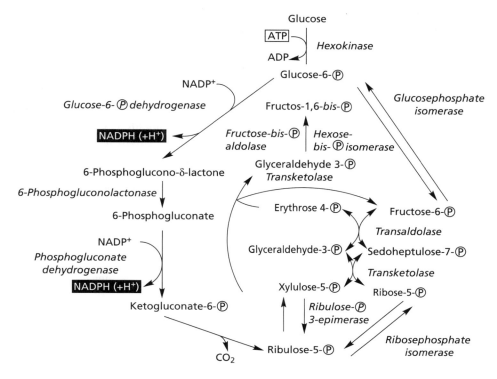

Fig. 9.6 The pentose phosphate pathway.

Unlike reduced NAD^+, reduced $NADP^+$ does not undergo oxidative phosphorylation to produce ATP and the main function of the pentose phosphate pathway is to provide reduced $NADP^+$ for tissues which have a specific demand for it, particularly those actively synthesising fatty acids. Almost one-third of the glucose metabolised by the liver may follow this pathway, and this figure may be exceeded in adipose tissue. Reduced $NADP^+$ can be converted to reduced NAD^+ via an energy-linked transhydrogenase and thus serve indirectly as a source of ATP.

Glycogen as an energy source

Glycogen is the major form of carbohydrate storage in animals and is present in most animal cells. It may form up to 8 per cent of the fresh liver and up to 1 per cent of fresh muscle. Most is present in the latter. The release of energy from glycogen in a utilisable form necessitates its breakdown to glucose, which is then degraded as previously described. Breakdown of glycogen within cells takes place through the action of inorganic phosphate and glycogen phosphorylase. This enzyme catalyses the rupture of the 1,4-glycosidic linkages of the glycogen (Chapter 2) and degradation begins at the non-reducing end of the chain. Glucose-1-phosphate molecules are released successively until a branch point is approached. A rearrangement of the molecule then takes place in the presence of oligotransferase, and a limit dextrin with a terminal 1,6-linked glucose is produced. Attack at the 1,6 linkage by an oligo-1,6-glucosidase releases free glucose, and glucose-1-phosphate is produced by further activity of the phosphorylase. The net effect of glycogen breakdown is the production of glucose-1-phosphate plus a little glucose. Glucose-1-phosphate is converted by phosphoglucomutase to glucose-6-phosphate, which enters the Embden–Meyerhof or the pentose phosphate pathway, as does the residual glucose. The production of glucose-6-phosphate from glycogen does not involve expenditure of ATP except for that used in the conversion of residual glucose to glucose-6-phosphate. Energy production from glycogen is thus slightly more efficient than it is from glucose.

Propionic acid as an energy source

In ruminant animals, considerable amounts of propionate are produced from carbohydrate breakdown in the rumen. The acid then passes across the rumen wall, where a little is converted to lactate. The remainder is carried to the liver, where it is changed into glucose. The first stage in this process is conversion to succinyl-coenzyme A (Fig. 9.7).

This enters the tricarboxylic acid cycle and is converted to malate (Fig. 9.5), and the equivalent of three moles of ATP are produced. The malate is transported into the cytosol where it is converted to oxalacetate and then phosphoenolpyruvate:

Malate   Oxalacetate   Phosphoenolpyruvate

The phosphoenolpyruvate may then be converted to fructose diphosphate by reversal of steps 10, 9, 8, 7 and 5 in the glycolytic sequence shown in Fig. 9.4. This is then converted to fructose-6-phosphate by hexosediphosphatase and then to glucose-6-phosphate by the reverse of step 2 and finally to glucose by glucose-6-phosphatase. The glucose may be used eventually to provide energy and we may prepare an energy balance sheet:

	Moles ATP	
	+	−
2 moles propionate to 2 moles succinyl-coenzyme A		6
2 moles succinyl-coenzyme A to 2 moles malate	6	
2 moles malate to 2 moles phosphoenolpyruvate	6	2
2 moles phosphoenolpyruvate to 1 mole glucose		8
1 mole glucose to carbon dioxide and water	38	
Total	50	16

There is thus a net gain of 17 moles of ATP per mole of propionic acid.

Small amounts of propionate are present in the peripheral blood supply. They may arise because of incomplete removal by the liver or from oxidation of fatty acids with an odd number of carbon atoms. Such propionate could conceivably be used directly for energy production. The pathway would be

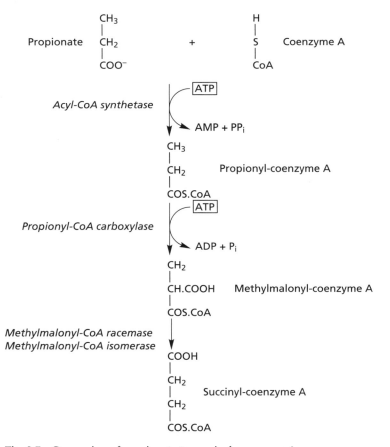

Fig. 9.7 Conversion of propionate to succinyl-coenzyme A.

the same as that described as far as phosphoenolpyruvate. This would then follow the Embden–Meyerhof pathway via pyruvate, acetyl-coenzyme A and the tricarboxylic acid cycle. The balance sheet for this process is:

	Moles ATP +	Moles ATP −
1 mole propionate to 1 mole succinyl-coenzyme A		3
1 mole succinyl-coenzyme A to 1 mole malate	3	
1 mole malate to 1 mole phosphoenolpyruvate	3	1
1 mole phosphoenolpyruvate to 1 mole acetyl-coenzyme A	4	
1 mole acetyl-coenzyme A to carbon dioxide and water	12	
Total	22	4
Net gain of ATP per mole propionic acid	18	

The pathway is thus marginally more efficient than that via glucose.

Butyric acid as an energy source

Butyric acid produced in the rumen is converted to β-hydroxybutyrate (D-3-hydroxybutyrate) in its passage across the ruminal and omasal walls. The pathway for this conversion is shown in Fig. 9.8.

The D-3-hydroxybutyrate may then be used as a source of energy by a number of tissues, notably skeletal and heart muscle. In non-ruminant, but not in ruminant, animals, utilisation by the brain increases markedly under conditions of glucose shortage. The reactions involved in energy production are shown in Fig. 9.9.

The acetyl-coenzyme A is metabolised via the tricarboxylic acid cycle. We may calculate the energy released from butyrate by the synthetase pathway as follows:

	Moles ATP +	Moles ATP −
1 mole butyrate to 1 mole D-3-hydroxybutyrate	5	5
1 mole D-3-hydroxybutyrate to 2 moles acetyl-CoA	3	2
2 moles acetyl-CoA to carbon dioxide and water	24	
Total	32	7
Net gain of ATP per mole butyric acid	**25**	

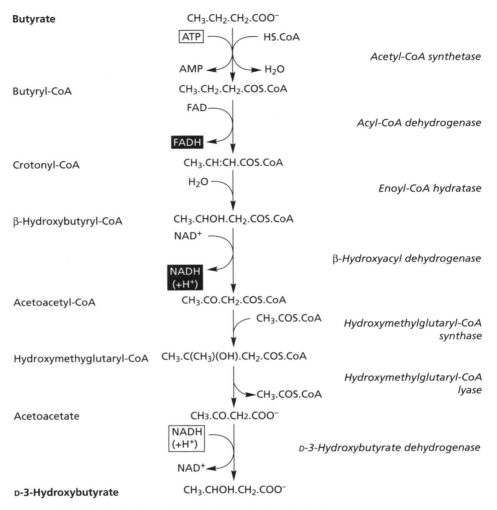

Fig. 9.8 Production of 3-hydroxybutyrate from butyrate.

If the change of acetoacetate to acetoacetyl-coenzyme A takes place via the succinyl-coenzyme A pathway, there is a saving of two moles of ATP and the net gain per mole of butyric acid is equivalent to 27 high-energy phosphate bonds. However, the energy cost of producing succinyl-coenzyme A has to be taken into account and this pathway is then slightly less efficient than the other.

Acetic acid as an energy source

Acetic acid is the major product of carbohydrate digestion in the ruminant and is the only volatile fatty acid present in the peripheral blood in significant amounts. It is used by a wide variety of tissues as a source of energy.

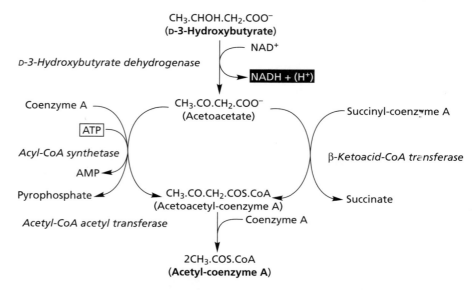

Fig. 9.9 Formation of acetyl-coenzyme A from D-3-hydroxybutyrate.

The initial reaction in this case is conversion of acetate to acetyl-coenzyme A in the presence of acetyl-coenzyme A synthetase:

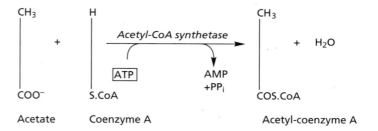

The formation of the acetyl-coenzyme A takes place in the cytosol whereas the oxidation via the tricarboxylic acid cycle is confined to the mitochondrial matrix. The acetyl-coenzyme A is unable to cross the mitochondrial wall and has to be complexed with carnitine to achieve this. Within the mitochondrial matrix the complex is broken down, releasing acetyl-coenzyme A, which then enters the tricarboxylic acid cycle and is oxidised yielding 12 moles of ATP per mole. Since two high-energy phosphate bonds are used in the initial synthetase-mediated reaction, the net yield of ATP is 10 moles per mole of acetate.

Fat as an energy source

The store of triacylglycerol in the body is mobilised to provide energy by the action of lipases, which catalyse the production of glycerol and fatty acids.

The glycerol is glycogenic and enters the glycolytic pathway (see Fig. 9.4) as dihydroxyacetone phosphate, produced as shown in the following reactions:

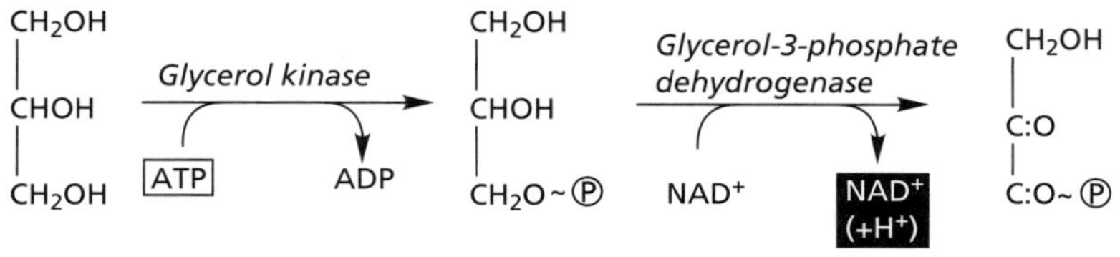

Glycerol | L-Glycerol 3-phosphate | Dihydroxyacetone phosphate

Glucose may then be produced by the reverse of the aldolase reaction to give fructose-1,6-diphosphate which is then converted to glucose by the action of hexose diphosphatase, glucose-6-phosphate isomerase and glucose-6-phosphatase. If the glucose is used to produce energy we may assess the efficiency of glycerol as an energy source:

	Moles ATP	
	+	−
2 moles glycerol to 2 moles dihydroxyacetone phosphate	6	2
2 moles dihydroxyacetone phosphate to 1 mole glucose		
1 mole glucose to carbon dioxide and water	38	
Total	44	2
Net yield of ATP per mole of glycerol	**21**	

On the other hand the dihydroxyacetone phosphate may enter the glycolytic pathway and be metabolised via pyruvate and the tricarboxylic acid cycle to carbon dioxide and water, with energy being released. The efficiency of glycerol as an energy source under these circumstances may be assessed as follows:

	Moles ATP	
	+	−
1 mole glycerol to 1 mole dihydroxyacetone phosphate	3	1
1 mole dihydroxyacetone phosphate to 1 mole pyruvate	5	
1 mole pyruvate to carbon dioxide and water	15	
Total	23	1
Net yield of ATP per mole of glycerol	**22**	

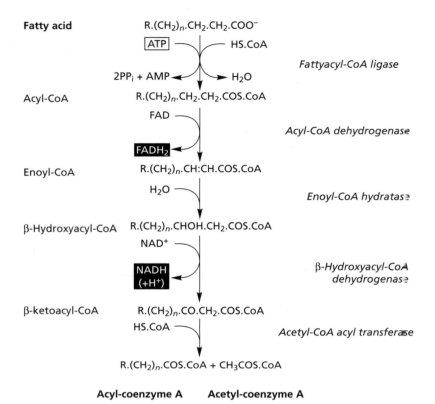

Fig. 9.10 Oxidation of a fatty acid to acetyl-coenzyme A.

The major part of the energy derived from fat is provided by the fatty acids. The major pathway for their degradation is that of β-oxidation, which results in a progressive shortening of the carbon chain by removal of two carbon atoms at a time. The first stage of β-oxidation is the reaction of the fatty acid with coenzyme A in the presence of ATP and fattyacyl-CoA ligase to give an acyl-coenzyme A. This occurs in the cytosol and the fattyacyl-CoA is then transferred into the mitochondria as a complex with carnitine and is there regenerated. It then undergoes a series of reactions to give an acyl-coenzyme A with two less carbon atoms than the original and a mole of acetyl-CoA is released. The pathway is illustrated in Fig. 9.10.

During the splitting off of the two-carbon acetyl-coenzyme A the equivalent of five moles of ATP is produced. The remaining acylcoenzyme undergoes the same series of reactions and the process continues until the carbon chain has been completely converted to acetyl-coenzyme A. This enters the tricarboxylic acid cycle and is oxidised to carbon dioxide and water, each mole of acetyl-CoA so metabolised giving 12 moles of ATP. Since the initial ligase reaction is only necessary once for each molecule, more ATP is produced for the same expenditure of energy by the oxidation of long rather than short chain acids. The oxidation of the 16-carbon palmitate is illustrated in Fig. 9.11.

The energy production in this sequence may be summarised as follows:

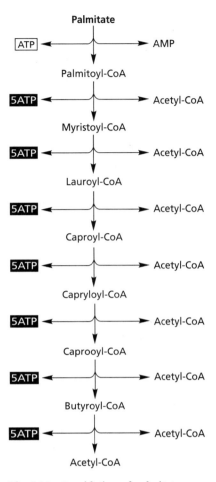

Fig. 9.11 β-oxidation of palmitate.

	Moles ATP	
	+	−
1 mole palmitate to palmitoyl-CoA		2
1 mole palmitoyl-CoA to 8 moles acetyl-CoA	35	
8 moles acetyl-CoA to carbon dioxide and water	96	
Total	131	2
Net gain of ATP per mole of palmitate	**129**	

Amino acids as sources of energy

When amino acids are available in excess of the animal's requirements, or
when the animal is forced to catabolise body tissues to maintain essential

body processes, amino acids may be broken down to provide energy. Amino acid degradation takes place in all animal tissues but mainly in the liver; the kidney shows considerable activity also. Muscular tissue is relatively inactive.

The first stage in the oxidative degradation of amino acids is the removal of the amino group by one or other of two main pathways, oxidative deamination and transamination. In transamination the amino group is transferred to the α-carbon atom of a keto acid, usually α-ketoglutarate, resulting in the production of another keto acid and glutamate. The reactions are catalysed by enzymes known as aminotransferases. The reaction for aspartate may be represented as:

$$
\underset{\text{Aspartate}}{\begin{array}{c} NH_3 \\ | \\ CH.COO^- \\ | \\ CH_2.COO^- \end{array}}
\; + \;
\underset{\text{α-Ketoglutarate}}{\begin{array}{c} O \\ \| \\ C.COO^- \\ | \\ CH_2 \\ | \\ CH_2.COO^- \end{array}}
\xrightleftharpoons{\text{Aspartate aminotransferase}}
\underset{\text{Oxalacetate}}{\begin{array}{c} O \\ \| \\ C.COO^- \\ | \\ CH_2.COO^- \end{array}}
\; + \;
\underset{\text{Glutamate}}{\begin{array}{c} NH_3^+ \\ | \\ CH.COO^- \\ | \\ CH_2 \\ | \\ CH_2.COO^- \end{array}}
$$

The glutamate so formed, as well as that which becomes available from the digestive tract and from protein breakdown in the tissues, may undergo oxidative deamination in the presence of glutamate dehydrogenase:

$$
\underset{\text{Glutamate}}{\begin{array}{c} NH_3^+ \\ | \\ CH.COO^- \\ | \\ CH_2 \\ | \\ CH_2.COO^- \end{array}}
\xrightarrow[\text{Glutamate dehydrogenase}]{NAD^+ \; H_2O \quad NADH \, (+H^+)}
\underset{\text{α-Ketoglutarate}}{\begin{array}{c} O \\ \| \\ C.COO^- \\ | \\ CH_2 \\ | \\ CH_2.COO^- \end{array}}
\; + \; NH_4^+
$$

The α-ketoglutarate may then be used in further transaminations and the reduced coenzyme is oxidised by oxidative phosphorylation (p. 203). Glutamate is the only amino acid in mammalian tissue which undergoes oxidative deamination at an appreciable rate. The initial transaminations which give rise to it are therefore of major importance when amino acids are being used as sources of energy. Flavin-linked D- and L-amino acid oxidases, which catalyse the production of keto acids and ammonia, do exist but are of minor importance only. The final product of amino acid degradation is acetyl-coenzyme A, which is then processed via the tricarboxylic acid cycle to yield energy. The acetyl-CoA may be produced directly (as in the case of tryptophan, leucine and isoleucine), via pyruvate (alanine, glycine, serine, threonine and cysteine) or via acetoacetyl-CoA (phenylalanine, tyrosine, leucine, lysine and tryptophan). Other amino acids are degraded by pathways of varying complexity to give products such as α-ketoglutarate oxalacetate, fumarate and succinyl-CoA, which enter the tricarboxylic acid cycle and yield acetyl-CoA via phosphoenolpyruvate (p. 207).

One of the consequences of amino acid catabolism is the production of ammonia, which is highly toxic. Some of this may be used in amination for amino acid synthesis in the body. In this case ammonia reacts with α-keto-glutarate to give glutamate, which is then used for the syntheses. The reaction is the reverse of oxidative deamination except that $NADP^+$ takes the place of NAD^+. Most of the ammonia is excreted from the body as urea in mammals and uric acid in birds.

Deamination of amino acids occurs in all the organs of the body but primarily in the liver. In most other tissues the ammonia is converted to glutamine or alanine (in muscle) before being transported to the liver and regenerated.

In mammals the ammonia is then converted into urea. This is a two-stage process requiring a supply of energy in the form of ATP. The first stage is the formation of carbamoyl phosphate from carbon dioxide and ammonia in the presence of carbamoyl phosphate synthetase:

$$CO_2 + NH_4^+ + H_2O \xrightarrow[\substack{2ATP}]{\substack{Carbamoyl\ phosphate \\ synthetase}} NH_2-\overset{\overset{\displaystyle O}{\|}}{C}.O\sim\textcircled{P}$$

$$2ADP + P_i$$

The carbamoyl phosphate then reacts with ornithine to start a cycle of reactions resulting in the production of urea (Fig. 9.12).

The aspartate entering the cycle is produced by reaction of glutamate with oxalacetate, the former being produced from α-ketoglutarate plus ammonia released by deamination of an amino acid. The oxalacetate is derived from the fumarate, released in the production of arginine from arginosuccinate, which enters the tricarboxylic acid cycle and is converted to malate and then oxalacetate. We then have a second associated cycle linking the urea and the tricarboxylic acid cycles, which may be visualised as shown in Fig. 9.13.

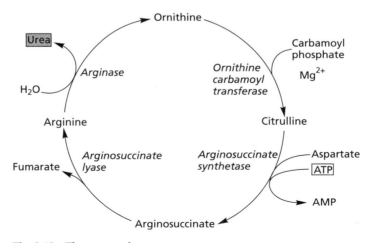

Fig. 9.12 The urea cycle.

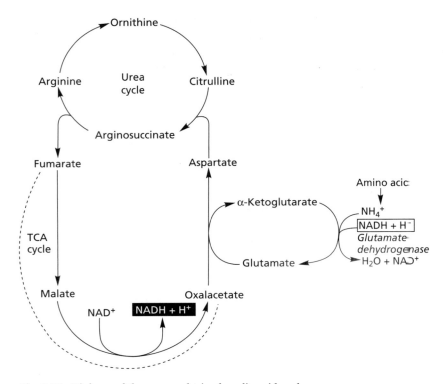

Fig. 9.13 Linkage of the urea and tricarboxylic acid cycles.

In ruminant animals ammonia is absorbed directly from the rumen and undergoes the same series of reactions. Most of the urea is excreted but a certain amount, depending upon the nitrogen status of the animal, is recycled via the saliva and directly across the rumen wall.

In assessing the efficiency of energy production from amino acids, the energy needed for urea synthesis must be set against that obtained by oxidation of the carbon skeleton of the acid. If we take aspartate as an example, this is first converted to oxalacetate and glutamate by reaction with α-ketoglutarate. The oxalacetate is oxidised via the phosphoenolpyruvate pathway and the tricarboxylic acid cycle. The glutamate is deaminated to regenerate α-ketoglutarate and the ammonia released is converted to urea. A balance sheet may be prepared:

	Moles ATP	
	+	−
2 moles aspartate to glutamate + oxalacetate	0	0
2 moles glutamate to α-ketoglutarate + ammonia	6	0
2 moles ammonia to glutamine	0	2
1 mole ammonia to carbamoyl phosphate	0	2

1 mole citrulline to arginosuccinate	0	2
1 mole malate to oxalacetate	3	0
1 mole ammonia to aspartate	0	3
2 moles oxalacetate to carbon dioxide and water	30	0
Total	39	9
Net gain from 2 moles of aspartate		30
Net gain from 1 mole of aspartate		**15**

In birds, excretion of ammonia is by way of uric acid. This involves incorporation of ammonia into glutamine by reaction with glutamate:

$$NH_3^+$$
$$|$$
$$CH.COO^- \quad + \quad NH_4^+ \quad \xrightarrow[\boxed{ATP}]{\textit{Glutamine synthetase}} \quad ADP + P_i \quad \begin{array}{c} NH_3^+ \\ | \\ CH.COO^- \end{array}$$

| Glutamate | | | | Glutamine |

Glutamine is then involved in a series of reactions with ribose-5-phosphate, glycine and aspartate to give inosinic acid, which contains a purine nucleus. This series of reactions may be represented as:

$$2\ Glutamine\ +\ Glycine\ +\ Aspartate \xrightarrow{\boxed{8\ ATP}} Inosinic\ acid\ +\ Fumarate$$

The ribose-5-phosphate residue is then removed giving hypoxanthine, which undergoes two xanthine oxidase mediated oxidations to give xanthine and then uric acid (Fig. 9.14).

Table 9.1 Comparison of the efficiency of certain nutrients as sources of energy as ATP[a]

Nutrient	ΔG (kJ/mole)	Moles ATP/mole nutrient	Moles ATP/100 g nutrient	Heat of combustion/ mole ATP (kJ)
Glucose	2870	38	21.2 (4)	75.5 (1)
Propionic acid	1528	17	22.9 (3)	89.9 (5)
Acetic acid	874	10	16.7 (5)	87.4 (4)
Butyric acid	2184	26	38.5 (2)	84.0 (3)
Aspartic acid	1568	15	11.4 (6)	104.5 (6)
Tripalmitin	32025	409	50.7 (1)	78.3 (2)

[a] Figures in parentheses show rank order.

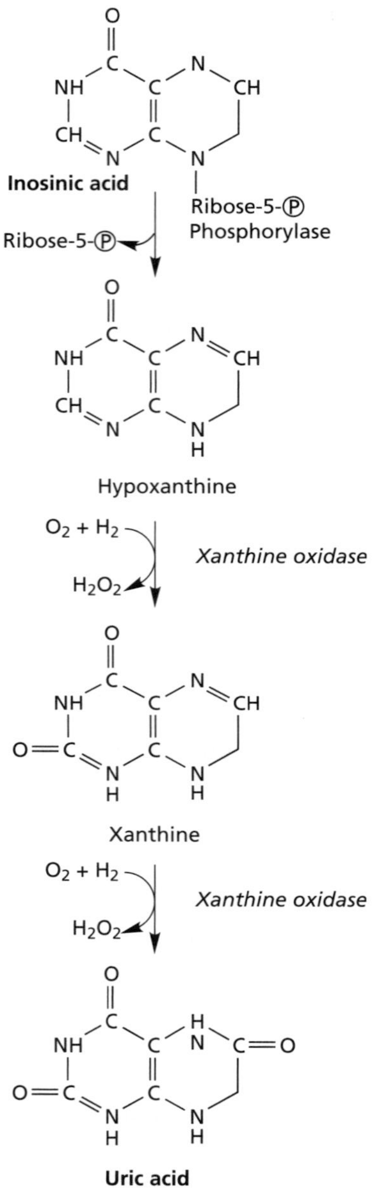

Fig. 9.14 Conversion of inosinic acid to uric acid.

Elimination of two moles of ammonia results in a net loss of 5 moles of ATP and, in addition, 2 moles of glutamate, 1 mole glycine and 1 mole aspartate are used up and 1 mole fumarate is produced.

The efficiency with which the nutrients are used as sources of energy is summarised in Table 9.1.

The unweighted average cost of producing a mole of ATP is here 86.3 MJ. The generally accepted figure is 85.4 MJ.

9.2 Protein synthesis

Proteins are synthesised from amino acids, which become available either as the end-products of digestion or as the result of synthetic processes within the body. Direct amination may take place as in the case of α-ketoglutarate, which yields glutamate:

α-Ketoglutarate Glutamate

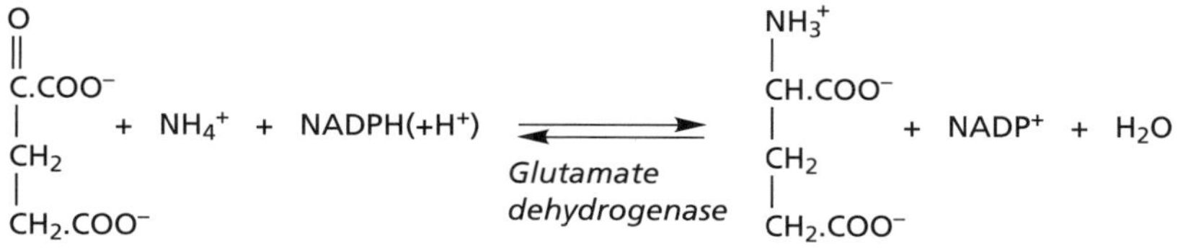

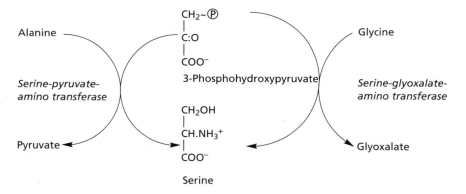

The glutamate may undergo further amination to give glutamine but, more importantly, may undergo transamination reactions with various keto acids to give amino acids as shown in Fig. 9.15.

Amino acids other than glutamate may undergo such transaminations to produce new amino acids. Thus both alanine and glycine react with phosphohydroxypyruvate to give serine:

Serine

Glutamate is the source material of proline, which contains a five-membered ring structure. The synthesis of proline takes place in two stages and requires energy in the form of reduced NAD^+ and $NADP^+$:

Glutamate Δ'-Pyrroline 5-carboxylate Proline

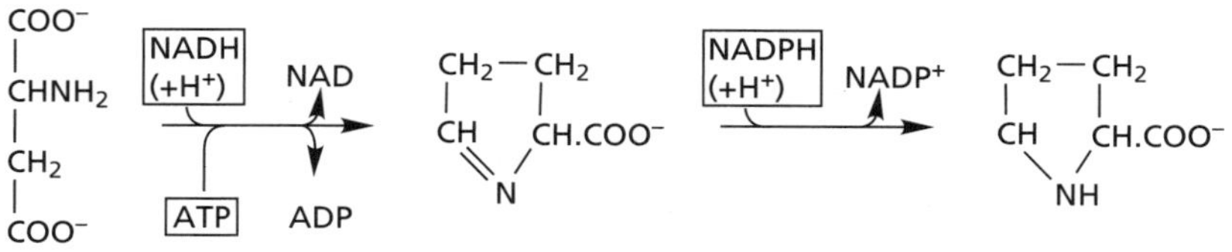

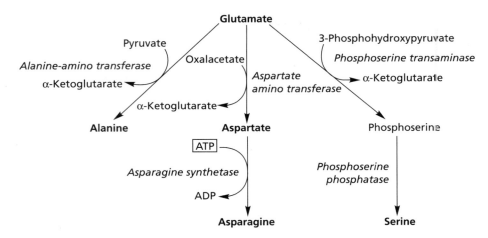

Fig. 9.15 Amino acid synthesis from glutamate.

Amino acids may also be formed by reaction of keto acids with ammonium salts or urea; arginine, as we have already seen, may be synthesised during urea formation.

Not all amino acids are capable of being synthesised in the body and others are not synthesised at sufficient speed to satisfy the needs of the body. Both these groups have to be supplied to the animal. Such amino acids are known as essential or indispensable amino acids (Chapter 4). The words 'essential' and 'indispensable', as used here, do not imply that yet other amino acids are not required for the well-being of the animal but simply that a supply of them in the diet is unnecessary. All the 25 amino acids normally found in the body are physiological essentials; some 10 or 11 are dietary essentials. As would be expected, the actual list of indispensable amino acids differs from species to species. In cattle and sheep, bacterial synthesis of amino acids in the rumen renders the inclusion of any specific amino acids in the diet unnecessary except under conditions of intensive production, as with high-yielding dairy cows or small animals making high weight gains.

The amino acids absorbed from the gut into the bloodstream are transferred into the cells. This requires a supply of energy since the concentration of amino acids in the cell may be up to 100 times that in the blood and transfer into the cell has to take place against a very considerable concentration gradient. A continuous exchange takes place between the blood and cellular amino acids but not between the free amino acids and those of the tissue proteins. The tissue proteins themselves undergo breakdown and resynthesis, their stability varying with different tissues. Thus liver protein has a half-life of seven days whereas collagen is so stable as to be considered almost completely inert.

The process of protein synthesis may be divided conveniently into four stages, these being activation of individual amino acids, initiation of peptide chain formation, chain elongation and chain termination.

Activation

The first step is enzymatic and requires the presence of ATP to give complexes:

Amino acid + $\boxed{\text{ATP}}$ + enzyme ⟶ Amino-acyl-AMP-enzyme +
pyrophosphate

The amino-acyl group is then coupled to a molecule of transfer RNA (tRNA):

Aminoacyl-AMP-enzyme + tRNA ⟶ Amino-acyl-tRNA + AMP + enzyme

Both reactions are catalysed by a single aminoacyl synthetase, Mg^{2+} dependent and specific for the amino acid and the tRNA. The synthetases discriminate between the 20 naturally occurring amino acids but the specificity is not absolute. The tRNA molecule is composed of a strand of nucleotides (Chapter 4), which exhibits considerable folding stabilised by hydrogen bonding. At one end of the chain is a final base arrangement of −C−C−A−OH, i.e. cytidine–cytidine–adenosine. The amino acid is attached to the ribose of the terminal adenosine. The other end of the chain frequently terminates in a guanine nucleotide. We may visualise a typical tRNA molecule as shown in Fig. 9.16.

There is at least one tRNA for each amino acid but only one amino acid for a given tRNA. Since the terminal regions of the various tRNA species are so similar, their specificity must reside within some arrangement in the

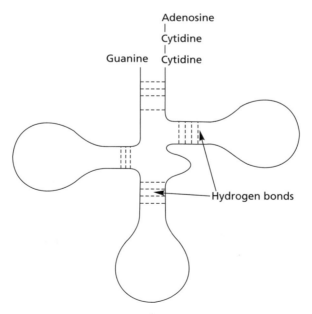

Fig. 9.16 Diagrammatic representation of a tRNA molecule.

Table 9.2 Examples of known codons on mRNA

Codon	Amino acid
UUU	Phenylalanine
UUA	Leucine
UCC	Serine
UCA	Serine
CCC	Proline
CGA	Arginine
AGC	Serine
AGA	Arginine
UGA	Stop

Note: A=adenine, C=cytosine, G=guanine, U=uracil.

interior of the molecules. This arrangement consists of a sequence of three bases, the nature and arrangement of which are specific for a particular amino acid.

When the amino acid has been coupled to the tRNA it is carried to one of the sites of protein synthesis, the ribosomes. These form part of the structures known as polysomes in which several ribosomes are linked by a strand of messenger RNA (mRNA). It is the sequence of bases on this mRNA strand, originally transcribed from the nuclear DNA, that dictates the amino acid sequence in the primary structure of the protein to be synthesised. A particular amino acid will be placed at the mRNA surface at a point having a specific arrangement of three bases, i.e. there is a base triplet code, known as a codon, for each amino acid. The tRNA carrying the specific amino acid for a particular codon will have a complementary arrangement of three bases known as an anti-codon. There are 64 possible base triplet combinations and 61 of these have been shown to code for the 20 amino acids involved in protein synthesis. There is thus more than one codon per amino acid and the code is said to be degenerate. However, any one codon codes for one amino acid only and so, although degenerate, the code is not ambiguous. The codons for individual amino acids have been elucidated and examples are given in Table 9.2.

Initiation of peptide chain formation

The ribosomes of higher animals consist of two subunits designated 40S (Svedberg units) and 60S, according to their sedimentation characteristics in the ultracentrifuge. These combine to form a functional 80S ribosome.

Initiation of peptide formation involves attachment of the smaller subunit to a tRNA and the mRNA. The first tRNA codes for methionine and is placed at an AUG codon (n) at the end of the mRNA chain (Phase 1 of Fig. 9.17).

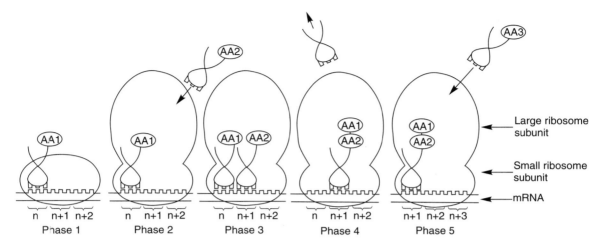

Fig. 9.17 Diagrammatic representation of the sequence of events occuring in the ribosome during the process of polypeptide synthesis.

The larger subunit becomes attached to form the complete ribosome, which is then ready to accept the next tRNA–amino acid complex at codon n + 1 (Phase 2). The insertion of each amino acid residue requires the expenditure of one high-energy phosphate bond, as GTP.

Chain elongation

The requisite amino acid (AA2) is then placed at codon n + 1 by its specific tRNA as shown in Phase 3. A peptide bond is then formed between AA1 (methionine) and AA2, and the tRNA for methionine is simultaneously ejected (Phase 4). The ribosome and the mRNA then move relative to one another and codon n + 1 is placed at the position previously occupied by codon n and codon n+2 moves into that previously occupied by n + 1, as shown in Phase 5. The process is then repeated with AA3 being placed at n + 2, followed by formation of a peptide bond and movement of the ribosome to expose codon n + 3. This continues until the chain is complete; each movement requires the expenditure of one high-energy bond in the form of GTP.

Termination

Chain elongation continues until a codon is reached which does not code for any amino acid, i.e. UAA, UAG or UGA. It then ceases and the formed peptide chain is liberated by hydrolysis, which requires the rupture of one high-energy phosphate bond in GTP. The methionine residue is then removed enzymatically.

The polypeptide is the primary structure of the protein. The chain then becomes arranged in a secondary spiral form stabilised by hydrogen bonding.

The tertiary structure involves extensive coiling and folding of the chain and is stabilised by hydrogen bonding, salt linkages and sulphur bridges. The quaternary structure involves polymerisation of these basic units (Chapter 4).

The mRNA forms a small proportion of the cell's RNA and has but a transient existence. In some microorganisms it may function as a template in synthesis 10–20 times only; in mammalian tissues its active life may be much longer and in some cases it may persist for several days.

The mechanism of protein synthesis discussed above does not involve addition of amino acids to preformed peptides; synthesis starts with an amino acid and a polypeptide chain is synthesised by the successive addition of single amino acids. Unless all the amino acids required to synthesise the peptide are present at the right time, synthesis will not take place and the amino acids which are present are removed and may be catabolised. Considerable wastage of amino acids may thus take place if an incomplete mixture is presented for synthesis.

Energy cost

During protein synthesis, energy is provided by hydrolysis of ATP and GTP, the production of each mole requiring the expenditure of 85.4 kJ by the body. If we make certain assumptions, an estimate of the energetic efficiency of protein synthesis may be made. Let us assume that the average gram molecular weight of amino acids in a given protein is 100. In such a protein the number of amino acids is large, say n, and the number of peptide bonds will be $n - 1$ but may be taken for all practical purposes as n. We may now construct an energy balance sheet as follows:

	Energy expended (kJ)	Energy stored (kJ)
100 g amino acid	2437	
2 moles ATP (activation)	170.8	
1 mole GTP (initiation)	85.4	
1 mole GTP (elongation)	85.4	
1 mole GTP (termination)	85.4	
100 g protein		2437
	2864	2437

Energetic efficiency $= 2437/2864 = 0.85$

NB The calculation assumes synchronous provision of the required amino acids; efficiency is probably much less under normal conditions.

Genetic engineering

Recombinant DNA technology allows the splitting off of a DNA fragment containing a gene of interest and its linkage to a DNA molecule capable of self-replication. This can then be propagated by introduction into a living cell and the characteristic for which the gene encodes conferred upon that cell. Early successes of the technique involved the cloning and expression in *E. coli* of a gene encoding human insulin, and of the gene encoding rat growth hormone in mice. Such genetically modified animals are known as transgenic animals.

The introduction of genes which alter the inherent biochemical pathways in organisms may be of critical significance for the nutrition of animals, since it could confer upon them the ability, which they did not previously possess, of producing essential nutrients. Thus:

1. The amino acid cysteine is essential for wool growth in sheep, which need a source of methionine to synthesise it. Certain bacteria have the ability to synthesise cysteine, the pathway involving the action of two enzymes, serine transacetylase and O-acetylserine sulphhydrylase. The genes encoding the enzymes have been successfully introduced into sheep, which then express the appropriate pathways but, so far, in inappropriate tissues only.
2. Genes for the biosynthesis of the essential amino acids threonine and lysine from aspartate have been successfully introduced into mouse cells as a preliminary to their introduction into the pig genome.
3. Transgenic mice have been produced which show pancreatic cellulase activity. The potential importance of this achievement for improving digestion in monogastric animals is self-evident.
4. Gene transfer has already been used to introduce cellulase activity into hind gut bacteria. If the ability to show cellulase activity under highly acid conditions could be conferred upon rumen microorganisms, it could have a significant effect in ameliorating the detrimental effects of high-level concentrate feeding on fibre digestion and forage intake. The transgenic organisms would of course have to be able to compete with the native rumen flora if they were to be successful in practice.

9.3 Fat synthesis

The glycerides (triacylglycerols) of the depot fat are derived from preformed glycerides or may be synthesised in the body from fattyacyl CoAs and L-glycerol-3-phosphate. This can take place in most tissues but is confined mostly to the liver and adipose tissue.

Synthesis of fattyacyl-CoAs

It is generally considered that there are three systems of fatty acid synthesis. The first, which is highly active, is centred in the cytosol and results, mainly, in the production of palmitate from acetyl-coenzyme A or butyryl-coenzyme A. Nearly all other fatty acids are produced by the modification of this acid.

The second occurs chiefly in the endoplasmic reticulum and to a minor extent in the mitochondria. It involves elongation of fatty acid chains by two-carbon addition, with malonyl CoA as donor. The third system, confined to the endoplasmic reticulum, brings about desaturation of preformed fatty acids.

Cytosolic synthesis of palmitate

In non-ruminant animals, acetyl-coenzyme A is produced in the mitochondria by oxidative degradation of amino and fatty acids, but mainly by oxidative decarboxylation of pyruvate produced from glucose, by glycolysis (p. 206). It must then be transported into the cytosol for the synthesis to take place. However, the mitochondrial membrane is impervious to acetyl-CoA, which has to be complexed with carnitine or changed to citrate in order that it may be transported into the cytosol. Regeneration of acetyl-CoA then takes place, e.g.:

$$\text{Citrate} + \text{Coenzyme A} \xrightarrow[\boxed{\text{ATP}} \quad \text{ADP}]{\textit{ATP-citrate lyase}} \text{Acetyl-CoA} + \text{Oxalacetate}$$

In the ruminant animal, acetate is absorbed directly from the gut and is changed to acetyl-CoA, in the presence of acetyl-CoA synthetase (p. 214), in the cytosol. This is the major source of acetyl-CoA in ruminants, in which ATP-citrate lyase activity is greatly reduced, and passage of mitochondrial acetyl-CoA to the cytosol is limited.

The system is active in liver, kidney, brain, lungs, mammary gland and adipose tissue. The requirements of the system are reduced $NADP^+$, ATP, HCO_3^- as a source of carbon dioxide, and manganese ions. The first stage is the transformation of acetyl-coenzyme A to malonyl-coenzyme A:

$$\begin{array}{c}
\text{CH}_3 \\
| \\
\text{COS.CoA}
\end{array}
+ \boxed{\text{ATP}} + \text{H}_2\text{O} + \text{CO}_2 \xrightarrow[\text{Mn}^{2+}]{\textit{Acetyl-CoA carboxylase}}
\begin{array}{c}
\text{COOH} \\
| \\
\text{CH}_2 \\
| \\
\text{COS.CoA}
\end{array}
+ \text{ADP} + \text{P}_i$$

The malonyl-coenzyme A then reacts with acyl-carrier protein (ACP), in the presence of malonyl-CoA-ACP transacylase, to give the malonyl-ACP complex. Acetyl-coenzyme A is then coupled with ACP in the presence of acetyl-CoA-ACP transacylase and this reacts with the malonyl-ACP, the chain length being increased by two carbon atoms to give the butyryl-ACP complex. The reactions involved are shown in Fig. 9.18.

The butyryl-ACP complex then reacts with malonyl-ACP complex, resulting in further elongation of the chain by two carbon atoms to give caproyl-ACP. Chain elongation takes place by successive reactions of the fattyacyl-ACP complexes with malonyl-coenzyme A until the palmitoyl-ACP complex is produced, when it ceases. Palmitic acid is liberated by the action of a specific deacylase. We may write the overall reaction as:

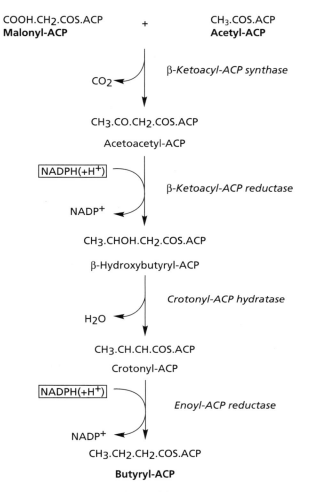

COOH.CH$_2$.COS.ACP　　　+　　　CH$_3$.COS.ACP
Malonyl-ACP　　　　　　　　　　**Acetyl-ACP**

CO$_2$　　　β-*Ketoacyl-ACP synthase*

CH$_3$.CO.CH$_2$.COS.ACP

Acetoacetyl-ACP

NADPH(+H$^+$)　　　β-*Ketoacyl-ACP reductase*

NADP$^+$

CH$_3$.CHOH.CH$_2$.COS.ACP

β-Hydroxybutyryl-ACP

Crotonyl-ACP hydratase

H$_2$O

CH$_3$.CH.CH.COS.ACP

Crotonyl-ACP

NADPH(+H$^+$)　　　*Enoyl-ACP reductase*

NADP$^+$

CH$_3$.CH$_2$.CH$_2$.COS.ACP

Butyryl-ACP

Fig. 9.18 Cytosolic synthesis of fatty acids.

CH$_3$.COS.CoA　+　7COO$^-$.(CH$_2$)COS.CoA　+　14NADPH(+H$^+$)

Acetyl-CoA　　　　Malonyl-CoA

CH$_3$.(CH$_2$)$_{14}$.COO$^-$　+　7CO$_2$　+　14NADP$^+$　+　6H$_2$O　+　8HS.CoA

Palmitate　　　　　　　　　　　　　　　　　　Coenzyme A

In assessing the energy requirements of the process, the energy cost of producing malonyl-coenzyme A from acetyl-coenzyme A must be taken into account and we may write:

$$8Ac\text{-}CoA \; + \; 7ATP \; + \; 14NADPH(+H^+)$$

$$C_{15}H_{31}.COO^- \; + \; NADP^+ \; + \; 7ADP \; + \; 6H_2O \; + \; 7P_i$$

The mammary gland contains deacylases specific for short and medium chain acyl complexes and acids of these chain lengths appear in milk fat.

Chain elongation

This system involves the incorporation of two carbon units into medium and long chain fatty acids and requires ATP and reduced $NADP^+$. The pathway is illustrated in Fig. 9.19.

The preferred substrate is palmitoyl-CoA and in most tissues the product is stearate. Long chain saturated acids with 18, 20, 22 and 24 carbon atoms are synthesised in the brain, where they are required as components of the brain lipids.

A mitochondrial system for elongation of fatty acid chains, using acetyl-CoA as the two-carbon donor, does exist but has limited activity with acyl-CoA substrates with 16 or more carbon atoms and is probably concerned with the lengthening of shorter chains.

Desaturation of preformed fatty acids

Double bonds may be introduced into fatty acid chains by the action of fattyacyl-CoA desaturases present in the endoplasmic reticulum. Thus palmitoleic and oleic acids are produced from the corresponding saturated acids by a Δ^9-desaturase system, steroyl CoA desaturase, which introduces a double bond between carbon atoms 9 and 10:

$$O_2 \qquad H_2O$$

$$CH_3.(CH_2)_6.CH_2.CH_2.(CH_2)_6.COO^- \longrightarrow \qquad \longrightarrow CH_3.(CH_2)_6.CH:CH.(CH_2)_6.COO^-$$

$$\boxed{\begin{array}{c} NADH \\ (+H^+) \end{array}} \qquad NAD^+$$

Palmitate   Palmitoleate

The system is confined to acids with a chain length of 15 or greater.

Mammalian cells also contain Δ^6 and Δ^5 desaturases but do not have systems capable of introducing double bonds beyond carbon atom 9. As a result it is not possible for mammalian tissues to synthesise either linoleic acid ($18:2^{\Delta9,12}$) or α-linolenic acid ($18:3^{\Delta9,12,15}$). These have to be provided in the diet and are referred to as essential fatty acids (EFA). Once they have been

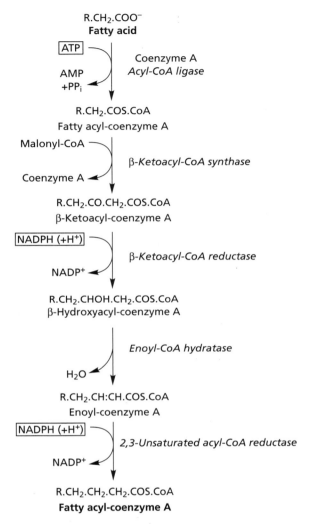

R.CH$_2$.COO$^-$
Fatty acid

ATP

AMP
+PP$_i$

Coenzyme A
Acyl-CoA ligase

R.CH$_2$.COS.CoA
Fatty acyl-coenzyme A

Malonyl-CoA

Coenzyme A

β-*Ketoacyl-CoA synthase*

R.CH$_2$.CO.CH$_2$.COS.CoA
β-Ketoacyl-coenzyme A

NADPH (+H$^+$)

NADP$^+$

β-*Ketoacyl-CoA reductase*

R.CH$_2$.CHOH.CH$_2$.COS.CoA
β-Hydroxyacyl-coenzyme A

Enoyl-CoA hydratase

H$_2$O

R.CH$_2$.CH:CH.COS.CoA
Enoyl-coenzyme A

NADPH (+H$^+$)

NADP$^+$

2,3-Unsaturated acyl-CoA reductase

R.CH$_2$.CH$_2$.CH$_2$.COS.CoA
Fatty acyl-coenzyme A

Fig. 9.19 Elongation of a fatty acid chain.

ingested, a range of acids, including γ-linolenic, arachidonic, eicosapentaenoic and docosahexanoic, can be synthesised from them by successive chain elongation and Δ^6 and/or Δ^5 desaturations (see Fig. 3.1).

Synthesis of L-glycerol-3-phosphate

The usual precursor is dihydroxyacetone phosphate produced by the aldolase reaction of the glycolytic pathway. This is reduced by the NAD-linked glycerol-3-phosphate dehydrogenase:

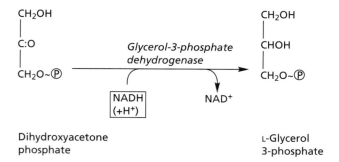

Dihydroxyacetone phosphate → L-Glycerol 3-phosphate (Glycerol-3-phosphate dehydrogenase; NADH (+H⁺) → NAD⁺)

It may also be formed from free glycerol, absorbed from the gut or arising from catalysis of triacylglycerols, in the presence of glycerol kinase. The reaction requires the expenditure of a high-energy bond as ATP. White adipose tissue does not contain significant amounts of glycerol kinase so that most glycerol phosphate is made available by glycolysis.

Synthesis of triacylglycerols

The first stage is the acylation, in the presence of glycerol-3-phosphate acyltransferase, of the free alcohol groups of the glycerol-3-phosphate by two molecules of fattyacyl-CoA to yield a phosphatidic acid:

$R_1.COS.CoA$ $HS.CoA$ $R_2.COS.CoA$ $HS.CoA$

$$
\begin{array}{ccc}
CH_2OH & CH_2.OCO.R_1 & CH_2.OCO.R_1 \\
| & | & | \\
CHOH \longrightarrow & CHOH \longrightarrow & CHOCO.R_2 \\
| & | & | \\
CH_2O{\sim}\textcircled{P} & CH_2O{\sim}\textcircled{P} & CH_2O{-}\textcircled{P}
\end{array}
$$

L-Glycerol-3-phosphate   Monoacylglycerol-3-phosphate   Diacylglycerol-3-phosphate

The reaction occurs preferentially with acids containing 16 and 18 carbon atoms. The phosphatidic acid is then hydrolysed to give a diacylglycerol, which reacts with a third fattyacyl-CoA to give a triacylglycerol:

$$
\begin{array}{ccc}
CH_2OCO.R_1 & CH_2OCO.R_1 & CH_2OCO.R_1 \\
| & | & | \\
CHOCO.R_2 \longrightarrow & CHOCO.R_2 \longrightarrow & CHOCO.R_2 \\
| & | & | \\
CH_2O{\sim}\textcircled{P} & CH_2OH & CH_2OCO.R_3
\end{array}
$$

Phosphatidate phosphatase ($H_2O \to P_i$)   Diacylglycerol acyltransferase ($R_3.COS.CoA \to HS.CoA$)

Diacylglycerol-3-phosphate   Diacylglycerol   Triacylglycerol

Direct synthesis of triacylglycerols from 2-monoacylglycerols, arising from lipid digestion in the intestine, takes place in the intestinal mucosa of higher animals.

Energy cost

The efficiency of fat synthesis may be calculated from the pathways described. The calculation for the synthesis of tripalmitin by the cytoplasmic system would be:

	Energy expended (kJ)	Energy stored (kJ)
8 moles acetate	6996.0	
8 moles acetate to acetyl-CoA	1366.4	
7 moles acetyl-CoA to malonyl-CoA	597.8	
7 additions of malonyl-CoA	3348.3	
Energy for 1 mole of palmitate	12308.5	
Energy for 3 moles of palmitate	36925.5	
$\frac{1}{2}$ mole glucose	1435	
$\frac{1}{2}$ mole glucose to dihydroxyacetone phosphate	85.4	
1 mole dihydroxyacetone phosphate to 1 mole L-glycerol 3-phosphate	256.2	
Energy for 1 mole L-glycerol 3-phosphate	1776.6	
Total energy for 1 mole of tripalmitin	38702.1	
Energy stored in 1 mole of tripalmitin		32025.0
Efficiency of synthesis = 32025/38702.1 = 0.83		

9.4 Carbohydrate synthesis

Glucose

Glucose is important as a source of energy:

- especially in nervous tissue and in erythrocytes,
- as the source of glycerol 3-phosphate for fat synthesis,
- for the maintenance of blood sugar levels,
- for the maintenance of glycogen reserves especially in the liver and muscle,
- as the source of oxalacetate to allow the oxidation of acetyl CoA,
- to allow clearing of lactate and glycerol,
- as the source of other carbohydrates.

In non-ruminant animals, glucose becomes available as the result of the digestion of dietary carbohydrate. When this source is inadequate glucose may be synthesised from a variety of non-carbohydrate sources, mainly lactate, glycerol and glucogenic amino acids (gluconeogenesis). In the ruminant animal, little or no glucose becomes available from dietary sources. The demands for glucose do not, however, differ significantly from those of the non-ruminant. As a result ruminant animals have developed highly efficient glucose conservation and gluconeogenic mechanisms. As in the non-ruminant, glucose may be synthesised from glycerol, lactate and the glucogenic amino acids, but the major source is propionate (about 90 per cent of which is used for glucose synthesis).

Lactate as a source of glucose

The first step in the production of glucose from lactate is its conversion to pyruvate by lactate dehydrogenase:

$$
\begin{array}{ccc}
CH_3 & & CH_3 \\
| & \text{\textit{Lactate dehydrogenase}} & | \\
CHOH & \xrightarrow{\hspace{3cm}} & C{:}O \\
| & \quad NAD^+ \qquad NADH & | \\
COO^- & (+H^+) & COO^- \\
\text{Lactate} & & \text{Pyruvate}
\end{array}
$$

Glucose cannot be produced from pyruvate by a simple reversal of the glycolytic sequence since three of the reactions, namely,

■ the conversion of glucose to glucose 6-phosphate,
■ the conversion of fructose 6-phosphate to fructose 1,6-bisphosphate,
■ the conversion of phosphoenolpyruvate to pyruvate,

are so highly exergonic as to be irreversible under normal cell conditions.

In gluconeogenesis, the conversion of pyruvate to phosphoenolpyruvate is achieved by a two-stage process involving:

■ pyruvate carboxylase, which converts pyruvate to oxalacetate (in non-ruminant animals the enzyme is located in the mitosol and conversion of pyruvate to oxalacetate can only occur in this location),
■ the phosphoenolpyruvate carboxykinase mediated change of oxalacetate to phosphoenolpyruvate.

The latter may take place in the mitosol and the phosphoenolpyruvate then passes into the cytosol. Alternatively oxalacetate may be transferred into the cytosol by means of the glutamate–aspartate shuttle and converted to phosphoenolpyruvate by cytosolic phosphoenolpyruvate carboxykinase. In the ruminant animal, pyruvate carboxylase is located in the cytosol as well as the mitosol, and pyruvate can be changed to phosphoenolpyruvate entirely in the cytosol.

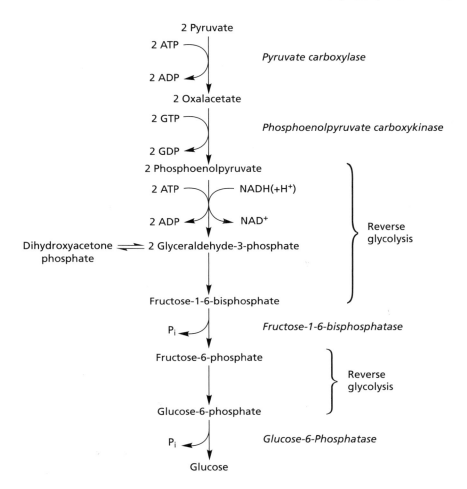

Fig. 9.20 Gluconeogenesis from pyruvate.

In the cytosol, the phosphoenolpyruvate is converted to fructose 1,6-bis-phosphate by the reverse of stages 10–5 of the glycolytic sequence (Fig. 9.4). This is then changed to fructose-6-phosphate by fructose-1,6-bisphosphatase and then to glucose-6-phosphate by the reverse of stage 2. The final change to glucose is achieved by glucose-6-phosphatase. We may then write the whole process as shown in Fig. 9.20.

Amino acids as sources of glucose

Catabolism of amino acids, except for leucine and lysine, results in net synthesis of TCA cycle intermediates or the production of pyruvate (p. 218). Conversion of the latter into glucose then takes place as shown in Fig. 9.20. The TCA cycle intermediates enter the cycle and are converted into malate, which crosses into the cytosol, where it is converted into oxalacetate by malate dehydrogenase:

COO⁻   COO⁻
 | *Malate dehydrogenase* |
CHOH   ─────────────────────────────────►   C:O
 | NAD⁺   NADH (+H⁺) |
CH₂   CH₂
 | |
COO⁻   COO⁻

Malate   Oxalacetate

The oxalacetate enters the glucogenic pathway.

Glycerol as a source of glucose

Glycerol is first phosphorylated, by glycerol kinase, to glycerol 3-phosphate. This is oxidised, by glycerol 3-phosphate dehydrogenase, to dihydroxyacetone phosphate, which enters the glucogenic pathway.

Propionate as a source of glucose

Propionate is first changed to succinyl-CoA (Fig. 9.7). This enters the TCA cycle and is converted to malate and leaves the mitosol as such. Conversion to glucose then takes place via the usual route through oxalacetate and phosphoenolpyruvate.

We may represent the major pathways of gluconeogenesis as shown in Fig. 9.21.

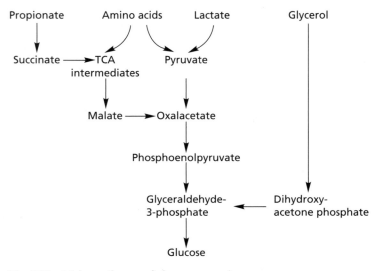

Fig. 9.21 Major pathways of gluconeogenesis.

Table 9.3 Energy cost
of gluconeogenesis

Source	Input	Energy cost (kJ)	Energy retained (kJ)	Efficiency
Aspartate	2 moles + 10ATP	3968	2870	0.72
Propionate	2 moles + 4ATP	3248	2870	0.88
Lactate	2 moles + 6ATP	3246	2870	0.88
Glycerol	2 moles − 4ATP	2982	2870	0.96

Consideration of the energy changes involved in the various metabolic pathways shows that it involves a considerable demand for energy (Table 9.3).

In the non-ruminant animal, amino acids and lactate form the major sources for gluconeogenesis; whereas in the ruminant animal, propionate is pre-eminent. Under normal feeding conditions propionate probably provides about 70 per cent of the glucose requirement of ruminant animals but becomes of lesser importance as the level of feeding falls. Under starvation conditions its contribution approaches zero and glycerol becomes of major significance in gluconeogenesis.

Glycogen synthesis

Glycogen is a complex polysaccharide made up of condensed glucose residues (Chapter 2) and has the ability to add on further glucose units when these are available in the body. The actual source material for glycogen formation is uridine diphosphate glucose (UDPG), which is produced from a variety of sources as shown in Fig. 9.22.

Glycogen is produced by reaction of the uridine diphosphate glucose with primer molecules, the most active of which is glycogen itself. Molecules with as few as four glucose residues may serve as primers, but the reaction rate is slow. As the complexity of the primer increases so does the reaction rate. The synthesis involves reaction of uridine diphosphate glucose with the fourth hydroxyl group of the non-reducing end of the primer chain in the presence of glycogen synthase:

$$\text{UDP-glucose} + (\text{glucose})_n \xrightarrow{\textit{Glycogen synthase}} (\text{Glucose})_{n+1} + \text{UDP}$$

The 1,6 linkages responsible for the branching within the glycogen molecule are formed by transfer of a terminal oligosaccharide fragment of six or seven glucose residues from the end of the glycogen chain to a 6-hydroxyl group in a glucose residue in the interior of the chain. This takes place in the presence of branching enzyme, or more correctly, amylo-(1,4–1,6) transglycosylase.

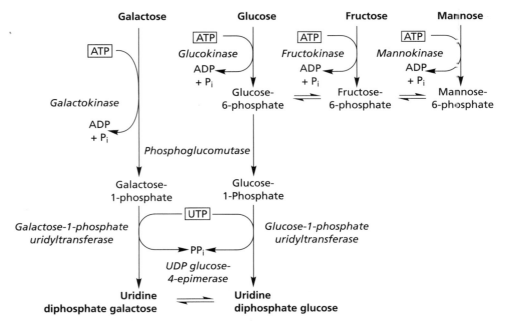

Fig. 9.22 Formation of uridine diphosphate glucose.

Lactose synthesis

Lactose, the sugar of milk, is produced in large quantities in the mammary glands of lactating animals. It is formed by condensation of one glucose and one galactose molecule. A supply of glucose is readily available but the galactose has to be synthesised, virtually in its entirety, from glucose, and this involves a configurational change at carbon atom 4. The glucose is first converted to glucose-1-phosphate and then to uridine diphosphate glucose, from which uridine diphosphate galactose is produced by the action of UDP galactose-4-epimerase, as shown in Fig. 9.23.

Lactose is then formed by the action of the UDP-D-galactose with glucose in the presence of the lactose synthetase system:

$$\text{UDP-galactose} + \text{D-glucose} \xrightarrow{\textit{Lactose synthetase}} \text{UDP} + \text{lactose}$$

The synthetase system is a complex of the enzyme galactosyl transferase with α-lactalbumin. The enzyme catalyses the attachment of galactose to protein containing a carbohydrate residue; α-lactalbumin alters the specificity of the enzyme so that it catalyses the linkage of galactose to glucose. The enzyme is present in the non-lactating gland but is feebly active only. With the onset of lactation, α-lactalbumin is produced in the gland and in its presence the enzyme becomes highly active in catalysing the linkage.

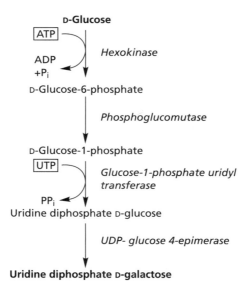

Fig. 9.23 Conversion of glucose to uridine diphosphate galactose.

The energetic efficiency of lactose synthesis may be assessed as follows:

	Energy expended (kJ)	Energy stored (kJ)
2 moles glucose	5606	
2 moles glucose to 2 moles glucose-1-phosphate	170.8	
1 mole glucose-1-phosphate to uridine diphosphate galactose	85.4	
Energy required for 1 mole lactose	5862.2	
Energy retained in 1 mole lactose		5648.4
Energetic efficiency = 5648.4/5862.2 = 0.96		

Estimates of the energetic efficiency of the synthetic processes described, although interesting, should not be given too much weight since their validity depends upon a number of factors including assumptions of complete coupling and ideal conditions, as well as availability of source materials.

9.5 Control of metabolism

An organism must adjust to constantly changing internal and external environments. In order to achieve this, efficient intercellular communication has to be established. Such communication is vested in two distinct but integrated systems. The first is the nervous system with a fixed physical framework; the second is the endocrine system utilising hormones which are transported from the secreting glands to various target tissues. The integration of the two systems is well illustrated by two examples. Thus vasopressin is synthesised in the hypothalamus and is transported along nerve fibres to the pituitary gland, from which it is secreted; on the other hand certain hormones such as insulin and adrenocorticotropic hormone (ACTH) have receptor sites within the brain. Hormones do not initiate processes but exert their control by regulating existing processes. This they do by influencing the rates of synthesis and breakdown of enzymes and by affecting the efficiency of enzyme catalysis and the permeability of cell membranes.

At the cellular level, chemical processes must take place at rates consistent with the needs of the whole organism. This is achieved by control of enzyme activity, which depends upon:

■ The amount of enzyme available, which is the resultant of synthesis and breakdown.

■ The presence of the enzyme in an inactive or proenzyme form Action of the enzyme then depends upon the presence of certain proteolytic agents which expose or create active sites on the proenzyme. Typical are enzymes concerned in digestion and blood clotting.

■ The compartmentalisation of specific processes within the cytosol or organelles of the cell, brought about by the impermeability of membranes to the passage of certain metabolites. For example, fat synthesis takes place in the cytosol whereas fat oxidation is limited to the mitosol, and the futile cycling which would take place were the two processes to occur in the same locus is obviated. The impermeability may be overcome by shuttle systems which require cytoplasmic and organelle forms of the same catalytic activity. This provides a measure of fine control via the availability of substrate.

■ The presence of proteins, which bind to enzymes and inhibit their activity, and of substances which complex with, and render unavailable, metal ions essential for the activity of specific enzymes.

■ The presence of proteins which alter the specificity of the enzyme. The role of α-lactalbumin in the case of galactosyl transferase is a good example. The level of α-lactalbumin is under hormonal control and therefore so is lactose synthesis.

■ The operation of feedback inhibition. This is probably the most common regulatory mechanism operating in metabolism. In this case, the activity of an enzyme is inhibited by the presence of an end-product of the reaction or pathway. In the synthesis of valine from pyruvate, for example, the first step is the formation of acetolactate, catalysed by acetolactate synthetase. The activity of this enzyme and the rate of valine formation are reduced by the presence of valine. A similar situation arises when accumulation of an end-product affects the rate of a reaction or pathway by a simple mass action effect. For example, the rate of glucose breakdown via the Embden–Meyerhof pathway is controlled by the reaction

$$1,3\text{-diphosphoglycerate} + ADP + P_i \rightarrow 3\text{-phosphoglycerate} + ATP$$

When ATP is being consumed rapidly its breakdown ensures a plentiful supply of ADP and phosphoric acid; the reaction thus proceeds rapidly from left to right. If, on the other hand, ATP is not being used, the supply of ADP and inorganic phosphate is reduced and so is speed of the reaction.

Summary

1. Metabolism is the name given to the sequence or succession of chemical processes taking place in the living organism. It involves synthesis (anabolism) and breakdown (catabolism) of complex compounds.

2. Exergonic reactions release energy and endergonic reactions (syntheses) require an input of energy. The link between the two, which allows the energy to be used, is provided by high-energy phosphate compounds, the chief of which is adenosine triphosphate (ATP).

3. ATP is produced from a number of sources by substrate level and oxidative phosphorylation.

4. Glucose is oxidised via the glycolytic pathway and the tricarboxylic cycle.

5. Glycogen is the main form of carbohydrate storage in the animal body. It is first broken down to give glucose, which is metabolised as described.

6. Propionic, butyric and acetic acids are metabolised via the tricarboxylic acid cycle.

7. Fat is the major energy store in the body. Energy is released following a preliminary lipolysis followed by β-oxidation of the fatty acids. The glycerol is first changed to glucose before oxidation.

8. Amino acids may be used as sources of energy when in excess or when other forms of energy are in deficit. The process involves excretion of the nitrogen as urea and is relatively inefficient.

9. Protein synthesis requires activation of amino acids, initiation of chain formation, chain elongation and termination, all of which have an energy cost.

10. Fatty acid synthesis may be cytosolic and produces palmitic acid. Addition of two-carbon units to preformed long and medium chain fatty acids takes place in the microsomes. Desaturation of fatty acids occurs in the endoplasmic reticulum, but mammals are incapable of introducing double bonds beyond carbon atom 9 (see Chapter 3).

11. Lactose is synthesised from glucose and galactose in the mammary gland by a complex enzyme system which is α-lactalbumin-dependent.

12. Glucose is central in a large number of metabolic processes. In the monogastric animal, it is obtained by direct absorption from the small intestine. In ruminant animals this does not take place and highly efficient gluconeogenic and glucose-conserving mechanisms have been developed.

13. On a whole-body basis, metabolism is controlled by an integrated system consisting of a nervous system with a fixed physical structure and an endocrine system secreting hormones which travel to various target tissues. At the cellular level control is exerted by manipulation of the active enzyme supply and by feedback inhibition.

Further reading

Devlin T M (ed.) 1997 *Textbook of Biochemistry with Clinical Correlations*, 4th edn, New York, John Wiley & Sons.

Mathews C K and van Holde K E 1999 *Biochemistry*, 3rd edn, Redwood City, CA, Benjamin Cummings Publishing Co.

Murray R K, Granner D K, Mayes P A and Rodwell V W 1993 *Harper's Biochemistry*, 23rd edn, USA, Appleton and Large.

Stryer L 2000 *Biochemistry*, 4th edn, San Francisco, W H Freeman.

10 Evaluation of foods: digestibility

10.1 Measurement of digestibility

10.2 Special methods for measuring digestibility

10.3 Validity of digestibility coefficients

10.4 Digestion and digestibility in the various sections of the tract

10.5 Factors affecting digestibility

10.6 Alternative measures of the digestibility of foods

10.7 Availability of mineral elements

This chapter marks a change from qualitative to quantitative nutrition. Those chapters preceding it have shown which substances are required by animals, how these substances are supplied in foods and the manner in which they are utilised. This chapter and those immediately following are concerned with the assessment, first, of the quantities in which nutrients are supplied by foods, and second, of the quantities in which they are required by different classes of farm animals.

The potential value of a food for supplying a particular nutrient can be determined by chemical analysis, but the actual value of the food to the animal can be arrived at only after making allowances for the inevitable losses that occur during digestion, absorption and metabolism. The first tax imposed on a food is that represented by the part of it which is not absorbed and is excreted in the faeces.

The digestibility of a food is most accurately defined as that proportion which is not excreted in the faeces and which is, therefore, assumed to be absorbed by the animal. It is commonly expressed in terms of dry matter and as a coefficient or a percentage. For example, if a cow ate 9 kg of hay containing 8 kg of dry matter and excreted 3 kg of dry matter in its faeces, the digestibility of the hay dry matter would be:

$$\frac{8-3}{8} = 0.625 \quad \text{or} \quad \frac{8-3}{8} \times 100 = 62.5 \text{ per cent}$$

Coefficients could be calculated in the same way for each constituent of the dry matter. Although the proportion of the food not excreted in the faeces is commonly assumed to be equal to that which is absorbed from the digestive tract, there are objections to this assumption, which will be discussed later.

10.1　Measurement of digestibility

In a digestibility trial, the food under investigation is given to the animal in known amounts and the output of faeces measured. More than one animal is used, first because animals, even when of the same species, age and sex, differ slightly in their digestive ability, and second because replication allows more opportunity for detecting errors of measurement.

In trials with mammals, male animals are preferred to females because it is easier to collect faeces and urine separately with the male. They should be docile and in good health. Small animals are confined in metabolism cages, which separate faeces and urine by an arrangement of sieves, but larger animals such as cattle are fitted with harness and faeces collection bags made of rubber or of a similar impervious material. For females a special device channels faeces into the bag while diverting urine. Similar equipment can be used for sheep.

For poultry, the determination of digestibility is complicated by the fact that faeces and urine are voided from a single orifice, the cloaca. The compounds present in urine are mainly nitrogenous, and faeces and urine can be separated chemically if the nitrogenous compounds of urine can be separated from those of faeces. The separation is based either on the fact that most urine nitrogen is in the form of uric acid or that most faecal nitrogen is present as true protein. It is also possible to alter the fowl's anatomy by surgery so that faeces and urine are voided separately.

The food required for the trial should if possible be thoroughly mixed beforehand, to obtain uniform composition. It is then given to the animals for at least a week before collection of faeces begins (and preferably two weeks with ruminants) in order to accustom the animals to the diet and to clear from the tract the residues of previous foods. This preliminary period is followed by a period when food intake and faecal output are recorded. This experimental period is usually 5–14 days long, with longer periods being desirable as they give greater accuracy. With simple-stomached animals the faeces resulting from a particular input of food can be identified by adding an indigestible coloured substance such as ferric oxide or carmine to the first and last meals of the experimental period; the beginning and the end of faecal collection are then delayed until the dye appears in and disappears from the excreta. With ruminants this method is not successful because the dyed meal mixes with others in the rumen; instead an arbitrary time-lag of 24–48 hours is normally allowed for the passage of food residues, i.e. the measurement of faecal output begins 1–2 days after that of food intake, and continues for the same period after measurement of food intake has ended.

In all digestibility trials, and particularly those with ruminants, it is highly desirable that meals should be given at the same time each day and that the amounts of food eaten should not vary from day to day. When intake is irregular there is the possibility, for example, that if the last meal of the experimental period is unusually large the subsequent increase in faecal output may be delayed until after the end of faecal collection. In this situation the output of faeces resulting from the measured intake of food will be

Table 10.1 Results of a digestibility trial in which three sheep were fed on hay

	Dry matter (DM)	Organic matter	Crude protein	Ether extract	Acid-detergent fibre
(1) Analyses (g/kg DM)					
Hay	–	919	93	15	350
Faeces	–	870	110	15	317
(2) Nutrients (kg/day)					
Consumed	1.63	1.50	0.151	0.024	0.57
Excreted	0.76	0.66	0.084	0.011	0.24
Digested	0.87	0.84	0.067	0.013	0.33
(3) Digestibility coefficients	0.534	0.560	0.444	0.541	0.579
(4) Digestible nutrients (g/kg DM)	–	515	41	8	203

underestimated and digestibility overestimated. The trial is completed by analysing samples of the food used and the faeces collected.

Table 10.1 gives an example of a digestibility trial in which sheep were fed on hay for a preliminary period of 14 days followed by an experimental period of 10 days. The results are the means for three animals. The calculations may be summarised as follows:

1. The average quantity of hay dry matter consumed was 1.63 kg per head per day, and the average quantity of dry matter excreted in the faeces was 0.76 kg per head per day. Hay and faeces were analysed, with the results shown in Section 1 of Table 10.1.
2. From these analytical data and the quantities of dry matter consumed and excreted, the quantities of individual nutrients consumed, excreted and – by difference – digested were calculated, as shown in Section 2.
3. Digestibility coefficients were calculated by expressing the weights of nutrients digested as proportions of the weights consumed (e.g. 0.87/1.63 = 0.534); see Section 3.
4. Finally, the composition of the hay was calculated in terms of digestible nutrients (e.g. $919 \times 0.560 = 515$); see Section 4.

The general formula for the calculation of digestibility coefficients, which combines stages 3 and 4 above, is:

$$\frac{\text{Nutrient consumed} - \text{Nutrient in faeces}}{\text{Nutrient consumed}}$$

In this example the food in question was roughage and could be given to the animals as the sole item of diet. Concentrated foods, however, may cause digestive disturbances if given alone to ruminants, and their digestibility is often determined by giving them in combination with a roughage of known digestibility. Thus the hay in the example of Table 10.1 could have been used in a second trial in which the sheep also received 0.50 kg oats per day. If the dry matter content of the oats was 900 g/kg, daily dry matter intake would increase by 0.45 kg. If faecal output increased from 0.76 to 0.91 kg per day the digestibility of the dry matter in oats would be calculated as:

$$\frac{0.45-(0.91-0.76)}{0.45}=\frac{0.45-0.15}{0.45}=0.667$$

In this example the hay is designated as the basal food and the oats as the test food. The general formula for calculating the digestibility of the test food is:

$$\frac{\text{Nutrient in test food} - (\text{Nutrient in faeces} - \text{Nutrient in faeces from basal food})}{\text{Nutrient in test food}}$$

10.2 Special methods for measuring digestibility

Indicator methods

In some circumstances the lack of suitable equipment or the particular nature of the trial may make it impracticable to measure directly either food intake or faeces output, or both. For instance, when animals are fed as a group it is impossible to measure the intake of each individual. Digestibility can still be measured, however, if the food contains some substance which is known to be completely indigestible. If the concentrations of this indicator substance in the food and in small samples of the faeces of each animal are then determined, the ratio between these concentrations gives an estimate of digestibility. For example, if the concentration of the indicator increased from 10 g/kg in the food dry matter to 20 g/kg in the faeces, this would mean that half of the dry matter had been digested and absorbed. In equation form:

$$\text{Digestibility of DM} = \frac{\text{g indicator/kg faeces DM} - \text{g indicator/kg food DM}}{\text{g indicator/kg faeces DM}}$$

The indicator may be a natural constituent of the food or be a chemical mixed into it. It is difficult to mix chemicals with foods like hay, but an indigestible constituent such as lignin may be used. Other indicators in use today are fractions of the food known as indigestible acid-detergent fibre and acid-insoluble ash (the latter being mainly silica), and also some naturally occurring *n*-alkanes of long chain length (C_{25}–C_{35}). The indicator most commonly added to foods is chromium in the form of chromic oxide, Cr_2O_3. Chromic oxide is very insoluble and hence indigestible; moreover, chromium is unlikely to be present as a major natural constituent of foods. For non-ruminants, titanium dioxide may be added to foods as an indicator.

Chromic oxide may be used as an indicator in a different way, to estimate faeces output rather than digestibility. In this application the marker is given for 10–15 days in fixed amounts (e.g. administered in a gelatin capsule) and once its excretion is assumed to have stabilised its concentration in faeces samples is determined. Faeces dry matter output (kg/day) is calculated as follows:

Marker dose (g per day) /Marker concentration in faeces DM (g per kg)

For example, if an animal was given 10 g of chromic oxide per day and the concentration of the marker was found to be 4 g/kg faeces DM, faeces output would be calculated as 10/4 = 2.5 kg DM/day. If food intake was known, dry matter digestibility could be calculated in the usual way.

Measuring the digestibility of herbage eaten by grazing animals presents a special problem. In theory, natural herbage constituents like lignin can be employed as indicators. In practice, this application of indicator methods is complicated by the difficulty of obtaining samples of the food (i.e. pasture herbage) which are truly representative of that consumed. Animals graze selectively, preferring young plant to old, and leaf to stem, and a sample of the sward cut with a mower is therefore likely to contain a higher concentration of fibrous constituents (including lignin) than the herbage harvested by the animal. One way of obtaining representative samples is to use an animal with an oesophageal fistula (an opening from the lumen of the oesophagus to the skin surface). When this is closed by a plug, food passes normally between mouth and stomach; when the plug is temporarily removed, herbage consumed can be collected in a bag hung below the fistula. Samples of grazed herbage obtained in this way can then be analysed, together with samples of faeces, for the indicator.

Laboratory methods of estimating digestibility

Since digestibility trials are laborious to perform, there have been numerous attempts made to determine the digestibility of foods by reproducing in the laboratory the reactions which take place in the alimentary tract of the animal. Digestion in non-ruminants is not easily simulated in its entirety, but the digestibility of food protein may be determined from its susceptibility to attack *in vitro* by pepsin and hydrochloric acid. It is also possible to collect digestive tract secretions via cannulae and to use them to digest foods *in vitro*.

The digestibility of foods for ruminants can be measured quite accurately in the laboratory by treating them first with rumen liquor and then with pepsin. During the first stage of this so-called 'two-stage *in vitro*' method a finely ground sample of the food is incubated for 48 hours with buffered rumen liquor in a tube under anaerobic conditions. In the second stage, the bacteria are killed by acidifying with hydrochloric acid to pH 2 and are then digested (together with some undigested food protein) by incubating them with pepsin for a further 48 hours. The insoluble residue is filtered off, dried and ignited, and its organic matter subtracted from that present in the food to provide an estimate of digestible organic matter. Digestibility determined *in vitro* is generally slightly lower than that determined *in vivo*, and corrective equations are required to relate one measure to the other; an example is illustrated in Fig. 10.1*a*.

Until superseded by other methods (see below) this technique was used routinely in the analysis of farm roughages for advisory purposes and for determining the digestibility of small samples such as those available to the plant breeder. A further application is found in estimating the digestibility of grazed pasture herbage when this is collected from an animal with an oesophageal fistula as described above.

The rumen liquor used in the first stage of this laboratory procedure may vary in its fermentative characteristics according to the diet of the animal from which it is obtained. In an attempt to obtain more repeatable estimates of digestibility, rumen liquor is sometimes replaced by fungal cellulase preparations. Figure 10.1*b* shows how incubation with pepsin followed by incubation with cellulase can be used to estimate the digestibility of forages for

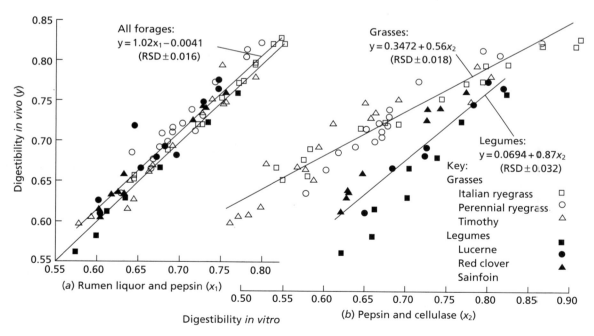

All forages:
$y = 1.02x_1 - 0.0041$
(RSD ± 0.016)

Grasses:
$y = 0.3472 + 0.56x_2$
(RSD ± 0.018)

Legumes:
$y = 0.0694 + 0.87x_2$
(RSD ± 0.032)

Digestibility *in vivo* (y)

(a) Rumen liquor and pepsin (x_1)

Digestibility *in vitro*

(b) Pepsin and cellulase (x_2)

Key:
Grasses
Italian ryegrass □
Perennial ryegrass ○
Timothy △
Legumes
Lucerne ■
Red clover ●
Sainfoin ▲

Fig. 10.1 Laboratory methods for estimating the dry matter digestibility of forages. (a) Incubation in rumen liquor followed by digestion with pepsin. (b) Digestion with pepsin followed by digestion with cellulase. (Adapted from Terry R A *et al.* 1978 *Journal of the British Grassland Society* **22**: 13.)

sheep. The relationship in Fig. 10.1*b*, however, is less close than that found for the same forages between digestibility estimated by fermentation with rumen liquor and that determined in sheep (Fig. 10.1*a*). The pepsin–cellulase method required separate prediction equations for grasses and legumes, and the residual standard deviations for these equations were larger than that of the single equation used for the rumen liquor–pepsin method.

Rumen liquor is also used in another method for the laboratory assessment of the digestibility of the foods of ruminants, but in this method the quantity or proportion of food digested is estimated indirectly from the volume of gas produced by the fermentation. Gas production in the rumen, and hence in the test tube, is proportional to the quantity of food fermented. About half of the gas is carbon dioxide arising from the neutralisation of acids by buffers, and the rest is a mixture of methane and carbon dioxide arising from the fermentation of carbohydrates and proteins to volatile fatty acids. The advantage of this method over the laboratory methods discussed above is that it can be readily applied to large numbers of food samples, especially if gas production is recorded automatically. One disadvantage is that gas production reflects only one aspect of rumen fermentation, namely volatile fatty acid production, and not the synthesis of microbial biomass. Thus gas production measurements need to be related to the quantity of nutrients remaining after fermentation.

The fistulated animals used to provide rumen liquor for the estimation of digestibility *in vitro* may also be used in a different way to assess rapidly the digestibility of small samples of food. The food samples (3–5 g of dry matter) are placed in small bags made of permeable synthetic fabric of standard pore size (400–1600 μm^2), which are inserted into the rumen through the cannula and incubated there for 24–48 hours. After each bag is withdrawn it is washed and dried to determine the quantity of food dry matter remaining as undigested material. This technique is known as the determination of digestibility *in sacco* (see Fig. 10.2). There are problems in its use, arising particularly from the need to select an appropriate period of incubation.

Laboratory methods of estimating digestibility can also be used to study the time course of digestion in the rumen. For example, Fig. 10.2 suggests rapid digestion at first, followed by a slower phase. This is consistent with the rapid breakdown of easily solubilised nutrients, such as sugars, followed by the slower degradation of cellulose and other complex carbohydrates. The time course of digestion can be described by fitting equations to data of the type shown in Fig. 10.2; this procedure is explained in Chapter 13.

Over the last decade there has been a change in the routine evaluation of foods for advisory purposes (e.g. evaluating a silage sample from a farm) from methods requiring rumen liqour obtained from fistulated animals to those requiring only laboratory apparatus. In Britain the preferred technique is near-infrared reflectance spectroscopy (see p. 12), which is claimed to provide more accurate predictions of digestibility than the 'two-stage *in vitro*' method when applied to defined groups of feeds, such as grass silages. This method is quickly executed and requires minimal sample preparation, but the equipment required is expensive, and the chemical nature of the food constituents contributing to the near-infrared spectrum has yet to be established.

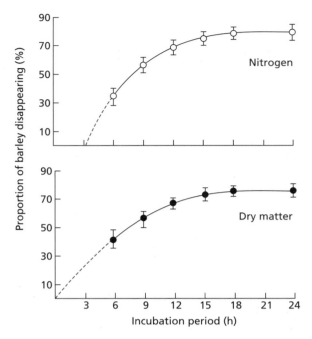

Fig. 10.2 Disappearance of dry matter and nitrogen from samples of barley incubated in artificial fibre bags in the rumen of sheep. Vertical lines indicate variation between replicate samples, expressed as standard deviations. (Adapted from Mehrez A Z and Ørskov E R 1977 *Journal of Agriculture Science, Cambridge* **88**: 645.)

10.3 Validity of digestibility coefficients

The assumption that the proportion of food digested and absorbed can be determined by subtracting the part excreted in the faeces is open to question on two counts. The first is that in ruminants the methane arising from the fermentation of carbohydrates is lost by eructation, and not absorbed. This loss leads to overestimation of the digestible carbohydrate and digestible energy content of ruminant foods.

More serious errors are introduced by the fact that, as discussed in Chapter 8, not all the faeces are actual undigested food residues. Part of the faecal material is contributed by enzymes and other substances secreted into the gut and not reabsorbed, and by cellular material abraded from the lining of the gut. Thus if a pig, for example, is fed on a nitrogen-free diet it continues to excrete nitrogen in the faeces. Since this nitrogen is derived from the body and not directly from the food, it is known as the *metabolic faecal nitrogen*; the amounts in which it is excreted are approximately proportional to the animal's dry matter intake. Faeces also contain appreciable quantities of ether-extractable substances and minerals which are of metabolic origin. Some of the ash fraction of faeces is contributed by mineral elements secreted into the gut, because the faeces serve as the route of true excretion of certain minerals, particularly calcium.

The excretion in faeces of substances not arising directly from the food leads to underestimation of the proportion of the food actually absorbed by the animal. The values obtained in digestibility trials are therefore called *apparent digestibility* coefficients to distinguish them from the coefficients of *true digestibility*. In practice, true digestibility coefficients are difficult to determine, because the fractions of the faeces attributable to the food and to the animal are in most cases indistinguishable from one another. Apparent coefficients of digestibility of organic constituents of foods are satisfactory for many purposes, and they do represent the net result of the ingestion of food. Apparent coefficients for some mineral elements, however, may be quite meaningless (see p. 259).

10.4 Digestion and digestibility in the various sections of the tract

As explained in Chapter 8, nutrients may be absorbed from several parts of the digestive tract. Even in non-ruminants, absorption occurs in two distinctly different parts, the small and large intestines, and in ruminants volatile fatty acids are absorbed from the rumen. A food constituent that is digested (and absorbed) at one site may give rise to nutrients differing quite considerably from those resulting from its digestion at another site, and the nutritive value to the animal of that constituent will therefore depend not only on the extent to which it is digested (i.e. its digestibility) but also on the site of digestion. For example, a carbohydrate such as starch may be fermented in the rumen to volatile fatty acids (and methane) or digested in the small intestine to glucose.

The digestibility of foods in successive sections of the digestive tract is most conveniently measured by the use of cannulated animals, prepared as described on p. 165. An example of the use of cannulated animals to measure digestion in the successive portions of the digestive tract of sheep is shown in Table 10.2. The sheep were cannulated at the duodenum and terminal ileum, thus allowing digestion to be partitioned between the stomach, the small intestine and the large intestine. The overall digestibility of the organic matter of the pelleted grass (0.78) was considerably lower than that

Table 10.2 Digestion of chopped or ground and pelleted dried grass in successive portions of the alimentary tract of sheep (After Beever D E, Coelho da Silva J P, Prescott J H D and Armstrong D G 1972 *British Journal of Nutrition* **28**: 347)

Food constituent:	Organic matter		Cellulose	
Form of grass:	Chopped	Pelleted	Chopped	Pelleted
Proportion digested in:				
Stomach	0.52	0.45	0.80	0.56
Small intestine	0.27	0.20	0.02	−0.02
Large intestine	0.04	0.13	0.05	0.23
Overall	0.83	0.78	0.87	0.77

of the chopped grass (0.83), microbial digestion of the former being partially transferred from the stomach (i.e. the rumen) to the large intestine. These differences were even more apparent for cellulose digestion, to which the small intestine made a negligible contribution.

In Chapter 8 reference was made to the degradation of dietary protein in the rumen. The nitrogen of food protein entering the rumen may leave it in the same form if the protein escapes degradation, or, if the protein is degraded, the nitrogen may leave the rumen either as microbial protein, if it is re-utilised, or as ammonia if it is not. The fate of dietary protein (or nitrogen) in the ruminant may be determined by collecting digesta from successive portions of the digestive tract. Table 10.3 contrasts the digestion of the nitrogen in two types of ryegrass. Although the two grasses were similar in total nitrogen content (line 1), considerably more nitrogen was 'lost' from the stomach – presumably by absorption of ammonia – with the first grass (line 1–2). Conversely, with this grass less nitrogen was absorbed in the small intestine, expressed either as total nitrogen (line 2–3), protein (line 9) or amino acids (line 10). The difference between total nitrogen and protein nitrogen absorption reflects absorption of ammonia reaching the intestine, and that between protein and amino acid nitrogen would be largely due to the nucleic acid nitrogen of microbial protein. A further loss of nitrogen, again presumably as ammonia, occurred in the large intestine. The net outcome was that although the short-rotation ryegrass contained slightly less nitrogen, of slightly lower overall digestibility, it provided the sheep with about 25 per cent more in terms of absorbed amino acids than did the perennial ryegrass. With the latter, of the total nitrogen apparently absorbed (32 g/day), less than half (14.6 g/day) was in the form of amino acids.

Table 10.3 Protein digestion and absorption by sheep given 800 g organic matter per day from one of two types of ryegrass (Adapted from MacRae J C and Ulyatt M J 1974 *Journal of Agricultural Science, Cambridge* **82**: 309)

			Perennial ryegrass	Short-rotation ryegrass
Total N (g/day)	(1)	In feed	37.8	34.9
	(2)	At duodenum	27.8	31.7
	(3)	At terminal ileum	9.0	9.3
	(4)	In faeces	5.8	6.7
Proportion of feed N digested	(5)	In stomach	0.26	0.09
	(6)	In small intestine	0.50	0.64
	(7)	In large intestine	0.08	0.07
	(8)	Overall	0.84	0.80
Protein N absorbed (g/day)[a]	(9)	In small intestine	15.0	19.1
Amino acid N absorbed (g/day)	(10)	In small intestine	14.6	18.3

[a] Protein calculated as 6.25 (non-ammonia nitrogen).

Another technique for studying the digestion of defined portions of food in sections of the tract involves the use of small bags similar to those employed in studies of digestion in the rumen (p. 251). For this so-called 'mobile nylon bag technique' small samples of food (0.5–1.0 g) contained in bags are inserted into the gut via a cannula (e.g. into the duodenum) and later recovered via a second cannula (e.g. at the ileo-caecal junction). The loss of nutrients between the two sites is taken to be the portion digested and absorbed. This technique is commonly used in pigs; in this species, digestibility to the end of the small intestine (sometimes called 'ileal digestibility') is reckoned to give a more accurate measure of the nutritive value of a food than would digestibility in the whole tract. For example, suppose that with a food of high nutritive value all the lysine was released from proteins and absorbed by the pig before the digesta reached the end of the ileum (i.e. ileal digestibility of lysine = 100 per cent). However, microorganisms in the caecum and colon would synthesise more lysine, which would be excreted in microbial protein in faeces and would thus reduce the apparent digestibility of lysine in the whole tract to less than 100 per cent. The mobile nylon bag technique is also used in horses, in which the bags may be introduced into the stomach via a naso-gastric tube.

10.5 Factors affecting digestibility

Food composition

The digestibility of a food is closely related to its chemical composition, and a food like barley, which varies little in composition from one sample to another, will show little variation in digestibility. Other foods, particularly fresh or conserved herbages, are much less constant in composition and therefore vary more in digestibility. The fibre fraction of a food has the greatest influence on its digestibility, and both the amount and chemical composition of the fibre are important.

Modern methods of food analysis attempt to distinguish fractions of cell walls and cell contents. When forages are heated with neutral detergent solution, the cell contents dissolve and the cell walls remain as a residue called neutral-detergent fibre (see p. 6). The cell wall fraction may be further divided into hemicellulose (extractable with acid detergent solution) and cellulose plus lignin (i.e. acid-detergent fibre). Cell contents are almost completely digested (i.e. true digestibility = 100 per cent) although their apparent digestibility will be 10–15 per cent lower because of the excretion of metabolic products of the gut (see p. 252). The digestibility of cell walls is much more variable and depends on the degree of lignification, which in chemical terms is expressed as the lignin content of acid-detergent fibre. But cell wall digestibility also depends on the structure of plant tissues. Thus tropical grasses are generally less digestible than their temperate counterparts because their leaves contain more vascular bundles, hence more lignin, and because they have dense masses of cells that resist invasion by microorganisms.

The digestibility of foods may be reduced by deficiencies or excesses of nutrients or other constituents. Such effects are most commonly found in ruminants, in which, for example, deficiencies in rumen liquor of ammonia nitrogen or of sulphur will restrict microbial growth and thus reduce fibre digestibility. A dietary excess of lipids will also inhibit rumen microorganisms. The high silica content of some foods, particularly rice straw, reduces their digestibility. In foods for non-ruminants, constituents that bind to proteins and amino acids, such as tannins, will reduce their digestibility.

Ration composition

The digestibility of a food is influenced not only by its own composition but also by the composition of other foods consumed with it. Thus one might find that if a roughage (DM digestibility 0.6) and a concentrate (0.8) were given as equal parts of a mixed ration, the digestibility of the complete ration might differ from the expected value of 0.7. This *associative effect of foods* represents a serious objection to the determination of the digestibility of concentrates by difference as described on p. 248. Associative effects are usually negative (i.e. the digestibility of mixed rations is less than expected) and greatest when a low-quality roughage is supplemented with a starchy concentrate. In these circumstances, rapid fermentation of starch to volatile fatty acids depresses the rumen pH to 6 or less. The low pH inhibits cellulolytic microorganisms and fibre digestibility is depressed. The depression can be partly prevented by adding a buffering agent such as sodium bicarbonate to the diet, but in addition to its 'pH effect' starch seems to have a direct effect on cellulolysis.

Preparation of foods

The commonest treatments applied to foods are chopping, chaffing, crushing or grinding, and cooking. In order to obtain maximum digestibility cereal grains should be crushed for cattle and ground for pigs, or else they may pass through the gut intact. Sheep can often be relied upon to chew whole cereal grains, thereby obviating mechanical processing, but if grains are given with a roughage that passes rapidly through the gut, such as grass silage, they should be crushed for sheep (see Chapter 23).

Roughages are subjected to several processes of comminution: the mildest process, *chaffing*, has little direct effect on their digestibility but may reduce it indirectly by preventing the selection by animals of the more digestible components. The *wafering* of roughages, a process involving their compression into round or square section blocks, also has little effect on their digestibility. The most severe process, *fine grinding* (often followed by *pelleting*), has a marked effect on the manner in which roughages are digested and hence on their digestibility. Ground roughages pass through the rumen faster than long or chopped materials and their fibrous components may be less completely fermented (see Table 10.2). The grinding of roughages therefore reduces the digestibility of their crude fibre by as much as 20 percentage units, and of the dry matter as a whole by 5–15 percentage units. The depression is often greatest for roughages intrinsically low in digestibility, and may be exaggerated by the effect of level of feeding on digestibility since grinding increases the acceptability of roughages to ruminants (see Chapter 17).

Roughages such as the cereal straws, in which the cellulose is mixed or bound with a high proportion of lignin, may be treated chemically to separate the two components. The treatment processes and their effects are described in detail in Chapter 21. The chemicals used are mainly alkalis (sodium and ammonium hydroxides) and they improve the dry matter digestibility of cereal straws quite dramatically, from 0.4 to 0.5–0.7.

Foods are sometimes subjected to *heat treatment* to improve their digestibility. A traditional form of heat treatment is the boiling of potatoes for pigs, but heat may also be applied to other foods as steam or by microwave irradiation (a process known as 'micronisation'). Applied to the cereals, such treatments cause relatively small increases in digestibility, although sorghum appears to be more responsive than other grains. Heat treatments are most effective in improving digestibility when used for the specific purpose of inactivating the digestive enzyme inhibitors that are present in some feeds. The best examples of these inhibitors are found in protein concentrates (see Chapter 24). Potatoes, and root crops such as swedes (*Brassica napus*), possess trypsin inhibitors which are inactivated by heat treatment; the benefit of such treatment is not so much the improvement in digestibility as the transfer of protein digestion from fermentation in the caecum to normal enzyme digestion in the small intestine.

Enzyme supplementation of foods

As non-ruminants are ill-equipped to break down many constituents of foods, enzyme preparations (usually of fungal origin) may be added to foods in the hope of improving their digestibility. The most consistently successful enzyme addition has been that of β-glucanase to barley in diets for poultry. If β-glucans escape digestion they appear in the excreta in the form of gels that cause undesirable 'sticky droppings'; the glucans also interfere with the digestion of other dietary constituents, so their enzymic destruction causes a general improvement in digestibility. Further examples of the use of enzyme supplements are described in Chapter 25.

Animal factors

Digestibility is more a property of the food than of the animal consuming it, but this is not to say that a food given to different animals will also be digested to the same extent. The most important animal factor is the species of the consumer. Foods low in fibre are equally well digested by ruminants and non-ruminants, but foods high in fibre are better digested by ruminants. Apparent digestibility coefficients for protein are frequently higher for pigs because their excretion of metabolic faecal nitrogen is smaller than that of ruminants. Differences in digestive ability between sheep and cattle are small and of little importance with most diets; however, highly digestible foods – such as cereal grains – tend to be more efficiently digested by sheep, and poorly digested foods – such as low-quality roughages – tend to be better digested by cattle.

Level of feeding

An increase in the quantity of a food eaten by an animal generally causes a faster rate of passage of digesta. The food is then exposed to the action of digestive enzymes for a shorter period, and there may be a reduction in its digestibility. As one would expect, reductions in digestibility due to increased rates of passage are greatest for the slowly digested components of foods, namely the cell-wall components.

Level of feeding is often expressed in multiples of the quantity of food required by the animal for what is called body maintenance (i.e. the quantity which prevents the animal from losing weight but does not allow growth; see Chapter 14). This level is defined as unity; in ruminants fed to appetite, the level can rise to 2.0–2.5 times the maintenance level in growing or fattening animals and to 3.0–5.0 times maintenance in lactating animals. Increasing the level of feeding by one unit (e.g. from maintenance to twice that level) reduces the digestibility of roughages such as hay, silage and grazed pasture by only a small proportion (0.01–0.02). With diets for ruminants containing finer food particles (e.g. mixtures of roughages and concentrates) the reduction in digestibility per unit increase in feeding level is greater (0.02–0.03); this would mean, for example, that the digestibility of the dry matter in a typical dairy cow diet might fall from 0.75 at the maintenance level of feeding to 0.70 at 3.0 times the maintenance level. Falls of this magnitude may be due to exaggerated associative effects of foods, and there is some evidence that they can be prevented by raising the protein content of the diet. The greatest reductions in digestibility with increasing feeding level are found with ground and pelleted roughages and also some fibrous by-products (0.05 units per unit increase in level).

In non-ruminants, feeding levels rise to 2.0–3.0 times maintenance in poultry, 3.0–4.0 times maintenance in growing pigs and 4.0–6.0 times maintenance in lactating sows, but there is little effect of feeding level on digestibility of conventional (i.e. low-fibre) diets.

10.6 Alternative measures of the digestibility of foods

We have seen already that digestibility may be expressed in a variety of ways. Thus we have to distinguish between the digestibility of food dry matter or organic matter (or of the individual constituents of the organic matter), between apparent and true digestibility, and between digestibility in the whole tract and that in part of the tract (e.g. 'ileal' digestibility). In addition, measures of digestibility may be determined directly, in animals (digestibility *in vivo*), or indirectly, either in the laboratory (*in vitro*) or by the use of polyester bags (*in sacco*). In addition to these measures there are some associated terms derived from digestibility data, which are intended to provide a measure of the energy value of the food. One such measure is the total digestible nutrients (TDN) content of the food, which is calculated as the combined weight in 100 kg of the food of digestible crude protein and digestible carbohydrate (crude fibre and N-free extractives), plus 2.25 times the weight of digestible ether extract. The ether extract is multiplied by 2.25 because the energy value of fats is approximately two-and-a-quarter times that of carbohydrates. If the digestible protein plus carbohydrate of the hay in Table 10.1 is calculated as the digestible organic matter less the digestible ether extract (i.e. as $515 - 8 = 507$), the TDN content is calculated as $507 + 2.25 \times 8 = 525$ g/kg, or 52.5 kg/100 kg. The measure TDN was at one time used extensively in the USA but is now virtually obsolete.

Another derived measure of the energy content of foods is the concentration of digestible organic matter in the dry matter (DOMD). For the hay in Table 10.1 this is 515 g/kg, or 51.5 per cent. The percentage figure is sometimes called the 'D' value of the food. The energy equivalents of TDN and DOMD are explained in later chapters.

10.7 Availability of mineral elements

There are endogenous faecal losses of most minerals, particularly calcium, phosphorus, magnesium and iron. These may arise from secretions into the gut, from which the minerals are not reabsorbed, and may be quite large; for example, in ruminants the quantity of phosphorus secreted into the gut via the saliva is generally greater than the quantity present in the food. The faeces may also carry minerals that have been absorbed in excessive amounts and therefore have to be excreted (i.e. the gut is the route for true excretion as well as endogenous excretion). For these reasons apparent digestibility coefficients for such elements have little significance. The measure of importance is therefore true digestibility, which for minerals is commonly called 'availability'. To measure the availability of a mineral one must generally distinguish between the portion in the faeces which is unabsorbed material and the portion which represents material discharged into the gut from the tissues. This distinction may be made by labelling an element within the body, and hence the portion secreted into the gut, with a radioactive isotope.

Mineral elements in digesta exist in three forms, as metallic ions in solution, as constituents of metallo-organic complexes in solution or as constituents of insoluble substances. Those present in the first form are readily absorbed and those in the third form are not absorbed at all. The metallo-organic complexes, some of which are chelates (see p. 110), may in some cases be absorbed. For some minerals there are opportunities for conversion from one form to another, so broadly speaking the availability of an element will depend on the form in which it occurs in the food and on the extent to which conditions in the gut favour conversion from one form to another. Thus sodium and potassium, which occur in digesta almost entirely as ions, have an availability close to 1.0. At the other extreme, copper occurs almost entirely as soluble or insoluble complexes and its availability is generally less than 0.1. Phosphorus, to take a further example, is present in many foods as a constituent of phytic acid (see p. 120) and its availability depends on the presence of phytases – of microbial or animal origin – in the digestive tract. A potent factor controlling the interconversion of soluble and insoluble forms of mineral elements is the pH of the digesta. In addition, there may be specific agents which bind mineral elements and thus prevent their absorption. For example, calcium may be precipitated by oxalates and copper by sulphides.

The availability of minerals is commonly high in young animals fed on milk and milk products but declines as the diet changes to solid foods. An additional complication is that the absorption, and hence apparent availability, of some mineral elements is partly determined by the animal's need for them. Iron absorption, discussed in Chapter 8, is the clearest example of this effect, but in ruminants calcium absorption also appears to be determined by the animal's need.

While no attempt is made here or in Chapter 8 to provide a complete list of the factors that influence the availability of minerals, those mentioned serve to illustrate why no attempt is made in tables of food composition to give values of availability comparable to the digestibility coefficients commonly quoted for organic nutrients. Because the availability of the minerals in a particular food depends so much on the other constituents of the diet and on the animal to which it is given, average values for individual foods would be of little significance.

Summary

1. The digestibility (%) of nutrients in a food (D) is calculated from the weights of nutrient consumed (I) and excreted in faeces (F) by the formula $100(I-F)/I$.

2. In a typical digestibility trial the food under investigation is given to animals for 3–4 weeks and faeces are collected for the last 1–2 weeks.

3. If food intake and/or faeces output cannot be measured quantitatively, digestibility may be estimated indirectly, from the relative concentrations in food and faeces of an indigestible substance, known as an indicator.

4. Digestibility of foods may be estimated in the laboratory (in vitro) by incubating them

in rumen liquor and various enzyme preparations. Digestion in the rumen may be estimated from small samples of food placed in permeable bags in that organ (digestibility *in sacco*). Near-infrared reflectance spectroscopy is being used increasingly to estimate the digestibility of foods in farm advisory work.

5. If a nutrient in faeces is derived from the body tissues (i.e. not directly from the food), its true digestiblity (as opposed to apparent digestibility) will be underestimated.

6. Cannulae inserted surgically in the gut of animals allow digesta to be sampled from each sector and digestibility to be calculated for successive sectors.

7. Although the digestibility of a food is determined mainly by its chemical composition (especially its fibre content), it may be modified by supplementation, by physical or chemical processing and by exogenous enzymes.

8. The digestibility of a given food is also affected by the species of animal to which it is given and by the level of feeding.

9. The digestible nutrients contents of foods are a primary measure of their value to the animal as a source of energy (see Chapter 11).

10. The digestibility of mineral elements (commonly called mineral availability) is affected by secretion and excretion of minerals into the gut and by selective absorption.

Problems

10.1 A pig was given 2 kg DM/day as a food containing 150 g protein per kg. It excreted in its faeces 0.4 kg DM containing 175 g protein per kg. Calculate the digestibility of the protein.

10.2 As a check on the measurement of digestibility in problem 10.1, above, the food dry matter contained 10 g/kg of the indicator chromic oxide. The faeces dry matter contained 50 g/kg chromic oxide. Did the actual quantity of dry matter excreted in faeces agree with that estimated by the indicator?

10.3 Of the 2 kg DM given to the pig in problem 10.1, above, 0.3 kg came from soya bean meal, containing 450 g protein per kg, for which digestibility was known to be 85 per cent. The rest of the ration was a cereal; what was the digestibility of the cereal protein?

Further reading

Ammerman C B, Baker D H and Lewis A J (eds) 1995 *Bioavailability of Nutrients for Animals: Amino Acids, Minerals and Vitamins*, San Diego, Academic Press.

British Society of Animal Science 1997 *In vitro* techniques for measuring nutrient supply to ruminants. *Occasional Publication of the British Society of Animal Science*, No. 22, Edinburgh.

Givens D I and Deaville E R 1999 The current and future role of near infrared reflectance spectroscopy in animal nutrition: a review. *Australian Journal of Agricultural Research* **50**: 1131–45.

Mayes R W and Dove H 2000 Measurement of dietary nutrient intake in free-ranging mammalian herbivores. *Nutrition Research Reviews* **13**: 107–38.

Nutrition Society 1977 Methods for evaluating feeds for large farm animals. *Proceedings of the Nutrition Society* **36**: 169–225.

Schneider B H and Flatt W P 1975 *The Evaluation of Feeds through Digestibility Experiments*, Athens, GA, University of Georgia Press.

Van Soest P J 1994 *Nutritional Ecology of Ruminants*, 2nd edn, Corvallis, OR, O and B Books.

Wheeler J L and Mochrie R D (eds) 1981 *Forage Evaluation: Concepts and Techniques*, Melbourne, CSIRO and Lexington, KY, American Forage and Grassland Council.

11 Evaluation of foods: energy content of foods and the partition of food energy within the animal

11.1   Demand for energy

11.2   Supply of energy

11.3   Animal calorimetry: methods for measuring heat production and energy retention

11.4   Utilisation of metabolisable energy

The major organic nutrients are required by animals as materials for the construction of body tissues and the synthesis of such products as milk and eggs, and they are needed also as sources of energy for work done by the animal. A unifying feature of these diverse functions is that they all involve a transfer of energy, and this applies both when chemical energy is converted into mechanical or heat energy, as when nutrients are oxidised, and when chemical energy is converted from one form to another, as for example when body fat is synthesised from food carbohydrate. The ability of a food to supply energy is therefore of great importance in determining its nutritive value. The purpose of this chapter and the next is to discuss the fate of food energy in the animal body, the measurement of energy metabolism and the expression of the energy value of foods.

11.1   Demand for energy

An animal deprived of food continues to require energy for those functions of the body immediately necessary for life – for the mechanical work of essential muscular activity, for chemical work such as the movement of dissolved substances against concentration gradients, and for the synthesis of expended body constituents such as enzymes and hormones. In the starving animal the energy required for these purposes is obtained by the catabolism of the body's reserves, first of glycogen, then of fat and protein. In the fed animal the primary demand on the energy of the food is in meeting this requirement for body maintenance and so preventing the catabolism of the animal's tissues.

When the chemical energy of the food is used for muscular and chemical work involved in maintenance, the animal does no work on its surroundings and all the energy used is converted into heat. Energy so used is regarded as having been expended, since heat energy is useful to the animal only in maintaining body temperature. In a fasting animal the quantity of heat produced is equal to the energy of the tissue catabolised and when measured under specific conditions is known as the animal's basal metabolism. In Chapter 14 it will be shown how estimates of basal metabolism are used in assessing the maintenance energy requirements of animals.

Energy supplied by the food in excess of that needed for maintenance is used for the various forms of production. (More correctly, the nutrients represented by this energy are so used.) A young growing animal will store energy principally in the protein of its new tissues, whereas an adult will store relatively more energy in fat and a lactating animal will transfer food energy into the energy contained in milk constituents. Some other forms of production are the performance of muscular work and the formation of wool and eggs. No function, not even body maintenance, can be said to have absolute priority for food energy. A young animal receiving adequate protein but insufficient energy for maintenance may still store protein while drawing on its reserves of fat. Similarly, some wool growth continues to take place in animals with sub-maintenance intakes of energy and even in fasted animals.

11.2 Supply of energy

Gross energy of foods

The animal obtains energy from its food. The quantity of chemical energy present in a food is measured by converting it into heat energy and determining the heat produced. This conversion is carried out by oxidising the food by burning it; the quantity of heat resulting from the complete oxidation of unit weight of a food is known as the gross energy or heat of combustion of that food (see Box 11.1).

Some typical gross energy values are shown in Fig. 11.1. The primary determinant of the gross energy content of an organic substance is its degree of oxidation, as expressed in the ratio of carbon plus hydrogen to oxygen. All carbohydrates have similar ratios and all therefore have about the same gross energy content (about 17.5 MJ/kg DM). Triglyceride fats, however, have relatively less oxygen and therefore have a gross energy value that is much higher (about 39 MJ/kg DM) than that of carbohydrates. Individual members of the fatty acid series vary in gross energy content according to carbon chain length; thus the lower members of the series (the volatile fatty acids) are lower in energy content. Proteins have a higher gross energy value than carbohydrates because they contain

Box 11.1
Measurement of gross energy

Gross energy is measured in an apparatus known as a bomb calorimeter, which in its simplest form consists of a strong metal chamber (the bomb) resting in an insulated tank of water. The food sample is placed in the bomb and oxygen admitted under pressure. The temperature of the water is taken and the sample is then ignited electrically. The heat produced by the oxidation is absorbed by the bomb and the surrounding water, and when equilibrium is reached the temperature of the water is taken again. The quantity of heat produced is then calculated from the rise in temperature and the weights and specific heats of the water and the bomb. The bomb calorimeter can be used to determine the gross energy content of whole foods or of their constituents and of animal tissues and excretory products.

the additional oxidisable element, nitrogen (and also sulphur). Methane has a very high energy value because it consists solely of carbon and hydrogen.

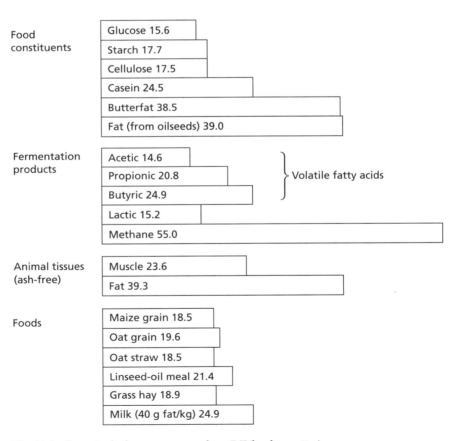

Fig. 11.1 Some typical gross energy values (MJ/kg dry matter).

In spite of these differences between food constituents, the predominance of the carbohydrates means that the foods of farm animals vary little in energy content. Only foods rich in fat, such as linseed-oil meal with 90 g ether extract/kg, have high values, and only those rich in ash, which has no calorific value, are much lower than average. Most common foods contain about 18.5 MJ/kg DM.

Of the gross energy of foods, not all is available and useful to the animal. Some energy is lost from the animal in the form of the solid, liquid and gaseous excretions; another fraction is lost as heat. These sources of energy loss are illustrated in Fig. 11.2. Their deduction from the gross energy content of the food gives rise to further descriptive categories of food energy; for example, gross energy less the energy content of faeces gives the category known as the digestible energy of the food (but which might be more accurately designated as the energy content of the digested nutrients). This and other categories will now be described.

Digestible energy

The apparently digestible energy of a food is the gross energy content of unit weight of the food less the gross energy content of the faeces resulting from the consumption of unit weight of that food.

In the example of a digestibility trial given in Table 10.1, the sheep ate 1.63 kg of hay DM having an energy content of 18.0 MJ/kg. Total energy intake was therefore 29.3 MJ/day. The 0.76 kg of faeces DM contained 18.7 MJ/kg, or a total of 14.2 MJ/day. The apparent digestibility of the energy of the hay was therefore $(29.3 - 14.2)/29.3 = 0.515$ and the digestible energy content of the hay DM was $18.0 \times 0.515 = 9.3$ MJ/kg.

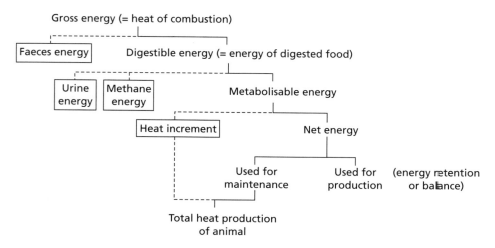

Fig. 11.2 The partition of food energy in the animal. Losses of energy are shown as the boxed items on the left.

Metabolisable energy

The animal suffers further losses of energy-containing substances in its urine and, particularly if it is a ruminant, in the combustible gases leaving the digestive tract. The metabolisable energy of a food (see Box 11.2) is the digestible energy less the energy lost in urine and combustible gases. The energy of urine is present in nitrogen-containing substances such as urea, hippuric acid, creatinine and allantoin, and also in such non-nitrogenous compounds as glucuronates and citric acid.

The combustible gases lost from the rumen consist almost entirely of methane. Methane production is closely related to food intake, and at the maintenance level of nutrition about 7–9 per cent of the gross energy of the food (about 11–13 per cent of the digestible energy) is lost as methane. At higher levels of feeding the proportion falls to 6–7 per cent of gross energy, the fall being most marked for highly digestible foods. With previously fermented foods, such as brewer's grains, methane production is low (3 per cent of gross energy).

When methane production cannot be measured directly it can be estimated as 8 per cent of gross energy intake. Another approximation allows the metabolisable energy values of ruminant foods to be calculated from digestible energy values by multiplying by 0.8. This implies that, on average, about 20 per cent of the energy apparently digested is excreted in the urine and as methane.

For poultry, metabolisable energy is measured more easily than digestible energy, because the faeces and urine are voided together. A rapid, standardised method has been developed for determining the metabolisable energy value of poultry foods. Cockerels are fasted (or fed only on a small quantity of glucose) until their digestive tract is empty, then force fed with a single meal of the food under investigation. Excreta are then collected until all residues of the single meal have been voided. At the same time the small quantities of excreta voided by fasted (or glucose-fed) birds are collected, as a measure of endogenous losses. The energy of these endogenous losses is deducted from the energy of the excreta of the fed birds, so the estimate of metabolisable energy obtained is a true rather than apparent value

Box 11.2
Measurement of metabolisable energy

The metabolisable energy value of a food is determined in a feeding trial similar to a digestibility trial, but in which urine and methane, as well as faeces, are collected. Metabolism cages for sheep and pigs incorporate a device for collecting urine. The urine of cattle is caught in rubber urinals attached below the abdomen for males and over the vulva for females, and is piped by gravity or suction to a collection vessel. An alternative method of collecting urine from females, which is commonly used for pigs, is to insert an inflatable catheter into the vagina. When methane production is measured the animal is usually kept in an airtight container known as a respiration chamber. The operation of such chambers is described in more detail later (p. 276).

(see p. 252); it is known as true metabolisable energy (TME), and is not directly comparable with measures of the metabolisable energy of foods obtained by other means.

Factors affecting the metabolisable energy values of foods

Table 11.1 shows metabolisable energy values for a number of foods. It is clear that, of the sources of energy loss so far considered, the faecal loss is by far the most important. Even for a food of high digestibility such as barley, twice as much energy is lost in the faeces as in the urine and methane. The main factors affecting the metabolisable energy value of a food are therefore those which influence its digestibility. These have been discussed earlier (Chapter 10), and the emphasis here will be on urine and methane losses.

The metabolisable energy value of a food will obviously vary according to the species of animal to which it is given, or more specifically, to the type of digestion to which it is subjected. Fermentative digestion, in the rumen or further along the gut, incurs losses of energy as methane. A lesser effect of the intervention of microorganisms in digestion is an increase in the losses of energy in either urine (as the breakdown products of the nucleic

Table 11.1 Metabolisable energy values of some typical foods[a]

Animal	Food	Gross energy	Energy lost in:			ME
			Faeces	Urine	Methane	
Fowl	Maize	18.4	2.2		–	16.2
	Wheat	18.1	2.8		–	15.3
	Barley	18.2	4.9		–	13.3
Pig	Maize	18.9	1.6	0.4	–	16.9
	Oats	19.4	5.5	0.6	–	13.3
	Barley	17.5	2.8	0.5	–	14.2
	Coconut cake meal	19.0	6.4	2.6	–	10.0
Sheep	Barley	18.5	3.0	0.6	2.0	12.9
	Dried ryegrass (young)	19.5	3.4	1.5	1.6	13.0
	Dried ryegrass (mature)	19.0	7.1	0.6	1.4	9.9
	Grass hay (young)	18.0	5.4	0.9	1.5	10.2
	Grass hay (mature)	17.9	7.6	0.5	1.4	8.4
	Grass silage	19.0	5.0	0.9	1.5	11.6
Cattle	Maize	18.9	2.8	0.8	1.3	14.0
	Barley	18.3	4.1	0.8	1.1	12.3
	Wheat bran	19.0	6.0	1.0	1.4	10.6
	Lucerne hay	18.3	8.2	1.0	1.3	7.8

[a] Uncorrected values – see text – expressed as MJ per kg food dry matter.

acids of bacteria that have been digested and absorbed) or faeces (as microorganisms grown in the hind gut and not digested). In general, losses of energy in methane and in urine are greater for ruminants than for non-ruminants, so foods such as concentrates, which are digested to the same extent in ruminants and non-ruminants, will have a higher metabolisable energy value for non-ruminants (*cf.* values for barley in Table 11.1). However, fibrous foods given to non-ruminants will also incur losses due to fermentative digestion in the hind gut. In ruminants, foods like silages that have been fermented prior to consumption by the animal will incur smaller energy losses in digestion but will already have incurred losses in the silo. Thus silages are said to contain less 'fermentable' metabolisable energy than comparable foods such as hays; this difference, however, is of greater importance to the protein nutrition of ruminants (see Chapters 8 and 13) than to energy nutrition.

A final comment on the effect of animal species is that differences between cattle and sheep in urine and methane losses of energy are small and of no significance.

The metabolisable energy value of a food will vary according to whether the amino acids it supplies are retained by the animal for protein synthesis, or are deaminated and their nitrogen excreted in the urine as urea. For this reason, metabolisable energy values are sometimes corrected to zero nitrogen balance by deducting for each 1 g of nitrogen retained, either 28 kJ (pigs), 31 kJ (ruminants) or 34 kJ (poultry). The factor most appropriate to each species of animal depends on the extent to which nitrogen is excreted as urea (gross energy 23 kJ/g nitrogen) and as other compounds (e.g. uric acid, 28 kJ/g nitrogen). If an animal is excreting more nitrogen in its urine than it is absorbing from its food (i.e. is in negative nitrogen balance: see Chapter 13), some of the urine nitrogen is not derived from the food, and in this case the metabolisable energy value must be subjected to a positive correction.

The manner in which the food is prepared may in some cases affect its metabolisable energy value. For ruminants, the grinding and pelleting of roughages leads to an increase in faecal losses of energy (see Table 10.2 for an example), but this may be partly offset by a reduction in methane production. For poultry, the grinding of cereals has no consistent effect on metabolisable energy values.

As discussed earlier (Chapter 10), increases in the level of feeding of ruminants may cause an appreciable reduction in the digestibility of their food, and hence in its metabolisable energy value. Increases in faecal energy loss may be partially offset by reductions in losses of energy in methane and urine. Nevertheless, for finely ground roughages and for mixed roughage and concentrate diets, metabolisable energy value is reduced by increases in level of feeding.

In theory it should be possible to prevent the production of methane in the rumen and thereby avoid losing 8 per cent of gross energy intake in this form. In practice it is possible to suppress methane production by adding antimicrobial drugs to the diet (one effective chemical is chloroform), but the consequences are not consistently favourable. Energy may be diverted to another gaseous by-product, hydrogen (see Fig. 8.3); furthermore, the rumen microorganisms may adapt to the presence of the drug and revert to the synthesis of methane. The coccidiostat monensin which has been widely used as a feed-borne growth stimulant for beef cattle, is considered to be a methane suppressant (see Chapter 26). However, there is some evidence that methanogenic microorganisms can adapt to this agent also.

Heat increment of foods

The ingestion of food by an animal is followed by losses of energy not only as the chemical energy of its solid, liquid and gaseous excreta but also as heat. Animals are continuously producing heat and losing it to their surroundings, either directly by radiation, conduction and convection, or indirectly by the evaporation of water. If a fasting animal is given food, within a few hours its heat production will increase above the level represented by basal metabolism. This increase is known as the heat increment of the food; it is quite marked in man after a large meal. The heat increment may be expressed in absolute terms (MJ/kg food DM), or relatively as a proportion of the gross or metabolisable energy. Unless the animal is in a particularly cold environment, this heat energy is of no value to it, and must be considered, like the energy of the excreta, as a tax on the energy of the food.

The causes of the heat increment are to be found in the processes of the digestion of foods and the metabolism of the nutrients derived from them. The act of eating, which includes chewing, swallowing and the secretion of saliva, requires muscular activity for which energy is supplied by the oxidation of nutrients; in ruminants chewing fibrous foods, the energy cost of eating is estimated to be 3–6 per cent of metabolisable energy intake. The energy cost of rumination, however, is much less than the cost of eating, and is estimated to be about 0.3 per cent of metabolisable energy intake. Ruminants also generate heat through the metabolism of their gut microorganisms; this is estimated to amount to about 7–8 per cent of metabolisable energy intake (or alternatively, as 0.6 kJ per kilojoule of methane produced).

More heat is produced when nutrients are metabolised. For example, it was shown in Chapter 9 that if glucose is oxidised in the formation of ATP, the efficiency of free energy capture is only about 0.69, 0.31 being lost as heat. Moreover, the efficiency will be even less if temporary storage of nutrients is required (e.g. glucose stored as glycogen) because more reactions are required. Similar inefficiency is apparent in the synthesis of the body's structural constituents. The linking of one amino acid to another, for example, requires the expenditure of four pyrophosphate 'high-energy' bonds, and if the ATP that provides these is obtained through glucose oxidation, about 2.5 MJ of energy will be released as heat for each kilogram of protein formed. Protein synthesis, it should be noted, occurs not only in growing animals but also in those kept at a maintenance level, in which protein synthesis is a part of the process of protein turnover (see p. 224). Protein metabolism is estimated to account for about 10 per cent of the animal's heat production. The animal also expends high-energy phosphate bonds to do the work involved in the movement of substances (e.g. Na^+ and K^+ ions) against concentration gradients. This so-called 'ion pumping' may also contribute 10 per cent of the animal's heat production. Heat is produced within the body in those regions with the most active metabolism. Thus it has been estimated that in ruminants, which have a large and metabolically active gut, as much as half of total heat production originates in the gut and liver.

We shall see later that the heat increment of foods varies considerably, according to the nature of the food, the type of animal consuming it and the processes for which nutrients are used.

Net energy and energy retention

The deduction of the heat increment of a food from its metabolisable energy gives the net energy value of the food. The net energy of a food is that energy which is available to the animal for useful purposes, i.e. for body maintenance and for the various forms of production (Fig. 11.2).

Net energy used for maintenance is mainly used to perform work within the body and will leave the animal as heat. That used for growth and fattening and for milk, egg or wool production is either stored in the body or leaves it as chemical energy, and the quantity so used is referred to as the animal's energy retention.

It is important to understand that of the heat lost by the animal only a part, the heat increment of the food, is truly waste energy which can be regarded as a direct tax on the food energy. The heat resulting from the energy used for body maintenance is considered to represent energy which has been used by the animal and degraded into a useless form during the process of utilisation.

11.3 Animal calorimetry: methods for measuring heat production and energy retention

Calorimetry means the measurement of heat. The partition of food energy depicted in Fig. 11.2 shows that if the metabolisable energy intake of an animal is known, the measurement of its total heat production will allow its energy retention to be calculated by difference (likewise, the measurement of energy retention will allow heat production to be calculated). In practice, measurement of either heat production or energy retention is used to establish the net energy value of a food.

The methods used to measure heat production and energy retention in animals can be quite complicated, both in principle and in practice. In the past, the complexity and cost of the apparatus required for animal calorimetry limited its use to a small number of nutritional research establishments. Improved funding of research has gradually removed this restriction, but even so, animal calorimetry remains a specialised topic and few nutritionists become involved in it. Nevertheless, the study of animal calorimetry on paper (as in this book) is valuable to all students of nutrition, because it reinforces their knowledge of the principles of energy metabolism. In the pages that follow, the principles of the methods used in animal calorimetry are explained within the main body of the text, and the apparatus employed is described in boxed sections of the text.

The heat production of animals can be measured by physical methods, by a procedure known as direct calorimetry. Alternatively, heat production can be estimated from the respiratory exchange of the animal; for this a respiration chamber is normally used and the approach is one of indirect calorimetry. Respiration chambers can also be used to estimate energy retention rather than heat production, by the procedure known as the carbon and nitrogen balance trial.

Direct calorimetry

Animals do not store heat, except for relatively short periods of time, and when measurements are made over periods of 24 hours or longer it is generally safe to assume that the quantity of heat lost from the animal is equal to the quantity produced. To determine the heat increment of a food, animals are given the food at two levels of metabolisable energy intake and their heat production is measured at each level. The heat increment is calculated as shown in Fig. 11.3. Two levels are needed because a part of the animal's heat production is contributed by its basal metabolism. An increase in food intake causes total heat production to rise, but the basal metabolism is assumed to remain the same. The increase in heat production is thus the heat increment of the extra food given.

In the example shown in Fig. 11.3, the food was given at levels supplying 40 and 100 MJ metabolisable energy. The increment of 60 MJ (BD on the figure) was associated with an increase in heat production, CD, of 24 MJ. The heat increment as a fraction was:

CD/BD or 24/60 = 0.4

It is also possible to make the lower level of intake zero, and to estimate the heat increment as the difference in heat production between the basal (or fasting) metabolism and that produced in the fed animal. In the example of Fig. 11.3 this method gives the heat increment as 16/40 = 0.4.

If a single food is being investigated, it may be given as the sole item of diet at both levels. If the food is one that would not normally be given alone, the lower level may be obtained by giving a basal ration and the higher level by the same basal ration plus some of the food under investigation. For example, the heat increment of barley eaten by sheep could be

Box 11.3
Animal calorimeters

Heat is lost from the body principally by radiation, conduction and convection from the body surface and by evaporation of water from the skin and lungs. The animal calorimeter is basically an airtight and insulated chamber. Evaporative losses of heat are measured by recording the volume of air drawn through the chamber and its moisture content on entry and exit. In most early calorimeters the sensible heat loss (i.e. that lost by radiation, conduction and convection) was taken up in water circulated through coils within the chamber; the quantity of heat removed from the chamber could then be computed from the rate of flow of the water and the difference between its entry and exit temperatures. In a more recent type, the gradient layer calorimeter, the quantity of heat is measured electrically as it passes through the wall of the chamber. This type of calorimeter lends itself well to automation, and both sensible and evaporative losses of heat can be recorded automatically. Most calorimeters incorporate apparatus for measuring respiratory exchange and can therefore be used for indirect calorimetry as well.

measured by feeding the sheep first on a basal ration of hay and then on an equal amount of the same hay plus some of the barley.

Animal calorimeters are expensive to build and the earlier types required much labour to operate them. Because of this most animal calorimetry today is carried out by the indirect method described below.

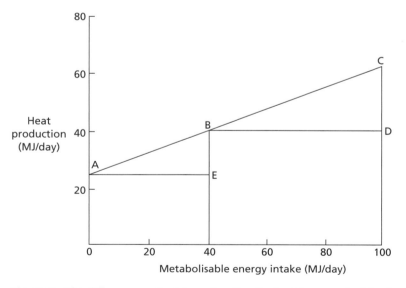

Fig. 11.3 The difference method for estimating the heat increment of foods. A is the basal metabolism and B and C represent heat production at metabolisable energy intakes of 40 and 100 MJ, respectively. For the sake of simplicity the relationship between heat production and metabolisable energy intake is shown here as being linear, i.e. ABC is a straight line; however, as explained later in the chapter, this is not usually the case.

Indirect calorimetry by the measurement of respiratory exchange

The substances that are oxidised in the body, and whose energy is therefore converted into heat, fall mainly into the three nutrient classes of carbohydrates, fats and proteins. The overall reaction for the oxidation of a carbohydrate such as glucose is:

$$C_6H_{12}O_6 + 6O_2 \rightarrow 6CO_2 + 6H_2O + 2.82 \text{ MJ} \qquad [11.1]$$

and for the oxidation of the typical fat, tripalmitin, is:

$$C_3H_5(OOC.C_{15}H_{31})_3 + 72.5O_2 \rightarrow 51CO_2 + 49H_2O + 32.02 \text{ MJ} \qquad [11.2]$$

One gram-molecule of oxygen occupies 22.4 litres at NTP. Thus in an animal obtaining all its energy by the oxidation of glucose, the utilisation of one litre of oxygen would lead to production of $2820/(6 \times 22.4) = 20.98$ kJ of heat; for mixtures of carbohydrates an average value is 21.12 kJ/litre. Such values are known as thermal equivalents of oxygen and are used in indirect calorimetry to estimate heat production from oxygen consumption. For an animal catabolising mixtures of fats alone, the thermal equivalent of oxygen is 19.61 kJ/litre (*cf.* 19.73 kJ/litre calculated for a single fat from Equation [11.2] above).

Animals do not normally obtain energy exclusively from either carbohydrate or fat. They oxidise a mixture of these (and of protein also), so that in order to apply the appropriate thermal equivalent when converting oxygen consumption to heat production it is necessary to know how much of the oxygen is used for each nutrient. The proportions are calculated from what is known as the respiratory quotient (RQ). This is the ratio between the volume of carbon dioxide produced by the animal and the volume of oxygen used. Since, under the same conditions of temperature and pressure, equal volumes of gases contain equal numbers of molecules, the RQ can be calculated from the molecules of carbon dioxide produced and oxygen used. From Equation [11.1] the RQ for carbohydrate is calculated as $6CO_2/6O_2 = 1$, and from Equation [11.2] that of the fat, tripalmitin, as $51CO_2/72.5O_2 = 0.70$. If the RQ of an animal is known, the proportions of fat and carbohydrate oxidised can then be determined from standard tables. For example, an RQ of 0.9 indicates the oxidation of a mixture of 67.5 per cent carbohydrate and 32.5 per cent fat, and the thermal equivalent of oxygen for such a mixture is 20.60 kJ/litre.

The mixture oxidised generally includes protein. The quantity of protein catabolised can be estimated from the output of nitrogen in the urine, 0.16 g of urinary N being excreted for each gram of protein. The heat of combustion of protein (i.e. the heat produced when it is completely oxidised) varies according to the amino acid proportions but averages 22.2 kJ/g. Protein, however, is incompletely oxidised in animals because the body cannot oxidise nitrogen, and the average amount of heat produced by the catabolism of 1 g of protein is 18.0 kJ. For each gram of protein oxidised, 0.77 litres of carbon dioxide is produced and 0.96 litres of oxygen used, giving an RQ of 0.8.

Heat is produced not only when organic nutrients are oxidised but also when they are used for the synthesis of tissue materials. It has been found, however, that the quantities of heat produced during these syntheses bear the same relation to the respiratory exchange as they do when the nutrients are completely oxidised.

The relationship between respiratory exchange and heat production is disturbed if the oxidation of carbohydrate and fat is incomplete. This situation arises in the metabolic disorder known as ketosis, when fatty acids are not completely oxidised to carbon dioxide and water, and carbon and hydrogen leave the body as ketones or ketone-like substances. Incomplete oxidation also occurs under normal conditions in ruminants, because an end-product of carbohydrate fermentation in the rumen is methane. In practice, heat production calculated from respiratory exchange in ruminants is corrected for this effect by the deduction of 2.42 kJ for each litre of methane.

The calculations explained above may be combined into a single equation called the Brouwer equation (after the Dutch scientist, E Brouwer):

$$HP = 16.18VO_2 + 5.16VCO_2 - 5.90N - 2.42CH_4 \qquad [11.3]$$

where: HP = heat production (kJ)
         VO_2 = oxygen consumption (litres)
         VCO_2 = carbon dioxide production (litres)
         N = urinary nitrogen excretion (g)
         CH_4 = methane production (litres)

For poultry the N coefficient is 1.20 (instead of 5.90), as poultry excrete nitrogen in the more oxidised form of uric acid, rather than as urea.

In some situations, which are discussed in more detail later, heat production has to be estimated from oxygen consumption alone. If a respiratory quotient of 0.82 and a thermal equivalent of 20.0 are assumed, departures from this RQ in the range 0.7–1.0 cause a maximum bias of no more than 3.5 per cent in the estimate of heat production. A further simplification is possible in respect of protein metabolism. The thermal equivalent of oxygen used for protein oxidation is 18.8 kJ/litre, not very different from the value of 20.0 assumed for carbohydrate and fat oxidation. If only a small proportion of the heat production is derived from protein oxidation it is unnecessary to assess it separately, and so urinary nitrogen output need not be measured.

An example of the calculation of heat production from respiratory exchange is shown in Table 11.2. If the Brouwer equation (Equation [11.3], above) had been applied to the respiratory exchange data of Table 11.2, heat production would have been estimated as 7858 kJ.

A recent significant change in studies of animal energetics has been that from studying the whole animal to measuring energy exchange in specific

Table 11.2 Calculation of the heat production of a calf from values for its respiratory exchange and urinary nitrogen excretion (After Blaxter K L, Graham N McC and Rook J A F 1955 *Journal of Agricultural Science, Cambridge* **45**: 10)

Results of the experiment (per 24 hours)		
Oxygen consumed, litre		392.0
Carbon dioxide produced, litre		310.7
Nitrogen excreted in urine, g		14.8
Heat from protein metabolism		
Protein oxidised, g	(14.8×6.25)	92.5
Heat produced, kJ	(92.5×18.0)	1665
Oxygen used, litre	(92.5×0.96)	88.8
Carbon dioxide produced, litre	(92.5×0.77)	71.2
Heat from carbohydrate and fat metabolism		
Oxygen used, litre	$(392.0 - 88.8)$	303.2
Carbon dioxide produced, litre	$(310.7 - 71.2)$	239.5
Non-protein respiratory quotient		0.79
Thermal equivalent of oxygen when RQ = 0.79, kJ/litre		20.0
Heat produced, kJ	(303.2×20.0)	6064
Total heat produced, kJ	$(1665 + 6064)$	**7729**

Box 11.4
Measuring respiratory exchange in respiration chambers

The apparatus most commonly used for farm animals is a respiration chamber. The simplest type of chamber, the closed-circuit type (Fig. 11.4*a*), consists of an airtight container for the animal together with vessels holding absorbents for carbon dioxide and water vapour. The chamber incorporates devices for feeding, watering and even milking the animal. The oxygen used by the animal is replaced from a metered supply. At the end of a trial period (24 hours) the carbon dioxide produced can be measured by weighing the absorbent; any methane produced can be measured by sampling and analysing the air in the chamber. The main disadvantage of the closed-circuit chamber is that large quantities of absorbents are required; thus for a cow, 100 kg of soda lime would be needed each day to absorb carbon dioxide and 250 kg of silica gel to absorb water vapour.

In the alternative, open-circuit type of chamber (Fig. 11.4*b*) air is drawn through the chamber at a metered rate and sampled for analysis on entry and exit. Thus carbon dioxide production, methane production and oxygen consumption can be estimated. As the differences in composition between inlet and outlet air must be kept small if conditions for the animal are to be kept normal, very accurate measures of gas flow and composition are required. Modern equipment, based on infrared analysers, meets this criterion, and open-circuit chambers have now largely replaced closed-circuit. With some chambers it is possible to alternate between closed- and open-circuit operation. Closed-circuit operation for perhaps 30 minutes, with no gas absorption, produces appreciable changes in the composition of the chamber air. Then a brief period (about 3 minutes) of open-circuit operation allows the chamber air to be flushed out through a flow meter and sampling device.

(a) Closed circuit

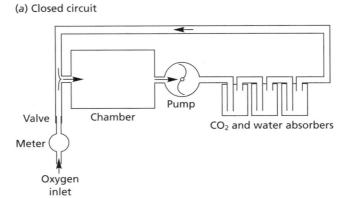

(b) Open circuit

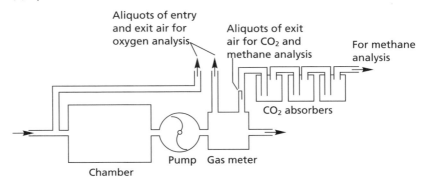

Fig. 11.4 Diagrams of respiration chambers.

organs or tissues. The basic methodology for such measurements is that catheters are inserted in blood vessels supplying and draining an organ, and the flow and composition of the blood are determined, in order to allow oxygen uptake and carbon dioxide production to be estimated. At the same time the uptake of metabolites, such as glucose, can be measured.

Box 11.5
Measuring respiratory exchange in unconfined animals

Respiratory exchange can be measured without an animal chamber if the subject is fitted with a face mask, which is then connected to either a closed or an open circuit for determining oxygen consumption alone or both oxygen consumption and carbon dioxide production. This method is suitable for short periods of measurement but cannot be used to estimate heat production when the animal is eating. For longer-term measurements of energy metabolism in unconfined (e.g. grazing) animals, heat production can be estimated with reasonable accuracy from carbon dioxide production alone. The latter is measured by infusing into the body fluids a source of radio-labelled carbon dioxide (^{14}C sodium bicarbonate) and sampling body fluids to determine the degree to which the labelled carbon dioxide is diluted by that produced by the animal.

Measurement of energy retention by the carbon and nitrogen balance technique

In respiration calorimetry, heat production is estimated, and energy retention is calculated as the difference between metabolisable energy intake and heat production (as in Table 11.2). An alternative approach is to estimate energy retention more directly and to calculate heat production by difference.

The main forms in which energy is stored by the growing and fattening animal are protein and fat, for the carbohydrate reserves of the body are small and relatively constant. The quantities of protein and fat stored can be estimated by carrying out a carbon and nitrogen balance trial; that is, by measuring the amounts of these elements entering and leaving the body and so, by difference, the amounts retained. The energy retained can then be calculated by multiplying the quantities of nutrients stored by their calorific values.

Both carbon and nitrogen enter the body only in the food, and nitrogen leaves it only in faeces and urine. Carbon, however, leaves the body also in methane and carbon dioxide, and the balance trial must therefore be carried out in a respiration chamber. The procedure for calculating energy retention from carbon and nitrogen balance data is best illustrated by considering an animal in which storage of both fat and protein is taking place. In such an animal, intakes of carbon and nitrogen will be greater than the quantities excreted, and the animal is said to be in positive balance with respect to these elements. The quantity of protein stored is calculated by multiplying the nitrogen balance by 1000/160 (= 6.25), for body protein is assumed to contain 160 g N/kg. It also contains 512 g C/kg, and the amount of carbon stored as protein can therefore be computed. The remaining carbon is stored as fat, which contains 746 g C/kg. Fat storage is therefore calculated by dividing the carbon balance, less that stored as protein, by 0.746. The energy present in the protein and fat stored is then calculated by using average calorific values for body tissue. These values vary from one species to another; for cattle and sheep those now recommended are 39.3 MJ/kg for fat and 23.6 MJ/kg for protein. An example of this method of calculating energy retention (and heat production) is shown in Table 11.3.

The advantages of the carbon and nitrogen balance technique are that no measure of oxygen consumption (or RQ) is required and that energy retention is subdivided into that stored as protein and that stored as fat.

Measuring energy retention by the comparative slaughter technique

Because calorimetric experiments require elaborate apparatus and can be conducted with only small numbers of animals, numerous attempts have been, and are still being, made to measure energy retention in other ways. In many feeding trials the animals' intakes of digestible or metabolisable

Table 11.3 Calculation of the energy retention and heat production of a sheep from its carbon and nitrogen balance (After Blaxter K L and Graham N McC 1955 *Journal of Agricultural Science, Cambridge* **46**: 292)

Results of the experiment (per 24 hours)	C (g)	N (g)	Energy (MJ)
Intake	684.5	41.67	28.41
Excretion in faeces	279.3	13.96	11.47
Excretion in urine	33.6	25.41	1.50
Excretion as methane	20.3	–	1.49
Excretion as CO_2	278.0	–	–
Balance	73.3	2.30	
Intake of metabolisable energy	–	–	13.95
Protein and fat storage			
Protein stored, g	(2.30×6.25)		14.4
Carbon stored as protein, g	(14.4×0.512)		7.4
Carbon stored as fat, g	$(73.3 - 7.4)$		65.9
Fat stored, g	$(65.9 \div 0.746)$		88.3
Energy retention and heat production			
Energy stored as protein, MJ	(14.4×23.6)		0.34
Energy stored as fat, MJ	(88.3×39.3)		3.47
Total energy retention, MJ	$(0.34 + 3.47)$		3.81
Heat production, MJ	$(13.95 - 3.81)$		10.14

energy can be measured satisfactorily but their energy retention can be estimated only from changes in liveweight. Weight changes, however, provide inaccurate estimates of energy retention, first because they may represent no more than changes in the contents of the gut or bladder, and second because the energy content of true tissue gain can vary over a wide range according to its proportions of bone, muscle and fat (see Chapter 14). These objections are only partly overcome in experiments where the energy retention is in the form of milk or eggs, whose energy is easily measured, for retention in these products is almost invariably accompanied by retention in other tissues (e.g. milking cows are normally increasing or decreasing in liveweight and in energy content).

Energy retention can, however, be measured in feeding trials if the energy content of the animal is estimated at the beginning and end of the experiment. In the comparative slaughter method, this is done by dividing the animals into two groups and slaughtering one (the sample slaughter group) at the beginning of trial. The energy content of the animals slaughtered is determined by bomb calorimetry, the samples used being taken either from the whole, minced body or from the tissues of the body after these have been separated by dissection. A relationship is then obtained between the liveweight of the animals and their energy content, and this is used to predict the initial energy content of animals in the second group. The latter are slaughtered at the end of the trial and treated in the same manner as those in the sample slaughter group, and their energy gains can then be calculated.

Table 11.4 shows an example of the use of the comparative slaughter technique in an experiment in which, for comparative purposes, respiration

Table 11.4 Use of the comparative slaughter technique to estimate the energy retention and heat production of poultry (After Fuller H L, Dale N M and Smith C F 1983 *Journal of Nutrition* **113**: 1403)

Details of birds	Initial	Final	Difference
Liveweight, g	2755	2823	68
Gross energy, kJ	27491	28170	679
Metabolisable energy intake, kJ		2255	
Heat production, kJ, (2255 − 679)		1576	
Heat production by respiration calorimetry, kJ		1548	

calorimetry was also employed. Cockerels were slaughtered either before or after a four-day period in respiration chambers. Their energy gain was calculated as the difference between initial and final body energy content, and deducted from the birds' known metabolisable energy intake to give an estimate of heat production, which agreed within 2 per cent with the estimate obtained by respiration calorimetry. Most comparative slaughter trials would last longer, and hence give a greater increment in energy balance, than that shown in Table 11.4. It is also worth noting that comparative slaughter trials often give lower estimates of energy retention than trials conducted with animals in calorimeters, possibly because the former allow more opportunity for animals to expend energy on muscular activity.

The comparative slaughter method requires no elaborate apparatus, but is obviously expensive and laborious when applied to larger animal species. The method becomes less costly if body composition, and hence energy content, can be measured in the living animal, or, failing that, in the whole, undissected carcass. Several chemical methods have been developed for estimating body composition *in vivo*. For most of them the principle employed is that the lean body mass of animals (i.e. empty body weight less weight of fat) is in several respects reasonably constant in composition. For example, in cattle 1 kg lean body mass contains 729 g water, 216 g protein and 53 g ash. This means that if the weight of water in the living animal can be measured, the weights of protein and ash can then be estimated; in addition, if total weight is known, weight of fat can be estimated by subtracting the lean body mass. In practice, total body water can be estimated by so-called 'dilution' techniques, in which a known quantity of a marker substance is injected into the animal, allowed to equilibrate with body water and its equilibrium concentration determined. The marker substances most commonly used are water containing the radioactive isotope of hydrogen, tritium, or its heavy isotope, deuterium. One difficulty with these techniques is that the markers mix not only with actual body water but also with water present in the gut (in ruminants as much as 30 per cent of total body water may be in gut contents). A second chemical method for estimating body composition *in vivo* is based on the constant concentration of potassium in the lean body mass.

The composition of a carcass is often estimated without dissection or chemical analysis from its specific gravity. Fat has an appreciably lower specific gravity than bone and muscle, and the fatter the carcass the lower will be its specific gravity. The specific gravity of a carcass is determined by weighing it in air and in water, but this method has technical difficulties (e.g. air trapped under water) that make it imprecise. Nevertheless, estimates of energy utilisation obtained by comparative slaughter and specific gravity measurements have been used in the USA to establish a complete cattle feeding system (see Chapter 12).

11.4 Utilisation of metabolisable energy

The general relationship between the metabolisable energy (ME) intake of an animal and its energy retention is shown in Fig. 11.5. When ME intake is zero (i.e. the animal is fasted), energy retention is negative; in this situation the animal uses its reserves to provide energy for the maintenance of its essential bodily functions, and this energy leaves the animal as heat. As ME intake increases, energy loss (i.e. negative retention) diminishes; when energy retention is zero, ME intake is sufficient to meet the animal's requirement for body maintenance. As ME intake increases further, the animal begins to retain energy, either in its body tissues or in products such as milk and eggs.

The slope of the line relating retention to intake is a measure of the efficiency of utilisation of ME. For example, if the ME intake of an animal was increased by 10 MJ and its retention increased by 7 MJ, the efficiency of utilisation of ME would be calculated as $7/10 = 0.7$. (Conversely, the heat increment would be calculated as $3/10 = 0.3$ of the ME.) Efficiency values like that of 0.7 calculated above are conventionally called 'k' factors, with

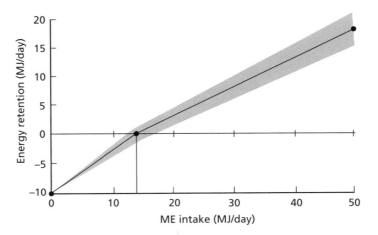

Fig. 11.5 Efficiency of utilisation of metabolisable energy (an example based on utilisation of metabolisable energy by the growing ruminant).

Box 11.6

The k factors commonly used to describe the efficiency of utilisation of metabolisable energy

k factor	Efficiency of utilisation for:
k_m	maintenance
k_p	protein deposition
k_f	fat deposition
k_g (or k_{pf})	growth in general
k_l	milk production (lactation)
k_c	foetal growth (the conceptus)
k_w	work (e.g. in draught animals)
k_{wool}	wool growth

the letter k carrying a subscript to indicate the function for which ME is being used. The commonly used k factors are shown in Box 11.6.

The term k_f has been used to denote both the specific efficiency of fat deposition (as indicated above) and the general efficiency of energy deposition in what were once called 'fattening' animals. Today, the preferred term for the latter use is k_g, as animals are now considered to grow rather than fatten.

In Fig. 11.5 the line relating energy retention to ME intake changes in slope at the maintenance level, becoming less steep and thus indicating a reduction in efficiency. Scientists argue as to whether there should be an abrupt bend (as in Fig. 11.5), or whether the relationship between retention and ME intake should be represented by a smooth curve. Conceptually, however, it is convenient to envisage a difference in the efficiency of utilisation of ME below and above maintenance. The apparently greater efficiency below maintenance is due to the fact that k_m is not an absolute measure of the utilisation of the energy of absorbed nutrients but is a relative measure of the efficiency with which nutrients obtained from food can be used to replace energy sources obtained from body reserves; this rather difficult concept is illustrated later.

An additional feature of Fig. 11.5 is the shaded area on either side of the lines. This is intended to indicate that the efficiency of utilisation of ME is quite variable. We shall see later that the principal causes of this variation in efficiency are, first, the nature of the chemical compounds in which the ME is contained (hence the nature of the food and the manner in which it is digested) and, second, the function for which these compounds are used by the animal.

Utilisation of metabolisable energy for maintenance

For maintenance purposes the animal oxidises the nutrients absorbed from its food principally to provide energy for work. If it is given no food, it obtains this energy mainly by the oxidation of body fat. When food is given, but in quantities insufficient to provide all the energy needed for maintenance, the task of providing ATP is partially transferred from the reserves of body fat to the nutrients absorbed. If the energy contained in these nutrients can be transferred to ATP as efficiently as can that in body fat, no extra heat will be produced by the animal apart from that associated specifically with the consumption, digestion and absorption of food. (Heat of fermentation comes into this category and also work of digestion, that is, heat arising from the energy used in the mastication of food and its propulsion through the gut, in the absorption of nutrients and in their transport to the tissues.)

The efficiency of free energy capture when body fats are oxidised and ATP is formed can be calculated from the reactions shown in Chapter 9 to be of the order of 0.67. For glucose, to take an example of a nutrient, the efficiency is similar, at about 0.70. One would therefore expect that glucose given to a fasting animal would be utilised without any increase in heat production, or in other words with apparent (calorimetric) efficiency of 1.0. Table 11.5 shows that this is approximately true. In sheep the efficiency is reduced through fermentation losses if the glucose passes into the rumen, but these losses are avoided if it is infused directly into the abomasum.

Table 11.5 also shows that dietary fat is used for maintenance with high energetic efficiency, as one would expect. When protein is used to provide energy for maintenance, however, there is an appreciable heat increment of about 0.2, which is in part attributable to the energy required for urea synthesis (see Chapter 9). In ruminants, energy for maintenance is absorbed largely in the form of volatile fatty acids. Experiments in which the pure acids have been infused singly into the rumens of fasting sheep have shown that there are differences between them in the efficiency with which their energy is utilised (Table 11.5). But when the acids are combined into mixtures representing the extremes likely to be found in the rumen, the efficiency of utilisation is uniform and high. Nevertheless the efficiency is still less than that for glucose, and this discrepancy, together with the energy lost through heat of fermentation in ruminants, leads one to expect that metabolisable energy will be utilised more efficiently for maintenance in those animals in which it is absorbed in the form of glucose than in ruminants.

Very few experiments have been carried out to determine the efficiency with which the metabolisable energy in foods is used for maintenance, and these few have been restricted almost entirely to ruminant animals fed on roughages. A selection of the results is given in Table 11.5.

Most of the metabolisable energy of the roughages would have been absorbed in the form of volatile fatty acids. Efficiency of utilisation is less, however, than for synthetic mixtures of these acids, since with whole foods heat losses are increased by heat of fermentation and by the energy used for work of digestion. In spite of this, the metabolisable energy in these foods was used with quite high efficiency.

Table 11.5 Efficiency of utilisation for maintenance of metabolisable energy in various nutrients and foods

	Ruminant	Pig, etc.[a]	Fowl
Food constituents			
Glucose	0.94 (1.00)[b]	0.95	0.89
Starch	0.80	0.88	0.97
Olive oil		0.97	0.95
Casein	0.70 (0.82)[b]	0.76	0.84
Fermentation products			
Acetic acid	0.59		
Propionic acid	0.86		
Butyric acid	0.76		
Mixture A[c]	0.87		
Mixture B[d]	0.86		
Concentrates			
Maize	0.80		
Balanced diets	0.70	0.85	0.90
Roughages			
Dried ryegrass (young)	0.78		
Dried ryegrass (mature)	0.74		
Meadow hay	0.70		
Lucerne hay	0.82		
Grass silages	0.65–0.71		

[a] Including dog and rat.
[b] Values in parentheses are from administration per duodenum.
[c] Mixture A: acetic 0.25, propionic 0.45, butyric 0.30.
[d] Mixture B: acetic 0.75, propionic 0.15, butyric 0.10.

Utilisation of metabolisable energy for productive purposes

Although energy may be stored by animals in a wide variety of products — in body fat, muscle, milk, eggs and wool — the energy of these products is contained mainly in fat and protein (only in milk is much energy stored as carbohydrate). The efficiency with which metabolisable energy is used for productive purposes therefore depends largely on the energetic efficiency of the metabolic pathways involved in the synthesis of fat and protein from absorbed nutrients. These pathways have been outlined in Chapter 9. In general the synthesis of either fat or protein is a more complicated process than its catabolism, in the same way that the construction of a building is more difficult than its demolition. Not only must the building materials be present in the right proportions, but they must arrive on the scene at the right time, and the absence of a particular material may prevent or seriously impair the whole process. Thus it was shown in Chapter 9 that fatty acid synthesis is dependent on a supply of reduced $NADP^+$. Because of the greater complexity of synthetic processes it is more difficult to estimate their theoretical efficiency.

Utilisation of metabolisable energy for growth

In Chapter 9, the synthesis of a triacylglycerol from acetate and glucose was shown to have a theoretical efficiency of 0.83. A higher value would be expected for fat synthesis from long chain fatty acids of dietary origin, but a lower value for the formation of fat from protein, as energy would be needed to synthesise the urea in which amino acid nitrogen would be excreted. In protein synthesis, the energy cost of linking amino acids together is relatively small and if these are present in the right proportions, the theoretical efficiency of protein synthesis is about 0.85 (Chapter 9). However, if some amino acids have to be synthesised while others undergo deamination, efficiency will be considerably less; as discussed later, the observed efficiency of protein synthesis is generally much lower than the observed efficiency of fat synthesis. The synthesis of lactose from glucose can be achieved with an efficiency of 0.96 (Chapter 9), but in the dairy cow the glucose so utilised will largely be formed from propionic acid, or possibly from amino acids (gluconeogenesis), and the efficiency of lactose synthesis will be lower.

The figures given above are all calculated from the appropriate metabolic pathways, and in relating them to the efficiency with which ME will be utilised it is important to remember that they will be reduced by those energy losses mentioned earlier (p. 270) which are directly referable to the consumption, digestion and absorption of food. Measuring in an animal the energetic efficiency with which a single body substance such as protein is synthesised is complicated by the fact mentioned previously that animals seldom store energy as single substances or even as single products. However, when growing animals are storing energy as protein and fat, it is possible to calculate, by the mathematical procedure known as regression analysis, the energy used for each of these processes. Analyses of this kind have been made for many calorimetric experiments carried out with pigs, and the results are shown in Table 11.6. The actual values are always lower than the theoretical, the discrepancies being relatively small for fat deposition but much larger for protein deposition. Thus for protein the value of about 0.5 estimated by calorimetry is much less than the theoretical value, given earlier, of 0.85. The main reason for this large discrepancy is believed to be that protein deposition is not just a matter of protein synthesis but is the resultant of two processes, synthesis and breakdown. Proteins in most body tissues are continuously being broken down and resynthesised, by reactions that generate heat, so this 'turnover' of protein reduces the calorimetric efficiency of protein deposition. One would expect that proteins in tissues with a high rate of turnover (such as those in the alimentary tract) would be deposited with particularly low calorimetric efficiency, whereas those with little or no turnover would be deposited more efficiently. Thus milk proteins, which leave the animal before they can be broken down, should be synthesised with high calorimetric efficiency. For pigs (and other non-ruminants) on normal diets, one would expect the efficiency of utilisation for growth as a whole to be intermediate between the efficiency values found for fat and protein deposition.

Table 11.6 Typical values for the efficiency of utilisation of metabolisable energy for growth in pigs

Form of energy storage	Origin of value	Substrate or diet	Efficiency
Fat (k_f)	Theoretical	Acetate + glucose	0.81
		Dietary fat	0.99
		Dietary protein	0.69
	Actual (calorimetric)	Normal diets	0.74
		Dietary fat	0.86
		Dietary carbohydrate	0.76
		Dietary protein	0.66
		Volatile fatty acids	0.65–0.71
Protein (k_p)	Theoretical	Amino acids	0.88
	Actual (calorimetric)	Amino acids	0.45–0.55
Protein plus fat (k_{pf})	Actual	Many diets (mean)	0.71
		Barley	0.6
		Maize	0.62
		Soya bean meal	0.48

In pigs, protein deposition commonly accounts for 20 per cent of total energy retention and the expected efficiency would therefore be 0.7. This value has been confirmed by many calorimetric experiments made with pigs.

In poultry, values for the efficiency of utilisation of ME for growth are similar to those found in pigs. They lie in the range 0.60–0.80 but for balanced diets are close to 0.70.

For ruminants the efficiency of utilisation of ME for growth is generally lower than that for pigs, and it is also more variable, as is shown in Table 11.7. When sheep and cattle are fed on diets similar to those commonly given to pigs (i.e. based largely on cereal-containing concentrates), the efficiency factor k_g rarely exceeds 0.62 and is therefore about 10 per cent lower than the average value for pigs suggested above, of 0.7. However, k_g values for ruminants are much lower when these animals are fed largely on forages and they are also much more variable. Thus the best forages, such as dried immature ryegrass, give k_g values >0.5, whereas poor-quality forages, such as wheat straw, give values as low as 0.2. We shall see later (Chapter 12) that the k_g values of forages for ruminants are related to their metabolisable energy concentrations. These k_g values are clearly far below both the theoretical and calorimetric values shown in Table 11.6. We now need to consider how they can be explained.

Heat of fermentation, as explained earlier (p. 270), accounts for part of the generally lower calorimetric efficiency of growth in ruminants; thus it would explain much of the 10 per cent difference in k_g between pigs and ruminants fed on similar, mainly concentrate diets. For a fuller explanation, however, we need to consider the end-products of digestion in ruminants. We have seen already (Chapter 8) that much of the energy of the products of digestion in ruminants is in the form of volatile fatty acids,

Table 11.7 Efficiency of utilisation of metabolisable energy in various nutrients and foods for growth and fattening of ruminants

Food constituents		Fermentation products	
Glucose	0.54 (0.72)[a]	Acetic acid	0.33–0.60
Sucrose	0.58	Propionic acid	0.56
Starch	0.64	Butyric acid	0.62
Cellulose	0.61	Mixture A[b]	0.58
Groundnut oil	0.58	Mixture B[c]	0.32
Mixed proteins	0.51	Lactic acid	0.75
Casein	0.50 (0.65)[a]	Ethanol	0.72
Concentrates		**Roughages**	
Barley	0.60	Dried ryegrass (young)	0.52
Oats	0.61	Dried ryegrass (mature)	0.34
Maize	0.62	Meadow hay	0.30
Groundnut meal	0.54	Lucerne hay	0.52
Soya bean meal	0.48	Grass silages	0.21–0.60
		Wheat straw	0.24
		Dried grass (chopped)	0.31
		Dried grass (pelleted)	0.46

[a] Values in parentheses are from administration per duodenum.
[b] Mixture A: acetic 0.25, propionic 0.45, butyric 0.30.
[c] Mixture B: acetic 0.75, propionic 0.15, butyric 0.10.

with relatively smaller amounts of energy in the forms of lipids, amino acids (from microbial and food proteins) and carbohydrates that have escaped rumen fermentation. Furthermore, the composition of the volatile fatty acid mixtures varies according to the nature of the diet, being dominated by acetic acid in the case of forage diets and by propionic acid in the case of concentrate diets. At an early stage in the study of energy metabolism of ruminants it was recognised that ME derived from poorly digested roughages, such as straws and low-quality hays, was utilised for growth with low efficiency (0.2–0.4, as illustrated in Table 11.7). At first, this poor efficiency was attributed to the energy costs of so-called 'work of digestion', which was considered to be the energy needed for mastication of fibrous foods and the propulsion of their undigested residues through the gut. Later, it seemed likely that the high acetic acid content of the volatile fatty acids arising from the digestion of such foods was responsible for their low k_g values. When contrasting mixtures of volatile fatty acids were infused into the rumen of fattening sheep, a large difference in k_g was observed (see Table 11.7). Subsequent experiments, however, in which less extreme mixtures of volatile fatty acids were used, in which the acid mixtures were better balanced with other nutrients (e.g. protein) and in which younger (growing rather than fattening) animals were used, showed much smaller effects of volatile fatty acid proportions on k_g. The current view is that acetic acid added to a well-balanced diet is used no less efficiently than propionic and butyric acids. Attention has therefore been directed again at 'work of digestion' as the cause of poor utilisation of ME from low-quality forages. As explained earlier (p. 271) the gut and the tissues associated with it that are served by the portal blood system (the so-called 'portal-drained viscera') have been shown to account for a high proportion – as much as 50 per cent – of the heat increment of feeding in ruminants, although an association between the heat produced from this source and the fibrousness of the diet has yet to be firmly established. It is significant, though, that treatment of forages to reduce their fibrousness, by grinding and pelleting them, improves the efficiency of utilisation of ME (Table 11.7).

Whatever the explanation for the poor utilisation of ME for growth by ruminants given low-quality forages, the practical problem remains. Ruminant production systems based on such feeds, as is the case in most of the tropical developing countries, are characterised by poor efficiency. Nevertheless, it is now recognised that efficiency can be improved by ensuring that the volatile fatty acids arising from digestion of poor-quality forages are balanced by supplements of other nutrients, especially protein and α-linked polysaccharides, which can escape rumen fermentation.

Utilisation of metabolisable energy for milk or egg production

These additional forms of animal production rarely occur alone, being usually accompanied by gains or losses of fat or protein from the body of the lactating mammal or laying bird. This means that estimates of the partial efficiency of milk or egg synthesis must usually be made by the mathematical partition of the metabolisable energy utilised. As rather complex diets are required for these syntheses, it is not possible to give efficiency coefficients for single nutrients as was done for maintenance and growth in Tables 11.5–11.7.

Over the past 40 years, many energy balance trials have been carried out with lactating cows, principally in the USA and the Netherlands similar trials have also been made with lactating sheep and goats. Analyses of the results of these trials have shown that the efficiency of utilisation of ME for the synthesis of milk in ruminants varies over a comparatively small range, from 0.56 for the poorest diets (7 MJ ME/kg DM) to 0.66 for the best diets (13 MJ ME/kg DM). For the diets commonly given to lactating ruminants, a constant value for k_l of 0.60 or 0.62 is often assumed. From the same analyses it has been calculated that when lactating animals are also building up a store of energy, probably mainly as fat, they can achieve this storage with almost the same efficiency (0.60) as that for milk synthesis. Thus the k_g or k_f factor for lactating ruminants is at the top end of the range of the factors for non-lactating ruminants (Table 11.7). Animals can use body reserves of energy to synthesise milk with an efficiency of about 0.84; this means that a cow which builds up her reserves towards the end of one lactation and then uses them for milk synthesis at the start of the next lactation has an overall efficiency of $0.60 \times 0.84 = 0.50$.

The greater efficiency of the lactating ruminant than of the growing or fattening animal is probably due in part to simpler forms of energy storage in milk, namely lactose and short chain fatty acids. Milk protein may be efficiently synthesised because it is rapidly removed from the body and is not required to take part in the turnover of amino acids experienced by most body proteins. In lactating sows, the efficiency of utilisation of ME for milk synthesis is about the same as the k_g factor in growing pigs (0.65–0.70).

The laying hen has been estimated to utilise metabolisable energy for egg synthesis with an efficiency in the range 0.60–0.80, and with a mean value of 0.69. For egg protein synthesis, efficiency is estimated to be 0.45–0.50 and for egg lipids, 0.75–0.80. Body tissue synthesis in laying hens is also achieved with high efficiency (0.75–0.80).

Other factors affecting the utilisation of metabolisable energy

Associative effects

In Chapter 10 it was explained that the digestibility of a food can vary according to the nature of the ration in which it is included. Associative effects of this kind have also been observed in the utilisation of metabolisable energy. In one experiment it was found that the metabolisable energy of maize meal was utilised for fattening with an apparent efficiency varying between 0.58 and 0.74 according to the nature of the basal ration to which it was added. In ruminants, such differences are likely to arise through variations in the effect of the food on the manner in which the whole ration is digested, and hence on the form in which metabolisable energy is absorbed. The implication is that values for the efficiency of utilisation of metabolisable energy for individual foods are of limited significance.

Balance of nutrients

The effect of the nutrient proportions of the food has been partly covered in earlier parts of this section. A fattening animal will tend to use metabolisable energy more efficiently if it is provided as carbohydrate than if it is provided as protein. Similarly, if a growing animal is provided with insufficient protein or with insufficient of a particular amino acid, it will tend to store energy as fat rather than as protein, and the efficiency with which it utilises metabolisable energy will probably be altered.

Deficiencies of minerals and vitamins, too, can interfere with energy utilisation. Thus a deficiency of phosphorus has been shown to reduce efficiency of utilisation by about 10 per cent in cattle. This effect is hardly surprising in view of the vital role of phosphorus in the energy-yielding reactions of intermediary metabolism.

Summary

1. The gross energy (GE) content of foods is measured as their heat of combustion in a bomb calorimeter. Typical values are (MJ/kg DM): carbohydrates 17.5, proteins 23.6, fats 39.3, average foods 18.5.

2. The digestible energy (DE) of a food is its GE minus the energy of the faeces (both expressed as MJ/kg food DM consumed).

3. The metabolisable energy (ME) of a food is its GE minus the energy of faeces, urine and methane (all expressed as MJ/kg food DM consumed).

4. Determination of ME in ruminants requires the use of a respiration chamber to collect methane. With poultry, which excrete faeces and urine together (but negligible methane), ME may be corrected for excreta of endogenous origin (measured in fasted birds) to provide estimates of true metabolisable energy (TME).

5. For a given animal species, losses of energy in urine and methane are relatively constant (amounting to about 20 per cent of DE), but may be affected by diet (e.g. high protein intake will increase urine energy loss).

6. In addition to their energy losses in excreta, animals also lose energy as heat. Part of their heat loss, called the heat increment of feeding, arises from work done by the animal in digesting and metabolising food. The remainder, called the basal or fasting metabolism, arises from the minimal work done by the animal for its essential bodily functions (i.e. body maintenance).

7. Deducting the heat increment of a food from its ME gives its net energy (NE) value. Net energy is used by the animal to meet its maintenance requirements and to form new body tissues (growth) or products (milk or eggs). The energy of these tissues and products is known as energy retention.

8. The heat loss of animals may be measured directly in an enclosed chamber called an animal calorimeter. Alternatively, it may be estimated from the animal's respiratory exchange (oxygen consumption and carbon dioxide production) in a respiration chamber.

9. Energy retention may be estimated indirectly, as ME intake minus heat production, or more directly from the animal's carbon and nitrogen retention in a respiration chamber. It may also be measured by the comparative slaughter and analysis of animals to determine their accretion of protein and fat.

10. Two vital statistics in the evaluation of foods for energy are the ME value and the efficiency of utilisation of ME. The latter is calculated as NE/ME and is called the 'k' factor.

11. The value of k varies according to the nature of the food, the species of animal and the function for which food energy is used by the animal.

12. Foods of high ME value tend to have higher k values because they are digested and metabolised with less expenditure of energy. Ruminants have lower k values than simple-stomached animals, mainly because of heat arising from microbial metabolism (heat of fermentation).

13. ME is used for maintenance with an efficiency (k_m) of 0.7–0.8. For tissue growth, efficiency (k_g) is 0.4–0.6 and depends on tissue proportions of protein and fat. Efficiency for protein retention (k_p) is lower (0.45–0.55) than that for fat synthesis (k_f, 0.65–0.85). For milk synthesis, efficiency (k_l) is about 0.6, and for egg production, 0.7.

Problems

11.1 A sheep consuming 1.2 kg/day of silage DM, containing 19 MJ GE/kg DM, excreted 5.0 MJ/day of energy in faeces, 1.36 MJ/day in urine and 1.80 MJ/day in methane. Calculate the ME value of the silage.

11.2 When the sheep in problem 11.1, above, was placed in a respiration chamber it used each day 536 litres of oxygen and excreted 429 litres of carbon dioxide, 45.8 litres of methane and 19 g of urinary nitrogen. Use the Brouwer equation to calculate its heat production. How much energy did it retain?

11.3 A pig in a respiration chamber stored 182.5 g of carbon and 10.4 g of nitrogen per day. Calculate its protein and fat deposition, and the energy content of its tissue growth.

11.4 A steer given daily 8 kg of maize silage DM, containing 11 MJ ME/kg DM, retained 12 MJ/day in its tissues. Its fasting metabolism was 42 MJ/day, and the efficiency of utilisation of ME for maintenance (k_m) was known to be 0.7. Calculate k_g, the efficiency of utilisation of the silage ME for growth.

Further reading

Blaxter K L 1967 *The Energy Metabolism of Ruminants*, London, Hutchinson.

Blaxter K L 1989 *Energy Metabolism in Animals and Man*, Cambridge University Press.

McLean J A and Tobin G 1987 *Animal and Human Calorimetry*, Cambridge University Press.

McCracken K, Unsworth E F and Wylie A R G 1998 *Energy Metabolism of Farm Animals*, Wallingford UK, CAB International (Proceedings of the 14th European Association for Animal Production Symposium: see also other volumes in this series).

Minson D J 1990 *Forage in Ruminant Nutrition*, New York, Academic Press.

Reid J T, White C D, Anrique R and Fortin A 1980 Nutritional energetics of livestock: some present boundaries of knowledge and future research needs. *Journal of Animal Science* **51**: 1393–415.

12 Evaluation of foods: systems for expressing the energy value of foods

12.1 Energy systems and energy models

12.2 Energy systems for ruminants

12.3 Energy systems for pigs and poultry

12.4 Energy systems for horses

12.5 Predicting the energy value of foods

For the farmer the essential steps in the scientific rationing of animals are, first, the assessment of their nutrient requirements, and second, the selection of foods which can supply these requirements. This balancing of demand and supply is made separately for each nutrient, and in many cases the nutrients given first consideration are those supplying energy. There are good reasons why energy should receive priority. In the first place the energy-supplying nutrients are those present in the food in greatest quantity; this means that if a diet has been devised to meet other nutrient requirements first and is then found to be deficient in energy, a major revision of its constituents will probably be needed. In contrast, a deficiency of a mineral or vitamin can often be rectified very simply by adding a small quantity of a concentrated source.

A further feature of the energy-containing nutrients that distinguishes them from others is the manner in which the performance of an animal, when measured as liveweight gain or milk or egg production, responds to changes in the quantities supplied. Whereas an animal with a given level of performance, a steer gaining 1 kg/day, for example, will respond eventually if the allowance of any one nutrient is reduced below the quantity required for this level of performance, increases in single nutrients above the requirement generally have little effect. Increasing the quantity of vitamin A supplied, for example, to twice the requirement is unlikely to affect the steer's liveweight gain (although it may increase its vitamin A reserves). But if energy intake alone is increased the animal will attempt to retain more energy, either partly as protein if nitrogen intake is adequate, or entirely as fat, and its liveweight gain will increase. In effect, energy intake is the pacemaker of production, as animals tend to show a continuous response to changes in the quantities supplied. If other nutrients are present in amounts only just sufficient for the animal's requirements, the response to an increase in energy intake is likely to be an undesirable one. Increasing the storage of body fat is likely to increase the need for minerals and vitamins associated with the enzyme systems involved in fat synthesis, and so to precipitate a deficiency of those substances. It is therefore important to maintain a correct balance between energy, the pacemaker, and other nutrients in the diet.

12.1 Energy systems and energy models

An energy system is essentially a set of rules relating the energy intake of an animal to its performance or productivity. The system can be used either to predict the animal's performance from a particular level of energy intake or to calculate the energy intake required to obtain a particular level of performance. The simplest energy systems consist of two sets of figures, one set for the energy requirements of animals and the other for the energy values of foods. Ideally, the two sets of figures will be expressed in the same units. For example, if an animal growing at the rate of 1 kg per day stores 15 MJ of energy, its energy requirement for growth will be stated as 15 MJ/kg gain. If the food to be used to produce this gain contains 5 MJ net energy per kilogram, the quantity required can readily be calculated as $15/5 = 3$ kg. In this example, both the animal's requirement for energy and the energy value of its food are expressed in terms of net energy, and the system used is described as a net energy system. We saw in the previous chapter, however, that the net energy value of a food is not a fixed value but varies according to the function for which it is required by the animal. As the metabolisable energy of the food is used more efficiently for maintenance than for growth (or for milk or egg production), a food will have at least two net energy values. For this reason it may be preferable to state the energy value of a food in units that are less variable, and in fact most energy systems employ metabolisable energy as the measure of the energy value of foods. When food energy is expressed as metabolisable energy and animal requirements are expressed as net energy, it is not possible to equate one with the other. To bring them together an additional element is needed for the system, which may be called an interface. The interface is essentially a method of calculating the equivalence of net and metabolisable energy for any specific rationing situation. Thus, in the example given above, if the energy value of the food was expressed as 10 MJ metabolisable energy per kilogram, the interface element required would be an estimate of k_g, the efficiency of utilisation of metabolisable energy for growth; if this were 0.5, the net energy value of the food would be calculated as

$$10 \times 0.5 = 5 \text{ MJ/kg}$$

as before.

The use of an interface allows energy systems to be more detailed and more complex than they would be if based solely on the energy input and output of animals. Thus it would be possible to subdivide energy input into the metabolisable energy provided by protein, by fat and by carbohydrates. Similarly, energy storage could be divided into that which is in the form of protein and that which is in the form of fat. The interface could now include the biochemical pathways linking nutrient intake and storage. The system would then have become what is commonly called a model. Although there is no sharp distinction between energy systems and energy models (and the systems can be seen as simple models), energy models are currently intended as scientific, rather than practical, tools. They can incorporate all the appropriate scientific information that is available and their operation is often valuable in identifying gaps in scientific knowledge. They are also interactive in the sense that they can be used not just to predict, say, energy retention and growth rate but also to show how energy storage might be divided among the various tissues and organs of the body. Although such complexity is not currently incorporated into energy systems, the increasing use of computers in the rationing of livestock (even on the farm) makes it easier for practical systems to become more detailed without becoming unworkable. Nevertheless, in this chapter the emphasis will be on energy systems rather than on energy models. A recent book edited by M K Theodorou and J France (see Further reading) describes current developments in both systems and models.

Before some actual energy systems are discussed, two further preliminary points must be made. The first is that the assessment of the energy value of a food requires a laborious and complicated procedure, which cannot be applied, for example, to samples of hay or silage brought by a farmer to an advisory chemist. For this reason an essential feature of most systems is a method for predicting energy value from some more easily measured characteristic of the food, such as its gross or digestible composition. Second, it is important to realise that energy systems for herbivores (ruminants and horses) are more complicated than those used for pigs or poultry. The main reasons for the greater complexity of systems for herbivores are the greater variety of foods involved and the wider range of options for digestion provided by the gut compartments.

12.2 Energy systems for ruminants

Early energy systems

Energy systems have a history going back to the first half of the nineteenth century but did not provide an adequate description of energy utilisation until methods of animal calorimetry came into use in the second half of that century. About 1900, H P Armsby at the University of Pennsylvania and O Kellner at the Möckern Experiment Station in Germany used the results of their calorimetric studies to devise energy systems based on the net energy value of foods. The systems differed in some respects, most evidently in the units used. Armsby expressed net energy in terms of calories (the unit ante-dating joules), but Kellner – believing that farmers would have difficulty in understanding calories – expressed net energy values of foods relative to the net energy value of that common food constituent, starch. For example, if the net energy value of barley was found to be 1.91 Mcal (megacalories) per kilogram, and that of starch, 2.36 Mcal/kg, then 1 kg of barley was stated to have a starch equivalent of 1.91/2.36 = 0.81 kg. Both Kellner's and Armsby's systems encountered difficulties caused by the differences in net energy values of foods for maintenance, growth, etc. and used approxima-tions to avoid these difficulties. Kellner's starch equivalent system was used (mainly in Europe) as the basis of practical rationing systems until the 1970s. A full description of the starch equivalent system was included in the first two editions of this book (1966 and 1973).

Armsby's net energy system was incorporated in what was at one time the standard reference work on the feeding of livestock in the USA, F B Morrison's *Feeds and Feeding*, but was not much used in practice. The pre-ferred system in the Americas was for many years the total digestible nutri-ents (TDN) system mentioned in Chapter 10. Although the TDN system, like Kellner's, employed units that were not units of energy, TDN values of foods can be fairly easily translated into metabolisable energy values.

The metabolisable energy system used in Britain

The energy system now used for ruminants in Britain was introduced in its original form in 1965 by the UK Agricultural Research Council. A revised version of this system was prepared by the Council in 1980 and published in a book entitled *The Nutrient Requirements of Ruminant Livestock*; this ver-sion is commonly referred to as the 'ARC 1980 system'. In 1990, the system was evaluated and modified by a UK national working party with member-ship drawn from research, advisory and commercial nutritionists. In 1993, a full description of the modified system was prepared by the Agricultural and Food Research Council and published by the Commonwealth Agricultural Bureaux (see Further reading). The description of the system given here is initially restricted to its essential features and its simplest forms of operation. The version of the system that is used in practice includes some modifications that will be discussed later. The system provides estimates of the energy requirements of six classes of livestock (cattle and sheep; growing, gestating and lactating), but the examples given in this chapter are restricted to grow-ing and lactating cattle. Energy requirements of the other classes of rumi-nants are referred to in later chapters.

In the ARC 1980 system, the energy values of foods are expressed in terms of metabolisable energy, and the metabolisable energy value of a ration is calculated by adding up the contributions of the foods making up the ration. The energy requirements of animals are expressed in absolute terms, as net energy. The essential feature of the interface is a series of equations to predict efficiency of utilisation of metabolisable energy for maintenance, growth and

Table 12.1 Efficiency of utilisation of metabolisable energy by ruminants for maintenance, growth and milk production

Metabolisability, q_m	0.4	0.5	0.6	0.7
Metabolisable energy concentration, MJ/kg DM	7.4	9.2	11.0	12.9
Maintenance, k_m	0.643	0.678	0.714	0.750
Growth and fattening, k_g	0.318	0.396	0.474	0.552
Lactation, k_l	0.560	0.595	0.630	0.665

Equations: $k_m = 0.35q_m + 0.503$
$\quad\quad\quad\quad k_g = 0.78q_m + 0.006$
$\quad\quad\quad\quad k_l = 0.35q_m + 0.420$

lactation (Table 12.1). The predictions are made from the metabolisable energy concentration of the diet, although this is expressed as the fraction ME/GE (sometimes called 'metabolisability') rather than as MJ/kg. Metabolisability may be converted to MJ ME/kg DM by multiplying it by 18.4, the mean concentration of gross energy in food dry matter (although this factor is too high for foods of high ash content and too low for those of high protein or lipid content). The efficiency values in Table 12.1 illustrate several points made earlier (in this chapter and Chapter 11). Although k_m and k_l vary with metabolisability (q_m), they vary much less than does k_g. To put this another way, for low-quality foods ($q_m = 0.4$), k_g is only 50 per cent of k_m, whereas for high-quality foods ($q_m = 0.7$) k_g is 74 per cent of k_m.

The system can be used in two ways, either to predict the performance of animals from a predetermined ration or to formulate a ration for a specific level of performance. The operation of the system is illustrated for beef cattle in Boxes 12.1 and 12.2. An example of ration formulation for dairy cows is given in Box 12.3. Ration formulation is often more difficult than performance prediction, because if the ME concentration of the diet is not known, the interface data cannot be calculated precisely. This means that if the procedure of Example (a) is carried out in reverse, provisional interface data have to be used, and the calculations have to be repeated until the correct match of diet and interface data is obtained. The repetition of the calculations (known as 'iteration') is easily carried out with a computer, but if a computer is not available, the approximate method illustrated in Example (b) can be used. For convenience, Example (b) starts at the endpoint of Example (a).

A new term required for Example (b) is k_{mp}, the average efficiency with which the metabolisable energy of a food is used for the combined functions of maintenance and production. This factor will be intermediate to k_m and k_p (in this case, k_g) and its value will vary with the production level of the animal. It is calculated from the relative requirements of net energy for maintenance and production, by the formula:

$$k_{mp} = (NE_m + NE_p)/(NE_m/k_m + NE_p/k_p)$$

Before leaving Example (b) we should look again at the equation for calculating k_{mp} from k_m and k_g. As mentioned above, the key factor in the calculation is the proportionality between the net energy requirement for

Box 12.1
Example (a): Prediction of performance in growing cattle

A 300 kg steer is to be fed on a ration of 4.5 kg of hay (containing 4 kg DM) and 2.2 kg of a cereal-based concentrate (2 kg DM).

Metabolisable energy supplied (MJ ME/day)

Hay (4 kg DM $\times$ 8 MJ/kg)	32
Concentrate (2 kg DM $\times$ 14 MJ/kg)	28
Total	60

Net energy required (MJ NE/day)

Maintenance (per day)	23
Liveweight gain (per kg)	15

Interface data

ME concentration of ration: (60/6)	10 MJ/kg DM
Metabolisability (q_m) : (10/18.4)	0.54
k_m (from Table 12.1)	0.692
k_g (from Table 12.1)	0.427

Partition of ME

Required for maintenance: (23/0.692)	33 MJ/day
Available for gain: (60–33)	27 MJ/day
Conversion to NE: (27 $\times$ 0.427)	11.5 MJ/day
Gain: (11.5/15)	0.77 kg/day

maintenance and the net energy requirement for production. The term $(NE_m + NE_p)/NE_m$ is sometimes known as the animal production level (APL). In this example it is $(23 + 11.5)/23 = 1.5$.

Ration formulation for dairy cows (Example (c)) may be as complicated as that for beef cattle, because – as shown in Table 12.1 – efficiency of utilisation of metabolisable energy for milk production (k_l) varies with M/D. In practice, some simplification is possible because the metabolisable energy concentration of dairy cow diets does not normally vary over such a wide range as does that of growing cattle, and it is therefore reasonable to assume constant values for both k_l and k_m. If k_m is assumed to be 0.72 and

Box 12.2

Example (b): Ration formulation for growing cattle

A steer of 300 kg is to be grown at a rate of 0.77 kg/day on a diet consisting of 3 kg of hay (i.e. less than in Example (a)) plus an unknown quantity of concentrate.

Net energy required (MJ NE/day)

Maintenance of 300 kg steer	23.0
Liveweight gain of 0.77 kg/day	11.5
Total	34.5

Feed and interface data

	Hay	Concentrate
Metabolisable energy (MJ/kg DM)	8	14
q_m (8/18.4; 14/18.4)	0.435	0.760
k_m (Table 12.1)	0.655	0.769
k_g (Table 12.1)	0.345	0.599
k_{mp} (see text for formula)	0.504	0.703
Net energy (ME × k_{mp}; MJ/kg DM)	4.03	9.84

Net energy supply (MJ NE/day)

From hay (3 kg; 2.7 kg DM): (2.7 × 4.03)	10.9
From concentrate: (34.5 − 10.9)	23.6
Concentrate required: (23.6/9.84)	2.4 kg DM
	2.7 kg concentrate

Thus the ration required is 3.0 kg of hay plus 2.7 kg of concentrate per day.

k_l 0.62, ration calculation becomes a matter of simple arithmetic. The gross energy content of milk, hence the net energy required for its production, varies with its fat content. In Example (c), the fat content is assumed to be 40 g per kg and the corresponding net energy requirement is 3.13 MJ per kg of milk.

Ration calculations for dairy cows are often complicated by changes in their energy reserves. If a cow is gaining weight and storing energy as fat, her requirements have to be considered in terms of three components: maintenance, milk synthesis and fattening. Conversely, if she is losing weight, then allowance must be made for the energy contributed from her fat reserves to milk synthesis or body maintenance.

Box 12.3
Example (c): Ration formulation for dairy cows

A 500 kg cow giving 20 kg of milk per day, containing 40 g of fat per kg, is to be fed on a diet containing 11 MJ ME/kg DM

Energy requirements (MJ/day)

	NE	k	ME
Maintenance: (500 kg)	37	/0.72	51
Milk: (20 kg × 3.13)	63	/0.61	101
Total			152
Feed required (kg DM/day): (152/11)	13.8		

Further refinements of the ARC 1980 system

All three components of the system, the energy content of foods, the energy requirements of animals and the interface linking them, can be given greater accuracy by introducing additional factors. The energy requirements of animals are discussed in later chapters; here we shall consider refinements allowing more accurate assessments of the animals' energy intake and their utilisation of metabolisable energy.

As the food intake of an animal increases, the metabolisability of the food energy declines (see p. 269). Food intake can be defined as the level of feeding, which is the metabolisable energy intake relative to that required for maintenance; thus in Example (a), above, the level of feeding is 60/33 = 1.82. (Note that the level of feeding is not exactly the same as the animal production level, as the former is calculated in metabolisable energy and the latter in net energy.) For growing cattle the level of feeding is commonly 2–2.5, but for lactating cattle it rises to 3–4. In the refined ARC 1980 system, the metabolisable energy requirement of lactating cows is increased by 1.8 per cent for each unit increase in the level of feeding above 1. Thus for a cow with a level of feeding of 3 (as in Example (c)) the initial estimate of ME requirement would be increased by $2 \times 1.8 = 3.6$ per cent to allow for the effects of increasing food intake on the ME content of the ration. The same correction is made for lactating ewes, but not for other classes of ruminants.

Increasing the level of feeding may also reduce the efficiency of utilisation of ME (i.e. reduce the k factors). Refinements of the system include corrections for this effect. For example, for growing cattle k_g is assumed to have its normal predicted value when the level of feeding is at twice the maintenance level, but if the level of feeding is greater than this, k_g is reduced, and if the level of feeding is less than twice the maintenance level, k_g is increased. For example, if cattle on a diet containing 10 MJ ME/kg

DM were fed at 2.5 times maintenance, k_g would be reduced from 0.43 to 0.39. The same correction is used for growing lambs.

A possible refinement of the ARC 1980 system, which has been advocated but not generally adopted, is to modify k_g values according to the nature of the diet. Table 12.1 shows that the k factors vary according to the metabolisability of the dietary energy and also with the function of the animal for which the energy is used. There is also evidence that when metabolisable energy is used for growth, the efficiency with which it is utilised varies with the nature of the diet. For example, when diets containing 11 MJ ME/kg DM are made up from either high-quality forage alone or poorer-quality forage plus concentrates, the diet containing concentrates will have a k_g value about 5 per cent higher than that of the diet of forage alone. Among the forages, there is evidence that k_g values are greater for first growths (i.e. spring growths) of temperate herbages than for later growths having the same metabolisable energy concentration, and are higher for temperate than for tropical forages. However, it is difficult to classify diets into different categories for the purpose of predicting k_g, and the single equations of Table 12.1 are used in the current refined version of the ARC 1980 system.

In its assessment of the ARC 1980 system, the 1990 UK working party referred to earlier concluded that the system generally underestimated the ME requirements of growing cattle, but not of dairy cows (or of other classes of ruminants). It recommended that energy requirements of growing cattle calculated by the system should be adjusted by the use of correction factors, which in effect increase the energy requirements for production (i.e. not for maintenance) by 10–15 per cent. The reason for the underestimation is not known; a likely explanation is that the ARC 1980 system underestimates the maintenance requirements of ruminants (a topic discussed in Chapter 14). Another possibility is that the positive correction applied to k_g at lower levels of feeding is too generous.

Although it seems undesirable to introduce an arbitrary correction factor to an otherwise logical system, it is important that the system should accurately predict the animal growth rates that are achieved in practice.

Alternative energy systems for ruminants

In 1990 the Australian Standing Committee on Agriculture introduced a set of feeding standards for ruminants that included an energy system based on the ARC 1980 system. The differences between the Australian and ARC systems reflect the differences in animal production systems between the two countries, the Australian system being intended mainly for use with grazing animals. It therefore has refined estimates of maintenance requirements that include the energy costs of grazing. The system also employs refined estimates of k_g for forage diets, which take into account the poorer utilisation of the metabolisable energy of tropical and later-growth temperate forages. The system also includes an estimate of the efficiency of utilisation of metabolisable energy for wool growth (k_{wool}, 0.18). Finally, the Australian system has no corrections to k_g for level of feeding, but for all classes of stock there is a positive correction to the maintenance requirement as total ME intake rises above the maintenance level; this correction raises the maintenance requirement by 10 per cent at a feeding level of twice maintenance.

In 1988 the European Association of Animal Production carried out a survey of the feeding systems in use in European countries. The report on energy systems for ruminants (see Further reading, under Van der Honing and Alderman) demonstrated the wide variety of systems in use in Europe, and it is not possible to describe all of them here. The Netherlands, Belgium, France, Germany, Switzerland, Italy and Austria have systems with many features in common, and the Dutch system will be described as an example. The metabolisable energy content of foods is calculated from digestible nutrients and is then converted to a net energy value. For growing animals, the basis for this conversion is that the animal production level (p. 297) is assumed to be constant, at 1.5, and hence that k_{mp}, as explained earlier in this chapter, has a unique value for a food of known metabolisable energy concentration. Each food can therefore be given a single net energy value for maintenance and production (NE_{mp}), but this is converted to a unit value by dividing it by the presumed NE_{mp} of barley (6.9 MJ/kg, or about 8 MJ/kg DM). For lactating cows a corresponding net energy value for maintenance and lactation is calculated by assuming that k_l is 0.60 when $ME/GE = 0.57$ (i.e. when $M/D = 10.5$ MJ/kg) and varies by 0.4 per unit change in ME/GE. For example, if $M/D = 11.5$ MJ/kg and $ME/GE = 0.62$, $k_l = 0.60 + (0.4 \times (0.6 - 0.57)) = 0.62$ and the net energy value of the food for lactation (called NE_l) $= 0.62 \times 11.5 = 7.1$ MJ/kg DM. The calculation is further complicated by reducing the predicted value of NE_l by 2.5 per cent, to allow for the normally high feeding level of dairy cows, and by converting NE_l to a unit value (barley again being assumed to contain 6.9 MJ NE/kg). In addition to being used for cows, NE_l values are used in the rationing of younger dairy animals (i.e. heifers being reared as dairy herd replacements).

The simplifying assumptions used in the Dutch scheme to calculate the net energy values of foods are therefore that (a) for growing animals, $APL = 1.5$, and (b) for growing heifers, $k_{mp} = k_l$. A third assumption, which is not stated but is implicit in the system, is that $k_m = k_l$. All three are liable to introduce inaccuracies and are allowed for by adjusting the net energy requirements of animals. For example, with regard to assumption (b) above, it is recognised that in heifers growing slowly (and therefore using most of their energy intake for maintenance) k_l is an underestimate of k_{mp}; whereas in heifers growing rapidly, k_l is an overestimate of k_{mp}. The net energy requirements of slow growing heifers have therefore been reduced and those of fast growing heifers increased, to cancel out the biases in the estimation of the net energy values of their foods.

The energy systems grouped with the Dutch system are similar to it in principle, but differ in detail. Thus the metabolisable energy values of foods may be calculated by different procedures, different corrections may be used for level of feeding and the units employed may be MJ rather than unit values. For example, in Switzerland the units are MJ of net energy.

Several countries in Scandinavia have net energy systems that were based on Kellner's starch equivalent system but now use Scandinavian feed units

(Norway and Finland) or fattening feed units (Denmark). Sweden, however, has a metabolisable energy system.

After using a system of total digestible nutrients (see p. 259) for many years, the USA changed to net energy systems for beef and dairy cattle, these being described in publications of the National Research Council (see Further reading). Metabolisable energy is calculated as $0.82 \times$ digestible energy and digestible energy is calculated as 4.409 Mcal (18.45 MJ) per kg TDN. For beef cattle, foods are given two net energy values, for maintenance (NE_m) and gain (NE_g), these being calculated from the metabolisable energy (ME) content of dry matter in each food by means of the following equations:

$$NE_m = 1.37ME - 0.138ME^2 + 0.0105ME^3 - 1.12$$

$$NE_g = 1.42ME - 0.174ME^2 + 0.0122ME^3 - 1.65$$

Where: ME = metabolisable energy content of dry matter, and all energy values are expressed in Mcal/kg.

These equations predict the NE_m and NE_g values of a food containing 11 MJ (2.63 Mcal) ME/kg DM to be 7.2 and 4.6 MJ/kgDM, respectively, which may be compared with the UK ARC values (derived from Table 12.1) of 7.9 and 5.2. For a poorer-quality food, containing 7.4 MJ (1.77 Mcal) ME/kg DM, the US values are 3.9 MJ NE_m and 1.6 MJ NE_g, and the corresponding UK values are 4.8 and 2.4 MJ, respectively. Thus the US values are considerably lower than those derived by the UK system, the difference being particularly large for foods low in ME content.

Net energy values for lactation (NE_l) for the US system have been calculated from TDN, digestible or metabolisable energy by an equation similar to that used in the Dutch system. For example, foods containing 10 or 12 MJ ME/kg DM are calculated to contain 6.0 and 7.1 MJ NE_l/kg DM in the American system and 5.8 and 7.2 MJ NE_l in the Dutch system. Net energy requirements for body maintenance and milk synthesis are expressed as NE_l, as they are in the Dutch and related European systems.

The US net energy systems for both beef and dairy cattle are currently being modified to incorporate the Cornell net carbohydrate and protein system referred to in Chapter 1. This system includes calculations of both the energy and protein values of feeds. The principle on which it operates is that if the quantity and characteristics (such as solubility) of each nutrient present in the food are known, the rate at which it is digested and the products of its digestion that are absorbed from the rumen and lower gut can be calculated. The energy and protein values of these products can then be summed to give an estimate of the values of the food as a whole. To be operated efficiently the system requires unusually detailed analytical data for the foods being used (see Chapter 1). Thus carbohydrates in foods are divided into two kinds of fibre (fermentable or not fermentable in the rumen) and two kinds of non-fibrous components (starch and sugars). Proteins are similarly divided into five fractions that vary in the rates at which they are degraded in the rumen. In the latest versions of the US National Research Council publications referred to above, the procedures of the Cornell system are offered as an alternative to simpler methods of calculating net energy values from metabolisable energy. The latest US system for dairy cows is outlined in Chapter 16.

Retrospect and prospect

The first edition of this book was published at a time (1966) when the older energy systems for ruminants, such as Kellner's starch equivalent system, were being re-evaluated in the light of newly acquired calorimetric data. In the former German Democratic Republic, Kellner's system had been re-modelled, and in Britain a new system had recently been introduced by the Agricultural Research Council. The hope at the time was that new information on energy utilisation would bring more accurate systems which might be less complicated and therefore more acceptable for international use. In the event this hope has not been fully realised. Greater accuracy may perhaps have been achieved, but at the expense of greater complexity and an increase in the number of systems in use.

The energy systems for ruminants which have been most improved since 1966 have been those for the lactating cow. As the reader will now appreciate, energy systems can accommodate the dairy cow better than they can growing and fattening cattle (or sheep). Efficiency of utilisation of metabolisable energy varies less in the cow than in the steer, and the energy content of the product, milk, is less variable and more predictable than is the case with liveweight gain. Various systems for dairy cows have been compared in Fig. 12.1, by taking a specimen diet and using the procedures of each system to predict first the energy value of the diet, and second, the quantities required for specified levels of production; the conclusion drawn is that systems for dairy cows do not differ greatly. However, a recent analysis of calorimetric data for dairy cows obtained over the last 20 years showed that the Australian system was better at predicting energy requirements than the UK, US and Dutch systems.

Figure 12.2 illustrates a different method for comparing energy systems, in this instance for beef cattle. A series of input–output data mainly for grass-fed beef cattle provided a basis for comparing actual output with that predicted from input by the use of the two systems. The conclusion drawn was that the US net energy system predicted the performance of the cattle more accurately than did the 1965 version of the UK Agricultural Research Council system.

It must be admitted, however, that comparisons of the kinds illustrated above are seldom sufficiently conclusive for the nutritionists of one country to discard their own system and adopt *in toto* that of another country.

A general change that has taken place in the use of energy systems is that the emphasis is now on performance prediction for foods eaten to appetite rather than on ration calculation. As the examples given earlier illustrate, the same figures are used for the two types of calculation; however,

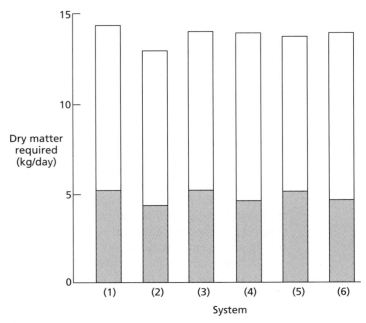

Key to systems:
(1) Metabolisable energy (UK)
(2) Scandinavian feed units (Denmark)
(3) Total digestible nutrients (USA)
(4) Energy feed units (former East Germany)
(5) Net energy for lactation (USA)
(6) Feed units for lactation (The Netherlands)

Fig. 12.1 Comparison of energy systems for predicting the requirements of dairy cows fed on a standard diet containing equal parts of roughages and concentrates. Requirements for maintenance alone are shown by the shaded areas, and for maintenance plus the production of 20 kg milk by the complete bars. (After Steg R and van der Honing Y 1979 *Report of the Dutch Institute for Cattle Feeding Research,* Hoorn, No. 49.)

performance prediction often requires estimates of voluntary food intake (see Chapter 17).

In the future, as energy systems are modified to incorporate new findings they are likely to become even more complex. As the need for simplicity of calculation is diminished by the increasing availability of computers, energy systems are becoming parts of much larger mathematical models of nutrient requirements that are capable of dealing simultaneously with energy, amino acids, vitamins and minerals (and with the interactions between them). Complex systems tend to obscure the principles upon which they are based. Students of animal nutrition should therefore pay particular attention to the principles of energy metabolism outlined in the previous chapter and at the same time should familiarise themselves with the energy systems currently used in their own countries.

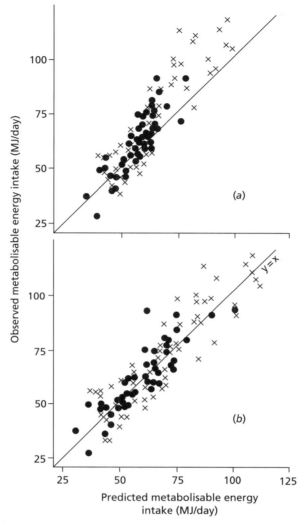

Fig. 12.2 Comparison of metabolisable energy intake observed in beef cattle with intake predicted from animal performance by two systems: (*a*) the metabolisable energy system of the UK Agricultural Research Council; (*b*) the net energy system of the US National Research Council. (From Joyce J P *et al.* 1975 *New Zealand Journal of Agricultural Research* **18**: 295.)

12.3 Energy systems for pigs and poultry

Energy systems for pigs and poultry are less complicated than those for ruminants, both in derivation and in application. One reason for this is that pigs and poultry, because they digest cellulose to a relatively small extent, are limited to a range of foods varying little in metabolisable energy concentration. It is also generally considered that in pigs and poultry the utilisation of metabolisable energy differs to a relatively small extent from one food to another. A further reason is that, unlike ruminants, pigs and poultry are less commonly rationed for a stated level of production. Poultry, in particular, are usually fed to appetite in the hope that they will achieve maximum rates of meat or egg production. Nevertheless, the energy concentration of foods and diets remains an important characteristic for poultry (and pigs) fed to appetite, because these animals tend to adjust their intakes to provide a constant energy intake (see Chapter 17). If the energy concentration of their diet is increased (for example, by adding fat) non-ruminant animals are likely to reduce their intake; if no adjustment is made in the concentration of amino acids and other nutrients, they are liable to suffer deficiencies. In the practical rationing of pigs and poultry, one component of energy systems, namely the energy concentration of foods, is quite important; whereas the other component, the energy requirements of the animals, is considered to be much less important.

Pigs

A net energy system for pigs was introduced at the time of the starch equivalent system for ruminants (by G Fingerling, Kellner's successor at the Möckern Experiment Station): revised versions of this, or of alternative net energy systems, have been used in a few European countries until the present day. In the last 10 years there have been further proposals for net energy systems for pigs, these being prompted by the use of greater ranges of foods for pigs, including fibrous by-product foods. As a measure of food energy, net energy will be superior to metabolisable energy (and digestible energy) if there are differences between foods in the efficiency of utilisation of metabolisable energy (i.e. the k factors). For many pig diets, the k factor for a single function is reasonably constant; for example, Table 11.6 shows that the efficiency for growth, k_g, is close to 0.71. Fibrous foods, however, provide exceptions, as the end-products of fermentative digestion are utilised less efficiently in the pig (just as they are in ruminants). The UK Agricultural Research Council, in its 1981 publication *The Nutrient Requirements of Pigs*, suggested that methane and heat loss associated with the metabolism of volatile fatty acids together reduced the net energy value of 'fermented' digestible energy to no more than two-thirds of that of digestible energy derived from absorption of nutrients from the small intestine.

It is also well recognised that when k_g is sub-divided into separate factors for deposition of protein (k_p) and fat (k_f), these factors are different. Thus separate values for k_p (0.56) and k_f (0.74) have been used to calculate the energy requirements for growth of pigs in both the UK and the USA. This difference has become important in pigs because of the breeding policy of selecting pigs to store less fat and more protein.

Despite this variation in k factors, most countries have based their energy systems for pigs on either digestible or metabolisable energy, although in some cases with modifications that acknowledge the lower net energy value of the metabolisable energy of fibrous foods. The system used in the UK is based on digestible energy, although metabolisable energy is a more common measure. It is argued that these two measures are closely related (e.g. $ME = 0.96DE$), although for fibrous foods that give rise to methane in the hind gut of the pig, the factor is 0.90 rather than 0.96.

Poultry

Net energy values of foods for poultry were determined 50–60 years ago in the USA by G S Fraps, who used the method of comparative slaughter to measure energy retention in young chickens. The figures obtained were called productive energy values, in order to emphasise that they were net energy values for growth, not maintenance. Fraps also devised methods for predicting the productive energy values of foods from their gross or digestible nutrients.

Fraps's productive energy values have never been widely employed and in most countries metabolisable energy has been used to express the energy value of poultry foods. As for pig foods, it has been argued that the metabolisable energy of poultry foods is used with reasonably constant efficiency. The recent expansion of calorimetric studies of poultry has yielded results that do not support this argument; in particular, metabolisable energy provided by dietary fat has been shown to be more efficiently utilised than that supplied by carbohydrate. There have therefore been further proposals for net energy systems for poultry, which would include equations for predicting net energy from metabolisable energy, but these have not been adopted (see Further reading). Metabolisable energy is very easily measured in poultry, since faeces and urine are voided together, and this is undoubtedly a strong factor favouring its retention in energy systems for poultry.

12.4 Energy systems for horses

The US National Research Council at one time expressed the energy value of foods for horses in terms of total digestible nutrients, but it now uses digestible energy in a system that is employed in the Americas, Australasia and parts of Europe (including the UK). In 1984 the French Institut National de la Recherche Agronomique introduced a system for horses that is based on net energy. Because of the digestive processes of horses, one would expect a poor correlation between digestible and net energy for the range of foods consumed by horses, which includes concentrates and forages such as poor hay. Thus hay would mainly be digested by microorganisms in the large intestine, with the consequent losses of energy as methane and heat arising from fermentation and the metabolism of volatile fatty acids. But a food such as maize would be mainly digested in the small intestine, with its energy being absorbed and metabolised as hexose sugars. For this reason the relationships between digestible, metabolisable and net energy in the net energy system for horses are much more variable than in corresponding systems for ruminants; for example, the factor k_m for horses ranges from 0.6 for fibrous roughages to 0.8 for starchy concentrates.

12.5 Predicting the energy value of foods

The precision of any energy system depends on the accuracy with which the energy content of individual foods or diets can be estimated. The metabolisable energy content of grass silage, for example, can vary from perhaps 8 to 12 MJ/kg DM according to the type of grass from which it is made and the silage-making practices employed. A farmer wishing to assess the energy value of his silage might use a reference book to identify its type and to provide an appropriate value for MJ ME/kg DM (see Appendix 2, Table A2.1 for examples). The assessment can be made more precisely, however, if the silage is subjected to chemical analysis. The analytical data can either be used to 'type' the silage more accurately, or better, they can be fitted into equations that have been devised to predict energy value from chemical composition. For example, the following equation allows the metabolisable energy value of silages to be predicted from their content of modified acid-detergent fibre (MADF, g/kg DM):

$$ME \; (MJ/kg \; DM) = 15.33 - 0.0152 \; MADF$$

An alternative to chemical analysis is the assessment of digestibility by fermentation *in vitro* (see p. 249) and the prediction of metabolisable energy value from the digestible organic matter content of the food. For roughages given to ruminants a simple formula that is commonly used is:

$$ME \; (MJ/kg \; DM) = 0.016 \; DOMD$$

where DOMD = g digestible organic matter per kg dry matter.

New methods of analysis, particularly near-infrared reflectance spectrometry (NIRS), are providing opportunities for predicting the digestibility and metabolisable energy value of foods in the laboratory (see Chapter 10). It must be emphasised that all laboratory methods of predicting energy value are dependent on the availability of reliable data obtained by conducting metabolism trials with animals. It is the systematic assessment of many food samples in such trials that provides the regression equations used to predict energy values from laboratory measures.

For cereal grains and other concentrated foods, prediction of energy value is easier than for silage because such foods are less variable in chemical composition; appropriate values can therefore be taken from tables. Nevertheless, there are occasions when the energy values of concentrates need to be predicted from their chemical composition or other characteristics. For example, many countries are introducing legislation to require compound feed manufacturers to declare the energy value (typically, the metabolisable energy value) of their products, and a simple method of checking such values is needed. In the UK, the following equations are used for predicting the energy content of compound feeds for poultry, pigs and ruminants.

Poultry:

$$ME \text{ (MJ/kg feed)} = 0.01551CP + 0.03431OIL + 0.01669STA + 0.01301SUG$$

Pigs:

$$DE \text{ (MJ/kg DM)} = 17.47 + 0.0079CP + 0.0158OIL - 0.0331ASH - 0.0140NDF$$

Ruminants:

$$ME \text{ (MJ/kg DM)} = 0.014DIGY + 0.025OIL$$

where:
CP = crude protein
OIL = lipid extracted with an organic solvent after acid hydrolysis
STA = starch
SUG = total sugars, expressed as sucrose
NDF = neutral-detergent fibre, determined by a method that includes treatment with amylase to remove starch
DIGY = digestibility estimated in the laboratory with neutral-detergent and enzyme treatment. Also known as NCGD (neutral-detergent cellulase plus gamanase digestibility).

and all the above measures are expressed as g/kg feed (poultry feeds) or g/kg dry matter (pig and ruminant feeds).

Note that the three equations differ in principle. The poultry equation assigns logically derived factors to each food constituent; for example, the factor for oil indicates that $0.03431/0.0393 = 0.87$ of the gross energy of lipids in poultry feeds is metabolisable. The pig equation starts with a constant value and adjusts it upwards for the protein and oil content of feeds, and downwards for ash and fibre. The ruminant equation is based on a biological, rather than chemical, evaluation of the feed, but includes an adjustment for oil.

Earlier equations were based on the original components of the proximate analysis (e.g. on crude fibre rather than neutral-detergent fibre). An example of an earlier equation for pig foods, which was based on a compilation of European studies made by the European Association for Animal Production, is:

$$ME \text{ (MJ/kg DM)} = 0.018CP + 0.0315OIL + 0.0163NFE - 0.0149CF$$

The precision of this and other equations can be assessed from their residual standard deviations (rsd). The ME equation based on proximate constituents (rsd 0.54) is less precise than the DE equation based on modern analytical components (rsd 0.33).

Summary

1. Energy systems have three components: (a) values for the energy requirements of animals, (b) values for the energy content of foods, and (c) information for connecting (a) and (b), called an interface.

2. In the UK metabolisable energy system for ruminants, animal requirements are expressed in terms of net energy (NE: e.g. the gross energy value of milk or body weight gain) and food energy is expressed as metabolisable energy (ME). The interface includes factors (k factors) for calculating the net energy of the food ME, which depend on the ME concentration in food dry matter (M/D) and the function for which the ME is being utilised, such as maintenance or growth.

3. Energy systems are used either to formulate rations for animals with a prescribed level of production or to predict the performance of animals given a prescribed ration.

4. The UK ME system for ruminants has been refined by including corrections for the tendency for increasing the level of feeding to reduce both the ME value of foods and the efficiency with which the ME is utilised (hence NE).

5. Australia has a system based on the UK ME system, but with refinements intended to cater especially for grazing animals.

6. Several countries of continental Europe have similar energy systems, in which a simplified interface is used to calculate NE from ME. The USA uses a net energy system in which foods are given NE values for maintenance, growth and milk production. This system is currently being modified to incorporate new knowledge of digestion in the ruminant.

7. When applied to dairy cows, the various energy systems give rather similar results, but for growing cattle (and sheep) the UK ME system predicts lower food requirements (or greater growth rates) than the US or European systems.

8. Energy systems for pigs and poultry are based on digestible or metabolisable energy, because with most foods k values are considered to be constant. However, systems for pigs are now beginning to include calculations of separate requirements for the protein and fat components of growth. In the case of poultry foods, ME is easily measured because faeces and urine are voided together.

9. Energy systems for horses are based on digestible or net energy.

10. In Britain there are standard equations for estimating the energy value of ruminant, pig or poultry feeds from the results of laboratory analyses or digestibility trials.

Problems

12.1 Female lambs weighing 30 kg are to be fed on a ration that will allow them to gain 0.2 kg/day. Their maintenance requirement is 2.4 MJ NE/day, and the energy content of their liveweight gains is 15.6 MJ/kg. The foods available are a forage containing 9 MJ ME/kg DM and a concentrate containing 13 MJ/kg DM. The lambs will be given 0.5 kg DM/day as forage. How much concentrate will they require?

12.2   A food intended for pigs was found to contain the following (g/kg DM):

Crude protein   160
Oil   50
Neutral-detergent fibre   150
Ash   60

(a)   Calculate its digestible energy value.
(b)   The food was given in an amount of 1.0 kg per head per day to pigs weighing 35 kg with a maintenance requirement of 7 MJ DE/day. The pigs made gains containing 210 g muscle DM (170 g protein) and 200 g fat per kg. Calculate (from data presented in Chapters 11 and 12) the daily weight gain of the pigs.

Further reading

Agnew R E and Yan T 2000 Impact of recent research on energy feeding systems for dairy cattle. *Livestock Production Science* **66**: 197–215.

Agricultural Research Council 1980 *The Nutrient Requirements of Ruminant Livestock*, Farnham Royal, Commonwealth Agricultural Bureaux.

Agricultural Research Council 1981 *The Nutrient Requirements of Pigs*, Farnham Royal, Commonwealth Agricultural Bureaux.

Agricultural and Food Research Council 1990 *Technical Committee on Responses to Nutrients, Report No. 4, Nutrient Requirements of Sows and Boars*. Wallingford, UK, CAB International. (See also *Nutrition Abstracts and Reviews, Series B* **60**: 383–406.)

Agricultural and Food Research Council 1993 *Energy and Protein Requirements of Ruminants* (an advisory manual prepared by the AFRC Technical Committee on Responses to Nutrients), Wallingford, UK, CAB International.

Alderman G 1985 Prediction of the energy value of compound feeds. In: Haresign W and Cole D J A (eds) *Recent Advances in Animal Nutrition – 1985*, London, Butterworth.

Henry Y, Vogt H and Zoiopoulos P E 1988 Feed evaluation and nutritional requirements: pigs and poultry. *Livestock Production Science* **19**: 299–354.

Morris T R and Freeman B M (eds) 1974 *Energy Requirements of Poultry*, Poultry Science Symposium No. 9, British Poultry Science Ltd.

National Academy of Sciences/National Research Council 1989 *Nutrient Requirements of Dairy Cattle*, 6th rev. edn, Washington, National Research Council.

National Academy of Sciences/National Research Council 1989 *Nutrient Requirements of Horses*, Washington, National Research Council.

National Academy of Sciences/National Research Council 2000 *Nutrient Requirements of Beef Cattle*, 7th rev. edn, Washington, National Research Council.

Pirgozliev V and Rose S P 1999 Net energy systems for poultry feeds: a quantitative review. *World's Poultry Science Journal* **55**: 23–36.

Theodorou M K and France J (eds) 2000 *Feeding Systems and Feed Evaluation Models*, Wallingford, UK, CAB International.

Van der Honing Y and Alderman G 1988 Feed evaluation and nutritional requirements: ruminants. *Livestock Production Science* **19**: 217–78.

Van Es A J H 1978 Feed evaluation for ruminants 1. The systems in use from May 1977 onwards in the Netherlands. *Livestock Production Science* **5**: 331–45.

Whittemore C T 1997 An analysis of methods for the utilisation of net energy concepts to improve the accuracy of feed evaluation in diets for pigs. *Animal Feed Science and Technology* **68**: 89–99.

Wiseman J and Cole D J A (eds) 1990 *Feedstuff Evaluation*, London, Butterworth.

Historical references

Armsby H P 1917 *The Nutrition of Farm Animals*, New York, Macmillan.

Kellner O 1926 *The Scientific Feeding of Farm Animals*, 2nd edn (translated by W Goodwin), London, Duckworth.

13 Evaluation of foods: protein

13.1   Crude protein

13.2   Digestible crude protein

13.3   Determination of endogenous nitrogen

13.4   Measures of protein quality for monogastric animals

13.5   Measures of food protein used in practice in the feeding of pigs and poultry

13.6   Measures of protein quality for ruminant animals

13.7   The UK metabolisable protein system

Proteins are made up of amino acids, the classification of which into indispensable (essential) and dispensable (non-essential) has already been described in Chapter 4, and in Section 9.2 under 'protein synthesis'. For food to be used with maximum efficiency, the animal must receive sufficient quantities of both the indispensable and dispensable amino acids to meet its metabolic demands. Simple-stomached animals such as pigs and poultry obtain these acids from the breakdown of food proteins during digestion and absorption. In the case of ruminant animals the situation is more complex.

Considerable degradation and synthesis of protein occur in the rumen, and the material which finally becomes available for digestion by the animal may differ considerably from that originally present in the food. Different approaches to the evaluation of protein sources are therefore necessary for ruminant and non-ruminant animals.

In the past, foods were evaluated as sources of protein by somewhat crude and simple methods which took no account of the species of animal for which the foods were intended. Certain of these methods are still sometimes used. These will be dealt with first and then the more refined methods of assessing protein quality for simple-stomached and ruminant animals will be discussed separately.

13.1   Crude protein (CP)

Most of the nitrogen required by the animal is used for protein synthesis. Most of the food nitrogen is also present as protein, and it is convenient and almost universal for the nitrogen requirements of animals and the nitrogen status of foods to be stated in terms of protein. Chemically, the protein content of a food is calculated from its nitrogen content determined by a modification of the classical Kjeldahl technique; this gives a figure which includes most forms of nitrogen although nitrites, nitrates and certain cyclic nitrogen compounds require special techniques for their recovery. Two assumptions are made in calculating the protein content from that of nitrogen: first, that all the nitrogen of the food is present as protein; and second, that all food protein contains 160 g N/kg. The nitrogen content of the food is then expressed in terms of crude protein (CP):

$$CP \ (g/kg) = gN/kg \times 1000/160$$

or more commonly:

$$CP \ (g/kg) = gN/kg \times 6.25$$

Both the assumptions are unsound. Different food proteins have different nitrogen contents and, therefore, different factors should be used in the conversion of nitrogen to protein for individual foods. Table 13.1 shows the nitrogen contents of a number of common proteins together with the appropriate nitrogen conversion factors.

Although fundamentally unsound, the use of an average conversion factor of 6.25 for all food proteins is justified in practice, since protein requirements of farm animals, expressed in terms of N × 6.25, are requirements for nitrogen and not protein *per se*. The publication of tables of protein contents based on true conversion factors could lead to considerable confusion and inefficiency in feeding. The assumption that the whole of the food nitrogen is present as protein is also false, since many simple nitrogenous compounds such as amides, amino acids, glycosides, alkaloids, ammonium salts and compound lipids may be present (Chapter 4). Quantitatively, only the amides and the amino acids are important and these are present in large amounts in only a few foods such as young grass, silage and immature root crops. About 95 per cent of the nitrogen of most mature seeds is present as true protein whereas leaves have 80–90 per cent, stems and roots 60 per

Table 13.1 Factors for converting nitrogen to crude protein (Adapted from Jones D B 1931 *USDA Circ.* 183)

Food protein	Nitrogen (g/kg)	Conversion factor
Cottonseed	188.7	5.30
Soya bean	175.1	5.71
Barley	171.5	5.83
Maize	160.0	6.25
Oats	171.5	5.83
Wheat	171.5	5.83
Egg	160.0	6.25
Meat	160.0	6.25
Milk	156.8	6.38

cent, and storage organs such as potatoes and carrots 30–40 per cent, in this form. In the diets of pigs and poultry, cereals and oilseeds predominate and they contain little non-protein nitrogen. Hence, in practice, there is little to be gained from attempting to differentiate between the two types of nitrogen, particularly as a considerable proportion of the non-protein fraction, amino acids and amides for example, may be utilised for amino acid synthesis by the animal. Tradition apart, there seems little justification for the use of the term 'crude protein' in nutrition. The expression of both animal requirements and the status of foods in terms of nitrogen would be more logical and would avoid confusion.

13.2 Digestible crude protein (DCP)

The crude protein figure provides a measure of the nitrogen present in the food but gives little indication of its value to the animal. Before the food becomes available to the animal it must undergo digestion, during which it is broken down to simpler substances which are absorbed into the body. The digestible protein in a food may be determined by digestibility trials in which nitrogen intake is measured along with the nitrogen voided in the faeces, as described in Chapter 10.

Digestibility coefficients based on collection and analysis of digesta from the terminal ileum are generally considered to give a more accurate measure of the nitrogen absorbed than do those based on the more usual faecal collection. Ileal collection eliminates the lower gut as a source of errors, and is justified since absorption from the large intestine is held to make little or no contribution to the protein status of the animal. Furthermore, a higher correlation has been demonstrated between daily liveweight gain in pigs and ileal rather than whole-tract digestibility ($r = 0.76$ versus 0.34), particularly with unusual protein sources.

It is assumed that the difference between the quantities of nitrogen in the food and faeces, or digesta in the case of ileal collection, represents the quantity absorbed in utilisable form by the body and that all nitrogen which appears in the faeces is of dietary origin. In the light of the fate of ruminal ammonia (Chapter 8) and the presence in faeces and digesta of nitrogen of metabolic or endogenous origin (Chapter 10), these assumptions are obviously untenable. The figures so obtained are for apparently digestible protein. Determination of the true digestibility must take account of the contribution of nitrogen of endogenous origin to that of the digesta. The endogenous nitrogen (EN) contains non-food substances entering the intestine, such as saliva, bile, gastric and pancreatic secretions, and cells sloughed off the mucous membrane of the gut. It may also contain significant amounts of bacterial nitrogen, which is not endogenous in origin. It may be divided into two fractions: that which is unrelated to the quality and the quantity of the dietary protein and is dependent only upon the quantity of dry matter passing through the gut and termed non-specific endogenous nitrogen (NSE); and that which is related to the quality and quantity of the dietary protein and termed specific endogenous nitrogen (SE).

As a result of the inclusion of the NSE, which is constant for a given level of dry matter intake, in the calculation of apparent digestibility, different values will be obtained for different levels of inclusion of dietary protein. It has been suggested that this anomaly may be overcome by eliminating NSE from the calculation to give a figure for 'standardised digestiblity' (SD):

$$SD = \frac{\text{ingested protein} - \text{excreted protein} - \text{non-specific endogenous protein}}{\text{ingested protein}}$$

Estimates of NSE may be obtained from trials carried out with protein-free diets.

The use of standardised digestibility figures overcomes the problem of multiple values and the systematic undervaluation of the digestibility of low-protein foods which is inherent in apparent digestibility figures. Standardised digestibility is sometimes referred to as true digestibility but it remains an apparent value and does not give a realistic picture of the true digestibility of the nitrogen. For this, the contribution of endogenous nitrogen to faecal or digesta nitrogen must be taken into account.

13.3 Determination of endogenous nitrogen

The magnitude of the endogenous nitrogen component may be estimated by one of the following procedures:

■ Measuring the nitrogen in digesta when protein-free diets are consumed, the assumption being that this is entirely endogenous in origin. The procedure is non-physiological, normal protein metabolism may be affected and nitrogen secretion into the gut reduced. Furthermore, the diet may lack the stimulatory effect of normal diets on endogenous secretions. However, the fibrous nature of protein-free diets may increase the contribution of that part of the endogenous component resulting from abrasion of the gut membrane. The technique cannot be used with ruminant animals as N-free diets inhibit rumen fermentation.

■ Measuring the nitrogen in digesta when completely digestible protein is consumed.

■ Measuring the recoveries of nitrogen, from digesta, when graded levels of protein are consumed. The recoveries are regressed on intake; extrapolation to zero intake is taken as the endogenous nitrogen loss. It is assumed that endogenous nitrogen loss is not related to protein intake. This is not a valid assumption and the estimates obtained by this method are, at best, minimum values.

■ Isotope dilution, in which the stable nitrogen isotope ^{15}N is introduced in known concentration in the food. The concentration in the faeces/digesta is then measured and the dilution with nitrogen of non-food (endogenous) origin is then calculated.

■ Treating dietary lysine with guanidine to give homoarginine. This is assumed to escape digestion in the gut and any lysine appearing in the digesta is assumed to be of endogenous origin. The relationship between lysine and the other amino acids is considered to be constant and is used in calculating their true digestibility.

■ Peptide alimentation of the alimentary canal with hydrolysed casein and separation of the exogenous and endogenous components on the basis of molecular weight, the endogenous being greater than 10 000.

The results may vary widely with the different techniques employed. Even two of the most favoured current techniques, that based on homoarginine and that using ^{15}N-labelled dietary protein, show average differences of about 5 per cent in the true digestibility values derived from them. Some figures obtained using ^{15}N labelling are given in Table 13.2.

As well as varying with the technique used, the magnitude of the endogenous nitrogen fraction is influenced by the type and amount of non-starch polysaccharide in the diet and the protein status of the experimental animals. Some workers consider that endogenous secretions are a component of the net value of the food and that digesta nitrogen should be set against dietary nitrogen whatever its origin. Others regard digestibility as purely a property of the food and consider that endogenous nitrogen should be eliminated from its calculation. The digestibility of the nitrogen of a complete diet could then be calculated additively from the digestibility values of the proteins of the foods making up the diet. Owing to the variability of the values for endogenous nitrogen, any assumed figure used to transform apparent to true digestibility values will be of doubtful validity, and so will the resulting coefficients. Most of the figures in current use are apparent values, but in feeding systems in which metabolic nitrogen loss is treated as part of the maintenance nitrogen demand, true rather than apparent coefficients have to be used.

Table 13.2 Apparent and true ileal protein digestibility values as determined by the ^{15}N-isotope dilution techniques (Adapted from Sauer W C and de Lange K 1992 Novel methods for determining protein and amino acid digestibilities in feedstuffs. In: Nissen S (ed.) *Modern Methods in Protein Nutrition and Metabolism*, London, Academic Press)

	Soya bean meal	Canola meal	Wheat	Barley
Digestibility %				
Apparent	83.8	66.0	80.0	69.5
True	97.5	84.1	99.0	94.2
Endogenous protein				
As g/kg dry matter intake	25.5	30.5	27.4	27.7
As percentage of total crude protein present at the distal ileum	84.6	53.5	94.5	81.1
As g/100 g crude protein intake	13.7	18.0	19.1	24.7

There is recent evidence that figures for the apparent ileal digestibility of lysine, and some other amino acids, overestimate availability to the animal, which reinforces the view that it is dangerous to equate digestibility with availability.

13.4 Measures of protein quality for monogastric animals

Digestible protein figures are not entirely satisfactory measures of the value of a protein to an animal, because the efficiency with which the absorbed protein is used differs considerably from one source to another. In order to allow for such differences, methods for evaluating proteins such as the protein efficiency ratio (PER), the net protein retention (NPR) and the gross protein value (GPV), which are based on the growth response of experimental animals to the protein under consideration, have been devised.

Protein efficiency ratio

This is defined as:

$$PER = \frac{\text{gain in body weight (g)}}{\text{protein consumed (g)}}$$

The rat is the usual experimental animal.

Net protein retention

This is calculated as:

$$NPR = \frac{\text{weight gain of TPG} - \text{weight loss of NPG}}{\text{weight of protein consumed}}$$

Where: TPG = group given the test protein
NPG = group on protein-free diet

Gross protein value

The liveweight gains of chicks receiving a basal diet containing 80 g crude protein/kg are compared with those of chicks receiving the basal diet plus 30 g/kg of a test protein, and of yet others receiving the basal diet plus 30 g/kg of casein. The extra liveweight gain per unit of supplementary test protein stated as a proportion of the extra liveweight gain per unit of supplementary casein is the GPV of the test protein, i.e.:

$$GPV = A/A°$$

where: A = g increased weight gain/g test protein
$A°$ = g increased weight gain/g casein

Nitrogen balance

Liveweight gains may not be related to protein stored and a more accurate evaluation of a protein may be obtained by using the results of nitrogen balance experiments. In such experiments, the nitrogen consumed in the food is measured, together with that voided in the faeces, urine and any other nitrogen-containing products such as milk, wool and eggs. When the nitrogen intake is equal to the output, the animal is said to be in nitrogen equilibrium; when intake exceeds output it is in positive balance, and when output exceeds intake it is in negative balance. Table 13.3 illustrates the calculation of a nitrogen balance for a 50 kg pig in positive nitrogen balance to the extent of 21.79 g/day.

Balance trials are susceptible to several sources of error:

■ inadequate adaptation of experimental animals to the diet and the environment,
■ collection and weighing of faeces and urine,
■ storage of faeces and urine,
■ preparation and sampling of faeces and urine for chemical analysis.

The results of balance trials have been shown to diverge considerably from those obtained with the comparative slaughter technique (Chapter 11). This too is liable to error, largely owing to the difficulty of animal selection for the initial slaughter group, which, if inadequate, can invalidate the comparison with the slaughter group. With larger animals, errors can arise from poor sampling of the carcasses owing to their physical size. They have then to be dissected into component parts for sampling, instead of the body sampling possible with smaller animals. In general, the balance technique

Table 13.3 Nitrogen balance for a Large White/Landrace pig on a diet of wheat, soya bean meal and herring meal (Adapted from Morgan C A and Whittemore C T, unpublished)

	Daily intake (g)	Daily output (g)
Food nitrogen	46.42	
Faecal nitrogen		4.99
Urinary nitrogen		19.64
Nitrogen retained by body		21.79
Totals	46.42	46.42
Balance		+21.79

gives higher estimates of nitrogen retention and most authorities agree that such estimates are not as reliable as those based on slaughter.

Biological value (BV)

This is a direct measure of the proportion of the food protein which can be utilised by the animal for synthesising body tissues and compounds, and may be defined as the proportion of the absorbed nitrogen which is retained by the body. A balance trial is conducted in which nitrogen intake and urinary and faecal excretions of nitrogen are measured along with the endogenous fractions in these two materials. The biological value is then calculated as:

$$BV = \frac{\text{N intake} - (\text{faecal N} - \text{MFN}) - (\text{urinary N} - \text{EUN})}{\text{N intake} - (\text{faecal N} - \text{MFN})}$$

Where: MFN = metabolic (endogenous) faecal N
EUN = endogenous urinary N

An example of such a calculation is given in Table 13.4.

The endogenous urinary nitrogen results from irreversible reactions involved in the breakdown and replacement of various protein secretions and structures within the body. Thus both the faecal and urinary endogenous fractions represent nitrogen which has been absorbed and utilised by the animal rather than nitrogen which cannot be so utilised. Their exclusion from the faecal and urinary values in the above formula gives a measure of the true biological value.

In determining biological value, as much as possible of the dietary protein should be provided by that under test. Protein intake must be sufficient to allow adequate nitrogen retention but must not be in excess of that required for maximum retention; if the latter level were exceeded, the general amino acid catabolism resulting would depress the estimate of biological value. For the same reason, sufficient non-protein nitrogenous nutrients must be given to prevent protein being catabolised to provide energy. The

Table 13.4 Calculation of biological value for maintenance and growth of the rat (Adapted from Mitchell H H 1942 *Journal of Biological Chemistry* **58**: 873)

Food consumed daily, g	6.0
Nitrogen in food, g/kg	10.43
Daily nitrogen intake, mg	62.6
Total nitrogen excreted daily in urine, mg	32.8
Endogenous nitrogen excreted daily in urine, mg	22.0
Total nitrogen excreted daily in faeces, mg	20.9
Metabolic faecal nitrogen excreted daily, mg	10.7

$$BV = \frac{62.6 - (20.9 - 10.7) - (32.8 - 22.0)}{62.6 - (20.9 - 10.7)}$$

$$= 0.79$$

Table 13.5 Biological values of the protein in various foods for maintenance and growth for the growing pig (Adapted from Armstrong D G and Mitchell H H 1955 *Journal of Animal Science* **14**: 53)

Food	BV
Milk	0.95–0.97
Fish meal	0.74–0.89
Soya bean meal	0.63–0.76
Cottonseed meal	0.63
Linseed meal	0.61
Maize	0.49–0.61
Barley	0.57–0.71
Peas	0.62–0.65

diet must also be adequate in other respects. Table 13.5 gives biological values for the proteins of some typical foods.

Such biological values are for the combined functions of maintenance, meaning the replacement of existing proteins, and growth (i.e. the formation of new tissues). Biological values for maintenance alone may be calculated from balance data. A linear relationship exists between nitrogen intake and balance below equilibrium, which may be represented by the following equation:

$$y = bx - a$$

Where: y = N balance
x = N absorbed
a = N loss at zero intake
  (all expressed as gN per basal kJ), and
b = the nitrogen balance index, i.e. that fraction of the absorbed nitrogen retained by the body and is equal to the BV for maintenance.

The product of BV and digestibility is termed the *net protein utilisation* (NPU) and is the proportion of the nitrogen intake retained by the animal.

The amino acids absorbed by the animal are required for the synthesis of body proteins. The efficiency with which this synthesis is effected depends partly on how closely the amino acid proportions of the absorbed mixture resemble those of the body proteins and partly on the extent to which the proportions can be modified. The biological value of a food protein therefore depends upon the number and kinds of amino acids present in the molecule: the closer the amino acid composition of the food protein approaches that of the body protein the higher will be its biological value. Animals have little ability to store amino acids in the free state and if an amino acid is not required immediately for protein synthesis it is readily broken down and either transformed into a non-essential amino acid or used as an energy source. Since essential amino acids cannot be effectively synthesised in the animal body, an imbalance of these in the diet leads to wastage. Food proteins with either a deficiency or an excess of any particular amino acid will tend to have low biological values.

If we consider two food proteins, one deficient in lysine and rich in methionine, and the other deficient in methionine but containing an excess of lysine, then, if these proteins are given separately to young pigs they will have low biological values because of the imbalance of these acids. If, however, the two proteins are given together then the mixture will be better balanced and have a higher value than either given alone. Such proteins complement each other. In practice, and for a similar reason, it often happens that a diet containing a large variety of proteins has a higher biological value than one containing a few proteins only. This also explains why biological values of individual foods cannot be applied when mixtures of foods are used, since clearly the biological value of a mixture is not simply a mean of the individual components. For the same reason it is impossible to predict the value of a protein as a supplement to a given diet from its biological value.

Animal proteins generally have higher biological values than plant proteins, although there are exceptions. Gelatine, for example, is deficient in several essential amino acids.

The amino acid composition of a given food protein will be relatively constant (Appendix Table 1.3) but that of the protein to be synthesised will vary considerably with the type of animal and the various functions it has to perform. For the normal growth of rats, pigs and chicks, for example, lysine, tryptophan, histidine, methionine, phenylalanine, leucine, isoleucine, threonine, valine and arginine are dietary essentials. Man does not require histidine whereas chicks need glycine, in addition to those acids required by the rat, to ensure optimum growth. On the other hand arginine is not a dietary essential for maintenance of the rat or pig.

The situation is further complicated by the fact that some amino acids can be replaced, at least in part, by others; for example, methionine can be partly replaced by cystine, and tyrosine can partly replace phenylalanine. In such cases the two amino acids are frequently considered together in assessing the animal's requirements. It is obvious that no single figure for biological value will suffice as a measure of the nutritive value of a food protein for different animals and different functions. The differences between the amino acid requirements of young pigs and those of laying hens, for example, are shown in Appendix Tables 9 and 10. The consequent need for multiple figures severely limits the use of the biological value concept in feeding practice.

The biological methods of evaluation reflect the content of the limiting amino acid in the protein. Changes in the levels of other amino acids will not affect the values until one or other of them becomes limiting Thus in milk protein, with an excess of lysine, change in lysine content will not affect the biological measures until it is reduced to such a level that it itself becomes the limiting acid. The biological measures are therefore of limited value in assessing the effects of certain processes, heat treatment for example, on nutritive value.

Since biological value is dependent primarily upon essential amino acid constitution, it would seem logical to assess the nutritive value of a protein by determining its indispensable amino acid constitution and then comparing this with the known amino acid requirements of a particular class of animal. Application of modern chromatographic techniques coupled with automated procedures allows relatively quick and convenient resolution of mixtures of amino acids. However, the acid hydrolysis used to produce such mixtures from protein destroys practically all the tryptophan and a considerable proportion of the cystine and methionine. Tryptophan has to be released by a separate alkaline hydrolysis and cystine and methionine have to be oxidized to cysteic acid and methionine sulphone to ensure their quantitative recovery. Losses of amino acids and the production of artefacts, which are greater with foods of high carbohydrate content, are reduced if the hydrolysis is carried out *in vacuo*. Evaluations of proteins in terms of each individual amino acid would be laborious and inconvenient and several attempts have been made to state the results of amino acid analyses in a more useful and convenient form.

Chemical score

In this concept, it is considered that the quality of a protein is decided by that amino acid which is in greatest deficit when compared with a standard. The standard generally used has been egg protein but many workers now use a defined amino acid mixture, the FAO Recommended Reference Amino Acid Pattern. The content of each of the essential amino acids of a protein is expressed as a proportion of that in the standard (the standard pattern ratio) and the lowest proportion taken as the score. In wheat protein, for example, the essential amino acid in greatest deficit is lysine. The contents of lysine in egg and wheat proteins are 72 and 27 g/kg respectively and the chemical score for wheat protein is therefore $27/72 = 0.37$. Scores correlate well with biological values for rats and human beings but not for poultry. They are useful for grouping proteins but suffer a serious disadvantage in that no account is taken of the deficiencies of acids other than that in greatest deficit.

The essential amino acid index (EAAI)

This is the geometric mean of the egg, or standard pattern, ratios of the indispensable amino acids. It has the advantage of predicting the effect of supplementation in combinations of proteins. On the other hand, it has the disadvantage that proteins of very different amino acid composition may have the same or a very similar index.

Both the chemical score and the essential amino acid index are based upon gross amino acid composition. It would be more logical to use figures for the acids available to the animal. Such figures may be obtained in several ways. Determinations of digestibility *in vivo* involve amino acid analyses of the food and faeces. The figures so obtained are suspect because the faeces contain varying amounts of amino acids not present in the food, mainly as a result of microbial activity in the large intestine. Thus net synthesis of methionine and lysine has been shown to take place in the hind gut of the pig. On the other hand cysteine, threonine and tryptophan disappear from the large intestine but make little or no contribution to the amino acid nutrition of the animal. This drawback may be overcome by determining apparent ileal digestibility, by measuring amino acids in the digesta at the terminal ileum instead of in the faeces. A more accurate estimate of the dietary amino acids absorbed is provided if the contribution of amino acids of endogenous origin to those found at the terminal ileum is used in the calculation of truly digestible amino acids. Digestibility trials *in vivo* are laborious and time consuming, require considerable technical resources and skill and are thus expensive. Determinations *in vitro* involve the action of one or, at most, a few enzymes and are not strictly comparable with the action *in vivo*, which entails a series of enzymes.

Biological assay of available amino acids

The available amino acid content of a food protein may be assayed by measuring the liveweight gain, food conversion efficiency or nitrogen retention of animals given the intact protein as a supplement to a diet deficient in the particular amino acid under investigation. The chick is the usual experimental animal and the response to the test material is compared with the response to supplements of pure amino acids. The method has been used successfully for lysine, methionine and cystine but, in addition to the usual disadvantages associated with biological methods – time, technical expertise and supply of suitable animals – there is the major problem of constructing diets deficient in specific amino acids but adequate in other respects.

Microbiological assay of essential amino acids

Certain microorganisms have amino acid requirements similar to those of higher animals and have been used for the evaluation of food proteins. The methods are based on measuring the growth of the microorganisms in culture media which include the protein under test. Best results have been obtained with *Streptococcus zymogenes* and *Tetrahymena pyriformis*. The former is used after an acid or enzymic predigestion of the food protein; estimates of the availability of lysine and methionine have agreed well with chick assays and measurements of NPU. *T. pyriformis* has intrinsic proteolytic activity and is used, for soluble proteins, without predigestion. An improved method, using predigestion with the enzyme pronase and measuring response in terms of the tetrahymanol content of the culture medium, has given results for available lysine, methionine and tryptophan which correlate well with those of biological assays. Tetrahymanol, the characteristic pentacyclic terpene synthesised by *T. pyriformis*, is determined by gas–liquid chromatography.

Chemical methods

It would be ideal if there were simple chemical procedures for determining the availability of amino acids, the results of which correlated well with those of accepted biological methods. The most widely used method is that for 'FDNB-reactive lysine', which was originally proposed by Dr K J Carpenter and has since undergone modification both by Carpenter and by other workers. The theoretical justification for the use of the method is that practically the only source of utilisable lysine in foods is that which has the epsilon-amino group free to react with various chemical reagents. The protein under test is allowed to react under alkaline conditions with fluoro-2,4-dinitrobenzene (FDNB) to give DNP-lysine, the concentration of which can be measured colorometrically. In practice, the method has been found to agree well with biological procedures for evaluating proteins as supplements to diets, such as those containing high proportions of cereals, in which lysine is limiting. The correlation has also been good with diets based largely on animal protein. With vegetable protein and diets containing high levels of carbohydrate the method is not so satisfactory, the results being too low owing to destruction of the coloured lysine derivative during acid hydrolysis. Various methods have been proposed to counter this but none has proved completely satisfactory. From the available evidence it seems clear that FDNB-available lysine overestimates the availability of the acid when applied to heat-treated meals, as illustrated in Table 13.6.

The gross protein value is probably the most commonly used biological method for evaluating proteins. Gross protein values measure the ability of proteins to supplement diets consisting largely of cereals and they correlate well with FDNB-reactive lysine figures as shown in Fig. 13.1.

Dye-binding methods

These have been widely used for estimating protein in such foods as cereals and milk. The methods are rapid and give reproducible results and attempts have been made to use them for measuring total basic amino acids and reactive lysine. The latter requires blocking of the epsilon-amino group to

Table 13.6 Mean amounts of total lysine, FDNB-reactive lysine and truly digestible lysine in heat-treated meal and bone meals (Adapted from Moughan P J 1991. In: Haresign W and Cole D J A (eds) *Recent Advances in Animal Nutrition*, London, Butterworth-Heinemann)

	Sample					
	1	2	3	4	5	6
Total lysine (g/100 g)	2.65	2.59	2.82	2.73	3.89	2.68
FDNB-available lysine (g/100 g)	2.17	1.91	2.53	2.32	2.57	2.11
Ileal truly digestible lysine (g/100 g)	1.72	1.75	1.93	1.97	2.88	2.03

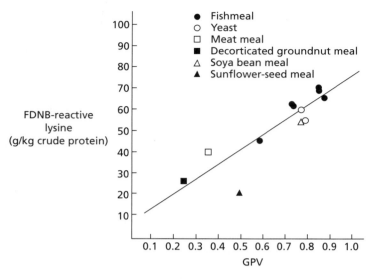

Fig. 13.1 Relationship between gross protein value and FDNB-reactive lysine. (After Carpenter K J and Woodham A A 1974 *British Journal of Nutrition* **32**: 647.)

prevent reaction with the dye. Orange G has been used, along with 2,4,6-trinitrobenzene sulphonic acid and propionic anhydride as blocking agents, and has proved effective for estimating the lysine content of cereals. It is less effective for fish and meat meals.

The increased use of synthetic lysine in foods presents a further problem. This amino acid has both amino groups available for reaction with FDNB. The resulting compound is soluble in ether and is not estimated by the Carpenter method.

Interpretation of amino acid assays

Several factors may be responsible for a lack of agreement between estimates of protein quality based on amino acid content and those made in animal experiments:

- Even small changes in the concentration of one or more amino acids may increase the amounts of others required to maintain growth rates.
- Certain acids, such as tryptophan and histidine may be toxic, albeit at concentrations far greater than normally occur in food proteins.
- Antagonisms may exist between specific acids, which prejudice their utilisation. Thus the addition of as little as 20 g/kg of leucine to a diet deficient in isoleucine may have deleterious effects on performance, and the arginine requirement of the rat may be increased by giving higher levels of lysine.
- Antinutritional factors (ANF) are frequently present in foods used primarily as protein sources. Chief among these are enzyme inhibitors, lectins,

polyphenols and certain non-protein amino acids. All are capable of lowering the absorption and/or the utilisation of amino acids by the animal but are not taken into account in evaluations of protein sources based on amino acid content. The nature and mode of action of anti-nutritional factors is dealt with in Chapter 24 with particular reference to their importance in individual foods.

■ There is considerable evidence that growing animals, such as young rats and chicks, do not fulfil their growth potential if the dietary nitrogen is entirely in the form of essential amino acids. Additional nitrogen is required and is best supplied as a mixture of non-essential amino acids; glutamate, alanine and ammonium citrate are effective sources also. Allowance must be made for these factors when a protein source is being evaluated on the basis of its content of one or more indispensable amino acids.

13.5 Measures of food protein used in practice in the feeding of pigs and poultry

The difficulty of assessing the value of a dietary protein is reflected in the variety of methods that have been proposed, all of which have considerable limitations. A crude protein figure gives a measure of the total nitrogen content and is useful since the digestibility of the proteins in foods commonly given to pigs and poultry is fairly constant. This is not true for new or unusual sources of protein and this approach is not a valid one.

The quality of dietary protein is indicated by stating the contents of all the essential amino acids or of those most likely to be in deficit. Practical pig and poultry diets are based largely on cereals, and assessment of foods as sources of protein for such animals is a matter of measuring their ability to supplement the amino acid deficiencies of the cereals. The main deficiencies in these cases are of lysine or methionine, so that the most useful measures of protein quality are those which reflect the available lysine or methionine content of the food. The determination of available lysine is now accepted as a routine procedure for protein foods in many laboratories.

The most recent method for evaluating dietary protein for growing pigs is based on the concept of 'ideal protein'. This is a modification of the chemical score with the amino acid pattern of the particular tissue protein serving as the reference pattern (Table 13.7).

If the main limiting amino acid was lysine at 50 g/kg CP, then the score would be $50/70 = 0.70$, and the ideal protein content would be 700 g/kg CP. A diet with 170 g/kg of this protein would then supply $170 \times 0.7 = 119$ g ideal protein/kg.

Ideal protein is conceived as providing the essential amino acids in the proportions required by the pig and of having the correct balance between the essential and non-essential acids. Once the balance of amino acids has been described in this way there is no need to tabulate requirements for

Table 13.7
Recommended balance
of amino acids (g/kg) in
ideal protein (Adapted
from Agricultural Research
Council 1981 *The Nutrient
Requirements of Pigs*,
Farnham Royal,
Commonwealth
Agricultural Bureaux)

Lysine	70	Leucine	70
Methionine + cystine	35[a]	Histidine	23
Threonine	42	Phenylalanine + tyrosine	67[b]
Tryptophan	10	Valine	49
Isoleucine	38	Non-essential amino acids	596

[a] At least half should be methionine.
[b] At least half should be phenylalanine.

individual amino acids. The statement of a requirement for ideal protein automatically fixes the requirement for each acid.

The concept of ideal protein makes no allowance for an excess of individual amino acids, which, in certain circumstances, could be detrimental. It is usual therefore to limit the concentration of any one acid to less than 1.2 times that of the reference protein. It is assumed in calculating the requirement for ideal protein that the proportions of amino acids made available to the metabolic body pool are the same as those in the dietary protein. This will not be so if the acids are not digested or absorbed to the same extent. Requirements are usually stated in terms of apparently digested ideal protein (ADIP) and a digestibility of 0.75 is assumed in transforming the ideal protein supply to an ADIP supply. As more information becomes available on ileal digestibility values these will probably be used instead of the single assumed figure. The ileal-digestible amino acid patterns of foods could then be compared with the balance in ideal protein and the scores calculated as before.

The balance of amino acids in the ideal protein applicable to maintenance, growth of lean tissue, pregnancy and lactation will be different, reflecting differences in the composition of the proteins synthesised. In the case of lactating sows and young fast-growing pigs, the maintenance requirement is small in relation to the total, and the composition of the product (body and milk protein) will dominate the demand for amino acids and consequently the composition of the ideal protein. In the case of pregnant sows and boars, the maintenance demand will be large in relation to that for production and will be dominant.

In practice, pig diets are formulated to ileal-digestible lysine, and the ileal-digestible contents of the other essential amino acids in the diet are maintained in the same proportion to lysine as they are in the ideal protein.

For poultry, evaluation of protein sources is based upon the amounts of the three major limiting amino acids, lysine, methionine and tryptophan. It is generally assumed that diets adequate in these acids will automatically provide sufficient amounts of the others.

13.6 Measures of protein quality for ruminant animals

Traditionally, proteins in foods for ruminant animals have been evaluated in terms of crude protein (CP) or digestible crude protein (DCP).

Realisation that the crude protein fraction contained variable amounts of non-protein nitrogen led to the use of true protein instead of crude protein but this was unsatisfactory since no allowance was made for the nutritive value of the non-protein nitrogen fraction. The concept of protein equivalent (PE), introduced in 1925 but now no longer used in this context, was an attempt to overcome this difficulty by allowing the non-protein nitrogen fraction half the nutritive value of the true protein. The term 'protein equivalent' is currently used in connection with foods containing urea. Such foods must by law be sold with a statement of their content of protein equivalent of urea. This means the amount of urea nitrogen multiplied by 6.25.

The use of DCP for evaluating food proteins for ruminants has been largely abandoned. This resulted from a growing awareness of the extensive degradative and synthetic activities of the microorganisms of the rumen (Chapter 8). Rumen microorganisms are responsible for providing the major part of the energy requirements of the host animal by transforming dietary carbohydrates to acetate, propionate and butyrate. In order to do this and to exploit the energy potential of the food fully, they must grow and multiply and this involves large-scale synthesis of microbial protein. The nitrogen for this is obtained, in the form of amino acids, peptides and ammonia, by breakdown of the nitrogen fraction of the food. Bacteria acting on the structural carbohydrate (SC) fraction of the diet use only ammonia, whereas those acting on the non-structural fraction (NSC) derive about 65 per cent of their nitrogen from amino acids and peptides, and the remainder from ammonia.

The microbial protein passes from the rumen, is digested in the small intestine, and so makes a contribution to satisfying the nitrogen requirements of the host animal. The magnitude of this contribution depends upon the speed and extent of microbial breakdown of the dietary nitrogen fraction, upon the efficiency of the transformation of the degraded material into microbial protein (nitrogenous compounds), the digestibility of the microbial protein and the biological value of the latter.

The degradative and synthetic processes taking place in the rumen are of major importance in the nitrogen economy of the host animal since they determine the nature of the amino acid mix made available for protein synthesis at tissue level. The series of changes undergone by dietary protein between mouth and body tissue in the ruminant animal is illustrated schematically in Fig. 13.2.

Satisfying the demands of the rumen microorganisms for readily available nitrogen is a major function of the diet and to this end a certain proportion of the nitrogen fraction must be degradable by the rumen microorganisms.

Current systems for the evaluation of food protein for ruminant animals involve determinations of the degradability of protein in the rumen, the synthesis of microbial protein, the digestion in the lower gut of both food and microbial proteins, and the efficiency of utilisation of absorbed amino acids. The methods used to determine these components of the system are described next, after which their use in the systems will be illustrated.

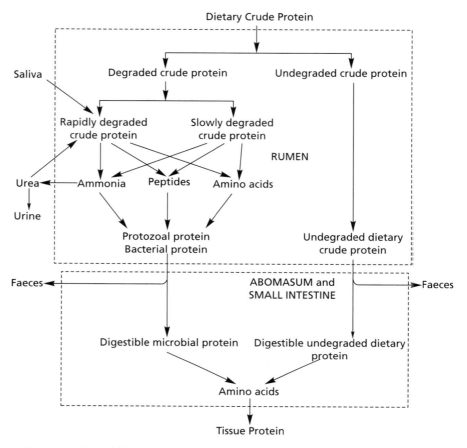

Fig. 13.2 Fate of dietary crude protein in the ruminant animal.

Degradability of the nitrogen fraction of the diet

Nitrogen fractions within the diet will vary in their susceptibility to breakdown, from immediately degraded to undegradable, and from 0 to 1 in the extent to which they are degraded in the rumen and digested when they reach the small intestine (Table 13.8).

Degradability will be affected by such factors as the surface area available for microbial attack and the protective action of other constituents as well as the physical and chemical nature of the protein. Claims have been made that the solubility of a protein is correlated with ease of breakdown but these do not survive critical examination. Thus casein, which is readily degraded in the rumen, is not readily soluble; whereas albumin, which is resistant to breakdown, is readily soluble. It has been suggested that a major factor affecting degradability is the amino acid sequence within the protein molecule. If this is so then the nature of the microbially produced rumen peptidases is of considerable importance and it seems doubtful whether any simple laboratory test for degradability is possible.

Table 13.8 Composition, rumen degradation and intestinal digestion of protein fractions (Adapted from Chalupa W and Sniffen C J 1994. In: Garnsworthy P C and Cole D J A (eds) *Recent Advances in Animal Nutrition,* University of Nottingham Press)

Fraction	Composition	Rumen degradation (% hour)	Intestinal digestibility (%)
A	Ammonia, nitrate, amino acids, peptides	Instantaneous	None reaches intestine
B1	Globulins, some albumins	200–300	100
B2	Most albumins, glutelins	5–15	100
B3	Prolamins, cell-wall proteins, denatured proteins	0.1–1.5	80
C	Maillard products, nitrogen bound to lignin, tannin-bound protein	0	0

The extent to which a nitrogen fraction is degraded in the rumen will depend upon its innate degradability and the time it spends in the rumen, i.e. upon rate of passage. As the rate of passage increases so the extent of ruminal breakdown is reduced.

Measurement of degradability in vivo

This involves the measurement of dietary nitrogen intake, the endogenous nitrogen (EN), the non-ammonia nitrogen (NAN) and microbial nitrogen (MN) of dietary origin passing the duodenum. Degradability (Dy) of nitrogen is then expressed as:

$$Dy = 1 - \frac{NAN - (MN + EN)}{\text{dietary N intake}}$$

This method requires accurate measurement of duodenal flow and microbial and endogenous nitrogen. The flow measurement, which requires the use of a dual-phase marker system, has a large coefficient of variation (between animals) and many published values must be suspect owing to the small number of animals used in their determination. Microbial nitrogen in duodenal nitrogen is usually identified by means of marker substances such as diaminopimelic acid (DAPA), aminoethylphosphoric acid (AEPA), ribonucleic acid and amino acids labelled with ^{35}S, ^{32}P and ^{15}N. The concentration of marker in the microorganisms is measured in a sample of rumen fluid. Different markers may give results which vary widely, sometimes by as much as 100 per cent. The assumption that the microorganisms isolated from rumen fluid are representative of those in the duodenum is of doubtful validity, since the latter include organisms normally adherent to food particles and/or the rumen epithelium. The endogenous fraction constitutes about 50–200 g/kg of the duodenal nitrogen but is difficult to quantify. A value of 150 g/kg is frequently assumed. Measurements of degradability are thus subject to possible errors owing to uncertainties in measuring duodenal flow and microbial and endogenous nitrogen, and are affected by dietary considerations such as level of feeding and the size and frequency of meals. It has been calculated that estimates of degradability may vary over a range of 0.3–0.35 owing to errors of determination alone. Despite its inadequacies this technique remains the only method currently available for providing an absolute measure of protein degradability and is the standard against which other methods have to be assessed.

Determination of degradability in sacco (or in situ)

This involves incubation of the food in synthetic fibre bags suspended in the rumen, as described in Chapter 10. The degradability is calculated as the difference between the nitrogen initially present in the bag and that present after incubation, stated as a proportion of the initial nitrogen:

$$\text{degradability} = \frac{\text{initial N} - \text{N after incubation}}{\text{initial N}}$$

When protein disappearance (p) is regressed on time, p increases but at a reducing rate. The relationship may be described by an equation of the following form:

$$p = a + b(1 - e^{-ct})$$

in which a, b and c are constants which may be fitted by an iterative least-squares procedure. The relationship is illustrated in Fig. 13.3.

In Fig. 13.3, a is the intercept on the y-axis and represents degradability at zero time. It is that part of the protein which is water-soluble and which is considered to be immediately degradable; b is the difference between a and the asymptote and represents that part of the protein which is degraded more slowly, c is the rate of disappearance of the potentially degradable fraction b, and t is the time of exposure. The extent of protein breakdown will depend upon the time for which the protein remains in the rumen (i.e. upon its rate of passage through the rumen). The effective degradability P may then be defined as:

$$P = a + [bc/(c + r)][1 - e^{-(1c + r)t}]$$

in which r is the rate of passage from the rumen to the abomasum (p. 335). As the time of incubation increases, the fraction of the protein remaining in

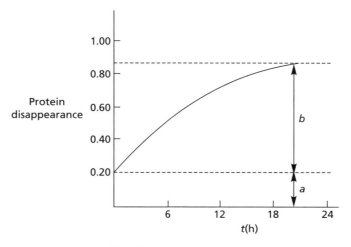

Fig. 13.3 Relationship of protein disappearance to time of incubation.

the rumen falls to zero, as does the rate of breakdown, and P may then be defined as:

$$P = a + bc/(c + r)$$

In this equation a is the immediately degradable protein and $bc/(c + r)$ the slowly degradable fraction.

If we assume a rate of passage of 0.05, with $a = 0.30$, $b = 0.70$ and $c = 0.02$, the effective degradability would be $0.3 + 0.7 \times 0.02/(0.02 + 0.05) = 0.50$. The protein available to the rumen microorganisms would then be $0.50 \times CP$.

The technique is subject to several inherent sources of error which must be controlled if reproducible results are to be obtained. Chief of these are sample size, bag size, porosity of the bag material and treatment of the bags following removal from the rumen. Ring tests (i.e. tests carried out at several laboratories) have shown unacceptably high inter-laboratory variability, indicating a need for strictly defined standard procedures which have to be rigidly adhered to if the results are to be universally applied in practice. A standardised procedure is given in the Agricultural and Food Research Council (1992) *Technical Committee on Responses to Nutrients, Report No. 9, Nutrient Requirements of Ruminant Animals: Protein.*

A basic assumption of this method is that disappearance of nitrogen from the bag, virtually reflecting solubility in rumen fluid, is synonymous with degradability. It has been known for some time that small amounts of food protein which are solubilised may leave the rumen without being degraded, and this must cast doubt on the veracity of values obtained using the technique. Even more serious in this context is the recent observation that acid-detergent insoluble nitrogen (ADIN), known to be undegradable, may disappear during incubation. A further complication is the presence in the bags of rumen bacteria, which contribute to the nitrogen of the contents.

Laboratory procedures for determining nitrogen degradability

Solubility in buffer solutions

Significant correlations have been demonstrated between the degradability values for the nitrogen fractions of foods and their solubility characteristics in a range of buffer solutions, including McDougall artificial saliva, borate–phosphate buffer and Wise Burroughs's buffer. When the methods are used for a range of foods, errors of prediction may be high but within food types predictions are improved sufficiently to allow the use of buffer solubility in the routine monitoring of concentrate foods.

When used in conjunction with methods for the fractionation of dietary nitrogen to predict degradability, solubility in buffer solution has shown good corrrelation with figures based on enzyme solubility (see below).

Solubility in enzyme solutions

Solubilisation of protein by purified enzymes from fungi and bacteria has been widely investigated as a means of estimating degradability. Different proteases have given varying results when compared with the *in sacco* technique. This is not unexpected in view of the fact that a single enzyme is being used to simulate the action of the complex multi-enzyme system of the rumen. As with the buffer solutions, accuracy of prediction is poor over a range of foods but improves when the technique is applied within food types. The most promising enzyme sources have been *Streptococcus griseus*, *Streptococcus bovis*, *Bacteroides amylophilus* and *Butyrovibrio* strain 7. *Streptococcus griseus* protease is the preferred source of enzyme for estimating degradability in the French 'protein digested in the intestine' (PDI) system. The method used incorporates corrections for different food types and the regression equation of degradability *in sacco* on enzyme solubility has a quoted residual standard deviation (rsd) of 0.025. Thus, estimates of degradability could be expected to be within ±0.05 of the *in sacco* value on 95 per cent of occasions.

Chemical analysis

A number of workers have shown significant correlations between crude protein content and degradability, reflecting the decreased proportion of the nitrogen fraction bound to fibre with increasing nitrogen content. The following equation for predicting the nitrogen degradability of grasses has been claimed to have an acceptable error of prediction:

$$Dy = (0.9 - 2.4r)(CP - 0.059NDF)/CP$$

where CP and NDF are as g/kg and r is the outflow rate per hour.

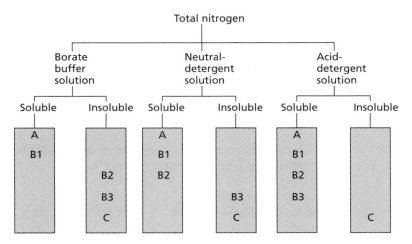

A = soluble in buffer; B1 = soluble in buffer and precipitated by trichloracetic acid;
B2 = insoluble in buffer but soluble in neutral- and acid-detergent solutions;
B3 = insoluble in buffer, insoluble in neutral-detergent solution but soluble
in acid-detergent solution; C = insoluble in buffer and both neutral- and acid-detergent solutions.

Fig. 13.4 Partitioning of dietary protein. (From Chalupa W and Sniffen C J 1994. In: Garnsworthy P C and Cole D J A (eds) *Recent Advances in Animal Nutrition*, University of Nottingham Press.)

In the Cornell net protein and carbohydrate system, food protein is separated into fractions based on a combination of borate–phosphate extractions and the detergent system of carbohydrate analysis of Goering and Van Soest (Fig. 13.4).

The protein is then allocated a degradability value based on the proportions of the various fractions present and their enzyme-determined degradability values (Table 13.8).

Near-infrared reflectance spectroscopy (NIRS)

NIRS measurements reflect the types and proportions of organic structures within a material. As such they are widely used for the routine analysis of foods and their nutritional evaluation (Chapter 1). The technique might therefore be expected to provide a solution to the problem of determining degradability. Current indications are that NIRS has the ability to estimate nitrogen degradability with a high degree of precision and considerable accuracy with R^2 values of 0.80–0.87 being claimed.

Rate of passage

The extent to which a protein is broken down in the rumen depends not only on its innate degradability but also upon the length of time for which it is exposed to breakdown and therefore upon its rate of passage through the rumen. It may be defined as:

$$RD = Kd/(Kd + Kp)$$

Where: RD = rumen degradability
Kd = rate of digestion
Kp = rate of passage

The rate of passage of food from the rumen is affected by a complex of food and animal factors. Passage is faster for:

■ smaller particles,
■ particles of higher density,
■ more highly hydrated particles,
■ more highly digested particles.

It will thus increase as digestion and rumination proceeds.

Rate of passage increases with increased dry matter intake and is thereby affected by a number of animal and environmental factors:

■ advancing pregnancy limits rumen fill and increases rate of passage,
■ lactation increases intake and rate of passage,
■ excessive body condition can reduce intake and rate of passage, and
■ high environmental temperatures will reduce intake and throughput.

The value for r may be determined by treatment of the protein with dichromate. The treatment renders the protein completely indigestible, there is no loss of chromium from the protein subsequent to treatment and particle size distribution is not affected. The rate of dilution of chromium in samples of rumen contents taken over a period of time (Chapter 8) can therefore provide an estimate of the rate of passage of the protein from the rumen.

The rate of passage of forages may be estimated from the dry matter intake (DMI) of the animal and the proportion of this provided by forage (F), by means of the following equation (Chalupa W et al. 1991; see Further reading):

$$Kp = 0.388 + 0.002 \times DMI + 0.0002 \times F$$

Where: Kp = passage rate (% per hour)
DMI = g DM consumed per kg metabolic bodyweight ($W^{0.75}$) per day
F = forage DM in DMI (%)

For concentrates:

$$Kp = Kp \text{ (forage)} \times 1.45 - 0.424$$

The predicted values are then adjusted for particle size using a factor, Af, which is based on effective NDF (ENDF):

Af (forages) = 100/(ENDF + 70)

Af (concentrates) = 100/(ENDF + 90)

Where: ENDF = proportion of NDF that is effective in meeting fibre requirements (%)

Kp = passage rate (per cent per hour)

Efficiency of nitrogen capture

The efficiency with which nitrogen is captured by the microorganisms of the rumen depends not only upon the speed and extent of breakdown but also upon the synchronous provision of a readily available, utilisable source of energy to fuel the synthesis of microbial protein. Failure to achieve this balance can result in too rapid and extensive a breakdown, and the synthetic powers of the rumen microorganisms may be overwhelmed. Wastage may then occur since the excess ammonia is absorbed and largely excreted as urea; some, though, is recycled via the rumen wall and contributes further to the nitrogen economy of the rumen. The extent of recycling has been estimated as about 70 per cent of nitrogen intake for diets of low protein content (about 5 g/kg) and as little as 11 per cent for foods with about 20 g/kg. The extent of recycling in a particular situation may be calculated using the following equation;

$$Y = 121.7 - 12.01X + 0.3235X^2$$

Where: Y = recycled urea nitrogen as a percentage of nitrogen intake

X = per cent crude protein in dry matter

That part of the food crude protein which is immediately degradable is unlikely to be as effective a source of nitrogen for the microorganisms as that which is more slowly degraded. It is generally considered that the slowly degraded nitrogen fraction is incorporated into microbial protein with an efficiency of 1.0, whereas that immediately degraded is less efficiently used. Estimates of the efficiency with which immediately degraded protein is incorporated vary, but 0.8 is a commonly used figure.

Yield of microbial protein

The yield of microbial protein which becomes available for digestion and absorption post-ruminally by the host has been related to the energy of the diet stated in terms of digestible organic matter (DOM), digestible organic matter digested in the rumen (DOMADR), total digestible nutrients (TDN), net energy for lactation (NE_L), metabolisable energy (ME), fermentable organic matter (FOM), fermentable metabolisable energy (FME) and rumen-degradable carbohydrate. The last three eliminate fat and products of fermentation, neither of which are considered to provide energy which can be utilised by the rumen microorganisms. The energy of fermentation products is significant in the case of silages and some distillery and brewery by-products. Hays are not considered to have undergone fermentation though they often contain measurable amounts of fermentation acids such as acetic and propionic. The routine measurement of the contribution of fermentation products to metabolisable energy in individual foods has not been a feasible proposition and assumed values are commonly used. The validity of such values is questionable in view of the variation in the magnitude and diversity of the fermentation products in individual foods.

A simple relationship between available energy and microbial protein yield cannot reflect the true position. It takes no account of:

■ The maintenance requirement of the microorganisms, which has been estimated to vary from 0.022 to 0.187 g carbohydrate/g bacteria per hour. When fermentation is slow, as with diets rich in structural carbohydrate, maintenance costs may be significant and estimates of microbial yield may be exaggerated.

■ The rumen environment. Lowering the rumen pH from 6.7 to 5.7 has been shown to halve microbial protein production, which may be significant when diets are rich in soluble carbohydrate and low in fibre with consequent production of lactic acid and ruminal acidosis.

■ Variation in the form of nitrogen required by the different types of microorganism. Thus the organisms splitting non-structural carbohydrates (NSC) are able to utilise peptide nitrogen and ammonia whereas those splitting structural carbohydrates are unable to use amino nitrogen and have to rely on ammonia as their source of nitrogen. It has been claimed that the yield of NSC-splitting bacteria is increased by almost 20 per cent when the proportion of peptides in the total NSC + peptides increases from 0 to 14 per cent. Above 14 per cent there is no further increase in yield.

The relationships used in predicting microbial protein from fermentable energy have high standard errors of estimate and should be used with caution. In addition, such relationships may be used to calculate yields of microbial protein only when the supply of energy is limiting. When the supply of protein to the microorganisms is limiting, this will determine the yield of microbial protein.

Sophisticated models attempt to relate microbial yield to the rate of carbohydrate fermentation and rate of passage, theoretical growth rate, the energy cost of bacterial maintenance and the form of nitrogen available to the rumen microorganisms. Many of the relationships involved in such calculations are based on laboratory characterisation of the food, and the value of the model will depend on the validity of the relationships between the laboratory determinations and the values used in the models.

True digestibility of protein

The microbial protein synthesised in the rumen may be protozoal or bacterial, the relative proportions depending upon conditions within that organ. Thus low rumen pH tends to reduce protozoal activity and stimulate that of certain bacteria. The mixture of bacterial and protozoal protein, along with dietary protein not degraded in the rumen, passes to the abomasum and small intestine. Here it is broken down to amino acids, which are then absorbed into the body. The digestibility of bacterial protein is lower (about 0.75) than that of the protozoal (about 0.90) and the overall digestibility of microbial protein will depend to some extent upon the rumen environment. However, protozoal protein constitutes some 5 to 15 per cent only of the total microbial protein flow from the rumen and its influence on the overall digestibility of microbial protein will be small. The composition of bacteria is variable but that shown in Table 13.9 is an acceptable approximation.

About 15 per cent of the total nitrogen is in the form of nucleic acids, about 25 per cent is cell wall protein and the remainder is true protein. Available evidence indicates that the digestibility of nucleic acid nitrogen is of the order of 0.8–0.9, that of microbial true protein 0.85–0.9. It is commonly assumed that the protein associated with the cell walls is completely indigestible. Most estimates of the true digestibility of microbial protein are of the order of 0.85–0.87, which is higher than might be expected in view of the proportions of the fractions present in the protein. Although nucleic acids are highly digestible their nitrogen is of no use to the animal, as after absorption they are totally excreted in the urine.

The digestibility of the undegraded dietary protein is a characteristic of the protein mix in the food and may vary considerably from diet to diet. The true digestibility of the undegraded dietary protein will vary with the proportion of the various protein fractions present. Thus amino acids, peptides, globulins, albumins and glutelins will be almost completely digested; prolamins, proteins associated with the cell walls and denatured proteins will have digestibility values of about 0.8; the protein of Maillard products and nitrogen bound to lignin will be completely indigestible. Some estimates of

Table 13.9 Composition of rumen bacteria

	g/kg dry matter
Crude protein	625
True protein	375
Cell wall protein	155
Nucleic acid (nitrogen $\times$ 6.25)	95
Carbohydrate	210
Fat	120
Ash	45

the true digestibility values of the various protein fractions are shown in Table 13.8.

Digestibility has been shown to be inversely related to the content of acid-detergent insoluble nitrogen (ADIN), which reflects that part of the food nitrogen which is closely bound to insoluble fibre. The digestible undegradable protein content (DUP) of a food is calculated thus:

$$DUP = 0.9 \, (\text{undegradable protein} - ADIN \times 6.25)$$

This equation is based on the assumptions that ADIN is indigestible and that the digestible fraction has a true digestibility of 0.9.

In the case of foods such as maize gluten and some distillery and brewery by-products, which have been heat treated under moist conditions, Maillard-type reactions (Chapter 4) may occur, resulting in an increase in the concentration of nitrogenous compounds insoluble in acid-detergent. Such 'acquired ADIN' does have a finite though low digestibility and the above equation is unreliable when used for such foods.

Efficiency of utilisation of absorbed amino acids

The mixture of amino acids of dietary origin absorbed from the small intestine (i.e. the truly digested amino acids) is utilised for the synthesis of tissue protein. The efficiency of this process, which depends upon the composition of the mix relative to that of the protein to be synthesised, is best represented by its true biological value. This will in turn depend upon the biological values of the digested undegraded dietary protein and the digested microbial protein, and upon the relative proportions of each contributing to the mix. In addition, it will vary with the primary function for which it is required. Microbial protein is thought to have a relatively constant biological value of about 0.8, whereas that of dietary origin will be variable and characteristic of the foods making up the diet. Prediction of such dietary values is extremely difficult, since the biological values of the individual proteins are no guide to their value in combinations. The variability of estimates with which truly digestible true protein presented for absorption is used for various functions is illustrated in Table 13.10.

The Agricultural and Food Research Council, in its 1992 publication (see Further reading), describes the efficiency of utilisation of truly digested true protein in terms of the limiting efficiency of use of an ideally balanced amino acid mixture (k_{aai}), taken as 0.85 under most conditions and to be 1.0 for maintenance. Relative values (RV) were then proposed for different functions:

Growth	0.7
Pregnancy	1.0
Lactation	0.8
Wool growth	0.3

Table 13.10 Comparison of estimates of the efficiency of utilisation of truly digested true protein made in some protein evaluation systems for ruminants

System[a]	Maintenance	Lactation	Growth	Wool/Hair
PDI		0.64	0.28–0.68	
CPFD	0.80	0.80	0.80	
DVE	0.67	0.64		
AAT-PBV		0.75		
AP	0.67	0.65	0.50	0.15
ADPLS	0.70	0.70	0.70	0.60
CNCPS	0.67	0.65	0.41–0.75	

[a] PDI, the French 'protein digested in the intestine' system; CPFD, the German 'crude protein flow at duodenum' system; DVE, the Dutch 'digestible protein in the intestine' system; AAT-PBV, the Nordic system; AP, the American 'absorbed true protein' system; ADPLS, the Australian 'apparently digested protein leaving the stomach' system; CNCPS, the Cornell 'net protein and carbohydrate system'.

The two factors were then combined to give the following working values (k_n):

Maintenance	k_{nm}	= 1.00
Growth	k_{ng}	= 0.59
Pregnancy	k_{nc}	= 0.85
Lactation	k_{nl}	= 0.68
Wool growth	k_{nw}	= 0.26

An alternative approach is to estimate the supply of essential amino acids made available to the tissues (i.e. those absorbed from the small intestine) and to relate this to the amino acid requirements of the animal. This approach needs information on the truly digestible amino acid content of the undegraded dietary and the microbial protein.

The essential amino acid content of ruminal microbial protein is frequently claimed to be relatively constant. In fact, large differences in the amino acid composition of samples of microbial protein have been shown to exist. Some figures, based on 441 samples from 35 experiments covering 61 diets, are given in Table 13.11.

There is evidence that the essential amino acid composition of undegradable dietary protein may differ significantly from that of the original dietary material and it has been suggested that estimates of its contribution to the amino acid supply should be based on the amino acid profile of the insoluble fraction of the dietary protein rather than that of the whole dietary protein. A comparison of the essential amino acid profiles of whole and insoluble proteins in some common foods is shown in Table 13.12.

A measure of the effect of assuming one or other of the measures of amino acid composition can be gained if we consider the following daily ration:

	DM (kg)	CP (g)	Methionine (g)		Lysine (g)	
			A	B	A	B
Maize silage	10.0	1100	9.9	9.9	19.8	24.2
Maize	5.0	490	7.8	6.9	10.8	8.3
Soya bean meal	3.0	1509	15.1	12.0	78.4	99.5
Total	18.0	3099	32.8	28.8	109.0	132.0
B as per cent of A				88		121

A, calculated by assuming the amino acid composition of the whole protein.
B, calculated by assuming the composition of the insoluble available fraction.

Amino acids are usually considered to have the same true digestibility as true protein, but there is evidence that the efficiency with which individual amino acids are metabolised will vary with the biological value of the protein and the amino acid supply relative to requirement.

A system for the quantitative nutrition of ruminant animals should embody the processes described,which requires that factors such as degradability, efficiency of nitrogen capture, microbial protein yield, digestibility of microbial protein, digestibility of dietary undegraded protein and the true biological value of the absorbed nitrogen or its essential amino acid content, be quantified.

Table 13.11 Amino acid composition of ruminal bacteria (g/100g of amino acids) (From Clark J H, Klusmeyer T H and Cameron M R 1992 *Journal of Dairy Science* **75**: 2304–23)

Amino acid	Mean	Minimum	Maximum	Standard deviation	Coefficient of variation (%)
Arginine	5.1	3.8	6.8	0.7	13.2
Histidine	2.0	1.2	3.6	0.4	21.3
Isoleucine	5.7	4.6	6.7	0.4	7.4
Leucine	8.1	5.3	9.7	0.8	10.3
Lysine	7.9	4.9	9.5	0.9	11.9
Methionine	2.6	1.1	4.9	0.7	25.6
Phenylalanine	5.1	4.4	6.3	0.3	6.4
Threonine	5.8	5.0	7.8	0.5	8.9
Valine	6.2	4.7	7.6	0.6	10.1
Alanine	7.5	5.0	8.6	0.6	7.3
Aspartic acid	12.2	10.9	13.5	0.6	4.8
Glutamic acid	13.1	11.6	14.4	0.7	5.3
Glycine	5.8	5.0	7.6	0.5	8.2
Proline	3.7	2.4	5.3	0.5	13.2
Serine	4.6	3.4	5.4	0.4	8.9
Tyrosine	4.9	3.9	7.7	0.6	13.2

Table 13.12 Amino acid composition of whole and insoluble protein in some common foods (g/100 g protein) (After Muscato T V, Sniffen C J, Krishnamoorthy U and Van Soest P J 1983 *Journal of Dairy Science* **66**: 2198–2207)

	Maize		Dried brewer's grains		Soya bean meal		Maize silage		Timothy hay	
	A	B	A	B	A	B	A	B	A	B
Methionine	1.6	1.4	1.4	0.6	1.0	0.8	0.9	0.9	0.7	0.9
Lysine	2.2	1.7	2.1	2.1	5.2	6.6	1.8	2.2	2.9	2.8
Histidine	2.2	1.9	1.7	1.3	1.7	1.3	0.9	0.9	1.1	1.1
Phenylalanine	4.2	4.5	4.4	3.9	4.1	3.9	3.0	3.1	3.4	3.4
Threonine	2.7	2.7	2.6	2.3	3.0	3.0	2.4	2.2	2.9	2.8
Leucine	10.3	11.6	8.1	6.8	6.2	6.2	6.5	6.9	5.2	5.4
Isoleucine	3.4	2.9	3.8	3.2	4.4	4.3	2.8	2.5	3.4	2.8
Valine	3.5	3.9	3.8	3.6	3.4	3.8	3.7	3.5	3.5	3.9
Arginine	3.7	2.7	3.6	3.0	6.0	7.8	1.7	2.2	3.2	3.0

A, dietary protein.
B, insoluble protein (aa in whole protein – aa in buffer-soluble protein – aa in cell wall protein).

13.7 The UK metabolisable protein system

The system is fully described in the 1992 Report of the Agricultural and Food Research Council's Technical Committee on Responses to Nutrients (see Further reading).

The microbial demand for protein is stated in terms of effective rumen degradable protein (ERDP) and foods have to be evaluated in the same terms. The ERDP of a food is calculated as:

$$\text{ERDP} = \text{CP} \times [0.8a + bc/(c + r)]$$

where a, b and c are the fitted parameters derived from the determination *in sacco* of the degradability of the food, 0.8 is the efficiency of capture of the nitrogen of the readily degradable fraction and r, the outflow rate, is varied as follows:

Animal	r
Cattle and sheep at low planes of nutrition	0.02
Calves, beef cattle, sheep and dairy cows (up to twice the maintenance level of feeding)	0.05
Dairy cows yielding more than 15 kg of milk per day	0.08

Alternatively, the following equation may be used for calculating r for levels of feeding (L) as multiples of ME for maintenance:

$$r = -0.02 + 0.14(1 - e^{-0.35L})$$

Account is thus taken of the differential capture of rapidly and slowly degradable proteins and the rate of passage through the rumen.

The demand for amino acids at tissue level is quantified in terms of truly digestible protein required to be absorbed from the small intestine and designated 'metabolisable protein' (MP) as described in Chapters 14, 15 and 16.

Microbial protein contributes towards satisfying this demand. The yield of microbial crude protein is related to the energy available to the rumen microorganisms in terms of fermentable metabolisable energy:

$$FME = ME - ME_{fat} - ME_{ferm}$$

ME_{ferm} is assumed to be 0.1ME for silages and 0.05ME for brewery and distillery by-products, and ME_{fat} is 35 MJ/kg. The assumption made here for silage must be suspect. The major fermentation product in well-made silages is lactate and there is evidence that several rumen bacteria, notably *Megasphaera elsdenii* (Table 8.3), are able to utilise lactate with the production of propionate. The microbial crude protein yield (g) is calculated as:

$$FME(MJ)y$$

Where: $y = 9$ at maintenance, 10 for growth and 11 for lactation; or alternatively:

$$y = 7 + 6(1 - e^{-0.35L})$$

The proportion of the microbial crude protein present as true protein is assumed to be 0.75 and the true digestibility to be 0.85, and the contribution of microbial protein (DMP) to the truly absorbed amino acids is:

$$DMP \ (g/kg \ DM) = FME \ (y \times 0.75 \times 0.85) = 0.6375(FME \ y)$$

When this contribution is taken into account, there remains a residual metabolisable protein requirement, which may be calculated as MP–DMP and which has to be satisfied by the truly digestible undegraded protein of the diet. The true digestibility of dietary undegraded protein is calculated on the assumption that the ADIN content is indigestible and that the remainder has a true digestibility of 0.9. The truly digestible undegraded true protein (DUP) is then:

$$DUP = 0.9[CP[1 - a - bc/(c + r)] - 6.25ADIN]$$

Where: a, b and c are the usual *in sacco* constants and DUP, CP and ADIN are g/kg DM.

The metabolisable protein supplied by the food may be calculated as: MP (g/kg DM) = DMP + DUP. An example of the evaluation of a protein source in these terms is given in Box 13.1.

Box 13.1
Evaluation of a protein source for a ruminant animal

Animal: lactating dairy cow, $y = 11$
A concentrate has:

CP (g/kg DM) = 550
EE (g/kg DM) = 20
ME (MJ/kg DM) = 12.5
$a = 0.2$
$b = 0.65$
$c = 0.06$
$r = 0.05$
ADIN (g/kg DM) = 0.20

Then

ERDP (g/kg DM) = 550[0.8 × 0.2 + (0.65 × 0.06/(0.06 + 0.05))] = 283
FME (MJ/kg DM) = 12.5 − (35 × 0.02) = 11.8
ERDP/FME = 283/11.8 = 23.98 > y and energy is limiting

Then

DMP (g) = 0.6375 (11.8 × 11) = 82.7
DUP (g/kg DM) = 0.9{[55091 − 0.2 − (0.65 × 0.06 / (0.06 + 0.05))] − 6.25 × 0.2} = 219.4
MP (g/kg) (DMP+ DUP) = 82.7 + 219.4 = 302.1

Then

ERDP = 283 g/kg DM
DUP = 219 g/kg DM

The metabolisable protein content of a food is of no use as a guide to the food's ability to satisfy the residual demand for metabolisable protein, since it includes a contribution from ERDP which has already been taken into account in the form of DMP. The protein content of foods is thus stated in terms of ERDP and DUP.

Summary

1. For food protein to be used with maximum efficiency, its constituent indispensable and dispensable amino acids must be present in the diet in sufficient quantities to meet the animal's metabolic demands.

2. Crude protein is a measure of food protein calculated as nitrogen content multiplied by 6.25. Digestible crude protein is calculated as 6.25 times the difference between nitrogen intake and nitrogen voided in the faeces.

3. Ileal digestibility is measured as the difference between nitrogen intake and nitrogen voided at the terminal ileum.

4. For true digestibility, the endogenous component of the faecal or ileal matter must be known.

5. Standardised digestibility takes account of part only of the endogenous nitrogen and remains an apparent figure.

6. Standardised methods ('protein efficiency ratio', 'net protein ratio' and 'gross protein value') based on the growth response of experimental animals are used to evaluate protein sources for monogastric animals.

7. Nitrogen balance is the difference between nitrogen intake and nitrogen output in the form of faeces, urine, milk and eggs. As an estimate of protein deposition in the animal's tissues it is subject to several errors and the results have been known to diverge from those of the comparative slaughter technique.

8. Biological value is stated as the proportion of the nitrogen intake which is actually retained and takes account of endogenous urinary and faecal nitrogen.

9. Chemical score and the essential amino acid index are based on the proportion of the main limiting amino acid of the protein in relation to that in a standard protein.

10. Assays of available amino acids may be made by measuring the liveweight gain or food conversion efficiency of animals given the intact protein as a supplement to a diet deficient in a particular essential amino acid.

11. Certain microorganisms have amino acid requirements similar to those of higher animals and have been used for protein evaluation.

12. Determination of available lysine by reaction of fluoro-4-dinitrobenzene with the reactive epsilon group has been shown to correlate well with gross protein values of animal foods and for supplements for high-cereal diets.

13. Dye-binding methods have proved satisfactory for cereal and milk products.

14. 'Ideal protein' is used for evaluating proteins for pigs. It is a modification of the chemical score technique with the amino acid pattern of a particular tissue or tissues serving as the reference pattern.

15. In practice, pig diets are frequently formulated to the commonly major limiting amino acids, lysine, cystine + methionine, threonine and tryptophan.

16. For poultry, evaluation of protein sources is based on the contents of the three major amino acids: lysine, methionine and tryptophan.

17. In evaluating protein sources for ruminant animals, account must be taken of:

- the degradability of the protein within the rumen,

- the rate of passage of the food through the rumen,

- the efficiency of capture of degraded protein by the rumen microbes,

- the yield of microbial protein,

- the true digestibility of the protein reaching the small intestine,

- the efficiency of utilisation of nitrogen absorbed from the small intestine.

18. The UK metabolisable protein system divides the requirement of an animal into that which is required for supplying the needs of the rumen microbes and that which is required at tissue level. After estimating the contribution of microbial protein to satisfying this demand, the requirement for undegraded dietary protein is calculated.

Further reading

Agricultural Research Council 1980 *The Nutrient Requirements of Ruminant Livestock*, Farnham Royal, UK, Commonwealth Agricultural Bureaux.

Agricultural Research Council 1984 *The Nutrient Requirements of Ruminant Livestock*, Supplement No. 1, Farnham Royal, UK, Commonwealth Agricultural Bureaux.

Agricultural and Food Research Council 1992 Technical Committee on Responses to Nutrients, *Report No. 9. Nutritive Requirements of Ruminant Animals: Protein*, Farnham Royal, UK, Commonwealth Agricultural Bureaux. (See also *Nutrition Abstracts and Reviews, Series B* **62**: 787–835.)

Chalupa W, 1991 Model generated protein degradation nutrition information. In: *Proceedings of the Cornell Nutrition Conference*, Ithaca, NY, p. 44.

Haresign W and Cole D J A (eds) 1988 *Recent Advances in Animal Nutrition*, London, Butterworth.

Haresign W and Cole D J A (eds) 1991 *Recent Advances in Animal Nutrition*, London, Butterworth/Heinemann.

Nissen S (ed.) 1992 *Modern Methods in Protein Nutrition and Metabolism*, London, Academic Press.

van Weerden I, van Weerden E J and Huisman J (eds) (1992) *Nutritive and Digestive Physiology in Monogastric Farm Animals*, Wageningen, Pudoc.

Wiseman J and Cole D J A (eds) 1990 *Feedstuff Evaluation*, London, Butterworth.

14 Feeding standards for maintenance and growth

14.1   Feeding standards for maintenance

14.2   Animal growth and nutrition: feeding standards for growth and wool production

14.3   Mineral and vitamin requirements for maintenance and growth

14.4   Nutritional control of growth

Statements of the amounts of nutrients required by animals are described by the general term 'feeding standards'. Two other terms used in the same context are 'nutrient requirement' and 'allowance'. Neither is strictly defined, but a rough distinction between them is that, if the requirement is a statement of what animals on average require for a particular function, the allowance is greater than this amount by a safety margin designed principally to allow for variations in requirement between individual animals.

Feeding standards may be expressed in quantities of nutrients or in dietary proportions. Thus the phosphorus requirement of a 50 kg pig might be stated as 11 g P/day or as 5 g P/kg of the diet. The former method of expression is used mainly for animals given exact quantities of foods, the latter for animals fed to appetite. Various units are used for feeding standards. For example, the energy requirements of ruminants may be stated in terms of net energy, metabolisable energy or feed units, and their protein requirements in terms of crude protein, digestible crude protein or metabolisable protein. It is obviously desirable that the units used in the standards should be the same as those used in the evaluation of foods. Standards may be given separately for each function of the animal or as overall figures for the combined functions. The requirements of dairy cows, for example, are often given separately for maintenance and for milk production, but those for growing chickens are for maintenance and growth combined. In some cases the requirements for single functions are not known; this is true particularly of vitamin and trace-element requirements.

As mentioned above, the translation of a requirement into an allowance that is to be used in feeding practice is accompanied by the addition of a safety factor. The justification for such safety factors is illustrated by the following example. Suppose that in cattle of 500 kg liveweight the requirement for energy for maintenance is found to range in individuals from 30 to 36 MJ net energy per day, with a mean value of 33 MJ. Although some of the variation may be caused by inaccuracies in the methods of measurement used, much of it will undoubtedly reflect real differences between animals. This being so, the adoption of the mean estimate of requirement, 33 MJ, as the allowance to be used in practice will result in some cattle being over- and others under-fed. Underfeeding is regarded as the greater evil, and so a safety factor may be added to the requirement when calculating the allowance to be recommended. This safety factor will be designed to ensure that no animals, or only those with an exceptionally high requirement, will be underfed. It may be an arbitrary addition or, better, one based mathematically on the expected variation between animals; the larger this variation, the greater should be the safety factor. Safety factors have been criticised on the grounds that overfeeding, say, 90 per cent of the population to ensure that the remaining 10 per cent are not grossly underfed is a wasteful procedure. They are justifiable for nutrients whose deficiency can cause severe disorders and even death, and for which the cost of oversupply is relatively low (e.g. magnesium). For nutrients supplying energy, however, safety factors are probably not justified. Oversupply of energy is likely to be expensive, and although animals will respond to the excess, an increase in rate of production may not be desirable. For example, animals might store excessive amounts of fat.

Variations between animals, and also between samples of a food, must always be borne in mind when applying feeding standards. Such variations mean that the application of standards to individual animals and individual samples of foods must inevitably be attended by inaccuracies. For this reason feeding standards should be considered as guides to feeding practice, not as inflexible rules; they do not replace the art of the stockman in the finer adjustment of food intake to animal performance. Feeding standards, however, are not restricted in application to feeding individual animals; they can be used on a larger, farm scale to calculate, for example, the total winter feed required by a dairy herd, and even on a national scale to assist in planning food imports.

Between 1960 and the mid-1980s, feeding standards in the UK were drawn up by research scientists under the aegis of the Agricultural Research Council, and then translated into practical manuals by extension workers of the Ministry of Agriculture and associated governmental and commercial organisations. Publications from these two sources are listed at the end of this chapter. In 1983, a single organisation for the UK, the Technical Committee on Responses to Nutrients, became responsible for both revising standards and producing practical manuals. Operating through working parties that included both research scientists and extension workers, it has produced reports published by the Commonwealth Agricultural Bureaux in *Nutrition Abstracts and Reviews*. Some of these reports are listed at the end of the appropriate chapters. The last of these reports, on nutrient requirements of goats, appeared in 1994. At the time of writing (2001) there is no UK organisation with responsibility for feeding standards, despite the continuing need for revision of current standards that arises from new research findings.

Many other countries have manuals of feeding standards. The standards used in the USA are prepared by committees of the National Research Council and published under the general heading *Nutrient Requirements of Domestic Animals*. Australia has a national committee to prepare and publish standards. In Europe, the Netherlands has a Central Feed Bureau to produce and revise standards, and Germany, France and the Scandinavian countries also have feeding standards. Some of the manuals produced by these countries are listed at the end of this chapter or in Appendix 2. Examples of their use are included in the text, but it is not possible to provide a comprehensive coverage of all the standards available.

The main users of feeding standards are the commercial companies that supply concentrate mixtures as complete diets for pigs and poultry and as supplements to the forages consumed by ruminants. Farmers' advisers and consultants also need them. Commercial feed companies tend to modify published standards for their own use, in order to meet the special needs of their customers. For example, the nutrient requirements of poultry tend to change from one generation to the next because of the speed of genetic selection and improvement, and national feeding standards fail to keep up with the changes. The multiplicity of feeding standards, the tendency for them to be frequently revised and the increasing use of computers in ration formulation are factors that have encouraged users to be more flexible in their selection of feeding standards. Also, the emphasis today is less on the setting of minimum requirements and more on describing the continuous relationships between nutrient intake and animal performance (hence the formation in the UK of the Technical Committee on Responses to Nutrients).

Feeding standards have been criticised on the grounds that there may be little opportunity to apply them on the farm. For example, grazing animals will eat what they can harvest, not a prescribed ration. However, if grazing animals fail to grow because of a suspected copper deficiency, a feeding standard for copper will be needed to confirm the deficiency of the element in the pasture herbage. In developing countries, standards are often difficult to apply because the foods needed to supplement local resources are not available. In developed countries there may be a tendency to oversupply nutrients that are freely available and cheap (e.g. calcium). However, the need to avoid oversupply (by applying standards) has recently been recognised in countries where the excretion of excess nutrients by livestock causes environmental pollution; an example is the pollution of soil and water in the Netherlands by phosphorus (and also nitrogen) compounds. In the near future, feeding standards are likely to prescribe maximum levels of nutrients as well as minimum levels.

14.1 Feeding standards for maintenance

An animal is in a state of maintenance when its body composition remains constant, when it does not give rise to any product such as milk and does not perform any work on its surroundings. As farm animals are rarely kept in this non-productive state, it might seem to be only of academic interest to determine nutrient requirements for maintenance; but the total requirements of several classes of animal, notably dairy cows, are arrived at factorially by summing requirements calculated separately for maintenance and for production. Consequently, a knowledge of the maintenance needs of animals is of practical as well as theoretical significance. The relative importance of requirements for maintenance is illustrated in Table 14.1, which shows the proportion of their total energy requirements used for this purpose by various classes of animal.

The animals exemplified in Table 14.1 are all highly productive, but less productive animals use proportionately more of their energy intake for maintenance. It may be calculated, for example, that cattle in Africa on average use about 85 per cent of their energy intake for maintenance.

Animals deprived of food are forced to draw upon their body reserves to meet their nutrient requirements for maintenance. We have seen already that a fasted animal must oxidise reserves of nutrients to provide the energy needed for such essential processes as respiration and the circulation of the blood. Since the energy so utilised leaves the body, as heat, the animal is then in a state of negative energy balance. The same holds true for other nutrients: an animal fed on a protein-free diet continues to lose nitrogen in its faeces and urine and is therefore in negative nitrogen balance. The purpose of a maintenance ration is to prevent this drain on the body tissues, and the maintenance requirement of a nutrient can therefore be defined as the quantity which must be supplied in the diet so that the animal experiences neither net gain nor net loss of that nutrient. The requirement for maintenance is thus the minimum quantity promoting zero balance. (The qualifi-

Table 14.1 Approximate proportions of the total energy requirements of animals which are contributed by their requirements for maintenance

	Requirement (MJ net energy) for:		Maintenance as a percentage of total
	Maintenance	Production	
Daily values			
Dairy cow weighing 500 kg and producing 20 kg milk	32	63	34
Steer weighing 300 kg and gaining 1 kg	23	16	59
Pig weighing 50 kg and gaining 0.75 kg	7	10	41
Broiler chicken weighing 1 kg and gaining 35 g	0.50	0.32	61
Annual values			
Dairy cow weighing 500 kg, producing a calf of 35 kg and 5000 kg milk	12 200	16 000	43
Sow weighing 200 kg, producing 16 piglets, each 1.5 kg at birth, and 750 kg milk	7 100	4 600	61
Hen weighing 2 kg, producing 250 eggs	190	95	67

cation 'minimum' is necessary, because if the animal is unable to store the nutrient in question, increasing the quantity supplied above that required for maintenance will still result in zero balance.)

Energy requirements for maintenance

Basal and fasting metabolism

It was explained earlier (p. 270) that energy expended for the maintenance of an animal is converted into heat and leaves the body in this form. The quantity of heat arising in this way is known as the animal's basal metabolism, and its measurement provides a direct estimate of the quantity of net energy that the animal must obtain from its food in order to meet the demand for maintenance. The measurement of basal metabolism is complicated by the fact that the heat produced by the animal does not come only from this source but comes also from the digestion and metabolism of food constituents (the heat increment of feeding), and from the voluntary muscular activity of the animal. Heat production may be further increased if the animal is kept in a cold environment (see p. 358).

When basal metabolism is measured, the complicating effect of the heat increment of feeding is removed by depriving the animal of food. The period of fast required for the digestion and metabolism of previous meals to be completed varies considerably with species. In man, an overnight fast is sufficient; but in ruminant animals, digestion, absorption and metabolism continue for several days after feeding ceases, and a fast of at least four days would be needed. The same period is recommended for the pig and two days for the fowl. There are a number of criteria for establishing whether the animal has reached the post-absorptive state. If heat production can be measured continuously, the most satisfactory indication is the decline in heat production to a steady, constant level. A second indication is given by the respiratory quotient (p. 274). In fasting, the oxidation mixture gradually changes from absorbed fat, carbohydrate and protein to body fat and some body protein. This replacement in the mixture of carbohydrate by fat is accompanied by a decline in the non-protein respiratory quotient, and when the theoretical value for fat (0.7) is reached it may be assumed that energy is being obtained only from body reserves. In ruminants, an additional indication that the post-absorptive state has been reached is a decline in methane production (and therefore digestive activity) to a very low level.

The contribution of voluntary muscular activity to heat production can be reduced to a low level when basal metabolism is measured in human subjects, but in farm animals the cooperation needed to obtain a state of complete relaxation can rarely be achieved. Fasting may limit activity, but even the small activity represented by standing as opposed to lying is sufficient to increase heat production. Consequently, the term 'fasting metabolism' is to be preferred to 'basal metabolism' in studies with farm animals, since strict basal conditions are unlikely to be observed.

A term used in conjunction with fasting metabolism is 'fasting catabolism'. This includes the relatively small quantities of energy lost by fasting animals in their urine.

Some typical values for fasting metabolism are given in Table 14.2. As one would expect, the values are greater for large than for small animals, but column 2 shows that per unit of liveweight, fasting metabolism is greater in small animals. At an early stage in the study of basal metabolism it was recognised that fasting heat production is more nearly proportional to the surface area of animals than to their weight, and it became customary to compare values for animals of different sizes by expressing them in relation to surface area (column 3 of Table 14.2). The surface area of animals is obviously difficult to measure and methods were therefore devised for predicting it from their bodyweight. The basis for such methods is that, in bodies of the same shape and of equal density, surface area is proportional to the two-thirds power of weight ($W^{0.67}$). The logical development of this approach was to omit the calculation of surface area and to express fasting metabolism in relation to $W^{0.67}$. When the link between fasting metabolism and bodyweight was examined further it was found that the closest relationship was between metabolism and $W^{0.73}$, not $W^{0.67}$. The function $W^{0.73}$ was used as a reference base for fasting metabolism of farm animals until 1964, when it was decided to round off the exponent to 0.75 (see column 4 of Table 14.2).

There has been considerable discussion of whether surface area or $W^{0.75}$ (often called *metabolic liveweight*) is the better base. This will not be repeated here, but is contained in the books listed at the end of the chapter. Mathematically there is nothing to choose between the two bases, for their relationships with fasting metabolism are equally close.

The fasting metabolism of adult animals ranging in size from mice to elephants was found by S. Brody to have an average value of 70 kcal/kg $W^{0.73}$ per day; the approximate equivalent is 0.27 MJ/kg $W^{0.75}$ per day. There are, however, considerable variations from species to species, as shown in Table 14.2. For example, cattle tend to have a fasting metabolism

Table 14.2 Some typical values for the fasting metabolism of adult animals of various species

Animal	Live-weight (kg)	Fasting metabolism (MJ/day)			
		Per animal (1)	Per kg liveweight (W) (2)	Per m² surface area (3)	Per kg $W^{0.75}$ (4)
Cow	500	34.1	0.068	7.0	0.32
Pig	70	7.5	0.107	5.1	0.31
Man	70	7.1	0.101	3.9	0.29
Sheep	50	4.3	0.086	3.6	0.23
Fowl	2	0.60	0.300	–	0.36
Rat	0.3	0.12	0.400	3.6	0.30

about 15 per cent higher than the interspecies mean, and sheep a fasting metabolism 15 per cent lower. There are also variations within species, notably those due to age and sex. Fasting metabolism per unit of metabolic liveweight is higher in young animals than in old, being for example 0.39 MJ/kg $W^{0.75}$ per day in a young calf, but only 0.32 MJ/kg $W^{0.75}$ in a mature cow; it is also 15 per cent higher in male cattle than in females or castrated males.

Other methods of estimating maintenance energy requirements

The quantity of energy required for maintenance is, by definition, that which promotes energy equilibrium (zero energy balance). This quantity can be estimated directly in fed, as opposed to fasted, animals if the energy content of their food is known and their energy balance can be measured. In theory the quantities of food given could be adjusted until the animals were in exact energy equilibrium, but in practice it is easier to allow them to make small gains or losses and then to use a model of the kind depicted in Fig. 11.5 to estimate the energy intake required for equilibrium. For example, suppose a 300 kg steer was given 3.3 kg DM/day in the form of a food with M/D = 11 MJ/kg DM and with k_f = 0.5. If the steer retained 2 MJ/day, its maintenance requirement for metabolisable energy would be calculated as:

$$(3.3 \times 11) - (2/0.5) = 32.3 \text{ MJ ME/day}$$

This approach can also be followed in feeding trials in which animals are not kept in calorimeters. The animals are given known quantities of food energy, and their liveweights and liveweight gains or losses are measured. The partition of energy intake between that used for maintenance and that used for liveweight gain can be made in two ways. The simpler method involves the use of known feeding standards for liveweight gain. The alternative is to analyse the figures for energy intake (I), liveweight (W) and liveweight gain (G) by solving equations of the form:

$$I = aW^{0.75} + bG$$

The coefficients a and b then provide estimates of the quantities of food energy used for maintenance and for each unit of liveweight gain, respectively. This form of analysis can be extended to animals with more than one type of production, such as dairy cows, by adding extra terms to the right-hand side of the equation.

The main objection to determining energy requirements for maintenance (and also for production) in this way is that liveweight changes may fail to give a correct measure of energy balance. It is possible, however, to put the method on a sounder 'energy' basis by using the comparative slaughter technique to estimate changes in the energy content of the animals.

Fasting metabolism as a basis for estimating maintenance requirements

The 'feeding trial' method of estimating maintenance requirements has the advantage of being applied to animals kept under normal farm conditions, rather than under the somewhat unnatural conditions represented by a fast in a calorimeter. It is often difficult to translate values for fasting metabolism into practical maintenance requirements. One factor to be taken into account is that animals on the farm commonly use more energy for voluntary muscular activity. Another factor is that productive livestock must operate with more intense metabolism than fasted animals and thereby incur a higher maintenance cost. Third, animals on the farm experience greater extremes of climate and may need to use energy specifically to maintain their normal body temperature. The first two factors are discussed below, and the effects of climate on energy requirements for maintenance are discussed on p. 358.

Estimates of the energy costs of various forms of voluntary muscular activity are shown in Table 14.3. In the first column the costs are given per unit of liveweight. The second column gives estimates of the number of units of activity likely to be undertaken, and the third column gives estimated daily costs for a 50 kg sheep. For example, if the sheep walks 3.0 km per day (line 3) and the unit cost is 2.6 kJ per kg liveweight per km, a 50 kg sheep will incur a cost of $2.6 \times 3.0 \times 50 = 390$ kJ per day. Table 14.3 includes a value for fasting metabolism, and it can be calculated that the net energy required by the sheep for maintenance will be increased by $100(390/4300) = 9$ per cent if it walks 3.0 km per day.

Some of the activities listed in Table 14.3 are likely to be carried out by all animals (standing, getting up and lying down, plus a minimal amount of locomotion), and their energy cost is always added to the fasting metabolism when calculating maintenance requirements. For example, in the ARC 1980 system (as modified by MAFF) this activity allowance is about 7 kJ per kg liveweight per day. For a 50 kg sheep this amounts to 350 kJ per day, or about 8 per cent of the fasting metabolism.

Table 14.3 Energy costs of physical activity in a 50 kg sheep

Activity	Cost per kg liveweight	Duration or frequency of activity	Cost per day (kJ)
Standing	0.4 kJ/h	9 h/day	180
Changing position (stand then lie)	0.26 kJ	6 times/day	78
Walking	2.6 kJ/km	5 km/day	650
Climbing	28 kJ/km	0.2 km/day	280
Eating	2.5 kJ/h	2–8 h/day	250–1000
Ruminating	2.0 kJ/h	8 h/day	800
(Fasting metabolism			4300)

The energy costs of eating (prehension, chewing and swallowing) and of rumination are included in the heat increment of feeding (i.e. they are taken into account in the estimation of the k factors). However, if animals are grazing, rather than having foods delivered to them, their energy costs of muscular activity will be much increased. Table 14.3 shows that if a 50 kg sheep has to walk 5 km and climb 0.2 km a day in search of food, and has to extend its eating time from 2 to 8 hours a day, its energy cost will be increased by $650 + 280 + 750 = 1680$ kJ per day, which is equivalent to nearly 40 per cent of its fasting metabolism. In general, grazing animals are likely to have maintenance requirements that are 25–50 per cent greater than those of housed animals, the actual increase depending on terrain and vegetation.

The efficiency with which metabolisable energy is used to meet the costs of muscular activity (k_w) is generally assumed to be the same as k_m.

Although fasting metabolism is measured under standardised conditions, it is known that the value obtained for a particular animal will depend on that animal's previous energy status. If an animal on a high plane of nutrition is suddenly fasted, its metabolism will be greater than that of a similar animal that has previously been kept on a lower plane. In one comparison made with 35kg lambs, those previously on a high plane had a fasting metabolism 20 per cent higher than those previously kept on a moderate plane. The same effect is demonstrated when one attempts to give animals just enough food to keep their weight constant (i.e. to keep them at maintenance). As time passes, the ration has to be progressively reduced to maintain the required equilibrium. The inference is that animals can adapt to low-level (maintenance) rations either by improving their efficiency of energy utilisation or – more likely – by reducing non-essential muscular activity. This means that if an animal's fasting metabolism is determined after a period of low-plane nutrition, as is commonly the case, then the value obtained, even when increased to allow for additional muscular activity, may well underestimate its maintenance requirement during periods of higher-plane feeding. This source of error is recognised in the Australian energy system for ruminants (see p. 300), maintenance requirements being increased as levels of energy intake rise.

The fasting metabolism of an animal when expressed per unit of metabolic weight may vary with the body composition of the animal. The metabolically active tissues are the internal organs and musculature, with fat depots presumably requiring less energy for maintenance. Thus a fat animal is likely to have a lower fasting metabolism than a thin animal of the same weight.

Present standards

The maintenance requirements of cattle in the ARC (1980) system, as modified by MAFF, are based on data for fasting metabolism (*F*, MJ/day). For steers and heifers the equation is:

$$F = 0.53(W/1.08)^{0.67}$$

Liveweight is reduced to an estimated fasted weight by dividing by 1.08 and metabolic weight is calculated with the power 0.67 rather than the more usual 0.75. An allowance for the minimal activity expected of housed animals is calculated as $0.0071W$ for growing cattle and $0.0091W$ for dairy cows. Thus the net energy requirement for maintenance of a 500 kg cow would be:

$$0.53(500/1.08)^{0.67} + 0.0091 \times 500 = 36.9 \text{ MJ/day}$$

If the cow's diet contained 11 MJ ME/kg DM, k_m (Table 12.1) would be 0.714, and the cow's requirement for metabolisable energy would be:

$$36.9/0.714 = 51.7 \text{ MJ ME/day}$$

As their fasting metabolism is greater, bulls are considered to have maintenance requirements 15 per cent higher than those of steers and heifers of the same weight. For sheep, the equation for predicting the maintenance requirement (as net energy) for ewes and wethers is:

$$0.226(W/1.08)^{0.75} + 0.007W$$

From this, a 60 kg ewe is calculated to require 5.0 MJ NE/day for maintenance, or if $k_m = 0.714$, its requirement could alternatively be expressed as 7.0 MJ ME/day. As with cattle, intact males (rams) are considered to require 15 per cent more energy for maintenance.

Australian estimates of the maintenance requirements of ruminants are 5–10 per cent higher than those given above, because they take into account the factor mentioned earlier (p. 355), that fasting metabolism underestimates the energy required for maintenance in productive animals. In the latest (2000) version of its publication *Nutrient Requirements of Beef Cattle* the US National Research Council has introduced a multifactorial method of calculating energy requirements for maintenance that includes allowances for climatic factors and previous nutritional status.

The energy requirements of pigs and poultry are normally stated for maintenance and production combined, although some theoretical standards for maintenance have been calculated. The UK Technical Committee on Responses to Nutrients gives the maintenance requirements of sows as 430 kJ ME/kg $W^{0.75}$ per day (17.5 MJ for a 140 kg sow), with the requirement of boars being 15 per cent higher, at 495 kJ ME/kg $W^{0.75}$ per day. For laying hens, the maintenance requirement is about 550 kJ ME/kg $W^{0.75}$ per day.

Finally, the maintenance energy requirement of horses (MJ digestible energy per day) is estimated by the US National Research Council from the equation:

$$4.08 + 0.879W$$

Thus a 400 kg horse would require 39.2 MJ DE per day, but this is increased by 43 per cent to allow for normal levels of activity (i.e. to 56.3

MJ). In the French net energy system, the maintenance requirement of horses is estimated as 351 kJ NE/kg $W^{0.75}$ per day. Stallions require 10 per cent more than this for maintenance, and all horses need a further 10–15 per cent as an activity increment. Thus a 400 kg mare would require 31.4 MJ NE per day, which becomes $1.67 \times 31.4 = 52.4$ MJ DE per day, or 58–60 MJ DE per day after allowing for activity. The US and French estimates are therefore in reasonable agreement.

The influence of climate on energy metabolism and on energy requirements for maintenance

The influence of climate on the nutrition of farm animals is not confined to energy requirements for maintenance but extends to other aspects of energy metabolism and also to nutrients other than those supplying energy. Nevertheless, the greatest influence of climate is on energy requirements, and in cold climates it is animals kept at or below the maintenance level that are most affected by climate.

Mammals and birds are *homeotherms*, which means that they attempt to keep their body temperature constant. Animals produce heat continuously and, if they are to maintain a constant body temperature, must lose it to their surroundings. The two main routes by which they may lose heat are first the so-called *sensible* loss, by radiation, conduction and convection from their body surface, and second the evaporative loss, of water from both their body surface and their lungs (2.52 MJ/kg water). The rate at which heat is lost depends in the first instance on the difference in temperature between the animal and its surroundings: for farm animals, the rectal temperature, which is slightly lower than the deep body temperature, lies in the range 36–43 °C. The rate of heat loss is also influenced by characteristics of the animal, such as the insulation provided by its tissues and coat, and by such characteristics of the environment as air velocity, relative humidity and solar radiation. In effect, the rate of heat loss is determined by a complex interaction of factors contributed by the animal and its environment.

The physics and physiology of heat loss from farm animals may be illustrated by the example of a pig kept under basal conditions, that is, fasting, resting and at a 'comfortable' air temperature of 22 °C (Fig. 14.1, solid lines). Its heat production (and loss) is 5 MJ per day, and its heat loss is divided approximately equally between the evaporative and sensible routes. If the air temperature is gradually reduced, the pig will be liable to lose heat more rapidly. It can counter this effect to some extent by reducing its evaporative loss and perhaps also by reducing blood flow (hence heat transfer) to the body surface. The latter response will reduce the skin temperature and presumably cause the pig to feel cold. As the fall in air temperature continues, a stage is reached when the pig can maintain its deep body temperature only by increasing its heat production, which it might do by the muscular activity of shivering. The environmental temperature below which heat production is increased is known as the *lower critical temperature* of the pig, and in this example is 20 °C.

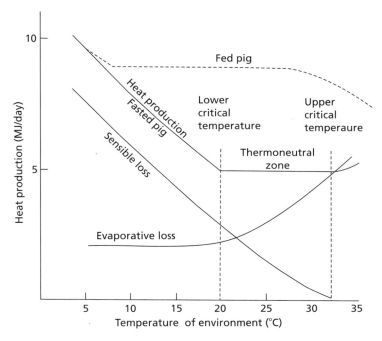

Fig. 14.1 Effect of environmental temperature on the heat production of a pig.

If the pig were fed rather than fasted, its heat production would be increased by the heat increment of feeding and its lower critical temperature would be reduced (see Fig. 14.1, broken line); in fact, the pig would not have to increase its heat production until the environmental temperature fell to 7 °C.

If this pig were subjected to an increasing temperature, it would have difficulty in losing heat by the sensible route and would need to increase its evaporative loss. Eventually, a temperature would be reached at which the pig would need to reduce its heat production, which it might do by restricting its muscular activity and also by reducing its food intake. The temperature above which animals must reduce their heat production is known as the *upper critical temperature*. The range between the lower and upper critical temperatures is known as the *thermoneutral zone*.

We can now move from the simple situation of the pig exposed to only one climatic variable, temperature, to other animals and other climates. Lower critical temperatures for animals kept in various circumstances are given in Table 14.4. Ruminants have a wider thermoneutral zone and a lower critical temperature than pigs (and poultry) because they have a greater capacity to regulate heat loss by evaporation. Their heat increment of feeding is also greater (i.e. lower k factors) than is that of non-ruminants. Also, ruminants produce heat at a steady rate throughout the day, whereas pigs tend to digest and metabolise their food quickly, and then experience cold when the heat increment has declined. Small animals are potentially more susceptible to cold because they are often less well insulated, but tend

Table 14.4 Some examples of the lower critical temperature (°C) of farm animals in various environments

Animal		State	Feeding or production level	Wind speed (km/h)	
				0	15
Sheep:	lamb	Newborn	–	28	34
	adult	Shorn	Fasted	31	35
		50 mm wool	Fasted	22	28
			Maintenance	7	18
			Ad libitum	−10	5
		100 mm wool	*Ad libitum*	−40	–
Cattle:	calf	Newborn	–	18	28
	beef	Store	Maintenance	−16	−3
		Growing (30 mm coat)	Gaining 0.8 kg/day	−32	−10
	cow	Dairy (20 mm coat)	Maintenance 30 l milk/day	−8	10
				−30	−20

				Floor	
				Straw	Concrete
Pig:	sow	Adult (160 kg)	Maintenance	22	–
	growing (40 kg)	Individual	High	14	19
		Group	High	7	13

to counteract this by having a higher basal metabolism per unit of body-weight. Nevertheless, the lower critical temperature of an adult sheep (50 kg) is higher than that of a cow (500 kg) kept in a similar environment (Table 14.4).

An animal's insulation depends on its subcutaneous fat (pig) or coat depth (sheep fleece, cattle hair, poultry feathers). Thus a sheep that has been shorn is particularly vulnerable to cold even in summer, and especially if its food intake is restricted. By disturbing the coat or fleece, wind reduces insulation and, as shown in Table 14.4, increases the critical temperature. Rain increases heat loss both by reducing insulation and through heat of vaporisation; its effects are not illustrated in Table 14.4, but in an adult sheep with a 50 mm coat, for example, 30 mm of rain per day can raise the critical temperature by 2–6 °C. For housed pigs, the floor surface is a determinant of its insulation, and pigs kept on straw have lower critical temperatures than those kept on bare concrete. Groups of pigs can huddle together, thereby reducing their combined surface area, their heat losses and their critical temperatures.

The farm animals that are most likely to suffer cold stress are the newborn lamb, calf and pig. They are small, their insulation may be low because of a lack of subcutaneous fat or a thin coat of hair or wool, and they are wet when born. If the newborn animal fails to suck sufficient milk from its mother, its heat increment of feeding will be low. The lamb and calf have a special tissue for generating heat which is known as brown adipose tissue. Deposits of 'brown fat' are located at strategic points of the young animal's body, such as the shoulders and abdomen. Fat globules are stored in actively metabolising cells, and the tissue is well supplied with blood. When the fat is metabolised, the oxidation is 'uncoupled', energy being released as heat instead of being captured as ATP. The heat so generated is carried to other parts of the body by the blood. The protective action of brown adipose tissue in young animals is limited by their small fat reserves, and it is essential that young lambs and calves should receive food (i.e. colostrum and milk) as soon after birth as possible.

A comparison of an animal's lower critical temperature with the environmental temperature tells us whether the animal is likely to need an additional source of energy to increase its heat production but does not tell us how the energy should be supplied or in what quantity. The alternative strategies for countering cold stress are, first, to make the environment warmer (e.g. by improving the insulation of buildings or reducing draughts), second, to allow the animal to increase heat production from its existing resources (e.g. by metabolising fat reserves), or third, to increase the animal's heat production by modifying its diet. The last would seem to be the preferable nutritional strategy.

In the case of housed animals, the increase in heat loss per 1 °C fall in temperature below the critical temperature is reasonably constant. A value of 18 kJ per kg $W^{0.75}$ per day has been quoted for adult pigs. Thus if a sow weighing 160 kg (45 kg $W^{0.75}$) were kept on a maintenance ration (19.4 MJ ME per day) at a temperature 5 °C below its critical temperature (22 °C), its daily heat loss would be potentially $45 \times 5 \times 18 = 4050$ kJ (4.05 MJ) greater than its heat production. This quantity is about 20 per cent of its maintenance requirement. When metabolisable energy is used to meet a deficit in heat production it is utilised with an efficiency of 100 per cent (i.e. k = 1), so that the sow would need an additional 4.05 MJ ME per day to ensure energy equilibrium. An alternative strategy would be to increase ME intake further, to a level at which the additional heat increment would meet the heat deficit. For example, if k_g was 0.7 the sow would need 13.5 MJ ME per day above its maintenance requirement, which would yield 4.05 MJ as heat and 9.45 MJ as retained energy.

Laying hens are normally fed to appetite and are therefore able to adjust their food and energy intake to produce the heat needed to maintain their body temperature. As the environmental temperature falls below 25 °C their daily metabolisable energy intake increases by about 22 kJ for each 1 °C fall. For a 1.8 kg bird this is equivalent to 14 kJ per kg $W^{0.75}$ (cf. 18 kJ for sows); the extra metabolisable energy consumed by hens has little effect on egg production and is therefore apparently used solely to generate heat. Calculations of this kind can be used by the farmer to determine whether it is more economical to provide extra heat or insulation for poultry buildings or extra food for the birds.

For housed ruminants the increases in heat loss induced by a 1 °C fall in environmental temperature are comparable (10–20 kJ per kg $W^{0.75}$ per day) to those for pigs and poultry, but are much greater (20–40 kJ) for ruminants kept out of doors and exposed to wind and rain. For ruminants it is possible to increase heat production by changing the quality of their rations. Forages of low metabolisable energy concentration have low k values and thus yield more heat per unit of metabolisable energy intake than concentrate-rich rations.

In a hot climate, the animal's problem is one of disposing of the heat it produces. We have seen already (Fig. 14.1) that as air temperature increases, less heat can be lost by the sensible route of radiation, conduction and convection, and more has to be lost by evaporation. The domestic species vary considerably in their ability to lose heat by the evaporation of water. Most of the mammals are poorly equipped with sweat glands and birds have none; cattle, however, are able to lose appreciable quantities of water and heat via these glands, especially tropical cattle (*Bos indicus*). Evaporation from the skin can be increased through surface water acquired by wallowing, but the major route by which water vapour is lost is the respiratory tract. The farmer can assist the animal to lose heat by providing shade, ventilation and possibly water sprays. However, if these and the animal's own heat loss mechanisms become overtaxed, the animal has to reduce its heat production, which it does by reducing its food and energy intake. This means that potentially high-producing livestock, such as dairy cows, are seriously handicapped in the tropics by their inability to maintain high levels of energy intake. Ruminants generally are handicapped in hot climates by their dependence on low-quality forages, the metabolisable energy of which is used with low efficiency, hence a high heat increment.

Protein requirements for maintenance

An animal placed on a nitrogen-free, but otherwise adequate, ration continues to lose nitrogen in its faeces and urine. The nitrogen of the faeces, as described earlier (Chapter 10), arises from the enzyme and cell residues of the digestive tract, and from microbial residues. If the animal continues to eat, it must continue to lose nitrogen in this fashion.

It is less obvious, perhaps, why an animal on a nitrogen-free diet should lose nitrogen in its urine. In part, this excretion represents nitrogen which has been incorporated into materials subsequently expended and which cannot be recovered by the body for reuse. Thus the creatine of muscles is eventually converted into creatinine, which is excreted in the urine. But the greater part of the nitrogen in the urine of animals not receiving food nitrogen is (in mammals) in the form of urea, the typical by-product of amino acid catabolism, which arises from the turnover of body proteins described in Chapter 11. The turnover rate of body proteins varies considerably from one tissue to another; proteins are replaced at intervals of hours or days in the intestine and liver, whereas in bone and nerve the interval may be one of months or even years. The amino acids released when body proteins are broken down form a pool from which the replacement proteins are synthesised; a particular amino acid molecule may therefore be present one day in a protein of the liver, for example, and the next day in muscle protein. In effect, the body proteins exchange amino acids among themselves. Like the synthesis of body proteins from absorbed amino acids (Chapter 13), however, this domestic traffic in amino acids is not completely efficient. Acids liberated from one protein may fail to be incorporated in another and are catabolised, their amino groups thus yielding the urea which is excreted in the urine.

When an animal is first placed on a nitrogen-free diet, the quantity of nitrogen in its urine may fall progressively for several days before stabilising at a lower level, and when nitrogen is reintroduced into the diet there is a similar lag in the re-establishment of equilibrium. This suggests that the animal possesses a protein reserve which can be drawn upon in times of scarcity of dietary nitrogen and restored in times of plenty. In times of scarcity, the tissues most readily depleted of protein are those in which the proteins are most labile, such as the liver. Depletion of liver nitrogen is accompanied by some reduction in enzyme activity, and the reserve protein is therefore envisaged as a 'working reserve' which consists of the cytoplasmic proteins themselves.

Once the reserve protein has been depleted, the urinary nitrogen excretion of an animal deprived of food nitrogen reaches a minimal and approximately constant level. (This level will be maintained only if energy intake is adequate, for if tissue proteins are catabolised specifically to provide energy, urinary nitrogen excretion will rise again.) The nitrogen excreted at this minimal level is known as the *endogenous urinary nitrogen* and represents the smallest loss of body nitrogen commensurate with the continuing existence of the animal. The endogenous urinary nitrogen excretion can therefore be used to estimate the nitrogen (or protein) required by the animal for maintenance. It is analogous to the basal metabolism, and in fact there is a relationship between the two. The proportionality commonly quoted is 2 mg endogenous urinary nitrogen per kcal basal metabolism (about 500 mg/MJ). For adult ruminants, however, the ratio is appreciably lower, at 300–400 mg endogenous urinary nitrogen per MJ of fasting metabolism. The reason for this is that ruminants on diets devoid of, or deficient in, nitrogen are capable of recycling urea to the rumen or large intestine, and nitrogen that would in a non-ruminant be excreted in the urine is therefore excreted as nitrogen of microbial residues. The *total* or *basal* endogenous nitrogen excretion (i.e. urinary plus faecal nitrogen) of ruminants is calculated as 350 mg N/kg $W^{0.75}$ per day and is equivalent to 1000–1500 mg/MJ fasting metabolism; it is therefore two to three times as great as in non-ruminants.

When nitrogen is reintroduced into the diet, the quantity of nitrogen excreted in the urine increases through the wastage of amino acids incurred in utilising the food protein. Urinary nitrogen in excess of the endogenous portion is known as the *exogenous urinary nitrogen*. This name implies that such nitrogen is of food origin as opposed to body origin; but, with the exception of the creatinine fraction of the endogenous portion, it is doubtful whether such a strict division is justified. It is more realistic to regard the so-called exogenous fraction as the extension of an existing loss of nitrogen rather than as an additional source of loss, for it probably reflects mainly an increase in turnover and wastage of amino acids.

The quantity of nitrogen or of protein required for maintenance is that which will balance the metabolic faecal and endogenous urinary losses of nitrogen (and also the small dermal losses of nitrogen occurring in scurf, hair and sweat). The two most commonly employed methods for estimating this quantity are analogous to those used for determining the quantity of energy needed for maintenance. The first, analogous to the determination of the fasting catabolism, involves measuring the animal's losses of nitrogen when it is fed on a nitrogen-free diet, and calculating the quantity of food nitrogen required to balance these losses. In the second method, the quantity of food nitrogen required is determined directly by finding the minimum intake which produces nitrogen equilibrium. This method is similar to that used for estimating maintenance energy requirements (p. 354).

Estimating protein requirements for maintenance of ruminants from endogenous nitrogen losses (the factorial method)

The starting point in the factorial calculation for ruminants is the basal endogenous nitrogen excretion. At 350 mg N/kg $W^{0.75}$ per day, for a 600 kg cow this would be 42.4 g/day. Dermal losses of nitrogen in hair and scurf would add 2.2 g N/day, giving a total of 44.6 g N, or 279 g protein. In the metabolisable protein evaluation system described in Chapter 13, protein absorbed from the intestine (metabolisable protein, MP) is considered to be utilised for maintenance with an efficiency of 1.0, so the MP requirement of the 600 kg cow for maintenance is also 279 g/day. This protein is likely to come from microbial protein (MCP). When allowance is made for the proportion of true protein in MCP (0.75) and for the true digestibility of this true protein (0.85), the requirement for MCP is:

$$279/(0.75 \times 0.85) = 438 \text{ g/day}$$

The quantity of MCP produced in the rumen depends on the quantity of organic matter fermented and hence on the quantity of fermentable metabolisable energy (FME). For cattle at the maintenance level the relationship is quantified as 9 g MCP per MJ FME. Thus if the cow's intake of FME was 55 MJ (which approximates to its energy requirement for maintenance), the quantity of MCP supplied would be $55 \times 9 = 495$ g. Thus MCP alone (i.e. without the contribution from the food of any digestible rumen-undegradable protein, DUP) would be more than sufficient to meet the requirement for maintenance. In fact, it is commonly the case with ruminants that a diet that meets the energy requirement for maintenance of the animal and the protein or nitrogen requirement of the rumen microorganisms will also meet the protein requirement of the animal for maintenance.

The final step in the calculation might be to estimate the minimum protein content of the cow's diet. As the microbial protein, and hence the protein requirement of the rumen microorganisms, is greater than the animal's requirement, the diet must contain sufficient effective rumen-degradable protein (ERDP) for the microorganisms. Ideally, the diet should contain protein of high degradability. If the diet contained 8 MJ FME/kg DM, the quantity required would be 55/8 = 6.9 kg DM. Thus the concentration of ERDP should be 495/6.9 = 72 g/kg DM. Low- to moderate-quality grass silage would meet this specification. At the low rumen outflow rate (0.02) expected at maintenance intake levels, the degradability of silage protein would be about 0.7, so the minimum protein content of the silage would need to be 72/0.7 = 102 g/kg DM. This food would also supply about 15 g/kg of DUP, although the animal would not require this for maintenance.

Other protein systems for ruminants (listed in Chapter 13) also base their estimates of maintenance requirements on endogenous losses of nitrogen but use different factors to translate endogenous losses into dietary requirements. However, the final outcome may not be so diverse. Thus in the French system (protein digestible in the intestine, PDI), the daily maintenance requirement of a 600 kg cow is stated as 395 g PDI, which is close to the figure calculated above, of 438 g metabolisable protein.

Alternative estimates of protein requirements for maintenance

For pigs and poultry, protein requirements are usually stated for maintenance and production together. It is, however, possible to calculate the requirements of these animals for maintenance alone from endogenous nitrogen losses. For example, the UK Agricultural Research Council calculated that the obligatory nitrogen loss of sows could be met by supplying only 0.90 g of digestible ideal protein (see Chapter 13) per kg $W^{0.75}$ per day (i.e. 36.6 g for a sow of 140 kg).

An alternative method of estimating protein requirements for maintenance is to feed animals on a range of diets varying in protein content and to measure their nitrogen balance. The lowest dietary protein level that promotes a nitrogen balance close to zero is the maintenance level. This method has been used to estimate the protein requirements of horses for maintenance as 2.4 g digestible crude protein per kg $W^{0.75}$ per day (e.g. 215 g/day for a 400 kg horse).

14.2 Animal growth and nutrition: feeding standards for growth and wool production

The simplest manifestation of growth in farm animals is their increase in size and weight. All animals start their lives as a single cell weighing practically nothing, then grow to reach mature weights that range from 2 kg for a laying hen to 1000 kg or more for a bull. The pattern by which animals grow from conception to maturity can be represented by a sigmoid (s-shaped) curve, as shown in Fig. 14.2. During the foetal period and from birth to puberty, the rate of growth accelerates; after puberty it decelerates and reaches a very low value as the mature weight is approached. In practice, features of the animal's environment, particularly nutrition, cause its growth to depart from the sigmoid curve. Periods of food scarcity (cold or dry seasons) may retard growth, or even cause the animal to lose weight, after which, periods of food abundance will allow the animal to grow rapidly. In general, animals kept under conditions of so-called 'intensive' husbandry will follow the growth curve illustrated in Fig. 14.2, whereas those kept under natural (extensive) conditions will follow interrupted curves, with their overall growth rate being less than that of the idealised pattern.

As animals grow, they do not simply increase in size and weight but also show what is termed *development*. By this we mean that the various parts of the animal, defined as anatomical components (e.g. legs), as organs (e.g. the liver) or as tissues (e.g. muscle) grow at different rates, so that the proportions of the animal change as it matures. For example, in cattle at birth the head is relatively large, forming 6.2 per cent of bodyweight of 40 kg, but when the calf has reached 100 kg the head forms only 4.5 per cent of bodyweight, and the proportion continues to decline until the animal reaches maturity. Sixty years ago Dr (later Sir) John Hammond, at Cambridge University, described the development of farm animals as a series of 'growth waves'. To take the major tissues as an example, in early life (including prenatal life), nerve and bone tissues have priority for available nutrients and grow rapidly. Later, muscle has priority and, finally, adipose tissue grows most rapidly. When growth is fast, the waves will overlap; for example, a fast-growing animal will begin forming substantial fat depots quite early in life, while muscle growth is still in progress.

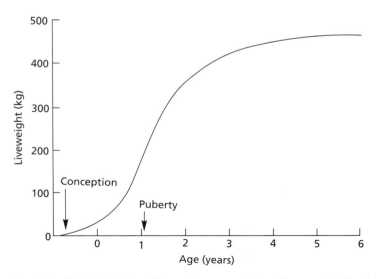

Fig. 14.2 The typical sigmoid growth curve as it would appear for the dairy cow.

Animal growth and animal nutrition interact with one another, in the sense that each can influence the other. The growth pattern of an animal determines its nutrient requirements. Conversely, by altering its nutrition, an animal's growth pattern can be modified. Another aspect of the interaction is that the growth pattern of animals determines the composition of the product of growth, meat, and so affects the consumer of the meat, man.

When feeding animals for meat production, the farmer is often aiming to produce carcasses that meet a particular specification for weight and composition. However, animals used for other purposes, such as reproduction or milk or egg production, may need to follow growth patterns that differ from those of meat animals. The main objective of this section of the chapter is to show how nutrient requirements for growth are determined and how they may vary according to the nature of the animal and the purpose for which it is kept. A secondary objective is to show how animal growth and development may be modified by control of nutrition. Although growth starts from conception, that which occurs *in utero* (or in the avian egg) forms a specialised subject that is covered in the next chapter. This section is concerned solely with post-natal growth.

Chemical growth

Although growth and development can be measured in terms of body parts, organs and tissues, the nutritionist's primary interest is in the growth of the chemical constituents of the body, as this determines the animal's nutrient requirements. The gross chemical constituents of the animal, protein, water and ash (together with small amounts of essential lipids such as phospholipids, and of carbohydrates such as glycogen), are combined in approximately constant proportions to form the *lean body mass* referred to earlier (p. 280). The animal also contains a variable proportion of storage lipid (fat determined as ether extract). The energy content of the body is almost entirely contributed by the energy contained in protein and lipid. In addition to these integral components, the body carries with it the extraneous and variable components of gut and bladder contents. The growth of all these components can be investigated by slaughtering and analysing animals at successive stages of growth. Figure 14.3 shows the results of the analysis of large numbers of data for the body composition of cattle, the weights of the body components being plotted against the animal's so-called 'empty bodyweight', which is liveweight less the extraneous gut and bladder contents.

Figure 14.3 shows that as the empty bodyweight increases, the weights of all chemical constituents increase, but at differing rates. Fat is deposited at an increasing rate and the lean body components (exemplified in Fig. 14.3 by protein) at decreasing rates. The energy content of the body follows a curve similar to that for fat content. The relationship of the weight of each component to empty bodyweight appears to be curvilinear. However, when all the weights are expressed as their logarithms, the relationships can be described by straight lines.

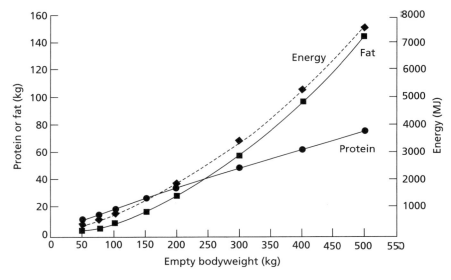

Fig. 14.3 Growth of protein, fat and energy in cattle. (Plotted from the data of the Agricultural Research Council 1980 *The Nutrient Requirements of Ruminant Livestock*, Farnham Royal, UK, Commonwealth Agricultural Bureaux.)

The equations for the logarithmic relationships are of the form:

$$\log y = \log b + a \log x$$

where y = the weight of the component and x = empty bodyweight. The algebraic form of this equation is:

$$y = bx^a$$

Such equations are known as *allometric equations* and were introduced by J S Huxley in 1932. The coefficient a is known as the growth coefficient and is a measure of the rate of growth of the part relative to the rate of growth of the whole animal. If it has a value greater than unity, the part is growing faster than the whole, and the part's contribution to the whole is increasing; the part in question is thus described as a late-maturing part of the body. Conversely, if its growth coefficient is less than unity, a part is said to be early-maturing.

Differentiation of the allometric equations allows the composition of gains in empty bodyweight to be determined for any particular bodyweight or range in bodyweight. Table 14.5 illustrates this procedure when applied to the data used to derive Fig. 14.3. The table shows that as the animal grows the composition of its gains changes in accordance with Hammond's growth waves. In early life the gains consist mainly of the water, protein and minerals (ash) required for growth of bone and muscle; later, the gains contain an increasing proportion of fat (and as a result, their energy content increases).

Table 14.5 Composition of gains in empty bodyweight (EBW) made by castrated male cattle of a medium-sized breed (Adapted from Agricultural Research Council 1980 *The Nutrient Requirements of Ruminant Livestock,* Farnham Royal, UK, Commonwealth Agricultural Bureaux)

Empty body weight (kg)	Protein (g/kg)	Fat (g/kg)	Energy (MJ/kg)
50	181	86	7.65
100	167	148	9.76
150	160	204	11.80
200	155	256	13.72
300	148	353	17.36
400	144	442	20.77
500	140	527	24.01

Equations:
$$\log_{10} \text{Protein} = 0.8893 \times \log_{10} \text{EBW} - 0.5037$$
$$\log_{10} \text{Fat} = 1.788 \times \log_{10} \text{EBW} - 2.657$$
$$\text{Energy} = 23.6 \times \text{Protein} + 39.3 \times \text{Fat}$$

Allometric equations can be used in other ways to study growth. The range of chemical components can be widened to include, for example, individual amino acids or mineral elements, with the analyses being used to define requirements for these nutrients. Allometric equations can also be applied to other body entities, such as particular organs or tissues.

Although the principal determinant of the composition of gains made by growing animals, and hence of their nutrient requirements for growth, is their bodyweight, there are other factors that can affect the composition of gains. Animal species is an obvious factor, for at a particular weight small species (species with a low mature weight) will be at a much more advanced stage of growth and maturity than large species. For example, at 60 kg liveweight, sheep have gains containing about 500 g/kg fat, whereas cattle have gains containing only about 75 g/kg. The composition of gains made by several species is shown in Table 14.6.

If small and large species differ in the composition of their gains, it seems likely that within a species, smaller breeds will make gains that differ in composition from those of the gains of larger breeds. Table 14.7 illustrates such an effect for cattle. It seems that the real determinant of the composition of gains is not absolute bodyweight, but bodyweight relative to the mature weight of the animal. This theory is supported by the effects of sex on composition of gains (also illustrated in Table 14.7). Females are smaller than males at maturity and therefore at a common weight make gains containing more fat and energy. Castrates tend to be intermediate between males and females.

A final factor influencing the composition of gains is the general growth rate of the animal. From the growth wave theory it seems likely that immature animals with limited nutrients available for growth, and therefore growing slowly, will use them for bone and muscle growth, whereas animals with more nutrients available will also store fat. Thus the fat content (g/kg) of the gain of pigs growing at 0.9 kg/day is likely to be greater than that of

Table 14.6 Percentage composition and energy content of the gains made by animals of various ages and liveweights (Adapted from Mitchell H H 1962 *Comparative Nutrition of Man and Domestic Animals*, Vol. 1, New York and London, Academic Press)

Animal	Live-weight (kg)	Age	Composition of gain (g/kg)				
			Water	Protein	Fat	Ash	Energy (MJ/kg)
Fowl	0.23	4.4 weeks	695	222	56	39	6.2
(White Leghorn	0.7	11.5 weeks	619	233	86	37	10.0
pullets – slow growth)	1.4	22.4 weeks	565	144	251	22	12.8
Sheep	9	1.2 months	579	153	248	22	13.9
(Shropshire ewes)	34	6.5 months	480	163	324	31	16.5
	59	19.9 months	251	158	528	63	20.8
Pig	23	–	390	127	460	29	21.0
(Duroc-Jersey	45	–	380	124	470	28	21.4
females	114	–	340	110	520	24	23.3
Cow	70	1.3 months	671	190	84	–	7.8
(Holstein heifers)	230	10.6 months	594	165	189	–	11.4
	450	32.4 months	552	209	187	–	12.3

pigs growing at 0.3 kg/day. This is commonly the case, although the effect is small in very immature animals or in genotypes selected for restricted fat deposition.

Energy requirements for growth

Ruminants

The UK Agricultural Research Council analysed data for large numbers of cattle and sheep that had been slaughtered at various weights and ages, and subjected to physical dissection and chemical analysis; some of the results of the analyses for cattle have been illustrated in Fig. 14.3 and Table 14.5. For cattle the energy content of the gains made by a castrate male of a medium-sized breed could be estimated with the following equation:

Table 14.7 Differences between breeds and sexes in the body composition of cattle of 300 kg empty bodyweight (Calculated from the data of Ayala H J 1974 PhD thesis, Cornell University, Ithaca, NY, USA)

Component	Breed	Sex		
		Male	Castrate	Female
Protein, g/kg	Aberdeen-Angus	172	161	150
	Holstein	186	187	167
Fat, g/kg	Aberdeen-Angus	190	227	314
	Holstein	136	172	213

$$\text{EV}_g = (4.1 + 0.0332W - 0.000\,009W^2)/(1 - 0.1475\Delta W)$$

where: EV_g = energy value of liveweight gain, MJ/kg
  W = liveweight, kg
  ΔW = liveweight gain, kg/day

The first bracketed term of the equation describes the increasing energy content of gain as cattle increase in size, and the second term provides a correction for the increase in energy content consequent upon faster liveweight gains. Thus a 100 kg animal gaining 0.5 kg/day is predicted to have gains containing 7.9 MJ/kg, whereas a 500 kg animal growing at the same rate has gains containing 19.9 MJ/kg. The corresponding values for animals growing at 1.0 kg/day are 8.6 and 21.6 MJ/kg, respectively.

To allow for the effects of breed and sex on the energy content of gains, the Agricultural Research Council introduced a simple 15 per cent correction factor. Thus for small breeds (maturing earlier) the value predicted by the above equation is increased by 15 per cent and for large breeds it is reduced by 15 per cent. Likewise, for females the value is increased by 15 per cent and for males (bulls) it is reduced by 15 per cent. Thus a 500 kg female of a small breed growing at 0.5 kg/day would be predicted to make gains of $19.9 \times 1.15 \times 1.15 = 26.3$ MJ/kg.

The sheep data examined by the Agricultural Research Council showed large effects of sex on the energy value of gains, but only small effects of breed (merinos tending to have more fat and hence more energy in their gains than other breeds), and no significant effect of rate of gain. The equations derived for the sexes were:

Castrates:	$\text{EV}_g = 4.4 + 0.35W$
Females:	$\text{EV}_g = 2.1 + 0.45W$
Males:	$\text{EV}_g = 2.5 + 0.35W$

These predict the energy content of the liveweight gains of 30 kg lambs, for example, to be 14.9 MJ/kg for castrates, 15.6 for females and 13.0 for males.

The above equations for cattle and sheep are used in the latest version of the 'ARC 1980' feeding standards for ruminants, which is the version recommended by the Technical Committee on Responses to Nutrients (see Further reading). The Australian Standing Committee on Agriculture, in the standards it proposed in 1990, used an ingenious method to allow the energy content of gains to be predicted by a single equation for both sheep and cattle, of any sex, of almost any breed and for any rate of gain. The basis of this method is to allocate to each class of animal a standard reference weight (SRW), which is defined as 'the liveweight that would be achieved by that animal when skeletal development is complete and the empty body contains 250 g fat per kg'. It is a mature working weight, not a maximal weight. Thus the SRW varies between breeds and also between sexes; for example, Friesian cattle are allocated SRW of 550 kg for females, 660 kg for castrates and 770 kg for males. For the prediction equation, the main variable is the animal's current weight as a proportion of its SRW; a second variable adjusts the energy content of gain for the rate of gain (this also being defined as a proportion of the SRW). Although the equation proved suitable for all sheep breeds and most breeds of cattle, it had to be modified for the large European breeds, such as the Charolais and Simmental. These are breeds that are capable of making gains unusually low in fat and, hence, in energy content. The US National Research Council has also adopted the concept of a standard reference weight for its feeding standards for cattle.

Pigs

In the UK, the Agricultural Research Council has devised energy feeding standards for growing pigs by using a factorial model. The model is designed principally for performance prediction (i.e. predicting growth rate for a given energy intake). The fixed factors in the model are an assumed maintenance requirement of 0.639 MJ ME/kg $W^{0.67}$/day and requirements for protein and fat deposition of 42.3 and 53.5 MJ/kg, respectively. (As protein and fat contain 23.7 and 39.6 MJ GE, the quantities above are equivalent to k_p and k_f values of 0.56 and 0.74, respectively.) The variable factors in the model are the nitrogen retention of the pig, and hence the partition of energy storage between protein and fat. For example, a 60 kg pig with an average capability for depositing protein is reckoned to retain 31 g/day at a maintenance level of feeding, plus 4.43 g for each 1 MJ ME in excess of its maintenance requirement. If its ME intake was 25 MJ/day, its protein and fat deposition would be predicted as:

ME intake	25.0 MJ/day
ME for maintenance ($0.639 \times 60^{0.67}$)	9.9 MJ/day
ME for production ($25.0 - 9.9$)	15.1 MJ/day
Protein deposited ($31 + 4.43 \times 15.1$)	98 g (0.1 kg)
ME used for protein (42.3×0.1)	4.23 MJ
ME used for fat ($15.1 - 4.23$)	10.87 MJ
Fat deposited ($10.87/53.5$)	0.203 kg

The weight of lean tissue deposited can be estimated from its known protein content (213 g/kg) to be $98/0.213 = 460$ g (0.46 kg). Total empty body gain is therefore $0.46 + 0.20 = 0.66$ kg/day. The energy content of gain is calculated to be $(0.1 \times 23.7) + (0.20 \times 39.6) = 10.3$ MJ, or $10.3/0.66 = 15.5$ MJ/kg.

For practical use, information of the kind provided by this model may be used to devise feeding scales for pigs that allow rapid growth without excessive fat deposition. The fastest growth is achieved by allowing the pigs to eat to appetite, and some types of pig can eat to appetite from birth until slaughter at 'bacon weight' (90 kg) without laying down too much fat. With some pigs, however, food intake must be restricted during growth from when they reach 45 kg, if a sufficiently lean carcass is to be obtained. Feeding scales have therefore been devised which stipulate the quantities of food or digestible energy to be given at different liveweights. A typical scale for a diet containing 13.5 MJ digestible energy per kg would rise gradually from 1.2 kg at a liveweight of 20 kg to 2.2 kg at 50 kg liveweight and 2.4 kg at 90 kg. Such restriction of intake slows down the rate of growth, but it reduces the fat content of the carcass and makes the carcass more acceptable to the bacon curer.

Poultry

With the exception of birds reared for breeding (see Chapter 15), growing poultry are normally fed to appetite, and feeding standards for them are therefore expressed, not as amounts of nutrients, but as the nutrient proportions of the diet (Appendix 2, Table A2.10).

As explained in Chapter 17, the quantities of food eaten by poultry are inversely related to the concentration of energy in their diets. This means that if the energy concentration of a diet is increased without change in the concentration of, for example, protein, and birds begin to eat less of the diet, then although their energy intake may remain approximately at the former level, their intake of protein will fall. The birds may then be deficient in protein. To generalise: a nutrient concentration that is adequate for a diet of low energy content may be inadequate for one richer in energy. It follows that feeding standards stated as nutrient concentrations are satisfactory only when applied to diets with a particular energy concentration. The standards of Appendix 2, Table A2.10 for chicks up to six weeks of age apply to diets containing 11.5 MJ metabolisable energy per kg and need to be adjusted for diets containing more or less energy. Some adjustments are discussed later in this chapter.

Horses

For the US digestible energy system, requirements for growth (MJ DE/kg) are estimated from the age and growth rate of horses by the equation:

$$(20.13 + 4.90x - 0.096x^2)y$$

where: x = age in months
y = growth rate in kg/day

For a horse 12 months old and gaining 0.65 kg per day this equation predicts the energy requirement for gain to be 42.3 MJ DE/kg. This is equivalent to a net energy value of $42.3/1.67 = 25.3$ MJ/kg. In the French net energy system, energy requirements for growth (MJ NE/kg) are estimated by means of an allometric equation which, like the equation above, varies requirements with age and rate of liveweight gain.

Protein requirements for growth

Ruminants

In the factorial system of estimating protein requirements described earlier (p. 364), the protein content of liveweight gain is added to the estimates of endogenous losses. For example, suppose a 20 kg lamb growing at 0.2 kg/day is estimated to have endogenous losses of nitrogen equivalent to 21 g protein per day, and suppose its liveweight gains are estimated to contain 140 g protein per kg. In addition, it will store about 6 g of protein per day in its fleece. Its net requirement for protein will therefore be:

$$21 + (0.2 \times 140) + 6 = 55 \text{ g/day}$$

To calculate the requirement for metabolisable protein, each of the net requirements must be divided by the appropriate efficiency factor. For maintenance the factor is 1.0, for liveweight gain it is 0.59 and for wool growth it is 0.26 (this last factor is low because of the unusual amino acid composition of wool protein; see p. 378). Thus the MCP requirement is:

$$21 + 28/0.59 + 6/0.26 = 92 \text{ g/day}$$

If the lamb's fermentable metabolisable energy intake is 7 MJ/day, this will yield $7 \times 10 = 70$ g microbial protein, which will meet $70 \times 0.75 \times 0.85 = 45$ g of the metabolisable protein requirement. The deficit in metabolisable protein is $92 - 45 = 47$ g/day, which must be met by digestible undegradable protein. Thus the food must supply 70 g of effective rumen-degradable protein, plus 47 g of digestible undegradable protein. If the lamb's food intake was 700 g dry matter per day, the concentrations needed would be 100 g/kg of ERDP plus 67 g/kg of DUP.

The previous example of the calculation of metabolisable protein requirement (p. 364) was for maintenance only, and the requirement could be met entirely by microbial protein from the rumen. In contrast, the fast-growing lamb of the current example has a requirement for protein that is high in relation to its energy requirements. Consequently, protein synthesised in the rumen is not sufficient to meet the animal's requirement, and there is therefore a substantial additional requirement for undegradable food protein. To keep the correct balance between rumen-degradable and undegradable protein it would probably be necessary to include in the diet a particularly good source of undegradable protein. If it was not possible to ensure the correct balance of the two categories of protein, it would be necessary to raise the total protein content of the diet, which might then supply sufficient undegradable protein but also supply a wasteful excess of degradable protein. Lambs (and calves) that are being suckled have a good balance of DUP and ERDP as they receive a good supply of the former as milk protein that bypasses the rumen by means of the oesophageal groove mechanism (p. 179), and a supply of ERDP from other foods such as pasture herbage.

Pigs and poultry

In addition to a general need for protein, non-ruminant animals have specific dietary requirements for ten or so essential (or indispensable) amino acids. During the past 30 years many experiments have been carried out to determine quantitative requirements for indispensable amino acids, and feeding standards expressed in terms of total protein are normally supplemented (or even completely replaced) by standards for all or some of these amino acids. Requirements may also be stated in terms of 'ideal protein' (protein containing indispensable amino acids in exactly the proportions required by the animal) as explained in Chapter 13.

It is possible to express protein requirements in terms of total protein alone in circumstances of animals being fed on a very limited range of foods of known amino acid composition. This is the case, for example, for growing pigs in the USA that are fed largely on maize and soya bean meal. But such a simplified approach cannot be sustained when wider ranges of foods and by-products are employed, and when diets must be formulated to give not only fast growth but also optimal carcass composition. Feed compounders therefore formulate pig diets to meet standards for at least three amino acids (lysine, methionine plus cystine and threonine). They will also take into account the availability of certain amino acids, assessed from digestibility to the terminal ileum (as explained in Chapter 10).

The requirements of pigs and poultry for indispensable amino acids

The requirement for an indispensable amino acid is assessed by giving diets containing different levels of the acid in question but equal levels of the remaining acids and measuring growth or nitrogen retention. Diets differing in their content of just one amino acid may be prepared from foods naturally deficient in it, to which are added graded amounts of the pure acid. Figure 14.4 shows the outcome of an experiment with chicks in which a diet low in lysine was supplemented in this way so as to give diets ranging in lysine content from 7 to 14 g/kg. The lysine requirement of the chick was concluded from this experiment to be 11 g/kg of the diet. In other experiments it has been found more convenient to use synthetic diets in which much or all of the nitrogen is in the form of the pure amino acids.

Standards for the indispensable amino acid requirements of chicks, turkey poults and young pigs have now been devised, and some of these are given in Appendix 2, Tables A2.9 and A2.10. At present they must be regarded only as approximate standards, for there are considerable complications to defining requirements for amino acids. Thus requirements are influenced by interactions among the indispensable amino acids themselves, between indispensable and dispensable acids and between amino acids and other nutrients. For chicks, the requirement for glycine is increased by low concentrations in the diet of methionine, arginine or B-complex vitamins. Interactions may be due to the conversion of one amino acid into another. If cystine, or its metabolically active form cysteine, is deficient in the diet it

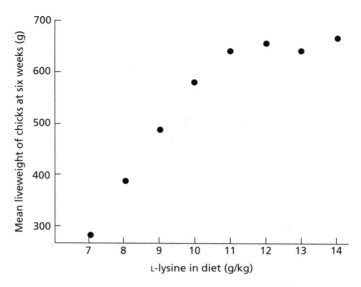

Fig. 14.4 Growth of chicks given diets containing different levels of lysine. (Plotted from the data of Edwards H M, Norris L C and Heuser G F 1956 *Poultry Science* **35**: 385.)

is synthesised by the animal from methionine. The requirement for methionine is therefore partially dependent on the cystine (or cysteine) content of the diet, and the two acids are usually considered together (i.e. the requirement is stated for methionine plus cystine). It should be noted, however, that the acids are not mutually interconvertible; methionine is not synthesised from cystine and therefore a part of the total requirement must always be met by methionine. Phenylalanine and the dispensable acid tyrosine have a similar relationship. In the chick, glycine and serine are interconvertible.

Further complications are introduced by relationships between amino acid requirements and the total protein content of the diet. If the latter is altered to compensate for a change in energy content, then the amino acid requirements will also change. For this reason, amino acid requirements are sometimes expressed as g amino acid/MJ of digestible or metabolisable energy.

The application in feeding practice of standards for as many as 10 or 11 amino acids is likely to be rather laborious. How necessary is it, then, to consider all of them when devising diets for pigs and poultry? In theory there is almost unlimited scope for adjusting the proportions of dietary constituents (including synthetic acids) until the indispensable amino acid contents of the diet exactly equal those demanded by the standards. In practice it is usually found that the amino acid mixtures provided by the diet are out of proportion, and hence inefficiently utilised, because one or two amino acids are very markedly deficient. In consequence, the degree of success achieved in applying the standards depends particularly on whether or not the requirements of these 'limiting' amino acids are met. By comparing requirements for amino acids with the amino acid contents of typical diets, it can be shown that for pigs the acid likely to be most deficient is lysine. For chicks the 'first-limiting' amino acid is most commonly methionine, although lysine and perhaps arginine may also be deficient.

In practice, therefore, it may be sufficient to ensure that diets for pigs and poultry contain, first, adequate total protein, and, second, adequate contents of those amino acids most likely to be deficient. When this simplification is made, however, it should be remembered that, for pigs at least, standards for total protein have often been set high in order to be adequate for diets containing low-quality protein, i.e. for diets most deficient in the more frequently limiting amino acids. If more attention is paid to these amino acids it seems likely that standards for total protein will be reduced, bringing into prominence other amino acids which are not at present regarded as likely to be near the borderline of adequacy.

Horses

The US National Research Council relates protein requirements for growth in horses to their energy intake; its recommendations are 12 g of digestible crude protein per MJ of digestible energy for young (weanling) horses and 10.8 g/MJ DE for yearlings. The French feeding system for horses uses a different method of calculation but appears to provide similar estimates of protein requirements for growth. Horses have specific requirements for indispensable amino acids, and the US standards state that the amino acid most likely to be limiting, lysine, should be supplied at 502 and 454 mg per MJ DE, for weanlings and yearlings, respectively.

Nutrient requirements for wool production

Wool fibres are grown in follicles formed by the invagination of the epidermis, a bulb at the base of the follicle being the site of very active cell division. As the blood supply to the epidermis is subject to large fluctuations (e.g. in response to environmental temperature) the metabolism of the follicles needs to be protected from variations in the supply of nutrients and oxygen. Thus the follicles have stores of energy in the forms of glycogen and glutamine, and they release this energy by aerobic or anaerobic glycolysis, which produces lactate and is less dependent on a supply of oxygen than is the tricarboxylic acid cycle (see Chapter 9). The wool fibres consist almost entirely of the protein wool keratin, which is characterised by its high content of the sulphur-containing amino acid cystine. Although cystine is not an indispensable amino acid, much of that in wool is synthesised from the indispensable acid methionine. Methionine is also used by the follicles to provide the polyamines needed in protein synthesis. Thus the wool follicle requires a good supply of the sulphur-containing amino acids.

The weight of wool produced by sheep varies considerably from one breed to another, and an average value is useful only as an example. For an example we shall take a fleece of 4 kg, this representing the annual production of a merino weighing 50 kg. Such a fleece would contain about 3 kg of actual wool fibres, the remaining 1 kg being wool wax, suint, dirt and water. Wool wax is produced by the sebaceous glands and consists mainly of esters of cholesterol and other alcohols, with the acids normally found in glycerides and other aliphatic acids. Suint, the secretion of the sudoriferous glands, is a mixture of inorganic salts, potassium soaps and potassium salts of lower fatty acids.

To grow in one year a fleece containing 3 kg protein the sheep would need to deposit a daily average (i.e. ignoring seasonal fluctuations in wool growth) of about 8 g protein or 1.3 g nitrogen. If this latter figure is compared with the 6.6 g nitrogen that a sheep of 50 kg might lose daily as endogenous nitrogen, it will be seen that in proportion to its requirement for maintenance the sheep's nitrogen requirement for wool growth is small. These figures, however, do not tell the whole story, since the efficiency with which absorbed amino acids are used for wool synthesis will be less than that with which they are used for maintenance. The efficiency with which food protein can be converted into wool is likely to depend on their respective proportions of cystine and methionine. Keratin contains 100–120 g/kg of these acids, compared with the 20–30 g/kg found in plant proteins and in the microbial proteins synthesised in the rumen, and so the biological value of food protein for wool growth is likely to be not greater than 0.3. In the metabolisable protein system the factor for conversion of metabolisable protein to wool protein is 0.26.

With regard to energy requirements for wool growth, a sheep producing a fleece of 4 kg would retain about 0.23 MJ/day. The efficiency with which metabolisable energy is used for wool production is estimated to be about 0.18. Thus the sheep would require $0.23/0.18 = 1.3$ MJ ME/day for wool growth, which may be compared with its maintenance requirement of approximately 6 MJ ME/day.

While nutrient requirements for wool production, even for protein, are quantitatively small, it must not be supposed that maximum wool growth will take place at a level of nutrition only slightly above the maintenance level. Wool growth reflects the general level of nutrition of the sheep. At sub-maintenance levels, when the sheep is losing weight, its wool continues to grow, although slowly. As the plane of nutrition improves and the sheep gains in weight, so wool growth too increases. In the ARC 1980 system, wool protein deposition (NP_w, g/day) is linked to protein deposited in other tissues (NP_g) by the equation:

$$NP_w = 3 + 0.1\, NP_g$$

The Australian Standing Committee on Agriculture has estimated the growth of clean dry wool in merinos to be in the range 0.5–0.9 g/MJ ME. There appears to be a maximum rate of wool growth, which varies from sheep to sheep within a range as great as 5–40 g/day.

The dependence of wool growth on the plane of nutrition (i.e. energy intake) of the sheep is due in part to the association between energy intake and the synthesis of microbial protein. Thus the real determinant of wool growth rate is likely to be the quantity of protein digested and absorbed in the small intestine of the sheep, and it has been shown, for example, that to achieve its maximum rate of wool growth a merino must absorb 120–150 g protein per day, containing 1.9 g cystine and 2.3 g methionine. In a ewe with a fermentable metabolisable energy intake of 10 MJ/day (i.e. twice its maintenance requirement for ME), only 100 g microbial protein would be synthesised per day and only $100 \times 0.75 \times 0.85 = 64$ g would be absorbed as amino acids. To achieve maximum growth of wool the sheep is therefore dependent on a good source of undegraded food protein. In practice this is likely to be supplied by the consumption of large quantities of protein in pasture herbage. This may be relatively highly degradable but can still supply much undegradable protein. For example, a ewe might consume 250 g protein/day, of which 0.3 (75 g) would be digestible undegradable protein. Nevertheless, wool growth in sheep is considerably increased by protein supplements protected from rumen degradation, such as formalin-treated casein. As would be anticipated, the most effective of such supplements are those rich in the sulphur-containing amino acids.

Wool quality is influenced by the nutrition of the sheep. High levels of nutrition increase the diameter of the fibres, and it is significant that the finer wools come from the nutritionally less favourable areas of land. Periods of starvation may cause an abrupt reduction in wool growth; this leaves a weak point in each fibre and is responsible for the fault in fleeces with the self-explanatory name of 'break'. An early sign of copper deficiency in sheep is a loss of 'crimp' or waviness in wool; this is accompanied by a general deterioration in quality, the wool losing its elasticity and its affinity for dyes. These effects are thought to be due to the involvement of copper in the enzyme system responsible for disulphide linkages in keratin. Copper deficiency also restricts the formation of the pigment of wool and hair, melanin. Zinc deficiency causes the production of brittle wool fibres. Many vitamin deficiencies affect wool and hair follicles, but with the exception of cyanocobalamin, these deficiencies are uncommon in sheep.

14.3 Mineral and vitamin requirements for maintenance and growth

This section is concerned with the general principles governing the determination of feeding standards for minerals and vitamins. No attempt is made to discuss individual nutrients, since to do so would result in duplicating the material of Chapters 5 and 6.

Minerals

Animals deprived of a dietary supply of mineral elements continue to excrete these nutrients. Those elements that occur in the body mainly as constituents of organic compounds, such as the iron of haemoglobin and the iodine of thyroxine, are released from these compounds when they are expended or 'worn out'. To a variable but often large extent the elements so liberated are reutilised, but reutilisation is never complete and a proportion of each mineral will be lost from the body in the faeces and urine and through the skin. Of those elements that occur in inorganic forms, such as calcium, sodium, potassium and magnesium, there are losses in the urine such as those arising from the maintenance of the acid–base balance of the animal, and losses in the faeces occurring through secretions into the gut which are not reabsorbed. Because they suffer all these endogenous losses, animals require minerals for maintenance.

Endogenous losses of minerals are often small, however, in relation to the mineral content of the body. A pig of 30 kg, whose body contains about 230 g calcium, suffers endogenous losses amounting to 0.9 g of the element per day, and thus needs to replace daily about 0.4 per cent of its body calcium. The same pig contains about 40 g sodium and needs to replace 0.036 g or 0.09 per cent each day. In contrast, the pig would need to replace about 0.7 per cent of its body nitrogen each day.

The approaches used in assessing requirements for minerals are the same as those used in determining standards for energy and protein. A 'theoretical' approach is provided by the factorial method, and 'practical' estimates of requirement by balance and growth trials. Since the mineral contents of foods are expressed as the total or gross amounts present, requirements are stated in the same terms. The standards must therefore make allowance for the differences in mineral availability that occur between different species and age classes of animal (see Chapter 10).

Factorial estimates of mineral requirements

The net requirement of a mineral element for maintenance plus growth is calculated as the sum of the endogenous losses and the quantity retained. To determine the dietary requirement, the net requirement is divided by an average value for availability (expressed as a decimal). For example a heifer of 300 kg liveweight gaining 0.5 kg/day might have endogenous losses of calcium of 5 g/day and retain 6 g/day: its net requirement would therefore be 11 g calcium/day. For an animal of this weight an average value for the availability of calcium would be about 0.68 and the animal's dietary requirement for calcium would therefore be 11/0.68 = 16 g/day.

The difficulties involved in the factorial approach to mineral requirements are the same as those associated with factorial estimates of protein requirement. Whereas the mineral composition of liveweight gains may readily (if laboriously) be determined by carcass analysis, the assessment of endogenous losses, and hence also of availability, is more difficult. Diets for ruminants which are completely free of an element are particularly difficult to prepare. Perhaps because of these difficulties of technique, theoretical estimates of mineral requirements often do not agree with practical estimates.

Growth and balance trials

When mineral requirements are assessed by comparing the effects on the animal of diets supplying different quantities of an element, the great problem is to establish a satisfactory criterion of adequacy. That intake which is sufficient to prevent clinical signs of deficiency may be insufficient to support maximum growth. Of the bone-forming elements, the intake which gives a maximum rate of liveweight gain may yet be inadequate if judged by the strength of bone produced. The position is further complicated by the mineral reserves of the animal. If these are large at the beginning of an experiment of short duration, they may be sufficient to allow normal health and growth even if the diet is inadequate, and it is therefore desirable that mineral balance should be determined directly or assessed indirectly by analysis of selected tissues. Even balance trials, however, may be difficult to interpret, since if the element is one for which the animal has great storage capacity, a dietary allowance that promotes less than maximum retention may still be quite adequate. In long-term experiments, then, such as those continuing for one or more annual cycles of the dairy cow, health and productivity alone may provide reliable indications of minimum mineral requirements. For growing animals, on the other hand, trials must usually be of shorter duration, and measurements of liveweight gain should be supplemented by measurements of mineral retention.

Present standards

The requirements for minerals given in the tables of Appendix 2 are based partly on factorial calculations and partly on the results of feeding trials. For all species, the elements that have been the subject of most investigation have been calcium and phosphorus; these being the elements most likely to be deficient in diets. In the case of ruminants, estimates of calcium and phosphorus requirements have changed markedly over the past 30 years as new information has been provided on endogenous losses and availability. For example, in 1965 the UK Agricultural Research Council stated the phosphorus requirement of a 600 kg cow producing 30 litres of milk per day to be 85 g/day. In 1980 it revised its estimate to 59 g P per day. In 1991 the AFRC Technical Committee on Responses to Nutrients estimated the phosphorus requirement of such a cow to be even lower (54 g/day) if it was fed on a high-energy diet (13 MJ ME/kg DM), but considerably higher (82 g/day) if it was fed on a diet of lower energy content (11 MJ ME/kg DM). The difference between diets is partly due to the fact that cattle on lower-energy diets have been found to secrete relatively more phosphorus in saliva and thus to suffer greater endogenous losses of the element.

Vitamins

There are no estimates of endogenous losses on which to base factorial estimates of vitamin requirements, and standards must therefore be derived from the results of feeding trials. As in the assessment of mineral requirements, it may be difficult in these trials to select the criteria by which the allowances compared are judged adequate or inadequate. The main criteria are again growth rate and freedom from signs of deficiency, deficiency being detected either by visual examination of the animals or by physiological tests such as determination of the vitamin levels in the blood. Vitamin storage may also be assessed, either from actual analyses of tissues or from such indirect evidence of tissue saturation as is provided by excretion of the vitamin in the urine. The difficulties involved in assessing requirements are illustrated in Table 14.8, which shows that the apparent requirement can vary widely according to the criterion of adequacy preferred.

In practice, vitamin allowances must be at least high enough to prevent signs of deficiency and not to restrict the rate of growth. Higher allowances, which promote storage or higher circulatory levels of the vitamin, are justifiable only if they can be shown in the long term to influence the health and productivity of animals in a way that does not become apparent in the short-term experiments by which requirements must often be assessed. Some storage can be justified, since in most animals there may be fluctuations in both the needs for the vitamins and the supply of them, and allowances are usually set at levels permitting the maintenance of stores.

Requirements for vitamins A and D, in older animals at least, are considered to be proportional to bodyweight. Those for vitamin E and the B group, however, which are more intimately concerned with metabolism, vary with food intake in general, or in some cases with intake of specific nutrients. Thus the requirement for thiamin, which is particularly concerned with carbohydrate metabolism, varies according to the relative importance of carbohydrate and fat in the diet. For similar reasons riboflavin requirements are increased by a high protein intake. Requirements also vary according to the extent to which B vitamins are synthesised in the alimentary tract. In herbivores, sufficient synthesis occurs to make the animal independent of dietary supplies. In pigs and poultry, considerable synthesis takes place in the lower gut, but the vitamins produced may fail to be absorbed: the contribution of the intestinal synthesis may then depend on whether the animals are free to practise coprophagy (the eating of faeces). A final point of importance concerns the availability of vitamins. Requirements are often determined from diets containing synthetic sources of the vitamins, whose availability may well be higher than that of the vitamins of natural foods. Although little is

Table 14.8 The vitamin A requirement of calves (From the data of Lewis J M and Wilson L T 1945 *Journal of Nutrition* **30**: 467)

Minimum requirement for:	Vitamin A (i.u./kg liveweight/day)
Prevention of night blindness	32
Optimal growth	64
Limited storage of vitamin A	250
Maximal blood level of vitamin A	500

known about vitamin availability, a well-documented instance of non-availability is provided by the nicotinic acid in cereals, some of which is in a bound form not available to pigs.

A further factor to be considered when formulating diets to provide requirements for vitamins is that many vitamins (as discussed in Chapter 5) are unstable and are partly destroyed during the preparation and storage of compound feeds.

14.4 Nutritional control of growth

The preceding sections of this chapter have shown how the growth patterns of animals determine their nutrient requirements. We must now consider the second aspect of the nutrition–growth interaction; how growth can be controlled by nutrition. The aims of any person controlling animal growth by nutrition are generally twofold: to use the nutritional resources available to achieve a high rate of growth, and to produce a carcass that meets the requirements of the consumer. The rate of growth of an animal is controlled by its nutrient intake, and especially by its energy intake. Energy was earlier described (p. 292) as the 'pacemaker' of animal production, and both natural (e.g. climatic) and imposed variations in the animal's energy supply will be reflected in its rate of growth. A rapid rate of growth is desirable because it minimises the 'overhead' cost of maintenance per unit of meat produced. In the animal production industries of developed countries, animal feeds are readily available although their use may be restricted by cost. In many developing countries supplies of feeds, particularly of high-energy concentrates, are often small or non-existent, and the situation often arises of inadequate supplies being spread over too many animals.

Until comparatively recently, fat was a highly prized component of meat. Until the late nineteenth century, vegetable oils were less readily available, and many consumers needed a high energy intake to fuel their manual work. The ideal meat animal was therefore one that fattened early in life, and it was provided by the genetic selection of the small, early-maturing breeds (such as Aberdeen-Angus cattle). We have seen already that animals that grow fast tend to have a greater amount of fat in each unit of gain, so the twin aims of rapid growth and a desirable carcass were compatible with each other. Over the past 20–30 years, however, fat has come to be regarded by many consumers as a less desirable component of the carcass, and the control of growth by nutritional means has been made difficult by the incompatibility of its aims; the danger is that animals fed to grow fast will become too fat. In current practice the danger is partly avoided by using breeds or selections of animals that are larger and later-maturing; thus the preferred beef animal today is one from the large European breeds that will grow rapidly and can be slaughtered when relatively immature, so giving a large but lean carcass. Another way to prevent excessive fat deposition is to treat animals with what are known as 'repartitioning' agents. These are hormones or related substances that alter the partition of energy deposition towards protein and away from fat. They include the sex hormones (both oestrogens and androgens), somatostatin (or 'growth hormone') and cimaterol (a β-adrenergic agonist). Growth hormone, in particular, is a powerful repartitioning agent, and the discovery of its effects has stimulated research into the use of genetic engineering to increase the animal's own production of the hormone. However, in many countries, including most of Europe, the use of these repartitioning agents is banned, mainly because of consumers' concern over their possible presence in meat. Nevertheless, their often dramatic effects serve to emphasise the scope for controlling growth by more acceptable methods, including nutrition.

The two aims in controlling growth may sometimes be combined into one, the maximising of lean tissue (or protein) deposition. To achieve this it is essential that the animal's protein supply should be optimal. In this and the preceding chapter, much emphasis has been put on the use of recently acquired knowledge to improve the protein nutrition of livestock. Some examples are the new systems introduced for calculating and meeting the protein requirements of ruminants and the concept of ideal protein for non-ruminants. To ensure the correct partition of energy between protein and fat it is essential that protein supply should match energy supply. With ruminants in particular, protein supply is often insufficient to meet the animal's potential for protein deposition, and some energy will be used for fat deposition. The energy : protein balance is set by the rumen, although it can be changed by supplying protein not degraded in the rumen but digestible in the lower gut (digestible undegradable protein, DUP; see Chapter 13). An extreme example of this is the feeding of over-fat lambs on straw (i.e. low energy) and fishmeal (high DUP). With such a diet, lambs can be 'remodelled' by causing them to use body fat to make up the dietary energy deficit while at the same time depositing protein.

Opportunities arise for controlling growth by nutrition through the phasing of nutrient intake; for example, food intake might be kept relatively low in early life and high thereafter, or vice versa. Classic experiments on the phasing of nutrient intake were conducted by Sir John Hammond and his colleagues, and their results provided the background to Hammond's growth wave theory, referred to earlier. The results of one of these experiments are illustrated in Table 14.9. Twenty pigs were grown to 90 kg liveweight with low or high rates of food intake giving low or high rates of growth. Some of the animals were changed from one rate to the other halfway through the experiment (high–low and low–high) and others were kept on the same rate throughout (high–high and low–low). The greatest differences in carcass composition between the pigs were in their dissectible fat content. The pigs that grew fast throughout (high–high) had a fat content 108 g/kg higher than those that grew slowly throughout (low–low), and there was a difference of the same order between the low–high and the high–low groups. Those with most fat and least muscle, the low–high group, fitted the growth wave theory, as a deficiency of nutrients in early life appeared to have restricted the muscle growth, which had the highest priority at that stage. However, if the results of the experiment are expressed in

Table 14.9 Composition (g/kg) of the carcasses of pigs grown at different rates (see text) and killed at 90 kg (After McMeekan C P 1940 *Journal of Agricultural Science, Cambridge* **30**: 511)

Growth rate: Age at slaughter (weeks):	High–high 20	High–low 28	Low–high 28	Low–low 46
Composition of whole carcass				
Bone	110	112	97	124
Muscle	403	449	363	491
Fat	383	334	441	275
Skin, etc.	105	106	99	110
Composition of fat-free carcass				
Bone	178	168	174	171
Muscle	653	674	649	677
Skin, etc.	170	160	177	152

a different way, as the fat-free carcass (see the second part of Table 14.9), it appears that tissues other than fat were not much affected by the treatments. The latter figures were calculated by F W H Elsley and his colleagues, who concluded that the main tissue affected by the phasing of nutrient intake was fat. Nevertheless, as we have already acknowledged, fat is important. For many years, bacon pigs had their food intake restricted so that they were fed according to a high–low sequence in order to limit fat deposition. More recently, the genotype of bacon pigs has been modified to provide animals that voluntarily restrict their intake as they approach bacon weight, and thus avoid excessive fat deposition.

Another example of the effects of phasing of nutrient intake is found in the phenomenon of compensatory growth. In many of the more primitive, or 'natural', systems of animal production a period of food shortage is followed by one of an abundant supply of nutrients (i.e. a low–high sequence). During the high phase, animals frequently grow very rapidly, and this compensatory growth may allow them to 'catch up' on animals that were not subjected to a low phase; an example of this is shown in Fig. 14.5. The cause of compensatory growth and also its effects on body composition are variable. Compensating animals frequently eat more per unit of bodyweight than others, and thus have more nutrients available for growth. Also, if they grow more in bone and muscle than in fat, their gains will contain less energy per unit weight, and each unit of energy intake will promote a greater gain in liveweight.

In conclusion, it is important to remember that the primary determinant of body composition, and hence of nutrient requirements for growth, is the weight of the animal, as is shown by the allometric equations. However, departures from the general allometric relationships, due to breed, sex, etc., are important in animal production because they affect nutrient requirements and also the composition and value of the carcass produced. Moreover, in modern production systems, farmers (and the scientists who assist them) are often attempting to bend or displace the allometric relationships in order to maximise their animals' rate of production of saleable

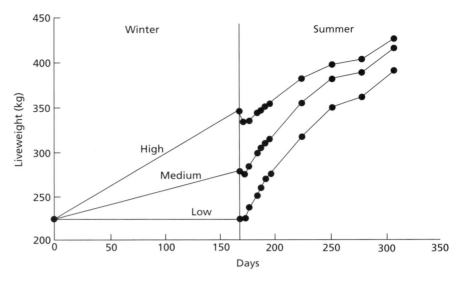

Fig. 14.5 Compensatory growth in cattle. Animals were kept on high, medium and low (maintenance) planes of nutrition during winter (days 0–168), then grazed together during the following summer (days 169–308). (Plotted from the data of Lawrence T L J and Pearce J 1964 *Journal of Agricultural Science, Cambridge* **63**: 5.)

meat (which today consists largely of lean muscle, hence protein). This aim can be achieved through genetic change, by the use (in some countries) of growth modifying agents and by control of nutrition.

Summary

1. Feeding standards are statements of the amounts of nutrients required by animals. They may be expressed as nutrient requirements or, with a safety factor added, as nutrient allowances. In Britain, standards were until recently set by the Agricultural (and Food) Research Council, in association with the Ministry of Agriculture, Fisheries and Food. Other countries have their own sets of standards.

2. Maintenance requirements for energy are generally estimated as the heat lost by an animal that is fasting and at rest (the so-called basal or fasting metabolism). Fasting metabolism is proportional to the so-called metabolic liveweight of the animal (typically $W^{0.73}$ or $W^{0.75}$). An average value for mammals is 70 kcal/kg $W^{0.73}$ (0.27 MJ/kg

$W^{0.75}$) per day, but there are variations due to age and sex. Maintenance energy requirements may also be estimated by means of feeding trials.

3. To translate fasting metabolism into maintenance requirement some allowance must be made for additional energy requirements for the muscular activity involved in searching for, harvesting and chewing food.

4. When animals are kept at low environmental temperatures their heat losses, hence their maintenance energy requirements, are increased. Lower and higher critical temperatures define the animal's thermoneutral range, which is affected by climatic conditions (sun, wind and rain), by the insulating properties of the animal and its

surroundings, and by food intake. Young animals are to some extent protected from cold by adipose tissue ('brown fat'), which is metabolised to produce heat.

5. Protein requirements for maintenance are estimated from the animal's endogenous urinary N excretion (i.e. N derived from body tissues and analogous to basal energy loss). There are also small losses of N in hair and scurf. In ruminants, the maintenance requirement for protein can generally be met by rumen microbial synthesis.

6. The growth of animals in weight generally follows a sigmoid curve against time. Individual tissues grow at different rates, and graphs of tissue weights against bodyweight show curvilinear relationships that can be described by so-called allometric equations. Typically, bone and muscle are early-maturing tissues, with fat being deposited later. Measurements of the chemical composition of gains made by growing animals provide estimates of the energy and protein requirements for growth. Requirements change as animals mature, and are also affected by their breed and sex.

7. In the UK ARC system for ruminants, equations and correction factors are used to estimate the energy and protein requirements for growth. Protein standards for pigs and poultry are more detailed in that they include estimates of requirements for specific amino acids.

8. In sheep, energy and protein requirements for wool growth are quantitatively small, but there are important qualitative requirements for sulphur-containing amino acids, copper and zinc. In practice, the rate of wool growth is often related to energy intake, because this determines the sheep's supply of microbial protein.

9. Net requirements for minerals for maintenance and growth are calculated like those for energy and protein, from endogenous losses and body composition data, and are translated into dietary requirements by the use of availability factors.

10. Requirements for vitamins are generally derived from feeding trials, which are complicated by difficulties in selecting the appropriate criterion for adequacy of supply.

11. Although growth patterns determine nutrient requirements, nutrition may influence the growth and body composition of animals. For example, a low plane of nutrition followed by a high plane may induce excessive fat deposition. The aims of any person controlling animal growth by nutrition are generally twofold: to use the nutritional resources available to achieve a high rate of growth, and to produce a carcass that meets the requirements of the consumer.

12. Because fat is no longer a desired component of meat, control of growth is often aimed at reducing fat and increasing muscle growth. In some countries, repartitioning agents (synthetic hormones) may be used to increase muscle growth.

Further reading

Alderman G 1993 *Energy and Protein Requirements of Ruminants*, Wallingford, UK, CAB International.

Australian Standing Committee on Agriculture 1990 *Feeding Standards for Australian Livestock: Ruminants*, Melbourne, CSIRO.

Black J L and Reis P J (eds) 1979 *Physiological and Environmental Limitations to Wool Growth*, Armidale, NSW, University of New England Press.

Hynd P I 2000 The nutritional biochemistry of wool and hair follicles. *Animal Science* **70**: 181–95.

Lawrence T L J and Fowler V R 1997 *Growth of Farm Animals*, Wallingford, UK, CAB International.

Leeson S and Summers J D 1997 *Commercial Poultry Nutrition*, 2nd edn, Guelph, Canada, University Books.

Mitchell H H 1962, 1964 *Comparative Nutrition of Man and the Domestic Animals* (2 vols), New York, Academic Press.

National Academy of Sciences / National Research Council 2000 *Nutrient Requirements of Beef Cattle*, 7th rev. edn, Washington, National Research Council.

O'Donovan P B 1984 Compensatory gain in cattle and sheep. *Nutrition Abstracts and Reviews, Series B* **54**: 389.

Pearson A M and Dutson T R (eds) 1991 *Growth Regulation in Farm Animals*, London, Elsevier.

Reid J T (ed.) 1968 *Body Composition in Animals and Man*, Washington, National Research Council, Publication no. 1598.

Theodorou M K and France J (eds) 2000 *Feeding Systems and Feed Evaluation Models*, Wallingford, UK, CAB International.

Historical reference

Brody S 1945 *Bioenergetics and Growth*, New York, Reinhold.

15 Feeding standards for reproduction

15.1 Nutrition and the initiation of reproductive ability

15.2 Plane of nutrition, fertility and fecundity

15.3 Egg production in poultry

15.4 Nutrition and the growth of the foetus

In reproducing animals, as in growing animals, there is an interaction between nutrition and production. Reproduction increases the animal's requirements for nutrients, but conversely, the nutrient supply of animals can influence their reproductive processes.

The influence of nutrition on reproduction begins early in the animal's life, as the plane of nutrition in young animals can affect the age at which they reach puberty. In mature animals, poor nutrition can reduce the production of ova and spermatozoa, so that the female either fails to conceive or produces fewer offspring than normal (i.e. litter size is reduced). In pregnancy, females have specific nutrient requirements for the maintenance and growth of the foetus(es).

The quantities of nutrients required for the production of ova and spermatozoa by mammals are small and of little significance. Thus the boar, which produces a large ejaculate of 150–250 ml, is estimated to require 0.4 MJ ME per ejaculate, which is equivalent to less than 2 per cent of its maintenance requirement for metabolisable energy (23.8 MJ/day). In birds, however, the quantities of nutrients required for egg production are large; these will be discussed in a special section of this chapter. Even in pregnancy, nutrient requirements for the growth of foetuses are in total relatively small. For example, a ewe producing twin lambs weighing a total of 7 kg at birth will deposit about 1.4 kg of protein in the foetuses and associated tissues (such as the placenta), which over a gestation period of 147 days is less than 10 g/day. In contrast, a growing sheep gaining 300 g in liveweight per day would deposit in its tissues about 50 g of protein per day. However, in assessing nutrient requirements for reproduction, several important features of reproductive processes must be borne in mind. The first of these is that reproduction is often not the sole productive process being carried out by the animal. In the case of cattle, for example, young females may be expected to conceive at 15–20 months of age, at about half of their mature bodyweight, and will have to continue growing while producing a calf. After calving, the female will be expected to conceive again within 2–3 months, at a time when she is producing large quantities of milk.

The second important feature of nutrient requirements for reproduction is that they vary considerably from one phase of the reproductive cycle to another. For example, the ewe conceiving twins will have very small protein requirements for foetal growth at the start of pregnancy, but by the last week of pregnancy she will be depositing as much as 30 g of protein per day in the foetuses. The final point to bear in mind is that reproduction is often an 'all or nothing' phenomenon, and the consequences of failure can be severe for the farmer. If a beef cow, kept solely to produce and rear calves, fails to conceive, her output will be zero and her owner will suffer a financial loss. Also, small litters – such as one lamb instead of two – can make the difference between profit and loss. It is therefore of great importance to ensure that reproduction in farm animals is not impaired by poor nutrition.

Research on the effects of nutrition on reproduction tends to make slow progress because the effects may be slow to manifest themselves and experiments must be of long duration. Also, the large random variability in reproduction causes statistical problems that can only be solved by using large numbers of animals (e.g. 80 sows per treatment to demonstrate a 5 per cent increase in litter size). Much recent research has been concerned with the mechanisms by which nutrition affects reproduction, and particularly with endocrinological processes. General undernutrition, and also the deficiency of a specific nutrient, can interfere with the synthesis of hormones involved in reproduction. In other cases, nutrition may affect the rate at which a hormone is destroyed by metabolism or may alter the sensitivity to a hormone of its target organ. Over the last 20 years the increasing use of techniques for superovulation of farm livestock and the culture *in vitro* of ova and embryos have both stimulated interest in, and provided opportunities for, the study of nutritional effects on reproduction.

This chapter will explain the general relationships between nutrition and reproduction at successive stages of the reproductive life of animals. Where appropriate, quantitative requirements for specific nutrients and energy will be presented and discussed. Much emphasis will be placed on the role of energy intake (or, in other words, the general plane of nutrition) in reproduction, because deficiencies or excesses of specific nutrients often affect reproduction through their influence on energy intake.

15.1 Nutrition and the initiation of reproductive ability

Puberty in cattle is markedly influenced by the level of nutrition at which animals have been reared. In general terms, the faster an animal grows, the earlier it reaches sexual maturity. In cattle, puberty occurs at a particular liveweight or body size rather than at a fixed age. This is illustrated in Table 15.1, which shows the effects of three planes of nutrition on the initiation of reproductive ability in dairy cattle. Although in both sexes there were considerable differences in age at puberty between the three treatments,

Table 15.1 Age and size at puberty of Holstein cattle reared on different planes of nutrition

Sex	Plane of nutrition (per cent of accepted standard for TDN)	At puberty		
		Age (weeks)	Weight (kg)	Height at withers (cm)
Females[a]	High (129)	37	270	108
	Medium (93)	49	271	113
	Low (61)	72	241	113
Males[b]	High (150)	37	292	116
	Medium (100)	43	262	116
	Low (66)	51	236	114

[a] From Sorenson A M *et al.* 1959 *Bulletin of the Cornell University Agricultural Experiment Station*, No. 936.
[b] From Bratton R W *et al.* 1959 *Bulletin of the Cornell University Agricultural Experiment Station*, No. 940.

differences in liveweight and in body size (as reflected in the measurement of height at withers) were much smaller. More recent experiments have shown that by liberal feeding with high-energy diets it is possible to induce puberty in cattle even earlier than the 37 weeks of Table 15.1. The attainment of puberty in sheep is complicated by their seasonal breeding pattern. Spring-born ewe lambs that are well nourished will reach puberty in the early autumn of the same year. Moderately fed lambs will also reach puberty in the same year, but later in the breeding season and at a lower liveweight. Poorly fed lambs will fail to come into oestrus until the following breeding season (i.e. at 18 months of age).

In pigs, on the other hand, high planes of nutrition do not advance puberty to any marked extent. The primary determinants of puberty in gilts are age (170–220 days), breed (crossbred gilts reach puberty about 20 days before purebreds) and the age at which gilt meets boar (a sudden meeting after about 165 days of age may induce first oestrus).

In practice, the factor which decides when an animal is to be first used for breeding is body size, and at puberty animals are usually considered to be too small for breeding. Thus although heifers of the larger dairy breeds may be capable of conceiving at seven months of age, they are not normally mated until they are at least 15 months old. The tendency today is for cattle, sheep and pigs of both sexes to be mated when relatively young, which means that in the female the nutrient demands of pregnancy are added to those of growth. Inadequate nutrition during pregnancy is liable to retard foetal growth and to delay the attainment of mature size by the mother. Incomplete skeletal development is particularly dangerous because it may lead to difficulties at parturition.

Rapid growth and the earlier attainment of a size appropriate to breeding has the economic advantage of reducing the non-productive part of the animal's life. With meat-producing animals a further advantage is that a high plane of nutrition in early life allows the selection for breeding purposes of the individuals which respond to liberal feeding most favourably in terms of growth, and which may therefore be expected to produce fast-growing offspring. But there are also some disadvantages of rapid growth in breeding stock, especially if there is excessive fat deposition. In dairy cattle, fatness in early life may prejudice the development of milk-secreting tissue, and there is also some evidence that rapid early growth reduces the useful life of cows. Over-fat gilts do not mate as readily as normal animals, and during pregnancy may suffer more embryonic mortality. The rearing of breeding stock is a matter requiring more long-term research; at present the best recommendation is that such animals should be fed at a plane of nutrition that allows rapid increase in size without excessive fat deposition.

15.2 Plane of nutrition, fertility and fecundity

Plane of nutrition of female animals

In female animals the primary determinant of fertility (i.e. whether or not the animal conceives) and fecundity (i.e. litter size) is the number of ova shed from the ovaries (the ovulation rate). In the cow the rate is normally one, in the ewe it is normally one–three (but may reach ten) and in the sow, 15–25. Not all ova are fertilised and survive to birth. For example, a flock of 100 ewes in a single oestrous cycle might produce 220 ova, of which 190 would be fertilised. Of the 190 embryos, 175 might survive for 15 days, to the point at which they become attached to the placenta, and of these, 170 might complete the full gestation period of 147 days. Thus the final 'lambing percentage' (lambs born per 100 ewes mated) would be 170.

It has long been recognised that female sheep gaining weight or fat reserves in the 3–4 weeks before mating are more likely to conceive and more likely to have twins or triplets than those in poorer condition. This has led to the practice of 'flushing' ewes by transferring them from a low to a high plane of nutrition (e.g. transferring them from hill pastures to lowland pastures or fodder crops) before mating. More recently it has been recognised that although a higher plane of nutrition may be valuable for previously undernourished ewes, animals kept continuously on a higher plane will also have high fertility and fecundity. The degree of fatness of breeding stock is commonly assessed by means of 'condition scoring', in which fat deposits along the back and tailhead are palpated and scored on a scale of 1 (low) to 5 (high). Table 15.2 demonstrates what are sometimes called the static and dynamic effects of plane of nutrition on condition score and on fertility and fecundity in ewes. The greatest difference in ovulation rate was between ewes that were in good condition at mating (condition score 3) and those in poorer condition at mating (score 1.5). In the former ewes there was little effect of the changes in condition and weight that occurred in the six weeks preceding mating, but in the latter ewes those

Table 15.2 Effects of body condition and of change in condition on the ovulation rate of Scottish Blackface ewes (After Gunn R G, Doney J M and Russel A J F 1969 *Journal of Agricultural Science, Cambridge* **73**: 289)

Condition score[a]		Liveweight[a]		Ovulation rate
(a)	(b)	(a)	(b)	
3.5 →	3.0	67 →	64	2.11
3.0 →	3.0	62 →	62	2.11
	3.0		61	2.00
2.5 ↗		60 ↗		
2.0 →	1.5	52 →	47	1.00
1.5 →	1.5	44 →	46	1.11
	1.5		49	1.38
1.0 ↗		39 ↗		

[a] Column (a) shows the condition score or weight six weeks before mating, and column (b) the corresponding value at mating.

whose condition had been improved (from score 1.0 to score 1.5) produced more ova than those whose condition had been reduced (from 2.0 to 1.5).

The gonadotrophic hormone, luteinising hormone, is needed for the maturation and release of oocytes, and a low plane of nutrition appears to reduce the frequency with which pulses of luteinising hormone are released. Thus flushing may improve ovulation rate by preventing this effect. Another endocrinological explanation for the effect of flushing is that a high plane of nutrition promotes a greater production of insulin, which encourages the uptake of glucose and the synthesis of steroid hormones by the ovary. There may also be a direct effect of nutrition on the ovary, as intravenous infusion of glucose in sheep increases the growth of oocytes. However, a delicate nutritional balance is required for sheep around the time of ovulation, as a high plane of nutrition can reduce the survival of oocytes and the embryos they become after fertilisation. One reason for this effect is that a high plane of nutrition stimulates the metabolism (i.e. destruction) of progesterone, the hormone required for the establishment and maintenance of pregnancy. Thus it is recommended that after the ewe has been mated, the flushing plane of nutrition should be reduced to about the maintenance level.

In cattle, a high plane of nutrition after mating is less damaging than in sheep (which is fortunate, as cattle are expected to conceive at the stage of lactation when they are being fed for a high milk yield); their progesterone supply may be augmented at this time by release of the hormone from mobilised fat reserves. Nevertheless, achieving reconception in cows two months after calving, at a time when the nutritional demands of lactation are high, is a problem for the dairy industry. In dairy cows in New York state, over a period of 17 years when milk yield increased by 33 per cent, the proportion of cows conceiving to their first insemination fell from 66 to 50 per cent. In Britain, conception rate in dairy cows appears to be declining by 0.33 per cent per year. Observations like these have led to the recommendation that cows should be regaining weight by the time they are inseminated or served by a bull, but it is often difficult to ensure a positive energy balance at this time, even in well-fed cows. In situations of food scarcity, as with cattle kept on natural pastures subject to drought, the recommendation is impossible to implement, and in such situations it is common to find that calving intervals are extended from the desired 12 months to as much as 24 months.

In the sow, flushing seems to have little effect on litter size, possibly because pigs are commonly kept at higher planes of nutrition than sheep; however, flushing the maiden gilt for ten days before mating can increase litter size.

Plane of nutrition of male animals

In mammals, the spermatozoa and ova and the secretions associated with them represent only very small quantities of matter. The average ejaculate of the bull, for example, contains 0.5 g of dry matter. It therefore seems reasonable to suppose that nutrient requirements for the production of spermatozoa and ova are likely to be inappreciable compared with the requirements for maintenance and for processes such as growth and lactation.

If this were so, one would expect that adult male animals kept only for semen production would require no more than a maintenance ration appropriate to their species and size. There is insufficient experimental evidence on which to base feeding standards for breeding males, but in practice such animals are given food well in excess of that required for maintenance in females of the same weight. There is no reliable evidence that high planes of nutrition are beneficial for male fertility, though it is recognised that underfeeding has deleterious effects (see below). The liberal feeding of males probably reflects the natural desire of farmers not to risk underfeeding and so jeopardise the reproductive performance of the whole herd or flock. Males, however, do have a higher fasting metabolism and therefore a higher energy requirement for maintenance than do females and castrates.

Effects of specific nutrients on fertility

Many nutrient deficiencies influence fertility indirectly, through their effects on the general metabolism of the animal. For example, phosphorus deficiency in grazing ruminants, which has often been associated with poor fertility, appears to affect reproduction because it restricts many metabolic processes, hence food intake and the general plane of nutrition. However, there is also some evidence that phosphorus deficiency has a direct effect on reproduction through suppressing oestrous cycles.

Protein deficiency, especially in ruminants, can be expected to influence reproduction because it reduces food intake. At the other extreme, it has recently been suggested that a supplement of digestible protein that is undegradable in the rumen (DUP; see Chapter 13) can increase the ovulation rate of both sheep and cattle. In Australia, lupin seed – which contains a high concentration of DUP – is used to flush ewes. In contrast, rumen-degradable protein (RDP) given in excessive amounts to dairy cows raises blood levels of ammonia (see Chapter 8), which in turn alters the ion fluxes in the endometrium and reduces embryo survival. In pigs, short-term deprivation of protein has no effect on fertility, but a prolonged protein deficiency – especially in younger animals – leads to reproductive failure.

Vitamin A deficiency must also be prolonged if it is to affect fertility; thus animals may suffer blindness before their reproductive organs are affected (by keratinisation of the vagina or degeneration of the testes). Recently β-carotene has been claimed to have a specific effect on fertility (i.e. independent of its role as a precursor of vitamin A), but the claim has yet to be adequately confirmed (see Chapter 6).

Vitamin E deficiency causes infertility in rats but there is no evidence that the vitamin plays an essential role in maintaining fertility in cattle and sheep. In pigs, however, vitamin E deficient diets have been reported to reduce reproductive performance. There is also evidence from experiments with mature fowls that a prolonged vitamin E deficiency causes sterility in the male and reproductive failure in the female; and in the male, sterility may become permanent through degenerative changes in the testes. In ruminants, a deficiency of the nutrient associated with vitamin E, selenium, reduces fertility, apparently by affecting fertilisation, and also the viability of spermatozoa. In a series of trials in selenium-deficient areas of New Zealand, a supplement of selenium increased the lambing percentage from 89 to 98, but vitamin E had little effect. However, selenium supplementation allied with vitamin E for dairy cows has been shown to reduce the incidence of retained placenta (the condition in which the placenta fails to be expelled after calving).

In non-ruminants, deficiencies of the B-group vitamins riboflavin and folic acid have been shown to reduce embryo survival.

In addition to selenium, the trace elements copper, molybdenum, iodine, manganese and zinc are important in influencing fertility. Herbage copper levels of less than 3 mg per kg of dry matter delay the return to oestrus and hence lengthen the calving interval in cattle. Copper deficiency can be induced by excessive molybdenum in the diet (see Chapter 6). An excess of molybdenum in the diet of young cattle delays and depresses oestrus, apparently because it reduces the secretion of luteinising hormone. Deficiencies of copper or iodine reduce egg production in poultry. A deficiency of manganese reduces fertility in sows by causing delayed or irregular oestrous cycles. Zinc deficiency is interesting because it affects reproduction through the male, in which it prevents spermatogenesis; zinc is a component of the enzyme thymidine kinase, which is required for spermatogenesis. In the female, zinc deficiency may possibly increase embryo mortality; thus it reduces hatchability in poultry.

15.3 Egg production in poultry

Rearing of hens

Birds intended for egg production are commonly fed to appetite during the rearing period, with the aim of achieving a liveweight of about 1.5 kg at 18–20 weeks of age, when egg production begins. Restriction of food intake during rearing reduces costs, but delays the onset of egg laying. Moreover, it is important that birds should be well grown at the onset of laying, because egg weight is related to bodyweight. Thus pullets are reared on diets high in metabolisable energy and protein (see Appendix 2, Table A.2.10).

In contrast to layers, pullets being reared as broiler breeders (i.e. meat-type birds) have to be restricted in food intake, or else they grow too fast and their subsequent egg production is much reduced. The recommended restriction to 40 per cent of appetite (i.e. their intake if unrestricted) leaves them hungry (and often difficult to manage).

Nutrient requirements of laying hens

Productive flocks of layers produce an average of about 250 eggs per bird per year; thus on an average day $250/365 = 70$ per cent of birds will lay an egg, hence the term '70 per cent production'. Their peak production, early in the laying period, reaches 90 per cent. Their eggs weigh on average 57 g and have the chemical composition shown in Table 15.3, and an energy value of about 375 kJ; this information can be used as the basis for a factorial calculation of the nutrient requirements of layers. At one time, laying hens were rationed according to a system in which they were given a certain amount of food per day for maintenance and a certain amount for the estimated egg production, but today they are almost invariably fed to appetite. Feeding standards for layers, as for other classes of poultry, are therefore expressed in terms of nutrient proportions rather than quantities. The requirements of layers are shown in Appendix 2, Table A.2.10.

Energy

A hen weighing 2.0 kg has a fasting metabolism of about 0.36 MJ/kg $W^{0.75}$ per day, or 0.60 MJ/day, and utilises metabolisable energy for maintenance and production with a combined efficiency of about 0.8. Its requirement for metabolisable energy for maintenance estimated from fasting metabolism would therefore be $0.60/0.8 = 0.75$ MJ/day, and for 90 per cent egg production, $0.375 \times 0.9/0.8 = 0.42$ MJ/day (total 1.17 MJ/day). Maintenance requirements increase as temperature falls; for example, 2 kg birds adapted to 25 °C would require an extra 0.018 MJ/day for each 1 °C fall in temperature below 25 °C. An alternative method of estimating energy requirements is to fit regression equations to data obtained under practical feeding conditions

Table 15.3 Average composition of the hen's egg

	Per kg whole egg	Per egg of 57 g	Proportion of nutrient in edible part of egg
Gross constituents, g			
Water	668	38.1	1.00
Protein	118	6.7	0.97
Lipid	100	5.7	0.99
Carbohydrate	8	0.5	1.00
Ash	107	6.1	0.04
Amino acids, g			
Arginine	7.2	0.41	0.97
Histidine	2.6	0.15	(assumed for all
Isoleucine	6.4	0.36	amino acids)
Leucine	10.1	0.57	
Lysine	7.9	0.45	
Methinonine	4.0	0.23	
Phenylalanine	6.0	0.34	
Threonine	5.5	0.31	
Tryptophan	2.2	0.13	
Valine	7.6	0.44	
Major minerals, g			
Calcium	37.3	2.13	0.01
Phosphorus	2.3	0.13	0.85
Sodium	1.2	0.066	1.00
Potassium	1.3	0.075	1.00
Magnesium	0.8	0.046	0.58
Trace elements, mg			
Copper	5.0	0.3	1.00
Iodine	0.3	0.02	
Iron	33	1.9	(traces of minor
Manganese	0.3	0.02	elements in shell)
Zinc	16	1.0	
Selenium	5.0	0.3	

for metabolisable energy intake and the weight, weight change and egg production of hens. Maintenance estimates derived in this way differentiate between hens of the smaller, white strains and the heavier, brown-feathered layers. Expressed per kg W these are 0.480 and 0.375 MJ ME per day, respectively, so 1.8 kg white and 2 kg brown birds would need 0.86 and 0.75 MJ ME per day (*cf.* 0.61 MJ estimated above, from fasting metabolism).

In commercial practice, laying hens are almost invariably fed to appetite; their metabolisable energy intake is commensurate with the estimates of requirement discussed above (about 1.0–1.2 MJ/day) and they gain 1–2 g/day. They adjust their food intake for the energy concentration of their diet, so that if the ME content of the diet is reduced they eat more of it, and if ME content is increased, they eat less (see Chapter 17). However, the adjustment of intake does not fully compensate for the change in energy content, and the diets of layers are therefore kept to a small range in energy content, cf 11–13 MJ ME/kg (12–14 MJ/kg DM). For Appendix 2, Table A2.10, a concentration of 11.1 MJ/kg (12.5 MJ/kg DM) has been assumed. Concentrations below 10 MJ/kg are likely to depress energy intake sufficiently to reduce egg production, and concentrations greater than 13 MJ/kg usually increase body weight gain rather than the number of eggs laid (although egg weight may be increased). The food intake of hens, like their maintenance requirement, is affected by the environmental temperature, and falls by 1–2 per cent for each 1 °C rise in temperature in the range 10–30 °C.

Protein

Laying hens weighing 1.8 kg and consuming 110 g/day of a diet containing 11.1 MJ ME/kg require a total protein concentration of about 160 g/kg of diet. The amino acid requirements of layers have not been as accurately defined as have those of chicks, because it is difficult to maintain a satisfactory level of egg production if dietary protein is given as mixtures of pure amino acids. For factorial calculations it is assumed that amino acids can be incorporated in egg proteins with an average efficiency of 0.83. For example, the lysine content of eggs is 7.9 mg/g (Table 15.3), so the hen's production requirement for absorbed (i.e. available) lysine is 7.9/0 83 = 9.5 mg/g of egg produced. An equation for predicting the available lysine requirement of hens is:

$$L = 9.5E + 60W$$

Where: L = available lysine (mg/day)
E = egg production (g/day)
W = bodyweight (kg)

Thus the term $60W$ estimates the maintenance requirement for available lysine. Similar equations are available for the other indispensable amino acids for which the requirements of layers have been quantified; in addition to lysine these include methionine (often the first limiting amino acid for layers), tryptophan and isoleucine. Glycine (or its substitute, serine) is apparently non-essential for layers.

When the amino acid levels of diets for layers are set, account needs to be taken of the fact that individual birds within a flock will be producing eggs at different rates, and their requirements will therefore differ. This is, of course, a problem with other livestock and other nutrients but it has received special attention in its application to layers. The amino acid levels of the diet can be raised to ensure that the requirements of, say, 95 per cent of the flock will be satisfied, but this will increase the cost of the diet.

Models have therefore been devised to compare the extra costs of raising amino acid levels with the extra returns expected from egg production.

As the egg production of hens falls during the period of laying from the peak of 90 per cent to perhaps 55 per cent, their requirements for amino acids (and other nutrients) decline. It is therefore possible to employ what is termed 'phase feeding' to reduce the protein content of their diet from an initial 170 g/kg to about 150 g/kg.

Mineral elements

The laying hen's requirement for calcium is two–three times greater than that of the non-layer, because of the large quantities of this element in the eggshell. The minimum requirement for maximum egg production is about 3 g/day, but maximum eggshell thickness is not achieved until calcium intake is increased to 3.8 g/day. The whole quantity of calcium required is commonly included in the mash (meal) or pellets, but if the hen is given a separate source of calcium, as grit, it is capable of adjusting its intake to its requirements. The hen's requirement for calcium fluctuates during the day, and for the 12–14 hours in which it is forming the egg shell, its requirement excedes the rate at which it can absorb calcium. The hen therefore has to use its reserves of calcium, although it attempts to maximise the quantity absorbed by selecting calcium-rich foods (when possible) and improving absorption from the gut. Phosphorus requirements are difficult to define because of uncertainties regarding the availability of phytate phosphorus; requirements are therefore often stated as non-phytin phosphorus or as the proportion of inorganic phosphorus to be added to the diet.

Other elements which are likely to be deficient in normal diets are sodium, chlorine, iron, iodine, manganese and zinc. Common salt is generally added to the diet of laying hens and is beneficial in counteracting cannibalism and feather pecking. The requirement of poultry for sodium is met by the provision of 3.8 g NaCl per kg of diet. In excessive amounts salt is definitely harmful, although adult birds can withstand 200 g/kg in the diet if adequate drinking water is available. The iron content of the egg is relatively high (see Table 15.3), and consequently the requirement of the laying hen is large compared with the requirement for maintenance. Excessive iron in the diet is, however, harmful and may give rise to rickets by rendering the phosphorus of the diet unavailable. Iodine and manganese are particularly important for breeding hens, since a deficiency of either leads to a reduction in the hatchability of eggs, and may also reduce the viability of the chicks after hatching. The requirements for manganese are influenced by breed differences as well as by the levels of calcium and phosphorus (and phytate) in the diet; this trace element is more likely to be deficient in diets predominantly rich in maize than in those based on wheat or oats. Zinc deficiency in the diet of laying hens adversely affects egg production and hatchability, and results in the production of weak chicks with a high mortality rate. In the past it is possible that the use of galvanised feeding and drinking troughs was an important source of this element.

Vitamins

An important feature of the vitamin requirements of laying hens is that the minimum amounts required to ensure maximum egg production may be insufficient to provide for the normal growth of the chick, both before and after hatching. Requirements for some vitamins are not yet known, but it appears that for most B vitamins the quantities needed for maximum hatchability are appreciably greater than those for egg production alone. For vitamins A and D this is not so.

The value of β-carotene as a source of vitamin A for poultry depends upon a number of factors, and it has been suggested that in practice this provitamin should be considered as having, on a weight basis, only 33 per cent of the value of vitamin A.

Regarding vitamin D, it should be remembered that D_3 (cholecalciferol) is about ten times as potent for poultry as D_2 (ergocalciferol).

15.4 Nutrition and the growth of the foetus

The role of the placenta

When the embryo becomes attached to the lining of the uterus, it produces tissues that link with those of the uterus to form the placenta; at this stage the embryo becomes a foetus. The functions of the placenta are to feed nutrients into the foetus and to remove excretory products from it, and it carries out these functions by having maternal and foetal blood vessels in close proximity to one another. Nutrients and metabolites pass from one circulation to the other by the same three types of process that operate in the gut (see Chapter 8). Simple diffusion (i.e. from high to low concentration) accounts for the transport of lactate, acetate, oxygen, carbon dioxide and urea; glucose also reaches the foetus by diffusion but at a faster rate than can be accounted for solely by a concentration gradient. Nutrients kept at a higher concentration in foetal blood than in maternal blood must reach the foetus by the energy-demanding process of active transport; amino acids, electrolytes (minerals) and water-soluble vitamins enter the foetus by this process. Large molecules, such as lipids and proteins, are generally unable to pass through the placenta, but in some species the process of pinocytosis allows the transfer to the foetus of immunoglobulins produced by the mother. In farm animals, however, immunoglobulins are generally transferred after birth, via the first-drawn milk, or colostrum (see Chapter 16).

The placenta grows during the first two-thirds of pregnancy. In animals carrying two or more foetuses, the placental tissues have to be divided among them, thereby reducing the nutrient supply to each individual foetus and posing the danger that unequal division may cause inadequate nourishment of some individuals and hence reduce their birthweight; this is often the origin of the underweight ('runt') piglets in a newborn litter. The efficiency of the placenta in providing nutrients to the foetus(es) is dependent on its permeability, and on the flow and nutrient concentrations of the maternal blood. Thus circumstances that reduce the blood flow to the uterus, as heat stress may do by diverting blood to the mother's exterior, will reduce foetal growth.

The main source of energy for the foetus is glucose. The placenta also utilises glucose as an energy source for its transport operations. The foetus uses most of its glucose supply to synthesise and maintain tissues consisting mainly of protein. Small excesses of glucose are used to synthesise energy stores of glycogen and lipid. In farm animals, however, these stores are small and amount to no more than 30 g of lipid per kg of bodyweight at birth. In ruminants, some acetate may also be used for lipid synthesis. In times of glucose scarcity, the foetus will use amino acids as an energy source, but the latter are required mainly for protein synthesis.

Nutrient requirements for foetal growth

The growth of the foetus is accompanied by the growth of the placenta and also by the fluid-containing sacs that surround the foetus. In addition, the uterus itself enlarges. The organs and tissues that grow with the foetus are known collectively as the adnexa. The quantities of nutrients deposited in the foetus can be determined by killing animals at birth and analysing them, and the time course of nutrient deposition during gestation can be determined by analysis of foetuses and adnexa obtained from animals slaughtered at successive stages of pregnancy. Figure 15.1 illustrates nutrient deposition in sheep foetuses alone and in foetuses plus adnexa (the gravid uterus); it is based on analyses made by the UK Agricultural Research Council of a number of experimental studies of the kind described above, and applies to a lamb weighing 4 kg at birth. Similar analyses have been made for cattle and pigs.

For the weight of the gravid uterus (uterus plus contents), and also for many of its components, the values shown in Fig. 15.1 are extremely small before about 63 days (9 weeks) of gestation, but then increase rapidly to full term, at 147 days (21 weeks). The increases are particularly rapid in the last one-third of pregnancy (i.e. from day 100 onwards). They are also faster for the foetus than for the gravid uterus, because the adnexa make their growth earlier, in mid-gestation. This is illustrated most clearly for the component sodium, most of which in mid-gestation is located in the foetal fluids. The main organic component is, of course, protein; fat forms a small proportion of the foetus, reaching only about 120 g, or 30 g/kg, at full term.

The patterns of growth shown in Fig. 15.1 can be described by what are known as Gompertz growth equations, which have the form:

$$\log Y = A - Be^{-Cx}$$

Where: Y = the weight of the foetus or a component of it
    A, B and C = constants
              x = day of gestation

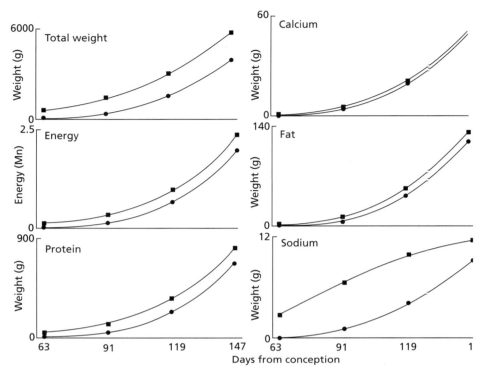

Fig. 15.1 Growth of the sheep foetus (●) and gravid uterus (■). (Plotted from the equations of the Agricultural Research Council 1980 *The Nutrient Requirements of Ruminant Livestock*, Farnham Royal, UK, Commonwealth Agricultural Bureaux.)

Differentiation of the Gompertz equations allows the calculation of nutrient deposition at successive stages of gestation, and some illustrative figures are shown in Table 15.4. For comparative purposes, Table 15.4 includes some estimates of the maintenance requirements of 40 kg ewes, expressed as net requirements. Until the last one-third of gestation, nutrient requirements for intra-uterine growth are so small relative to the ewe's maintenance requirements that they can be ignored. By the end of gestation, requirements for energy retention in the uterus are still small (17 per cent of the net maintenance requirement), but those for specific nutrients become more appreciable. For example, the intra-uterine requirement for calcium is over twice as great as the net maintenance requirement by the end of gestation.

Energy metabolism during gestation

If a pregnant animal is given a constant daily allowance of food, its heat production will rise towards the end of gestation. The increase is due mainly to the additional energy required by the foetus for both maintenance and

Table 15.4 Daily deposition of energy and selected nutrients in the uterus of a sheep at successive stages of gestation, and leading to the birth of a 4 kg lamb (Adapted from Agricultural Research Council 1980 *The Nutrient Requirements of Ruminant Livestock,* Farnham Royal, UK, Commonwealth Agricultural Bureaux)

Days (and weeks) from conception	Deposited in uterus (per day)			
	Energy (kJ)	Protein (g)	Calcium (g)	Phosphorus (g)
63 (9)	49.3	1.80	0.05	0.06
91 (13)	145.0	5.00	0.30	0.23
119 (17)	347.0	11.56	0.85	0.45
147 (21)	699.0	22.85	1.45	0.57
Ewe maintenance[a]	4000	33	0.64	0.56

[a] Approximate net daily requirement for maintenance of a 40 kg ewe.

growth. It has been found that metabolisable energy taken in by the mother in addition to her own maintenance requirement is utilised by the foetus with comparatively low efficiency (i.e. low k_c – see Chapter 11). For each additional MJ, only about 0.13 MJ is retained in the foetus, but the apparent k_c value of 0.13 is not directly comparable with the other k factors discussed in Chapter 11, because the heat production includes the basal heat production of the foetus as well as the heat produced during synthesis of foetal tissues. The efficiency with which metabolisable energy is used for the growth of the foetus (i.e. excluding the maintenance of the foetus) has recently been estimated as about 0.4. The demands of the foetus for maintenance and growth lead eventually to a considerable increase in the energy requirements of the mother. For example, for a ewe weighing 40 kg at the start of gestation, and requiring 6 MJ of metabolisable energy per day for maintenance, the total energy requirement will increase to about 11 MJ/day by the end of gestation. Thus the requirement for metabolisable energy in pregnancy is increased by far more than might be deduced from the storage of energy in the gravid uterus (Table 15.4).

Box 15.1
Mammary development

This takes place throughout pregnancy, but it is only in the later stages that it proceeds rapidly enough to make appreciable nutrient demands. Even then the quantities of nutrients laid down in the gland are quite small. In the heifer, for example, it has been shown by the analysis of animals slaughtered at various intervals during pregnancy that even in the last two weeks, when mammary growth is proceeding at its fastest rate, the quantity of protein deposited daily is no more than 45 g; this is only 20 per cent of the net protein requirement for maintenance, or 30 per cent of the protein deposited daily in the uterus at that stage of gestation.

Table 15.5 Digestible energy requirements (MJ/day) of a 140 kg sow for maintenance, foetal growth and maternal growth at successive stages of gestation (Adapted from Agricultural Research Council 1981 *The Nutrient Requirements of Pigs*, Farnham Royal, UK, Commonwealth Agricultural Bureaux)

Day of gestation	Maintenance requirement	Requirement for growth of:		Total requirement
		Foetuses, etc.	Maternal tissues[a]	
10	19.1	0.0	7.7	26.8
40	20.4	0.2	6.6	27.2
80	23.6	0.8	2.5	26.9
115	27.0	2.2	0.8	30.0

[a] Total gain in gestation of 20 kg.

Consequences of malnutrition in pregnancy

Malnutrition – meaning both inadequate and excessive intakes of nutrients – may affect pregnancy in several ways. The fertilised egg may die at an early stage (i.e. embryo loss), or later in pregnancy the foetus may develop incorrectly and die; it may then be resorbed *in utero*, expelled before full term

---

Box 15.2
Extra-uterine growth during gestation

The liveweight gains made by pregnant animals are often considerably greater than can be accounted for by the products of conception alone. For example, a litter of ten piglets and its associated membranes may weigh 18 kg at birth, but sows frequently gain over 50 kg during gestation. The difference represents the growth of the mother herself, and sows may in their own tissues deposit three to four times as much protein and five times as much calcium as is deposited in the products of conception. This pregnancy anabolism, as it is sometimes called, is obviously necessary in immature animals that are still growing, but it also occurs in older animals. Frequently, much of the weight gained during pregnancy is lost in the ensuing lactation.

If maternal growth (or regrowth) is regarded as an essential feature of gestation, then allowance must be made for it in feeding standards. Table 15.5 shows the effect on the energy requirements of 140 kg sows of providing for a maternal gain in gestation of 20 kg (most of this being made early in gestation, when requirements for foetal growth are small). Energy requirements for maternal growth are much larger than those for foetal growth, and as a result, total daily energy requirement does not change much over the course of pregnancy. The practical consequence of this is that sows are often given a flat rate of feed (about 2.0–2.7 kg of meal providing 26–34 MJ DE per day) throughout pregnancy.

Younger sows continue to grow lean tissue during pregnancy, so changes in bodyweight are not an accurate guide to changes in fat reserves (i.e. a sow gaining weight in pregnancy may still be losing fat). For this reason, the body condition score of sows may be included in models of energy requirements (see Agricultural and Food Research Council 1991, listed in Further reading).

A high plane of feeding, especially towards the end of gestation, is sometimes advocated as a means of increasing piglet birthweight and also the sow's mammary development; the greater energy reserves of the piglet and the greater milk yield of the sow should then increase the viability and growth of the litter. In practice, the improvements tend to be insignificant.

(abortion) or carried to full term (stillbirth). Less severe malnutrition may reduce the birthweight of the young, and the viability of small offspring may be diminished by their lack of strength or by their inadequate reserves (e.g. of fat). In some circumstances, it is the mother, not the foetus, that suffers from malnutrition. The foetus has a high priority for nutrients and if the mother has a low intake, her reserves will be used to meet the needs of the foetus. This priority is seen most strikingly in the case of iron, for the foetus can be adequately supplied with iron when the mother herself is anaemic. The protection thus afforded the foetus is not absolute, however, and in severe and prolonged deficiencies both foetus and mother will suffer. The degree of protection also varies from one nutrient to another, for although ewes as a result of an insufficient supply of energy may lose 15 kg of body substance during pregnancy and still give birth to normal lambs, an avitaminosis A that is without apparent effect on the ewe herself can lead to serious abnormalities in the young. The effects of underfeeding in pregnancy will also depend on the reserves of the mother, and particularly on the stage of pregnancy at which it occurs. In general, deficiencies are more serious the later they occur in pregnancy, but this rule is not invariable: vitamin A deficiency in early pregnancy, by interfering with the initial development of certain organs, can lead to abnormalities and even death in the young.

Effects on the young

Deficiencies of individual nutrients in pregnancy must be severe to cause the death of foetuses; protein and vitamin A are the nutrients most likely to be implicated, although deaths through iodine, calcium, riboflavin and pantothenic acid deficiencies have also been observed. Congenital deformities of nutritional origin often arise from vitamin A deficiency, which causes eye and bone malformations in particular. Iodine deficiency causes goitre in the unborn, and in pigs has been observed to result in a complete lack of hair in the young. Hairlessness can also be caused by an inadequate supply of riboflavin during pregnancy. Copper deficiency in the pregnant ewe leads to the condition of swayback in the lamb, as described earlier (p. 129).

In the early stages of pregnancy, when the nutritional demands of the embryo are still insignificant, the energy intake of the mother may influence embryo survival. In sheep and pigs both very low and very high intakes of energy at this stage may be damaging, especially in females in poor condition at mating. As discussed earlier (p. 393) the probable cause is a disturbance of the delicate hormone balance required at this time for implantation of the embryo. There is now evidence, some of which comes from the culture of embryos *in vitro*, that nutrition early in pregnancy can influence the later development of foetuses, for example by modifying the distribution of muscle fibres.

However, mid-pregnancy is a more critical period, because it is at this stage that nutrition affects the growth of the placenta. In the ewe, about two-thirds of the variation in lamb birthweight is associated with variation in placenta weight. In sheep a high plane of nutrition in mid-pregnancy *restricts* the growth of the placenta, possibly because a good supply of nutrients in blood can be transferred to the foetus with less than maximal placental development. Later in pregnancy, however, when the demands of the foetus are greater, its growth will be reduced because of the restricted growth of the placenta. The importance of this effect varies with the age and condition of the ewe; the practical recommendation is to feed older ewes and those in good body condition for a slight loss of weight in mid-pregnancy, but younger and thinner ewes should be fed to gain about 80 g/day. In pigs, a common recommendation is that feeding for the first two-thirds of pregnancy should be at about the maintenance level.

In the last one-third of pregnancy the requirements of the foetus(es) increase rapidly, and a low plane of nutrition at this time will restrict foetal growth. Ewes kept under natural conditions (e.g. mountain pastures) are frequently underfed in late pregnancy and thus lose weight: Table 15.6 illustrates the association between their weight losses and the birthweights of their lambs. Acute underfeeding in late pregnancy can stop foetal growth altogether, after which it may be impossible to restart it.

Young animals should be born with reserves of mineral elements, particularly iron and copper, and of vitamins A, D and E, because the milk, which may be the sole item of diet for a time after birth, is frequently poorly supplied with these nutrients. With regard to iron, it appears that if the mother is herself adequately supplied and is not anaemic, the administration of extra iron, whether in her food or by injection, will have no influence on the iron reserves of the newborn. If, however, the mother is anaemic these reserves – though not haemoglobin – may be reduced. The copper and fat-soluble vitamin reserves of the newborn are more susceptible to improvement through the nutrition of the mother.

Effects on the mother

The high priority of the foetus for nutrients may mean that the mother is the more severely affected by dietary deficiencies. The ability of the foetus to make the mother anaemic has already been mentioned; this situation is unusual in farm animals because their diets are normally well supplied with iron.

Table 15.6 The effects of energy intake during the last six weeks of pregnancy on the liveweight gains of ewes and on the birthweights of their twin lambs (Adapted from Gill J C and Thomson W 1954 *Journal of Agricultural Science, Cambridge* **45**: 229)

Group	Energy intake (MJ ME/day)	Liveweight change of ewes (kg)[a]	Birthweight of lambs (kg)
1	9.4	− 14.5	4.3
2	12.4	− 12.7	4.8
3	13.9	− 11.4	5.0
4	18.6	− 5.4	5.2

[a] From 6 weeks before to immediately after parturition.

The foetus has a high requirement for carbohydrate, and by virtue of its priority is able to maintain the sugar concentration of its own blood at a level higher than that of the mother. If the glucose supply of the mother is insufficient her blood glucose may fall considerably, to levels at which nerve tissues (which rely on carbohydrate for energy) are affected. This occurs in sheep in the condition known as pregnancy toxaemia, which is prevalent in ewes in the last month of pregnancy. Affected animals become dull and lethargic, lose their appetite and show nervous signs such as trembling and holding the head at an unusual angle; in animals showing these signs the mortality rate may be as high as 90 per cent. The disease occurs most frequently in ewes with more than one foetus – whence its alternative name of 'twin lamb disease' – and is most prevalent in times of food shortage and when the ewes are subjected to stress in the form of inclement weather or transportation. Loss of appetite is especially common among fat ewes. Blood samples from affected animals usually show, in addition to hypoglycaemia, a marked rise in ketone content and an increase in plasma free fatty acids. In the later stages of the disease the animal may suffer metabolic acidosis and renal failure.

There does not appear to be one single cause of pregnancy toxaemia. The main predisposing factors are undoubtedly the high requirement of the foetus for glucose and possibly a fall in the carbohydrate supply of the mother, which may arise through food shortage or through a decline in appetite in late pregnancy. One biochemical explanation for the disease hinges on the fact that the tricarboxylic acid cycle cannot function correctly without an adequate supply of oxalacetate, which is derived from glucose or such glucogenic substances as propionate, glycerol and certain amino acids. If the oxalacetate supply is curtailed, acetyl-CoA, which is derived from fats or from acetate arising through rumen fermentation, is unable to enter the cycle, and so follows an alternative pathway of metabolism which culminates in the formation of acetoacetate, β-hydroxybutyrate and acetone. In pregnancy toxaemia, the balance between metabolites needing to enter the cycle is upset by a reduction in glucose availability and an increase in acetyl-CoA production, the latter being caused by the animal having to metabolise its reserves of body fat. The clinical signs can thus be attributed both to hypoglycaemia and to the acidosis resulting from hyperketonaemia. An additional factor is that increased production of cortisol by the adrenal cortex in response to stress may reduce the utilisation of glucose; this possibility is supported by the fact that hyperketonaemia may continue after the blood glucose level has been restored to normal.

The disease has been treated by the injection of glucose, by feeding with substances likely to increase blood glucose levels, or by hormone therapy. Only moderate success has been achieved, however, and there is no doubt that the control of pregnancy toxaemia lies in the hands of the shepherd rather than the veterinary surgeon. The condition can be prevented by ensuring an adequate food supply in late pregnancy and by using foods which supply glucose or its precursors rather than acetate, i.e. concentrates rather than roughages.

Summary

1. Nutrition can affect both the fertility and the fecundity of farm animals, through its influence on age at puberty, on the production of ova and spermatozoa, and on the survival and growth of embryos and foetuses.

2. Cattle reach puberty at a specific skeletal size, rather than at a fixed age or weight. Pigs reach puberty at a fixed age, so nutrition has little effect. Well-nourished sheep reach puberty at about six months of age, but poorer animals will not reach it until the following breeding season (i.e. at 18 months of age).

3. Sheep in good or improving body condition produce more ova than those in poorer condition. This observation has led to the practice of 'flushing', by which the nutrition of ewes is improved for a few weeks before mating. In cattle, the demand for nutrients for lactation may interfere with re-conception two–three months after calving.

4. Deficiencies of protein, vitamins and minerals affect reproduction indirectly, through their effects on the general health of animals, but a few have more specific effects on reproduction. Thus zinc deficiency in males reduces the production of spermatozoa, and vitamin A deficiency causes congenital abnormalities in foetuses.

5. Egg production in poultry imposes large requirements for nutrients, especially amino acids and calcium, and these have been quantified in the form of detailed feeding standards.

6. For mammals, net requirements for the growth of the foetus increase exponentially during gestation and are quantitatively significant in the last one-third of this period. Nutrients are also required for the growth of the foetal membranes, uterus, mammary glands and (in some cases) the body reserves of the mother.

7. The placenta has a key role in ensuring that the foetus receives optimal supplies of nutrients, but overfeeding in mid-pregnancy restricts the size of the placenta and hence reduces birthweight. Underfeeding in late pregnancy also reduces birth weight.

8. Nutrient deficiencies in pregnancy may affect either or both the mother and the foetus. Many vitamin and mineral deficiencies (referred to in Chapters 5 and 6) are manifested in the foetus first. In the ewe, pregnancy toxaemia – which is basically a deficiency of glucose – is caused by the large demand for glucose by the foetus (often twin foetuses) and by a reduction in food intake in late gestation.

Further reading

Agricultural and Food Research Council 1991 *Nutrient Requirements of Sows and Boars*. Technical Committee on Responses to Nutrients, Report No. 4 (see also *Nutrition Abstracts and Reviews, Series B* **60**: 383–406).

Agricultural Research Council 1980 *The Nutrient Requirements of Ruminant Livestock*, Farnham Royal, UK, Commonwealth Agricultural Bureaux.

Agricultural Research Council 1981 *The Nutrient Requirements of Pigs*, Farnham Royal, UK, Commonwealth Agricultural Bureaux.

Agricultural Research Council 1984 *The Nutrient Requirements of Ruminant Livestock, Supplement No. 1*, Farnham Royal, UK, Commonwealth Agricultural Bureaux.

Blaxter K L 1967 *The Energy Metabolism of Ruminants*, London, Hutchinson.

Cupps P T 1991 *Reproduction in Domestic Animals*, New York, Academic Press.

Dunn T G and Moss G E 1992 Effects of nutrient deficiencies and excesses on reproductive efficiency of livestock. *Journal of Animal Science* **70**: 1580.

Mahan D C 1990 Mineral nutrition of the sow. *Journal of Animal Science* **68**: 573.

O'Callaghan D and Boland M P 1999 Nutritional effects on ovulation, embryo development and the establishment of pregnancy in ruminants. *Animal Science* **68**: 299–314.

Robinson J J, Sinclair K D and McEvoy T G 1999 Nutritional effects on foetal growth. *Animal Science* **68**: 315–31.

Rook J A F and Thomas P C (eds) 1983 *Nutritional Physiology of Farm Animals*, London, Longman.

Rose S P 1996 *Principles of Poultry Science*, Wallingford, UK, CAB International.

16 Lactation

16.1   Sources of milk constituents

16.2   Nutrient requirements of the lactating dairy cow

16.3   Nutrient requirements of the lactating dairy goat

16.4   Nutrient requirements of the lactating ewe

16.5   Nutrient requirements of the lactating sow

16.6   Nutrient requirements of the lactating mare

We are here concerned with the nutrient requirements for milk production, which involves a conversion of nutrients on a large scale and is a considerable biochemical and physiological achievement. A high-yielding dairy cow, for example, may in a single lactation produce five times as much dry matter in the form of milk as is present in her own body. The raw materials from which the milk constituents are derived, and the energy for the synthesis of certain of these in the mammary gland, are supplied by the food. The actual requirement for food depends upon the amount and composition of the milk being produced.

Qualitatively the milk of all species is similar in composition, although the detailed constitution of the various fractions such as protein and fat varies from species to species. Table 16.1 shows the typical composition of the milks of farm animals.

The major constituent of milk is water. Dissolved in the water are a wide range of inorganic elements, soluble nitrogenous substances such as amino acids, creatine, urea and the water-soluble protein albumin, lactose,

Table 16.1 The composition of milks of farm animals (g/kg)

	Fat	Protein	Lactose	Calcium	Phosphorus	Magnesium
Cow	37	34	48	1.2	0.9	0.12
Goat	45	33	41	1.3	1.1	0.20
Ewe	74	55	48	1.6	1.3	0.17
Sow	85	58	48	2.5	1.7	0.20
Mare	15	23	64	1.1	0.6	0.06

enzymes, water-soluble vitamins of the B complex and vitamin C. In colloidal suspension in this solution are inorganic substances, mostly compounds of calcium and phosphorus, and the protein casein. Dispersed throughout the aqueous phase is a suspension of minute milk fat globules. Triacylglycerols make up about 98 per cent of the fat phase, the remainder being composed of certain fat-associated substances such as phospholipids, cholesterol, the fat-soluble vitamins, pigments, traces of protein and heavy metals. The fat phase is usually referred to simply as 'fat' and the remaining constituents, other than water, are classified as 'solids-not-fat' or SNF.

16.1 Sources of the milk constituents

All or most of the major milk constituents are synthesised in the mammary gland from various precursors which are selectively absorbed from the blood. The gland also exerts this selective filtering action on certain proteins, minerals and vitamins, which are not elaborated by it but are simply transferred directly from the blood to the milk.

Milk proteins

About 95 per cent of the nitrogen in milk is in the form of protein, the remainder being present in substances such as urea, creatine, glucosamine and ammonia, which filter from the blood into the milk. In this respect milk functions as an alternative excretory outlet to urine. The protein fraction is dominated by the caseins. In cow's milk there are five of these, α_{s1}-, α_{s2}-, β-, κ- and γ-casein, which together contain about 78 per cent of the total milk nitrogen. The protein in next greatest amount is β-lactoglobulin. The remainder of the fraction is made up of small amounts of α-lactalbumin, serum albumin and the immune globulins, pseudo-globulin and euglobulin, all of which are absorbed directly from the blood.

Amino acids are absorbed by the mammary gland in quantities sufficient to account for the protein synthesised within it. Considerable interconversion of amino acids occurs before synthesis and certain amino acids are important as sources of others. Thus ornithine, which does not appear in milk protein, is absorbed and retained in large quantities by the mammary gland and has been shown to be a precursor of proline, glutamate and aspartate. Synthesis of the carbohydrate moieties of the proteins takes place in the mammary gland, as does phosphorylation of serine and threonine before their incorporation into the caseins.

Lactose

With the exception of traces of glucose, neutral and acid oligosaccharides and galactose, lactose is the only carbohydrate in milk. Chemically a molecule of lactose is produced by the union of one glucose and one galactose residue in the presence of an α-lactalbumin-dependent enzyme (Chapter 9). The galactose is derived almost entirely from glucose but a small part comes from acetate and glycerol. Virtually all the glucose is derived from the blood.

Milk fat

Apart from a minor fraction associated with the fat globule membrane, milk fat consists of a mixture of triacylglycerols containing a wide range of saturated and unsaturated fatty acids. The predominant saturated acid is palmitic and the unsaturated acids consist mainly of oleic with small contributions from linoleic and linolenic. The molar proportions of the fatty acids in the milk fats of different species are given in Table 16.2.

The fats are characterised by the presence of fatty acids of medium chain length ($8:0$ to $12:0$) which are specific to the mammary gland. Ruminant milk fats are further characterised by the presence of the low molecular weight butanoic ($4:0$) and hexanoic ($6:0$) acids, which here form 0.05 of the total fatty acids, on a molar basis. The milk fat of the mare also contains these acids, but the horse too is adapted for the fermentation of roughages with consequent production of volatile fatty acids in the gut. Table 16.3 gives a more comprehensive analysis of bovine milk fat and illustrates the wide range of fatty acids present. There are 22 in all, including those with odd numbers of carbon atoms and those with branched chains.

Table 16.2 Fatty acid composition (molar proportions) of milk lipids (From Bickerstaff R 1970 Uptake and metabolism of fat in the lactating mammary gland. In: Falconer I (ed.) *Lactation*, London, Butterworth)

Fatty acid	Cow	Goat	Sow	Mare
Saturated				
Butanoic	0.031	0.013	–	0.004
Hexanoic	0.019	0.028	–	0.009
Octanoic	0.008	0.083	–	0.026
Decanoic	0.020	0.129	0.002	0.055
Dodecanoic	0.039	0.036	0.003	0.056
Tetradecanoic	0.106	0.102	0.033	0.070
Hexadecanoic	0.281	0.245	0.303	0.161
Octadecanoic	0.085	0.098	0.040	0.029
Unsaturated				
Hexadecenoic	–	0.009	0.099	0.075
Octadecenoic	0.364	0.233	0.353	0.187
Octadecadienoic	0.037	0.018	0.130	0.076
Octadecatrienoic	–	–	0.025	0.161
Others	–	0.008	–	0.081

Table 16.3 Total fatty acid composition of bovine milk lipids (Adapted from Jensen R G, Ferrier A M and Lammi-Keefe C J 1991 *Journal of Dairy Science* **74**: 3228)

Fatty acid	Weight (%)	Fatty acid	Weight (%)
$C_{4:0}$	1.61	$C_{15:0}$	1.38
$C_{6:0}$	1.90	$C_{16:0}{}^{a}$	0.35
$C_{8:0}$	1.30	$C_{16:0}$	32.31
$C_{10:0}$	3.25	$C_{16:1}$	3.55
$C_{10:1}$	0.32	$C_{17:0}$	1.11
$C_{12:0}$	3.66	$C_{18:0}{}^{a}$	0.50
$C_{12:1}$	0.12	$C_{18:0}$	7.82
$C_{13:0}$	0.21	$C_{18:1}$	22.44
$C_{14:0}{}^{a}$	1.48	$C_{18:2}$	2.59
$C_{14:0}$	11.28	$C_{18:3}$	1.33
$C_{14:1}$	1.34	C_{20}	0.15

[a] Branched chain.

The fatty acids of milk fat are derived from two sources. The first is the chylomicrons and very low-density lipoproteins of the blood. The second is synthesis from acetate via the cytosolic malonyl-CoA pathway. In non-ruminant animals, the acetate is produced from glucose absorbed from the blood, whereas in ruminants it is derived from circulating blood acetate. Glucose also provides some of the glycerol moiety of the triacylglycerols, via the glycerol-3-phosphate pathway, as well as the reduced $NADP^{+}$ required for the cytosolic synthesis of the fatty acids. The major product of this pathway is palmitic acid. There is evidence that in non-ruminants this may undergo chain elongation but in ruminants this does not appear to be so. The 8 : 0 to 12 : 0 acids are produced by premature termination of the pathway. In the non-ruminant this is enzymatic, whereas in the ruminant it is an inherent property of the synthase system. Desaturation of acids takes place in both ruminant and non-ruminant mammary tissue, the products being 18 : 1 and 16 : 1 acids. The 4 : 0 and 6 : 0 acids are produced mainly by *de novo* synthesis from acetate in the mammary gland. However, D-3-hydroxybutyrate can be absorbed from the blood and used as a source of butyryl-CoA, which then joins the cytosolic pathway. Ruminant mammary gland tissue is unique in being able to esterify these acids into triacylglycerols. Acids with odd numbers of carbon atoms and those having branched chains are synthesised from propionate via methyl malonyl coenzyme A and the synthetase system.

About half the total acids of milk fat arise from blood lipids and the rest from *de novo* synthesis. Ultimately all the acids of milk fat originate in the products of digestion but not all do so directly. Some come from endogenous acetate and fatty acids after storage in the body.

That part of the glycerol moiety of the fat which is not derived from glucose originates in the acylglycerols of the blood.

Milk fat yield is influenced by the balance of fat synthesis and mobilisation. This is under hormonal control but depends upon the balance of glucogenic substances in the products of digestion. Thus a high proportion of propionate, glucose and amino acids stimulates fat deposition in adipose tissue and reduces the supply of fat precursors to the mammary gland.

Minerals

The inorganic elements of milk may be divided conveniently into two groups. The first is composed of the major elements calcium, phosphorus, sodium, magnesium and chlorine. The second group, the trace elements, contains some 25 elements whose presence in milk has been well authenticated; these include metals such as aluminium and tin; the metalloids boron, arsenic and silicon; and the halogens fluorine, bromine and iodine. Such substances are present in very small amounts and their presence in milk is coincidental with their presence in blood; nevertheless they may have an important bearing on the nutritive value of the milk and on the health and well-being of the suckled animal. The inorganic constituents of milk are absorbed directly from the blood by the mammary gland, which shows considerable selectivity; the gland is able to block the entry of some elements, such as selenium and fluorine, but allows the passage of others such as zinc and molybdenum. This selectivity may be a considerable disadvantage when it acts against elements whose presence at increased levels in milk would be desirable. Copper and iron, for example, are both elements important in haemoglobin formation and therefore for the nutrition of the young animal. Yet despite the fact that the levels of iron and copper in milk are never adequate, they cannot be raised by giving increased amounts to the lactating animal even when blood levels of these elements are so raised. The iron content of colostrum, the milk produced in the immediate *post partum* period, may be up to 15 times that of normal milk. However, during this time transfer of substances between blood and milk is abnormal.

Vitamins

Vitamins are not synthesised in the mammary gland and those present in milk are absorbed from the blood. Milk has considerable vitamin A potency owing to the presence of both vitamin A and β-carotene. The amounts of vitamins C and D present are very small and vitamins E and K occur only as traces. There is a large range of B vitamins in milk, including thiamin, riboflavin, nicotinic acid, B_6, pantothenic acid, biotin, folic acid, choline, B_{12} and inositol.

Figure 16.1 summarises the origins of the milk constituents.

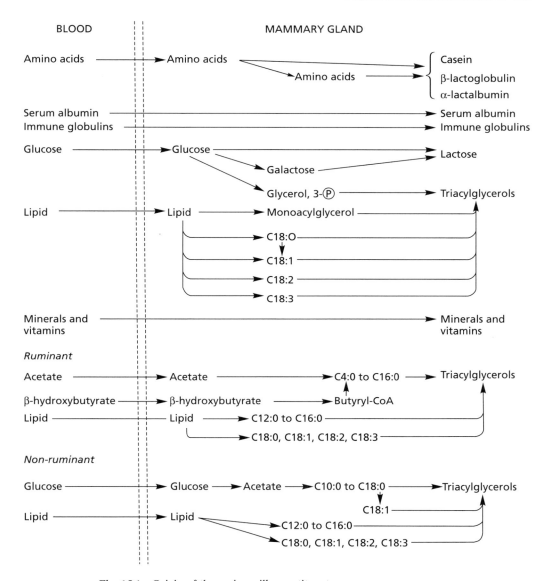

Fig. 16.1 Origin of the major milk constituents.

It will be clear from the foregoing that the mammary gland must be supplied with a wide range of materials if it is to perform its function of producing milk. In order for specific milk proteins to be synthesised, essential amino acids have to be available. In addition, a supply of non-essential amino acids or the raw materials for their synthesis must also be provided and non-specific milk proteins must be supplied as such.

Glucose and acetate are required for lactose and fat synthesis, and minerals and vitamins must be provided in quantities that allow the maintenance of normal levels of these milk constituents. The substances themselves, or the raw materials from which they are produced, have to be supplied either from the food or from the products of microbial activity in the alimentary canal.

16.2 Nutrient requirements of the lactating dairy cow

The nutrient requirements of the dairy cow for milk production depend upon the amount of milk being produced and upon its composition.

Milk yield

The yield of milk is decided primarily by the breed of the cow. Generally speaking, the order of yield for the main British dairy breeds is Holstein, Friesian, Ayrshire, Guernsey and Jersey (Table 16.4).

There are, however, considerable intra-breed variations with strain and individuality. Thus certain strains and individuals of a low-yielding breed may often outyield others of a higher-yielding breed. Old cows tend to have higher yields than younger animals but the main short-term factor affecting milk yield is the stage of lactation. Yield generally increases from parturition to about 35 days *post partum* and then falls regularly at the rate of about 2.5 per cent per week to the end of lactation. In individual cases, yield frequently reaches a peak earlier in lactation and the fall thereafter is much sharper.

As a result of these factors the yield of milk may vary over a very wide range. Fortunately, such variations present little difficulty in assessing the nutrient requirements of the cow since yield is easily and conveniently measured.

When estimates of yield are necessary for the long-term planning of the feeding of the lactating cow, several useful generalisations may be made which allow prediction of yield at a given stage of lactation. Thus peak yield may be calculated as one two-hundredth of the expected lactation yield or as 1.1 times the yield recorded two weeks *post partum*, e.g. a cow yielding 23 kg at this time could be expected to have a peak yield of 25 kg. The assumption of a weekly rate of decline from peak yield of 2.5 per cent per week is useful in predicting milk yield and also in monitoring deviations from normality

Table 16.4 Milk yields of the main British breeds of dairy cows (Adapted from *Dairy Facts & Figures* 1999)

Breed	Average lactation yield (kg)
Ayrshire	5940
Holstein–Friesian	6846
Shorthorn	5622
Guernsey	4898
Jersey	4611
All breeds	6755[a]

[a] Weighted for cow numbers.

during the progress of lactation. Such estimates are relatively imprecise and attempts to increase accuracy have resulted in highly sophisticated mathematical descriptions of the changes in yield with lactation.

In 1976, P D P Wood (1976: see Further reading) suggested that the yield of milk on any day *post partum* (*y*) may be calculated using equations of the following type:

$$y(n) = an^b e^{-cn}$$

where: n = week of lactation
   a = a positive scalar directly related to total milk production
   b = an index of the animal's capacity to utilise energy for milk production
   c = a decay rate

Values for a, b and c may be obtained from lactation yield data by a least-squares procedure. The constants for yield groups may then be used to predict week of peak yield (b/c) and daily yield in a given week of lactation. Typical figures for a cow with an expected lactation yield of 5500 kg would be:

$a = 26.69$
$b = 0.03996$
$c = 0.00942$

Week of peak yield would then be 4.24 (0.03996/0.00942) and predicted maximum daily yield 27.17 kg.

The Wood equation has been criticised in that it predicts zero yield at parturition which, in the light of the development of the secretory potential of the mammary gland at parturition, is obviously not so. A number of alternatives have been put forward. Most have been more complex and have proved to give less satisfactory performance. That of Emmans and Fisher 1986 in Fisher and Boormans (1986):

$$dY/dt = a\{\exp[-\exp(G_0 - bt)]\}[\exp(-ct)]$$

marginally outperforms the Wood equation, the proportion of variation explained (R^2) being 0.84 compared with 0.82, and dispersion around the regression line (residual standard deviation) being 1.74 compared with 1.86. In addition it does not predict zero yield at parturition and is more easily interpretable in biological terms. It does so, however, at the expense of increased complexity and the introduction of an extra term.

The Agricultural and Food Research Council in 1993 proposed the use of the Morant equation (1989):

$$Y(\text{kg/day}) = \exp\{a - b\ tl\ (1 + k\ tl) + c\ tl^2 + d/t\}$$

where: b, c and k = constants varying with parity and lactation yield
   a = the natural log of the expected yield at day 150 of lactation

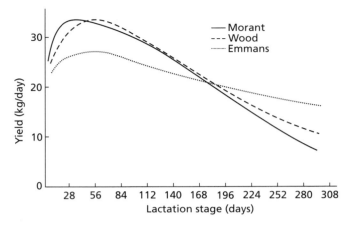

Fig. 16.2   Yield prediction curves for lactation yield of about 7000 kg.

$$t = \text{days since calving}$$
$$tl = (t - 150)/100$$

Figure 16.2 shows constructed yield curves for the lactation of a cow yielding about 7000 kg of milk in 48 weeks.

Milk composition

The composition of milk varies with a number of non-nutritional factors. Milking technique may have a profound effect on fat content and thus on total solids content since incomplete milking may leave a considerable volume of fat-rich milk in the udder. Unequal intervals between milkings may reduce yield and fat content when a single interval exceeds 16 hours and especially with high-yielding cows. Again diseases, particularly mastitis, may reduce the yield and compositional quality of milk. Lactose and potassium contents are lowered and those of sodium and chloride raised. Changes in fat content are erratic but crude protein shows little change. The net result, depending upon the severity of the infection, is a reduction in solids-not-fat and total solids content.

In a well-managed herd none of these factors should be of any importance. Certain variations in composition have to be accepted, however, since they are inevitable in a given herd. The factors responsible for these variations are breed, strain, individuality, age of cow and stage of lactation.

Effect of breed, strain within the breed and individuality on milk composition

There is a definite breed order in relation to milk quality, which is the reverse of that for milk yield. From Table 16.5 it can be seen that the Jersey

Table 16.5 Average values for the fat and protein contents of the milks of the main British breeds of dairy cows (Adapted from *Dairy Facts & Figures* 1999)

Breed	Fat (g/kg)	Protein (g/kg)
Ayrshire	40.8	33.4
Holstein–Friesian	40.4	32.8
Shorthorn	39.2	33.0
Guernsey	48.1	35.9
Jersey	54.8	38.7
All breeds	40.7[a]	32.9[a]

[a] Weighted for cow numbers.

produces the highest-quality milk and the high-yielding Holstein gives the poorest-quality product.

A more detailed though less recent report of the milks of the various breeds is given in Table 16.6.

It is interesting to note the difference in the constitution of the solids-not-fat of the different breeds; the milk of the high-yielding Friesian, for example, has a proportionately higher lactose and lower protein content than the milk from the lower-yielding Guernsey. Strain and individuality of the cows have an important effect on milk composition and many Friesian cows may average more than 40 g fat/kg and 89 g SNF/kg over a lactation, whereas some Channel Island cows may not match these figures. Thus, the ranges within the four breeds quoted in Table 16.6 were as given in Table 16.7.

Effect of age on milk composition

As the age of the cow increases so the quality of the milk produced becomes poorer. This is shown for Ayrshire cows in Table 16.8.

The regression of SNF content on age is linear and the decrease occurs almost equally in lactose and protein. Fat content, on the other hand, is relatively constant for the first four lactations and then decreases gradually with age. Studies on commercial herds indicate that over the first five lactations there is a linear decline in fat and solids-not-fat contents of about

Table 16.6 Average values for the detailed composition of the milk of four British dairy breeds (Adapted from Rook J A F 1961 *Dairy Science Abstracts* **23**: 251)

Constituent (g/kg)	Ayrshire	Friesian	Guernsey	Shorthorn
Fat	36.9	34.6	44.9	35.3
SNF	88.2	86.1	90.8	87.4
Protein	33.8	32.8	35.7	33.2
Lactose	45.7	44.6	46.2	45.1
Ash	7.0	7.5	7.7	7.6
Calcium	1.16	1.13	1.30	1.21
Phosphorus	0.93	0.90	1.02	0.96

Table 16.7 Within-breed variation in the composition of cow's milk (g/kg)[a] (Adapted from Rook J A F 1961 *Dairy Science Abstracts* **23**: 251)

Constituent (g/kg)	Ayrshire	Friesian	Guernsey	Shorthorn
Fat	35.7–38.7	33.2–37.2	43.1–49.0	33.7–38.1
SNF	86.5–89.4	84.0–87.5	88.2–93.0	85.7–89.0
Protein	33.0–34.7	32.0–34.4	33.9–37.3	31.6–34.2
Lactose	43.7–46.8	43.0–46.0	45.7–47.3	43.8–45.9

[a] Annual averages for individual herds.

2 and 4 g/kg, respectively. The age frequency distribution of a herd may profoundly affect the composition of the mixed herd milk.

Effect of stage of lactation on milk composition

Advancing lactation has a marked effect on the composition of milk, which is of poorest quality during that period when yield is at its highest. Both fat and solids-not-fat contents are low at this time and then improve gradually until the last three months of the lactation, when the improvement is more rapid. The changes are shown for Ayrshire cows in Fig. 16.3.

Solids-not-fat content fell during the first seven weeks of lactation, being the resultant of the fall in crude protein content from day 15 to day 45 (2.8 g/kg) and the rise in lactose content (0.4 g/kg) which took place during this period. Subsequently, the rise in protein content outweighed the fall in lactose content and the solids-not-fat content rose to the end of lactation. Fat content fell sharply in early lactation when yield was rising rapidly and then continued to fall more slowly until day 75 of lactation. Thereafter fat content rose slowly until day 195, after which the rise was much faster.

It will be obvious that assessment of milk composition is a more difficult task than assessment of milk yield, since there are five main variables to consider. Modern analytical methods allow for routine milk analysis on a

Table 16.8 Effect of age of cow on the composition of milk (g/kg) (After Waite R *et al.* 1956 *Journal of Dairy Research* **23**: 65)

Lactation	Fat	Solids-not-fat	Crude protein	Lactose
1	41.1	90.1	33.6	47.2
2	40.6	89.2	33.5	46.2
3	40.3	88.2	32.8	45.9
4	40.2	88.4	33.0	45.7
5	39.0	87.2	32.6	45.3
6	39.1	87.4	33.0	44.8
7	39.4	86.7	32.5	44.8
8	38.2	86.5	32.3	44.4
9	40.3	87.0	32.7	44.8
10	38.3	86.6	32.5	44.6
11	37.7	86.1	31.6	44.6

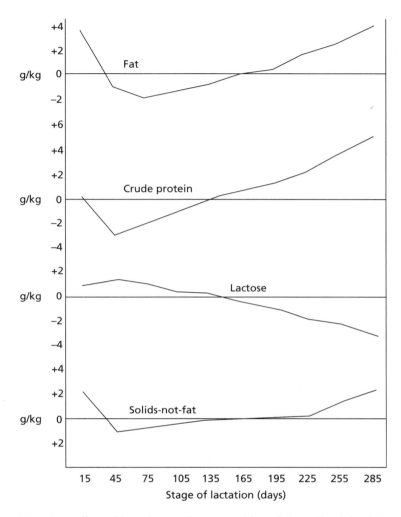

Fig. 16.3 Effect of lactation on the composition of the milk of the dairy cow. (After Waite R *et al.* 1956 *Journal of Dairy Research* **23**: 65.)

large scale and values for fat, lactose and protein contents of herd bulk milks are now readily available. When analytical results are not available, assumptions are often made concerning the quantitative relationships between constituents which allow composition to be predicted from the content of a single easily determined constituent, usually fat.

Energy requirements

Energy standards may be derived factorially. This involves an estimate of the gross energy value (EV_l) of the milk, which may be used along with the yield to estimate the net energy requirement for milk production.

Table 16.9 Calculation of the gross energy value of milk

Constituent	Content (g/kg)	Gross energy (MJ/kg)	Gross energy (MJ/kg milk)
Fat	40	38.12	1.52
Protein	34	24.52	0.83
Carbohydrate	47	16.54	0.73
Milk			3.13

The energy value of milk

Determination of the gross energy of milk involves either bomb calorimetry or a detailed chemical analysis; the amounts of fat, lactose and protein are then multiplied by their energy values and the products summed, as illustrated in Table 16.9.

In its 1980 publication *The Nutrient Requirements of Ruminant Livestock* the Agricultural Research Council proposed the following equation for calculating the energy value of milk:

$$EV_1 \text{ (MJ/kg)} = 1.509 + 0.0406F$$

in which F is fat content (g/kg). The equation has a standard error of estimate of 0.089 MJ/kg.

More accurate assessments may be obtained by including protein (P) and lactose (L) (both g/kg) in the prediction equation:

$$EV_1 \text{ (MJ/kg)} = 0.0384F + 0.0223P + 0.0199L - 0.108$$

The standard error of estimate is thus reduced to 0.035.

When compositional data are not available the energy values shown in Table 16.10, based on breed average fat and SNF values, may be used but with considerable reservations.

The net requirement for energy for milk production is the product of the energy value and the yield. The next step in the factorial estimate is the calculation of the amount of food energy required to provide the estimated net

Table 16.10 Energy values of the milks of the main British breeds of dairy cows

Breed	Energy value (MJ/kg)
Ayrshire	3.16
Holstein–Friesian	3.11
Guernsey	3.49
Jersey	3.79
SCM	3.14[a]

[a] Solids corrected milk with 40 g fat/kg and 89 g solids-not-fat/kg.

requirement. For this, the efficiency of utilisation of food energy for milk production must be known.

Efficiency of utilisation of food energy (ME) for milk production

From the calorimetric work of Forbes, Fries and Kellner an efficiency of utilisation of metabolisable energy for milk production (k_l) of about 0.70 is indicated. More recent estimates of k_l have varied widely from 0.50 to 0.81 but the majority cluster round about 0.60 to 0.65. There is considerable evidence that much of the variation is due to differences in the energy concentration of the diet. Van Es has suggested that the efficiency of utilisation of metabolisable energy for milk production is related to the metabolisability (q_m) of the diet, defined as the ME (MJ/kg DM) at the maintenance level as a proportion of the gross energy (MJ/kg DM). His implied relationships for (a) Dutch and (b) American data are:

$$\text{(a)} \quad k_l = 0.385 + 0.38\, q_m$$
$$\text{(b)} \quad k_l = 0.466 + 0.28\, q_m$$

where k_l is the efficiency of utilisation of ME for milk production at zero weight change. More recently AFRC (1993) has suggested that k_l is best calculated as:

$$k_l = 0.35\, q_m + 0.42$$

It has become common in deriving energy allowances to assume that the gross energy of the dry matter of all foods is constant at 18.4 MJ/kg and to transform this relationship to give:

$$k_l = 0.019\, M/D + 0.42$$

allowing k_l to be calculated from the energy concentration of the diet. For certain foods such as silage and high-fat dairy compounds, which frequently comprise a major part of the diets of lactating cows, the assumption is not valid and the equation based on metabolisability should be used. In such cases estimates of the gross energy of the foods are needed since routine bomb calorimetry is not feasible. Values calculated from the proximate composition are an acceptable alternative:

$$GE\ (MJ/kg) = 0.0226CP + 0.0407EE + 0.0192CF + 0.0177NFE$$

in which CP, EE, CF and NFE are stated as g/kg. When neither calorimetric nor compositional data are available, silages should be allocated a gross energy of 19.2 MJ/kg DM, high-oil compound foods 19.4 MJ/kg DM and all other foods a value of 18.4 MJ/kg DM. Efficiency of utilisation of ME for milk production will vary from 0.61 to 0.67 for diets with q_m 0.55 to 0.70. The range of energy concentrations encountered with diets for milk production is narrow and it is widely held that a single factor could be adopted without causing significant error; 0.62 is widely used.

The efficiency of utilisation of metabolisable energy is influenced by the level of protein in the diet. When protein content is inadequate body tissues are catabolised to make good the deficiency, a process which is wasteful of energy. When protein content is too high excess amino acids are used as a source of energy. Since protein is used relatively inefficiently for this purpose the overall efficiency of utilisation of metabolisable energy is reduced.

There is some evidence (Fig. 16.4) that the efficiency of utilisation of metabolisable energy for milk production is influenced by the proportion of acetate in the fatty acids produced during rumen fermentation.

It would appear that when the proportion of acetate is below 0.50, the cow is unable to synthesise sufficient of the lower and medium chain fatty acids which form a large part of milk fat; when the proportion of acetate in the total fatty acid production is greater than 0.65 efficiency is lowered owing to a shortage of propionate. This may limit the extent of gluconeogenesis from propionate, and gluconeogenic amino acids may be catabolised to make good the deficit. Gluconeogenesis from amino acids is a less efficient process and overall efficiency will be reduced. In addition, such catabolism may result in a shortage of amino acids for protein synthesis. The efficiency of utilisation of acetate depends on a supply of NADH, which in turn depends on a supply of glucose, which may be insufficient if gluconeogenesis is reduced.

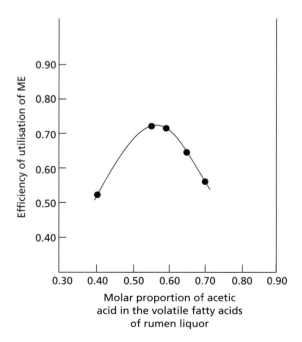

Fig. 16.4 Energetic efficiency of lactation. (After Blaxter K L 1967 *The Energy Metabolism of Ruminants*, London, Hutchinson, p. 259.)

There is other evidence that milk energy output is decreased when acetate forms less than 0.5–0.55 and propionate more than 0.35–0.45 of the rumen volatile fatty acids. This is thought to be the result of a reorientation of metabolism towards increased lipogenesis in adipose tissue and decreased lipogenesis in the mammary gland.

The inclusion of correction factors to allow for the effects of protein level and volatile fatty acid patterns on k_l is not justifiable, since the effects cannot be adequately quantified. At the same time, protein provides such a small proportion of the energy demand that any correction would be almost negligible. The effect of changes in the proportions of the volatile fatty acids within the normal range would be small and would in any case be partly taken into account by its association with energy concentration.

The requirement for metabolisable energy for milk production (M_l) may thus be calculated as:

$$M_l \; (\text{MJ/kg}) = EV_l/(0.35 \; q_m + 0.42)$$

The ME requirements of a cow producing milk having 40 g/kg fat, 34 g/kg protein and 48 g/kg lactose, and receiving a diet with $q_m = 0.60$, would be 5.0 MJ/kg.

Liveweight change in lactation

Lactating cows are usually gaining or losing bodyweight. A cow losing weight would be making reserves of energy available to maintain her level of milk production. If on the other hand she was gaining weight, some of the production ration would be diverted from milk production for this purpose.

Values quoted in the literature for the energy value of liveweight gain in dairy cows vary from 19 to 30 MJ/kg and appear to be related to body condition. In the Cornell net protein and carbohydrate system, the energy value of liveweight change is linearly related to body condition score and liveweight; in Australia a linear relationship with body condition score is used to predict the energy value of liveweight gain. Stage of lactation affects the energy value of liveweight change. The ratio of protein to fat in mobilised body tissue falls for about the first eight weeks of lactation and, as a result, the energy value of the tissue may be expected to increase.

AFRC (1993) postulated a value of 19 MJ/kg of liveweight change. This is very much lower than the value of 26 MJ/kg assumed earlier, and lower too than the mean value of about 6 Mcal (25 MJ)/kg quoted recently (2001) by the US National Research Council. The validity of a fixed value for the energy of body tissue is extremely doubtful but, for purposes of illustration, a preferred value of 25 MJ/kg has been assumed in the following treatment.

There is general agreement that the efficiency with which metabolisable energy is used for tissue deposition (k_g) in the lactating cow is higher than in the non-lactating animal. Values for k_g of up to 0.84 have been quoted but the majority of published work suggests that k_g is very similar to, but slightly lower than, k_l. For this reason the suggestion that $k_g = 0.95 \, k_l$ seems appropriate. Thus, for a cow gaining weight, each kilogram gained means that $25/(0.95 \times k_l)$ MJ of dietary metabolisable energy is unavailable for milk production, or that this amount of dietary metabolisable energy must be supplied in addition to that required for maintenance and milk production. In net energy terms, each kilogram of weight gain may be regarded as adding $25/0.95 = 26.3$ MJ to the lactation demand.

A recent estimate of the efficiency of utilisation of mobilised body tissue energy for lactation gave a mean of 0.74 with a range of 0.72–0.75. This is much lower than the figure of 0.84 used by AFRC (1993). If we accept the AFRC values, then, for each kilogram of body tissue mobilised, $25 \times 0.84 = 21$ MJ of energy is secreted as milk. This is equivalent to an additional 33.9 (21/0.62) MJ of dietary metabolisable energy for lactation, assuming a k_l of 0.62.

The use in the future of a variable energy value of liveweight change and perhaps a lower efficiency of utilisation must be given very serious consideration.

Energy requirements for maintenance

As well as the production of milk, the diet of the lactating cow must provide the energy for maintenance. This may be calculated (Chapter 15) as:

$$E_m \text{ (MJ/day)} = 0.53 \, (W/1.08)^{0.67} + 0.0091W$$

The activity allowance of $0.0091W$ applies to cows living indoors under normal loose housing conditions. It is based on certain assumptions of time spent standing, number of positional changes and distance walked and is valid only in situations in which these assumptions hold true (Chapter 14). Available evidence indicates that under such conditions a cow spends about 14 hours standing, stands up and lies down nine times and walks about 500 m, during a normal day's activity. The coefficient may then be calculated as shown in Box 16.1.

Box 16.1
Energy costs of physical activity

Activity	Energy cost	Energy expended (MJ/kg/day)
Standing (14 hours)	10 kJ/kg/day	0.0058
Positional changes (9)	0.26 kJ/kg	0.0023
Walking (500 m/day)	2.0 J/kg/m	0.0010
Total		0.0091

The subject of activity increments is dealt with more comprehensively in Chapter 14.

Efficiency of utilisation of dietary metabolisable energy for maintenance (k_m) may be calculated as:

$$k_m = 0.35\, q_m + 0.503$$

and the requirement for metabolisable energy for maintenance as:

$$M_m \text{ (MJ/day)} = [0.53\, (W/1.08)^{0.67} + 0.0091W]/(0.35\, q_m + 0.503)$$

There is some evidence that modern dairy cows have a higher maintenance requirement than that indicated by the above formula, probably owing to their greater genetic merit. Such animals have a greater proportion of body protein mass and there is increasing evidence that the basal metabolic rate – hence maintenance requirement – is dependent upon lean body mass rather than total liveweight. In addition, the increased intake and metabolism of nutrients needed to satisfy the demands of increased productivity may stimulate cardiac output, blood flow and oxygen consumption. Maintenance energy requirement has been shown to vary with the proportion of fibre in the diet, owing to the increased work of rumination and digestion, and an increased metabolic activity in the organs of the body.

In calculating the energy requirements of the dairy cow, cognisance must be taken of the decline in the efficiency of utilisation of metabolisable energy with increasing level of energy intake. In order to do this the calculated requirement has to be increased accordingly. The procedure, which involves the use of a correction factor, is best illustrated by an example, as shown in Box 16.2.

Box 16.2

Calculation of the metabolisable energy requirement of a 600 kg cow producing 30 kg/day of milk containing 40 g fat/kg and losing 0.4 kg liveweight/day on a diet having a q_m of 0.6

$E_m = 0.53 \times (600/1.08)^{0.67} + 0.0091 \times 600$	= 42.04 MJ/day
$k_m = 0.35 \times 0.6 + 0.503$	= 0.713
$E_l = 30(1.509 + 0.0406 \times 40)$	= 93.99 MJ/day
$k_l = 0.35 \times 0.6 + 0.42$	= 0.630
$M_l = 93.99/0.63$	= 149.19 MJ/day
$E_g = -0.4 \times 25$	= 10.0 MJ/day
Net energy spared by weight loss (10.0×0.84)	= 8.40 MJ/day
$M_m = 42.04/0.713$	= 58.96 MJ/day
$M_p = (93.99 - 8.40)/0.63$	= 135.86 MJ/day
Correction factor for level of feeding ($1 + 0.018 M_p/M_m$)	= 1.0415
$M_{mp} = (135.86 + 58.96) \times 1.0415$	= 202.91 MJ/day

It is common to use a safety margin in calculating allowances in an attempt to minimise the risks of underfeeding. A frequently used figure is 5 per cent, which would then give a revised figure of 213 MJ/day for the requirement of the cow in this calculation. There is evidence that the 5 per cent figure is inappropriate and that a safety margin closer to 15 per cent would be required to ensure that only 10 per cent of cows would be underfed. The high degree of variability implicit in the large safety margins is due partly to imperfect information on the parameters of the prediction model and also to the heterogeneity of the subject population. A safety margin has not been used in the box or in calculating the allowances given in the tables of Appendix 2.

Responses to increments of dietary energy

In experiments in which responses to the additions of energy to the diet have been measured in terms of milk yield, it has been found that only part of the theoretically expected increase in yield has been obtained. The discrepancy is the result of two factors:

■ Additions of concentrate foods to the diet bring about concomitant decreases in the roughage component, so that the increase in energy intake is less than is added in the supplement. The replacement rate, defined as change in forage intake per unit change in supplement intake, is greater for high-quality forages and at high intakes of supplement. In certain cases it may approach unity and the increase in the intake of dietary energy resulting from supplementation may be very small.
■ Energy consumed by the lactating animal over and above that required for maintenance is partitioned between milk production and body gain. Response to supplements of energy added to the diet is negatively curvilinear in the case of milk yield and positively curvilinear in the case of liveweight gain (Fig. 16.5).

Thus an increase in energy intake will result in an increase in milk yield together with a reduction in liveweight loss or an increase in liveweight gain. When these changes are considered along with the true increments of dietary energy, then responses approach the theoretical.

The ability of cows to divert part of their production ration for the growth of their own tissues, or alternatively to supplement the energy available for milk production by the breakdown of these tissues, varies considerably from one individual to another; cows of high yielding capacity use a higher proportion of a production ration for milk than those of lower potential. Within individuals the tendency is for the proportion of the energy of the increment which is used for milk production to decrease as the production ration is increased; in other words, the input of energy required per kilogram increases as the milk yield of the cow increases. Partitioning will also be influenced by the nature of the products of rumen fermentation, particularly the relative proportions of the volatile fatty acids, as discussed above.

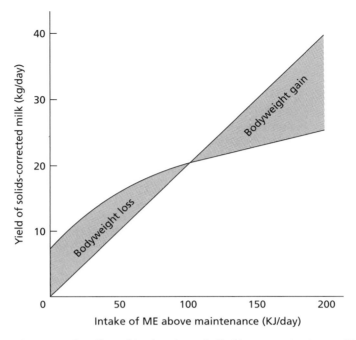

Fig. 16.5 The effect of intake of metabolisable energy (ME) on milk output and bodyweight change. (Adapted from Broster W and Thomas C 1981 *Recent Advances in Animal Nutrition*, In: Haresign W (ed.) London, Butterworth, pp. 49–69.)

Response, in terms of milk yield, to the addition of concentrate foods to a fixed ration is directly proportional to potential or current milk yield. This is well illustrated by the work of Blaxter, who allocated energy in excess of the level dictated by earlier standards to cows of different yield groups. He showed that the response ranged from 0.016 kg milk/MJ ME at a yield of 10 kg/day to 0.172 kg at 25 kg. A typical response to an increment of energy under such conditions would be 0.14 kg milk/MJ change in metabolisable energy intake (MEI), with responses of 0.003 and 0.01 kg/MJ MEI for yields of fat and solids-not-fat, respectively.

Under conditions of *ad libitum* feeding, response does not depend upon potential or current yield and a typical value would be 0.7 kg milk/kg of concentrate dry matter. If the true increase in dietary metabolisable energy is measured there is a trend for high-yielding cows to show greater responses.

The major determinant of total lactation yield is peak yield, whether this is achieved as a result of cow potential or feeding practice. Furthermore, responses to increments of energy decline as lactation progresses, and with low and medium planes of nutrition, elevation of energy intake in early lactation results in a 55 per cent residual effect in later lactation. Theoretically, therefore, allocation of high levels of concentrate supplements to early lactation in order to ensure maximum peak yield should result in higher lactation yields. Experimental evidence does not entirely support this expectation, since enhanced persistency may counterbalance a low peak yield. It would appear that when the plane of nutrition is low in relation to potential yield, the allocation of concentrates to early lactation improves lactation yield. When the plane of nutrition is generous in relation to potential then total input of concentrate foods over the lactation is more important than pattern of allocation.

For animals of high potential, it is very difficult to maintain generous levels of feeding in early lactation when intakes of dry matter are low. Such animals must be provided with high-energy concentrate foods and the finest-quality roughages if generous levels of feeding are to be achieved. A major problem at this time is to ensure that the ration does not cause rumen disorders and result in loss of appetite and the production of low-fat milk. The proportion of roughage should therefore not be allowed to fall below 35 per cent of the diet. Weight lost at this time needs to be replaced before the next lactation and this is usually achieved during late lactation and the dry period. In the light of the evidence of the high efficiency of utilisation of energy for body gain in the lactating cow, it may be that the most effective method of replacing lost tissue is by deliberate feeding in excess of requirement during late lactation. Accumulation of reserves is supported by the high efficiency of the use of such reserves for milk production.

Protein requirements

In the ruminant animal, dietary protein performs two functions:

- it must satisfy the nitrogen demands of the rumen microorganisms,
- it must supply the truly absorbable true protein required to satisfy the demand for amino acid nitrogen at tissue level.

Metabolisable protein requirement

The metabolisable protein requirement may be defined as the quantity of truly absorbable true protein required to satisfy the demand for amino acid nitrogen at tissue level. The net protein demand at tissue level is made up of the following:

1. A maintenance component, which may be calculated as 2.19 g/kg $W^{0.75}$ per day.
2. A dermal component resulting from the loss of hair and scurf, which may be calculated as 0.1125 g/kg $W^{0.75}$ per day.
3. A milk component calculated as milk crude protein $\times$ 0.95 g/kg milk produced. The factor of 0.95 is used because the non-protein nitrogen fraction of the milk, 5 per cent of the total nitrogen, is regarded as excretory material which has already been used by the body and has therefore formed part of a previously satisfied demand. When valid figures for protein content are not available, it may be calculated from fat

content (F) using regression equations such as that of Gaines and Overman:

$$\text{Protein (g/kg)} = 21.7 + 0.31F \text{ (g/kg)}$$

Alternatively, the protein contents of the milks of the main British breeds of dairy cows given in Table 16.5 may be used.

4. A component reflecting liveweight change (ΔW). Body tissue is assumed to contain 150 g protein per kilogram of empty bodyweight. Using a transformation factor of 1.09 this becomes $150/1.09 = 138$ g per kilogram of liveweight.

In order to calculate the quantities of metabolisable protein required to satisfy these net requirements, factors for the efficiency of utilisation of metabolisable protein for maintenance, dermal losses, lactation and growth and the conversion of mobilised body protein to milk protein are required. The following are recommended:

Maintenance = 1.0
Dermal   = 1.0
Growth   = 0.59
Lactation = 0.68

Amino acids arising from the mobilisation of body protein are utilised with the same efficiency as absorbed amino acids and liveweight loss has a sparing action on metabolisable protein requirement equal to its protein content, i.e. 138 g/kg. An example of the calculation of the metabolisable protein requirement of a lactating cow is given in Box 16.3.

Box 16.3

Calculation of the metabolisable protein requirement of a 600 kg cow producing 30 kg/day of milk containing 32 g CP/kg, and losing 0.4 kg W/day

	Net protein requirement (g/day)	Efficiency factor	Metabolisable protein requirement (g/day)
Maintenance	$2.19 \times 600^{0.75}$	1.00	265.5
Dermal loss	$0.1125 \times 600^{0.75}$	1.00	13.6
Milk	$32 \times 0.95 \times 30$	0.68	1341.2
Weight loss	-0.4×138	1.00	−55.2
Metabolisable protein	$(265.5 + 13.6 + 1341.2 - 55.2)$		1565.1

Effective rumen-degradable protein requirement

The protein requirements of the rumen microorganisms are stated in terms of effective rumen-degradable protein (ERDP). The requirement for effective rumen-degradable protein may be calculated relative to the dietary intake of fermentable metabolisable energy (FME) and is defined as:

$$\text{ERDP (g/day)} = \text{FME (MJ/day)} \times y$$

where y is the requirement of ERDP (g per MJ of FME) and varies with level of production.

The suggested values of y for different levels of animal performance are:

Animals at the maintenance level (M)	$y = 9$
Ewes in late pregnancy or lactation and lactating dairy cows	$y = 11$

Alternatively y may be calculated using the following equation:

$$y = 7 + 6(1 - e^{-0.35L})$$

where L is level of feeding relative to maintenance.

This equation is a mathematical convenience which smoothes the relationship between y and L and thus avoids boundary problems. It is not based on experimental data.

Microbial protein contributes to satisfying the demand for metabolisable protein but in the majority of cases, particularly at high levels of production, cannot completely satisfy the demand. The deficit has to be made good by the truly digestible undegradable true protein (DUP) of the diet.

Microbial crude protein (FME $\times$ 11) is assumed to contain 75 per cent of true protein (amino acids) and to have a true digestibility of 0.85. The contribution of microbial protein (MCP) to metabolisable protein (MP) is then MCP $\times$ 0.75 $\times$ 0.85 or 0.6375MCP. The requirement for truly digestible undegraded protein is then MP $-$ 0.6375MCP. An example of the calculation of the protein requirements of a dairy cow using an assumed figure for FME is given in Box 16.4.

Box 16.4

Calculation of the protein requirements of a 600 kg cow producing 30 kg of milk containing 32 g CP/kg, and losing 0.4 kg W/day

Metabolisable energy requirement (MJ/day)	202
Fermentable metabolisable energy of ration (MJ)	175
ERDP requirement (175 $\times$ 11) g/day	1925
Metabolisable protein requirement (g/day)	1565
Microbial protein contribution (1925 $\times$ 0.6375) g/day	1227
DUP requirement (1565 $-$ 1227) g/day	338

It is usual to use a safety margin of 5 per cent in converting requirements of protein to allowances but this is not suggested in this system. If safety margins are to be used they should be applied to the calculated requirement for ERDP and MP. However, in calculating the DUP allowance the uncorrected ERDP should be used to calculate truly digestible microbial true protein.

The application of the system to the evaluation of a ration as a protein source is straightforward. The MP, ERDP, DUP and FME contents of a ration are easily calculated from those of the constituents. The latter allows calculation of the requirement for ERDP, which may then be compared with that supplied by the ration. If ERDP/FME is equal to or greater than 11 (i.e. energy is limiting) then the DUP requirement may be calculated as described above and this can be compared with the supply (see Box 16.5).

The application of the system to the formulation of rations is rather more complicated. The ration must first be formulated to meet the requirement for metabolisable energy. This allows the fermentable metabolisable energy content to be calculated. Only then may the requirement for ERDP be calculated. The ERDP and DUP status of the ration has then to be assessed and

Box 16.5

Evaluation of a ration ($q_m = 0.6$) for a 600 kg cow yielding 30 kg milk with 40 g/kg fat and 35 g/kg protein, and losing 0.4 kg W/day

Requirement for metabolisable protein (MP g/day):

Maintenance	$= 2.3 \times 600^{0.75}$	$=$	279
Milk	$= 35 \times 0.95 \times 30/0.68$	$=$	1467
Weight change	$= -0.4 \times 138$	$=$	-55
Total		$=$	1691

Ration:

Food	kg	DM (kg)	ME (MJ)	FME (MJ)	ERDP (g)	DUP (g)
Silage	35	8.05	84.5	62.8	886	185
Maize gluten feed	2	1.80	22.9	20.7	234	74
Compound	8.6	7.67	99.1	95.0	858	345
Total		17.52	206.5	178.5	1978	604

ERDP/FME $= 1978/178.5 = 11.08$ and energy is limiting
MCP $= 178.5 \times 11 = 1964$ g/day
MCP contribution to metabolisable protein demand $= 1964 \times 0.6375 = 1252$ g/day
DUP requirement $= 1691 - 1252 = 439$ g/day

Then:

	Requirement	Supplied by ration
ME (MJ/day)	206	206.5
ERDP (g/day)	1964	1987
DUP (g/day)	439	604
MP (g/day)	1691	1856

brought into balance. This is most simply done by formulating a supplement having the same FME concentration as the basal ration and the necessary ERDP and DUP concentrations. This is not an uncomplicated procedure!

If the ideal blend of degradable and undegradable protein is not achieved the requirement may be considerably increased. The attainment of such a blend may allow the use of cheaper sources of protein or justify the use of expensive protein or the industrial processing needed to modify the degradability of unsuitable dietary protein.

Estimates of protein requirements may also be made from the results of feeding trials. In these, diets are used which are accepted as satisfactory in all respects other than protein, and the minimum intake of protein adequate for maximum production is determined. Such experiments have to be of a long-term nature since even on deficient diets production may be maintained owing to the cow's ability to utilise her body tissues. This will result in a negative nitrogen balance and studies of such balances are often carried out to supplement the main feeding trial. In treating the results of feeding trials, an allowance is made for protein required for maintenance and the residue equated to milk production. Estimates of the digestible protein requirement based on such trials have varied from 1.75 times that present in milk to as little as 1.25 times in more recent work. These low levels only apply where the content of crude protein in the diet is of the order of 160 g/kg; where the content is reduced to about 120 g CP/kg the requirement for milk production rises.

Mineral requirements

AFRC Technical Committee Report No. 6 (TCORN 6) proposed that the net daily requirement for calcium for maintenance (g/day) of the dairy cow may be calculated as:

$$0.0079W + 0.66DMI - 0.74$$

where DMI is defined as $MEI/18.4q_m$. For a 600 kg cow consuming 170 MJ ME on a diet with q_m of 0.6, DMI would be 15.4. This gives a net requirement of 23.6 mg/kg W per day.

Box 16.6

Alternative systems for estimating the energy and protein requirements of dairy cows

The examples given in this chapter are based on the Agricultural and Food Research Council systems employed in the UK. At the time of writing (February 2001) the US National Research Council is about to publish a new and much enlarged edition of its handbook *Nutrient Requirements of Dairy Cattle*, which will include new procedures for calculating requirements that are based on the Cornell net carbohydrate and protein system. From a preview of this publication (available on the internet at http/www.nap.edu) the new procedures can be seen as being complicated and thus employable only by the use of the computer program that is to be provided with the handbook. From the tables, energy requirements of lactating cows appear to be similar to those calculated by the AFRC system, but protein requirements are higher in the NRC system.

For the net phosphorus requirement the TCORN calculation is:

$$1.6(0.693DMI - 0.06)$$

For a 600 kg cow consuming 15.4 kg of dry matter, this suggests a net requirement of 28.2 mg/kg W/day. In addition to the requirements for maintenance, calcium and phosphorus must be provided for milk production. TCORN 6 suggested 1.2 and 0.9 g/kg for the calcium and phosphorus contents of milk. Availability values for calcium and phosphorus are 0.68 and 0.58.

TCORN 6 suggested that for diets with q_m greater than 0.7, maintenance requirements for phosphorus should be calculated using 1.0 instead of 1.6 in the above formula, and that the availability factor should be changed to 0.7. Such a sharp change can give rise to anomalies in calculating requirements. Thus a phosphorus requirement of 72 g/day would be changed to 57 g/day for a change of q_m from 0.69 to 0.71, which we find unacceptable. In the light of these observations and the fact that our previously used standards have proved satisfactory over many years, these were preferred, and they have been used, along with the TCORN methods of calculating net requirements, for calculating the allowances in the tables of Appendix 2.

The results of feeding trials suggest that allowances of calcium and phosphorus considerably lower than those indicated by factorial calculation can be given for long periods with no ill effects. Thus 25–28 g of calcium and 25 g of phosphorus per day have proved adequate for cows producing 4540 kg of milk per annum over four lactations, which implies a dietary requirement of 1.10 to 1.32 g calcium and 1.10 g phosphorus per kg of milk. The requirements given in Appendix 2, Table A2.3, have been derived by factorial calculation and are probably slightly higher than the minimum requirement but are considered necessary to ensure a normal life span and satisfactory reproduction. Balance experiments have shown that even very liberal allowances of calcium and phosphorus are frequently inadequate to meet the needs of the cow for these elements during the early part of the lactation. In the later stages and in the dry period, storage of calcium and phosphorus takes place. Figure 16.6, for example, shows the cumulative weekly calcium and phosphorus balances throughout a 47-week lactation for a mature Ayrshire cow producing 5000 kg of milk.

Despite the negative balances which occurred over considerable periods early in lactation, there was a net positive balance over the lactation and dry period as a whole. It has therefore become normal practice to consider the complete lactation in assessing calcium and phosphorus requirements; early negative balances are regarded as normal, since no ill effects are evident as long as subsequent replenishment of body reserves takes place, and daily requirements are formulated on the basis of total production over the lactation. However, although the lactation approach is satisfactory in many cases, considerable trouble may arise if the allowances used are too low. When a shortage is serious, progressive weakening and breaking of the bones may result and, in less severe cases, a premature drying off, which reduces yield and shortens the productive life of the cow. There seems to

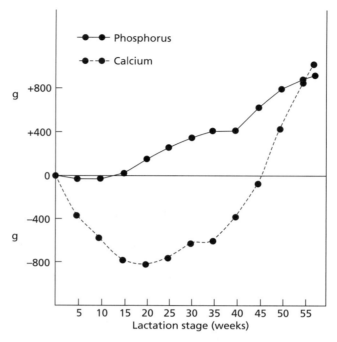

Fig. 16.6 Cumulative balances of calcium and phosphorus during lactation (47 weeks) and dry period. (Adapted from Ellenberger H B, Newlander J A and Jones C H 1931 *Bulletin of the Vermont Agricultural Experimental Station*, p. 331.)

be little reason why requirements should not be based on weekly yield measurements.

In diets which are deficient in phosphorus, the ratio of calcium to phosphorus can be important. With practical diets, evidence for the importance of the ratio is conflicting and in the absence of definitive evidence the best approach would be to keep the Ca : P ratio between 1 : 1 and 2 : 1.

In calculating magnesium allowances, a net daily requirement for maintenance of 3 mg/kg W may be assumed along with a concentration of 0.125 g/kg in the milk. Availability of dietary magnesium is very low at about 0.17.

Lactating cows are usually given a sodium chloride supplement. This is done by adding the salt to the food or by allowing continuous access to salt licks. The primary need is for sodium rather than chloride, which is more plentiful in normal diets. A deficiency manifests itself in a loss of appetite, rough coat, listlessness, loss of weight and a drop in milk production. Salt hunger and low levels of sodium in plasma and urine may occur in high-yielding cows after as little as three weeks if diets are unsupplemented. Loss of appetite, weight and production may take about a year to appear. The net requirement for sodium is about 8 mg/kg per day for maintenance plus 0.60 g/kg of milk. It is usually recommended that 28 g of sodium chloride per day should be provided in addition to that in the food, or that 15 kg/t of sodium chloride should be added to the concentrate ration.

Vitamin requirements

Vitamins are required by the lactating animal to allow proper functioning of the physiological processes of milk production and as constituents of the milk itself. It has yet to be shown that there is a dietary requirement for vitamins specifically for lactation but they have a role in the synthesis of milk constituents, as for instance biotin has in the synthesis of the milk fat. Most of the evidence points to the conclusion that as long as levels of vitamins in the diet are sufficient for maintenance, normal growth and reproduction then no further allowance for lactation need be made. However, normal levels of vitamins in the milk must be maintained and sufficient amounts have to be given to allow for this. The B vitamins are an exception, since an adequate supply becomes available as a result of microbial synthesis in the rumen. Maintenance of normal vitamin levels in milk is particularly important where milk is the sole source of vitamins for the young animal, as for example with the young pig and the suckled calf.

Winter milk has a vitamin A potency of about 2000 iu/kg. Apart from the almost colourless vitamin A, milk contains variable amounts of the precursor β-carotene. This is a red pigment, yellow in dilute solution as in milk, to which it imparts a rich creamy colour. The vitamin A potency of milk varies widely, being particularly sensitive to changes in dietary levels even though only about 3 per cent of the intake finds its way into the milk. Thus green foods are excellent sources of the provitamin, as is shown by the deep yellow colour of milk produced by grazing cows. Some breeds, e.g. the Jersey and the Guernsey, have higher ratios of β-carotene to vitamin A in their milks, which consequently have a deeper yellow colour. For example, feeding the Channel Islands breeds with vitamin A in excess of levels adequate for reproduction may increase the potency of the milk by up to 20 times but has no effect on the yield or gross composition of the milk. Considerable storage of vitamin A takes place in the body and these reserves may be tapped to maintain levels in the milk. Since the newborn animal normally has small reserves it is almost entirely dependent upon milk for its supply and it is essential to feed the nursing mother during pregnancy and lactation so as to maintain the potency of the milk. No problem arises with cattle and sheep which are given early access to green food but great care is required where this is not so, as for example in winter-calving suckler herds. The daily requirement of the lactating cow is about 99 iu/kg W or 30 μg/kg W.

There is some evidence that there may be a requirement for β-carotene itself, quite distinct from its function as a provitamin (see Chapters 5 and 15).

When lactating dairy cows are kept on diets deficient in vitamin D and irradiation is prevented, deficiency symptoms appear, showing that the vitamin is essential for normal health. There is no evidence, however, for a requirement greater than that which supports maintenance and reproduction. The vitamin D potency of milk is largely influenced by the extent of the cow's exposure to sunlight, and large dietary intakes are necessary for small increases in the concentration in the milk. Administration of the vitamin has little effect in ameliorating the negative balances of calcium and phosphorus which occur in early lactation, but very heavy doses

(20 000 000 iu/day) for three to five days *pre partum* and one day *post partum* have been claimed to provide some control of milk fever (p. 118). The daily requirement of the lactating cow is about 10 iu/kg W.

Dietary intakes of the B vitamins are of no significance in ruminant animals because of ruminal synthesis. A physiological requirement, in addition to that needed for maintaining normal levels in milk, does exist for many of them owing to their involvement in the complicated enzyme systems responsible for the synthesis of milk.

Effects of limitation of food intake on milk production

There is a great deal of evidence to show that reduction of food intake has a profound effect upon both the yield and the composition of milk.

When cows are kept without food the yield drops to very low levels of about 0.5 kg per milking within three days. At the same time the solids-not-fat and fat contents rise to about twice their previous levels, the increases being due to a concentration resulting from reduced yield. Less severe limitations reduce yields to a lesser extent; the solids-not-fat content is lowered but the effect on fat content is variable.

Limitation of the energy part of the diet has a greater effect on the solids-not-fat content than does limitation of the protein, although it is the protein fraction which is reduced in both cases. Lactose concentration shows little change, as would be expected of the major determinant of the osmotic pressure of milk. Most of the fall in protein content is probably due to increased gluconeogenesis from amino acids, owing to a reduced propionate supply on low-energy diets. As a result, the supply of amino acids to the mammary gland is reduced and so is protein synthesis. It is also conceivable that the low energy supply limits microbial protein synthesis in the rumen, and thus the amino acids available to the mammary gland. Throughout the winter feeding period in the UK there is a decline in milk yield and solids-not-fat content in most herds, the rate of decline being most marked in the later period. The traditional pattern is for both yield and solids-not-fat content to increase when the cows are allowed access to spring pasture. It has been shown experimentally that where levels of winter feeding are high, such increases do not take place and indeed the opposite effect may be produced. It would appear, therefore, that winter feeding of dairy cows is frequently inadequate.

Table 16.11 Comparison of fat contents of milks produced on various diets with those produced on a diet containing 5.4 kg of hay + concentrates (After Balch C C *et al.* 1954 *Journal of Dairy Research* **21**: 172)

Diet	Change in milk fat (g/kg)
3.6 kg hay + concentrates	−11.6
3.6 kg ground hay + concentrates	−17.2

The change to pasture feeding in spring is frequently accompanied by a fall in the fat content of the milk. Spring pastures have a low content of crude fibre and a high content of soluble carbohydrates; other diets having similar characteristics also bring about a decline in milk fat. These diets may have a low ratio of roughage to concentrate or the roughage that is present may be very finely ground. The effect is well illustrated in Table 16.11.

The drop in fat content becomes more obvious as the proportion of roughage in the diet falls below 400 g/kg DM, and below 100 g/kg DM herd mean fat content may be below 20 g/kg. Available data suggest that the measure of fibre most highly correlated with fat content is acid-detergent fibre (ADF), which consists of cellulose, lignin, acid-detergent insoluble nitrogen and acid insoluble ash. Generally, the ADF content of the diet of the dairy cow should be maintained above 190 g/kg DM but this may not be possible if energy requirements at times of highest yields are to be met. It is at such times that the danger of producing low-fat milk is greatest. Fineness of grinding (particle size), particularly of the roughage part of the diet, has an important influence on milk fat content and it has been suggested that the minimum allowable chop length is about 7 mm. More information on the influence of such factors as the buffering capacity of foods and frequency of feeding in ameliorating or accentuating the effects of low-fibre and finely ground diets is required before definitive statements can be made on optimum figures for either.

Lowered fat contents are usually accompanied by changes in the fatty acid composition with a decrease in the saturated acids and an increase in the unsaturated acids, particularly the 9-octadecenoic. The changes in fat content and composition are associated with changes in rumen fermentation patterns. Low-fibre diets fail to stimulate salivary secretion and hence diminish the buffering power of the rumen liquor. Such diets are often fermented rapidly, giving rise to pronounced peaks of acid production and very low pH values. As a result, the activity of the cellulolytic fibre-digesting microorganisms is inhibited and that of various starch utilisers encouraged. These changes are reflected in changes in the balance of volatile fatty acids (VFA) in the rumen. On high-fibre diets the molar proportions of the VFA would be about 0.70 acetic acid, 0.18 propionic acid and 0.12 butyric acid, together with a number of higher acids present in small amounts only. If the fibre content of the diet is reduced and that of the concentrates increased, the proportion of acetic acid falls and in extreme cases may be less than 0.4. This fall is usually accompanied by a decrease in butyric acid and an increase in propionic acid, which may be more than 0.45 of the total acids present: the concentration of valeric acid may also increase. Diets containing high proportions of carbohydrate which has been treated to increase its availability, as with the starch in flaked maize, are particularly effective in decreasing the ratio of acetate to propionate and reducing fat content (Table 16.12).

Experiments in which volatile fatty acids have been infused into the rumen of lactating cows have shown that acetate and butyrate increase milk fat content whereas propionate reduces it.

It has been suggested that if the ratio of acetic to propionic acid in the rumen contents falls below 3 : 1 then milks of low fat content will be

Table 16.12 Relation of dietary cooked starch to milk fat content and to rumen volatile fatty acids (After Ensor W L *et al.* 1959 *Journal of Dairy Science* **42**: 189)

Diet	Percentage change in fat content	Molar proportions of total rumen volatile fatty acids	
		Acetic acid	Propionic acid
12.4–14.5 kg ground pelleted hay	0	0.68	0.20
12.4 kg ground pelleted hay + 1.8 kg ground maize	−13	0.62	0.25
12.4 kg ground pelleted hay + 1.8 kg cooked maize	−53	0.54	0.31

produced. Other workers have claimed that the most important determinant of milk fat content is the balance of glucogenic to non-glucogenic VFA in rumen contents and that this is conveniently defined as the non-glucogenic ratio (NGR):

$$NGR = (A + 2B + V)/P + V$$

where A, P, B and V are molar proportions of acetate, propionate, butyrate and valerate in the rumen contents. If the ratio falls below 3 the danger of low-fat milks being produced is increased.

There is a tendency for dietary fat to be regarded simply as a source of energy. It has been shown, however, that when fat is replaced by an isocaloric amount of starch in the diet of the lactating cow then milk yields may be lowered. There is also some evidence that diets having 50–70 g/kg DM of ether extract produce more milk than those containing less than 40 g/kg. Most forages and food grains have low contents of lipids, of the order of 15–40 g/kg, and the fat content of the diet is usually boosted by that of supplementary compound foods. The provision of high levels of dietary fat is particularly important with high-yielding cows, for which the constraints of intake make it difficult to provide adequate energy. In addition, increased fat content at the expense of the starch fraction can help to correct the low-fat syndrome found with diets low in fibre and high in starch. This is the result of increased incorporation of long chain fatty acids of the dietary fat into the milk fat whereas *de novo* synthesis of short chain acids is decreased. Protein concentration but not total protein secretion in the milk may be reduced. Unfortunately, added fat tends to impair fermentation and digestion of plant cell wall constituents in the rumen and may depress intake. The effect is greatest with roughages of high quality and the effect in the case of low-quality materials may be negligible. In this respect unsaturated oils are much less desirable than the more saturated fats. Thus cod-liver oil given at about 200 g/day can reduce fat content by as much as 25 per cent and herring oil has a similar effect. It is generally considered that not more than 0.5 kg of fat should be added to the daily ration of a lactating cow. Higher levels may be used if the fat is protected so that its deleterious effects on the rumen are avoided and normal hydrogenation, solubilisation and absorption take place. Prills (pellets) of calcium salts of the fatty acids have proved to be an effective form of protected fat. Fatty acids in this form are less available and therefore less toxic to the microorganisms of the rumen than are the free fatty acids produced by fat hydrolysis. In addition, the release of free fatty acids into the rumen following upon the ingestion of fat results in the fixation of calcium, thus denied to the rumen microbes for which it is essential. This does not occur with the calcium soaps and it has been claimed that up to 15 per cent of the metabolisable energy requirement may be supplied by dietary fat when such products are used.

The nature of the dietary fat can have a profound effect on the composition of milk fat. Diets rich in acids up to palmitic generally increase the proportions of these acids in milk fat at the expense of the C_{18} acids. Dietary fats rich in saturated and unsaturated acids result in increased yields of oleic and stearic acids with associated decreases in shorter chain acids, particularly palmitic. The secretion of linoleic and linolenic acids is not affected owing to the extensive hydrogenation which occurs in the rumen. There are some indications that the polyunsaturated C_{18} acids can affect the acetate : propionate ratio in the rumen. Soya bean oil, for example, is rich in linoleic acid and can markedly reduce the ratio of acetate to propionate. Diets which reduce milk fat content increase the activity in adipose tissue of enzymes involved in fatty acid and triacylglycerol synthesis. At the same time a less pronounced reduction in the activity of such enzymes occurs in mammary tissue. As a result the amount of acetate available for milk fat synthesis is reduced owing to its use in fatty acid synthesis in the adipose tissue. Stimulation of triacylglycerol synthesis would have a similar effect on the level of plasma free fatty acids. Liver synthesis of low-density lipoproteins and the supply of these to the mammary gland for fat synthesis would thus be reduced. Finally, the concurrent changes in the mammary gland would cause the reduced supply of precursors to be used less efficiently for milk fat synthesis. Intravenous infusions of glucose reduce plasma glyceride concentrations, lending support to the view that the effect of low-fat producing diets is due to the increased glucogenic nature of the acid mixture absorbed from the rumen on such diets.

Changes in dietary carbohydrate which reduce milk fat content tend to increase protein content if dietary protein supply is adequate. The effect may require two–three weeks to manifest itself and be of the order of 8 g protein/kg milk. It is probable that the increased propionic acid production on such diets has a sparing effect on certain glucogenic amino acids such as glutamate, and more of these are then available to the mammary gland for protein synthesis. The increased intake of energy *per se*, which usually occurs on such diets, would have the same effect.

Reductions in dietary protein level may decrease milk yield and, almost invariably, non-protein nitrogen content. Milk protein is little affected until intake of protein falls below 60 per cent of requirement. This is probably due to an insufficiency of essential amino acids, primarily methionine, followed by threonine and tryptophan.

16.3   Nutrient requirements of the lactating dairy goat

In addition to the dairy cow, the goat is also used for the commercial production of milk for human consumption. As in the case of the dairy cow, nutrient requirements depend upon the amount of milk being produced and upon its composition.

Milk yield

Yield varies with breed (Table 16.13).

These yields are about 20 per cent less than those given in the last edition and until recently the only definitive values for the UK. A lactation normally lasts for about ten months, during which time up to 1350 kg of milk may be produced. In the short term, the major factor affecting yield will be stage of lactation, with the pattern of milk production being similar to that for the dairy cow (p. 418). Peak yield usually occurs at about the sixth week *post partum* and is maintained for the next four weeks before declining at the rate of about 2.5–3 per cent per week. Estimates of yield may be made using the following equation:

$$Y(kg/day) = A \; exp[B(1 + n'/2)n' + Cn'^2 - 1.01/n)]$$

n = days *post partum*
$n' = (n - 150)/100$
A = the yield at day 150 of lactation
B = persistency in the form of change in yield at day 150
C = curvature in the decline in yield in the period after the peak is achieved

Figure 16.7 shows a plot of the estimated yields for a lactation yield of about 750 kg, with $A = 3.47$, $B = -0.618$ and $C = -0.0707$.

Table 16.13 Total lactation milk yields of dairy goats in England and Wales (Adapted from AFRC 1994 *The Nutrition of Goats*, Technical Committee on Responses to Nutrients Report No. 10, Wallingford, UK, CAB International)

Breed	Lactation milk yield (kg)
Anglo-Nubian	681
Saanen	904
British Saanen	970
Toggenburg	672
British Toggenburg	1090
British Alpine	953
Golden Guernsey	820

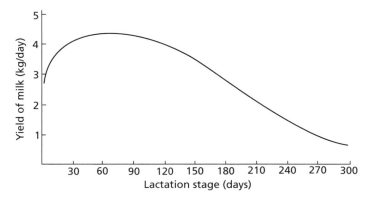

Fig. 16.7 Yield prediction for a goat producing about 750 kg milk/lactation.

Milk composition

The composition of milk is primarily affected by the breed of the goat. Representative figures for several breeds are given in Table 16.14. As with the dairy cow, breeds producing the larger yields tend to give milk of poorer quality.

Stage of lactation also affects the composition of the milk. Fat and solids-not-fat content fall to a minimum at about four months *post partum*, rise for the succeeding three months and then increase slowly until the close of the lactation.

Fat content has been shown to decrease with increasing parity and is lower for animals with high lactation yields.

The nutrient requirements of the lactating goat may be derived factorially from estimates of the requirements for maintenance, milk production and liveweight change.

Table 16.14 Composition (g/kg) of the milks of dairy goats in England and Wales

Breed	Fat	Protein	Lactose	Energy value (MJ/kg)
Anglo-Nubian	46.5	35.5	43.4	3.36
Saanen	35.1	28.8	44.8	2.79
British Saanen	37.6	29.2	42.8	2.86
Toggenburg	37.1	28.6	45.8	2.87
British Toggenburg	37.3	29.6	43.8	2.87
British Alpine	41.1	31.1	43.3	3.05
Golden Guernsey	41.9	31.7	43.3	3.09

Energy requirement

Maintenance

Estimates of the fasting metabolism for lactating goats show considerable variation, but a mean value of 315 kJ/kg $W^{0.75}$ per day, which closely resembles that of 319 kJ/kg $W^{0.75}$ per day for dairy cattle, is an acceptable estimate. This is the basic energy requirement. It should be increased by about 10 per cent for animals kept indoors, 20 per cent for grazing animals under lowland conditions and 30 per cent for those on the hill, to allow for increased activity. In very dry or mountainous situations these may have to be increased to as much as 100 per cent.

The efficiency of utilisation of metabolisable energy for maintenance (k_m) may be calculated as:

$$k_m = 0.35 q_m + 0.503$$

For a diet of $q_m = 0.6$, $k_m = 0.713$, and the basic requirement for metabolisable energy for maintenance for a 60 kg goat is then $0.315 \times 60^{0.75}/0.713 = 9.52$ MJ/day. For a goat on hill land it should be increased to 12.4 MJ/day.

Milk production

The net requirement of energy for milk production is the gross energy of the milk produced. This will depend upon the yield and the energy value (MJ/kg) of the milk.

In its 1997 publication (see Further reading) the Agricultural and Food Research Council proposed the use of the following equation for estimating the energy value of goat's milk:

$$EV_l \ (MJ/kg) = 0.0376 \times Fat \ (g/kg) + 0.0209 \times Protein \ (g/kg) + 0.948$$

When the protein content of the milk is not known, EV_l may be calculated as:

$$EV_l (MJ/kg) = 1.509 + 0.0406 \times Fat$$

When no data are available, a value of 3.25 MJ/kg or breed values based upon the figures in Table 16.14 may be adopted.

The efficiency of utilisation of metabolisable energy for milk production (k_l) may be calculated as:

$$k_l = 0.35 q_m + 0.42$$

The metabolisable energy requirement for the production of 1 kg of milk in the case of an Anglo-Nubian goat receiving a diet of q_m of 0.6 would then be $3.36/0.63 = 5.33$ MJ/day. The metabolisable energy requirement of a 60 kg goat of this breed producing 3 kg of milk at pasture would be:

$$(0.315 \times 1.2 \times 60^{0.75}/0.71) + (3 \times 3.36/0.63) = 27.5 \text{ MJ/day}$$

Lactating goats are almost always gaining or losing weight owing to the need for mobilisation of body tissue to make good the difference between energy input and output. In the situation where a goat is being fed to gain weight, extra energy has to be provided. AFRC (1997) suggests a value of 23.9 MJ/kg of liveweight change for the EV_g of lactating goats. This is equivalent to a dietary metabolisable energy contribution of 32.4 MJ ($23.9 \times 0.84/0.62$). The efficiency of utilisation of metabolisable energy for gain may be taken as 0.95 k_l. For 1 kg of liveweight gain, a dietary intake of metabolisable energy of 41 MJ ($23.9/(0.62 \times 0.95)$) would be required.

As with cattle a level of feeding correction should be used when calculating energy requirements.

Protein requirement

Maintenance

The net requirement for nitrogen for maintenance is that required to replace the endogenous urinary loss (EUN) plus part of the metabolic faecal nitrogen loss (MFN) plus loss of nitrogen in hair and scurf (D). The total is termed the basal endogenous nitrogen (BEN) and is of the order of 0.35 g/kg $W^{0.75}$/day, which translates to a metabolisable protein demand of 2.19 g/kg $W^{0.75}$. Metabolisable protein is assumed to be used for maintenance with an efficiency of 1.0.

Lactation

Estimates of the crude protein contents of the milks of modern dairy goats (Table 16.14) range from 28.6 to 35.5 g/kg. Taking an average value of 30.6 gives an 'acceptable' true protein content of 27.5 g/kg, assuming that 10 per cent of the milk nitrogen is non-protein nitrogen. When the information is available, a preferred figure should be used instead: for example, 32 g/kg for the Anglo-Nubian breed. The efficiency of utilisation of metabolisable protein for milk production is 0.68 and the metabolisable protein requirement for milk production is then $32/0.68 = 47$ g/kg.

When the lactating goat is losing weight, the mobilised tissue contributes nitrogen to the production of milk. Efficiency of mobilisation is taken to be 1.0 and the mobilised nitrogen is used with an efficiency of 0.68. The protein content of liveweight change (ΔW) in the lactating goat has been estimated as 138 g/kg. Liveweight loss thus contributes $138 \times 0.68 = 93.8$ g metabolisable protein per kilogram to the production of milk.

When the lactating goat is gaining weight there is a requirement in excess of that required for maintenance and lactation. Taking the protein content of liveweight gain to be 138 g/kg and the efficiency of utilisation of metabolisable protein for gain to be 0.59 gives an extra dietary requirement of 234 g/kg of metabolisable protein.

The metabolisable protein requirement, where Y is milk yield in kilograms, is then:

$$\text{MP (g/day)} = 2.19 \times W^{0.75} + 27.5Y/0.68 + 138 \times \Delta W/0.59$$

when ΔW is positive and

$$\text{MP (g/day)} = 2.19 \times W^{0.75} + 27.5Y/0.68 - 138 \times \Delta W \times 0.68$$

when ΔW is negative.

The daily requirement for degradable protein (ERDP) is given by (FME × 11) g, where FME (MJ/day) is the fermentable metabolisable energy intake. The contribution of dietary ERDP to satisfying metabolisable protein demand may be calculated, as for the cow, as $0.6375 \times \text{ERDP}$.

The daily requirement for digestible undegraded protein (DUP) is calculated as:

$$\text{DUP (g/day)} = \text{MP} - 0.6375 \times \text{ERDP}$$

Box 16.7

Calculation of the requirements of a 60 kg Anglo-Nubian goat producing 5 kg/day milk at pasture and losing 50 g W/day

M_m (MJ/day) $= (60^{0.75} \times 0.315 \times 1.2)/0.713$	$=$ 11.43
M_l (MJ/day) $= 5 \times 3.36/0.63$	$=$ 26.67
M_g (MJ/day) $= -0.05 \times 32.4$	$=$ -1.62
$M_m + M_p$ (MJ/day)	$=$ 36.47
Feeding level correction $= 1 + 0.018 \times M_p/M_m$	$=$ 1.0394
M_{mp} (MJ/day) $= 36.47 \times 1.0394$	$=$ 37.91
FME (MJ/day)	$=$ 32.68
ERDP (MJ/day) $= 32.68 \times 11$	$=$ 359.5
MP (g/day) $= (2.19 \times 60^{0.75}) + (35.5 \times 0.9 \times 5/0.68) - (0.05 \times 138)$	$=$ 275.8
Microbial protein supply (g/day) $= 359.5 \times 0.6375$	$=$ 229.2
DUP (g/day) $= 275.8 - 229.2$	$=$ 46.5
Ca (g/day) $= (2.37 + 5 \times 1.3)/0.51$	$=$ 17.4
P (g/day) $= (3.71 + 5 \times 0.9)/0.55$	$=$ 14.9
Mg (g/day) $= (50 \times 0.0035 + 5 \times 0.13)/0.17$	$=$ 4.9

Requirements for calcium, phosphorus and magnesium

Endogenous losses (net requirements for maintenance) may be calculated as:

$$\text{Calcium (g/day)} = 0.623 \times \text{DMI} + 0.228$$

$$\text{Phosphorus (g/day)} = 1.6(0.693 \times \text{DMI} - 0.06)$$

where DMI = metabolisable energy intake/$18.4q_m$ (NB: this is not a prediction of potential intake.)

$$\text{Magnesium (g/day)} = 0.0035W$$

Milk may be taken to contain 1.3 g calcium, 0.9 g phosphorus and 0.13 g magnesium per kilogram.

There is a dearth of information on the availability of dietary minerals for the goat and it is suggested that the values used for cattle, 0.51, 0.55 and 0.17 for calcium, phosphorus and magnesium, respectively, should be adopted.

An example of the calculation of the nutrient requirements of a goat with an assumed dry matter intake of 3.43 kg/day (37.91/18.4 × 0.6) is given in Box 16.7.

16.4 Nutrient requirements of the lactating ewe

The lactation of the ewe usually lasts from 12 to 20 weeks, although individuals show very considerable variations. Stage of lactation has a pronounced

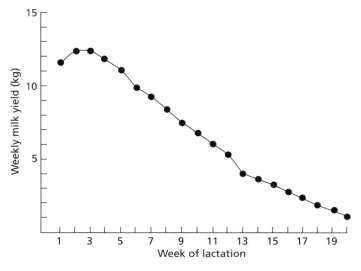

Fig. 16.8 Effect of stage of lactation on milk yield of the ewe. (From Wallace L R 1948 *Journal of Agricultural Science, Cambridge* **38**: 93.)

Table 16.15 Lactation (12 weeks) yields of different breeds of sheep

Breed	Yield (kg)	
	Twins	Singles
Romney Marsh	148	115
Cheviot	–	91
Border Leicester/Cheviot	211	124
Suffolk	145	94
Hampshire Down	79	75
Scottish Blackface	142	102
Finnish Landrace/Scottish Blackface	206	133

effect on milk yield, which is maximal at the second–third week and then falls steadily, as shown for Suffolk ewes in Fig. 16.8.

It has been calculated that about 38 per cent of the total yield is obtained in the first four weeks of lactation, 30 per cent in the succeeding four weeks, 21 per cent in the next four weeks and 11 per cent in the final four weeks. Comparison of the milk yields of different breeds is difficult, since the data have been obtained under widely differing climatic conditions, levels of feeding and sampling techniques. They indicate, however, that differences do exist (Table 16.15) and that within breed differences are frequently large.

Animals suckling more than one lamb produce more milk than those suckling single lambs. The higher yield is probably due to higher frequency of suckling and greater emptying of the udder, indicating that a single lamb is incapable of removing sufficient milk from the udder to allow the full milking potential of the ewe to be fulfilled.

Data on the composition of the milk of the ewe are relatively few. Such factors as sampling techniques, stage of lactation and milking intervals all affect composition and figures are not strictly comparable and show considerable variation. Thus, published figures for breed average fat and protein contents vary from 50 to 100 g/kg and from 40 to 70 g/kg, respectively. Breed differences in composition are illustrated by the figures given in Table 16 16.

The composition of ewe's milk is affected by stage of lactation as shown in Fig. 16.9. The changes are similar to those in the dairy cow if allowance is made for the different length of lactation.

Table 16.16 Effect of breed on the composition of ewe's milk (g/kg) (After Neidig R E and Iddings E J 1919 *Journal of Agricultural Research* **17**: 19)

Breed	Fat	Solids-not-fat
Hampshire	71	97.5
Cotswold	77	95.2
Shropshire	81	96.7
Southdown	75	103.6

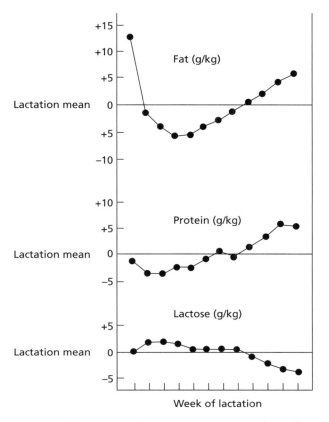

Fig. 16.9 Effect of stage of lactation on milk composition of Finnish Landrace × Blackface ewes. (Adapted from Peart J N *et al.* 1972 *Journal of Agricultural Science, Cambridge* **79**: 303.)

The nutrient requirements of the lactating ewe may be derived factorially from estimates of the requirements for maintenance, milk production and liveweight change.

Energy requirements

The net requirement for maintenance (E_m) of housed ewes may be calculated as:

$$E_m \ (MJ/day) = 0.226(W/1.08)^{0.75} + 0.0096W$$

where W is liveweight (kg).

The activity increment of $0.0096W$ will be greater for ewes kept outdoors. For grazing sheep under lowland conditions a value of 0.0109 would appear to be reasonable, and for hill sheep, 0.0196.

The energy value of ewe's milk is given by:

$$EV_l \ (MJ/kg) = 0.0328F + 0.0025D + 2.20$$

in which F = fat content (g/kg) and D = day of lactation. Alternatively, a value of 4.6 MJ/kg may be assumed when no information on composition is available.

Estimates of the energy value of mobilised body tissue in the lactating ewe have varied from 17 to 68 MJ/kg, being particularly high and variable in early lactation. The figure of 68 MJ/kg is not acceptable and can only be explained in terms of simultaneous loss of fat accompanied by storage of water. In the absence of definitive evidence, it is proposed that, by analogy with the lactating cow, a figure of 25 MJ/kg should be taken. Each kilogram of mobilised tissue contributes $25 \times 0.84 = 21.0$ MJ net energy as milk and each kilogram of liveweight gain adds an additional $25/0.95 = 26 \, 3$ MJ to the net lactation requirement of the animal for energy.

The relevant efficiency factors for calculating metabolisable energy requirements are:

$$k_m = 0.35q_m + 0.503$$
$$k_l = 0.35q_m + 0.42$$

Protein requirements

The requirement for metabolisable protein is made up of that for maintenance, calculated as 2.19 g/kg $W^{0.75}$, plus that for milk production calculated as $49/0.68 = 72$ g/kg milk, plus that for wool growth taken as 20 4 g/day, plus or minus an adjustment for liveweight change (ΔW), calculated as -119 g/kg ΔW when ΔW is negative, and $+140$ g/kg ΔW when ΔW is positive.

The requirement for effective degradable protein is given by ERDP (g/day) = FME (MJ/day) $\times 11$. The requirement for truly digested undegradable protein is:

$$DUP \ (g/day) = 2.19W^{0.75} + 49Y/0.68 + 20.4 - 119\Delta W - 0.6375ERDP$$

when the animal is losing weight, and

$$DUP \ (g/day) = 2.19W^{0.75} + 49Y/0.68 + 20.4 + 140\Delta W - 0.6375ERDP$$

when the animal is gaining weight, where Y is milk yield in kilograms per day and ΔW is liveweight change in kilograms per day.

Calcium, phosphorus and magnesium

As with the lactating dairy cow, the TCORN methods of calculating net requirements, along with our previously used figures for availability, have been used in calculating the values given in the tables of Appendix 2.

Table 16.17 Suggested milk yields for calculating nutrient allowances for lactating ewes

Type of ewe	Number of lambs	Milk yield			
		12 weeks (kg)	Day 1–28 (kg/day)	Day 29–56 (kg/day)	Day 57–85 (kg/day)
Hill	One	86	1.21	1.09	0.75
	Two	130	1.90	1.63	1.11
Lowland	One	140	2.00	1.80	1.20
	Two	190	2.90	2.31	1.56

The net daily maintenance requirement for calcium, phosphorus and magnesium may be calculated as:

$$Ca \ (g/day) = 0.623 \times DMI + 0.228$$

$$P \ (g/day) = 1.6(0.693 \times DMI) - 0.06$$

$$Mg \ (g/day) = 0.03W$$

The net requirements for milk production (g/kg) are 1.2 g calcium, 1.3 g phosphorus and 0.17 g magnesium. Absorbability values are 0.51, 0.55 and 0.17 for calcium, phosphorus and magnesium, respectively.

For the purposes of calculating nutrient allowances, the milk yields given in Table 16.17 may be adopted.

Box 16.8

Calculation of the nutrient requirements of a 75 kg lowland ewe in the fourth week of lactation, suckling two lambs, receiving a diet of $q_m = 0.6$ and losing 100 g W/day

$E_m \ (MJ/day) = 0.226 \times (75/1.08)^{0.75} + 0.0106 \times 75$	=	6.25
$k_m = 0.35 \times 0.6 + 0.503$	=	0.713
$M_m \ (MJ/day)$	=	8.77
$E_l \ (MJ/day) = 2.31 \times 4.6$	=	10.63
$k_l = 0.35 \times 0.6 + 0.42$	=	0.63
$M_l \ (MJ/day)$	=	16.87
$M_g \ (MJ/day) = -0.1 \times 21.0/0.63$	=	−3.33
$M_p (MJ/day) = 16.87 - 3.33$	=	13.54
Level of feeding correction $= 1 + 0.018 M_p/M_m$	=	1.0278
$M_{mp} \ (MJ/day) = 1.0278(M_m + M_p)$	=	22.93
FME (MJ/day)	=	20.3
ERDP (g/day) $= 20.3 \times 11$	=	223.3
MP (g/day) $= (2.19 \times 75^{0.75}) + (2.31 \times 72) + (-0.1 \times 119) + 20.4$	=	230.6
DUP (g/day) $= 230.6 - (223.3 \times 0.6375)$	=	88.2

An example of the calculation of the nutrient requirements of a lactating ewe is given in Box 16.8.

The allowances given in the tables of Appendix 2 do not include a safety margin.

Ewes which have been severely undernourished in pregnancy show a more rapid decline in milk production during the subsequent lactation than adequately nourished animals. This accords with independent observations of reduced metabolic capacity in ewes severely undernourished during pregnancy. Where restriction is less severe and ewes come to parturition in lean condition, they have been shown to milk as well when adequately nourished as do ewes with better condition scores at lambing. Ewes will not maintain high milk production at the expense of body reserves and even relatively small restrictions of intake depress milk production. Severe restriction of nutrient intake (to provide maintenance only) may reduce milk production by as much as 50 per cent in two–three days. If restriction is continued beyond the time when peak yield is normally achieved then recovery of yield may not be accomplished even if subsequent intake is raised.

16.5   Nutrient requirements of the lactating sow

In most breeding units, lactation lasts for three–six weeks and many litters are weaned at three–four weeks of age. Maximum yield of milk occurs at about four weeks and production falls gradually thereafter, as shown in Table 16.18.

Fat content rises to the third week and then falls to the end of lactation. Solids-not-fat content is at a minimum at the third week and then rises to the end of lactation owing mainly to a rise in protein content.

Milk yield also varies with breed, age and litter size. It increases with the number of piglets suckled, although yield per piglet decreases as shown in Table 16.19.

Modern hybrid sows of higher liveweight are generally considered to give higher yields than those shown in the tables. Various equations, similar to those for dairy cows, have been proposed for predicting the yields for such sows. It is suggested that the following provides acceptable estimates:

Table 16.18 Variation in yield and composition of sow's milk with stage of lactation (After Elsley F W H 1970 Nutrition and lactation in the sow. In: Falconer I R (ed.) *Lactation*, London, Butterworth, p 398)

	Week							
	1	2	3	4	5	6	7	8
Daily yield, kg	5.10	6.51	7.12	7.18	6.95	6.59	5.70	4.89
Fat, g/kg	82.6	83.2	88.4	85.8	83.3	75.2	73.6	73.1
SNF, g/kg	115.2	113.2	111.8	114.1	117.3	120.5	126.1	129.9
Protein, g/kg	57.6	54.0	53.1	55.0	59.2	62.3	68.3	73.4
Lactose, g/kg	49.9	51.5	50.8	50.8	49.0	48.6	47.5	45.6
Ash, g/kg	7.7	7.7	7.9	8.3	9.1	9.6	10.3	10.9

Table 16.19 Effect of litter size on milk yield in the sow (From Elsley F W H 1970 Nutrition and lactation in the sow. In: Falconer I R (ed.) *Lactation*, London, Butterworth, p. 396)

	Number of pigs								
	4	5	6	7	8	9	10	11	12
Daily milk yield, kg:									
per litter	4.0	4.8	5.2	5.8	6.6	7.0	7.6	8.2	8.6
per pig	1.0	1.0	0.9	0.9	0.9	0.8	0.8	0.7	0.7

$$Y \text{ (kg/day)} = a \times e^{-ct} \times u$$

where a is a scalar, t is day of lactation, u is $e^{-\exp}(G - B_1 t)$ and describes the degree of maturity of the mammary gland at parturition, and e^{-ct} describes the rate of decline of secretory capacity.

Estimates of yield based on $a = 18$ and 24, $c = 0.025$, $G = 0.5$ and $B_1 = 0.1$ are given in Table 16.20.

Table 16.20 Estimates of the milk yields of sows (kg/day)

	Day of lactation			
a	7	14	21	28
18	6.8	8.4	8.7	8.1
24	9.1	11.3	11.6	10.8

Energy requirements

The net requirement of the lactating sow for dietary energy is the sum of that expended in maintenance plus the gross energy of the milk less the contribution made by mobilised body tissue. The maintenance requirement for metabolisable energy is 0.439 MJ/kg $W^{0.75}$ per day. The gross energy content of the milk of the sow may be assumed to be 5.2 MJ/kg. Efficiency of utilisation of dietary metabolisable energy for milk production is 0.65 and the requirement for milk production is $(5.2/0.65) = 8.0$ MJ ME/kg. Body tissue that is mobilised to provide energy for milk production is assumed to contain 0.85 of fat and thus to have a gross energy of $39.4 \times 0.85 = 33.5$ MJ/kg. The efficiency of conversion of the energy of mobilised body tissue to milk energy is 0.85. Each kilogram of mobilised body tissue thus represents a contribution equivalent to $(33.5 \times 0.85/0.65) = 43.8$ MJ to the dietary metabolisable energy supply. The energy requirements of pigs are usually stated in terms of digestible energy (DE) and the ratio of ME : DE can be assumed to be 0.96. Thus a 200 kg sow producing 10 kg of milk and losing 0.2 kg/day would have a metabolisable energy requirement of:

$$(200^{0.75} \times 0.439) + (10 \times 8.0) - (0.2 \times 43.8) = 94.6 \text{ MJ/day}$$

and a digestible energy requirement of 94.6/0.96 = 98.5 MJ.

The validity of the results of such factorial calculations depends upon the validity of the assumptions upon which the calculations are based. The critical assumptions in this case are that the yield of milk is correct and that the maintenance requirement of the lactating sow is equal to that of a non-lactating sow. There is evidence that the maintenance requirement is considerably greater and it has been suggested that, instead of treating maintenance separately, a gross efficiency of energy conversion into milk of about 0.45 should be used in calculating energy requirements. This would give an estimate of:

$$[(10 \times 5.2) - (0.2 \times 33.5 \times 0.85)/0.45] = 103 \text{ MJ DE/day}$$

A similar approach is implicit in the derivation of equations for the regression of milk yield upon energy intake, such as that given below:

$$\text{Milk yield (MJ/day)} = 0.0015 + 0.47\text{FI}$$

where FI is intake of digestible energy in MJ.

The levels of intake indicated by these factorial calculations are higher than those that have been used in the recent past. The latter arose from the belief that it was desirable to feed for very high weight gains during pregnancy and to rely heavily on the mobilisation of body reserves to supply the energy requirement during lactation. High intakes in pregnancy can improve yield in the subsequent lactation but appetite in lactation is reduced. It has been claimed that the reduction is of the order of 0.5 kg for each increase of 0.5 kg in pregnancy feeding. As a result weight loss in lactation may be excessive, leading to prolonged intervals between weaning and conception. The maximum tolerable weight loss in lactation is of the order of 10–12 kg. It is now considered that the greatest efficiency of energy utilisation is achieved by giving adequate amounts of energy during lactation after restricted intake during pregnancy. A true gain in sow bodyweight of 12–15 kg over a complete reproductive cycle (about 25–27 kg during pregnancy) appears to give optimal reproductive performance and provides body reserves for lactation. When levels of food intake during lactation are designed to meet requirement, sows lose less weight and produce more milk than animals on a lower nutritional plane. On the other hand, piglets reared by sows receiving the higher intakes have failed to show significant advantage. This is partly because the lower-yielding sows produced richer milk and partly because the increased yields were obtained after the third week, when creep feed was being eaten and the piglets on the low-yielding sows ate more of this feed than the others. Consumption of creep feed is variable and cannot always be relied upon to make good the deficiencies in the milk yield of the sow; for safety it would be preferable to feed the piglets via the milk rather than by the more efficient direct creep feeding. There is evidence that low levels of energy during lactation have a cumulative effect and that considerable reduction of milk yield and depletion of subcutaneous fat may occur over three lactations. Various techniques including frequent feeding, wet feeding, pelleting, high-energy diets and avoidance of over-fatness at farrowing are used to ensure satisfactory intakes in lactation. It is important to appreciate the constraints of intake when diets are being formulated. Intakes may be as low as 4 kg/day at high temperatures but may reach 7–8 kg in cool conditions. Under normal circumstances intakes are rarely higher than 6.5 kg and are more commonly about 5 kg.

Protein requirements

The lactating sow is a highly efficient converter of protein into milk. Most of the available evidence indicates an apparent digestibility of protein in excess of 0.80 and an overall efficiency of protein utilisation for body gain and milk production of 0.7. Estimates of the gross efficiency of conversion of dietary protein to milk protein vary from 0.30 to 0.45. The net protein requirement for a lactating animal is made up of that for maintenance plus that secreted in the milk. Estimates of protein requirements for a lactation may be based on a figure for gross efficiency of protein utilisation, usually 0.40. A sow producing 10 kg of milk/day would secrete $(10 \times 57) = 570$ g of protein per day in her milk; the crude protein required in the daily ration would then be $(570/0.40) = 1425$ g.

A meal given at 7.5 kg per day would thus need to have a crude protein content of 190 g/kg. There is experimental evidence that raising the crude protein content of the sow's diet from 140 to 160 or 190 g/kg does not increase milk yield.

Current standards in the UK are derived factorially. Maintenance requirement is taken to be 0.45 g/kg W and milk to contain 57 g/kg of protein. Coupled with an efficiency of utilisation of digested protein of 0.7, and a digestibility of protein of 0.8, this gives a crude protein requirement (g/day) of:

$$(0.45W + 57Y)/(0.7 \times 0.8)$$

in which Y is milk yield (kg/day).

For the 200 kg sow (see above) the requirement for crude protein is

$$(200 \times 0.45 + 10 \times 57)/(0.70 \times 0.8) = 1178 \text{ g/day}$$

Dietary protein is the sole source of essential amino acids for the pig. The essential amino acid composition of sow's milk protein is as shown in Table 16.21, along with requirements based on an assumed digestibility of 0.8 and an assumed biological value of 0.70.

Evidence is accumulating that the high protein levels usually recommended for the diets of lactating sows may be required only because of inadequate

Table 16.21 Amino acid requirements of sows for milk production

Amino acid	g/kg of milk	Requirement (g/kg of milk)
Histidine	1.57	2.80
Isoleucine	2.34	4.18
Leucine	4.88	8.71
Lysine	4.32	7.71
Methionine	0.99	1.77
Methionine + cystine	1.88	3.36
Phenylalanine	2.26	4.04
Phenylalanine + tyrosine	4.82	8.61
Threonine	2.37	4.23
Valine	3.05	5.45
Tryptophan	0.75	1.34

protein quality. It has been demonstrated that the biological value of barley can be raised from about 0.56 to 0.72 by supplementation with 20 g of lysine per sow per day. This is to all intents and purposes the same as the figure of 0.73 for barley/fishmeal diets. In the light of this information it has been suggested that diets containing as little as 120 g crude protein per kilogram may be adequate for milk production, as long as lysine levels are adequate and intake is not less than 5 kg/day for a sow suckling eight piglets. Such a low level of dietary crude protein would indicate a gross efficiency of conversion of dietary to milk protein of 0.63, which is highly optimistic.

Mineral requirements

There is no evidence to suggest that minerals other than calcium and phosphorus have to be provided in the diet of lactating sows at levels above those necessary for normal reproduction. Balance experiments indicate that the gross efficiency of utilisation of calcium and phosphorus for lactation is about 0.47 and 0.5, respectively. Table 16.1 shows that the milk of the sow contains 2.5 g/kg of calcium and 1.7 g/kg of phosphorus. A sow producing 10 kg of milk per day would thus be secreting 25 g of calcium and 17 g of phosphorus. Obligatory losses (g/100 kg W) may be assumed to be 3.2 g calcium and 2 g phosphorus. The 200 kg sow (see above) would have the following requirements:

$$(3.2 \times 2 + 10 \times 2.5)/0.47 = 66.8 \text{ g Ca/day}$$

$$(2 \times 2 + 10 \times 1.7)/0.5 = 42.0 \text{ g P/day}$$

A meal given at 7.5 kg/day would need to contain 8.9 g Ca/kg and 5.6 g P/kg.

Vitamin requirements

Little information is available concerning the vitamin requirements for lactation in the sow. Those given in the tables of Appendix 2 are the same as for the pregnant sow. The assumption is made that levels which allow normal reproduction and maintenance are adequate for lactation.

16.6 Nutrient requirements of the lactating mare

The nutrient requirements of the mare for milk production depend upon the amount of milk being produced and upon its composition.

Milk yield

The yield of milk varies with breed, age, stage of lactation, with nutrient intake in the late stages of pregnancy, with current nutrient intake and with water availability. Some typical milk yields for mares of different liveweights are given in Table 16.22.

Yield for brood mares (kg/day) may be predicted as 3 per cent of liveweight during the first three months of lactation and 2 per cent of liveweight for months 4–6. For ponies the corresponding percentage values are 4 and 3.

Milk composition

Fat content rises to a maximum at about the twelfth day of lactation and then remains at about the same level for the rest of the lactation. Protein content declines sharply to day 12 of the lactation and then declines gradually to the end of lactation. Lactose shows a very slight rise as the lactation proceeds. The gross energy of the milk falls sharply in early lactation but from about day 12 shows little change from a value of about 2 MJ/kg.

Table 16.22 Typical milk yields (kg/day) of mares of various bodyweights during weeks 1–25 of lactation (Adapted from Frape D 1998 *Equine Nutrition and Feeding*, 2nd edn, Oxford, Blackwell Science)

Weeks	1–2	4–5	5–12	20–25
Quarter horse (500 kg)	10	14	10	–
TB, standardbred (494 kg)	12–16	14–16	18	–
Dutch saddlebred horse (600 kg)	14	16	19	11
French draft (726 kg)	20	25	27	–

Energy requirement

Energy standards may be derived factorially. This involves an estimate of the gross energy value (EV_1 (MJ/kg)) of the milk which, along with the yield (Y (kg)), may be used to estimate the net energy requirement for milk production. The US National Research Council (NRC, 1989: see Further reading) assumes the gross energy content of mare's milk to be 1.987 MJ/kg, which is the net requirement for milk production.

Digestible energy (DE) is assumed to be converted to milk energy with an efficiency of 0.6 and the requirement for milk production (DE_l) is then 1.987Y/0.6 MJ.

In the French net energy system for horses, the gross energy of milk is assumed to be 2.3 MJ/kg and the net requirement for milk production is 2.3Y MJ.

In addition to the requirement for the production of milk, the diet of the lactating mare must provide the energy for maintenance. NRC (1989) proposes that for mares of less than 500 kg liveweight, the DE required for maintenance (DE_m) may be calculated as:

$$DE_m \ (MJ/day) = 4.185(1.4 + 0.03W)$$

which includes an allowance for normal activity.

Mares of greater liveweight are considered to be less active and DE_m for these animals is calculated as:

$$DE_m \ (MJ/day) = 4.185(1.82 + 0.0383W + 0.000015W^2)$$

In the French net energy system, the requirement for maintenance is calculated as 0.351 MJ/kg $W^{0.75}$ with factors used to allow for activity which depend upon the type of horse and whether the animal is at work. Thus, for a riding horse at work the factor would be 1.05 at rest and 1.10 at work; for a draft animal at work the factors would be 1.0 and 1.05.

According to the NRC system, the energy requirement of a mare of 500 kg liveweight producing 15 kg of milk per day would be:

$$DE \ (MJ/day) = 4.185(1.4 + 0.03W) + (15 \times 1.987/0.6) = 68.63 + 49.68$$
$$= 118.3$$

The energy requirement of the same animal according to the French system would be:

$$NE \ (MJ/day) = (0.351 \times 500^{0.75}) + (15 \times 2.3) = 37.11 + 34.50 = 71.6$$

The French system provides tables of net energy values of some 149 foods for maintenance which are considered applicable to lactation as well. The values are stated in terms of UFC (unité fourragère cheval); one UFC is equal to the net energy of standard barley, equal to 9.42 MJ/kg.

Protein requirements

Milk production

The net protein requirement for milk production is the protein content of the milk. Estimates vary and stage of·lactation has a significant effect. For this reason NRC (1989) allocates figures of 21 and 18 g protein per kg for milk in early and late lactation, respectively. According to their own figures it would have been more logical to use 27 g/kg in the first month of lactation, 22 g/kg in the second month and 18 g/kg thereafter. Efficiency of conversion of digestible crude protein (DCP) is assumed to be 0.65, giving protein requirements of 32 and 28 g of digestible crude protein/kg in early and late lactation.

In the French system, the net requirement is taken to be 24 g/kg for the first three months of lactation and 21 thereafter. Protein values of foods are stated in terms of MADC (matières azotés digestibles cheval). The MADC is the protein truly digested in the small and large intestine and is calculated as $k \times DCP$, where $k = 1.0$, 0.9, 0.85, 0.80 and 0.70 for concentrate foods, green forages, grass hay, barley straw and grass silage, respectively.

Maintenance

NRC uses a value of 2.8 g DCP/kg $W^{0.75}$ per day for the protein requirement for maintenance. The French take this value to represent the net protein requirement. The DCP requirements would then be 2.8, 3.1, 3.3, 3.5 and 4.0 g DCP/kg $W^{0.75}$ for the five types of food above.

The amino acid balance in the protein is not considered to be significant by either NRC or INRA but diets are assumed to contain 3.5 g lysine/kg.

Mineral requirements

Maintenance

Endogenous losses of the major minerals, which represent the net requirements for maintenance, are:

Calcium = 20 mg/kg W/day
Phosphorus = 10 mg/kg W/day
Magnesium = 6 mg/kg W/day

Milk production

The mineral composition of mare's milk varies with stage of lactation and for the purpose of estimating dietary requirements the following values are acceptable:

	Months 1–3 of lactation	Months 4–6 of lactation
Calcium (g/kg)	1.2	0.8
Phosphorus (g/kg)	0.75	0.50
Magnesium (g/kg)	0.09	0.045

Efficiency of absorption is 0.5 for calcium, 0.45 for phosphorus and 0.40 for magnesium.

Vitamin requirements

Only limited data are available on the vitamin requirements of horses and definitive statements are not possible at this time. Table 16.23 gives some tentative minimum levels considered to be sufficient to allow normal functioning of the animal, and maximum levels designed to avoid toxicity.

The calculation by the NRC system of the requirements for a 500 kg mare producing 15 kg of milk is illustrated in Box 16.9.

Table 16.23 Suggested minimum and maximum allowances of some vitamins for lactating mares (After National Academy of Sciences/National Research Council 1989 *Nutrient Requirements of Horses,* 5th rev. edn, Washington, NRC)

	Minimum	Maximum
Vitamin A, iu/kg W/day	30–60	300
Vitamin D, iu/kg W/day	10	44
Vitamin E, iu/kg W/day	2	20
Vitamin C, mg/kg W/day	Unknown	20

Box 16.9

Calculation of the mutrient requirements of a 500 kg mare producing 15 kg of milk in early lactation

DE_m (MJ/day) = 4.184(1.4 + 0.03 × 500)	=	68.6
DE_l (MJ/day) = 15 × 1.9876	=	49.7
DE_{ml} (MJ/day) = 68.6 + 49.7	=	118.3
DCP_m (g/day) = 2.8 × $500^{0.75}$	=	296
DCP_l (g/day) = 15 × 32	=	480
DCP_{ml} (g/day) = 296 + 480	=	776
Calcium (g/day) = (0.02 × 500 + 1.2 × 15)/0.5	=	56
Phosphorus (g/day) = (0.01 × 500 + 0.75 × 15)/0.45	=	36
Magnesium (g/day) = (0.006 × 500 + 0.09 × 15)/0.40	=	11

Summary

1. Lactation is a major physiological and biochemical undertaking.

2. The net requirements for milk production are dependent upon the amounts of nutrients secreted in the milk, and therefore upon milk yield and composition. In addition to a production requirement, a lactating animal may have an increased maintenance requirement.

3. If the efficiency with which a dietary nutrient is used to provide the net requirements is known, then the dietary requirement may be calculated as the net requirement divided by the efficiency of dietary nutrient utilisation.

4. Qualitatively the milks of farm animals are almost identical in composition, consisting of water, fat, protein, lactose and mineral matter (dominated by calcium and phosphorus).

5. The constituents of milk are either absorbed directly from the blood or are synthesised in the mammary gland from raw materials absorbed from the blood. In either case they originate in the food of the animal.

6. The yield and composition of milk vary with species, breed, strain within the breed, age and stage of lactation.

7. Values for the efficiency of utilisation of metabolisable energy for maintenance and for milk production are related to the energy concentration of the diet and are very similar.

8. Concentrate foods added to the diet rarely bring about the expected response in milk yield, because they cause a concomitant decrease in roughage intake, which is greater for high-than for low-quality roughages. Additionally, the extra concentrate increases the proportion of propionate in the rumen volatile fatty acids and the response is negatively curvilinear in the case of milk yield and positively so for liveweight gain.

9. High energy intakes must include a certain level of roughage in the diet if an acceptable rumen fermentation is to be maintained and problems of acidosis, reduced intake and low-fat milk are to be avoided.

10. Reduction of food intake has a profound effect on both the yield and composition of milk. In extreme cases yield may be reduced to very low levels in a matter of days. Limitation of the energy of the diet leads to a reduction in protein content, probably because of diversion of amino acids to gluconeogenesis.

11. The production of low-fat milk is usually associated with a lack of coarse roughage in the diet and is accentuated by high levels of readily available energy foods. It is generally accompanied by a high proportion of propionate in the rumen volatile fatty acids.

12. Maintenance energy requirements of lactating ewes and goats are related to metabolic body-weight, with an additional allowance for activity, which is least for animals kept indoors and greatest for those on the hill in harsh climatic conditions. Their yield of milk will depend mainly upon breed, stage of lactation and litter size. Energy and protein content are chiefly a reflection of stage of lactation and breed.

13. Protein requirements for the ruminant animals are stated in terms of effective rumen-degradable protein and metabolisable protein.

14. The fat content of sow's milk rises to about the third week of lactation and falls thereafter. Protein falls in early lactation and then rises towards the end.

15. Milk yield in the sow varies with age, litter size and breed, with modern hybrid animals being especially productive. Energy requirements may be based on the conversion of dietary energy to milk energy with an efficiency of about 0.45.

16. Sows rely heavily upon the mobilisation of body reserves to satisfy the demands of milk production. Current thinking favours high dietary energy intakes at this time to minimise weight loss. Energy inputs during pregnancy are restricted to avoid loss of appetite in the subsequent lactation.

17. Dietary protein is the sole source of essential amino acids for the sow. Diets are usually formulated to ileal digestible lysine with the proportions of the other acids to lysine being kept the same as those in the ideal protein.

18. The milk yield of the mare is related to bodyweight and stage of lactation.

19. The gross energy of mare's milk falls to about day 12 of lactation and shows little change from about 2 MJ/kg thereafter. Energy requirements are stated as digestible energy (US NRC) or as net energy. In the French net energy system, the unit used is the UFC (unité fouragère cheval), equal to 1 kg of standard barley (9.42 MJ/kg).

20. The mare's protein requirement for maintenance is related to metabolic liveweight and breed. Production requirement depends primarily on stage of lactation.

21. Dietary protein requirements are stated in terms of digestible crude protein (NRC) or as MADC (matières azotés digestibles cheval), the protein truly digested in the intestine.

Further reading

Agnew R E and Yan T 2000 Impact of recent research on energy feeding systems for dairy cattle. *Livestock Production Science* **66**: 197–215.

Agricultural and Food Research Council 1990 Technical Committee on Responses to Nutrients, Report No. 4. *Nutrient Requirements of Sows and Boars*. Farnham Royal, UK, Commonwealth Agricultural Bureaux. (See also *Nutrition Abstracts and Reviews, Series, B* **60**: 383–406.)

Agricultural and Food Research Council 1992 Technical Committee on Responses to Nutrients, Report No. 5. *Nutrient Requirements of Ruminant Animals: Energy*, Farnham Royal, UK, Commonwealth Agricultural Bureaux. (See also *Nutrition Abstracts and Reviews, Series B* **60**: 729–804.)

Agricultural and Food Research Council 1992 Technical Committee on Responses to Nutrients, Report No. 9. *Nutrient Requirements of Ruminant Animals: Protein*, Farnham Royal, UK, Commonwealth Agricultural Bureaux. (See also *Nutrition Abstracts and Reviews, Series B* **62**: 787–835.)

Agricultural and Food Research Council 1993 *Energy and Protein Requirements of Ruminants*, AFRC Technical Committee on Responses to Nutrients, Wallingford, UK, CAB International.

Agricultural and Food Research Council 1998 Technical Committee on Responses to Nutrients, Report No. 10. *The Nutrition of Goats*, Wallingford, UK, CAB International.

Agricultural Research Council 1980 *The Nutrient Requirements of Ruminant Livestock*, Farnham Royal, UK, Commonwealth Agricultural Bureaux.

Agricultural Research Council 1981 *The Nutrient Requirements of Pigs*, Farnham Royal, UK, Commonwealth Agricultural Bureaux.

Agricultural Research Council 1984 *The Nutrient Requirements of Ruminant Livestock, Supplement No. 1*, Farnham Royal, UK, Commonwealth Agricultural Bureaux.

Emmans G C and Fisher C in Fisher C and Boorman K N (eds) 1986 *Nutrient Requirements of Poultry and Nutritional Research*, London, Butterworth, pp 9–39.

Falconer I R (ed.) 1998 *Lactation*, London, Butterworth.

Frape D 1998 *Equine Nutrition and Feeding*, 2nd edn, Oxford, Blackwell Science.

Morand-Fehr P (ed.) 1991 *Goat Nutrition*, Wageningen, Pudoc.

Morant S V and Gnanasakthy A 1989 A new approach to the mathematical formulation of lactation curves. *Animal Production* **49**: 151–62.

National Research Council 1989 National Academy of Sciences/National Research Council. *Nutrient Requirements of Horses*, 5th rev. edn, Washington, NRC.

Rook J A F and Thomas P C (eds) 1983 *Nutritional Physiology of Farm Animals*, London, Longman.

Theodorou M K and France J (eds) 2000 *Feeding Systems and Feed Evaluation Models*, Wallingford, UK, CAB International.

Wood P D P 1976 Algebraic models of lactation curves for milk, fat and protein production with estimates of seasonal variation. *Animal Production* **22**: 35–40.

17 Voluntary intake of food

17.1   Food intake in monogastric animals

17.2   Food intake in ruminants

17.3   Prediction of food intake

In previous chapters attention has been concentrated on the energy and nutrient requirements of farm animals for maintenance and various productive processes. An additional important factor that must be considered is the quantity of food that an animal can consume in a given period of time. The more food an animal consumes each day, the greater will be the opportunity for increasing its daily production. An increase in production that is obtained by higher food intakes is usually associated with an increase in overall efficiency of the production process, since maintenance costs are decreased proportionately as productivity rises. There are, however, certain exceptions to the generalisation; for example, with some breeds of bacon pigs, excessive intakes of food lead to very fat carcasses, which are unacceptable to the consumer and therefore economically undesirable.

Feeding is a complex activity which includes such actions as the search for food, recognition of food and movement towards it, sensory appraisal of food, the initiation of eating and ingestion. In the alimentary tract the food is digested and the nutrients are then absorbed and metabolised. All these movements and processes can influence food intake on a short-term basis. In addition it is necessary to consider why, in most mature animals, bodyweight is maintained more or less constant over long periods of time, even if food is available *ad libitum*. Thus the concepts of short- and long-term control of food intake must be considered, the former being concerned with the initiation and cessation of individual meals and the latter with the maintenance of a long-term energy balance. Although many of these control systems are thought to be similar in all classes of farm animals, there are important differences between species that depend mainly on the structure and function of their digestive tracts.

Control mechanisms for the food intake of farm animals can be envisaged as operating at three levels. At the metabolic level, concentrations of nutrients, metabolites or hormones may stimulate the nervous system to cause the animal to start or stop feeding. At the level of the digestive system, the quantities of digesta may determine whether or not the animal ingests more food. Finally, external influences such as climatic variables or the ease with which food can be ingested will also influence food intake. In monogastric animals fed on concentrated foods and protected from adverse environmental features, control is primarily at the metabolic level. As diets become more fibrous, control shifts to the digestive system, and in ruminants that are not grazing this is usually the operative level. For grazing ruminants, environmental factors assume greater importance, and may be the determinants of intake. The first part of this chapter is concerned with monogastric animals and the second with ruminants (including grazing animals).

17.1 Food intake in monogastric animals

Control centres in the central nervous system

Feeding in mammals and birds is controlled by centres in the hypothalamus, situated beneath the cerebrum in the brain. It was originally proposed that there were two centres of activity. The first of these was the feeding centre (lateral hypothalamus), which caused the animal to eat food unless inhibited by the second, the satiety centre (ventromedial hypothalamus), which received signals from the body as a result of consumption of food. Quite simply it was considered that the animal would continue to eat unless the satiety centre received signals which inhibited the activity of the feeding centre. There is little doubt that this is an oversimplification and, although the hypothalamus does play an important role in intake regulation, it is now believed that other areas of the central nervous system are also involved.

Short-term regulation

Chemostatic theories

The release of nutrients from foods in the digestive tract, their absorption and passage via the portal system to the liver, and their presence in circulating blood all provide opportunities for nutrients to signal their presence to the satiety centre of the hypothalamus. According to chemostatic theory, the rise in concentration at these sites of some critical substance sends a signal to the brain to cause the animal to stop eating, and a fall in concentration causes the animal to start eating. Earlier studies concentrated on blood glucose concentration. Small doses of insulin, which lower the concentration of glucose in blood, cause an animal to feel hungry and hence to start eating. It is also known that blood glucose level rises after a meal and then falls slowly. It has been suggested that gluco-receptors, perhaps located in the hypothalamus itself, monitor either the absolute concentration of glucose in blood or the difference in concentration between arterial and venous blood. More recent studies have investigated receptors located closer to the point of origin of glucose (and other nutrients), in the gut and liver, which could evoke a faster response to the ingestion of food. Thus glucose infused into the intestine or the hepatic portal system causes a greater reduction in intake than glucose infused into the peripheral circulation.

Glucose in the duodenum has been shown to generate signals that are transmitted neurally and may cause the flow of digesta from the stomach to be retarded, hence reducing food intake. Another possible means of communication between the gut and the brain is provided by the peptide hormone cholecystokinin; this is released into the gut when digestive products such as amino acids and fatty acids reach the duodenum (see p. 169) and is known to act on the hypothalamus. In the liver, glucose and other nutrients that are oxidised in that organ are believed to send signals via the vagus nerve that eventually reach the hypothalamus.

The short-term control of intake in the fowl does not seem to be influenced to the same extent by blood glucose or other nutrients and it appears that signals are received directly from the crop, as is explained later.

Thermostatic theory

This theory proposes that animals eat to keep warm and stop eating to prevent hyperthermia. Heat is produced during the digestion and metabolism of food and it is considered that this heat increment could provide one of the signals used in the short-term regulation of food intake. It has been established that there are thermoreceptors, sensitive to changes in heat, present in the anterior hypothalamus and also peripherally in the skin, but experiments involving the heating or cooling of the hypothalamus have not induced consistent changes in food intake. However, at a practical level (as mentioned in Chapter 14) a high environmental temperature causes animals to reduce their heat production by eating less.

Long-term regulation

The long-term preservation of a relatively constant bodyweight, combined with an animal's desire to return to that bodyweight if it is altered by starvation or forced feeding, implies that some agent associated with energy storage acts as a signal for the long-term regulation of food intake. One suggestion is that this might be fat deposition. Studies with poultry tend to support this lipostatic theory of regulation. Cockerels forced to eat twice their normal intake of food deposited fat in the abdomen and liver. When force-feeding was stopped the birds fasted for 6–10 days, and when voluntary feeding recommenced food intake was low. It was evident from these studies that the birds lost weight when force-feeding ceased, and tissue fat concentration decreased to levels approaching normal after 23 days. In pigs, it would appear that any feedback mechanism from body fat to the controlling centres of feeding is not as sensitive as that in poultry and other animals. This insensitivity may have arisen through early genetic selection for rapid weight gain, when excessive carcass fat was not considered, as it is today, an undesirable characteristic. The natural propensity of the modern pig to fatten is usually counteracted in practice by restricted feeding and also by selection of pigs with a smaller appetite.

The role of fat depots in the regulation of food intake has recently been confirmed by the discovery of the hormone leptin. This polypeptide is secreted by white adipose tissue and acts on the hypothalamus, where it stimulates or suppresses the release of neuropeptides. These reduce food intake, and increase thermogenesis and physical activity. Its discovery arose from the study of mice that were hyperphagic and therefore obese, and which lacked the gene for leptin. When these mice were treated with leptin, their food intake was reduced to normal levels. Further experiments with species including farm animals have shown that injecting leptin into individuals with no genetic deficiency of the hormone has less dramatic effects than in the obese mice. Nevertheless, leptin is regarded as a key element in the 'feedback' control of food intake.

Sensory appraisal

The senses of sight, smell, touch and taste play an important role in stimulating appetite in man, and in influencing the quantity of food ingested at any one meal. It is a common assumption that animals share the same attitudes to food as ourselves but it is now generally accepted that the senses play a less important role in food intake in farm animals than they do in man.

The term 'palatability' is used to describe the degree of readiness with which a particular food is selected and eaten, but palatability and food intake are not synonymous. Palatability involves only the senses of smell, touch and taste. Most domestic animals exhibit sniffing behaviour, but the extent to which the sense of smell is necessary in order to locate and select foods is difficult to measure. A variety of aromatic substances, such as dill, aniseed, coriander and fenugreek, are frequently added to animal foods. The inference is that the odour from these spices makes the food more attractive and hence increases intake. Although transitory increases in food intake may occur, the effects of these additives have yet to be convincingly demonstrated to be long-lasting in terms of overall increased food intake.

Similarly, with the sense of taste, most animals show preferences for certain foods when presented with a choice. Typical is the preference of young pigs for sucrose solutions rather than water. The fowl is indifferent to solutions of the common sugars but finds xylose objectionable, and will not ingest salt solutions in concentrations beyond the capacity of its excretory system. Every species studied has shown considerable individual variability; for example, in a litter of pigs tested with saccharin solutions of different concentrations, some animals preferred high levels of the sweetener whereas others rejected them.

Physiological factors

The classical experiments of E F Adolph in 1947 demonstrated that, when the diets of rats were diluted with inert materials to produce a wide range of energy concentration, the animals were able to adjust the amount of food eaten so that their energy intake remained constant. This concept that 'animals eat for calories' has also been shown to apply to poultry and other non-ruminant farm animals. The manner in which chicks respond to diets of differing energy content is illustrated in Table 17.1, in which a normal diet containing 8.95 MJ productive energy (or about 13.2 MJ metabolisable energy) per kg was 'diluted' with increasing proportions of a low-energy constituent, oat hulls. The most diluted diet had an energy concentration which was only half that of the original and much lower than the range normally experienced by chicks. The chicks responded by eating up to 25 per cent more food, but even so energy intake declined by up to 29 per cent. If the energy content of a diet is increased by the addition of a concentrated source of energy such as fat, chicks respond in the opposite way. They eat less, but the reduction in intake may be insufficient to prevent a rise in energy intake. Where extensive diet dilution is carried out, by using low-digestibility materials, the ability to adjust the intake may be overcome because gastrointestinal capacity becomes a limiting factor. The crop appears to be concerned with intake in the fowl, since cropectomised birds eat less than normal when feeding time is restricted. Inflation or introduction of inert materials into the crop is known to cause a decrease in food intake. In mammals, distension and tension receptors have been identified in the oesophagus, stomach, duodenum and small intestine. Distension in these areas of the tract increases the activity in the vagus nerve and in the satiety centre of the hypothalamus.

Experiments similar to that shown in Table 17.1 have been made with pigs and show the same picture. In the mid-range of energy concentration pigs can compensate for variations in concentration, but with low-energy diets (e.g. containing 9 MJ DE/kg) compensation is incomplete and energy intake is reduced. Conversely, if the energy content of diets is increased to

Table 17.1 The effects of reducing the energy content of the diet on the food and energy intakes of chicks and on their growth (After Hill F W and Dansky L M 1954 *Poultry Science*, **33**: 112)

	Diet no.				
	1	2	3	4	5
Energy content of diet					
Productive energy, MJ/kg	8.95	7.91	6.82	5.73	4.64
Metabolisable energy, MJ/kg	13.18	11.59	10.21	8.91	7.45
Metabolisable energy (% of diet no. 1)	100	88	78	68	57
Performance of chicks to 11 weeks of age (% of result for diet no. 1)					
Total food intake	100	101	113	117	125
Total metabolisable energy intake	100	90	88	80	71
Liveweight gain	100	99	102	98	98
Fat content of carcass at 11 weeks of age (% of dry matter)					
(Male chicks only)	26.8	23.2	21.1	18.1	16.1

15 MJ DE/kg by the addition of fat, pigs fail to reduce their food intake proportionately, and energy intake increases. For pigs, as for poultry, it has been suggested that energy intake with low-energy diets is restricted by gut distension. It has also been suggested that high-energy diets tend to be 'overeaten' because they fail to provide sufficient bulk.

The general relationship between food intake and energy requirement suggests that, as with energy, intake should vary not directly with liveweight but with metabolic liveweight ($W^{0.75}$). This relationship is generally held to exist although it may vary with the physiological state of the animal. For example, lactation is usually associated with a marked increase in food intake, and in the rat at the peak of lactation, food intake may be nearly three times that of a non-lactating animal. In sows, the smaller the amount of food given during pregnancy, the greater is the amount consumed during lactation. If fed *ad libitum* on normal diets, pregnant sows overeat and become too fat, but their energy intake can be reduced by incorporating fibrous by-product foods, such as sugar beet pulp, in their diet.

It would seem reasonable to assume that intake increases with exercise, and studies with rats have shown that there is a linear relationship between food intake and the duration of exercise. However, the limited information on farm animals suggests that exercise reduces intake in the short term, possibly because of fatigue.

Nutritional deficiencies

Utilisation by the tissues of the absorbed products of digestion depends upon the efficient functioning of the many enzymes and coenzymes of the various metabolic pathways, and dietary deficiencies of indispensable amino acids, vitamins and minerals are likely to affect the intake of food. In poultry, severe deficiencies of amino acids reduce food intake whereas moderate deficiencies, insufficient to affect growth markedly, increase intake. When hens are given a diet containing high concentrations of calcium (30 g/kg), intake is about 25 per cent greater on egg-forming than on non-egg-forming days. This large variation does not occur when low-calcium diets are given with calcium being provided separately as calcareous grit. It would appear that laying hens 'eat for calcium'; the hormone controlling their calcium intake appears to be oestradiol. The effects of specific deficiencies of trace elements, especially cobalt, copper, zinc and manganese, and also vitamins such as retinol, cholecalciferol, thiamin and B_{12} on appetite have already been dealt with in Chapters 5 and 6. Mammals do not need to develop a calcium appetite, because they regulate their calcium supply by varying absorption.

Choice feeding

Animals have precise nutritional requirements, but under natural conditions are faced with a wide variety of foods to choose from, some of which are nutritionally inadequate.

The domestic rat and mouse are known to regulate their intakes of foods to satisfy, as far as the properties of the foods allow, their requirements for energy, protein and certain other nutrients. In studies with farm animals attention has concentrated on poultry, and it has been demonstrated that the domestic fowl has specific appetites for calcium (as discussed above) and also for phosphorus, zinc, thiamin and various amino acids.

It is possible to feed poultry on two separate dietary components, such as a whole cereal grain (e.g. whole wheat) and a balancer food that contains relatively high levels of amino acids, vitamins and minerals. This practice is known as choice feeding; it allows the birds to balance the energy : protein ratio of their overall diet. The balancer food is formulated so that equal proportions of the two foods are expected to be eaten, and because milling, mixing and pelleting costs are avoided for the whole cereal, total feeding costs can be reduced. The choice feeding system has been used successfully with large flocks of growing turkeys, but has not proved consistently successful with broiler chickens and adult laying hens. The theory that poultry have a control system that allows them to choose suitable amounts of different foods to satisfy their nutritional requirements is regarded as being too simplistic, and other factors such as the physical form of ingredients, composition of the food, trough position and previous experience are also likely to be involved.

As early as 1915, T M Evvard, in the USA, reported that pigs were able to choose a satisfactory diet when given access to several foods, and that as they grew they changed the proportions of the foods selected to maintain a suitable balance between protein and energy. The pigs required a period of familiarisation with the foods in order to be able to associate some of their properties with their nutritional effects. Recent investigations have confirmed the ability of pigs to select a diet of suitable protein content when given a choice between pairs of foods differing in protein content. Table 17.2 shows that when offered pairs of foods ranging in protein content from 125 to 267 g/kg, four out of six groups of pigs (groups 2–5) were able to select the proportions giving the appropriate protein concentration for their overall diet of about 200 g/kg. The first group was offered foods that were both

Table 17.2 Diet selection of young pigs offered a choice of foods differing in crude protein content (Adapted from the data of Kyriazakis I, Emmans G C and Whittemore C T 1990 *Animal Production* **51**: 189)

Group no.	Protein content (g/kg) of food		Food intake (g/day)	Proportions of foods 1 and 2	Protein content of overall diet (g/kg)
	1	2			
1	125	174	1106	29:71	160
2	125	213	1013	6:94	208
3	125	267	1055	44:56	204
4	174	213	1028	31:69	202
5	174	267	1076	66:34	205
6	213	267	1054	98:2	218

too low in protein to allow the correct selection, and the last group was given foods that both had protein concentrations higher than the desired level. Related experiments have shown that choice-fed pigs reduce the protein content of their selected diet as their protein requirements decline with increasing liveweight. Furthermore, pigs with the genetic capacity to grow more lean tissue (intact males of breeds selected for bacon production) select diets higher in protein content.

17.2 Food intake in ruminants

Although food intake can be controlled at the metabolic level in ruminants, the signals are likely to be different from those in monogastric animals. The amount of glucose absorbed from the digestive tract of the ruminant is relatively small and blood glucose levels show little relation to feeding behaviour. It would therefore seem unlikely that a glucostatic mechanism of intake control could apply to ruminants. A more likely chemostatic mechanism might involve the volatile fatty acids absorbed from the rumen. Intraruminal infusions of acetate and propionate have been shown to depress intake of concentrate diets by ruminants, and it is suggested that receptors for acetate and propionate occur on the luminal side of the reticulo-rumen. Infusions of these acids into the hepatic portal vein also reduce intake, apparently via signals sent from the liver to the hypothalamus. Butyrate seems to have less effect on intake than acetate or propionate, probably because butyrate is normally metabolised to acetoacetate and β-hydroxybutyrate by the rumen epithelium. With diets consisting mainly of roughages, infusions of volatile fatty acids have had less definite effects on intake. As mentioned above, in ruminants on such diets control of intake appears to be exercised at the level of the digestive system, and features of the food have an important influence on intake.

Food characteristics that determine intake

Ruminants are adapted to the utilisation of what may be termed 'bulky' foods, but may nevertheless have difficulty in processing such foods. Rumination and fermentation are relatively slow processes, and fibrous foods may have to spend a long time in the digestive tract for their digestible components to be extracted. If foods and their indigestible residues are detained in the digestive tract, the animal's throughput – hence its daily intake – will be reduced. In many feeding situations intake seems to be restricted by the capacity of the rumen, with stretch and tension receptors in the rumen wall signalling the degree of 'fill' to the brain, but what constitutes the maximum – and hence critical – 'fill' of the rumen is uncertain. The notion that voluminous, 'bulky' foods, such as hay and straw, will fill the rumen to a greater degree than concentrates has received some support, although after being chewed, the voluminous foods are not as 'bulky' as they are in the trough. Another concept of rumen fill is that animals eat to maintain a constant amount of dry matter in the rumen; again there is some experimental evidence for this, although there are some foods (e.g. some types of silage) that do not promote as great a fill of dry matter as other foods. There is also doubt about the contribution of water to rumen fill. Water-filled balloons in the rumen, which reduce its effective volume, will reduce food intake, but water added to food does not have this effect. However, there is some evidence that foods with a particularly high content (about 900 g/kg) of water bound within plant tissues promote a lesser dry matter intake than comparable foods of lower water content.

It has long been recognised that in ruminants there is a positive relationship between the digestibility of foods and their intake. A simple and early example of this relationship is shown in Fig. 17.1, which is derived from experiments in which sheep were fed to appetite on various roughages as the sole item of diet. Intake more than doubled as the energy digestibility of the food increased from 0.4 (oat hay; the lowest point on the graph) to 0.8 (artificially dried grass; the highest point).

A range in digestibility like that in Fig. 17.1 can also be created by taking a single roughage and supplementing it with increasing proportions of highly digestible concentrates, and in these circumstances there is also a positive relationship between intake and digestibility in ruminants (which contrasts with the *negative* relationship shown for a monogastric species in Table 17.1). The effect on intake of adding a concentrate supplement to a roughage depends on the digestibility of that roughage. If its digestibility is low (e.g. cereal straw with dry matter digestibility of 0.4) total intake will be increased more than if its digestibility is high (e.g. young grassland herbage, 0.8). To put this another way, concentrate added to roughage of

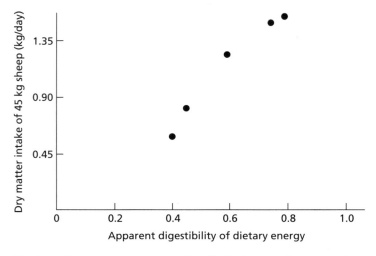

Fig. 17.1 Food consumption and digestibility in sheep fed on roughages. (After Blaxter K L, Wainman F W and Wilson R S 1961 *Animal Production* **3**: 51.)

low digestibility tends to be eaten in addition to the roughage, but when added to roughage of high digestibility it tends to replace the roughage.

The examination of relationships of the kind illustrated in Fig. 17.1 has shown that intake is actually more closely related to the *rate of digestion* of diets than to digestibility *per se*, although the two last measures are generally related to one another. In other words, foods that are digested rapidly, and are also of high digestibility, promote high intakes. The faster the rate of digestion, the more rapidly is the digestive tract emptied, and the more space is made available for the next meal. The primary chemical component of foods that determines their rate of digestion is neutral-detergent fibre (NDF), which is itself a measure of cell-wall content; thus there is a negative relationship between the NDF content of foods and the rate at which they are digested.

One consequence of the relationships just described is that foods that are equal in digestibility but differ in NDF (or cell-wall) content will promote different intakes. An example is provided by the two families of pasture plants: grasses and legumes. At equal digestibility, legumes contain less cell wall (hence more cell contents) and are consumed in quantities about 20 per cent greater than grasses. Another difference between legumes and grasses is that in the former lignification is restricted to the vascular bundles, whereas in grasses the lignin is more widely distributed and has a greater inhibitory effect on rate of digestion.

Although rate of digestion and intake are related to the concentration of cell walls in ruminant foods, the physical form of the cell walls also affects intake. The mechanical grinding of roughages partially destroys the structural organisation of cell walls, thereby accelerating their breakdown in the rumen and increasing food intake. This effect is illustrated in Table 17.3, which shows also that the increase in intake due to grinding and pelleting was achieved despite a *reduction* in digestibility. The fine particles produced in ground roughages pass rapidly out of the rumen, leaving room for more food but allowing some digestible material to escape undigested; this may be digested in the small intestine or – if cell wall – by fermentation in the caecum. We shall see later (Chapter 21) that chemical treatments of

Table 17.3 The effects on dry matter intake and digestibility of grinding and pelleting roughage-based diets for sheep and cattle (mean values for three diets) (Adapted from Greenhalgh J F D and Reid G W 1973 *Animal Production* **16**: 223)

Measures	Species	Form of roughage[a]		Percentage difference
		Long	Pelleted	
Intake	Sheep	56.8	82.4	+45
(g/kg $W^{0.75}$ per day)	Cattle	81.8	90.7	+11
Digestibility	Sheep	0.672	0.586	−13
	Cattle	0.699	0.569	−19

[a] The diets were dried grass, barn-dried hay and a mixture of 60 per cent hay and 40 per cent barley. For each diet the roughage was either ground and pelleted or left unprocessed ('long').

forages that disrupt the cell wall structure cause large increases in intake. Another example of the influence of cell-wall structure comes from the comparison of intakes of the leaves and stems of pasture plants. Although the two components may be equal in digestibility, the cell walls in leaves are more easily broken down, so animals given leaves eat about 40 per cent more dry matter per day than those offered stems.

The breakdown of food particles in the rumen and its effect on intake has been the subject of much recent research. In practice it is possible, as explained above, to overcome the resistance of plant cell walls by mechanical or chemical treatment, but the processes involved are expensive, may have undesirable side-effects (e.g. mechanical treatment reducing digestibility), and cannot be applied to grazed forages. In the longer term the aim is to identify new forage species or breed new varieties that are more rapidly broken down in the rumen.

Nutrient deficiencies that reduce the activities of rumen microorganisms are liable to reduce food intake. The most common is protein or nitrogen deficiency, which may be corrected by supplementation with rumen-degradable protein or with a simple source of nitrogen such as urea. However supplements of undegradable protein may also increase the intake of low-protein forages, either because the nitrogen of protein digested post-ruminally is partially recycled to the rumen, or because of the effect of protein on the animal's tissue metabolism. Other nutrients whose deficiencies are liable to restrict food intake in ruminants are sulphur, phosphorus, sodium and cobalt.

There are some foods that are eaten in lesser quantity than would be expected from their digestibility or cell-wall content. These include some types of silage, particularly those with a high content of fermentation acids or those that have been badly fermented and hence have a high ammonia content (see Chapter 20). The physical form of the food may also be involved, as fine chopping of the silage – or of the grass from which the silage is made – will increase intake, possibly because it prevents the formation in the rumen of a dense mat of fibrous material. Sheep are more likely than cattle to have sub-normal intakes from silages and respond more to having their foods (including silages) chopped or ground.

Foods with intakes less than expected may also be labelled as 'unpalatable'. As discussed earlier (p. 467) the concept of palatability is not easily defined, but there is some evidence, obtained with low-quality roughages, that if some food is put directly into the rumen (via a cannula), consumption by mouth is not decreased proportionately, and the animal therefore digests more than it would voluntarily consume. Generally, however, palatability is not thought to be an important factor determining intake, except where the food is protected against consumption (e.g. by spines) or contaminated in some way (e.g. by excreta).

Animal factors affecting intake in ruminants

If the capacity of the rumen is a critical factor in determining the food intake of ruminants, then circumstances that change the relationship between the size of the rumen and the size of the whole animal are likely to affect intake. As ruminants of a given species grow, their food intake follows approximately the proportionality to metabolic bodyweight ($W^{0.75}$) referred to earlier. However, cattle have a greater intake per unit of metabolic bodyweight than sheep; for example, a growing steer (300 kg) on a diet containing 11 MJ ME/kg DM will consume about 90 g DM per kg $W^{0.75}$ per day (6.3 kg DM per animal per day) whereas a growing lamb (40 kg) will consume only 60 g DM per kg $W^{0.75}$ per day (0.96 kg per animal). Intake seems to be related to fasting heat production, which is itself related to metabolic bodyweight (see Chapter 14) but is smaller per unit of metabolic bodyweight in sheep than in cattle. When animals become excessively fat, their intake tends to stabilise, or in other words, not to increase as body weight continues to increase. This may be due to abdominal fat deposits reducing the volume of the rumen, but may also be a metabolic effect (i.e. lipostatic limitation of intake determined by the secretion of leptin). Conversely, in very lean animals intake per unit of metabolic bodyweight tends to be high. This effect is seen in animals showing compensatory growth after a period of food restriction (p. 385); it is also to be seen in ruminants chronically short of food, as in some developing countries, where animals appear to be constructed of skin and bone enclosing a large rumen.

In pregnant animals, two opposing effects influence food intake. The increased need for nutrients for foetal development causes intake to rise. In the later stages of pregnancy the effective volume of the abdominal cavity is reduced as the foetus increases in size, and so is the space available for expansion of the rumen during feeding. As a result, intake will be depressed, especially if the diet consists predominantly of roughage.

The increased intake in ruminants with the onset of lactation is well known. This increase is mainly physiological in origin although there may also be a physical effect from the reduction in fat deposits in the abdominal cavity. There is a noticeable lag in the response of food intake to the increased energy demand of lactation. In early lactation, the dairy cow loses weight that is replaced at a later phase of lactation when milk yields are falling while intakes of DM remain high. These changes are illustrated in Fig. 17.2 for lactating and non-lactating identical twin, Jersey crossbred cows fed exclusively on fresh pasture herbage over a 36-week period. The intakes of gross energy by the lactating cows were about 50 per cent higher than those of the non-lactating animals.

Environmental factors affecting the food intake of ruminants

Grazing ruminants

The intake of ruminants in their natural habitat (i.e. at pasture) is influenced not only by the chemical composition and digestibility (or rate of digestion) of the pasture herbage but also by its physical structure and distribution. The grazing animal has to be able to harvest sufficient herbage to meet its needs without undue expenditure of energy. Its intake is determined

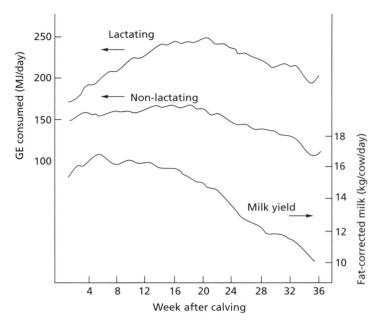

Fig. 17.2 Intake of gross energy and changes in milk production in lactating and non-lactating cows. (After Hutton J B 1963 *Proceedings, New Zealand Society for Animal Production* **23**: 39.)

by three factors, bite size (the quantity of dry matter harvested at each bite), bite rate (number of bites per minute) and grazing time. For example, a grazing dairy cow (600 kg) has a maximum bite size of 0.6 g DM, grazes at 60 bites per minute and hence harvests at 36 g DM per minute, or 2.16 kg per hour. To achieve a reasonable intake of 16 kg DM per day it will therefore need to graze for 16/2.16 = 7.4 hours per day. Cows normally graze for about eight hours per day, but sometimes for as much as 10 hours per day. In this instance the cow should be able to achieve the necessary intake in a grazing time within the eight-hour period. For the cow to achieve maximum bite size and bite rate the herbage must be suitably distributed. In general, relatively short (12–15 cm) and dense swards allow maximum bite size for cattle. Tall, spindly plants such as many tropical grasses restrict bite size because the animal is unable to get a mouthful of herbage at each bite. Low plant density (e.g. less than 1500 kg DM/ha for sheep) is also a restricting factor, and may be exacerbated by the animal's desire to graze selectively. Animals tend to prefer rapidly digested leaf to slowly digested stem, and also prefer green to dead material. Some plants may be rejected because they are rendered unpalatable by protective spines or by contamination with excreta.

In a good grazing environment with short, dense swards of highly digestible herbage, ruminants will consume as much dry matter as when food is offered to them in a trough, but in more difficult environments they often fail to achieve the intake of food that they are capable of digesting and metabolising.

Other environmental factors

Environmental temperature influences the intake of ruminants as it does of monogastric species. At temperatures below the thermoneutral zone (p. 358), intake is increased, and at temperatures above the thermoneutral zone, intake is reduced. Well-fed ruminants have a broad thermoneutral zone, extending to quite low critical temperatures at the lower end (see Table 14.4). However, at the upper end (i.e. in hot climates) temperature can exert a strong influence on intake, especially in more productive animals with a high nutrient demand. For example, for temperate (i.e. *Bos taurus*) breeds of cattle it has been estimated that intake falls by 2 per cent for every 1 °C rise in average daily temperature above 25 °C.

Another feature of the environment that has an effect on intake is day length. This effect is most evident in deer, which reduce their food intake very severely as day length declines; where short days coincide with a shortage of food, the effect is a survival mechanism to ensure that the limited supplies of food last through the critical period. Sheep also reduce their intake as days get shorter, but to a much lesser extent than do deer. Cattle seem not to be affected by day length.

Ill-health can reduce the intake of both ruminants and non-ruminants. Contrary to popular opinion, infestations of gastrointestinal parasites tend to reduce intake, presumably because the interference with digestive function overrides any metabolic stimulus arising from a reduction in the absorption of nutrients. There is also evidence that stimulation of the animal's immune system, as happens with parasite infestations, may be responsible for a reduction in food intake. Infestations with external parasites, such as ticks, also reduce intake.

17.3 Prediction of food intake

For both monogastric and ruminant species it is often necessary to be able to predict intake. Animals are commonly fed to appetite, and it is not possible to predict their performance by the use of feeding standards without an estimate of intake. For pigs and poultry, prediction is relatively simple as it is based mainly on the characteristics of the animals involved (but can become more complex in monogastric animals fed on fibrous foods). For ruminants, prediction is much more difficult, as many food variables may have to be taken into account.

For approximate predictions of intake there are some 'rules of thumb'. Thus the daily dry matter intake of beef cattle is often assumed to be about 22 g/kg liveweight, whereas that of dairy cows is higher, at about 28 g/kg in early lactation and 32 g/kg at peak intake (see Fig. 17.2).

An example of prediction equations for intake that may be used in conjunction with feeding standards is provided by the UK Agricultural Research

Council's Technical Committee on Responses to Nutrients, which constructed a series of equations for predicting the intake of grass silage by cattle (both dairy and beef) fed on silage and concentrates. The following is an example of the equations derived for beef cattle:

$$SDMI = 24.96 - 0.5397CDMI + 0.1080SDM - 0.0264AN + 0.0458DOMD$$

where: SDMI = silage dry matter intake (g/kg $W^{0.75}$ per day)
         CDMI = concentrate dry matter intake (g/kg $W^{0.75}$ per day)
          SDM = silage dry matter content (g/kg)
           AN = silage ammonia N content (g/kg total N)
         DOMD = digestible organic matter in silage dry matter (g/kg)

Thus this equation estimates that silage dry matter intake will be reduced by about 0.54 g for each 1 g of concentrate dry matter consumed. In addition, intake of the silage is related to three measures of quality, two positively (dry matter content and digestible organic matter content) and one negatively (ammonia nitrogen as a proportion of total nitrogen content). Other measures of silage quality employed in the equations were nitrogen content, butyric acid content and pH. The dairy cow equations included as predictors the yields of milk and of milk constituents, and also the week of lactation. Despite this degree of complexity, the most precise of the equations could account for only 60–70 per cent of the variation in intake, so their accuracy of prediction was limited.

The Australian Standing Committee on Agriculture has adopted a computer-based model (called GRAZFEED) to predict the intake of grazing ruminants. The animal factors in this model include the animal's current weight in proportion to its so-called 'standard reference weight' (SRW; see p. 371), body condition (i.e. fatness) and stage of lactation. The food factors include herbage digestibility and any supplementary foods. Environmental factors in the model are the features of the pasture that determine the structure of the sward, and there are also adjustments of intake for climatic factors. Expressing an animal's liveweight as a proportion of its SRW has a large influence on predicted intake. For example, a 400 kg steer that has reached its SRW (400 kg) is predicted to eat 5.9 kg per day of dry matter from good-quality pasture (dry matter digestibility 0.7), whereas a 400 kg animal of a larger breed that is only halfway towards its SRW (800 kg), and is therefore still growing, is predicted to eat 10.1 kg dry matter per day.

For the Cornell net carbohydrate and protein system for predicting the requirements of cattle, the intake of lactating cows is predicted from the net energy content of the food and the milk yield and milk fat content of the animal. Predicted values are then adjusted for six other factors: age, breed and body fat content of the cattle, the inclusion in the diet of any additives, the environmental temperature and another component of the environment called 'mud'.

Summary

1. Control mechanisms for food intake operate at the metabolic level (e.g. metabolite concentrations signalling the need to start or stop feeding) and at the level of the digestive system (e.g. signals from the degree of 'fill' of the rumen). In addition, intake is influenced by features of the environment such as temperature.

2. In monogastric animals control is mainly metabolic. In the short term, blood concentrations of glucose and other nutrients may control intake via the hypothalamus. In the longer term, fat depots send signals to the brain by the hormone messenger leptin.

3. The palatability of foods is not a major determinant of intake in farm animals.

4. When the energy content of foods is changed, animals attempt to keep their energy intake constant by changing food intake, but there are limits to the compensatory mechanism. Deficiencies of some specific nutrients, such as P, Ca, Cu and thiamin, reduce food intake. When given a choice of foods, animals are often able to select a diet with nutrient concentrations close to their needs.

5. Food intake in ruminants fed on high-energy diets is controlled metabolically, but in those fed on forages it is limited by the rate at which food can be digested in the rumen. Forages with a high content of cell walls (or neutral-detergent fibre) are digested slowly, are low in digestibility and promote low intakes. Disrupting the cell walls of forages by mechanical or chemical treatment markedly increases intake.

6. Intake is related to metabolic weight ($W^{0.75}$). In late pregnancy it is restricted by the foetus reducing rumen volume, but it increases in lactation.

7. When animals are grazing their intake may be restricted by the distribution of herbage (e.g. short herbage restricting bite size). Grazing animals prefer leaves to stems and may reject herbage that is dead or contaminated with excreta.

8. Low environmental temperatures cause increases in intake and high temperatures cause decreases.

9. For both monogastric and ruminant animals there are equations for predicting intake from food composition and features of the animal and its environment.

Further reading

Agricultural and Food Research Council 1991 Technical Committee on Responses to Nutrients, Report No. 8, *Voluntary Intake of Cattle*, Wallingford, UK, CAB International (see also *Nutrition Abstracts and Reviews, Series B* **61**: 815–23).

Australian Standing Committee on Agriculture 1990 *Feeding Standards for Australian Livestock: Ruminants*, Melbourne, CSIRO.

Campling R C and Lean I J 1983 Food characteristics that limit voluntary intake. In: Rook J A F and Thomas P C (eds) *Nutritional Physiology of Farm Animals*, London, Longman.

Dulphy J P and Demarquilly C 1983 Voluntary feed consumption as an attribute of feeds. In: Robards G E and Packham R G (eds) *Feed Information and Animal Production*, Farnham Royal, UK, Commonwealth Agricultural Bureaux.

Forbes J M 1995 *Voluntary Food Intake and Diet Selection in Farm Animals*, Wallingford, UK, CAB International.

Forbes J M, Varley M A and Lawrence T L J (eds) 1989 *The Voluntary Food Intake of Pigs*, Occasional Publication of the British Society of Animal Production, No. 13.

Hacker J B (ed.) 1982 *Nutritional Limits to Animal Production from Pastures*, Farnham Royal, UK, Commonwealth Agricultural Bureaux.

Hacker J B and Ternouth J H (eds) 1989 *The Nutrition of Herbivores*, Sydney, Academic Press.

Houseknecht K L, Baile C A, Matteri R L and Spurlock M E 1998 The biology of leptin: a review. *Journal of Animal Science* **76**: 1405–20.

Sykes A H 1983 Food intake and its control. In: Freeman B M (ed.) *Physiology and Biochemistry of the Domestic Fowl*, London, Academic Press.

18 Animal nutrition and the consumers of animal products

18.1   Comparative nutrition

18.2   The contribution of animal products to human requirements

18.3   Objections to the use of animal products

18.4   Future trends in the consumption of animal products

Animal products – meat (including fish), milk, milk products and eggs – are used principally as food for man, although some may also be used to feed other animals. In developed countries, the by-products of the meat industry are used to feed companion animals (cats and dogs) and may also be incorporated in the diets of farm animals (e.g. as meat and bone meal). Milk by-products are used to make milk replacers for young animals reared away from their dams. In this chapter we shall review the use of animal products as food for man and companion animals; their use for feeding farm animals is discussed in Chapter 24.

18.1   Comparative nutrition

The nutrient requirements of farm animals have been discussed in earlier chapters. Here, some selected values from these chapters are compared with the nutrient requirements of human beings and their companion, the

Box 18.1
Alternative uses of animals

Although animals are used mainly as sources of food, it is well to remember that they provide other benefits for man. Skins, wool and hair are utilised to make clothing and other fabrics, such as carpets. In many countries, animals are much used for draught purposes, to carry loads and pull farm machinery. As the alternatives to these uses are mainly derived from fossil fuels, there are clear incentives to maintain an interest in these animal products. Animals are also used to provide vaccines and other medical products for man; the range of these products is currently being extended by the application of genetic engineering.

Table 18.1 Comparative nutrient requirements of man and some domestic animals for growing animals of 30 kg liveweight[a]

Source	Man[a]	Dog[b]	Pig[c]	Sheep[d]
Energy, MJ/day	10	15	21	10
Protein, g/day	45	250	300	180
Lysine, g/day	2	8	17	–
Methionine & cystine, g/day	0.8	5	8	–
Calcium, g/day	0.7	10	15	3.4
Phosphorus, g/day	0.7	8	10	3.3
Copper, mg/day	1.0	7	8	2.7
Thiamin, mg/day	0.7	1.0	2.2	–
Riboflavin, mg/day	1.0	2.2	4.5	–
Vitamin A, iu/day	1500	5000	2000	1000
Vitamin D, iu/day	400	500	1500	180

Sources:
[a] Mainly UK estimated average requirements (Garrow J S and James W P T (eds) 1993 *Human Nutrition and Dietetics*, 9th edn, Churchill Livingstone, London).
[b] American Association of Feed Control Officials.
[c] Appendix 2, Table A.2.9.1 (1.5 kg of food per day).
[d] Appendix 2, Tables A.2.7.1 and A.2.7.2 (150 g/day of liveweight gain).

dog. Table 18.1 summarises the requirements of 30 kg growing individuals of the species, sheep, pig, dog and man, and Table 18.2 gives values for three of these species in a state of maintenance.

For growing animals (Table 18.1) requirements vary markedly between species, with fast-growing species having greater requirements. For example, the fast-growing pig needs six–seven times as much protein and 15–20 times as much calcium per day as a teenage human being of the same weight. The requirements of the growing dog are also high, and those of the slower-growing

Table 18.2 Comparative nutrient requirements of man and some domestic animals for animals maintained at 50 kg liveweight[c]

Source	Man[a]	Dog[b]	Sheep[d]
Energy, MJ/day	10	8.4	6.5
Protein, g/day	40	48	100
Lysine, g/day	3.6	3.6	–
Methionine & cystine, g/day	0.7	2.5	–
Calcium, g/day	0.5	3.4	1.1
Phosphorus, g/day	0.5	2.9	1.3
Copper, mg/day	1	4.2	3
Thiamin, mg/day	0.7	0.6	–
Riboflavin, mg/day	1	1.3	–
Vitamin A, iu/day	1500	2850	1800
Vitamin D, iu/day	100	285	300

[a, b, d] see Table 18.1 for comparative sources.
[c] Pigs are not included in this table because they are not normally maintained at 50 kg liveweight.

sheep are intermediate to those of the pig and man. For adult animals requiring nutrients for maintenance only (Table 18.2), requirements are again generally higher for the domestic animals than for man, but the differences are smaller than for growing animals. Indeed, for energy, man's requirement for maintenance is greater than that of the other species.

These figures demonstrate that although human beings tend to be concerned with their nutrient intakes, their domestic animals have relatively greater needs for nutrients, especially if they are to grow or produce rapidly.

18.2 The contribution of animal products to human requirements

After comparing the nutrient requirements of man with those of domestic animals, we must now assess the quantities of nutrients supplied by the latter species to human diets. The contribution of animal products to world food supplies is summarised in Table 18.3. In total, animal products provide about one-sixth of energy supplies and one-third of protein supplies; meat is the major contributor, followed by milk and milk products. The figures for individual countries and world regions differ considerably from the world averages (Table 18.4). Thus in Europe and North America, meat consumption is 30–40 times greater than in the countries of the Indian subcontinent, although the discrepancy for milk consumption is not as great.

When the figures in Tables 18.3 and 18.4 are translated into nutrient intakes they show, first, that over the world as a whole, foods derived from animals provide each person with about 1.9 MJ of energy and 28 g of protein per day (Table 18.5). These quantities are equivalent to about 16 per cent of man's energy intake and 34 per cent of protein intake. In India, animal products

Table 18.3 Contribution of various food groups to world food supplies (FAO)

Food group	Energy (%)	Protein (%)
Cereals	49	43
Roots, tubers and pulses	10	10
Nuts, oils, vegetable fats	8	4
Sugar and sugar products	9	2
Vegetables and fruits	8	7
All plant products	84	66
Meat	7	15
Eggs	1	2
Fish	1	5
Milk	5	11
Other	2	1
All animal products	16	34

Table 18.4 Meat and milk consumption in selected countries and world regions (kg per head per year) (FAO)

Country or region	Meat	Milk
USA	115.5	247.0
Argentina	91.1	179.5
Europe	83.1	228.7
UK	71.6	232.7
Burundi	3.7	7.8
Bangladesh	2.4	11.9
India	3.6	54.2
Sri Lanka	2.4	31.3

supply only 7 per cent of man's energy intake and 15 per cent of protein, whereas in the USA the corresponding figures are 28 and 64 per cent. The most important contribution of animal products to man's requirements is that of protein. In Europe, this contribution rises from the world average figure of 28 g/day to just over 50 g/day; in Africa, it falls to slightly less than 10 g/day. For individual countries, the range is even greater, from 3.4 g/day in Burundi and Mozambique to 75.1 g/day in France. The figures for meat consumption in the UK and USA provide an interesting comparison. Meat in the UK provides about the same quantity of energy as meat in the USA, but much less protein; this difference is presumably a reflection of a lower content of fat in meat in the USA, and is likely to be diminishing as the fat content of meat in the UK continues to decline.

The main factor determining the intake of animal protein is the wealth of the human population, but this factor is modulated by additional factors, especially the availability of alternative sources of protein and the religious

Table 18.5 Contribution of animal products to human diets (FAO, 1998)

	Energy (MJ/day)			Protein (g/day)		
	Meat	Milk	Total[a]	Meat	Milk	Total[a]
World	0.87	0.59	1.90	12.6	7.0	27.6
Developed countries	1.41	1.40	3.62	25.9	17.3	55.5
Developing countries	0.72	0.37	1.41	8.8	4.2	19.7
Africa	0.30	0.29	0.74	5.6	3.6	12.5
Ghana	0.13	0.01	0.36	3.5	0.2	11.4
India	0.08	0.60	0.76	1.7	6.2	10.0
Europe	1.45	1.51	3.83	24.1	18.2	53.8
UK	1.94	1.61	4.38	26.3	19.3	54.9
USA	1.86	1.77	4.33	40.5	22.8	72.8

[a] Includes eggs and fish.

beliefs and social customs of consumers. Thus there are parts of the world, such as arctic and desert areas, where crop production is not feasible, and the human population is largely dependent on animals for a protein supply. In arctic areas, Eskimos eat fish and eat animals that also live on fish. In desert areas, nomadic people survive on the products of animals such as the camel, which can live on the sparse natural vegetation.

The consumption of pigmeat is prohibited by several of the major religious groups (Muslims, Jews and Hindus), who regard pigs as 'unclean' animals, probably because of their ability to infect man with certain parasites (discussed later). Hindus and many Muslims do not eat beef. However, sheep meat and chicken are not commonly proscribed, except by people who choose to be vegetarians. The consumption of milk, milk products and eggs is subject to fewer religious and social restrictions, although the extreme vegetarians, known as vegans, exclude these from their diets. In many parts of the world fresh milk is not consumed by adults, who then lose the ability to secrete the digestive enzyme lactase and hence to digest lactose. They are said to be 'lactose intolerant', and if they subsequently ingest foods containing lactose it is fermented in the large intestine and causes a digestive upset.

These restrictions clearly influence global and national patterns of consumption of animal products. For example, in India, where pigmeat and beef are generally not eaten, meat consumption is very low and milk, milk products and eggs supply a high proportion of animal protein intake. In the USA, which has a wealthy population whose eating habits are not so much determined by religious beliefs, both meat and milk are consumed in large amounts. Meat consumption is high also in countries with a well-developed pastoral agriculture, such as Australia and Argentina. European countries follow close behind.

Within the world's poorer, developing countries, there is a close relationship between social class (i.e. wealth) and the consumption of animal products, with the consumption patterns of richer people approaching those of the developed countries. In the developed countries, however, this relationship is much less marked, as even poorer people can afford meat and milk products. Table 18.6 illustrates the consumption pattern for social classes in Britain, and shows that the gap in the consumption of meat

Table 18.6 Effect of income on the consumption of meat and milk in the UK (g/head/week) (Ministry of Agriculture, Fisheries and Food 1998 *National Survey of Domestic Food Consumption and Expenditure*).© Crown copyright. Reproduced by permission of the controller of Her Majesty's Stationery Office

| | Gross weekly income of head of household (£) | | | | Old-age pensioners |
	>640	330–640	160–330	<160	
Meat	918	898	963	775	1071
Milk	2020	1912	1919	2012	2487

between the richest and poorest families is only about 15 per cent. The type of meat consumed may vary between social classes, however, with richer people eating more steaks and fewer hamburgers, hence more protein and less fat. Milk consumption shows no systematic change with social class. It seems surprising that the highest figures for both meat and milk should be those for pensioners, but this may be due to the fact that the survey is restricted to foods consumed in the home, and pensioners may eat less outside their homes.

In developed countries, the patterns of consumption of animal products are liable to become ever more confused by the growing awareness of consumers of moral objections to, and possible health risks from, such foods. In Britain, vegetarians are thought to form only about 7 per cent of the population, but there are additional categories of people who claim to eat only a little meat, or to eat only 'white' meats (chicken and pork) and not 'red' meat (beef). Others reject the white meats because they dislike the intensive farming methods used to produce them. Concern for human health, which is discussed later, centres on the avoidance of the saturated fats found in many animal products.

Over the world as a whole, as people and countries get richer they tend to increase their consumption of animal products, but this eventually reaches a plateau. The level of the plateau is not necessarily the same for all: thus meat consumption in Britain seems to have levelled off at about 70 kg per head per year, whereas the average figure for Europe as a whole is about 80 kg, and for the USA, 115 kg. Consumer preference for animal products may be partly based on their supposed superior nutritional value, but is probably more strongly determined by their organoleptic characteristics (i.e. their taste and texture). Wholly vegetable diets tend to be bland and unexciting, and meat and the other animal products are used to add variety. Improved methods of preservation of animal products, such as refrigeration, heat processing and canning or vacuum sealing, have made it easier for people to enjoy a continuous supply of these products. However, the improved availability of foods in general also means that people can experiment with exotic ingredients and culinary techniques that – like animal products – add variety to their diet. Thus people eventually reach a point on the scale of affluence at which they no longer need or desire to increase their consumption of animal products.

The phrase 'need or desire . . . animal products' introduces the question of their essentiality in human nutrition. Do we really need these foods or do we just like or prefer them? The continuing successful existence of the vegetarians, and – more particularly – the vegans, among us demonstrates the non-essentiality of animal products for man; all the nutrients required by man can be met by foods not of animal origin. There are, however, several major nutritional advantages in meeting man's requirements partly from animal, rather than entirely from plant, sources. The first is that animal products supply nutrients in proportions closer to those required by man. This is best illustrated by the essential amino acids. Table 18.1 shows that a growing child requires 2 g of lysine and 45 g of total protein a day, a ratio of 4.4 g lysine per 100 g protein. For rice and wheat proteins the lysine : protein ratio is much lower (2.8 and 3.1, respectively), so these cereals need to be balanced in the diet by a lysine-rich protein source. A 'good quality' plant protein such as that of the soya bean has a lysine : protein ratio of 6.4, but animal proteins in milk and beef have even more favourable ratios of 8.2 and 9.1, respectively. Thus animal proteins are valuable for supplementing the proteins of staple foods such as cereals by supplying lysine and other essential amino acids, and this is particularly important for growing children, for whom amino acid requirements are most critical. If lysine requirements have to be met with cereal proteins, protein intake has to be high and much of it is wasted.

There is one essential nutrient, vitamin B_{12} (cyanocobalamin), which is synthesised by microorganisms and present in animal products but virtually absent from plant-derived foods. Vegans, in particular, have to ensure that they have a supply of this nutrient from a supplementary source such as yeast (see Chapter 5). Animal products are also good sources of other vitamins, especially vitamin A, thiamin, riboflavin and niacin.

Another advantage of animal-derived foods for man is that their nutrients are more accessible for digestion than those of plant-derived foods. Plant cell walls impede digestion in the stomach and small intestine and, although they may be digested in the large intestine, the consequent release of nutrients may be too late to allow efficient absorption. Some minerals in plant tissues are bound in compounds that resist digestion, an example being phosphorus in phytates (see pp. 120 and 623). Animal products are good sources of the minerals iron, calcium and zinc.

18.3 Objections to the use of animal products

Ethical and environmental objections

Objections to the use of animals to feed human beings arise first from ethical considerations, a full discussion of which is beyond the scope of this book. The primary argument, in brief, is that man has no right to exploit other animal species. The objections to using animals are lessened if they are not killed (i.e. kept for milk or egg production), and increased if they are kept under unnatural and perhaps harmful conditions. A second type of ethical argument is that plant-derived foods should not be diverted to animal feeding when they could be used directly to feed human populations that may be short of food. Until recently (e.g. the nineteenth century in Britain and other early developing countries) this argument did not apply because farm animals were used as scavengers to convert plants and plant by-products inedible by man into human foods (and even today, the world's crop by-products contribute about 500 Mt of dry matter per year as animal feeds). The argument has gathered force, however, with the increasing use of cereal and other grain crops for animal feeding. In developed countries, other than those with a predominantly pastoral agriculture, around 70 per cent of the cereals grown are used to feed livestock, and even in developing countries (including those with food shortages) considerable areas of land are used to grow crops for animal feeding. Over the world as a whole, cereal usage as animal feed amounts to 115 kg per person per year; the range across countries is from 4 kg in India and sub-Saharan Africa to 600 kg in the USA.

Objections to the use of animals to provide human food are also made on environmental grounds. Over-grazing can destroy plant communities; demand for additional grazing can cause deforestation; the excreta of intensively kept livestock cause pollution problems; methane from ruminants contributes to global warming. These raise complex issues which again cannot be fully explored in this book.

The direct, nutritional objections to animal-derived foods arise mainly from two sources. First, farm animals may harbour organisms such as pathogenic bacteria and intestinal parasites that may be transmissible to man through the consumption of animal products. Second, some of the supposedly valuable nutrients in animal products – fats, in particular – have been implicated in the causation of certain diseases of man.

Nutrition and human health

The *Merck Veterinary Manual* lists 150 diseases transmissible from animals to man (known collectively as zoonoses), but the majority of these are transferred by contact or bites, and/or are carried by wild animals. The food-borne diseases of man that arise from farm animals form a relatively small – but nevertheless important – group and are summarised in Box 18.2. The infection of man with these diseases can be minimised by various means, the first of which involves their restriction in, or elimination from, animals. One example is the regular use of anthelmintics to restrict intestinal parasites, and another is the slaughter of infected stock to restrict or eradicate bovine tuberculosis. The milk-borne disease tuberculosis also provides an example of another means of controlling infection, this being the treatment of animal products before they are consumed. The pasteurisation (heat treatment) of milk is designed to kill tuberculosis bacilli and other bacteria. Attention to hygiene in slaughterhouses and food stores, and appropriate cooking of meat, are also important in the control of zoonoses.

Antibiotics in feeds (see Chapter 25) have been used in intensive livestock systems to restrict infections, but their routine administration is now prohibited or discouraged because of the danger of producing antibiotic-resistant organisms.

Of the diseases listed, those regarded today as being the most important in developed countries are the enteric infections from *Campylobacter*, *E. coli* and *Salmonella* organisms. Although cases of 'food poisoning' have always occurred, people today are now less tolerant of them, both mentally (food is expected to be safe) and perhaps physically (as a generally cleaner environment has prevented the development of resistance to the organisms responsible).

Box 18.2
Some important diseases transmissible in food from farm animals to man

Disease and causative organism		Farm animals involved
(a) Bacteria		
Brucellosis: (undulant fever)	*Brucella abortus*	Cattle
	B. melitensis	Goats, sheep
Campylobacter enteritis:	*Campylobacter* spp.	All farm animals
Clostridial diseases:	*Clostridium botulinum*	Domestic animals
	C. perfringens	
Coliform infections:	*Escherichia coli*	Poultry, pigs and cattle
Listeriosis:	*Listeria monocytogenes*	All farm animals
Salmonellosis:	*Salmonella* spp.	Poultry, pigs, cattle and horses
Tuberculosis:	*Mycobacterium bovis*	Cattle
(b) Protozoa		
Sarcocytosis: (cyst formation)	*Sarcocystis suihominis*	Pigs and cattle
Cryptosporidiosis: (diarrhoea)	*Cryptosporidium parvum*	Cattle
(c) Cestodes, trematodes and nematodes		
Fascioliasis: (liver fluke)	*Fasciola hepatica*	Cattle and sheep
Tapeworms:	*Echinococcus and Taenia* spp.	Cattle, sheep and pigs
Trichinosis: (cyst formation)	*Trichinella spiralis*	Pigs

Source: The Merck Veterinary Manual, 8th edn, 1998 New Jersey, Merck and Co., Inc.

Not listed in Box 18.2 is the possible (or perhaps probable) zoonosis arising from bovine spongiform encephalopathy (BSE). The 1998 *Merck Veterinary Manual* states 'There is no evidence that the transmissible spongiform encephalopathies of man are acquired from animals', but the occurrence of so-called 'new variant' Creutzfeld–Jacob disease in man that has coincided with an epidemic of 'mad cow disease' in cattle and related species has inevitably cast doubt on this statement and has led to the introduction of special control measures to prevent the transmission of BSE from cattle and to man, and also to eradicate it from domestic animals (see also Chapter 24).

The chief chemical constituents of animal products that are implicated in diseases of man are fats in general and saturated fatty acids in particular. The diseases with which they are associated are those of the circulatory system that are characterised by damage to the arterial walls (atherogenesis) and the formation of blood clots (thrombogenesis). When arteries are damaged, fibrous plaques containing lipids are formed, and these may break away to form clots. If clots form in the blood vessels and impede the blood supply to the heart muscle they cause what is commonly called 'a coronary' (i.e. coronary heart disease); if they block the vessels supplying the brain they cause 'a stroke'; if they block the vessels of the lungs they cause a pulmonary embolism. These conditions are frequently fatal, and if the victim survives, he or she may be severely handicapped. Similar conditions can be caused by the rupture of damaged blood vessels.

The link between fatty deposits in the circulatory system and dietary fats is the lipid transport system that employs lipoproteins (see Chapters 3 and 9). The lipoproteins occur in various forms, which are defined by their density (see Table 3.6, p. 50), and the concentrations of these forms in blood are used to assess the risk of heart attacks and strokes. High risk factors are high concentrations of low-density lipoproteins (LDL) and very low-density lipoproteins (VLDL). Conversely, high concentrations of high-density lipoproteins (HDL) indicate a low risk. A high concentration of blood cholesterol, which is a constituent of lipoproteins, is also regarded as a high-risk indicator. The significance of these indicators is a matter of continuing research and debate. Thus some authorities consider cholesterol level to be a poor indicator. The LDL are now thought to be dangerous when present as small particles, a state characterised by a raised concentration of triacylglycerols.

As mentioned earlier, it is the saturated fatty acids (SFA) of foods that are regarded as the cause of a high-risk pattern of blood lipoproteins; octadecanoic (stearic, C_{18}), tetradecanoic (myristic, C_{14}) and any *trans* acids are considered to be the most damaging. With increasing consumption of SFA, blood levels of cholesterol and LDL are raised. Conversely, the polyunsaturated fatty acids (PUFA) are judged to be beneficial, although the various 'families' of PUFA differ in their effects; the omega-6 PUFA (which occur mainly in plant lipids) reduce the blood concentration of LDL, and the omega-3 PUFA (from fish lipids) reduce VLDL. In between the SFA and PUFA are the monounsaturated fatty acids (MUFA), such as octadecenoic (oleic, 18:1), which are regarded as neutral or possibly beneficial to blood lipoproteins.

As the association between lipid consumption and cardiovascular disease has been exposed and explored, many countries have produced nutritional guidelines that are intended to induce people to reduce their intake of fat and especially of SFA. A common recommendation is that fat should provide no more than 30 per cent of total energy intake, and that this fat should be divided equally among SFA, MUFA and PUFA (i.e. each supplying 10 per cent of energy intake). A less extreme proposal is that the ratio of PUFA to SFA (called the P : S ratio) should be 0.5–0.8.

In Britain, fat intake has fallen over the last ten years, from 42 to about 38 per cent of energy intake, but further changes are needed to meet the guidelines. In the USA, fat consumption has already fallen further, to 33 per cent of energy intake, and the P : S ratio is now 0.4–0.5. Although consumers can and do reduce their total intake of fat, they have more difficulty in modifying the proportions in their diet of the three main types of fatty acids. Only plant lipids have the 10 : 10 : 10 ratio suggested above. The fats of terrestrial animals have a predominance of saturated fatty acids. Thus in milk fat the ratio SFA : MUFA : PUFA is 8.5 : 3.3 : 0.3, and in meat, 8.3 : 8.3 : 2.0.

Box 18.3
Evidence for the association between dietary lipids and cardiovascular disease in man

The evidence comes mainly from the analysis of statistics for cardiovascular disease and diet composition obtained either from national populations or from smaller groups within countries. In addition to these *retrospective* studies there have also been *prospective* investigations, in which the diet and health of selected groups of people have been studied for long periods. Finally, there have been some *intervention* trials in which the diets of groups of people have been altered – for example, by reducing lipid consumption – and the consequences determined by comparing their health with that of control groups. Laboratory animals do not provide a satisfactory model for cardiovascular disease; closely controlled experimentation is therefore difficult, although pigs are used in some studies.

A common finding from the retrospective studies has been that communities with a high intake of SFA (e.g. SFA providing 15–25 per cent of energy intake) have a high incidence of coronary heart disease (CHD). Nevertheless, there are anomalies in the data, the most commonly quoted being the higher incidence of CHD in Britain than in France despite very similar intakes of lipid and SFA. A comparison of vegetarians and meat eaters in Britain showed that the former had lower levels of blood cholesterol (4.88 versus 5.31 m.mol/litre) and were 24 per cent less likely to die from CHD. Over the quarter-century that has passed since dietary fat was first linked to CHD, many developed countries have shown a fall in the incidence of CHD. On the other hand, the richer inhabitants of developing countries, who can afford the fat-rich diets of developed countries, are now experiencing CHD.

The figures given immediately above demonstrate the difficulty – perhaps even the impossibility – of meeting the guidelines for fat consumption with a diet containing a high proportion of animal products. The preferred strategy of those who wish to meet the guidelines seems to be a reduction in intake of animal fat but no reduction in consumption of the other constituents of animal products. In other words, people tend to maintain their consumption of meat and milk (and their derivatives), but to select against the fat in these foods. Selection can be exercised by switching from high-fat meat to that containing less fat in total and less SFA in particular; this is one reason for the continuing replacement of beef by chicken. Fat may be trimmed from joints of meat and replaced as a cooking aid by vegetable oils. Much of the milk consumed in liquid form has its fat content reduced to around 20 g/kg (i.e. half the 'natural' content): in Britain fat-reduced milks account for more than 50 per cent of liquid consumption. Of milk products, butter has to a large extent been replaced by spreads based on vegetable oils (although all the milk fat produced in Britain is still consumed in some form or another!).

Animal nutritionists, in association with animal breeders, have responded to the challenge of maintaining the acceptability of animal-derived foods by modifying their lipid constituents. As discussed earlier (Chapter 14), animals are selected for leanness, are fed to give maximal growth of muscle and are slaughtered when immature (hence having less fat). With pigs and poultry it is possible to modify the constitution of body fats via their diet; for example, the proportions of omega n-3 and omega n-6 PUFA can be changed. Ruminants tend to deposit saturated fat because the unsaturated lipids of their plant diet are hydrogenated in the rumen (Chapter 8). It is possible to introduce unsaturated fatty acids into either their body or their milk fats by protecting dietary lipids against intervention by rumen microorganisms, but the physical properties of the resulting fats are inevitably altered and may render them unacceptable to the consumer. For example, milk fat enriched with unsaturated fatty acids is more susceptible to oxidation and hence rancidity.

The fat content of cow's milk can be reduced by feeding the cow on an extreme type of diet (typically, a diet low in fibre; see Chapter 16), but the reduction is achieved only by upsetting the normal metabolism of the animal. This raises the question of whether it is morally acceptable to disadvantage the animal in order to meet the perceived needs of its consumer. Pigs selected to be ultra-lean have metabolic problems, and all pigs rely on subcutaneous fat to provide insulation against a cold environment.

Comparisons of national populations have shown some association between the consumption of meat (especially red meat, such as beef and lamb) and the incidence of cancer of the bowel, breast and prostate gland. However, comparisons made within populations (e.g. between vegetarians and meat eaters) have shown no such association, and the wholesale condemnation of meat, or other animal products, as being responsible for cancer is therefore not justified.

The place of animal fats – and particularly the fats of ruminants – in the diets of man has been given a new dimension by the discovery that one particular fatty acid, known popularly as conjugated linoleic acid (CLA) or more precisely as *cis*-9, *trans*-11 octadecadienoic acid, has a beneficial role in the body. This acid has been shown to be anti-atherogenic and anti-carcinogenic, and also to limit obesity and stimulate immune function. The CLA is produced in the rumen as an intermediate in the bacterial hydrogenation of unsaturated fatty acids present in the diet (hence its alternative name of rumenic acid), but it may also be synthesised in animal tissues. It is therefore present in both milk and meat from ruminants. Ruminants given foods that contain relatively high concentrations of unsaturated fatty acids, such as young pasture herbage, produce fats with particularly high contents of CLA.

18.4 Future trends in the consumption of animal products

Despite the arguments advanced against meat consumption – on ethical, environmental and health grounds – world demand for all types of meat is predicted to increase steadily over the next 20 years or so (Table 18.7). For meat in total, consumption per person per year in the developed countries is predicted to continue to rise slowly, by 0.2 per cent per year, but for the developing countries the corresponding figure is much greater, at 1.6 per cent per year. World demand on an absolute basis (i.e. allowing for population growth) is predicted to increase more rapidly, by 0.6 per cent per year in developed countries and by 4.1 per cent per year in developing countries. There are some interesting differences in the projections for individual animal species; for example, the demand for and production of

Table 18.7 Current and future demand for meat in the world and selected subunits (kg/head/year) (Adapted from Rosegrant M W, Leach N and Gerpacio R V 1999 Alternative futures for world cereal and meat consumption. *Proceedings of the Nutrition Society* 58: 219–34)

Meat	World	Developed countries	Developing countries	China	India	Sub-Saharan Africa
	1993/2020	1993/2020	1993/2020	1993/2020	1993/2020	1993/2020
Beef	9.8/10.7	25.2/25.8	5.2/7.4	2.1/4.4	2.6/4.0	4.1/5.5
Pig	13.7/15.8	29.4/29.4	9.0/12.8	24.5/43.1	0.4/0.7	1.2/1.5
Sheep + goat	1.8/2.1	2.8/3.2	1.5/1.9	1.2/1.4	0.7/0.8	1.6/1.8
Poultry	8.5/10.7	20.3/24.7	5.0/7.7	5.0/10.7	0.5/0.9	1.9/2.4
All	33.9/39.3	77.7/83.0	20.8/29.7	32.8/59.6	4.3/6.5	8.8/11.1

pigmeat in the developed countries is predicted to grow more slowly than that for poultrymeat.

It is possible that the arguments against the consumption of meat have yet to make their full impact on consumers. However, it has been calculated that if consumers in the developed countries (i.e. those most likely to be influenced by anti-meat arguments) were to reduce their meat consumption by half over the interval 1993–2020, world demand for meat would still increase, by about 1.5 per cent per year. Moreover, additional projections show that, because of the adverse economic effects on world agriculture of a reduced demand for meat in the developed countries, the partial switch from animal to vegetable foods would not increase the world supply of food per person per day by any significant amount.

Summary

1. The nutrient requirements of man may be compared with those of domestic animals, exemplified by the dog, pig and sheep. Requirements of growing animals differ considerably between species, mainly because of differences in growth rate, but those for adult maintenance are less discordant.

2. Foods derived from animals provide one-third of the protein and one-sixth of the energy supplied to the world's human population. Figures for individual countries vary widely, with the richest consuming seven times as much animal protein per head as the poorest. Although wealth is the main determinant of meat and milk consumption, environment and religion are important influences.

3. Human beings have no specific need for animal products, but these products are better sources of some important nutrients (e.g. iron and vitamin B_{12}) than are plant products.

4. Animal products present hazards to consumers from food-borne diseases and from the saturated fats they contain. The latter problem is being reduced by advice to consumers and by modifying the concentration and constitution of fats in animal products.

5. Meat consumption, both in total and per head of the world population, is predicted to increase over the next 20 years, but by more in developing countries than in developed.

Further reading

Bradford G E 1999 Contributions of animal agriculture to meeting global human food demand. *Livestock Production Science* **59**: 95–112.

Garrow J S and James W P T (eds) 1993 *Human Nutrition and Dietetics*, 9th edn, London, Churchill Livingstone.

Lichtenstein A H *et al*. 1998 Dietary fat consumption and health. *Nutrition Research Reviews* **56**: S3–S19.

Ministry of Agriculture, Fisheries and Food (UK) 1998 *National Food Survey 1998*, London, MAFF.

Nutrition Society (UK) 1999 Symposium: Meat or wheat for the next millenium? *Proceedings of the Nutrition Society* **58**: 209–75.

19 Grass and forage crops

19.1   Pastures and grazing animals

19.2   Grasses

19.3   Legumes

19.4   Other forages

19.1   Pastures and grazing animals

From a nutritional point of view, the simplest pastures are those created by sowing a single species of pasture plant, which may be a grass such as Italian ryegrass or a legume such as lucerne (alfalfa). These pastures have a limited life (1–4 years) and are described as *temporary pastures*. Sown *permanent pastures* normally include several species of both grasses and legumes. *Natural grasslands* have an even greater spread of species, which will include herbs and shrubs (some of which may be regarded as weeds). In this chapter we shall describe first the nutritional value of grasses (i.e. species of the Gramineae), then the value of legumes (Leguminoseae). The chapter is concluded with a description of special crops grown in association with pastures.

The nutrition of the grazing animal is different from that of housed livestock, for several reasons. In the first place, its diet is variable. As indicated above, the variability may be due to the botanical composition of pastures. There is also variability over time, for different species may grow at different times of the year, and even a single species will change in composition and nutritive value as it grows to maturity. A second important feature of the grazing animal is that it has to spend time and energy in harvesting its food. Thus a cow consuming 15 kg of dry matter a day from a pasture yielding 1000 kg of dry matter per hectare must harvest an area of 150 m^2 per day with a mowing apparatus only 0.01 m wide. The diet of the grazing animal is therefore difficult to evaluate; moreover, even if its nutritive value is known, the opportunities for correcting its deficiencies may be limited.

At its best, pasture herbage is a food of high nutritive value for ruminants. For example, young, leafy herbage from a perennial ryegrass pasture may contain 12 MJ of metabolisable energy and 200 g crude protein per kg DM, and be eaten in quantities sufficient for the needs of a high-producing dairy cow, for maintenance and the production of 25 kg of milk per day. Unfortunately, the nutritive value of herbage often falls far below its optimum; soil and climate may be unsuitable for growing the most nutritious pasture plants, and the grazing management system employed may fail to ensure that herbage is consistently grazed at the stage of growth when its nutritive value is at its highest. In many grazing situations the herbage barely supplies the maintenance needs of the animal. Achieving the optimal nutrition of grazing animals is a continuing challenge for nutritional scientists and farmers.

19.2 Grasses

Pattern of growth

In cold and temperate climates, grass starts to grow in the spring when soil temperatures reach 4–6 °C. From then on the pattern of growth is very much the same irrespective of species or cultivar. There is a rapid production of leaf followed by an increase in the growth of the stem, leading to the ultimate emergence of the flowering head and, finally, to the formation of the seed. As the grass grows in the spring the concentration of DM in the crop increases, at first slowly then more rapidly as the stems grow and the ears emerge and, finally, more slowly as the ears begin to ripen.

In hot climates, soil temperature is likely to be high enough to allow grass growth throughout the year, but growth is commonly restricted by lack of water. Where the climate is characterised by clearly defined wet and dry seasons, grass growth is very rapid during a warm wet season, but as the soil dries out the herbage matures and dies, leaving a feed resource that is sometimes described as 'standing hay'. Even in wetter conditions, plants in deep shade at the base of the sward may die, thus giving senescent herbage of inferior nutritive value.

The rate at which grass grows is dependent upon the environment, the nutrients available and the amount of leaf within the sward which is intercepting light. Immediately after harvesting there is a period of slow regrowth followed by an accelerated rate and finally a period of decreasing growth as the herbage matures. As grass swards increase in leaf area, the photosynthetic capacity of successive newly expanded leaves is progressively reduced because of the increasing shade in which they develop. The rate at which regrowth occurs depends upon the maturity of the crop at the time of harvesting. If the grass is young and leafy it recovers more quickly and starts regrowth earlier than when mature herbage is harvested. Typical growth rates for temperate pastures in the spring are 40–100 kg DM per hectare per day.

Chemical composition

The composition of the dry matter of pasture grass is very variable; for example, the crude protein content may range from as little as 30 g/kg in very mature herbage to over 300 g/kg in young, heavily fertilised grass. The fibre content is, broadly, related inversely to the crude protein content, and the acid-detergent fibre may range from 200 to over 450 g/kg in very mature moorland species of grasses.

The moisture content of grass is of particular importance when a crop is being harvested for conservation; it is high in very young material, usually 750–850 g/kg, and falls as the plants mature to about 650 g/kg. Weather conditions, however, greatly influence the moisture content.

The composition of the dry matter is dependent on the relative proportions of cell walls and cell contents. The cell walls consist of cellulose and hemicelluloses, reinforced with lignin. The cellulose content is generally within the range of 200–300 g/kg DM and that of hemicelluloses may vary from 100 to 300 g/kg DM. The concentrations of both these polysaccharide components increase with maturity; so also does that of lignin, which reduces the digestibility of the polysaccharides.

The cell contents include the water-soluble carbohydrates and much of the protein. The water-soluble carbohydrates of grasses include fructans and the sugars glucose, fructose, sucrose, raffinose and stachyose (see Table 19.1). In temperate grasses the storage carbohydrate fructan is the most abundant of the soluble carbohydrates and is found mainly in the stem. Grasses of tropical and subtropical origin accumulate starches, instead of fructans, in their vegetative tissues and these are stored primarily in the leaves. The water-soluble carbohydrate concentration of grasses is very variable, ranging from as little as 25 g/kg DM in some tropical species to over 300 g/kg DM in some cultivars of ryegrass.

Proteins are the main nitrogenous compounds in herbage, with true protein accounting for about 80 per cent of total nitrogen. Although the total protein content decreases with maturity the relative proportions of amino acids do not alter greatly. Similarly, the amino acid composition of proteins varies little among grass species. This is not surprising as up to half of the cellular protein in grasses is in the form of a single enzyme, ribulose bisphosphate carboxylase, which plays an important role in the photosynthetic

Table 19.1 Composition of the dry matter of a sample of Italian ryegrass cut at a young leafy stage (g/kg)

Proximate composition		Carbohydrates		Nitrogenous components		Other constituents	
Crude protein	190	Glucose	16	Total N	30	Lignin	52
Ether extract	45	Fructose	13	Protein N	27		
Crude fibre	208	Sucrose	45	Non-protein N	3		
Nitrogen-free		Oligosaccharides[a]	19				
extractives	449	Fructans	70				
Ash	108	Galactan	9				
		Araban	29				
		Xylan	63				
		Cellulose	202				

[a] Excluding sucrose.

fixation of carbon dioxide. Grass proteins are particularly rich in the amino acid arginine, and also contain appreciable amounts of glutamic acid and lysine. They have high biological values for growth compared with seed proteins, with methionine being the first and isoleucine the second limiting amino acid for growth. However, because of the considerable metabolism of amino acids in the rumen, this factor is of little significance to ruminant animals. In general, the protein contents of tropical grasses tend to be lower than those of temperate species.

The amino acid composition of forages is clearly important when the feeds are used as protein sources for non-ruminants. For ruminants, however, the most important characteristics of forage proteins are their rumen degradability and their overall digestibility. In immature forages, both measures are commonly very high (0.7–0.8), but they decline as the forages mature (and their total protein content declines). In mature herbage, a significant proportion of the protein may be indigestible because it is bound to fibre (acid-detergent insoluble nitrogen, ADIN; see p. 340).

The non-protein nitrogenous fraction of herbage varies with the physiological state of the plant. Generally, the more favourable the growth conditions, the higher is the non-protein N content as well as the total nitrogen value, and as the plants mature the contents of both decrease. The main components of the non-protein N fraction are amino acids, and amides such as glutamine and asparagine, which are concerned in protein synthesis; nitrates may also be present, and these may be toxic for grazing animals (see p. 505).

The lipid content of grasses, as determined in the ether extract fraction, is comparatively low and rarely exceeds 60 g/kg DM. The components of this group include triacylglycerols, glycolipids, waxes, phospholipids and sterols. Triacylglycerols occur only in small amounts and the major components are galactolipids, which constitute about 60 per cent of the total lipid content. Linolenic acid is the main fatty acid, comprising between 60 and 75 per cent of the total fatty acids present, with linoleic and palmitic acids the next most abundant.

The mineral content of pasture is very variable, depending upon the species, stage of growth, soil type, cultivation conditions and fertiliser application; an indication of the normal range in content of some essential elements is given in Table 19.2.

Green herbage is an exceptionally rich source of β-carotene, a precursor of vitamin A, and the dry matter of the young green crop may contain as much as 550 mg/kg. Herbage of this type supplies about 100 times the vitamin A requirement of a grazing cow when eaten in normal quantities.

It has generally been considered that growing plants do not contain vitamin D, although precursors are usually present. Studies suggest, however, that vitamin D may be present in herbage but in relatively small amounts. The greater vitamin D content of mature herbage than of young material may be caused in part by the presence of dead leaves in which vitamin D_2 has been produced from irradiated ergosterol.

Most green forage crops are good sources of vitamin E and of many of the B vitamins, especially riboflavin.

Table 19.2 Ranges of essential mineral contents of temperate pasture grasses

Element	Low	Normal	High
g/kg DM			
Potassium	<12	15–30	>35
Calcium	<2.0	2.5–5.0	>6.0
Phosphorus	<2.0	2.0–3.5	>4.0
Sulphur	<2.0	2.0–3.5	>4.0
Magnesium	<1.0	1.2–2.0	>2.5
mg/kg DM			
Iron	<45	50–150	>200
Manganese	<30	40–200	>250
Zinc	<10	15–50	>75
Copper	<3.0	4.0–8.0	>10
Molybdenum	<0.40	0.5–3.0	>5.0
Cobalt	<0.06	0.08–0.25	>0.30
Selenium	<0.02	0.03–0.20	>0.25

Factors influencing the nutritive value of herbage

Stage of growth

Stage of growth is the most important factor influencing the composition and nutritive value of pasture herbage. As plants grow there is a greater need for fibrous tissues to maintain their structure, and therefore the main structural carbohydrates (cellulose and hemicelluloses) and lignin increase, and the concentration of protein decreases; there is therefore an inverse relationship between the protein and fibre contents in a given species, although this relationship can be upset by the application of nitrogenous fertilisers.

The variations in chemical composition of two species of grasses, perennial ryegrass (*Lolium perenne*) and purple moor grass (*Molinea caerulea*), at three stages of growth are shown in Table 19.3. In addition to the changes in organic components, changes also occur in the mineral or ash constituents. The total ash content decreases as the plant matures. This is reflected in the calcium content, which follows a similar pattern to that of the total ash in grasses. The magnesium content is generally high in the early spring, but falls off sharply; during the summer it rises, reaching high values in the autumn.

The digestibility of the organic matter is one of the main factors determining the nutritive value of forage, and this may be as high as 0.85 in young spring pasture grass and as low as 0.45 in winter forage. The basic determinant of forage digestibility is the plant anatomy. Plant cell contents, being mainly soluble carbohydrates and proteins, are almost completely digestible, but cell walls vary in digestibility according to their degree of reinforcement with lignin. Thus digestibility decreases as plants increase in maturity, but the relationship is complicated by there being a spring period of up to a month during which the herbage digestibility remains fairly

Table 19.3 Composition and nutritional value of *Lolium perenne* and *Molinea caerulea* grown in Scotland and harvested at three stages of maturity (After Armstrong R H, Common T G and Smith H K 1986 *Grass and Forage Science* **41**: 53–60)

	Lolium perenne			*Molinia caerulea*[a]		
	24 May	7 July	4 Aug	3 July	16 Aug	6 Oct
Proportion of leaf	0.63	0.29	0.27	0.95	0.87	0.78
DM, g/kg	165	338	300	374	468	428
Components of DM[b]						
g/kg DM						
CP	143	69	48	149	97	59
Ash	88	68	74	27	29	20
ADF	227	316	347	327	414	435
Lignin	16	41	49	52	95	95
Organic matter						
Digestibility[c]	0.80	0.68	0.59	0.67	0.54	0.48
Intake[c],						
g/kg $W^{0.75}$/day	73	56	39	70	44	28

[a] Harvested from a hill sward containing about 80 per cent *M. caerulea*.
[b] CP = crude protein; ADF = acid detergent fibre.
[c] Determined using sheep.

constant. This period has been described as the 'plateau'. The end of this period is associated in some plant species with ear emergence, after which digestibility of organic matter may decrease abruptly. In grasses grown in the UK, this rate of decrease is about 0.004 units per day.

Digestibility of grasses is also influenced by leaf/stem ratios. Techniques involving fermentation *in vitro* have enabled the digestibility of different fractions of plants to be determined. In very young grass the stem is more digestible than the leaf, but whereas with advancing maturity the digestibility of the leaf fraction decreases very slowly, that of the stem fraction falls rapidly. As plants mature, the stem comprises an increasing proportion of the total herbage and hence has a much greater influence on the digestibility of the whole plant than the leaf.

The decrease in digestibility with stage of growth is also reflected in the metabolisable and net energy values of grasses as illustrated in Table 19.4. The low net energy value of mature herbage is not only due to a low organic matter digestibility but is also associated with a high concentration of cellulose. The digestion of this polysaccharide in the rumen and the metabolism of the end-products give rise to a high heat increment (see Chapter 11).

Species

The Gramineae form a very large family that has been subdivided into 28 tribes, of which the six largest contain most pasture grasses of economic

Table 19.4 Nutritional value of four cuts of perennial ryegrass (After Armstrong D G 1960 *Proceedings of the Eighth International Grassland Congress,* p 485)

	Cut 1 Young leafy	Cut 2 Late leafy	Cut 3 Ear emergence	Cut 4 Full seed
Constituents, g/kg DM				
Crude protein	186	153	138	97
Crude fibre	212	248	258	312
Cellulose	253	284	299	356
Energy, MJ/kg DM[a]				
Metabolisable	13.1	12.2	11.6	8.9
Net (for maintenance)	10.3	9.3	8.8	7.3
Net (for liveweight gain)	6.9	6.9	5.6	3.8

[a] Determined using mature sheep.

importance. All six of these tribes have a wide distribution but their importance in any particular region is determined largely by the temperature and to a lesser extent by the rainfall. The distribution of these tribes, and some agriculturally important members of them, are shown in Table 19.5.

In temperate areas having a reasonably uniform distribution of rainfall, grasses grow and mature relatively slowly and can thus be utilised at an early stage of growth when their nutritive value is high. In warmer climates, however, grasses mature more rapidly, their protein and phosphorus contents

Table 19.5 The main tribes of the Gramineae, their areas of major importance, and some examples of their grass-land species

Tribe	Major areas	Examples
Agrosteae	All temperate	*Agrostis* spp. – bent grasses *Phleum pratense* – timothy
Aveneae	Cold and temperate	*Holcus lanatus* – Yorkshire fog *Danthonia pilosa* – tussock grass
Festuceae	Temperate, particularly USA	*Lolium* spp. – ryegrasses *Festuca* spp. – fescues *Bromus* spp. – bromes

Eragrosteae	Tropical and
	    warm temperate
		Eragrostis curvula – weeping
		    love grass

Andropogoneae	Tropics, particularly SE Asia	*Andropogon gayanus* – gamba *Hyparrhenia rufa* – Jaragua

Paniceae	Tropics and
	    subtropics
		Digitaria decumbens – pangola
		    grass
		Panicum maximum – Guinea grass
		Paspalum dilatatum – dallis
		    grass
		Pennisetum purpureum – elephant
		    grass

Table 19.6 Composition (g/kg DM) and ME values (MJ/kg DM) of three cuts of Jaragua grass (*Hyparrhenia rufa*) grown in Brazil (Adapted from Gohl B 1981 *Tropical Feeds*, FAO, Rome)

Stage of growth	Composition				ME
	Crude protein	Crude fibre	Ether extract	Ash	
Vegetative	92	289	26	149	8.4
Full bloom	35	314	19	136	7.0
Milk stage	28	337	15	115	6.5

falling to very low levels, and their fibre content rising (see Table 19.6). In the wet tropics the herbage available is commonly fibrous but lush (i.e. high in water content); in drier areas the mature herbage becomes desiccated and is grazed as 'standing hay'. In both cases digestibility is low, typical values for tropical herbage being 0.1–0.15 units lower than for temperate herbage. The differences in composition between temperate and tropical grasses are not only a result of climate. Temperate species of grasses belong to the C_3 category of plants, in which the three-carbon compound phosphoglycerate is an important intermediate in the photosynthetic fixation of carbon dioxide. Most tropical grasses have a C_4 pathway of photosynthesis, in which carbon dioxide is first fixed in a reaction involving the four-carbon compound oxalacetate. The low protein contents often found in tropical grasses are an inherent characteristic of C_4 plant metabolism, which is associated with survival under conditions of low soil fertility. Another feature of tropical species is their storage of carbohydrates as starch, rather than as fructans.

A further factor of nutritional importance is that the anatomy of the leaves of tropical grasses differs from that of temperate grasses. In tropical grasses there are more vascular bundles and thick-walled bundle sheaths, hence more lignin, and in the central tissue of the leaf the mesophyll cells are more densely packed than those in temperate grasses. Thus in tropical grasses, intercellular air spaces represent only 3–12 per cent of leaf volume compared with 10–35 per cent in temperate species. This may partly explain why tropical grasses have a higher tensile strength than temperate ones, a feature which retards both mechanical and microbial degradation in the rumen. The consequences of this are lower digestibility of tropical grasses and lower voluntary dry matter intakes.

The selection of pasture species and cultivars is based on such agronomic characters as persistency and productivity, but nutritive value is also taken into account. Cultivars within a species generally differ to only a small degree in nutritive value, if the comparison is made at the same stage of growth, but differences between comparable species may be larger. A classical example for the temperate grasses is the difference between British cultivars of perennial ryegrass (*Lolium perenne*) and of cocksfoot or orchard grass (*Dactylis glomerata*). At the same stage of growth cocksfoot has a lower concentration of soluble carbohydrates and is lower in dry matter digestibility than ryegrass.

Pasture species may also influence the value of the herbage through their growth habit. As discussed in Chapter 17, tall and spindly (as opposed to short dense) swards are difficult for the animal to harvest efficiently, so its intake may be reduced.

In Britain, perennial ryegrass is the most important species of sown pastures, but Italian ryegrass (*Lolium multiflorum*), timothy (*Phleum pratense*), cocksfoot and the fescues (*Festuca* spp.) are also common. In older pastures these are accompanied by weed species, particularly meadowgrass (*Poa pratensis*), Yorkshire fog (*Holcus lanatus*) and the bents (*Agrostis* spp.). In the uplands, however, some of these weed species, such as the bents, along with other species such as mat grass (*Nardus stricta*) and purple moor grass (*Molinia caerulea*) (see Table 19.3), are valued constituents of the sward.

Soils and fertiliser treatment

The type of soil may influence the composition of the pasture, especially its mineral content. Plants normally react to a mineral deficiency in the soil either by limiting their growth or by reducing the concentration of the element in their tissues, or more usually by both. In addition, deficiencies of mineral elements may affect the utilisation of herbage; thus in sheep, sulphur deficiency reduces the digestibility of herbage. The most common mineral deficiencies of grass herbage are those of phosphorus, magnesium, copper and cobalt.

The acidity of the soil is an important factor that can influence, in particular, the uptake of many trace elements by plants. Both manganese and cobalt are poorly absorbed by plants from calcareous soils, whereas low molybdenum levels of herbage are usually associated with acid soils. The induced copper deficiency known as teart (see p. 130), associated with high herbage molybdenum levels, generally occurs on pasture grown on soils derived from Lower Lias clay or limestone.

Liberal dressings of fertilisers can markedly affect the mineral content of plants, and it is also known that the application of nitrogenous fertilisers increases leaf area as well as rate of photosynthesis. As a consequence, the crude protein content, and frequently the amide and nitrate contents, are increased.

Application of nitrogenous fertilisers also depresses the water-soluble carbohydrate content of temperate grasses, which may have an adverse effect on fermentation if the crop is preserved as silage (see p. 515). Fertilisers may also affect, indirectly, the nutritive value of a sward by altering the botanical composition. For example, legumes do not thrive on a lime-deficient soil, while heavy dressings of nitrogen encourage growth of grasses and at the same time depress clover growth.

Grazing system

In many traditional grazing systems, animals are kept on the same area of pasture throughout the year (continuous grazing). In such systems, the ideal 'stocking rate' (i.e. animals per unit area) is one that maintains a perfect balance between the growth of new herbage and its harvesting by animals; in this situation the animal is presented with a constant supply of young (and therefore nutritious) herbage. In practice, this ideal situation is rarely achieved. If the rate of growth exceeds the rate of harvesting, herbage accumulates and matures, thus reducing the nutritive value of the material on offer. However, an increase in the amount of herbage on offer allows the animals to graze selectively, and they are able to compensate to some extent for the general fall in nutritive value by selecting plants, or parts of plants, that are higher in nutritive value than the rest. For example, leaf may be selected in preference to stem. Such selective grazing is particularly important with pastures containing a wide variety of plants (e.g. shrubs and trees as well as grasses and herbaceous legumes). If the rate of harvesting exceeds the rate of growth of the pasture, the 'grazing pressure' on the sward is said to increase. Selection by animals is reduced, and the pasture plants may be so denuded of foliage that their root reserves are depleted and they fail to regrow. Both under- and over-grazing of pastures may change their botanical composition and therefore the nutritive value of their herbage.

In rotational grazing systems, pastures are grazed for short periods at a high stocking rate and grazing pressure; animals harvest most of the herbage on offer, and the pastures are then rested for longer periods of recovery. For example, a farmer might divide his pastures into 28 paddocks and graze each for one day, thus allowing 27 days for regrowth. Once a rotation has been established, animals should receive herbage of reasonably constant nutritive value. In these systems the balance between herbage grown and herbage harvested may also be controlled by varying the numbers of animals or paddocks (e.g. some paddocks may be cut for silage).

Pastures are not necessarily harvested by grazing. In some areas, grass may be cut by machine and transported to housed animals. This practice of 'zero grazing' allows greater control over the diet of grass-fed livestock.

Other factors affecting the nutritive value of grass

Such factors as climate and season may influence the nutritive value of pasture. The concentration of sugars and fructans, for example, can be influenced markedly by the amount of sunshine received by the plant. Generally on a dull cloudy day the soluble carbohydrate content of grass will be lower than on a fine sunny day. Rainfall can affect the mineral composition of pasture herbage. Calcium, for example, tends to accumulate in plants during periods of drought but to be present in smaller concentration when the soil moisture is high; on the other hand phosphorus appears to be present in higher concentrations when the rainfall is high.

The net energy value of autumn grass is often lower than that of spring grass, even when the two are cut at the same stage of growth and are equal in digestibility or metabolisable energy concentration. For example the dry matter of spring grass containing 11 MJ ME/kg DM was found to supply 5.2 MJ net energy/kg for fattening, whereas later growths of the same metabolisable energy content supplied only 4.3 MJ net energy/kg. Late-season herbages are often lower in soluble carbohydrate content and higher in protein content than early-season herbages; there is some evidence that this combination leads to poor trapping of herbage nitrogen in the rumen, hence increased absorption of ammonia and reduced production of microbial protein.

Nutritional disorders associated with grasses

Nitrate poisoning

As mentioned earlier, nitrate can accumulate in grasses. Nitrate *per se* is relatively non-toxic to animals. The toxic effect in ruminants is caused by the reduction of nitrate to nitrite in the rumen. Nitrite, but not nitrate, oxidises the ferrous iron of haemoglobin to the ferric state, producing a brown pigment, methaemoglobin, which is incapable of transporting oxygen to the body tissues. Toxic signs include trembling, staggering, rapid respiration and death. It has been reported that toxicity may occur in animals grazing herbage containing more than 0.7 g nitrate-N/kg DM, although the lethal concentration is much higher than this. Some authorities have quoted a lethal figure for nitrate-N of 2.2 g/kg DM, whereas others have suggested a value far in excess of this. A sudden intake of nitrate may be particularly dangerous; experimentally this may be brought about by drenching, but may occur in practice when herbage that is normally non-toxic is eaten unusually quickly. Nitrate is sometimes less toxic if the diet also contains soluble carbohydrates. The nitrate content of grasses varies with species, variety and manuring, although the amount present is generally directly related to the crude protein content.

Mycotoxicoses

Several related disorders of grazing animals have as their primary cause the invasion of grasses by fungi, which produce substances known as mycotoxins. The condition known as 'ryegrass staggers' occurs in ruminants and horses grazing perennial ryegrass in New Zealand, Australia, North America and (occasionally) Europe. Affected animals are uncoordinated and may collapse, although mortality is low. The endophytic fungus involved, *Acremonium loliae*, produces the neurotoxic alkaloid lolitrem B. The tropical pasture plant *Paspalum* is associated with the related condition, paspalum staggers, but the fungus involved in this case is *Claviceps paspali*, which is related to the ergot fungus (see p. 576). An ergot alkaloid, ergovaline, is also produced by the fungus *Acremonium coenophialum*, which infects the grass tall fescue. Ergovaline causes the condition known as 'fescue foot', in which grazing animals become lame and develop gangrene at their extremities. The last mycotoxin to be considered is sporidesmin, which is produced by the pasture fungus *Pithomyces chartarum*. In Australasia it causes liver damage in grazing animals, which leads to the release into the circulation of bile and the chlorophyll breakdown product phylloerythrin. These compounds cause jaundice and photosensitisation of the skin, hence the name of the disorder, 'facial exzema'.

Control of these mycotoxicoses is based on the use of fungicides on pasture, on the selection of resistant plants and animals, and on the modification of grazing practices to avoid consumption of infected herbage (e.g. avoiding the grazing of infected dead plant material close to the ground).

Grass sickness

Horses grazing pasture after a dry period can develop this condition, for which the scientific name is equine dysautonomia. It is characterised by muscular tremors, difficulty or inability to swallow, regurgitation of stomach contents, abnormal stance, impaction of the colon and loss of weight. These signs are the results of damage to the autonomic nervous system. This damage extends to the nerve cells in the autonomic ganglia and in the enteric nervous system, the plexus of neurons within the intestinal wall. The exact cause of grass sickness is not known but there is evidence that a botulinum toxin is involved, this arising from the organism *Clostridium botulinum* type C in the soil or on the herbage. Large amounts of the toxin have been found in the gut of animals suffering acute signs of the condition, and antibodies have been isolated in chronically ill animals. It is also thought that the animal's susceptibility to the toxin is affected by its immune defences in the gut mucosa, and is also influenced by a nutritional trigger involving the biochemical status of the grass.

19.3 Legumes

Pasture and field crop legumes

The family Leguminoseae contains about 18 000 species, which are valued for their ability to grow in a symbiotic relationship with nitrogen-fixing bacteria and for their drought resistance. The commonest legumes found in pastures are the clovers (*Trifolium* spp.), the main representatives being red clover (*T. pratense*) and white clover (*T. repens*) in the cooler and wetter regions such as Europe and New Zealand, and subterranean clover (*T. subterraneum*) in drier areas such as southern Australia.

Nutritionally, the clovers are superior to grasses in protein and mineral content (particularly calcium, phosphorus, magnesium, copper and cobalt), and their nutritive value falls less with age. Studies with white clover have shown that the rates of reduction in particle size and of movement of particulate matter from the rumen are more rapid than with grass. Sheep and cattle, offered white clover as fresh forage, consumed 20 per cent more dry matter than from grass of the same metabolisable energy content. Similar high voluntary intakes of dry matter have been obtained from red clover and other legumes.

The sugars present in clovers are similar to those found in grasses, the main sugar being sucrose. Fructans are generally absent, but starch is present and concentrations of this polysaccharide as high as 50 g/kg DM have been reported in the dried leaves of red clover.

Many tropical pastures are deficient in indigenous legumes, but determined efforts are being made to introduce them. In Australia, for example, the South American legume centro (*Centrosema pubescens*) has been introduced into pastures of the wet tropical type, and siratro (*Macroptilium atropurpureum*) has been bred from Mexican cultivars for use in drier areas. Because tropical legumes are similar anatomically to temperate legumes, they differ much less in digestibility than do tropical and temperate grasses.

Lucerne or alfalfa (*Medicago sativa*) also occurs in pastures, but like many other legumes is more commonly grown on its own. It is found in warm temperate areas and in many tropical and subtropical countries. The protein content is comparatively high and declines only slowly with maturity (Table 19.7). Lucerne grown in the UK tends to be high in fibre, particularly the stem, and at the late flowering stage crude fibre may be as high as 500 g/kg DM. Lucerne cultivars are distinguished by the time of flowering, and for UK conditions early flowering types are recommended. These varieties usually flower in the second week of June, but to obtain a cut with acceptable digestibility, the crop should be first harvested at the early bud stage (end of May), when the expected digestible organic matter (DOM) content would be 620–640 g/kg DM, and subsequently cut at six–eight-week intervals to give DOM values of 560–600 g/kg DM.

In Britain, the small area of lucerne grown is harvested mainly for silage or for artificial drying (see Chapter 21), but in other parts of the world, notably the USA (where it is known as alfalfa), the crop is also used for grazing.

Berseem or Egyptian clover (*Trifolium alexandrinum*) is an important legume grown in the Mediterranean area and India. It is valued for its rapid growth in the cooler winter season in the subtropics and for its good recovery after cutting or grazing. It has a nutritive value very similar to that of lucerne.

Sainfoin (*Onobrychis viciifolia*) is a legume of less economic importance than lucerne, and in the UK is confined to a few main areas in the south. In common with most green forages the leaf is richer than the stem in crude protein, ether extract and minerals, especially calcium. Changes which occur in the composition of the plant are mainly due to variation in stem

Table 19.7 Composition of the dry matter of lucerne (Adapted from MAFF 1975 *Energy Allowances and Feeding Systems for Ruminants*, Technical Bulletin, No. 33, HMSO, London, p 70)

	Pre-bud	In-bud	Early flower
Crude fibre, g/kg	220	282	300
Ash, g/kg	120	82	100
Crude protein, g/kg	253	205	171
Digestible organic matter, g/kg	670	620	540
Metabolisable energy, MJ/kg	10.2	9.4	8.2

composition and leaf–stem ratio. The crude protein content in the dry matter may vary from 240 g/kg at the early flowering stage to 140 g/kg at full flower. Corresponding crude fibre values at similar stages of growth may be 140 and 270 g/kg DM.

Peas (*Pisum sativum*), beans (*Vicia faba*) and vetches (*Vicia sativa*) are sometimes grown as green fodder crops, and when cut at the early flowering stage are similar in nutritive value to other legumes.

Leguminous trees and shrubs

Animals grazing mature grass pastures are often able to supplement their diet by consuming the foliage of trees and shrubs, many of which are legumes. The collective term applied to food obtained in this way is 'browse'; in addition to being harvested by animals, browse may be cut and carried to housed livestock. The foliage of leguminous trees is high in protein (200–300 g/kg DM) and minerals, but is also high in fibre (500–600 g neutral-detergent fibre per kg DM). Foliage of some species also has a high concentration of condensed (i.e. non-hydrolysable) tannins, whose role in nutrition is problematical. Low–moderate concentrations precipitate soluble plant proteins and thus protect them against digestion in the rumen (see Chapter 8), but if the proteins are too firmly bound to the tannins they are not digested in the small intestine. For foliage with a high concentration of condensed tannins (>100 g/kg DM), digestibility (especially of protein) tends to be low. Tannins and possibly other constituents reduce the palatability of browse, so its nutritional value may be as a food reserve, to be consumed when grass herbage is no longer available.

One of the best-known browse species is leucaena (*Leucaena leucocephala*), also known as ipil-ipil, which is a valuable tropical legume that is cultivated extensively in many parts of the world, notably in SE Asia, Latin America and the West Indies. It is a valuable source of protein and minerals, and is also rich in β-carotene. However, it also contains the toxic amino acid mimosine (see below). Other examples of legume forages are gliricidia (*Gliricidia sepium*), sesbania (*Sesbania sesban*) and acacia (*Acacia angustissima*), all of which are cultivated in Africa.

Nutritional disorders associated with legumes

A disorder that is frequently encountered in cattle and sheep grazing on legume-dominated pastures is bloat. The most serious problems are associated with clovers and lucerne. The primary cause of bloat is the retention of the fermentation gases in a stable foam, preventing their elimination by eructation. Soluble leaf proteins are thought to play the major role in the formation of the foam (see p. 186). Legumes that contain significant concentrations of condensed tannins (>20 g/kg), such as sainfoin, are unlikely to cause bloat, probably because of the ability of tannins to precipitate soluble proteins.

Associated with grazing sheep on pure lucerne is a sudden death syndrome termed 'redgut'. This is thought to be caused by the rapid passage of highly digestible forage through the rumen that causes increased fermentation in the large intestine. In early weaning experiments in New Zealand, 1–3 per cent of lambs grazing lucerne died of this disorder. The incidence was reduced when the lucerne was supplemented with meadow hay.

A large number of species of plants are known to contain compounds which have oestrogenic activity. Pasture plants containing these phytoestrogens are mainly of the species *Trifolium subterraneum* (subterranean clover). *T. pratense* (red clover), *Medicago sativa* (lucerne) and *M. truncatula* (barrel medic). The oestrogens in *Trifolium* sp. are mainly isoflavones, whereas those in *Medicago* sp. are usually coumestans.

Naturally occurring isoflavones and coumestans have relatively weak oestrogenic activity, but this activity can be increased as a result of metabolism in the rumen. For example, the main isoflavone in subterranean clover, formononetin, is converted to equol in the rumen. Some plants, e.g. *T. repens* (white clover), are normally non-oestrogenic but when infected with fungi can produce high concentrations of coumestan.

The consumption of oestrogenic pasture plants by sheep leads to severe infertility and post-natal death in lambs. The infertility can persist for long periods after the ewes have been taken off the oestrogenic pastures. The main cause of the infertility is a cystic glandular hyperplasia of the uterus, which causes an increased flow of mucus and hence poor sperm penetration to the oviduct. A 'temporary infertility' may occur in ewes grazing oestrogenic pastures at the time of mating. Fertility is restored when sheep are moved to other pastures. Improved cultivars of subterranean clover containing lower contents of oestrogenic substances are now grown in Australia.

Cattle grazing oestrogenic pastures do not appear to suffer the severe infertility problems that affect sheep.

Leucaena contains the toxic amino acid, mimosine. In the rumen this is converted to dihydroxypyridine (DHP), a compound with goitrogenic properties (see p. 135). Ruminants consuming large quantities of leucaena may suffer weight loss, thyroid dysfunction and loss of hair or wool (alopecia). In countries with a natural population of leucaena, grazing animals possess a rumen microorganism capable of destroying DHP (*Synergistes jonesii*). In Australia, the ill effects of introduced leucaena have been ameliorated by inoculating grazing animals with rumen contents taken from animals in Hawaii.

19.4 Other forages

Cereals

Cereals are sometimes grown as green forage crops, either alone or mixed with legumes. Like the grain, the forage is rich in carbohydrate and low in protein, its nutritive value depending mainly on the stage of growth when

Table 19.8 Composition of whole barley at different stages of growth (Adapted from Edwards R A, Donaldson E and MacGregor A W 1968 *Journal of the Science of Food and Agriculture* **19**: 656–60; MacGregor A W and Edwards R A 1968 *Journal of the Science of Food and Agriculture* **19**: 661–6)

	Stage of growth[a]						
	1	2	3	4	5	6	7
Dry matter g/kg	191	200	258	293	353	387	425
Components of DM, g/kg							
Crude protein	103	87	72	67	56	60	66
Crude fibre	313	321	286	253	204	195	254
WSC [b]	193	200	265	326	255	185	86
Fructose	60	50	41	31	29	31	22
Glucose	60	60	42	29	28	20	11
Sucrose	19	15	23	33	21	20	4
Fructans	31	33	72	128	122	66	23
Starch	3	3	4	10	185	348	413
Calcium	5.1	5.5	5.2	4.1	3.5	2.3	1.9
Phosphorus	2.3	2.0	1.8	1.7	1.6	1.7	2.1
D Value [c]	62	54	58	61	63	61	55

[a] Stage of growth: 1 = heading completed, 2 = flowering, 3 = watery kernels, 4 = milky kernels, 5 = early mealy ripe, 6 = late mealy ripe, 7 = ripe for cutting.
[b] Water-soluble carbohydrates.
[c] Digestible organic matter as a percentage of the dry matter (*in vitro*).

harvested (see Table 19.8). The crude protein content of the cereal at the grazing stage is generally within the range of 60–120 g/kg DM. At the time of ear formation the concentration of crude fibre falls as a result of the great increase in starch, which tends to maintain the digestibility value.

Sugarcane

Sugarcane (*Saccharum officinarum*) is a tropical or subtropical perennial grass that grows to a height of 4.5–6.0 m or more. The crop is processed for its sugar, which leaves two by-products: molasses and a fibrous residue termed bagasse. Sugarcane molasses is a high-energy, low-protein food similar in composition to the molasses obtained as a by-product from sugar beet (see p. 555). Bagasse is a high-fibre, low-protein product of very low digestibility which is sometimes mixed with the cane molasses for cattle feeding. The unmolassed bagasse has a digestibility of about 0.28, but this can be dramatically increased to about 0.55 by short-term (5–15 minute) treatment with wet steam at 200 °C. Steam-treated bagasse, supplemented with urea, has been shown to be suitable as a maintenance feed for beef cows.

In some countries, the whole sugarcane crop is used as forage for ruminants. The whole crop on a dry matter basis has an ME value of about 9 MJ/kg and a low crude protein content of about 40 g/kg.

Brassicas

The genus *Brassica* comprises some 40 species, of which the following are of agricultural importance: kales, cabbages, rapes, turnips and swedes. Some of the brassicas are grown primarily as root crops, and these will be discussed in Chapter 22.

Kales (Brassica oleracea)

The kales include a very wide variety of plant types, which range from short leafy plants, 30 cm high, to types 2 m tall with stems strong enough to be used in building. The commonest short type is thousandhead kale (var. *fruticosa*) and the commonest tall type is marrowstem kale (var. *acephala*), which is known in Australasia as 'chou moelier'. They are grown in temperate parts of the world to provide green fodder during winter, but in drier areas they may also be used to supplement summer grazing.

Kales are low in dry matter content (about 140 g/kg), which is rich in protein (about 150 g/kg), water-soluble carbohydrates (200–250 g/kg) and calcium (10–20 g/kg), and their digestibility is generally high. The woody stems of marrowstem kale are lower in digestibility than the rest of the plant and may be rejected by animals.

Rapes

The rapes grown in Britain are usually swede-rapes (*B. napus*) although turnip-rapes (*B. campestris*) also occur. Rape may be included in seed mixtures for pastures, in order to provide forage until the grasses become established. The nutritive value of rapes is similar to that of the kales.

Cabbages (B. oleracea, *var. capitata*)

These are grown for both human and animal consumption and range in type from 'open-leaved' to 'drumhead'. All have a low proportion of stem and hence are less fibrous than either kales or rapes.

Toxicity of brassica forage crops

All brassicas, whether grown as forage, root or oilseed crops, contain goitrogenic substances (see p. 135). In the forage crops these are mainly of the thiocyanate type, which interferes with the uptake of iodine by the thyroid gland and whose effects can be overcome by increasing the iodine content of the diet. All animals grazing on forage brassicas may develop goitre to some extent, but the most serious effects are found in lambs born to ewes which have grazed on brassicas during pregnancy; these lambs may be born dead or deformed. It has been suggested (but not adequately confirmed) that cows grazing on kales may secrete sufficient goitrogen in their milk to cause goitre in children drinking it.

Forage brassicas may also cause a haemolytic anaemia in ruminants, in extreme cases of which the haemoglobin content of the blood falls to only one-third of its normal value and the red cells are destroyed so rapidly that haemoglobin appears in the urine (haemoglobinuria). The condition is due to the presence in brassicas of the unusual amino acid S-methylcysteine sulphoxide, which in the rumen is reduced to dimethyl disulphide:

$$2CH_3 - \overset{\overset{\displaystyle O}{\|}}{S}.CH_2CHNH_2COOH + H_2 \longrightarrow CH_3.S.S.CH_3 + 2CH_3CO.COOH + 2NH_3$$

| S-methylcysteine sulphoxide | Dimethyl disulphide | Pyruvic acid |

The dimethyl disulphide is known to damage the red cells. Green brassicas contain 10–20 g S-methylcysteine sulphoxide per kg DM. The condition is best avoided by ensuring that kale or rape contributes no more than one-third of the animal's total dry matter intake.

Green tops

Mangel, fodder beet, sugar beet, turnip and swede tops may all be used for feeding farm animals. Care is required in feeding with mangel, fodder and sugar beet tops, since they contain a toxic ingredient which may lead to extensive scouring and distress and, in extreme cases, death. The risk appears to be reduced by allowing the leaves to wilt. The toxicity has been attributed to oxalic acid and its salts, which are supposed to be reduced or removed by wilting. A recent study casts some doubt on this theory, since the oxalate content of the leaves is practically unaffected by wilting. It is possible that the toxic substances are not oxalates but other factors which are destroyed during wilting.

Swede and turnip tops are safe for feeding, and may have a crude protein content in the dry matter as high as 200 g/kg, the digestibility of the organic matter being about 0.70; like kale, rape and cabbage, they may cause haemolytic anaemia in ruminants.

Sugar beet tops generally contain the upper part of the root as well as the green leaves and are more digestible, about 0.77. All these green tops are excellent sources of β-carotene.

Summary

1. Temporary pastures often consist of a single, sown plant species, whereas permanent pastures include several species of grasses and legumes. Natural grasslands consist of a wide variety of plants, including shrub species.

2. As grazing animals are free to select their diet, their nutrient intake is generally unknown. They generally have to work hard to harvest their food.

3. Grasses grow from green leafy material to mature stems and flowering heads, and their pattern of growth is reflected in their chemical composition. Fructans and other soluble carbohydrates decline from about 250 to 25 g/kg DM, and protein from 250 to 50 g/kg DM. Fibre, in the form of cellulose + hemicellulose, increases to as much 600 g/kg DM. If not restricted by soil supplies, grasses are good sources of minerals (and also of vitamins, except vitamin D). Digestibility and metabolisable energy value of grasses decrease as they mature.

4. Although growth stage is the main determinant of the nutritive value of grasses, there are relatively small differences among temperate species but large differences between temperate and tropical species. The latter differences are due to plant anatomy and photosynthetic pathways. Variations in soils and fertiliser applications have major effects on minerals in grasses, but may also affect other nutrients; thus fertiliser N increases plant protein content.

5. Grazing systems that allow selection by animals tend to cause the accumulation of mature herbage of low nutritive value.

6. Nutritional disorders associated with grass pastures include nitrate poisoning, various mycotoxicoses and (in horses) grass sickness.

7. In comparison with grasses, pasture legumes contain more protein and minerals, and starch in place of fructans. They are more-rapidly digested than grasses and therefore promote higher intakes. Clovers are grown with grasses, but other legumes, such as lucerne (alfalfa) are often grown alone. Shrub and tree legumes, such as leucaena, provide browse for grazing animals.

8. The disorder bloat, in which gas accumulates in the rumen of grazing animals, is caused by legumes such as the clovers that contain high concentrations of soluble leaf proteins and low concentrations of the condensed tannins which precipitate proteins.

9. Other forage crops include immature cereals, sugarcane (and its fibrous by-product, bagasse), various brassica species and the green tops of root crops. Brassicas such as kale contain goitrogens and also the unusual amino acid S-methyl-cysteine sulphoxide, which causes haemolytic anaemia in ruminants.

Further reading

Acamovic T, Stewart C S and Topps J H (eds) 1997 Legume forages and indigenous browse for ruminants in the semi-arid tropics. *Animal Feed Science and Technology* **69**: 1–287.

Barry T N and Blaney B J 1987 Secondary compounds of forages. In: Hacker J B and Ternouth J H (eds) *The Nutrition of Herbivores*, Sydney, Academic Press.

Butler G W and Bailey R W (eds) 1973 *Chemistry and Biochemistry of Herbage*, Vols 1–3, London, Academic Press.

Fahey G C, Collins M, Mertens D R and Moser L E 1994 *Forage Quality, Evaluation and Utilization*, American Society of Agronomy, Lexington, KY.

Hopkins A (ed.) 2000 *Grass, Its Production and Utilisation*, 3rd edn, Oxford, Blackwell Science.

Jones D I H and Wilson A D 1987 Nutritive quality of forage. In: Hacker J B and Ternouth J H (eds) *The Nutrition of Herbivores*, Sydney, Academic Press.

Leng R A 1990 Factors affecting the utilisation of 'poor quality' forages by ruminants particularly under tropical conditions. *Nutrition Research Reviews* **3**: 277–303.

Mello J P F and Devendra C (eds) 1995 *Tropical Legumes in Animal Nutrition*, Wallingford, UK, CAB International.

Minson D J 1990 *Forage in Ruminant Nutrition*, New York, Academic Press.

Thomas C, Reeve A and Fisher G E J (eds) 1991 *Milk from Grass*, 2nd edn, Reading, British Grassland Society.

Whiteman P C 1980 *Tropical Pasture Science*, Oxford University Press.

20 Silage

20.1   Silage, ensilage and silos

20.2   Role of plant enzymes in ensilage

20.3   Role of microorganisms in ensilage

20.4   Nutrient losses in ensilage

20.5   Classification of silages

20.6   Nutritive value of silages

20.1   Silage, ensilage and silos

Silage is the material produced by the controlled fermentation of a crop of high moisture content. Ensilage is the name given to the process and the container, if used, is called the silo. Almost any crop can be preserved as silage, but the commonest are grasses, legumes and whole cereals, especially wheat and maize.

The first essential objective in preserving crops by natural fermentation is the achievement of anaerobic conditions. In practice this is done by chopping the crop during harvesting, by rapid filling of the silo and by adequate consolidation and sealing. The main aim of sealing is to prevent re-entry and circulation of air during storage. Where oxygen is in contact with herbage for any period of time, aerobic microbial activity occurs and the material decays to a useless, inedible and frequently toxic product.

The second essential objective is to discourage the activities of undesirable microorganisms such as clostridia and enterobacteria, which produce objectionable fermentation products. These microorganisms can be inhibited either by encouraging the growth of lactic acid bacteria or by using chemical additives. Lactic acid bacteria ferment the naturally occurring sugars (mainly glucose and fructose) in the crop to a mixture of acids, but predominantly lactic. The acids produced increase the hydrogen ion concentration to a level at which the undesirable bacteria are inhibited. The critical pH at which inhibition occurs varies with the dry matter content of the crop ensiled. The attainment of the critical pH is more difficult with crops of high buffering capacity. Legumes are more highly buffered than grasses and are consequently more difficult to ensile satisfactorily. With grass crops having a dry matter content of about 200 g/kg the achievement of a pH of about 4.0 will normally preserve the crop satisfactorily, as long as the silo remains airtight and is free from penetration by rain. Wet crops are very difficult to ensile satisfactorily and should either be prewilted under good weather conditions or treated with a suitable additive. Similarly, crops low in water-soluble carbohydrates, or those which are highly buffered, must also be treated with an effective additive before ensiling.

The types of silo in which the farmer may choose to ferment his crop are very varied, ranging from small plastic bags to large cylindrical towers built of concrete, steel or wood. In recent years the amount of silage conserved as big bales, usually weighing from 0.5 to 0.75 tonne and encased in plastic bags or wrapped in plastic film, has increased dramatically. Provided the bags are well sealed and not punctured during storage, this method of conserving grass is satisfactory. The development of effective chopper balers has increased the efficiency of the technique, and improved the preservation and nutritional quality of the silage. Currently, about 20–25 per cent of UK silages are made by this method, but the commonest silo used is still of the clamp or bunker type. This generally consists of three solid walls some 2–3 metres in height and often built beneath a Dutch barn to protect the silage from the weather. When full, the surface of these silos is covered with plastic sheeting and weighted with some suitable material such as tyres or bales of straw.

20.2 Role of plant enzymes in ensilage

Immediately after cutting the crop and during the early stages of ensiling, chemical changes occur as a result of the activity of enzymes present in the plant tissue. The processes of respiration and proteolysis are of particular importance in influencing the nutritional value of the final product.

Respiration

Respiration may be defined as the oxidative degradation of organic compounds to yield usable energy. In higher plants, as in animals, oxygen is the terminal electron acceptor. Carbohydrates are the major respiratory source and the substrate for oxidation is usually a hexose sugar, which undergoes glycolysis and subsequent oxidation via the tricarboxylic acid cycle to carbon dioxide and water. In the harvested plant, biosynthetic reactions are limited and virtually all the energy in the hexose is converted into heat. In the isolated plant this heat energy would be dissipated into the atmosphere, but in the silo the heat is retained in the mass of herbage, causing an increase in temperature. This loss of soluble carbohydrates, through respiration, is a wasteful process and may result in such a depletion of substrate that subsequent fermentation may be adversely affected. Plant respiration will continue in the silo as long as both oxygen and a supply of substrate are available. The simplest method of limiting respiration is to achieve anaerobic conditions in the silo as rapidly as possible.

Proteolysis

In fresh herbage, 75–90 per cent of the total nitrogen is present as protein. After harvesting, rapid proteolysis (hydrolysis of peptide bonds) occurs and, after a few days of wilting in the field, the protein content may be reduced by as much as 50 per cent. The extent of protein degradation varies with plant species, DM content and temperature. Once the material is ensiled, proteolysis continues but the activity declines as the pH falls. The products of proteolysis are amino acids and peptides of varying chain length.

Further breakdown of amino acids occurs as a result of plant enzyme action, although this is considered to be limited. Most destruction of amino acids in silage is brought about by microbial activity rather than by plant enzymes.

20.3 Role of microorganisms in ensilage

Aerobic fungi and bacteria are the dominant microorganisms on fresh herbage, but as anaerobic conditions develop in the silo they are replaced by bacteria able to grow in the absence of oxygen. These include lactic acid bacteria, clostridia and enterobacteria.

Lactic acid bacteria

The lactic acid bacteria, which are facultative anaerobes (able to grow in the presence or absence of oxygen), are normally present on growing crops in small numbers but usually multiply rapidly after harvesting, particularly if the crop is chopped or lacerated. They can be divided into two categories: the homofermentative bacteria (e.g. *Lactobacillus plantarum*, *Pediococcus pentosaceus* and *Enterococcus faecalis*) and the heterofermentative bacteria (e.g. *Lactobacillus brevis* and *Leuconostoc mesenteroides*). When the crop is ensiled, the lactic acid bacteria continue to increase, fermenting the water-soluble carbohydrates in the crop (see p. 497) to organic acids, mainly lactic, which reduce the pH value. The homofermentative lactic acid bacteria are more efficient at producing lactic acid from hexose sugars than are the heterofermentative organisms (Table 20.1). During ensilage, some hydrolysis of hemicelluloses also occurs, liberating pentoses, which may be fermented to lactic and acetic acids by most types of lactic acid bacteria.

Clostridia

Clostridia are present on crops but the main source in silage is soil contamination. They occur in the form of spores and only grow under strict anaerobic conditions. They can be divided into two major groups: saccharolytic clostridia and proteolytic clostridia. The saccharolytic bacteria (e.g.

Table 20.1 Some fermentation pathways in ensilage

Lactic acid bacteria

Homofermentative:

Glucose→2 Lactic acid

Fructose→2 Lactic acid

Pentose→Lactic acid + Acetic acid

Heterofermentative:

Glucose→Lactic acid + Ethanol + CO_2

3 Fructose→Lactic acid + 2 Mannitol + Acetic acid + CO_2

Pentose→Lactic acid + Acetic acid

Clostridia

Saccharolytic:

2 Lactic acid→Butyric acid + $2CO_2$ + $2H_2$

Proteolytic:

Deamination

Glutamic acid→Acetic acid + Pyruvic acid + NH_3

Lysine→Acetic acid + Butyric acid + $2NH_3$

Decarboxylation

Arginine→Putrescine + CO_2

Glutamic acid→γ-Aminobutyric acid + CO_2

Histidine→Histamine + CO_2

Lysine→Cadaverine + CO_2

Oxidation/reduction (Stickland)

Alanine + 2 Glycine→3 Acetic acid + $3NH_3$ + CO_2

Enterobacteria

Glucose→Acetic acid + Ethanol + $2CO_2$ + $2H_2$

Clostridium butyricum and *Clostridium tyrobutyricum*) ferment lactic acid and residual water-soluble carbohydrates to butyric acid, resulting in a rise in the pH. Proteolytic clostridia (e.g. *Clostridium bifermentans* and *Clostridium sporogenes*) ferment mainly amino acids to a variety of products including acetic and butyric acids, amines and ammonia (Table 20.1).

Clostridia grow best at pH 7.0–7.4. They cannot tolerate acid conditions and a pH of 4.2 is usually considered to be low enough to inhibit their growth. The nature of the acids responsible for lowering the pH is important, with undissociated organic acids being the most potent. Clostridia are particularly sensitive to water availability and require very wet conditions for active growth. With very wet crops (i.e. those with a DM concentration of about 150 g/kg), even the achievement of a pH value as low as 4 may not inhibit their activity. Growth of clostridia is severely restricted if the dry matter of the ensiled material is above 300 g/kg, but complete inhibition may require considerably higher figures, perhaps as much as 400 g/kg.

As well as being inimical to preservation, clostridia may be a health hazard to animals on diets based on contaminated silage. A number of cases of botulism, caused by *Clostridium botulinum*, have been reported in horses and cattle receiving such diets. The condition is uncommon in the UK but is more so in the USA and the Baltic countries. It has also been claimed to be the causative organism of grass sickness of horses, although a nutritional trigger has been postulated.

Listeria

Low dry matter silages contaminated with soil may also contain *Listeria monocytogenes*, a bacterium known to be responsible for several diseases such as meningoencephalitis, anterior uveitis, and placentitis and subsequent abortion in cattle. A number of fatalities in ruminant animals (mainly sheep) on silage diets have been confirmed as listerioses. Listeriosis in horses is uncommon, probably because silage is rarely given to horses.

Meningoencaphalitis, abortion and fatal septicaemias of foals have been reported following infection with *L. monocytogenes*. The organism has not been isolated from silages with a pH below 4.7. Big bale silages are particularly at risk, owing to their high surface area to volume ratio and consequent increased susceptibility to the ingress of air and mould infestation.

Enterobacteria

The enterobacteria associated with silage, sometimes described as 'acetic acid bacteria' or 'coliform bacteria', are usually present in very low numbers on crops. Unlike the clostridia, they are facultative anaerobes, and consequently compete with the lactic acid bacteria for the water-soluble carbohydrates. They ferment these to a mixture of products including acetic acid, ethanol and hydrogen. Like the clostridia, they can decarboxylate and deaminate amino acids with consequent production of large concentrations of ammonia. The optimum pH for the growth of these organisms is about 7.0, and they are usually active only in the early stages of fermentation, when the pH is favourable for their growth. Examples of species commonly found in silage are *Escherichia coli* and *Erwinia herbicola*.

Bacillus species

These species are widespread contaminants of harvested grass, but their growth in well-preserved silage is limited by the development of lactic acid. However, they are abundant in silage which has undergone aerobic deterioration. Bovine abortions have been attributed to the consumption of silage contaminated with *Bacillus lichiniformis*.

Fungi

Fungi, which are present in soil and on vegetation, grow either as single cells, the yeasts, or as multicellular filamentous colonies, the moulds. Yeasts associated with silages include species of *Candida, Saccharomyces* and *Torulopsis*. They play a particularly important role in the deterioration of silage when it is exposed to air.

The majority of moulds are strict aerobes and are active on the surface layers of silage. A wide variety have been isolated from different types of silage, particularly when aerobic decomposition has occurred. Many of them are capable of producing mycotoxins. Some of the commoner mycotoxergenic fungi found in silage are shown, along with their associated mycotoxins, in Table 20.2.

There have been relatively few cases in which mycotoxicosis has been definitively identified as the cause of disease in animals consuming silage, or in human beings handling it. In part, this may be because of the ability of the rumen microbes to metabolise some of the toxins, as is the case with zearelenone, ochratoxin and some of the trichothecenes. In addition, the ruminant appears to be able to metabolise ingested trichothecenes with considerable efficiency. Failure to prove a causative effect does not preclude the occurrence of disease at sub-clinical levels. Many of the proven effects of the mycotoxins on fertility, and the immune and nervous systems, are long-term and may only show after continued exposure. The potential for deleterious effects on animals consuming spoiled silage and workers handling it does exist, as does the danger of transfer of the toxins into the human food chain. Silage which has undergone aerobic deterioration owing to mould infestation should not be given to animals, and workers handling such material should do so with great care.

The achievement of adequate compaction and sealing during the making of silage should reduce aerobic deterioration to a minimum as long as access of air during storage is denied. This may be a particular problem with silage made in big bales. Even small perforations of the plastic may be very serious owing to the large surface area to volume ratio in the bales. Birds can be a particular problem in this respect but netting of the bale stacks can provide an effective deterrent. The bagged, as distinct from the wrapped

Table 20.2 Some fungi, together with their mycotoxins, found in silage

Fungus	Mycotoxin
Penicillium roqueforti	Roquefortine A, B & C; PR toxin; microfinolic acid; penicillic acid
Byssochlamys nivea	Patulin
Paecilomyces viriotii	Patulin
Aspergillus clavatus	Patulin; cytochanasin E; tryptoquinolins
Aspergillus fumigatus	Fumiclavines A and C; fumitoxins A, B & C; gliotoxin, several tremorgens
Aspergillus flavus	Aflatoxins; cyclopiazonic acid
Fusarium culmorum	Dioxynivalenol; T2 toxin; HT toxin; zearelenone
Fusarium crookwellense	Zearelenone

silages, are also susceptible at the neck, where fungal infestation results from an inadequate tie.

In most cases deterioration occurs after opening of the silo and during feeding. Good management of the exposed surface to limit access of air is essential in the case of clamp and bunker silos at this time. In this context, the width of the face and the rate of usage of the silage will be the important factors to be taken into account. Silages made with additives which restrict fermentation, and which have raised pH and significant amounts of residual water-soluble carbohydrate, are particularly susceptible. Some measure of protection against fungal infestation may be provided by additives containing organic acids or their salts.

20.4 Nutrient losses in ensilage

Field losses

With crops cut and ensiled the same day, nutrient losses are negligible and even over a 24-hour wilting period, losses of dry matter of not more than 1 or 2 per cent may be expected. Over periods of wilting longer than 48 hours, considerable losses of nutrients can occur depending upon weather conditions. Dry matter losses as high as 6 per cent after five days and 10 per cent after eight days of wilting in the field have been reported. The main nutrients affected are the water-soluble carbohydates and proteins, which are hydrolysed to amino acids.

Oxidation losses

These result from the action of plant and microbial enzymes on substrates, such as sugars, in the presence of oxygen, with the concomitant formation of carbon dioxide and water. In a silo which has been rapidly filled and sealed, the oxygen trapped within the plant tissues is of little significance and causes dry matter losses of about 1 per cent only. Continuous exposure of herbage to oxygen, as sometimes occurs on the sides and upper surface of ensiled herbage, leads to the formation of inedible composted material. Measurements of this as surface waste can be misleading, since losses of dry matter of up to 75 per cent can occur in its formation, and the visible waste is a small part only of the lost material. Such losses may be particularly important when silage is made in 'big bales', owing to the high surface area to weight ratio and the vulnerability of the bales to air penetration.

Fermentation losses

Although considerable biochemical changes occur during fermentation, especially to the soluble carbohydrates and proteins, overall dry matter and energy losses arising from the activities of lactic acid bacteria are low. Dry matter losses can be expected to be less than 5 per cent and gross energy losses, because of the formation of high-energy compounds such as ethanol, are even less. In clostridial and enterobacterial fermentations, nutrient losses will be much higher than in lactic acid bacterial fermentations because of the evolution of the gases carbon dioxide, hydrogen and ammonia.

Effluent losses

In most silos, free drainage occurs and the liquid, or effluent, carries with it soluble nutrients. The amount of effluent produced depends largely upon the initial moisture content of the crop. Several equations, based on the dry matter of the ensiled material, have been suggested for estimating effluent loss. Typical is that of Bastiman:

$$V_n = 767 - 5.34D + 0.009\ 36D^2$$

Where: V_n = volume of effluent produced (litre/t of herbage)
D = dry matter content of herbage (kg/t)

As well as dry matter, factors such as type of silo, degree of consolidation and the nature and pretreatment of the crop will all affect effluent loss, but it will obviously be increased if the silo is left uncovered so that rain enters. Effluent contains sugars, soluble nitrogenous compounds, minerals and fermentation acids, all of which are valuable nutrients. Ensiling crops with a dry matter content of about 150 g/kg may result in effluent dry matter losses as high as 10 per cent. With crops wilted to about 300 g/kg DM there may be little or no effluent loss.

Table 20.3 Classification of silages

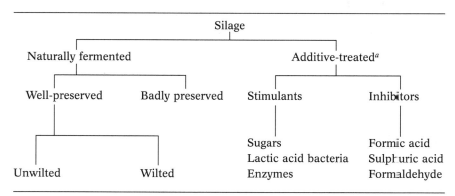

^{a} Only a few commonly used examples are given.

20.5 Classification of silages

Silages may be classified into two main categories, naturally fermented and additive-treated. These may be further subdivided as shown in Table 20.3.

Naturally fermented silages

Well-preserved unwilted silages

In this type of silage, commonly made from grasses and whole cereal crops, lactic acid bacteria have dominated the fermentation. A typical composition is shown in Table 20.4.

These silages are characterised by having low pH values, usually between 3.7 and 4.2, and containing high concentrations of lactic acid. In grass silages, the lactic acid contents generally lie in the range 80–120 g/kg DM, although higher amounts may be present if silages are made from wet crops rich in water-soluble carbohydrates. In maize silages, lactic acid contents are usually much lower than those of well-preserved grass silages because of the higher DM content and lower buffering properties of the original crop.

The silages usually contain small amounts of acetic acid and may also contain traces of propionic and butyric acids. Variable amounts of ethanol and mannitol derived from the activities of lactic acid bacteria and yeasts are present. Very little water-soluble carbohydrates remain after fermentation, usually less than 20 g/kg DM.

Table 20.4 Typical composition of well-preserved silages made from perennial ryegrass[a] and maize (Adapted from Donaldson E and Edwards R A 1976 *Journal of the Science of Food and Agriculture* **27**: 536–44; Wilkinson J M and Phipps R H 1979 *Journal of the Science of Food and Agriculture, Cambridge* **92**: 485–91)

	Grass silages		Maize silage
	Unwilted	Wilted	
DM, g/kg	186	316	285
pH	3.9	4.2	3.9
Total N, g/kg DM	23.0	22.8	15.0
Protein N, g/kg TN[b]	235	289	545
Ammonia N, g/kg TN	78	79	63
WSC[c], g/kg DM	10	47	16
Starch, g/kg DM	–	–	206
Acetic acid, g/kg DM	36	24	26
Butyric acid, g/kg DM	1.4	0.6	0
Lactic acid, g/kg DM	102	59	53
Ethanol, g/kg DM	12	6.4	<10

[a] Both ryegrass silages made from the same ryegrass source.
[b] TN = Total nitrogen.
[c] WSC = Water-soluble carbohydrates.

The nitrogenous components of well-preserved silages are mainly in a soluble non-protein form in contrast to those present in fresh forage crops, where most of the total nitrogen (TN) is present as protein. Some deamination of amino acids may occur during fermentation, but this activity is likely to be low and consequently the ammonia content of these silages will also be low, usually less than 100 g NH_3-N/kg TN.

Because of the extensive changes to the water-soluble carbohydrates resulting in the formation of high-energy compounds such as ethanol (gross energy = 29.8 MJ/kg), the gross energy concentrations of these silages are higher than those of the parent material.

Well-preserved wilted silages

Wilting a crop prior to ensiling restricts fermentation increasingly as dry matter content increases. In such wilted silages, clostridial and enterobacterial activities are normally minimal although some growth of lactic acid bacteria occurs, even in herbage wilted to dry matter contents as high as 500 g/kg. With very dry silages of this type, anaerobic storage in bunker silos is difficult and tower silos are preferred because there is less risk of air penetration. For bunker-type silos, a more normal aim is to prewilt the crop to a dry matter content of 280–320 g/kg. The composition of a typical wilted grass silage is shown in Table 20.4. In general, fermentation is restricted as dry matter content increases, and this is reflected in higher pH and soluble carbohydrate values, and in lower levels of fermentation acids. Wilting does not prevent proteolysis occurring, but if it is carried out rapidly, under good weather conditions, deamination of amino acids will be reduced. The gross energy contents are normally similar to those of the parent material.

Badly preserved silages

The term 'badly preserved silages' refers to silages in which either clostridia or enterobacteria, or both, have dominated the fermentation. It does not include those silages which have deteriorated as a result of oxidation. Such aerobically deteriorated material is liable to be toxic and should never be offered to animals (see above).

Badly preserved silages are frequently produced from crops which are either ensiled at too high a moisture content or which contain low levels of water-soluble carbohydrates. They may also be produced if the ensiled forage is deficient in lactic acid bacteria. The composition of two typical badly preserved silages, one made from cocksfoot and one from lucerne, both low in dry matter and water-soluble carbohydrates, is shown in Table 20.5.

In general, silages of this type are characterised by having high pH values, usually within the range 5.0 to 7.0. The main fermentation acid present is either acetic or butyric. Lactic acid and residual water-soluble carbohydrates are present in low concentrations or are absent. The ammonia-N levels are usually above 200 g/kg TN. This ammonia, which is derived from

Table 20.5 The composition of two badly preserved silages made from either cocksfoot or lucerne (Adapted from McDonald P, Henderson A R and Heron S J E 1991 *The Biochemistry of Silage*, Marlow, UK, Chalcombe Publications, p 271)

	Cocksfoot	Lucerne
pH	5.4	7.0
DM, g/kg	162	131
Total N, g/kg DM	37	46
Protein N, g/kg TN	302	260
Ammonia N, g/kg TN	323	292
WSC, g/kg DM	4	nil
Acetic acid, g/kg DM	37	114
Butyric acid, g/kg DM	36	8
Lactic acid, g/kg DM	1	13

the catabolism of amino acids, is accompanied by other degradation products such as amines and various keto acids and fatty acids (see Table 20.1).

Additive-treated silages

Silage additives can be classified into two main types: *fermentation stimulants*, such as sugar-rich materials, inoculants and enzymes, which encourage the development of lactic acid bacteria; and *fermentation inhibitors* such as acids and formalin, which partially or completely inhibit microbial growth.

Fermentation stimulants

Molasses, which is a by-product of the sugar beet and sugarcane industries (see p. 525), was one of the earliest silage additives to be used as a source of sugars. The by-product has a water-soluble carbohydrate content of about 700 g/kg DM and the additive has been shown to increase the dry matter and lactic acid contents, and to reduce the pH and ammonia levels in treated silages.

It was originally considered that, provided the general principles of silage making were adhered to, the natural lactic acid bacterial population on an ensiled crop would be sufficient to ensure a satisfactory fermentation. However, it is now known that growing crops are often poor sources of lactic acid bacteria and that some strains of these organisms are not ideally suited for ensiling purposes. A number of commercial inoculants containing freeze-dried cultures of homofermentative lactic acid bacteria are now available and some of these have proved effective in improving silage fermentation. Successful control of fermentation, using these inoculants, depends upon a number of factors including the inoculation rate, which should be at least 10^5 (but preferably 10^6) colony-forming units (cfu)/g fresh crop, and the presence of an adequate level of fermentable carbohydrates. The rapid domination of the fermentation by homolactic bacteria ensures

Table 20.6 Composition and nutritive value of grass silages inoculated with lactic acid bacteria compared with an untreated control (Adapted from Henderson A R, Seale D R, Anderson D H and Heron S J E 1990 In: *Proceedings of the Eurobac Conference*, Uppsala, August 1986, pp 93–8)

	Untreated	Inoculated[a]
DM, g/kg	168	181
pH	4.6	4.1
Total N, g/kg DM	33	32
Protein N, g/kg TN	386	407
Ammonia N, g/kg TN	130	88
WSC, g/kg DM	0	20
Acetic acid, g/kg DM	46	30
Butyric acid, g/kg DM	5	5
Lactic acid, g/kg DM	59	84
Ethanol, g/kg DM	13	7
DM digestibility[b]	0.74	0.77
ME, MJ/kg DM	11.4	12.5
Silage DM intake, g/day[b]	681	792
Liveweight gain, g/day[b]	71	129

[a] *Lactobacillus plantarum + Pediococcus pentosaceus* (10^6 colony-forming units/g).
[b] Using lambs.

the most efficient use of the water-soluble carbohydrates and, when levels of these in the crop are critical, increases the chances of producing a well-preserved silage. An illustration of the beneficial effects of a mixture of two homofermentative strains of lactic acid bacteria on the fermentation of an ensiled ryegrass crop is shown in Table 20.6.

When compared with the untreated control silage, the inoculated material had a lower pH, higher concentrations of water-soluble carbohydrates and lactic acid, and lower concentrations of acetic acid and ethanol.

Some commercial silage additives now contain enzymes along with an inoculum of suitable strains of lactic acid bacteria. The enzymes are usually cellulases and hemicellulases, which degrade the cell walls of plants thus releasing sugars, which are then available for fermentation by the lactic acid bacteria. The enzymes appear to be most effective when added to young herbage ensiled at low dry matter content.

Fermentation inhibitors

A large number of chemical compounds have been tested as potential fermentation inhibitors, but very few have been accepted for commercial use. One of the earliest was a mixture of mineral acids proposed by A I Virtanen, the technique being referred to as the AIV process. The acids, usually hydrochloric and sulphuric, were added to the herbage during ensiling in sufficent quantity to lower the pH value below 4.0. This process was for many years very popular in the Scandinavian countries and, if carried

Table 20.7 Composition and nutritive value of grass silages treated with two different additives (Adapted from Henderson A R, McDonald P and Anderson D H 1982 *Animal Feed Science and Technology* **7**: 303–14)

	Untreated	Formic acid[a]	Sulphuric acid + formalin[b]
DM, g/kg	181	184	176
pH	3.8	3.7	4.0
Total N, g/kg DM	27	23	25
Protein N, g/kg TN	400	490	509
Ammonia N, g/kg TN	65	49	44
WSC, g/kg DM	7	84	81
Acetic acid, g/kg DM	34	15	25
Butyric acid, g/kg DM	0.02	0.03	0.21
Lactic acid, g/kg DM	98	44	64
Ethanol, g/kg DM	7	9	18
DM digestibility[c]	0.74	0.74	0.72
ME, MJ/kg DM	12.1	11.3	10.3
Silage DM intake, g/day[c]	1020	1106	1020
Liveweight gain, g/day[c]	200	231	236

[a] 3.4 litres/tonne of 85 per cent formic acid.
[b] 4.6 litres/tonne containing 15 per cent sulphuric acid and 23 per cent formaldehyde.
[c] Using sheep.

out effectively, is a very efficient method of conserving nutrients. In recent years, however, formic acid has largely replaced mineral acids in Scandinavia and this organic acid, which is less corrosive than mineral acids, has also been accepted as an additive in many other countries. In the UK, it is commonly applied in the form of an aqueous solution of ammonium tetraformate. The recommended application rate varies from 2.5 to 5 litres per tonne of fresh crop, depending on the dry matter content of the crop. Complete inhibition of microbial growth does not take place, some lactic acid fermentation occurring. The beneficial effects of formic acid on the fermentation characteristics of crops low in water-soluble carbohydrates, such as legumes and grasses, have been well established and are well shown in Table 20.7.

Formalin, a 40 per cent solution of formaldehyde in water, has been used as a fermentation inhibitor. It is applied either on its own or more effectively with an acid such as sulphuric or formic. Typical results of a sulphuric acid/formalin mixture applied to ryegrass at the rate of 4.6 litres/tonne are shown in Table 20.7; the formaldehyde combines with the protein, protecting it from hydrolysis by plant enzymes and microorganisms in the silo. The acid in the mixture acts as a fermentation inhibitor, preventing, in particular, the development of undesirable bacteria in the silage. Formalin is now used only in conjunction with formic acid, which has proved to be more effective than sulphuric acid. In Europe, the use of formaldehyde as an additive has been banned owing to concern about its carcinogenic properties.

20.6   Nutritive value of silages

The nutritional value of a silage depends first upon the species and stage of growth of the harvested crop (factors which have already been discussed in the previous chapter), and second upon the changes resulting from the activities of plant enzymes and microorganisms during the harvesting and storage period.

Energy

Gross energy

Owing to the production, during ensilage, of high-energy compounds, silages tend to have higher gross energy values than the materials from which they were made. The magnitude of the increase will depend upon the degree of fermentation which has occurred during ensiling. This is well illustrated by the data in Table 20.8.

In each case, ensilage has resulted in an increase in gross energy but this is smaller for the prewilted and additive-treated materials.

Metabolisable energy (ME)

The energy requirements of ruminant animals and the energy values of foods are currently expressed in the UK in terms of metabolisable energy (Chapters 11 and 12).

Faecal losses of energy when silages are consumed by animals usually lie between 25 and 35 per cent and are mainly dependent upon the nature of the source material. The effects of the ensiling process on energy losses in the faeces are generally considered to be small. However, when the prewilting period is prolonged, and especially when accompanied by adverse conditions such as high winds and intermittent rain, they may be increased by as much

Table 20.8 Gross energy contents (MJ/kg DM) of silages and the grasses from which they were made

	No.	Gross energy	Standard deviation (+/−)	Increase (%)
Grasses	18	18.3	0.68	
Lactate silages	18	20.0	1.06	9.0
Grasses	7	18.4	0.45	
Wilted silages	7	19.1	0.40	3.8
Grasses	7	18.7	0.45	
Additive-treated silages	7	20.0	0.95	7.0

as 8 per cent. It has been claimed that the use of formic acid as an additive in silage making has reduced faecal losses of energy by up to 7 per cent but most work shows the reduction to be in the range of 0–2 per cent. Formaldehyde, used as an additive at the rate of 6.4 litres/tonne of fresh grass, has been shown to reduce the digestibility of energy by 4–5 percentage units.

Losses of energy in urine, on silage diets, are variable, ranging from 3 to 7 per cent of the gross energy. For most silages a value of about 5 per cent would appear to be acceptable. The major determinant of urinary energy loss appears to be the intake of nitrogen. Following an intake of silage, the concentration of ammonia in rumen contents rises sharply. It is generally considered that, when the concentration exceeds 150m g/litre, the rumen microorganisms cannot fully utilise the nitrogen and the excess is lost as urea. It is important that a readily available source of energy is provided along with the silage at this time if the rumen microorganisms are to make full use of the ammonia and thus reduce potential loss of energy and nitrogen to the animal.

The most comprehensive studies involving energy losses as methane are those carried out at the Rowett Research Institute (see Further reading). For 48 silages, mean energy loss as methane was 7.7 per cent of the gross energy with a standard deviation of 0.71.

In practice, the metabolisable energy contents of silages vary widely and it would be very unwise to rely on an average figure for silages of any one class and even more so for individual materials. For 2321 samples of farm silages analysed at the Scottish Agricultural Colleges in 1998, the range of metabolisable energy contents was 7.0–13.0 MJ/kg DM.

Routine determination of the ME of silages is frequently based on determination of organic matter digestibility and an assumed relationship between digestible organic matter and metabolisable energy. A comparison of some common methods of estimating the organic matter digestibility of silage is given in Table 20.9.

Table 20.9 Comparison of several laboratory methods for predicting the organic matter digestibility *in vivo* of grass silage. (After Barber *et al.* 1990 *Animal Feed Science and Technology* **28**: 115–28)

Method	Validation statistics[a]			
	R^2	SEP	Slope	Bias
NIRS (8-term)	0.76	2.6	0.93	−0.79
In vitro OMD[b]	0.64	3.6	0.89	1.85
Pepsin–cellulase	0.40	4.7	0.71	2.33
Modified acid-detergent fibre	0.20	5.1	0.52	−0.59
Acetyl bromide lignin	0.14	5.3	0.48	1.18

[a] R^2, proportion of variation explained by the regression equation; SEP, standard error of prediction (%).
[b] OMD, organic matter digestibility.

The following is typical of the type of relationship with digestible organic matter in dry matter (DOMD), which is used for calculating metabolisable energy:

$$\text{ME (MJ/kg DM)} = \text{DOMD (g/kg DM)} \times 0.016$$

which assumes an energy loss in urine and methane of 15 per cent of the gross energy, assumed to be 18.4 MJ/kg DM. For silages this factor is likely to underestimate metabolisable energy owing to the higher gross energy content of their dry matter.

Net energy

The net energy value of silages, like that of other foods, is related to metabolisable energy by efficiency factors (k factors: see p. 282). For maintenance, the efficiency of utilisation of metabolisable energy in good-quality silages (ME/GE $q_m = 0.56-0.70$) has been found to vary over a narrow range of $k_m = 0.68-0.71$. As with other forages, efficiency of utilisation of silage metabolisable energy for growth and fattening is lower and more variable ($k_g = 0.21-0.61$; approximate mean 0.45). Calorimetric measurements of the efficiency of utilisation of silage metabolisable energy for lactation are few in number, but lie in the range 0.53–0.58, and are therefore lower than the commonly assumed value of 0.62.

In practice, most estimates of the productive potential of silages are based on k values calculated from the metabolisable energy concentration of the dry matter of the diet. A typical equation is that currently used in the UK for estimating k_f:

$$k_f = 0.78 q_m = 0.006$$

This is based on calorimetric data and is derived from a general and not a silage population. Frequently, an assumed value of 18.4 MJ/kg DM is used for GE in calculating q_m. This is acceptable for the general run of foods but is clearly inappropriate for the majority of silages. Thus for a silage with a metabolisable energy content of 10.5 MJ/kg DM, the calculated value for k_f would be 0.45 assuming a gross energy content of 18.4, but 0.42 assuming a value of 19.6, which would be more appropriate for silage. Estimates of the productive potential of silages based on experimentally determined or book metabolisable energy values frequently appear to be optimistic and this may, in part, be due to a failure to use an appropriate figure for GE in such equations.

Protein

The process of ensilage results in proteolysis and an increase in the proportion of ammonia-N and free α-amino-N in the silage compared with the

Table 20.10 Variation in rumen degradability of the protein of silages (Weddell J R, Scottish Agricultural Colleges)

Material	No.	Mean degradability	Standard deviation	Coefficient of variation (%)	95% confidence limits
Pit, 1st cut	374	0.73	0.05	6.9	0.63–0.83
Pit, 2nd cut	172	0.71	0.04	5.6	0.63–0.79
Bale	15	0.74	0.05	6.8	0.64–0.84

original material. The proportion of potentially degradable nitrogen is reduced whereas the undegradable nitrogen, that associated with the cell walls, remains almost unchanged. Overall, there is an increase in the nitrogen which is readily available to the rumen microbes and which may be regarded as rapidly degraded.

Prewilting, and the use of formic acid or formaldehyde as an additive in ensilage, will reduce the extent of proteolysis and give silages with lower contents of both ammonia-N and free α-amino acid-N, and will reduce the rate of degradation of the nitrogen fraction. The use of bacterial inoculants and enzyme preparations as additives will increase proteolysis and result in silages of higher effective degradability. The reduction in degradation rate will be beneficial with regard to nitrogen capture by the rumen microbes and will reduce energy losses in urea. In the case of formaldehyde, the beneficial effect in reducing degradability is accompanied by a lowered digestibility in the lower gut. This may be significant when high levels of the additive are used.

Nitrogen degradability values for silages show great variability, as illustrated in Table 20.10, and the use of tabulated values to characterise individual samples may lead to unacceptable errors in ration formulation.

The highly degradable nature of the nitrogen in most silages points to the need for adequate supplementation of silage-based diets with a readily available supply of carbohydrate, so that the rumen microbes can cope with the rapid influx of ammonia following an intake of silage and so maximise the synthesis of microbial protein and minimise the loss of both nitrogen and energy. Silages, despite their high gross energy contents, are generally considered to be poor sources of energy for the microbes. Silage-based diets must, therefore, contain a supplemental source of energy to maximise the utilisation of the nitrogen of the diet. The addition of starch and certain sugars has been shown to increase microbial protein synthesis in the rumen. A similar increase has been obtained by supplementing silage diets with soya bean meal, presumably by making amino nitrogen available to rumen microorganisms otherwise dependent upon ammonia as their main source of nitrogen.

In general, the degradability of silage nitrogen is high, with most figures in the range of 75–85 per cent. Effective rumen degradability figures for various silages at different outflow rates are given in Appendix 2, Table A.2.2. As previously indicated, such figures should only be used when valid estimates of degradability are not available, e.g. those based on NIRS (Chapter 13).

Dry matter intake

The major determinant of nutrient intake on a silage diet is not the concentrations of the various nutrients the silage contains but the amount of it that is eaten. A number of factors have been shown to affect dry matter intake (Chapter 17). We are here concerned with those silage characteristics which affect intake, and particularly those which may be used to give valid prediction of its potential intake. Among the characteristics which have been shown to be related to intake are:

■ pH,
■ buffering capacity,
■ lactate (g/kg DM),
■ acetate (g/kg DM),
■ butyrate (g/kg DM),
■ ammonia nitrogen (g/kg TN),
■ digestible organic matter (g/kg DM),
■ rate of digestion,
■ fibre (g/kg DM).

A number of the factors interact. Thus the organic acids contribute to the acidity and to the buffering capacity of the silage and their effects on intake will be reflected in the effect of these two parameters. Similarly, high fibre concentrations, reflecting the maturity of the crop, are associated with poor digestibility, slow rates of digestion and low intakes. However, if oxidation losses are high there will be an increase in fibre content. Thus we may be faced with similar levels of fibre of a very different kind; one due to maturity of the crop and the other due to a concentration effect. The effect on dry matter intake will be different in each case. The complicated nature of the problem is further highlighted by pH. Here the relationship with intake is not linear. Low pH reduces intake but high pH also does so, the latter reflecting the influence of the concentration of ammonia-N in decreasing intake while increasing pH.

There is general acceptance of dry matter, digestible organic matter and ammonia nitrogen as the major determinants of silage dry matter intake and it behoves the silage maker to optimise fermentation and to minimise:

■ secondary fermentation,
■ clostridial activity,
■ aerobic deterioration,

if satisfactory intakes of his product are to be achieved.

For the ration formulator, the problem is to predict how much silage dry matter a particular animal will consume, so that its contribution to the animal's nutrient requirements can be calculated. The traditional approach to the problem has been to use regression equations based on

Table 20.11 Comparison of the accuracy of some methods of predicting silage dry matter intake (expressed as the standard error of cross validation, g/kg metabolic liveweight per day) (After Offer N W *et al.* 1998 *Animal Science* **66**: 357–67)

Method	Lambs	Cows
1 Best equation based on traditional laboratory determinations (Basal)	0.81	7.3
2 Basal + fermentation values (HPLC[a])	0.81	7.3
3 Basal + fermentation values (ET[b])	0.75	5.9
4 NIRS on silage DM	0.52	5.1
5 NIRS on wet silage	0.56	2.5

[a] HPLC, by high-pressure liquid chromatography.
[b] ET, by electrometric titration.

one or more of the parameters listed above. Typical is the following equation (Lewis 1981) predicting the potential intake of silage dry matter (I) by dairy cattle:

$$I \text{ (g/kg } W^{0.75}/\text{day)} = (0.103 \times \text{DM}) + (0.0516 \times \text{DOMD}) \\ - (0.05 \times \text{NH}_3 - \text{N}) + 45$$

Where: DM = dry matter as g/kg
DOMD = digestible organic matter (g/kg DM)
$\text{NH}_3 - \text{N}$ = ammonia nitrogen (g/kg total nitrogen)

Such equations have worked well in practice, but published work would suggest that the use of NIRS for predicting intake significantly improves prediction. A comparison of some methods of prediction is given in Table 20.11.

Similar findings have been obtained by other workers, who quote a standard error of prediction of 4.7 g/kg $W^{0.75}$ and an R^2 value of 0.79 from using an NIRS technique for estimating silage dry matter intake in beef cattle.

Such estimates of silage dry matter intake have to be modified in the light of various factors such as the class of animal to which it is to be offered, the level of concentrate supplementation, the level of production, feeding technique and environmental factors. For example, the equation for calculating I above is expanded for use with mixed diets as follows:

$$\text{SDMI(kg/day)} = (1.0681 - 0.002\,447 IC - 0.00337 C^2 - 10.9) W^{0.75}/1000 \\ + 0.001\,75 Y^2$$

Where: C = concentrate dry matter intake (g/kg $W^{0.75}$/day)
Y = milk yield (kg/day)

The UK Agricultural and Food Research Council proposed in 1992 the following equation for predicting the intake of silage by cattle:

$$SDMI(kg/day) = -3.74 - 0.387C + 1.48(F + P) + 0.0066W_n$$
$$+ 0.0136 \ (DOMD)$$

Where: $F + P$ = yield of fat and protein (kg/day)

W_n = weight of the cow in the week under consideration

C = concentrate dry matter (kg/day)

$DOMD$ = digestible organic matter (g/kg DM)

The equation is claimed to be more accurate overall than the Lewis equation.

Summary

1. Silage is the material produced by controlled fermentation of a crop of high moisture content. Ensilage is the name given to the process.

2. Ensilage is succesful if the activities of undesirable organisms and the production of objectionable fermentation products are minimised, which is achieved by encouraging the growth of lactic acid bacteria, which lower the pH of the mass.

3. Stability of the preserved mass is maintained by compaction and by efficient sealing of the silo.

4. In recent years, preservation in big bales has increased greatly at the expense of the more common bunker silo.

5. As a result of the activity of plant enzymes, water-soluble carbohydrates in the preserving mass are dissipated as carbon dioxide and water; temperature rises and preservation may be put at risk owing to a shortage of substrate for lactic acid production.

6. Plant enzymes also bring about proteolysis and the production of amino acids and peptides from the protein of the original material.

7. The most efficient preservation is achieved by the action of heterofermentative lactic acid bacteria.

8. Saccharolytic clostridia break lactic acid down to acetic and butyric acids, causing a rise in pH and putting preservation at risk. Proteolytic clostridia break down amino acids with the production of a number of unpleasant substances. *Clostridium botulinum* is the causative organism of botulism and has been associated with grass sickness in horses.

9. There has been an increasing appreciation in recent years of the importance of fungi in ensilage owing to their action on cell-wall structures. A number of the fungi present in silage, particularly aerobically spoiled material, are capable of producing mycotoxins, which can have very unpleasant effects on farm animals and human beings.

10. Losses of nutrients arising from normal fermentation are minimal. Most losses occur as a result of aerobic deterioration before the material is ensiled, or as result of ingress of air during storage or after opening the silo.

11. Significant losses of dry matter may occur as a result of loss of effluent when crops of very high moisture content are ensiled.

12. Additives are used in silage making if there is any doubt as to whether natural fermentation is capable of ensuring satisfactory preservation, or routinely as an insurance against poor preservation. They may be stimulants, which provide substrate for lactic acid production, or boost the population of desirable bacteria in the material

to be ensiled. Fermentation inhibitors are used to render the environment inimical to the development of undesirable microorganisms.

13. The metabolisable energy contents of silages are highly variable and difficult to predict satisfactorily by laboratory means. Currently, near-infrared reflectance appears to be the best single predictor.

14. Nitrogen degradability values are highly variable, and tabulated values are little more than a rough guide.

15. Silage dry matter intakes tend to be low and are strongly influenced by a number of factors, particularly pH, concentration of organic acids, buffering capacity and ammonia nitrogen content.

Further reading

Lewis M 1981 *Proceedings of the 6th Silage Conference*. In: Harkess R D and Castle M E (eds) Edinburgh, pp 35–6.

McDonald P, Henderson A R and Heron S J E 1991 *The Biochemistry of Silage*, 2nd edn, Marlow, UK, Chalcombe Publications.

Merry R J, Jones R and Theodorou M K 2000 The conservation of grass. In: Hopkins A (ed.) *Grass: Its Production and Utilization*, 3rd edn, Oxford, Blackwell Science.

Murdoch J C 1989 The Conservation of Grass. In: Holmes W (ed.) *Grass, Its Production and Utilisation*, 2nd edn, Oxford, Blackwell Scientific Publications.

Nash M J 1985 *Crop Conservation and Storage*, Oxford, Pergamon Press.

Stark B A and Wilkinson J M (eds) 1987 *Developments in Silage*, Marlow, UK, Chalcombe Publications.

21 Hay, artificially dried forages, straws and chaff

21.1   Hay

21.2   Artificially dried forages

21.3   Straws and related by-products

21.1   Hay

The traditional method of conserving green crops is that of haymaking, the success of which until fairly recently was entirely dependent upon the chance selection of a period of fine weather. The introduction of rapid drying techniques using field machinery and barn drying equipment has, however, considerably improved the efficiency of the process, and reduced the need to be dependent upon the weather. Although in many countries of Western Europe, including Britain, ensilage has overtaken haymaking as the preferred mode of conservation of green forages, haymaking is generally the more common process.

The aim in haymaking is to reduce the moisture content of the green crop to a level low enough to inhibit the action of plant and microbial enzymes. The moisture content of a green crop depends on many factors, but may range from about 650 to 850 g/kg, tending to fall as the plant matures. In order that a green crop may be stored satisfactorily in a stack or bale, the moisture content must be reduced to 150–200 g/kg. The custom of cutting the crop in a mature state when the moisture content is at its lowest is clearly a sensible procedure for rapid drying and maximum yield, but unfortunately the more mature the herbage, the poorer is the nutritive value (see Chapter 18).

Chemical changes and losses during drying

Chemical changes resulting in losses of valuable nutrients inevitably arise during the drying process. The magnitude of these losses depends to a large extent upon the speed of drying. In the field the loss of water from the swath is governed by the natural biological resistance of the leaf and swath to water loss, the prevailing weather conditions and swath microclimate, and the mechanical treatment of the crop during harvesting and conditioning. The losses of nutrients during haymaking arise from the action of plant and microbial enzymes, chemical oxidation, leaching and mechanical damage.

Action of plant enzymes

In warm, dry, windy weather the wet herbage, if properly handled and mechanically agitated, will dry rapidly and losses arising from plant enzyme activity will be small. The main changes involve the soluble carbohydrates and nitrogenous components. In the early stages of the drying process changes in individual components of the water-soluble carbohydrates occur such as the formation of fructose from hydrolysis of fructans. During extended periods of drying there is a considerable loss of hexoses as a result of respiration and this loss leads to an increase in the concentration of other constituents in the plant, especially the cell-wall components, which are reflected in the neutral-detergent fibre content. In the freshly cut crop, proteases present in the plant cells rapidly hydrolyse the proteins to peptides and amino acids, hydrolysis being followed by some degradation of specific amino acids. The effects of wilting ryegrass under ideal dry conditions and in a poor humid environment are compared in Table 21.1.

A number of devices and methods of treatment are used to speed up the drying process in the field and these include 'conditioners', such as crushers, rollers and crimpers, which break the cellular structure of the plant and allow the air to penetrate the swath more easily. A more traditional method, which is still practised in some parts of the world, notably Switzerland, Italy, Germany and Scandinavia, is to make hay on racks, frames or tripods. Table 21.2 shows a comparison in composition and nutritive value between hay made by 'tripoding' and the traditional 'field curing' method. The difference between these two methods is reflected in the crude fibre, digestible crude protein, digestible organic matter and metabolisable energy values.

Table 21.1 Changes in nitrogenous components of ryegrass/clover during the early stages of field drying (From Carpintero M C, Henderson A R and McDonald P 1979 *Grass and Forage Science* **34**: 311)

	Dry matter (DM) (g/kg)	Water-soluble carbohydrates (g/kg DM)	Total N (g/kg DM)	N components (g/kg total N)		
				Protein N	Non-protein N	Ammonia N
Fresh grass	173	213	26.6	925	75	1.2
Wilted 6 h (dry conditions)	349	215	28.2	876	124	1.1
Wilted 48 h (dry conditions)	462	203	28.9	835	165	2.6
Wilted 48 h (humid conditions)	199	211	29.9	753	247	2.6
Wilted 144 h (humid conditions)	375	175	31.0	690	310	26.4

Table 21.2 Composition (g/kg DM) and nutritive value of perennial ryegrass and of hay made from it by two methods in south-east Scotland

	Fresh grass	Tripod hay	Field-cured hay
Organic matter	932	908	925
Crude fibre	269	324	362
Crude protein	128	121	99
Digestible crude protein	81	72	47
Digestible organic matter	711	614	547
Metabolisable energy^a, MJ/kg DM	10.7	9.2	8.2

^{a} Calculated from digestible organic matter.

Action of microorganisms

If drying is prolonged because of bad weather conditions, changes brought about by the activity of bacteria and fungi may occur. Bacterial fermentation takes place in cut herbage that has lain in the field for a few days, and leads to the production of small quantities of acetic and propionic acids. Mouldy hay is unpalatable and may be harmful to farm animals and man because of the presence of mycotoxins. Such hay may also contain actinomycetes, which are responsible for the allergic disease affecting man known as 'farmer's lung'.

Oxidation

When herbage is dried in the field, a certain amount of oxidation occurs. The visual effects of this can be seen in the pigments, many of which are destroyed. The provitamin, carotene, is an important compound affected and may be reduced from 150 to 200 mg/kg in the dry matter of the fresh herbage to as little as 2–20 mg/kg in the hay. Rapid drying of the crop by tripoding or barn drying conserves the carotene more efficiently, and losses as low as 18 per cent in barn-dried hay have been reported. On the other hand, sunlight has a beneficial effect on the vitamin D content of hay because of irradiation of the ergosterol present in the green plants.

Leaching

Losses due to leaching by rain mainly affect the crop after it has been partly dried. Leaching causes a loss of soluble minerals, sugars and nitrogenous constituents, hence a concentration of cell-wall components, which is reflected in a higher fibre content. Rain may prolong the enzyme action within the cells, thus causing greater losses of soluble nutrients, and may also encourage the growth of moulds.

Mechanical damage

During the drying process the leaves lose moisture more rapidly than the stems, so becoming brittle and easily shattered by handling. Excessive mechanical handling is liable to cause a loss of this leafy material, and since the leaves at the hay stage are richer in digestible nutrients than the stems, the resultant hay may be of low feeding value. Leaf loss during haymaking is particularly likely to occur with legumes such as lucerne. There are a number of modern machines available which reduce the losses caused by leaf shattering. If the herbage is bruised or flattened, the drying rates of stems and leaves differ less. Baling the crop in the field at a moisture content of 300–400 g/kg, and subsequent drying by artificial ventilation, will reduce mechanical losses considerably.

Stage of growth

The stage of growth at the time of cutting is the most important crop factor determining the nutritive value of the conserved product. The later the date of cutting the larger will be the yield, the lower the digestibility and net energy value, and the lower the voluntary intake of dry matter by animals. It follows that if their drying conditions are similar, hays made from early cut crops will be of higher nutritive value than hays made from mature crops.

Plant species

The differences in chemical composition between species have already been discussed in Chapter 19. Hays made from legumes are generally richer in protein and minerals than grass hay. Pure clover swards are not commonly grown for making into hay in the UK, although many 'grass' hays contain a certain amount of clover. Lucerne or alfalfa (*Medicago sativa*) is a very important legume which is grown as a hay crop in many countries. The value of lucerne hay lies in its relatively high content of crude protein, which may be as high as 200 g/kg DM if it is made from a crop cut in the early bloom stage.

Cereals are sometimes cut green and made into hay, and this usually takes place when the grain is at the 'milky' stage. The nutritive values of cereal hays cut at this stage of growth are similar to those of hays made from mature grass, although the protein content is generally lower. Table 21.3 shows the composition of a number of hays made from different species. These are representative values and give no indication of the ranges in nutritive value. If extremes are considered, it is possible to produce hay of excellent quality with a digestible crude protein content of 115 g/kg DM and an ME value in excess of 10 MJ/kg DM (see Table 21.4). On the other hand poor-quality hay made from mature herbage harvested under bad weather conditions may have a negative digestible crude protein content and an ME value below 7 MJ/kg DM; material of this type is no better in feeding value than oat straw.

Table 21.3 Composition and nutritive value of hays (After Watson S J and Nash M J 1960 *The Conservation of Grass and Forage Crops*, Edinburgh, Oliver and Boyd)

	No. of samples	Dry matter basis			
		Crude fibre (g/kg)	Crude protein (g/kg)	Digestible crude protein (g/kg)	Metabolisable energy [a] (MJ/kg)
Grasses					
Meadow	686	298	113	67	8.8
Mixed grass	68	301	114	63	8.6
Cocksfoot (orchard grass)	17	356	82	42	8.0
Fescue	22	315	90	48	8.6
Ryegrass	39	305	96	48	8.9
Timothy	218	341	77	36	8.2
Legumes					
Clover	284	319	143	89	8.6
Lucerne	474	322	165	118	8.3
Vetches	28	277	213	163	9.1
Soya bean	42	366	156	101	7.8
Cereals					
Barley	19	265	93	52	8.6
Oat	48	329	80	41	8.5
Wheat	20	268	82	44	7.8

[a] Calculated from TDN values.

Changes during storage

The chemical changes and losses associated with haymaking do not completely cease when hay is stored in the stack or barn. The stored crop may contain 100–300 g/kg moisture. At the higher moisture levels chemical

Table 21.4 Composition (g/kg) and nutritive value of the dry matter of 47 grass hays [a] made during 1963–1965 in England and Wales (Adapted from *ADAS Science Arm Report* 1972, London, MAFF, HMSO, p 95)

	Range	Mean	SD
Ash	57–117	80	±11
Crude fibre	274–412	335	±31
Crude protein	63–167	96	±22
Digestible crude protein [b]	21–115	51	–
Digestible organic matter	391–711	563	±61
Metabolisable energy, MJ/kg	5.7–11.5	8.5	±1.1

[a] Mainly ryegrass but including some timothy/meadow fescue hays.
[b] Estimated from crude protein.

changes brought about by the action of plant enzymes and microorganisms are likely to occur.

Respiration ceases at about 40 °C, but the action of thermophilic bacteria may go on up to about 72 °C. Above this temperature chemical oxidation can cause further heating. The heat tends to accumulate in hay stored in bulk and eventually combustion may occur.

Prolonged heating during storage can have an adverse effect on the proteins of hay. New linkages are formed within and between peptide chains. Some of these linkages are resistant to hydrolysis by proteases, thereby reducing the solubility and digestibility of the proteins.

Susceptibility of proteins to heat damage is greatly enhanced in the presence of sugars, the damage being done by Maillard-type reactions (see p. 63). Temperature has an important effect on the reaction rate, which is 9000 times faster at 70 °C than at 10 °C. The amino acid lysine is particularly susceptible to reactions of this type. The products are colourless at first, but eventually turn brown; the dark brown colour of overheated hays, and other foods, can be attributed mainly to Maillard reactions.

Losses of carotene during storage depend to a large extent on the temperature. Below 5 °C little or no loss is likely to occur, whereas in warm weather losses may be considerable.

The changes that take place during storage increase the proportion of cell-wall constituents and reduce nutritive value.

Overall losses

The overall losses during haymaking can be appreciable under poor weather conditions. In a study on six commercial farms carried out over a three-year-period in north-east England by the Agricultural Development and Advisory Service (ADAS), losses of nutrients were measured between harvesting and feeding. Total dry matter losses averaged 19.3 per cent, made up of 13.7 per cent field loss and 5.6 per cent loss in the bale. The losses of digestible organic matter and digestible crude protein were both about 27 per cent.

Hay preservatives

The main objective in using hay preservatives is to allow hay to be stored at moisture levels which, in the absence of the preservative, would result in severe deterioration through moulding. The chemical preservatives tested have included propionic acid and its less volatile derivative ammonium bis-propionate. In the laboratory, propionic acid applied at a rate of 10 g/kg water (i.e. 3 kg/t of forage with a moisture content of 300 g/kg), will prevent moulding and preserve the hay. However, in the field it is difficult to apply the preservative uniformly, and a higher application rate, of 12 kg propionic acid per tonne of forage with 300 g/kg of moisture, is recommended. Hays with moisture contents as high as 400–500 g/kg can, after propionate treatment, be stored satisfactorily, provided the additive is both applied in sufficient quantity and distributed uniformly. Lactic acid producing bacteria have also been tested as preservatives of hay, but seem to have been much less effective than when used to preserve silage (see Chapter 20).

More recently, the success of anhydrous ammonia treatment of straw (see p. 546) has stimulated studies on treatment of hay with this gas. Anhydrous ammonia injected into plastic-covered stacks of bales of moist hay has increased the stability, under aerobic and anaerobic conditions, and has improved the nutritional value of the hay (but with some danger of the formation of toxins: see below).

21.2 Artificially dried forages

The process of artificial drying is a very efficient, though expensive, method of conserving forage crops. In northern Europe, grass and grass–clover mixtures are the commonest crops dried by this method, whereas in the USA lucerne (alfalfa) is the primary crop that is dehydrated. In Europe, artificially dried forage accounts for less than 0.5 per cent of all conserved forage dry matter. Even in the USA, which produces about 1 Mt of artificially dried forage a year, the proportion of all conserved forage is less than 1 per cent.

The drying is brought about by allowing the herbage to meet hot gases at a temperature that varies with the type of drier used. In the 'low-temperature' type of equipment, the hot gases are usually at a temperature of about 150 °C and the drying time varies from about 20 to 50 minutes depending upon the drier design and the moisture content of the crop. With 'high-temperature' driers, the temperature of the gases is initially within the range of 500–1000 °C and the time taken to pass through the drier varies from about 0.5 to 2 minutes.

In both processes, the temperature and time of drying are very carefully controlled so that the forage is never completely desiccated, and the final product usually contains about 50–100 g/kg of moisture. As long as some moisture remains in the material, the temperature of the herbage is unlikely to exceed 100 °C. It is obvious, however, that if the material is left in contact with the hot gases too long it will be charred or even completely incinerated.

After drying, the forage may be processed to suit the class of animal for which it is intended. For pigs and poultry, it is usually hammer milled and stored either as meal or as pellets. For ruminants the dried forage can be used as a 'long roughage' or more commonly packaged into different forms described as pellets, cobs or wafers. A pellet is a package made in a rotary-die press from milled dried forage; a cob is made in a rotary-die press from chopped and dried forage, while a wafer is usually made in a piston-type machine from either milled or chopped dried forage.

More widespread use of artificial drying of forage crops is limited by the high initial capital and high running costs of the process. To produce a tonne of dried grass from material with an initial dry matter content of 200 g/kg requires 300 litres of fuel oil.

Nutritive value

As a conservation technique, artificial drying is extremely efficient. Dry matter losses from mechanical handling and drying are together unlikely to exceed 10 per cent, and the nutritive value of the dried product is therefore close to that of the fresh crop. A survey of artificially dried forages in England showed that perennial ryegrass cut early in the growing season gave dried forage containing 200 g protein and 270 g acid-detergent fibre per kg dry matter, and 11.5 MJ metabolisable energy per kg dry matter. Later-season material was similar in protein content but contained slightly more fibre (290 g/kg DM) and less metabolisable energy (10.3 MJ/kg DM).

Some oxidation of carotene may occur, especially during long periods of storage of dried forage exposed to light and air; dried grass meal can lose as much as half its carotene during seven months storage under ordinary commercial conditions. A high-quality meal should have a carotene content of about 250 mg/kg, although under exceptional conditions carotene contents as high as 450 mg/kg have been obtained. Because irradiation of sterols cannot take place during the rapid drying process, the vitamin D content of dried forage will be very low. Originally most of the dried material produced from very young leafy herbage was used for pig and poultry diets. For this market, which is still an important one, the product is sold mainly for its content of protein, carotene and xanthophyll (the pigment mainly responsible for egg yolk colour). Dried grass and lucerne are also included in diets for horses and rabbits.

Dried grass made from young herbage is also used in the diets of ruminant animals, mainly as a replacement for the cereal and protein concentrates given with silage and hay. A particular feature of dried grass/silage combinations is the high total dry matter intake which can be obtained. Although there is some evidence that the apparent digestibility of the crude protein might be reduced slightly during drying, the disadvantages of this are more than offset by the increased amino acid supply to the animal because of the larger amount of protein which escapes degradation in the rumen and is digested in the small intestine. In a study with sheep the proportion of the total amino acid nitrogen intake that was apparently absorbed in the small intestine was 0.41 for fresh grass and 0.51 for the same grass after drying.

In the USA, considerable amounts of lucerne (alfalfa) are artificially dried and sold as a high-vitamin feed supplement for broilers. As the production of this is seasonal, large volumes of dried meal must be stored for periods of six months or more, and unless precautions are taken losses of carotene, xanthophyll and vitamin E are liable to occur as a result of oxidation. Since the rate of loss is a function of temperature, large volumes were stored under refrigeration in the past. More recently the dried product has been stored satisfactorily under inert gas, which virtually eliminates oxidative losses up to the time of removal from storage. Many manufacturers also add antioxidants to protect the product from the time it is removed from the inert gas until it is used.

21.3 Straws and related by-products

Straws consist of stems and leaves of plants after the removal of the ripe seeds by threshing, and are produced from most cereal crops and from some legumes. Chaff consists of the husk or glumes of the seed which are separated from the grain during threshing. Modern combine harvesters put out straw and chaff together, but older methods of threshing (e.g. hand threshing) yield the two by-products separately. A by-product similar in composition to straw is sugarcane bagasse, but this has been described earlier (Chapter 19). Other fibrous by-products of cereals are referred to in Chapter 23. All the straws and related by-products are extremely fibrous, most have a high content of lignin, and all are of low nutritive value. Their high fibre content restricts their use to that as food for ruminants.

On a world scale the total production of straws and related materials is sufficient to meet the maintenance needs of all ruminant livestock. However, the actual usage of straw as a food for livestock varies a great deal from one part of the world to another. Much straw is produced in regions such as the North American prairies that have relatively few livestock, and the cost of transporting a bulky, low-value food to other regions is too great to permit its export. Other parts of the world, such as Europe, produce a great deal of straw, but are also well supplied with better-quality forages. But in many tropical and subtropical countries that cannot afford to use land for forage production, straw is the essential basal food for ruminant livestock

Maize, wheat and rice crops are the main sources of world straw supplies, but in the UK barley provides much of the 15–20 per cent of total straw production that is used for animal feeding. When oats were grown for horses, oat straw was preferred for animal feeding, but the decline in the area of this crop has made its straw less important.

Barley and oat straw

Of the cereal straws, oat straw used to be popular in many areas of the UK as a bulky food for store cattle, along with roots and concentrates, and in limited quantities as a source of fibre for dairy cows. With the increasing use of barley grain as a major concentrate food for farm animals, especially in northern Europe, large quantities of barley straw are available and attention has concentrated in recent years on methods of trying to improve the nutritional value of this low-grade material.

The composition of both barley and oat straw may vary, although this is influenced more by stage of maturity of the crop at harvesting and environment than by the cultivar grown. The crude protein content of the dry matter of both straws is low, usually between 20 and 50 g/kg, with the higher values obtained in crops grown under cold and wet conditions where they do not mature completely. The rumen degradability of the protein is relatively low (<0.4), and of the undegradable protein much is likely to be indigestible. A major component of the dry matter is the fibre, which contains a relatively high proportion of lignin. In a chemical examination of the dry matter of barley straw, it was shown that this consisted of about 400–450 g/kg of cellulose, 300–500 g/kg of hemicellulose and 80–120 g/kg of lignin.

The digestibility of the organic matter of these straws rarely exceeds 0.5 and the metabolisable energy value is about 7 MJ/kg DM or in the case of winter barley cultivars, less than this. Of the ash fractions, silica is the main component and straws generally are poor sources of essential mineral elements, as can be seen from the results of a comparison between hays and barley straws shown in Table 21.5.

Apart from the low digestibility of these cereal straws, a major disadvantage is the low intake obtained when they are given to ruminant animals. Whereas a cow will consume up to 10 kg of medium-quality hay, it will eat only about 5 kg of straw. Improvements in both digestibility and intake can be obtained by the addition of nitrogen in the form of protein or urea.

Maize straw

Maize straw, or corn stover, has a higher nutrient content and is more digestible than most other straws. It has a crude protein content of about 60 g/kg DM and a metabolisable energy value of about 9 MJ/kg DM. In North America, corn stover is frequently used as a major part of the diet for dry, pregnant beef cows. The animals may be turned into the cornfields after the grain has been harvested or the stover may be chopped, ensiled and consumed in a similar way to maize silage. Alternatively, the stover, after drying in the field, can be stacked or harvested as large round bales.

Table 21.5 Some mineral contents of hay and barley straw obtained from 50 farms in southeast Scotland 1964–1966 (Adapted from Mackenzie E J and Purves D 1967 *Edinburgh School Agricultural Experimental Work*, p 23)

Hay		Barley straw	
Range	Mean	Range	Mean
g/kg DM			
Ca 3.0–6.3	4.6	1.5–4.5	3.1
Mg 0.6–1.4	1.1	0.3–0.6	0.5
Na 0.2–1.9	1.0	0.1–1.0	0.5
mg/kg DM			
Cu 1.5–10.0	6.0	0.6–4.0	2.4
Mn 30–150	80	1.8–22.0	12.1
Fe 30–120	106	18–170	78

Rice straw

In many of the intensive rice-growing areas of the world, particularly Asia, this straw is used as a food for farm animals. Its protein content and metabolisable energy value are similar to those of spring barley straw. It has an exceptionally high ash content, about 170 g/kg DM, which consists mainly of silica. The lignin content of this straw, about 60–70 g/kg DM, is, however, lower than that of other cereal straws. In contrast to other straws, the stems are more digestible than the leaves.

Wheat and rye straws

Until recently, wheat and rye straws were considered to be so poor in nutritional value that their use as foods for farm animals was not recommended. However, recent research has demonstrated that the digestibility of most cereal straws can be markedly improved with alkali treatment (see below).

Legume straws

The straws of beans and peas are richer in protein, calcium and magnesium than the cereal straws, and if properly harvested are useful roughage foods for ruminant animals. Because of their thick fibrous stems they are more difficult to dry than cereal straws and frequently become mouldy during storage.

Alkali treatment of straws and other forages

When straw is exposed to an alkali, the ester linkages between lignin and the cell-wall polysaccharides, cellulose and hemicelluloses, are hydrolysed, thereby causing the carbohydrates to become more available to the microorganisms in the rumen. This effect was first used to improve the digestibility of straw in Germany in the early 1900s. In the Beckmann process, straw was soaked for one–two days in a dilute solution (15–30 g/litre) of sodium hydroxide, then washed exhaustively to remove excess alkali. This process increased dry matter digestibility from 0.4 to 0.5–0.7, despite the fact that the washing removed a considerable proportion of the soluble – and presumably more digestible – constituents of the straw. In the process currently employed, chopped or milled straw is sprayed in a mixer with a small volume of concentrated sodium hydroxide (typically 170 litre/t of straw of a solution of 300 g/litre of NaOH, supplying 50 kg NaOH). The product is not washed and the alkali forms sodium carbonate, which gives the product a pH of 10–11. This process is slightly less effective in improving digestibility than is the Beckmann process, but gives a product that may be mixed with other foods and may also be pelleted.

An alternative alkali to sodium hydroxide is ammonia, which may be applied to straw in the anhydrous form or as a concentrated solution. As both forms are volatile the process has to be carried out in a sealed container, which may be formed by wrapping a stack of straw bales in plastic sheeting. As ammonia is a weaker alkali than sodium hydroxide it reacts slowly with the straw; the time required for treatment ranges from one day, if heat is applied to raise the temperature to 85 °C, to one month at winter temperatures. The ammonia is added at 30–35 kg/t of straw, and when the stack is exposed to the air about two-thirds of this is lost by volatilisation. The remainder is bound to the straw and raises its crude protein content by about 50 g/kg. In addition to this advantage over sodium hydroxide, ammonia does not leave a residue of sodium (which increases the water intake of animals).

Both sodium hydroxide and ammonia have been used on a wide range of low-quality forages, including straws, husks and hays. Table 21.6 summarises the results of 44 experiments in which animals were fed on diets containing a high proportion of forage, either untreated or treated with sodium hydroxide or ammonia. It should be noted that, in addition to improving digestibility, the alkali treatments caused increases in intake.

A danger arising from ammonia treatment is that it may sometimes cause the production of toxic imidazoles, which arise from reactions between ammonia and sugars. Forages containing more sugars than do straws, such as hays, are more likely to form imidazoles, and their production is encouraged by high temperatures. The toxins cause a form of dementia, which in cattle is sometimes called 'bovine bonkers'.

A chemical for treatment of forages that is easier to handle, and often cheaper, than ammonia is urea. When exposed to the enzyme urease, urea is hydrolysed to yield ammonia:

$$NH_2-CO-NH_2 + H_2O \rightarrow 2NH_3 + CO_2$$

Table 21.6 Performance of cattle and sheep fed on diets containing treated and untreated roughages[a] (Adapted from Greenhalgh J F D 1983 *Agricultural Progress* **58**: 11)

Species	Alkali:	NaOH		NH$_3$	
	Treatment:	−	+	−	+
Cattle	No. of experiments	[17	]	[10	]
	Roughage in diet, %	[64	]	[61	]
	Digestibility[b]	0.56	0.64	0.58	0.63
	Intake, kg/day[b]	7.2	8.1	6.8	7.8
	Liveweight gain, kg/day	0.62	0.82	0.40	0.71
Sheep	No. of experiments	[10	]	[7	]
	Roughage in diet, %	[66	]	[65	]
	Digestibility[b]	0.57	0.65	0.52	0.62
	Intake, g/day[b]	994	1259	1156	1147
	Liveweight gain, g/day	39	126	73	99

[a] The roughages used were mainly wheat and barley straws but included rice straw, and also maize by-products such as cobs and stalks. Some of the roughages were ground and pelleted.
[b] Of dry matter in total diet.

Straw normally carries bacteria that secrete the necessary urease; it is important that the straw should be wet enough (about 300 g water per kg) to allow the hydrolysis to take place. After the application of urea the straw is sealed in the same way as for treatment with ammonia. Urea ammoniation of straw has proved reasonably effective in improving its nutritive value, but is not as consistently effective as ammonia or sodium hydroxide.

Urea can also be used simply as a supplement to straw (i.e. be added at the time of feeding). Table 21.7 compares the effects of adding urea at this time (in this experimental situation, by intra-ruminal infusion) with adding it to the straw a month beforehand to generate ammonia. Both of these treatments, when applied separately (columns 2 and 3 versus 1), improved intake and digestibility, and reduced the weight loss of the sheep, but the double treatment (column 4) gave the best results, possibly because the urea-ammoniated straw (column 3) did not maintain a sufficient concentration of ammonia in the rumen.

In addition to treating straw and hay, alkalis may be applied to whole-crop cereal forages. In Chapter 20 the ensiling of cereals cut when immature was mentioned. If they are cut later, the grain forms a greater proportion (e.g. 60 per cent) of the crop and its dry matter content is greater. Ammonia or sodium hydroxide raises the digestibility of the straw in the crop and also acts as a preservative by preventing mould growth.

Other chemicals that have been used effectively to improve the digestibility of straw include alkaline hydrogen peroxide and mineral acids, but these are probably too expensive for practical use.

Supplementation of straws

Chemical treatments of straws have attracted a great deal of research interest, but their use in practice is somewhat limited. Countries that would benefit most from them have neither sufficient resources of chemicals nor the technology needed to apply them. The key to improvements in the utilisation of straw is supplementation. The first type of supplement required for

Table 21.7 A comparison of urea used to ammoniate rice straw with urea used as a dietary supplement for sheep (Adapted from Djajanegara A and Doyle P T 1989 *Animal Feed Science and Technology* **27**: 17)

Straw treatment:[a] Supplement:[b]	None None (1)	None Urea (2)	Urea None (3)	Urea Urea (4)
Dry matter digestibility	0.42	0.48	0.54	0.55
Dry matter intake, g/day	682	951	931	1114
Liveweight change, g/day	−138	−20	−10	38
Rumen ammonia-N, mg/litre	12	104	57	203

[a] Straw sprayed with 1 litre/kg of a solution containing 60 g urea, then sealed for 28 days.
[b] 11.5 g urea and 2.35 g sodium sulphate per kg dry matter consumed, given by continuous intra-ruminal infusion.

straw is one that provides adequate supplies of nutrients for the rumen microorganisms, the critical nutrients being nitrogen and sulphur, and perhaps phosphorus, sodium and cobalt. In the case of the nitrogen supplement, the aim is to provide a reasonably constant rumen ammonia concentration, and if the nitrogen is in a soluble and rapidly degradable form, such as urea, the supplement needs to be taken in frequently, in small quantities. Supplementary nutrients can be supplied as a solution sprayed on to the straw. Alternatively, they may be added to a small amount of concentrate food, or offered to animals as feedblocks or licks. None of these methods is consistently effective, but the first ensures that straw and supplement are consumed together.

The second type of supplement required for straw is one that provides the animal with additional protein that is not degraded in the rumen (digestible undegradable protein, see Chapter 13). This often stimulates intake (see Chapter 17) and ensures a proper balance between the protein and energy supplied to the animal's tissues. However, supplements of this second type, which are often based on fishmeal or specially treated vegetable protein sources, are in limited supply and expensive, and are therefore unsuitable for widespread use in developing countries. In those countries there is now much interest in supplementing straw with locally available materials, especially forages from protein-rich leguminous shrubs such as *Leucaena leucocephala* and *Gliricidia sepium* (see Chapter 19). These are valuable as sources of minerals and vitamins, but are less effective as sources of digestible undegradable protein.

Summary

1. Hay is made mainly by the sun-drying of grass and other forage crops. After the crop has been cut, its treatment in the field is intended to minimise losses of valuable nutrients caused by the action of plant respiration, by microorganisms, by oxidation, by leaching and by mechanical damage.

2. The nutritive value of hay is determined by the growth stage and plant species of the parent crop, by field losses of nutrients and by changes taking place during storage (which can be reduced by the use of chemical preservatives). Even under good conditions overall loss of dry matter may be about 20 per cent.

3. Artificially dried forages are higher in nutritive value than hays; they are expensive to produce and may be given to non-ruminant livestock as sources of minerals and vitamins.

4. Straws, mainly from cereals, are of low nutritive value but are important foods for ruminants, especially in developing countries.

5. The digestibility and intake of straws may be improved by treatment with sodium hydroxide or ammonia (as a gas or derived from urea). Such treatment is expensive, and the more appropriate way of making the best use of straws is to supplement them, especially with a source of rumen-degradable protein.

Further reading

Ho Y H, Wong H K, Abdullah N and Tajuddin Z A (eds) 1991 *Recent Advances on the Nutrition of Herbivores*, Kuala Lumpur, Malaysian Society of Animal Production.

Nash M J 1985 *Crop Conservation and Storage*, Oxford, Pergamon Press.

Staniforth A R 1979 *Cereal Straw*, Oxford, Clarendon Press.

Sullivan J T 1973 Drying and storing herbage as hay. In: Butler G W and Bailey R W (eds) *Chemistry and Biochemistry of Herbage* Vol. 3, London, Academic Press.

Sundstöl F and Owen E (eds) 1984 *Straw and Other Fibrous By-products as Feed*, Amsterdam, Elsevier.

22 Roots, tubers and related by-products

22.1 Roots

22.2 Tubers

The most important root crops used in the feeding of farm animals are turnips, swedes (or rutabagas), mangels (or mangolds) and fodder beet. Sugar beet is another important root crop, but is grown primarily for its sugar content and is normally not given as such to animals. However, the two by-products of the sugar extraction industry, sugar beet pulp and molasses (from both sugar beet and sugarcane), are important and nutritionally valuable animal foods.

The main tubers are potatoes, cassava and sweet potatoes, the last two being tropical crops.

Box 22.1

Declining importance of some root and tuber crops

The main fodder roots (turnips, swedes and fodder beet) were in the past valued for their ability to produce heavy crops in unfavourable environments, but are less popular today. The total (world) areas devoted to these crops were as follows (million hectares):

	1989	1999
Turnips	0.198	0.139
Swedes	0.055	0.051
Fodder beet	0.441	0.189

In contrast, the crops grown mainly for human consumption, but partly used to feed animals, have maintained their popularity, their current world areas (million hectares) being: potatoes 18.0, sweet potatoes 9.1, cassava 16.6 and sugar beet 6.7.

22.1 Roots

The main characteristics of roots are their high moisture content (750–940 g/kg) and low crude fibre content (40–130 g/kg DM). The organic matter of roots consists mainly of sugars (500–750 g/kg DM) and is of high digestibility (about 0.80–0.87). Roots are generally low in crude protein content although like most other crops this component can be influenced by the application of nitrogenous fertilisers. The degradability of protein in the rumen is high, at about 0.80 to 0.85.

The composition is influenced by season and the variation may be quite large; low DM roots are produced in a wet season and relatively high DM roots in a hot dry season. The composition also varies with size, the large root having lower DM and fibre contents and being of higher digestibility than the small root. Winter hardiness is associated with higher DM content and keeping quality. In the past, root crops have been considered as an alternative to silage in ruminant diets but their value as cereal replacements is now recognised. Roots are not a popular food for pigs and poultry because of their bulky nature, although those which have higher DM contents, such as fodder beet, are given to pigs. The root crops listed in Table 22.1 are poor sources of vitamins.

Roots are often stored in clamps during winter; during this period losses of dry matter of up to 10 per cent are not uncommon.

Swedes and turnips

Swedes (*Brassica napus*), which were introduced into Britain from Sweden about 200 years ago, and turnips (*Brassica campestris*) are chemically very similar, although turnips generally contain less DM than swedes (Table 22.1). Of the two types of turnip that are grown, the yellow-fleshed cultivars are of higher DM content than the white-fleshed cultivars. The ME value of swedes is usually higher than that of turnips, i.e. about 13 and 11 MJ/kg DM, respectively (Table 22.1). The main sugars present are glucose and fructose.

Both swedes and turnips are liable to taint milk if given to dairy cows at or just before milking time. The volatile compound responsible for the taint is absorbed from the air by the milk and is not passed through the cow.

Mangels, fodder beet and sugar beet

These three crops are all members of the same species, *Beta vulgaris*, and for convenience they are generally classified according to their dry matter content. Mangels are the lowest in DM content, richest in crude protein and lowest in sugar content of the three types. Fodder beet can be regarded as lying between mangels and sugar beet in terms of DM and sugar content, while sugar beet is richest in DM and sugar content, though poorest

Table 22.1 Typical composition and nutritional values of roots, root by-products and tubers (dry matter basis)

	Dry matter (g/kg)	Organic matter (g/kg)	Crude protein (g/kg)	Crude fibre (g/kg)	Rumen degradability of protein	ME[a] (MJ/kg)
Roots						
Swedes	120	942	108	100	0.85	12.8
Turnips	80	922	122	111	0.85	11.2
Mangels	110	933	100	58	0.85	12.4
Fodder beet	185	925	62	53	0.85	11.8
Sugar beet	230	970	48	48	–	13.7
Root by-products						
Beet molasses	750	931	40	0	0.80	12.0
Sugar beet pulp (molassed)	860	918	110	132	0.70	12.5
Tubers						
Cassava	370	970	35	43	0.80	12.8
Potatoes	210	957	110	38	0.85	13.3
Sweet potatoes	320	966	39	38	–	12.7

[a] For ruminants.

in crude protein. On a DM basis the ME values range from about 12 to 14 MJ/kg, the higher values applying to sugar beet. The main sugar present in these roots is sucrose.

Mangels

'Low dry matter' mangels have a DM content of 90–120 g/kg. 'Medium dry matter' mangels have a DM content of 120–150 g/kg. This group is usually smaller in size than the low dry matter group, but usually develops fairly large tops.

It is customary to store mangels for a few weeks after lifting, since freshly lifted mangels may have a slightly purgative effect. The toxic effect is associated with the nitrate present, which on storage is converted into asparagine. Unlike turnips and swedes, mangels do not cause milk taints when given to dairy cows.

Fodder beet

'Medium dry matter' fodder beet contains 140–180 g/kg of dry matter, whereas the 'high dry matter' varieties may contain up to 220 g/kg. In addition to varietal type, the DM content is also influenced by the stage of growth at harvesting and environmental conditions. Fodder beet is a poor source of protein (Table 22.1).

Fodder beet is a popular food in Denmark and the Netherlands for dairy cattle and young ruminants. Care is required in feeding cattle on high dry matter fodder beet, since excessive intakes may cause digestive upsets, hypocalcaemia and even death. The digestive disturbances are probably associated with the high sugar content of the root.

The use of fodder beet as the bulk ration for feeding pigs has given satisfactory results, but experiments have shown that the fattening period is slightly longer than when sugar beet is used. The digestibility of the organic matter of fodder beet is very high (about 0.90).

Sugar beet

Most sugar beet is grown for commercial sugar production, though it is sometimes given to animals, especially cows and pigs. Because of its hardness the beet should be pulped or chopped before feeding.

After extraction of the sugar at the sugar beet factory, two valuable by-products are obtained which are given to farm animals. These are sugar beet pulp and beet molasses.

Sugar beet pulp

The sugar beet on arrival at the factory is washed, sliced and soaked in water, which removes most of the sugars. After extraction of the sugar the residue is called sugar beet pulp. The water content of this product is 800–850 g/kg and the pulp may be sold in the fresh state for feeding farm animals, but because of transport difficulties it is frequently dried to a moisture content of 100 g/kg. Since the extraction process removes the water-soluble nutrients, the dried residue consists mainly of cell-wall polysaccharides, and consequently the crude fibre content is relatively high (about 200 g/kg DM); the crude protein and phosphorus contents are low, the former being about 100 g/kg DM.

Most sugar beet pulp is now sold after drying and the addition of molasses; the molasses provides about 20 per cent of the dry matter and raises the water-soluble carbohydrate (i.e. sugar) content from 200 to 300 g/kg DM. Molassed sugar beet pulp is extensively used as a food for dairy cows and is also given to fattening cattle and sheep. Originally, the pulp was considered to be too fibrous a food for pigs, but recent studies have shown that the digestibility of beet pulp fibre is high, at about 0.80–0.85, even for young pigs. It is recommended that up to 15 per cent molassed sugar beet pulp may be included in the diet of growing pigs and up to 20 per cent in the diets of finishers and sows. The product is not a suitable food for poultry.

Sugar beet pulp may be combined with other by-products such as distiller's grains (see Chapter 23).

Beet molasses

After crystallisation and separation of the sugar from the water extract, a thick black liquid termed beet molasses remains. This product contains 700–750 g/kg of DM, of which about 500 g consists of sugars. The molasses dry matter has a crude protein content of only 20–40 g/kg, most of this being in the form of non-protein nitrogenous compounds, including the amine betaine, which is responsible for the 'fishy' aroma associated with the extraction process.

Beet molasses is a laxative food and is normally given to animals in small quantities. Usually molasses is added to beet pulp (as mentioned above) or to other foods that include bran, grains, malt culms, spent hops and sphagnum moss. Molasses is also used, generally at levels of 5–10 per cent, in the manufacture of compound cubes and pellets. The molasses not only improves the palatability of the product but also acts as a binding agent. An additional use of molasses is as a constituent (and binding agent) in the compressed feed blocks that are used as protein, mineral and vitamin supplements for ruminants. Since beet molasses is a rich and relatively cheap source of soluble sugars it is sometimes used as an additive in silage making.

Molasses is used as the feedstock for a number of industrial fermentations. When the sugars have been fermented and the fermentation products removed, there is a residue rich in nitrogen-containing substances and ash. This is partially dried to give a material known as condensed molasses solubles (CMS), which contains about 350 g of crude protein per kg DM. The CMS is then mixed with molasses in the ratio 20 : 80 to give a protein-fortified food.

Cane molasses

Although sugarcane is not a root or tuber crop, one of its major by-products, cane molasses, is a similar food to beet molasses, and its use is therefore described in this chapter. In the tropical and subtropical countries where sugarcane is grown, there is often a serious shortage of sugar- or starch-containing animal foods, so cane molasses is a valuable resource. In addition to its uses as a carrier for supplements such as urea (see Chapter 24) and as a conditioner of compound foods, cane molasses is used as a source of supplementary energy for forage diets, and even as the main component of diets. In Cuba, for example, diets for beef cattle have been based on cane molasses offered *ad libitum* and providing 500–800 g/kg of total dry matter intake, together with some forage and a supplement of protein (preferably protein of low rumen degradability, as in fishmeal). Such diets can give liveweight gains of about 1 kg/day, but sometimes induce a condition in cattle known as molasses toxicity. This is characterised by uncoordination and blindness caused by deterioration of the brain similar to that seen in cerebro-cortical necrosis (see p. 92). Its origins lie in the unusual rumen fermentation of molasses diets, which gives rise to volatile fatty acid mixtures rich in butyrate and poor in propionate. Like cerebro-cortical necrosis, the toxicity may be due to thiamin deficiency, but it may also be due to a shortage of glucose caused by the poor production of propionate in the rumen. The condition is best avoided by ensuring that animals have sufficient good-quality forage available.

22.2 Tubers

Tubers differ from the root crops in containing either starch or fructan instead of sucrose or glucose as the main storage carbohydrate. They have higher dry matter and lower fibre contents (Table 22.1) and consequently are more suitable than roots for feeding pigs and poultry.

Potatoes

In potatoes (*Solanum tuberosum*) the main component is starch. The starch content of the dry matter is about 700 g/kg; this carbohydrate is present in the form of granules that vary in size depending upon the variety. The sugar content in the dry matter of mature, freshly lifted potatoes rarely exceeds 50 g/kg although values in excess of this figure in stored potatoes have been obtained. The amount present is affected by the temperature of storage and values as high as 300 g/kg have been reported for potatoes stored at −1 °C.

The crude protein content of the dry matter ranges from about 90 to 123 g/kg with a mean value of about 110 g/kg. About half of this is in the form of non-protein nitrogenous compounds. One of these compounds is the alkaloid solanidine, which occurs free and also in combination as the glyco-alkaloids chaconine and solanine. Solanidine and its derivatives are toxic to animals, causing gastroenteritis. The alkaloid levels may be high in potatoes exposed to light. Associated with light exposure is greening due to the production of chlorophyll. Green potatoes should be regarded as suspect. Removal of the eyes and peel, in which the solanidine is concentrated, will reduce the toxicity, but this is not a practical proposition in feeding farm animals Young shoots, even if white, are also likely to be rich in solanidine and these should be removed and discarded before feeding. Immature potatoes have been found to contain more solanidine than mature tubers. The risk of toxicity is reduced considerably if potatoes are steamed or otherwise cooked, the water in which the tubers have been boiled being discarded. Ensiling also destroys some of the toxin so that inclusion of slightly greened potatoes with grass should be acceptable. Ruminants are more resistant to toxicity than monogastric animals, presumably because of partial destruction of the toxin in the rumen.

The crude fibre content of potatoes is very low, usually less than 40 g/kg DM, which makes them particularly suitable for pigs and poultry. However, the protein in uncooked potatoes is frequently poorly digested by these animals, and protein digestibility coefficients as low as 0.23 have been reported for pigs. In similar trials with cooked potatoes, digestibility coefficients for protein generally exceed 0.70. Potatoes contain a protease inhibitor that reduces the digestibility not only of potato protein but also of protein in other components of the diet. The inhibitor is destroyed by heating; it is normal practice to cook potatoes for pigs and poultry, although cooking is unnecessary for ruminants, presumably because the inhibitor is destroyed in the rumen. For pigs and poultry the ME value of cooked potatoes is similar to that of maize, about 14–15 MJ/kg DM.

Potatoes are a poor source of minerals, except for the abundant element potassium, the calcium content being particularly low. The phosphorus content is rather higher, since this element is an integral part of the potato starch molecule, but some 20 per cent of it is in the form of phytates (see p. 120).

During the storage of potatoes considerable changes in composition may occur. The main change is a conversion of some of the starch to sugar and the oxidation of this sugar, with the production of carbon dioxide during respiration. The respiration rate increases with an increase in temperature. There may also be a loss of water from the tubers during storage.

Dried potatoes

The difficulty of storing potatoes satisfactorily for any prolonged period of time has led to a number of processing methods. Several methods of drying are used. In one method the cooked potatoes are passed through heated rollers to produce dried potato flakes. In another method sliced tubers are dried directly in flue gases; the resultant potato slices are frequently ground to a meal before marketing. The products are valuable concentrate foods for all classes of animal.

Potato processing waste

This product is the dried residue obtained in the processing of potatoes for canning and chipping for human consumption. It consists of peel and small pieces of tubers which have been coarsely ground and dried. It is of variable composition, the crude fibre content ranging from about 30 to 70 g/kg DM and the crude protein from 70 to 140 g/kg DM. Provided the product is free from soil contamination it is a useful food in small quantities for pigs, poultry and ruminants.

Cassava

Cassava (*Manihot esculenta*), also known as manioc, is a tropical shrubby perennial plant which produces tubers at the base of the stem. The chemical composition of these tubers varies with maturity, cultivar and growing conditions. Cassava tubers are used for the production of tapioca starch for human consumption although the tuber is also given to cattle, pigs and poultry. Its ME value is similar to that of potatoes but it has higher dry matter and lower crude protein contents (Table 22.1).

Cassava plants and tubers are to a certain degree poisonous since they contain varying proportions of two cyanogenetic glucosides (linamarin and lotaustralin), which readily break down to give hydrocyanic acid (see p. 21). In all cases care must be taken in the use of the tuber and wherever the plant is grown, indigenous methods of removing the glucosides have been devised. Such treatments include boiling, or grating and squeezing, or grinding to a powder and then pressing.

Large amounts of dehydrated cassava meal have recently become available for feeding farm animals. The meal can be used as a partial cereal grain replacer, provided the protein deficiency is rectified.

Cassava pomace is the residue from the extraction of starch from cassava tubers. Because of its high crude fibre content (about 270 g/kg DM) its use in the diets of non-ruminant animals should be restricted.

Sweet potatoes

The sweet potato (*Ipomoea batatas*) is a very important tropical plant whose tubers are widely grown for human consumption and as a commercial source of starch. The tubers are of similar nutritional value to ordinary potatoes although of much higher dry matter and lower crude protein contents (Table 22.1). Fresh tubers, surplus to requirements, are often cut into small pieces, sun-dried and then ground to produce a sweet potato meal, a high-energy food of low protein content. Sun-drying does not destroy the trypsin inhibitors believed to be present in the tubers, and levels in the diets of farm animals are usually restricted.

Summary

1. The fodder root crops of temperate climates (turnips, swedes, mangels and fodder beet) are characterised by their high water content (750–940 g/kg), low crude fibre content (40–130 g/kg DM) and high sugar content (500–750 g/kg DM). Digestibility and ME value are high (11–13 MJ/kg DM). They are mainly given to ruminants.

2. Sugar beet, grown mainly for extraction of sugar, yields two by-products for animal feeding. Sugar beet pulp is similar in composition to cereals, except for the replacement of starch by a high fibre content (200 g/kg DM). Beet molasses (and also sugarcane molasses) contain sugars and only 20–40 g protein per kg (mainly non-protein nitrogen). Excessive amounts of molasses in

ruminant diets give rise to butyric acid in the rumen, and may be toxic.

3. Potato tubers are rich in starch and low in fibre, their ME value for pigs and poultry being 14–15 MJ/kg DM. Potato processing waste (mainly peel) is higher in fibre than whole potatoes

4. The tropical plant, cassava, has tubers rich in starch but low in protein. Dried cassava meal and by-products from starch extraction are used in animal feeds, but must be prepared in such a way that cyanogenetic glucosides are inactivated.

5. Sweet potatoes may be dried for animal feeding, but contain a trypsin inhibitor.

Further reading

Barber W P and Lonsdale C R 1980 By-products from cereal, sugarbeet and potato processing. In: Ørskov E R (ed.) *By-products and Wastes in Animal Feeding*, Reading, British Society of Animal Production Occasional Symposium Publication No. 3.

Göhl B 1981 *Tropical Feeds*, Rome, FAO.

Greenhalgh J F D, McNaughton I H and Thow R F (eds) 1977 *Brassica Fodder Crops*, Edinburgh, Scottish Agricultural Development Council.

Kelly P 1983 Sugar beet pulp – a review. *Animal Feed Science and Technology* **8**: 1–18.

Nash M J 1985 *Crop Conservation and Storage*, Oxford, Pergamon Press.

Oke O L 1978 Problems in the use of cassava as animal feed. *Animal Feed Science and Technology* **3**: 345–80.

Preston T R and Leng R A 1986 *Matching Livestock Production Systems to Available Resources*, International Livestock Centre for Africa, Addis Ababa, Ethiopia.

23 Cereal grains and cereal by-products

23.1   The nutrient composition of grains

23.2   Barley

23.3   Maize

23.4   Oats

23.5   Wheat

23.6   Other cereals

23.7   Cereal processing

23.1   The nutrient composition of grains

The name 'cereal' is given to those members of the Gramineae which are cultivated for their seeds. Cereal grains are essentially carbohydrate concentrates, the main component of the dry matter being starch, which is concentrated in the endosperm (Fig. 23.1). The dry matter content of the grain depends on the harvesting method and storage conditions but is generally within the range of 800–900 g/kg.

Of the nitrogenous components 85–90 per cent are in the form of proteins. The proteins occur in all tissues of cereal grains, but higher concentrations are found in the embryo and aleurone layer than in the starchy endosperm, pericarp and testa. Within the endosperm, the concentration of protein increases from the centre to the periphery. The total content of protein in the grain is very variable; expressed as crude protein it normally ranges from 80 to 120 g/kg DM, although some cultivars of wheat contain as much as 220 g/kg DM. Cereal proteins are deficient in certain indispensable amino acids, particularly lysine and methionine. It has been shown that the value of cereal proteins for promoting growth in young chicks is in the order oats > barley > maize or wheat. The high relative value of oat protein for growth has been attributed to its slightly higher lysine content. This is demonstrated in Fig. 23.2, which compares the main limiting amino acid components of a number of cereal grains.

The lipid content of cereal grains varies with species. Wheat, barley, rye and rice contain 10–30 g/kg DM, sorghum 30–40 g/kg DM and maize and oats

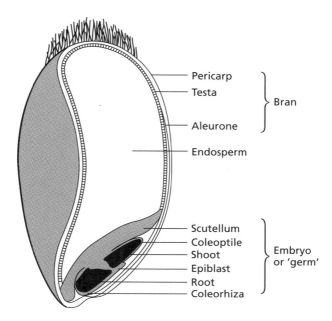

Fig. 23.1 Longitudinal section of caryopsis (grain) of wheat.

40–60 g/kg DM. The embryo or 'germ' contains more oil than the endosperm; in wheat, for example, the embryo has 100–170 g/kg of DM of oil, while the endosperm contains only 10–20 g/kg DM. The embryo of rice is exceptionally rich in oil, containing as much as 350 g/kg DM. Cereal oils are unsaturated, the main acids being linoleic and oleic, and because of this they tend to become rancid quickly, and also produce a soft body fat in pigs and poultry.

The crude fibre content of the harvested grains is highest in those such as oats or rice which contain a husk or hull formed from the fused glumes (palea and lemma), and is lowest in the 'naked' grains, wheat and maize. The husk has a diluent effect on the grain as a whole and reduces the energy value proportionally. Of the grains as harvested, oats has the lowest metabolisable energy value and maize has the highest, the respective values (MJ/kg DM) for poultry being around 12 and 16 and for ruminants 12 and 14.

Starch occurs in the endosperm of the grain in the form of granules, whose size and shape vary with different species. Cereal starches consist of about 25 per cent amylose and 75 per cent amylopectin, although 'waxy' starches contain greater proportions of amylopectin.

The cereals are all deficient in calcium, containing less than 1 g/kg DM. The phosphorus content is higher, being 3–5 g/kg DM, but part of this is present as phytic acid (see p. 120), which is concentrated in the aleurone layer. Cereal phytates have the property of being able to bind dietary calcium and probably magnesium thus preventing their absorption from the gut; oat phytates are more effective in this respect than barley, rye or wheat phytates. The cereal grains are deficient in vitamin D and, with the exception of yellow maize, in provitamins A. They are good sources of vitamin E

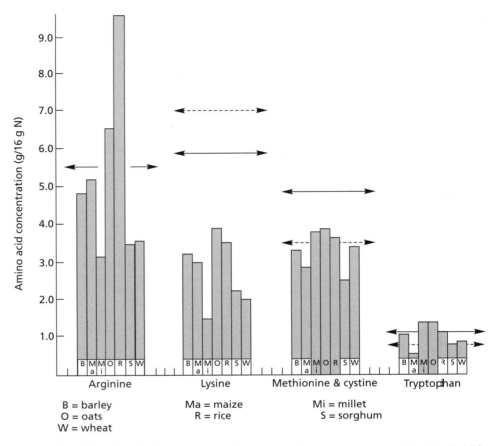

Fig. 23.2 Main limiting indispensable amino acids of cereal grains (g/16 g N) (Straight lines indicate requirements for chicks: dotted lines for growing pigs.)

and thiamin, but have a low content of riboflavin. Most of the vitamins are concentrated in the aleurone layer and the germ of the grain.

Calves, pigs and poultry depend upon cereal grains for their main source of energy, and at certain stages of growth as much as 90 per cent of their diet may consist of cereals and cereal by-products. Cereals generally form a lower proportion of the total diet of ruminants, although they are the major component of the concentrate ration.

23.2 Barley

Barley (*Hordeum sativum*) has always been a popular grain in the feeding of farm animals, especially pigs. In most varieties of barley the kernel is surrounded by a hull, which forms about 10–14 per cent of the weight of the grain.

Table 23.1 Composition and nutritive value of whole-grain barley samples of parent variety NP 113, and of mutant varieties Notch 1 and Notch 2 (After Balaravi S P *et al.* 1976 *Journal of the Science of Food and Agriculture* **27**: 545)

	NP 113	Notch 1	Notch 2
Protein, g/kg	117	157	146
Lysine, g/16 g N	3.88	4.00	3.96
Starch, g/kg	662	396	414
Crude fibre, g/kg	70	104	128
BV[a]	0.76	0.86	0.88
NPU[a]	0.66	0.68	0.73

[a] With rats.

The metabolisable energy value (MJ/kg DM) is about 13.3 for ruminants and 13.2 for poultry, and the digestible energy value for pigs is 15.4. The crude protein content of barley grain ranges from about 60 to 160 g/kg DM with an average value of about 120 g/kg DM. As with all cereal grains the protein is of low quality, being particularly deficient in the amino acid lysine. High-lysine mutants of barley have been produced by plant breeders and the superior nutritional value of two such mutants – Notch 1 and Notch 2 – is shown in Table 23.1. Unfortunately with many of these mutants the yields of grain are much lower (about 30 per cent) than from parent varieties, and the starch contents may be reduced.

The lipid content of barley grain is low, usually less than 25 g/kg DM. The range in dry matter composition of 179 samples of barley grain harvested in Wales is given in Table 23.2.

In many parts of the world, and in particular in the UK, barley forms the main concentrate in the diets of pigs and ruminants. In the 'barley beef' system of cattle feeding, beef cattle are fattened on concentrate diets consisting of about 85 per cent bruised barley without the use of roughages. In this process the barley is usually treated so that the husk is kept in one piece and at the same time the endosperm is exposed, the best results being obtained by rolling grain at a moisture content of 160–180 g/kg. Storage of high-moisture barley of this type can present a problem because of the possibility of mould growth. Satisfactory preservation of the moist grain can be obtained if it is stored anaerobically. An additional or alternative safeguard is to treat the grain with a mould inhibitor such as propionic acid (see p. 541). Certain hazards, such as rumen acidosis (rapid fermentation of the starch to lactic acid, resulting in depression of digestion of fibre and food intake) and bloat (see p. 186), can be encountered with high-concentrate diets given to ruminants and it is necessary to introduce this type of feeding gradually over a period of time. It is important that a protein concentrate with added vitamins A and D and minerals be used to supplement high-cereal diets of this type.

Barley should always have the awns removed before it is offered to poultry, otherwise digestive upsets may occur.

Table 23.2 Dry matter composition of 171 oat and 179 barley grain samples grown in Wales, 1961–1963 (After Morgan D E 1967 and 1968 *Journal of the Science of Food and Agriculture* **18**: 21; **19**: 393)

	Oats			Barley		
	Range	Mean	Coeff. of variation[a]	Range	Mean	Coeff. of variation[a]
Proximate constitutents, g/kg						
Crude protein	72–145	107	13.4	66–153	108	15.7
Crude fibre	80–179	125	13.6	38–73	56	12.5
Ether extract	9–80	52	20.2	11–32	19	15.8
Ash	22–41	31	8.7	17–42	25	12.4
Major mineral elements, g/kg						
Ca	0.7–1.8	1.1	18.2	0.5–1.6	0.8	25.6
Mg	1.0–1.8	1.3	13.1	0.9–1.6	1.2	8.3
K	3.1–6.5	4.7	17.0	3.5–6.3	4.9	12.2
Na	0.04–0.6	0.2	47.8	0.06–0.4	0.2	41.2
P	2.9–5.9	3.8	10.5	2.6–5.2	3.8	11.8
Cl	0.4–1.8	0.9	33.3	0.8–2.2	1.4	21.4
Trace elements, mg/kg						
Cu	3.0–8.2	4.7	17.0	3.5–19.8	6.6	27.3
Co	0.02–0.17	0.05	53.0	0.02–0.18	0.07	48.6
Mn	22–79	45	28.9	5–47	16	31.3
Zn	21–70	37	27.0	19–77	37	77.0
Energy, MJ/kg						
Metabolisable energy for poultry	9.5–14.4	12.1	7.2	12.9–15.0	14.1	2.6

[a] Standard deviation as % of mean.

Barley by-products

By-products of the brewing industry (Fig. 23.3)

In brewing, barley is first soaked and allowed to germinate. During this process, which is allowed to continue for about six days, a complete enzyme system for hydrolysing starch to dextrins and maltose develops. Although the enzymatic reactions have been initiated in this germination or 'malting' process, the main conversion of the starch in the grain to maltose and other sugars takes place during the next process, described as 'mashing'. After germination but before mashing, the grain or malt is dried, care being taken not to inactivate the enzymes. The sprouts are removed and are sold as malt culms or coombs. The dried malt is crushed, and small amounts of other cereals such as maize or rice may be added. Water is sprayed on to the mixture and the temperature of the mash increased to about 65 °C.

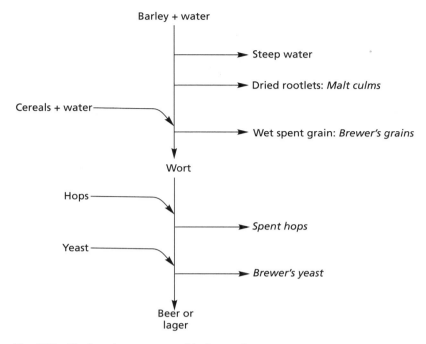

Fig. 23.3 The brewing process and its by-products.

The object of mashing is to provide suitable conditions for the action of enzymes on the proteins and starch, the latter being converted to dextrins, maltose and small amounts of other sugars. After the mashing process is completed the sugary liquid or 'wort' is drained off, leaving 'brewer's grains' as a residue. Brewer's grains are sold wet or dried as food for farm animals.

The wort is next boiled with hops, which give it a characteristic flavour and aroma; the hops are then filtered off and after drying are sold as spent hops. The wort is then fermented in an open vessel with yeast for a number of days, during which time most of the sugars are converted to alcohol and carbon dioxide. The yeast is filtered off, dried and sold as brewer's yeast.

The by-products obtained from the brewing process are therefore malt culms, brewer's grains, spent hops and brewer's yeast.

Malt culms

Malt culms consist of the plumule and radicle of barley, and are relatively rich in crude protein (about 280 g/kg DM). They are also produced as a by-product of the distilling industry (see below). They are not a high-energy food, however, and because of their fibrous nature their use is generally restricted to the feeding of ruminants and horses. The quality of protein in malt culms is poor and this, together with the high fibre content, limits their use for pigs to diets for pregnant sows, or at low levels in finishing diets. Malt culms have a bitter flavour owing to the presence of the amino acid asparagine, which forms about one-third of the crude protein. However, when mixed with other foods they are accepted readily by cattle and have been included in concentrate mixes at levels up to 500 g/kg. A related by-product is malt residual pellets, which comprise malt culms and other malt screenings. These have lower fibre and higher starch contents than plain malt culms, giving a slightly higher ME for ruminants (11.5 versus 11 MJ/kg DM) and lower crude protein (220 versus 280 g/kg DM).

Brewer's grains

Brewer's grains or 'draff' consist of the insoluble residue left after removal of the wort. In addition to the insoluble barley residue this product may contain maize and rice residues and, because of this, the composition of the product can be very variable, as is illustrated in Table 23.3.

The fresh brewer's grains contain about 700–760 g water/kg and may be given to cattle, sheep and horses in this fresh state or alternatively preserved as silage. More brewer's grains are produced in summer than winter, therefore ensilage is a popular form of storage for winter feeding. The wet product can be dried to about 100 g water/kg and sold as dried brewer's grains. The rumen degradability of the protein of the dried product is about 0.6 compared with about 0.8 in the original barley. Brewer's grains are a concentrated source of digestible fibre, and energy losses from the rumen as methane are lower than with high-starch feeds. They are high in phosphorus but low in other minerals. Brewer's grains have always been a popular food for dairy cows, but they are of little value for poultry. They are not very suitable for pigs except for pregnant sows, which have a large intake capacity and an active hind gut fermentation which enables them to utilise this material.

Table 23.3 The nutritional value of fresh brewer's grains[a] (Adapted from Barber W P and Lonsdale C R 1980 Occasional Publication No. 3, British Society of Animal Production, Reading, pp 61–9)

	Mean	Range
Dry matter, g/kg	263	244–300
Crude protein g/kg DM	234	184–262
Crude fibre, g/kg DM	176	155–204
Ether extract, g/kg DM	77	61–99
Total ash, g/kg DM	41	36–45
Digestible organic matter, g/kg DM[b]	594	552–643
Metabolisable energy, MJ/kg DM[c]	11.2	10.5–12.0
Digestible crude protein, g/kg DM[c]	185	139–213

[a] Results for seven samples selected from widely different sources in the UK.
[b] In vitro.
[c] Measured in sheep.

Spent hops

Dried spent hops are a fibrous product and comparable to poor hay in nutritive value, but are less palatable, probably because of their bitter flavour. This product is rarely used as a food for animals today, most of it being sold for use as fertiliser.

Dried brewer's yeast

Dried yeast is a protein-rich concentrate containing about 420 g crude protein/kg. It is highly digestible and may be used for all classes of farm animals. The protein is of fairly high nutritive value and is specially favoured for feeding pigs and poultry. It is a valuable source of many of the B group of vitamins, is relatively rich in phosphorus but has a low calcium content. Other yeasts are now available as protein concentrates and these will be described in Chapter 24.

By-products of the distilling industry

In distilling, the soluble materials may be extracted, as in brewing, or the whole mass fermented, the alcohol then being distilled off. The residue after filtration is sold as wet or dried distiller's grains. In Scotland, whisky distilleries are either 'malt' or 'grain' types (Fig. 23.4). The former use barley malt alone, whereas the latter use a mixture of cereals, which may include barley, maize, wheat and oats.

Distiller's grains (draff)

The composition of distiller's grains obviously will depend on the starting materials and can vary widely (Table 23.4). Malt distiller's grains are less variable in composition. Grain distiller's grains have a higher energy content than malt distiller's grains but have a lower content of some minerals. In general distiller's grains are low in soluble minerals, sodium and potassium, and also in calcium and magnesium. As with brewer's grains, distiller's grains are a useful feed for dairy cows and are often ensiled for winter feeding. Most of the lipid in the original grain is retained in this by-product and it has a high content of unsaturated fatty acids, which reduces microbial digestibility of fibre in the rumen and depresses intake. Digestibility and intake can be improved by the addition of calcium carbonate, which forms insoluble calcium soaps of the unsaturated fatty acids, thereby overcoming their effects on the rumen microbes. The low dry matter and high fibre contents restrict the inclusion of distiller's grains in pig diets to those for pregnant sows. The production of dried distiller's grains (distiller's light grains) has now ceased owing to the high costs of drying.

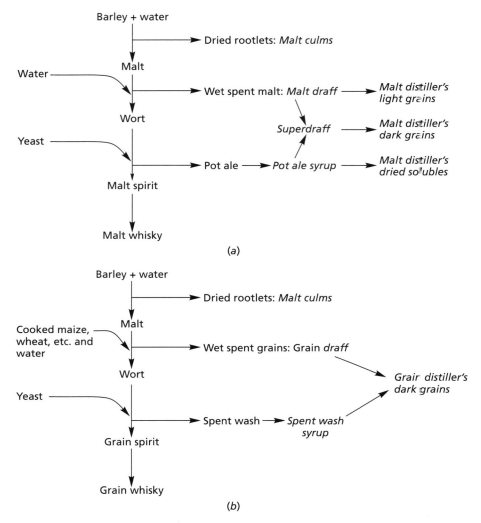

Fig. 23.4 (a) The malt distilling process and its by-products. (b) The grain distilling process and its by-products.

Distiller's solubles

After distilling off the alcohol, the liquor ('spent wash' in grain distilleries and 'pot ale' in malt distilleries) remaining in the whisky still is evaporated and then spray dried to produce a light brown powder of variable composition known as 'distiller's solubles'. Lime is added to aid drying so that it is rich in calcium. Only small amounts of dried solubles are produced because of the high cost of drying and their use is usually restricted to low levels in diets for pigs and poultry. They are a valuable source of the B group of vitamins and, although the protein content is high, the heating during drying reduces the availability of the amino acids. The dried solubles are reported to stimulate the activity of the rumen microflora. A cruder preparation of pot ale is the condensed form 'pot ale syrup', which contains about 300–500 g DM/kg and 350 g crude protein/kg DM and has a metabolisable energy value for ruminants of 14.2 MJ/kg DM. Much of the crude protein is in the form of breakdown products such as peptides and amino acids and is virtually all degradable in the rumen. However, the heat treatment results in the protein being of poor quality for pigs and it should be offered with good-quality protein sources, such as fishmeal. The mineral content of pot ale syrup is unbalanced with low calcium and sodium but high phosphorus and potassium contents. It should be given with caution to sheep because of its high content of copper, present as a contaminant from copper stills.

Distiller's dark grains

The spent wash or pot ale is often mixed (in proportions of 1:2–1:4) with the distiller's grains and dried together to yield a material sold as 'distiller's dried grains with solubles' or 'dark grains' (Table 23.4). The mixture from malt distilleries is referred to as 'malt' or 'barley dark grains', whereas that from grain distilleries is referred to as 'grain', 'wheat' or 'maize dark grains'. They are usually pelleted. Grain dark grains have the higher energy value owing to higher fat and protein contents and a lower fibre content. Dark grains generally are a balanced feed for ruminants but the degradability of

Table 23.4 The nutritive value of distiller's grains (Adapted from Black H *et al.* 1991 *Distillery By-products as Feeds for Livestock,* Scottish Agricultural College, Aberdeen)	Malt distiller's grains	Grain distiller's grains (supergrains)	Malt distiller's dark grains	Grain distiller's dark grains[a]
Dry matter, g/kg	230 (270[b])	260	900	890 (890)
Crude protein, g/kg DM	198	320	275	340 (317)
Ether extract, g/kg DM	125	106	35	69 (110)
Crude fibre, g/kg DM	174	160	121	89 (91)
Ash, g/kg DM	36	16	60	52 (46)
Metabolisable energy (ruminants), MJ/kg DM	11.1 (10.8[b])	13.0	12.2	13.5 (14.0)
Degradability of crude protein	0.80	0.80	0.70	0.70 (0.70)
Digestible energy (pigs[c]), MJ/kg DM	11.7	12.0	10.0	10.5
Lysine, g/kg DM	5.8	6.9	8.9	7.8

[a] Values are for wheat-based (maize-based) dark grains.
[b] Values for ensiled grains are given in parentheses.
[c] Values are applicable for growing/finishing pigs, corrected for less efficient use of products of fermentation; values for dry sows are likely to be 10–15 per cent higher.

the protein may vary according to the drying process. Additionally, the quality of the undegradable protein may be poor as a result of heat damage. This factor also limits the value of the protein for pigs, although with appropriate supplementation with amino acids, the grains can form up to 15 per cent of the diets for growing and finishing pigs. Dry sows can be fed at higher levels. Like other distillery by-products, dark grains are a good source of phosphorus and the copper content may be high, especially in malt distiller's dark grains. Some distilleries sell the mixed draff and pot ale syrup in the fresh state, a material known as 'superdraff'.

By-products of the pearl barley industry

In the preparation of pearl barley for human consumption, the bran coat is removed and the kernel is polished to produce a white shiny grain. During this process three by-products, described as coarse, medium and fine dust, are produced and these are frequently mixed and sold as barley feed. Barley feed contains about 140 g crude protein/kg DM and about 100 g crude fibre/kg DM. The amount of this product available in the UK is very small.

23.3 Maize

A number of different types of maize (*Zea mays*) exist and the grain appears in a variety of colours, yellow, white or red. Yellow maize contains a pigment, cryptoxanthin, which is a precursor of vitamin A. In the USA, where it is known as corn, large amounts of this cereal are grown, the yellow varieties being preferred for animal feeding. The pigmented grain tends to colour the carcass fat, which in the UK is not considered desirable, so that white maize varieties are preferred here for fattening animals. However, the pigment is useful in diets for laying hens, where it contributes to the production of the orange coloration of egg yolk.

Maize, like the other cereal grains, has certain limitations as a food for farm animals. Though an excellent source of digestible energy it is low in protein, and the proteins present are of poor quality (Fig. 23.2). Maize contains about 730 g starch/kg DM, is very low in fibre and has a high metabolisable energy value. The starch in maize is more slowly digested in the rumen than that of other grains, and at high levels of feeding a proportion of the starch passes into the small intestine, where it is digested and absorbed as glucose. This may have advantages in conditions such as ketosis (p. 275). When the starch is cooked during processing it is readily fermented in the rumen. The oil content of maize varies from 40 to 60 g/kg DM and is high in linoleic acid, which is an important factor in the diet controlling the egg size of hens. However, it tends to produce a soft body fat.

The crude protein content of maize is very variable and generally ranges from about 90 to 140 g/kg DM, although varieties have been developed recently containing even higher contents. In the USA, the tendency has been to develop hybrid varieties of lower protein content.

The maize kernel contains two main types of protein. Zein, occurring in the endosperm, is quantitatively the more important, but this protein is deficient in the indispensable amino acids tryptophan and lysine (see Fig. 23.2). The other protein, maize glutelin, occurring in lesser amounts in the endosperm and also in the germ, is a better source of these two amino acids. Varieties of maize have been produced with amino acid components different from those present in normal maize. One such variety is Opaque-2, which has a high lysine content. The difference between this variety and normal maize is primarily attributed to the zein : glutelin ratio.

Opaque-2 has been reported to be nutritionally superior to normal maize for the rat, pig, man and chick, but only in methionine-supplemented diets. A newer variety, Floury-2, has increased contents of both methionine and lysine and has been shown in studies with chicks to be superior to Opaque-2 maize in diets not supplemented with methionine.

Maize by-products

In the manufacture of starch and glucose from maize, a number of by-products are obtained which are suitable for feeding farm animals (Fig. 23.5).

The cleaned maize is soaked in a dilute acid solution and is then coarsely ground. The maize germ floats to the surface and is removed for further

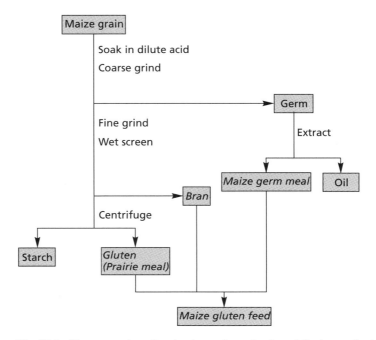

Fig. 23.5 The processing of maize to produce starch and the by-products formed.

processing. The de-germed grain is then finely ground and the bran is separated by wet screening. The remaining liquid consists of a suspension of starch and protein (gluten), which are separated by centrifugation. The process gives rise to three by-products – germ, bran and gluten.

The germ is very rich in oil, which is valued by the human food industry, and most is extracted before producing the germ meal. Maize germ meal is a variable product, depending on the degree of separation at milling (starch and fibre content) and the extraction process – screw versus solvent (oil content). In a recent study of five samples of the meal, the starch, neutral-detergent fibre and oil contents were in the ranges 435–570, 220–572 and 35–127 g/kg DM, respectively. In the UK the Feeding Stuffs Regulations require the declaration of oil and protein content. The protein has a good amino acid balance. The oil in high oil content meals can oxidise rapidly if they are not stored under the appropriate conditions. Maize gluten (prairie) meal has a very high protein content (up to about 700 g/kg DM) and is high in pigments so it is valued in poultry diets. The three by-products (germ, bran and gluten) are frequently mixed together and sold as maize gluten feed. This food has a variable protein content, normally in the range 200–250 g/kg DM, of which about 0.6 is degraded in the rumen. Dark brown material indicates heat damage, which will decrease the digestibility of the protein. Maize gluten feed has a crude fibre content of about 80 g/kg DM, and metabolisable energy values of about 9 and 12.5 MJ/kg DM for poultry and ruminants, respectively, and a digestible energy value for pigs of 14.5 MJ/kg DM. Since it is a milled product the fibre will not have the same effect as a long roughage in ruminant diets. Nevertheless, maize gluten feed has been used as a substantial proportion of the concentrate feed of dairy cows. It is generally limited to about 10–20 per cent in pig foods.

23.4 Oats

The oat (*Avena sativa*) has always been a favourite cereal for ruminant animals and horses, but has been less popular in pig and poultry feeding because of its comparatively high fibre content and low energy value.

The nutritive value of oats depends to a large extent on the proportion of kernel (groat) to hull. The proportion of hull in the whole grain depends upon the variety, environment and season, and can vary from 23 to 35 per cent (average 27 per cent). Oats of high hull content are richer in crude fibre and have a lower metabolisable energy value than low-hulled oats.

The crude protein content, which ranges from 70 to 150 g/kg DM, is increased by the application of nitrogenous fertilisers. Oat proteins are of poor quality and are deficient in the essential amino acids methionine, histidine and tryptophan, the amount of each of these acids in oat protein being generally below 20 g/kg. The lysine content is also low but is slightly higher than that of the other cereal proteins. Glutamic acid is the most abundant amino acid of oat protein, which may contain up to 200 g/kg.

The oil content of oats is higher than that of most of the other cereal grains, and about 60 per cent of it is present in the endosperm. As mentioned earlier, the oil is rich in unsaturated fatty acids and has a softening effect on the body fat. The range in dry matter composition of 171 samples of oat grain harvested in Wales is shown in Table 23.2.

The husk of a variant of oats, naked oats (*Avena nuda*), is removed easily during threshing, leaving the kernel. Originally the yield and nutritional quality were low but an improved variety, Rhiannon, was subsequently developed by the Welsh Plant Breeding Station (now the Institute for Grassland and Environmental Research). Naked oats have about 130–140 g crude protein, 6 g lysine and 100 g oil/kg DM.

Oat by-products

During the commercial preparation of oatmeal for human consumption, a number of by-products are obtained which are available for animal feeding. When the oats are received at the mill they contain a number of foreign grains, mainly other cereals and weed seeds, which are removed as cockle before processing. The cleaned oats are then stabilised by steaming to inactivate the enzyme lipase, which is located almost entirely in the pericarp of the kernel. After stabilisation, the oats are kiln dried before passing on to the huller, which removes the husks. The kernels are then brushed or scoured to detach the fine hairs that cover much of their surface.

The main by-products of oatmeal milling are oat husks or hulls, oat dust and meal seeds. The hulls form the main by-product, about 70 per cent of the total, and the commercial product consists of the true husks with a variable proportion, up to 10 per cent, of kernel material. *Oat hulls* are of very low feeding value, being little better than oat straw. Their crude protein content is so low (about 30 g/kg DM) that in digestibility studies negative digestibility coefficients for nitrogen are likely to be obtained, as the amount of metabolic nitrogen excreted is greater than that digested from them. The crude fibre content is usually between 350 and 380 g/kg DM, which makes the by-product valueless as food for animals other than ruminants.

Oat dust is rich in kernel material and includes the kernel hairs removed from the grain during brushing. It has a protein content of about 100 g/kg DM. *Meal seeds* consist of slivers of husk and fragments of kernels in approximately equal proportions.

Oat hulls may be combined with oat dust in the proportion in which they come from the mill (4 to 1) to produce a product sold as 'oat feed'. This material is rather better in feeding value than the hulls alone but the digestibility of the protein is still low. In the UK oat feed should not, by legal definition, contain more than 270 g crude fibre/kg. An alternative use for the hulls is in the brewing industry, where they are often added to the malt to assist in the drainage of wort from the mash tun.

The dehusked oats themselves (kernels or groats) are of high nutritive value, containing about 180 g crude protein/kg DM and less than 30 g crude fibre/kg DM. The groats are generally too expensive to give to farm animals, and are ground into oatmeal after removal of the tips. The tips are mixed with any residues which accumulate during the flow of the oats during milling and the product is designated 'flowmeal'. Flowmeal can be a very valuable food since it may contain the germ; most of this by-product, however, is absorbed by the compound trade.

Cooked oatflakes, although expensive, are a useful ingredient in the diet of very young piglets.

23.5 Wheat

Grain of wheat (*Triticum aestivum*) is very variable in composition. The crude protein content, for example, may range from 60 to 220 g/kg DM, though it is normally between 80 and 140 g/kg DM. Climate and soil fertility as well as variety influence the protein content. The amount and properties of the proteins present in wheat are very important in deciding the quality of the grain for flour production. The most important proteins present in the endosperm are a prolamin (gliadin) and a glutelin (glutenin). The mixture of proteins present in the endosperm is often referred to as 'gluten'. The amino acid composition of these two proteins differs, glutenin containing about three times as much lysine as that present in gliadin. The main amino acids present in wheat gluten are the dispensable acids glutamic acid (330 g/kg) and proline (120 g/kg). Wheat glutens vary in properties and it is mainly the properties of the gluten which decide whether the flour is suitable for bread or biscuit making. All glutens possess the property of elasticity. Strong glutens are preferred for bread making, and form a dough which traps the gases produced during yeast fermentation.

This property of gluten is considered to be the main reason why finely ground wheat is unpalatable when given in any quantity to animals. Wheat, especially if finely milled, forms a pasty mass in the mouth and rumen and this may lead to digestive upsets. Poultry are less susceptible, although wheat with a high gluten content should not be given since a doughy mass may accumulate in the crop. Newly harvested wheat is apparently more harmful in this respect than wheat which has been stored for some time.

Wheat by-products

The wheat grain consists of about 82 per cent endosperm, 15 per cent bran or seed coat and 3 per cent germ. In modern flour milling, the object is to separate the endosperm from the bran and germ. The wheat after careful cleaning and conditioning is blended into a suitable mix (grist) depending upon the type of flour required, and is passed through a series of rollers arranged in pairs. The first pair have a tearing action and release the bran coat from the endosperm. The rollers gradually break up the kernels and at the end of the various stages the flour is removed by sieving. The proportion of flour obtained from the original grain is known as the extraction rate. The mechanical limitations of milling are such that in practice about 75 per cent is the limit of white flour extraction; higher extraction rates result in the inclusion of bran and germ with the flour. In the UK, wholemeal and brown flour are frequently made by adding all, or some of, the milling by-products respectively to the straight-run white flour. Alternatively, the whole grain may be ground between stones to form a coarse wholemeal.

In the production of white flour, the extraction rate varies in different countries but in the UK is about 74 per cent. The remaining 26 per cent constitutes the residues or 'offals'. Before roller milling replaced stone milling, many different grades of wheat offals were sold. The names of these varied in different parts of the country and even from mill to mill. Some names simply indicate the quality of the by-product or the stage of the process at which they arose, for example middlings and thirds. In modern roller milling, the offals may be sold complete as straight-run wheat feed or as three separate products – germ, fine wheat feed (shorts in the USA; pollard in Australia) and coarse wheat feed or bran.

The germ or embryo is very rich in protein (*ca* 250 g/kg DM), low in fibre and an excellent source of thiamin and vitamin E. It may be collected separately or may be allowed to flow on to the fine wheat feed by-product.

Fine wheat feed varies considerably in composition depending on the original grist and the extraction rate. The crude protein content is generally within the range 160–210 g/kg DM and the crude fibre content about 40–100 g/kg DM. Fine wheat feed can be safely used for all classes of farm animals and levels up to 30 per cent can be used satisfactorily in diets for finishing pigs.

Coarse wheat feed, or bran, contains more fibre and less protein than fine wheat feed and has always been a popular food for horses. It is not considered to be a suitable food for pigs and poultry because of its high fibre content. However, very little bran is now available for feeding animals as most of it is used in the preparation of breakfast cereals.

23.6 Other cereals

Rice

Rice (*Oryza sativa*), the main cereal crop of eastern and southern Asia, requires a sub-tropical or warm temperate climate and little is grown in Europe north of latitude 49°.

Rice, when threshed, has a thick fibrous husk or hull like that of oats, and in this state is known as rough rice. The hull amounts to some 20 per cent of the total weight and is rich in silica. The hull is easily removed to leave a product known as brown rice. Brown rice is still invested in the bran, which may be removed with the aleurone layer and the germ by skinning and polishing, thus producing polished rice.

Rough rice may be used as a food for ruminants and horses, but brown rice is preferable for pigs and compares favourably with maize in protein and energy value. Most rice, however, is used for human consumption and little is available in the United Kingdom for farm animals.

The two main by-products obtained from rice milling are the hulls and rice meal. The hulls are high in fibre content and can contain up to 210 g/kg DM of silica. They also have sharp edges which may irritate the intestine, and should never be given to animals. Rice meal or rice bran comprises the pericarp, the aleurone layer, the germ and some of the endosperm, and is a valuable product containing about 120–145 g crude protein/kg DM and 110–180 g oil/kg DM. The oil is particularly unsaturated and may become rancid very quickly; if it is removed a product, extracted rice bran, of better keeping quality is obtained. The amounts of oil, crude protein and crude fibre in rice meal sold in the United Kingdom must be declared.

In the preparation of starch from rice, a product known as rice sludge or rice slump is left as a residue. The dried product has a crude protein content of about 280 g/kg DM and low crude fibre and oil contents, and is suitable for ruminants and pigs.

Rye

The use of rye (*Secale cereale*) in the United Kingdom is relatively small and little is grown for feeding farm animals. Rye grain is very similar to wheat in composition although rye protein has higher lysine and lower tryptophan contents than wheat protein. It is regarded as being the least palatable of the cereal grains. It is also liable to cause digestive upsets and should always be given with care and in restricted amounts.

Rye contaminated with ergot (*Claviceps purpurea*) may be dangerous to animals. This fungus contains a mixture of alkaloids, of which ergotamine and ergometrine are the most important and, in view of their action on uterine muscle, have been implicated as a cause of abortion in cattle consuming ergot-infested rye. However, it is not certain that the quantity of ergotamine could be sufficient to cause abortion. More importantly, chronic poisoning by the alkaloids causes injury to the epithelium of the capillaries, reducing blood flow and resulting in coldness and insensitivity of the extremities. Subsequently lameness and necrotic lesions occur in the feet, tail and ears of mammals and the comb, tongue and wattle of birds.

Like wheat, rye should be crushed or coarsely ground for feeding to animals. Rye is not commonly given to poultry. Studies with chicks have shown that rye contains at least two detrimental factors, an appetite-depressing factor located primarily in the bran, and a growth-depressing factor found in all parts of the grain.

Most of the rye grown in the United Kingdom is used for the production of rye breads and speciality products for human consumption. Some is used for brewing and distilling. The offals from the production of rye malt are rye bran and rye malt culms, but these are available in such small amounts in the United Kingdom as to be of little importance.

Triticale

Triticale is a hybrid cereal derived from crossing wheat with rye. Its name is derived from a combination of the two generic terms for the parent cereals (*Triticum* and *Secale*). The objective in crossing the two cereals was to combine the desirable characteristics of wheat such as grain quality, productivity and disease resistance with the vigour and hardiness of rye.

Triticale is grown commercially in central and northern Europe, North America and South America, mainly for animal feeding. Its composition is very variable; Hungarian strains, for example, can range in crude protein content from 110 to 185 g/kg DM. Recent strains of triticale are at least equal in protein content to wheat and the quality of protein in the hybrid is better than that in wheat because of its higher proportion of lysine and sulphur-containing amino acids. However, it is deficient in tryptophan. As with rye, triticale is subject to ergot infestation. Studies using this hybrid have demonstrated increased liver abscesses in steers when compared with sorghum diets. Triticale contains trypsin inhibitors and alkyl resorcinols and both of these have been implicated in problems of poor palatability and performance in pigs. It is generally recommended that triticale be limited to 50 per cent of the grain in the diets of farm animals.

Millet

The name 'millet' is frequently applied to several species of cereals which produce small grains and are widely cultivated in the tropics and warm temperate regions of the world.

The most important members of this group include *Pennisetum americanum* (pearl or bulrush millet), *Panicum miliaceum* (proso or broomcorn millet), *Setaria italica* (foxtail or Italian millet), *Eleusine coracana* (finger or birdsfoot millet), *Paspalum scorbiculatum* (kodo or ditch millet) and *Echinochloa crusgalli* (Japanese or barnyard millet).

The composition of millet is very variable, the crude protein content being generally within the range 100–120 g/kg DM, the ether extract 20–50 g/kg DM and the crude fibre 20–90 g/kg DM.

Millet has a nutritive value very similar to that of oats, and contains a high content of indigestible fibre owing to the presence of hulls, which are not removed by ordinary harvesting methods. Millet is a small seed and is usually ground for feeding animals other than poultry.

Sorghum

Sorghum (*Sorghum bicolor*) is the main food grain in Africa and parts of India and China. This cereal is also grown in the southern parts of the United States, where it is the second most important feed grain. It is more drought-resistant than maize, for which it is used as a replacement. There are many different types of sorghum (e.g. dari, milo) and, as they vary in grain size, typical analyses can be misleading when applied to individual samples.

The kernel of sorghum is very similar to that of maize, although smaller in size. It generally contains rather more protein but less oil than maize and has no pigmenting xanthophylls. Dark varieties contain tannins, which reduce protein digestibility. Whole sorghum grains can be given to sheep and poultry but are usually ground for other animals. Care is needed in the grinding process as this may produce a fine powder, which is pasty and unpalatable.

Cereal grain screenings

Grain screenings are the residues from the preparation, storage and shipment of cereal products and comprise broken pieces of grain, small grains and the dust from the outer layers of the grain. Their properties and nutritional quality vary widely according to such factors as type and method of processing of grain and collection point in the system. Screenings from the individual cereals may be sold separately or in combination. Problems associated with their use include the presence of weed seeds and, in old products, mycotoxins and rancidity of oils.

23.7 Cereal processing

The processing of cereals for use as animal foods, by simple techniques such as rolling or grinding, has been common practice for many years. More recently a range of other techniques have become available and these can be classified into two basic types – 'hot processes', in which heat is either applied or created during the treatment process, and 'cold processes', in which the temperature of the grain is not increased significantly. The hot treatments include steam flaking, micronisation, roasting and hot pelleting. Steam flaking is often carried out on maize by cooking the grain first with steam, then passing it through rollers to produce a thin flake, which is then dried. Flaked maize is considered to be more acceptable to animals and is of slightly higher digestibility than the unprocessed grain. Steaming and flaking are also known to increase the proportion of propionic acid in the volatile fatty acids in the rumen. Whereas about 75 per cent of the starch of ground maize is digested in the rumen, this is increased to about 95 per cent following steaming and flaking. Even larger effects have been recorded with sorghum (ground: 42 per cent *versus* steam processed: 91 per cent). Conversely, the starch of ground barley is well digested in the rumen as is that of ground wheat.

The term 'micronisation', in the context of grain processing, is used specifically to describe cooking by radiant heat followed by crushing in a roller mill. In this process the starch granules swell, fracture and gelatinise, thus making them more available to enzyme attack in the digestive tract. For poultry, hot (steam) pelleting appears to be superior to cold pelleting as measured by growth rate and feed conversion efficiency. Steam-processed or pressure-cooked sorghum grains also appear to be better utilised than unprocessed sorghum by chicks. Cooked cereals are also desirable in the diet of young pigs, which have a limited capacity to digest raw starch.

The 'cold processes' include grinding, rolling, cracking or crimping, cold pelleting, and addition of organic acids or alkalis. Grinding of cereals is essential for maximum performance of poultry kept under intensive conditions as they do not have access to grit, which is used in the gizzard to break down grains. Pigs are rather poor chewers of food and when given whole cereal grains a high proportion passes through the gut undigested. Therefore, pigs are usually given ground grains, and those receiving ground barley generally perform better than those given crimped barley. Efficient rolling to produce flattened grains is also successful. The degree of grinding should not be excessive as grinding beyond a certain fineness does not improve digestibility or performance and can precipitate problems with regard to the health of the pigs. Fine grinding can produce dusty material which can be inhaled and can cause irritation to the eyes and may induce vomiting. In addition feeding with finely ground cereals, particularly maize, is associated with ulceration of the oesophageal region of the stomach. Here the fine particles result in very fluid stomach contents and pepsin and acid are transported very easily within the stomach and reach the unprotected oesophageal region. The increased fluidity may also result in regurgitation of duodenal digesta. With wheat there is the additional problem of fine grinding producing an unpalatable pasty mass in the mouth. However, coarse grinding of wheat results in poor digestibility and efficient rolling of the grain produces the best product, but this can be difficult to achieve. Owing to their high fibre content, fine grinding of oats produces the best response.

In the case of horses, starch digestibility from cereals over the whole gut is high irrespective of the cereal source. However, this masks the extent of digestion in the different segments of the gut. It is important that the starch is mainly digested pre-caecally, since starch that reaches the large intestine will be fermented by the microbial population to produce lactic acid. This reduces the pH of the digesta, irritates the gut wall, and causes the death of the normal fibre-digesting bacteria with the release of endotoxins, which are absorbed into the bloodstream and can cause laminitis. Opportunity for starch digestion in the small intestine of the horse is limited by a rapid transit of digesta and limited α-amylase activity. Thus methods of processing cereals to ensure rapid digestion of starch in the small intestine are desirable. In the case of oats the size of the grain is appropriate for effective disruption by chewing and the starch granules are small and easily digested, so oats can be given whole to horses. Barley and wheat grains are small and escape efficient breakage during chewing and they need to be physically broken in order to expose the endosperm. Maize grains, on the other hand, are large, but tend to be hard, so they also require breakage. Bruising or coarse grinding of cereal grains rather than complete disintegration by fine grinding is more desirable since the latter does not give a further increase in starch digestion and creates an unpalatable, pasty product which is less stable in storage. Heat treatment by steam flaking, micronisation or extrusion increases the susceptibility of the starch to digestion in the small intestine and reduces the amount reaching the large intestine.

It is generally accepted that for cattle, barley grain should be coarse ground or rolled. When consumed whole a large proportion of the grains escape chewing and digestion, the proportion depending on the age of the animal and the forage : grain ratio in the diet. In high-grain diets, such as those used for 'barley beef' production, more of the whole grain is digested and the proportion appearing in the faeces can drop significantly, especially with light barley. In such cases the benefit from processing must be weighed against its cost. Studies with grains in nylon bags suspended in the rumen of cattle have shown that only minor damage to the seed coat is required for efficient digestion by the microflora. Therefore, the degree of processing should be minimal, as this avoids acidic conditions in the rumen. With sheep, however, because they masticate their feed so well, there is generally no advantage in processing grains. This is illustrated in Table 23 5, which shows the results of a study with early weaned lambs given whole grains or ground and pelleted grains. There were no marked effects of grinding and pelleting on liveweight gain and food conversion efficiency, although there were differences between cereals. Although processing had no effect on nitrogen digestibility, it significantly depressed the organic matter digestibility of barley and increased that of wheat. Furthermore, whole grain feeding had two additional advantages over pelleted grain. First, high levels of pelleted barley for lambs resulted in unacceptably soft fat, which was due to the deposition of increased amounts of branched chain and odd-numbered fatty acids. This occurred as a result of a failure of the liver to metabolise the increased propionate, leading to its direct incorporation into fatty acids (odd-numbered chains) and the utilisation of methyl malonate (branched chains). Second, the occurrence of rumenitis in lambs fed on high-concentrate diets was less pronounced with whole barley than with rolled and pelleted barley. This may be related to the lower rumen pH with processed barley. One exception to the use of whole grain in sheep diets is that it should not be given to ewes as a supplement to silage. Here a significant proportion of the grains would escape digestion.

Table 23.5 Performance and digestibility of early weaned lambs given four cereals (Adapted from Ørskov E R, Fraser C and Gordon J G 1974 *British Journal of Nutrition* **32**: 59; Ørskov E R, Fraser C and McHattie I 1974 *Animal Production* **18**: 85)

Cereal	Processing	Liveweight gain (g/day)	Feed conversion efficiency (kg feed/kg gain)	Digestibility	
				Organic matter	Nitrogen
Barley	Whole, loose	340	2.75	0.81	0.72
	Ground, pelleted	347	2.79	0.77	0.66
Maize	Whole, loose	345	2.52	0.84	0.75
	Ground, pelleted	346	2.62	0.82	0.69
Oats	Whole, loose	241	3.07	0.70	0.78
	Ground, pelleted	238	3.33	0.68	0.77
Wheat	Whole, loose	303	2.97	0.83	0.71
	Ground, pelleted	323	2.56	0.87	0.76

Organic acids, such as propionic acid, are sometimes added to high-moisture grain, especially barley, as a mould inhibitor. Unless the acid is effectively distributed, patches of mouldy grain may present a health hazard. Certain *Fusarium* species have been associated with such mouldy grain and these are known to produce metabolites such as zearalenone, which has oestrogenic activity and can cause vulvovaginitis and the characteristic splay-leg syndrome in pigs. The required rate of application of the acid increases with the moisture content of the grain. With grain containing up to 250 g moisture/kg (750 g dry matter/kg), barley can be treated and stored whole and processed at the time of feeding. Alternatively, the grain can be processed and the acid applied before sealing in a pit. This technique generally requires a higher rate of application (by about 10 per cent) of acid. With higher moisture contents of 350–450 g/kg, the grain can be crimped and the propionic acid, in combination with other short chain fatty acids and formaldehyde, is applied before storage of the grain in a sealed pit. The latter two methods result in a product that does not require further processing before being given to cattle.

Chemical treatment with sodium hydroxide, in granular form or solution, has been used as an alternative to mechanical treatment (e.g. rolling) of barley and other cereal grains. The intention is to soften the husk but not to expose the endosperm to rapid fermentation in the rumen, which would create excessively acid conditions. Originally, in practice it proved difficult to achieve these objectives by sodium hydroxide treatment of barley. Recently a method of treating barley and wheat grains has been developed using a mixer wagon. Sodium hydroxide and water (depending on the moisture content of the grain) are added, followed by mixing and then spreading the treated grain to allow it to cool. It is then heaped up and is ready for feeding after four days. Treated wheat grain is considered to have beneficial effects on dairy cows, especially in counteracting very acid or highly buffered silages, in addition to any effects of reducing the rate of fermentation in the rumen.

Satisfactory liveweight gains have been achieved with ammonia-treated whole moist barley and wheat given to beef cattle. Use of ammonia preserves the grain and increases its crude protein (rumen-degradable protein) content in addition to eliminating the need to dry and process the grain. Ammonia-treated grain was eaten at a slower rate than processed grain, which is advantageous in that it reduces the chance of the development of acid conditions in the rumen. Although some whole grain passed through the animal undigested, this had no effect on performance. Treating grain with ammonia requires specialised equipment. Recently a more simple method of treatment has been adopted which uses a solution of urea and employs the natural urease enzyme activity of the grain to release ammonia, which dissolves in the moisture on the grain. This method is easily applied on farm and retains all the advantages of ammonia treatment outlined above. Grain is harvested at 250–300 g moisture/kg and a concentrated urea solution is applied using an auger at the rate of 30 g urea/kg DM for wheat or 40 g/kg DM for barley. The grain is then sealed in a clamp and left for four weeks, during which the urea is converted to ammonia, which attacks the seed coat.

Summary

1. Cereals are the principal sources of energy in diets for non-ruminants and in concentrates for ruminants.

2. They are characterised by high starch and low fibre contents.

3. The crude protein content of cereals ranges from 80 to 120 g/kg dry matter.

4. The calcium content is low and the phosphorus content is moderate, but has a reduced availability to non-ruminants, much of it being in the form of phytate phosphorus.

5. There are many cereal by-products from the human food-processing industry that are useful as foods for animals. The by-products of the brewing industry are malt culms, brewer's grains, spent hops and dried brewer's yeast. The main by-products of the distilling industry are distiller's grains, pot ale syrup and dried distiller's dark grains. Maize by-products from starch production comprise maize germ meal, prairie meal and maize gluten feed. Oat milling by-products include oat hulls, oat feed and flowmeal. Wheat bran and wheat feed are by-products of flour milling.

6. There are several methods of processing of cereals and the extent of or need for processing varies between the grains and the target animals. Generally, the outer layer of the grain requires to be disrupted in some way in order to ensure efficient digestion of the starchy endosperm. Inapproproiate processing can lead to problems such as gastric ulcers in pigs and acidosis in the hind gut of horses. Poultry which have access to grit, and sheep, can be given whole grains.

7. Organic acids are used to preserve high-moisture grain.

8. Chemical treatment with sodium hydroxide, ammonia or urea can be used as an alternative to physical processing of grains for ruminants.

Further reading

Black H, Edwards S, Kay M and Thomas S 1991 *Distillery By-products as Feeds for Livestock*, Aberdeen, Scottish Agricultural College.

Church D C 1984 *Livestock Feeds and Feeding*, 2nd edn, Corvallis, OR, O and B Books.

Duffus C M and Slaughter J C 1980 *Seeds and Their Uses*, Chichester, John Wiley & Sons.

Gõhl B 1981 *Tropical Feeds*, Rome, FAO.

Hoseney R C, Varriano-Marston E and Dendy D A V 1981 Sorghum and millets. In: Pomeranz Y (ed.) *Advances in Cereal Science and Technology*, Vol. 4, St Paul, MN, American Association of Cereal Chemists.

Kent N L 1983 *Technology of Cereals*, 3rd edn, Oxford, Pergamon Press.

Ørskov E R 1981 Recent advances in the understanding of cereal processing for ruminants. In: Haresign W and Cole D J A (eds) *Recent Developments in Ruminant Nutrition*, London, Butterworth.

Symposium 1999 Premium grains for livestock. *Australian Journal of Agricultural Research* **50**: 629–908.

24 Protein concentrates

24.1   Oilseed cakes and meals

24.2   Oilseed residues of minor importance

24.3   Leguminous seeds

24.4   Animal protein concentrates

24.5   Milk products

24.6   Single-cell protein

24.7   Synthetic amino acids

24.8   Non-protein nitrogen compounds as protein sources

24.1   Oilseed cakes and meals

Oilseed cakes and meals are the residues remaining after removal of the greater part of the oil from oilseeds. The residues are rich in protein (200–500 g/kg) and most are valuable foods for farm animals. The total usage of these products by the food-compounding industry in the UK in 1999 was of the order of 2.7 million tonnes. This is a considerable fall in usage from the figure of 3.5 million tonnes quoted for 1991 in the last edition.

Soya bean meal made up about 38 per cent of the total, compared with 42 per cent in 1991, and rapeseed, at 23 per cent, was again the next largest contributor. Some more recent figures indicate an increasing importance of rapeseed meal relative to soya bean meal. An increased contribution was made by sunflower meal (18 per cent), and whole seeds at 3 per cent. The remaining 21 per cent encompassed a wide range of less well-known products such as sheanut, sesame and guar as well as the more familiar products such as linseed and cottonseed meals. The pattern of usage in recent years is illustrated in Table 24.1.

Most oilseed residues are of tropical origin; they include groundnut, cottonseed, linseed and soya bean. Some seeds such as castor bean yield residues unsuitable for animal consumption because they contain toxic substances.

Two main processes are used for removing oil from oilseeds. One employs pressure to force out the oil, and the other uses an organic solvent, usually hexane, to dissolve the oil from the seed. Some seeds such as groundnut, cottonseed and sunflower have a thick coat or husk, rich in fibre and of low digestibility, which lowers the nutritive value of the material. It may be

Table 24.1 Usage of oilseed products by the UK feed-compounding industry, 1996–1999

Raw material	Usage per year (million tonnes)[a]			
	1996	1997	1998	1999
Soya bean meal	1.025	1.013	1.026	1.022
	(37.5)	(39.0)	(40.7)	(37.9)
Rapeseed meal	0.567	0.542	0.526	0.616
	(20.8)	(20.9)	(20.9)	(22.8)
Sunflower meal	0.520	0.537	0.450	0.490
	(19.0)	(20.7)	(17.9)	(18.2)
Whole oilseeds	0.073	0.072	0.070	0.082
	(2.7)	(2.8)	(2.8)	(3.0)
Others	0.544	0.434	0.448	0.489
	(19.9)	(16.7)	(17.8)	(18.1)
Total	2.731	2.597	2.520	2.698

[a] Figures in parenthesis are percentages of the total.

completely or partially removed by cracking and riddling, a process known as decortication. The effect of decortication of cottonseed upon the nutritive value of the cake derived from it is shown in Table 24.2.

Removal of the husk lowers the crude fibre content and has an important effect in improving the apparent digestibility of the other constituents. As a result, the nutritive value of the decorticated cake is raised significantly above that of the undecorticated. The latter is suitable for feeding adult ruminants only. In this class of animal it may have a particular role in maintaining the fibre levels of the diets. Undecorticated cakes are rarely produced nowadays but partial decortication is widely practised.

Table 24.2 Composition and nutritive value of cottonseed cakes

	Composition (g/kg)					
	Dry matter	Crude protein	Ether extract	N-free extractives	Crude fibre	Ash
Undecorticated	880	231	55	400	248	66
Decorticated	900	457	89	293	87	74

	Digestibility				Metabolisable energy (MJ/kg DM)
	Crude protein	Ether extract	N-free extractives	Crude fibre	
Undecorticated	0.77	0.94	0.54	0.20	8.5
Decorticated	0.86	0.94	0.67	0.28	12.3

In the press process, the seed from which oil is to be removed is cracked and crushed to produce flakes about 0.25 mm thick, which are cooked at temperatures up to 104 °C for 15–20 minutes. The temperature is then raised to about 110–115 °C until the moisture content is reduced to about 30 g/kg. The material is then passed through a perforated horizontal cylinder in which revolves a screw of variable pitch that gives pressures up to 40 MN/m². The residue from screw pressing usually has an oil content between 25 and 40 g/kg. The cylindrical presses used for extraction are called expellers and the method of extraction is usually referred to as the *expeller process*.

Only material with an oil content of less than 350 g/kg is suitable for *solvent extraction*. If material of higher oil content is to be so treated it first undergoes a modified screw pressing to lower the oil content to a suitable level. The first stage in solvent extraction is flaking; after this the solvent is allowed to percolate through the flakes, or a process of steeping may be used. The oil content of the residual material is usually below 10 g/kg and it still contains some solvent, which is removed by heating. Some meals may benefit from being heated and advantage is taken of the evaporation of the solvent to do this; soya bean meal, for example, is toasted at this stage in its production.

About 950 g/kg of the nitrogen in oilseed meals is present as true protein, which has an apparent digestibility of 0.75–0.90 and is of good quality. When biological value is used as the criterion for judging protein quality, that of the oilseed proteins is considerably higher than that of the cereals (Table 24.3).

Some of the oilseed proteins approach animal proteins like fishmeal and meat meal in quality, though as a class they are not as good. Certainly they are of poorer quality than the better animal proteins such as those of milk and eggs. The figures for protein efficiency ratio and gross protein value confirm the good quality of the oilseed proteins but their chemical scores are low. This means that they have a badly balanced amino acid constitution, having a large deficit of at least one essential amino acid. In general,

Table 24.3 Nutritive value of some food proteins

Source	Biological value (rat)	Chemical score	Protein efficiency ratio (rat)	Gross protein value (chick)
Oats	0.65	0.46		
Wheat	0.67	0.37	1.5	
Maize	0.55	0.28	1.2	
Cottonseed meal	0.80	0.37	2.0	0.77
Groundnut meal	0.58	0.24	1.7	0.48
Soya bean meal	0.75	0.49	2.3	0.79
White-fish meal	0.77			1.02
Milk	0.85	0.69	2.8	0.90
Whole egg	0.95	1.00	3.8	

oilseed proteins have a low cystine and methionine content, and a variable but usually low lysine content. As a result, they cannot provide adequate supplementation of the cereal proteins with which they are commonly combined and they should be used in conjunction with an animal protein when given to simple-stomached animals. The quality of the protein in a particular oilseed is relatively constant but that of the cake or meal derived from it may vary depending upon the conditions employed for the removal of the oil. The high temperatures and pressures of the expeller process may denature the protein and reduce its digestibility, with a consequent lowering of its nutritive value. For ruminant animals, such a denaturation may be beneficial owing to an associated reduction in degradability. The high temperatures and pressures also allow control of deleterious substances such as gossypol and goitrin. Solvent extraction does not involve pressing, temperatures are comparatively low and the protein value of the meals is almost the same as that of the original seed.

The oilseed cakes may make a significant contribution to the energy content of the diet, particularly when the oil content is high. This will depend upon the process employed and its efficiency. Expeller soya bean meal may have an oil content of 66 g/kg DM and a metabolisable energy concentration of 14 MJ/kg DM, whereas solvent-extracted meal has 17 g oil and 12.3 MJ of metabolisable energy per kilogram dry matter. Digestive disturbances may result from uncontrolled use of cakes rich in oil and, if the oil is unsaturated, milk or body fat may be soft and carcass quality lowered.

The oilseed meals usually have a high phosphorus content, which tends to aggravate their generally low calcium content. They may provide useful amounts of the B vitamins but are poor sources of carotene and vitamin E.

Soya bean meal

Soya beans contain from 160 to 210 g/kg of oil and are normally solvent-extracted; the residual meal has an oil content of about 10 g/kg. The meal is generally regarded as one of the best sources of protein available to animals, and in 1999 was still the major protein source used in animal feeding in the UK. The protein contains all the essential amino acids but the concentrations of cystine and methionine are suboptimal. Methionine is the first limiting amino acid and may be particularly important in high-energy diets.

Soya bean meal contains a number of toxic, stimulatory and inhibitory substances including allergenic, goitrogenic and anticoagulant factors. Of particular importance in nutrition are the protease inhibitors, of which six have been identified. Two of these, the Kunitz anti-trypsin factor and the Bowman–Birk chymotrypsin inhibitor, are of practical significance. The protease inhibitors are partly responsible for the growth-retarding property of raw soya beans or unheated soya bean meal. The retardation has been attributed to inhibition of protein digestion but there is evidence that pancreatic hyperactivity results in increased production of trypsin and chymotrypsin. The consequent loss of cystine and methionine accentuates the marginal status of soya bean meal with regard to these acids and results in induced deficiencies of both. The Kunitz factor, but not the Bowman–Birk, is inactivated (30–40 per cent) by human gastric juice *in vivo* at pH 1.5–2.0. However, the pH of the stomach is above 2.0 most of the time. The Bowman–Birk factor is inactivated in its passage through the intestine of the chick.

Another substance contributing to the growth retardation is a haemagglutenin which is capable of agglutinating red blood cells in rats, rabbits and human beings but not in sheep and calves. The toxic agent belongs to a group of compounds known as lectins. These are proteins capable of recognising and binding reversibly to the carbohydrate moieties of glycoproteins on the surface of cell membranes, and have been shown to be responsible for impaired growth and death in test animals. They exert their toxicity by binding to the epithelial cells lining the small intestine, disrupting the brush border and reducing the efficiency of the absorption of nutrients. Lectins have also been shown to cause inhibition of brush border hydrolases. It has been suggested that lectins impair the body's defence system against bacterial infection and there is an increased tendency for invasion of the body by the gut microflora. Others have postulated that lectins encourage colonisation of the small intestine by coliforms. About half the growth-retarding effect of soya bean meal in monogastric animals has been attributed to the lectin content. Their toxicity varies, being extreme in the castor bean but relatively mild in the soya bean. The inhibitors are inactivated by heating, which accounts for the preference shown for toasted meals for simple-stomached animals. For ruminant animals, the inhibitors are not important and toasting is unnecessary. The process of toasting must be carefully controlled since overheating will reduce the availability of lysine and arginine and reduce the value of the protein.

Provided the meal has been properly prepared (has less than a specified maximum urease activity of 0.4 mg N/g per minute) it forms a very valuable food for farm animals. However, if soya bean meal is used as the major protein food for simple-stomached animals certain problems arise. The meal is a poor source of B vitamins, and these must be provided either as a supplement or in the form of an animal protein such as fishmeal. If such supplementation is not practised sows may produce weak litters which grow slowly because of reduced milk yields, and older pigs show lack of coordination and failure to walk. On such diets, breeding hens produce eggs of poor hatchability, giving chicks of poor quality; such chicks may have an increased susceptibility to haemorrhages owing to a shortage of vitamin K. Soya bean meal is a better source of calcium and phosphorus than the cereal grains, but when it replaces animal protein foods, adjustments must be made in the diet, particularly for rapidly growing animals and laying hens. As long as adequate supplementation is practised it may form up to 400 kg/t of poultry diets and 250 kg/t of pig diets.

Soyabean meal contains about 1 g/kg of genistein, which has oestrogenic properties and a potency of 4.44×10^{-6} times that of diethyl-stilboestrol. The effect of this constituent on growth rate has not been elucidated.

The oil in the soya bean has a laxative effect and may cause soft body fat to be produced. The extracted meal does not contain sufficient oil to cause this problem but it should be borne in mind in view of the increasing tendency to use full fat soya bean products in dietary formulations, especially for pigs. The full fat products are produced by batch pressure cooking or extrusion of the whole bean. The extruded product has a higher metabolisable energy content but this advantage is nullified if the products are ground and pelleted.

Soya protein concentrates are produced by solvent extraction and removal of insoluble carbohydrate. Protein concentration is of the order of 70 per cent. During processing, antigenic and antinutritional substances are removed and the materials are suitable for inclusion in calf milk replacers and baby pig diets.

Cottonseed meal

The protein of cottonseed meal is of good quality but has the common disadvantage of oilseed proteins of having a low content of cystine, methionine and lysine, the last named being the first limiting amino acid. The calcium content is low and since the calcium to phosphorus ratio is about 1 : 6, deficiencies of calcium may easily arise. It is a good though variable source of thiamin but a poor source of carotene.

When cottonseed meal is used as a protein source for young, pregnant or nursing pigs, or young or laying poultry, it needs to be supplemented with fishmeal or meat and bonemeal to make good a shortage of essential amino acids and calcium. A supplement of vitamins A and D should also be provided. Pigs and poultry do not readily accept the meal, mainly because of its dry, dusty nature. No such difficulty is encountered with lactating cows, although complications may arise when large amounts are given since the milk fat tends to become hard and firm. Butter made from such fat is often difficult to churn and tends to develop tallowy taints. Another factor to be considered when using cottonseed meal is its costive action. This is not normally a problem and may indeed be beneficial in diets containing large amounts of laxative constituents.

Cottonseeds may contain from 0.3 to 20 g/kg DM of a yellow pigment known as gossypol and concentrations of 4–17 g/kg DM have been quoted for the kernels. Gossypol is a polyphenolic aldehyde (alkanal), which is an antioxidant and polymerisation inhibitor toxic to simple-stomached animals. The general symptoms of gossypol toxicity are depressed appetite, loss of weight, laboured breathing and cardiac irregularity. Death is usually associated with a reduced oxygen-carrying capacity of the blood, haemolytic effects on the erythrocytes and circulatory failure. Post-mortem examinations usually show extensive oedema in the body cavities, indicating an effect on membrane permeability. Although acute toxicity is uncommon, ingestion of small amounts over a prolonged period can be lethal. In the past it has been considered important to distinguish between free (soluble in 70–30 v/v aqueous acetone) and bound gossypol, since only the former was considered to be physiologically active. It is now considered that some of the bound material is active, but this does not negate the general thesis that it is the free gossypol content which determines the toxic potency of the material. The free gossypol content of cottonseed meal decreases during processing and varies according to the methods used. Screw-pressed materials have 200–500 mg free gossypol/kg, prepressed solvent-extracted meals 200–700 and solvent-extracted 1000–5000 mg/kg. Processing conditions have to be carefully controlled to prevent loss of protein quality owing to the binding of gossypol to lysine at high temperatures. Fortunately, the shearing effect of the screw press in the expeller process is an efficient gossypol inactivator at temperatures which do not reduce protein quality.

It is generally considered that pig and poultry diets should not contain more than 100 mg free gossypol/kg and that inclusions of cottonseed meal should be between 50 and 100 kg/t. Particular care is required with laying hens since comparatively low levels of the meal may cause an olive-green discoloration of the yolk in storage. An associated pink discoloration of the albumen is now considered to be due to cyclopropenoids and not gossypol as was once thought. Treatment with ferrous sulphate can ameliorate the effects of gossypol, with doses ranging from 1 to 4 parts to 1 part gossypol. Ruminant animals do not show ill effects even when they consume large quantities of cottonseed meal.

In the UK the free gossypol content of foods is strictly controlled by law. Straight foods, except cotton cake or meal, must not contain more than 20 mg/kg and the same limit applies to complete foods for laying hens and piglets. For poultry and calves, the limit is 100, for pigs 60 and for cattle, sheep and goats 500. Cotton cakes and meals are allowed to contain up to 1200 mg free gossypol/kg. The concentrations refer to foods with a moisture content of 120 g/kg.

Coconut meal

The oil content of coconut meal varies from 25 to 65 g/kg, meals of higher oil content being very useful in the preparation of high-energy diets. However, they suffer the disadvantage of being susceptible to becoming rancid in store. The protein is low in lysine and histidine and this together with the generally high fibre content of about 120 g/kg limits the use of the meal for simple-stomached animals. It is usually recommended that it should form less than 25 kg/t of pig diets and less than 50 kg/t of poultry diets. When low-fibre coconut meals are available for simple-stomached animals they have to be supplemented with animal proteins to make good their amino acid deficiences. Neither protein quality nor fibre content is limiting for ruminant animals and coconut meal provides an acceptable and very useful protein supplement. In diets for dairy cows it is claimed to increase milk fat content. Some published work has shown increased butterfat contents in the milks of cows given supplements of coconut meal but the basal diet had a fat content of about 10 g/kg only. The milk fats produced on diets containing considerable amounts of coconut meal are firm and excellent for butter making.

Coconut meal has the valuable property of absorbing up to half its own weight of molasses and as a result is popular in compounding.

Palm kernel meal

This food has a comparatively low content of protein and the balance of amino acids is poor. The first limiting amino acid is lysine. The ratio of calcium to phosphorus is more favourable than in many other oilseed residues. The meal is dry and gritty, especially the solvent-extracted product, and is not readily eaten; it is therefore used in mixtures with more acceptable foods. Attempts to use it mixed with molasses, as molassed palm kernel cake, have not been successful. It has a reputation for increasing the fat content of milk and its chief use is for dairy cows. Palm kernel meal has been described as being balanced for milk production but in fact contains too high a proportion of protein to energy.

Palm kernel meal is not used widely in pig and poultry diets. This is partly because it is unpalatable and partly because of its high fibre content (150 g/kg DM), which reduces its digestibility for such animals. The highest level of palm kernel meal recommended in the diet of simple-stomached animals is about 50 kg/t.

Linseed meal

Linseed meal is unique among the oilseed residues in that it contains from 30 to 100 g/kg of mucilage. This is almost completely indigestible by non-ruminant animals but can be broken down by the microbial population of the rumen. It is readily dispersible in water, forming a viscous slime. Immature linseed contains a small amount of a cyanogenetic glycoside, linamarin, and an associated enzyme, linase, which is capable of hydrolysing it with the evolution of hydrogen cyanide. This is extremely toxic Death results from combination of the cyanide with cytochrome oxidase, leading to an immediate cessation of cellular respiration and anoxia. Low-temperature removal of oil may produce a meal in which unchanged linamarin and linase persist; such meals have proven toxic when given as a gruel since cyanide production begins as soon as the water is added. The meals are safe if given in the dry state, since the pH of the stomach contents of the pig is sufficiently low to inactivate linase. Normal processing conditions destroy linase and most of the linamarin and the resultant meals are quite safe. In ruminant animals the hydrogen cyanide formed by linase action is absorbed into the blood very slowly and this, coupled with its rapid detoxification in the liver and excretion via the kidney and lungs, ensures that it never reaches toxic levels in the blood. In the UK, linseed cake or meal must, by law, contain less than 350 mg of hydrocyanic acid per kilogram of food with a moisture content of 120 g/kg.

It has been reported that linseed meal has a protective action against selenium poisoning. The protein of linseed meal is of poorer quality than those of soya bean or cottonseed meals, having lower methionine and lysine contents. Linseed meal has only a moderate calcium content but is rich in phosphorus, part of which is present as phytate. It is a useful source of thiamin, riboflavin, nicotinamide, pantothenic acid and choline.

Linseed meal has a very good reputation as a food for ruminant animals which is not easy to justify on the basis of its proximate analysis. Part of the reputation may be the result of the ability of the mucilage to absorb large amounts of water, resulting in an increase in the bulk of the meal; this may increase retention time in the rumen and give a better opportunity for microbial digestion. The lubricating character of the mucilage also protects the gut wall against mechanical damage and, together with the bulkiness, regulates excretion and is claimed to prevent constipation without causing looseness. Linseed meal given to fattening animals results in faster gains than other vegetable protein supplements making the same protein contribution; cattle attain a very good sleek appearance though the body fat may be soft. The meal is readily eaten by dairy cows but tends to produce a soft milk fat which is susceptible to the development of oxidative rancidity.

Linseed meal is an excellent protein food for pigs as long as it is given with an animal protein supplement to make good its deficiencies of methionine, lysine and calcium. This is particularly important with diets containing large amounts of maize.

Linseed meal is not a satisfactory food for inclusion in poultry diets. Retardation of chick growth has been reported on diets containing 50 kg/t of linseed meal and deaths in turkey poults with 100 kg/t of diet. These adverse effects can be avoided by autoclaving the meal or by increasing the levels of vitamin B_6 in the diet; the untreated meal is thought to contain an unidentified anti-pyridoxine factor. Some workers consider the adverse effects of the meal to be due to the mucilage, since this collects as a gummy mass on the beak, causing necrosis and malformation and reducing the bird's ability to eat. Pelleting or coarse granulation can overcome this trouble. If linseed meal has to be included in poultry diets the levels should not exceed 30 kg/t.

Rapeseed meal

The 1981 world production of rapeseed was estimated at about 11.5 million tonnes, making it fifth in order of importance of the oil-producing seed crops. The current picture is of increasing production, particularly in Western Europe, where growing the crop is actively encouraged by the European Union. In 1983 rapeseed accounted for 37 per cent of oilseeds crushed in the UK. In 1991 the figure was 60 per cent. Production in the UK in 1992 was 1.25 million tonnes compared with 402 000 tonnes in 1980–1982. By 1999 production had reached 1.74 million tonnes, but in 2000 this fell back to 1.129 million tonnes. Extraction of the oil, by a prepress solvent extraction procedure, leaves a residue containing about 400 g protein/kg DM. It contains more fibre (140 g/kg DM) than soya bean meal and its metabolisable energy value is lower, about 7.4 MJ/kg DM for poultry, 11.5 for pigs and 11.1 for ruminants. Both protein content and digestibility are lower than for soya bean meal but the balance of essential amino acids compares favourably, the rapeseed meal having less lysine but more methionine. The balance of calcium and phosphorus is satisfactory and it contains a higher phosphorus content than other oilseed residues.

In the past, the use of rapeseed meals produced from rape (*Brassica napus*) grown in Europe was restricted, particularly for pigs and poultry, by the presence of glucosinolates (thioglucosides) accompanied by a thioglucosidase known as myrosinase. Under a variety of conditions these may give rise to isothiocyanates, organic thiocyanates, nitriles and 5-vinyloxazolidine-2-thione (goitrin). They exhibit a variety of toxic effects manifested as goitres, and liver and kidney poisoning. Their presence is not serious for ruminant animals although there is some evidence of reduced intake, minor liver damage and reduced volatile fatty acid production when the toxins have been administered orally. With pigs, meals with high glucosinolate levels may reduce food intake, growth and carcass quality. Piglets have shown poor survival rates and enlarged thyroids when maternal diets included high levels of glucosinolates.

Rapeseeds contain erucic acid, which has been known to cause heart lesions in experimental animals, but it is unlikely to be a problem for farm animals since it partitions with the oil during extraction.

A certain measure of control of the goitrogenic activity of meals is achieved by manipulating the pretreatment of the seed before extraction so as to ensure the earliest possible destruction of myrosinase. Such control is partial only, since bacterial thioglucosidases produced in the gut may hydrolyse residual glucosinolate in the meal.

Glucosinolates are no longer a problem when rapeseed meals are consumed as the meals are now produced from varieties which have greatly reduced contents of glucosinolates and erucic acid. The potential for using low-glucosinolate meals in pig and poultry diets is considerable, but account must be taken of the need for amino acid supplementation, especially with lysine, when high levels of inclusion are practised. It must be borne in mind that antinutritional factors are still present, even if at reduced levels, in the low-glucosinolate meals. This is of particular importance with the early weaned pig, where reduction in intake may be significant, and with breeding animals owing to the possible effect on the foetus.

Rapeseed meals frequently contain tannins. These are polyphenolic compounds which complex with proteins and carbohydrates to form enzyme-resistant substrates with a consequent lowering of digestibility. This may result also from the combination of the tannins with digestive enzymes with a consequent loss of activity. Tannins may cause damage to the intestinal mucosa and are known to interfere with iron absorption.

Evidence on permissible levels of inclusion of rapeseed meals in the diet is conflicting; some acceptable values for different animals are in Box 24.1.

The highly variable responses to dietary inclusions of rapeseed meals require that, in practice, such figures are applied with caution. Some authorities, for example, consider that rapeseed meals should not be used in starter or sow diets and that levels in grower and finisher diets should not exceed 50 and 100 kg/t, respectively.

Chick growth may be adversely affected if inclusion rates exceed 50 kg/t. The eggs of some hens producing brown-shelled eggs are susceptible to the development of fishy taints when rapeseed meals are included in the diet.

Box 24.1
Best estimates of permissible levels of inclusion of rapeseed meal in pig diets (kg/t)

	High-glucosinolate rapeseed meal	Low-glucosinolate rapeseed meal	
		UK	Canada
Starting pigs (7–15 kg W)	40	50	80
Growing pigs (15–45 kg W)	50	100	120
Finishing pigs (>45 kg W)	80	150	150
Gilts	0	100	120
Sows	30	120	120

This is because of the inability of these birds to oxidise trimethylamine produced from the polyphenolic choline ester sinapine.

For ruminant animals rapeseed meal may be used as the sole source of dietary protein.

In the UK, whole foods must, by law, contain less than 1000 mg 5-vinyloxazolidine thione per kg except those for laying hens, which must contain less than 500 mg/kg. Levels of isothiocyanates in whole foods are strictly prescribed, and rape cakes or meals must contain less than 4000 mg allylthiocanate per kilogram referred to a moisture content of 120 kg/t.

Groundnut meal

The seeds of the groundnut are borne in pods, usually is pairs or threes. The seeds contain 250–300 g/kg of crude protein and 350–600 g/kg of lipid material. The pod or husk is largely fibrous. Groundnut meal is now usually made from the kernels and only occasionally is the whole pod used as the source of an undecorticated meal. The most common method of extraction is screw pressing, giving a meal with 50–100 g/kg of oil. Lower oil levels can only be achieved by solvent extraction, but this has to be preceded by screw pressing to reduce the initially high oil content. The composition of the meal will depend on the raw material and the method of extraction used.

The protein of groundnut meal has suboptimal amounts of cystine and methionine, although the first limiting amino acid is lysine. When the meal is used in high-cereal diets, adequate supplementation with animal protein is necessary. This also ensures that the deficiencies of vitamin B_{12} and calcium are made good. Such supplementation is particularly important for young fast-growing animals such as pigs and poultry. The palatability of the meal for pigs is high, but it should not form more than 25 per cent of the diet as it tends to produce a soft body fat and may have a troublesome laxative action. This also limits its use for lactating cows, for whom it otherwise forms an excellent and acceptable protein source. It has been reported that both a growth factor and an antitrypsin factor occur in groundnut meal. The latter has antiplasmin activity and so shortens bleeding time; it is destroyed by heating.

In 1961 reports appeared which implicated certain batches of groundnut meal in the poisoning of turkey poults and ducklings. The toxic factor was shown to be a metabolite of the fungus *Aspergillus flavus* and was named aflatoxin. This is now known to be a mixture of four compounds designated aflatoxins B_1, G_1, B_2 and G_2, of which B_1 is the most toxic. There are considerable species differences in the susceptibility to these toxins; turkey poults and ducklings are highly susceptible, calves and pigs are susceptible, whereas mice and sheep are classed as resistant. Young animals are more susceptible than adults of the same species. A common feature in affected animals is liver damage with marked bile duct proliferation, liver necrosis and, in many cases, hepatic tumours. In fact, aflatoxin has been shown to be a potent liver toxin and a very active carcinogen.

There are several reports of deaths in calves under six months of age when fed on contaminated groundnut meal. Older cattle are more resistant but cases of deaths in store cattle, and loss of appetite and reduced milk yield in cows, have been reported. Deaths have occurred in six-month-old steers given 1 mg/kg of aflatoxin B_1 in their diet for a period of 133 days and liveweight gains were generally significantly reduced. Administration of 0.2 mg/kg of aflatoxin B_1 in the diets of Ayrshire calves significantly reduced liveweight gains. Experiments on the inclusion of 150–200 kg/t of toxic groundnut in dairy cow diets have shown significant falls in milk yield. A metabolite of aflatoxin known as aflatoxin M_1, which causes liver damage in ducklings, has been shown to be present in the milk of cows fed on toxic meals. The effect of aflatoxin on human beings has not been clearly established. There have been no reports of clinical poisoning in sheep.

Aflatoxins are relatively stable to heat, and methods of eliminating them from meals are elaborate. The best method of control is suitable storage to prevent mould growth, although aflatoxins may be produced in the growing crop. In the UK, prescribed maximum limits for aflatoxins in foods for animals are laid down by law (see Box 24.2).

The stringency of these standards together with the difficulty of ensuring that groundnut products comply with them has been a major factor contributing to the decline in the usage of groundnut products in the UK compounding industry, where they now occupy a minor position.

Sunflower seed meal

The meal is produced when the oil is removed from the seed by hydraulic pressure or solvent extraction. The hulls are usually partially, rather than completely, removed, but the resulting high-fibre meals (up to 420 g/kg DM) are readily accepted by older animals provided they are finely ground. As a result of the variation in the extent of dehulling before extraction, the meals vary widely in composition and nutritive value.

Solvent-extracted meals contain, on average, about 220 g crude fibre and 430 g crude protein per kilogram of dry matter and have metabolisable energy contents of 8.1, 10.4 and 11.2 MJ/kg DM for poultry, cattle and sheep, respectively. For pigs, a digestible energy content of 10.6 MJ/kg DM would

Box 24.2

UK prescribed limits for aflatoxins in foods for animals (mg aflatoxin B_1/kg referred to a moisture content of 120 g/kg)

Straight feeding stuffs:	0.05
Except: groundnut, copra, palm kernel, cottonseed, babasu, maize and products derived from the processing thereof	0.02
Complete feeding stuffs for cattle, sheep and goats (except dairy animals, calves, lambs and kids)	0.05
Complete feeding stuffs for pigs and poultry (except piglets and chicks)	0.02
Other complete feeding stuffs	0.01
Complementary feeding stuffs for cattle, sheep and goats (except complementary feeding stuffs for dairy animals, calves and lambs)	0.05
Complementary feeding stuffs for pigs and poultry (except young animals)	0.03
Other complementary feeding stuffs	0.005

be appropriate. Expeller meals have higher fat, lower crude fibre and crude protein contents and metabolisable energy values of about 13 MJ/kg DM for cattle. Sunflower oil has a high content of polyunsaturated fatty acids and may cause soft body fat in pigs, particularly when expeller meals are given. The oil is very susceptible to oxidation and the meals have a short shelf life owing to the development of rancidity, which renders them unpalatable. The meals are useful sources of protein which is low in lysine, the main limiting amino acid, but has about twice as much methionine as soya protein.

Maximum rates of inclusion in diets are 200 kg/t for adult cattle, 150 kg/t for adult sheep, 25 kg/t for growing pigs, 50 kg/t for finishing pigs and 100 kg/t for sows. They are not recommended for calves, lambs or young pigs. For adult poultry, sunflower meals may be included at 100 kg/t of the diet but they are not recommended for young birds.

Sesame seed meal

The meals currently available may be produced by hydraulic pressing or solvent extraction. The former have the lower protein content (about 400 g/kg DM compared with 500 g/kg DM for the solvent-extracted material), but have oil contents of more than 100 g/kg DM compared with 20 g/kg DM for the extracted meal and make a significantly greater contribution to the energy of the diet.

The protein is rich in leucine, arginine and methionine but is relatively low in lysine. The meal therefore needs to be combined with foods rich in lysine when given to pigs and poultry. For ruminants, the protein has a degradability of 0.65–0.75 depending upon the rate of passage through the rumen.

The residual oil of the cake or meal is highly unsaturated and may result in soft body and milk fat if consumed in excessive amounts and may also impart a disagreeable flavour to milk. The oil rapidly becomes rancid and unpalatable and meals containing it have been implicated in cases of vitamin E deficiency. The meal has a high content of phytic acid, which makes much of its phosphorus unavailable: rations containing the meal may also need extra supplementation with calcium.

The hulls of sesame seeds contain oxalates and it is essential that meals should be completely decorticated in order to avoid toxicities.

Meals in good condition are palatable but have a laxative action. The diets of young ruminants should not contain more than 50 kg/t, whereas the maximum inclusion rate for adults is 100–150 kg/t. The meal should not be given to young pigs or poultry but may be included at up to 50 kg/t for adults of these species.

24.2 Oilseed residues of minor importance

A number of less well-known meals are available in relatively small amounts. These are usually of poor feeding value and low palatability and frequently contain toxic material. Their use is thus limited to certain classes of animal at low rates of inclusion.

Extracted cocoa bean meal

The meals contain varying concentrations of theobromine, an alkaloid which is lethal to chickens and toxic to other monogastric animals and young ruminants. The materials are therefore suitable for adult ruminants only. They have a protein content of about 150 g/kg DM with an apparent digestibility for cattle of about 0.40. Quoted metabolisable energy values range from 5.3 MJ/kg DM for cattle to 9.0 for sheep. The suggested maximum dietary inclusion rate for adult ruminants is 30 kg/t of the total ration. Owing to the risk of a positive drug test, cocoa products should not be given to racing horses.

Shea nut meal

The meals have protein contents of about 190 g/kg DM but their digestibility (cattle) is very low at about 0.12. The energy value for adult ruminants is usually considered to be about half that of barley. They contain an alkaloid, saponine, which causes injury to the digestive mucosa and haemolytic changes in the blood. It is particularly toxic to pigs but can be tolerated by adult ruminants. The meals are bitter to the taste and of low palatability. They may be included at a maximum inclusion rate of 100 kg/t of the total ration of adult beef cattle and 50 kg/t for other adult ruminants. The product should carry a guarantee of saponine content if its use is to be considered.

24.3 Leguminous seeds

The Leguminosae are a large family of plants with about 12 000 recognised species. Four tribes within the family are of particular importance since they include all the common peas and beans. The Hedysareae contain the groundnut; the Vicieae include the genera *Vicia, Cicer, Pisum, Lens* and *Lathyrus*; the Genisteae contain the genus *Lupinus*; the Phaseoleae include the genera *Phaseolus, Dolichos* and *Glycine*.

The combined usage of peas and beans by UK compounders increased from 105 700 tonnes in 1996 to 174 500 tonnes in 1999, with peas forming about 80 per cent of the total.

Many leguminous plants are toxic to animals. Consumption of certain species within the genus *Lathyrus*, such as the Indian pea (*L. sativus*), may result in the condition known as lathyrism. This may take the form of osteolathyrism caused by the presence in the seed of β-L-glutamylamino-propionitrile or neurolathyrism, which is the form affecting human beings and of which β-*N*-oxalyl-L-α,β-diaminopropionic acid is thought to be the causative agent. The former is characterised by bone deformations and weakness in connective tissue. The latter is characterised by retardation of sexual development and increasing paralysis, which may prove fatal if the larynx is affected.

Vicia faba (broad bean) may cause the condition in man known as favism. This is characterised by haemolytic anaemia, and affected individuals suffer nausea, shortness of breath, abdominal pain, fevers and sometimes renal failure. The disease occurs in individuals with a genetic deficiency of glucose-6-phosphate dehydrogenase in their erythrocytes. It has been suggested that the causative agents are divicine and isouramil, pyrimidine derivatives present in the seed.

Phaseolus lunatus beans (lima bean, Java bean) contain the cyanogenetic glucoside phaseolunatin, which is extremely toxic when hydrolysed. The glucoside is present, but in small quantities only, in cultivated varieties of *P. lunatus* such as the 'butter bean'. A number of species including *Cicer ensiformis, Dolichus biflorus, D. lablab, P. lunatus, P. vulgaris* and *P. communis* are known to contain lectins, which are toxic upon oral ingestion. Unfamiliar leguminous seeds should be treated with caution until they have been proved to be safe.

Beans

Beans belong mainly to the Vicieae and Phaseoleae and are used as food for human beings and animals all over the world. The chief member of the Vicieae is *Vicia faba*, known as the broad bean, horse bean and Windsor bean. The most numerous genus in the Phaseoleae is *Phaseolus*, and the best, known species is *P. vulgaris*, with a number of varieties known as kidney, field, garden and haricot beans. There are a large number of other *Phaseolus* species which are locally important as sources of food, as are several other genera such as *Vigna, Dolichus* and *Canavalia*. Nutritionally the species are very similar, being good sources of protein with a high lysine content; they are good sources of energy and phosphorus but have low contents of calcium. Beans have little or no carotene or vitamin C but may contain significant amounts of thiamin, nicotinamide and riboflavin.

There are a large number of varieties of field bean, which fall into two classes: winter and spring. The winter varieties outyield the spring varieties, the yield levels in the UK being about 3.4 and 3.0 tonnes/hectare, respectively. Spring varieties usually have a higher protein content than winter varieties, about 270 compared with 230 g/kg DM, and a lower fibre content, about 69 compared with 78 g/kg DM. These fibre levels are higher than in the common cereals, except for oats. The ether extract concentration in both winter and spring beans is low, about 13 g/kg DM, but it has a high proportion of linoleic and linolenic acids. The mineral composition of beans is similar to that of the cereals and the oilseed residues, with high phosphorus and low calcium concentrations. They contain little or no sodium or chlorine and are poor sources of manganese.

Beans are regarded primarily as sources of protein of relatively good quality. This is a reflection of the amino acid composition characterised by a high lysine content similar to that of fishmeal protein; levels of cystine and methionine are lower than in common animal and vegetable proteins.

As well as being a good source of protein, beans make a significant contribution to the energy economy of the animal, having a metabolisable energy content of 13.5 MJ/kg DM for ruminants and 12.0 MJ/kg DM for poultry and a digestible energy content of 13.3 MJ/kg DM for pigs.

Beans are used in the diets of all the major classes of farm animals. Levels in the diets of calves up to three months of age are usually of the order of 150 kg/t but can be increased considerably thereafter. Mixtures containing 250 kg/t have been used quite satisfactorily for intensively fed steers. The beans are usually cracked, kibbled or coarsely ground for feeding but it would appear that whole beans are quite satisfactory for older ruminants, which rapidly adapt to chewing them. Dairy cow concentrates may contain 150–200 kg/t of beans and recent work has shown that levels of up to 350 kg/t may be used with no loss of milk yield.

For pigs, beans are usually ground to pass a 3 mm screen and are used in sow, weaner and fattening diets; it is not usual to include them in creep feeds. Although there is no objective evidence in its support, the view is widely held that newly harvested beans should be allowed to mature for several weeks before being given to pigs. The usual rate of inclusion of beans is 50–150 kg/t and should not exceed 200 kg/t.

Cystine and methionine are important in the diets of poultry owing to the demands of feathering, and beans may be expected to be of limited value for poultry. Traditionally, they have been little used in poultry diets but present evidence indicates that field beans can be used to replace soya bean meal as long as adequate methionine supplementation is practised.

Many species of beans, including the field bean, have been shown to contain antitryptic factors (see p. 326). In beans grown in the UK, the latter have not been shown to have any significant effect on performance.

Peas

The peas grown as a source of protein for animals in the UK (*Pisum sativum*) belong to the Vicieae. Other species, such as the chick pea (*Cicer arietinum*) in India, are important locally. Peas are basically similar to beans but have lower contents of crude protein (260 g/kg DM) and crude fibre (< 60 g/kg DM). The oil content is slightly higher than that of beans but the degree of saturation is similar. Like beans, peas are regarded primarily as a source of protein. That of peas has the better balance of amino acids, having higher contents of lysine, methionine and cystine. However, methionine is still the main limiting amino acid. Peas make a significant contribution to the energy intake of the animal, having a metabolisable energy content of 13.4 MJ/kg DM for ruminants and 12.7 MJ/kg for poultry, and a digestible energy content of 15.0 MJ/kg DM for pigs.

The maximum rate of inclusion of peas in ruminant diets may be as high as 400 kg/t, but problems of mixing and cubing limit their inclusion in pelleted foods to a maximum of 200 kg/t. They are particularly useful in that they are able to replace soya bean meal in pig and poultry diets, whereas beans are largely confined to ruminant diets.

Lupin seed meal

Lupin seed meal is made by grinding the whole seeds. It is a useful European-grown source of protein. There are three species of lupin, distinguished by the colour of the flowers. Those of *Lupinus albus* are white, those of *L. angustifolius* blue and those of *L. luteus* yellow. Within species, there are sweet and bitter varieties. The latter contain from 10 to 20 g/kg of toxic alkaloids such as lupinin and angustifolin and should not be given to animals: even sweet varieties may contain low levels of alkaloids. For safety, the alkaloid content must be less than 0.6 g/kg.

The seed coat is fibrous and its inclusion in the meal adversely affects its digestibility, especially for young monogastric animals. White varieties have the lowest fibre and highest oil and protein contents and meals made from them are more valuable for pigs and poultry than are the blue and yellow varieties. A typical meal will have a metabolisable energy content of 11.5 MJ/kg DM for poultry and 13.2 MJ/kg DM for ruminants, and a digestible energy content for pigs of 17.3 MJ/kg DM. The amino acid pattern is not well balanced and diets containing significant quantities of the meal may require supplementation with methionine, which is the first limiting amino acid.

Maximum levels of inclusion are 150 kg/t for ruminant diets, 100 kg/t for those of adult poultry and pigs, and 50 kg/t for growing pigs and broilers.

Because of rapid oxidation of the oil, the meal has to be used immediately or an antioxidant incorporated.

24.4 Animal protein concentrates

These materials are given to animals in much smaller amounts than the foods so far discussed, because they are not used primarily as sources of protein *per se* but to make good deficiencies of certain indispensable amino acids from which non-ruminants may suffer when they are fed on all-vegetable-protein diets. In addition, they often make a significant contribution to the animal's mineral nutrition as well as supplying various vitamins of the B complex. A further reason why these products are given in limited quantities to farm animals is that they are expensive, which makes their large-scale use uneconomic. Total usage of animal substances in the compound industry in the UK in 1992 was 409 000 tonnes but had fallen to 177 700 tonnes in 1999. At the earlier date over half was meat and bone-meal, the rest being fishmeal with minor contributions from other sources such as blood meal and feather meal. In 1999, some 2400 tonnes only of meat and bone meal were used.

Production of animal protein foods in the UK is controlled by the Processed Animal Protein Order 1989. This states that producers must be registered and lays down procedures for the sampling and testing of foods for *Salmonella* bacteria. Samples must be taken on each day that a consignment leaves the premises, and submitted for testing. When food produced on the premises is incorporated in a feeding stuff for livestock or poultry kept on those premises then, on each day immediately before any supplies of any such processed animal protein are incorporated in such feeding stuff, a sample must be taken and submitted for testing. It also lays down procedures in the event of a sample failing a test.

Although *Salmonella* infection from meat and bone meal continues to be a matter of concern, the greater concern at the present time is that of infection of animal (and human) consumers with forms of spongiform encephalopathy. Under current UK legislation (the Bovine Spongiform Encephalopathy (No. 2) Order 1996 (SI 1996 No. 3163)) mammalian meat and bone meal cannot be given to ruminant or non-ruminant animals; mammalian protein other than meat and bone meal may be given to non-ruminant animals but not to ruminant animals. Blood meal, tallow, fishmeal and poultry offal meal may be given to both classes of animal.

A European Council decision of 4 December 2000, which remains in force but under review, defined 'Processed animal proteins' as meat and bone meal, meat meal, bone meal, blood meal, dried plasma and other blood products, hydrolysed proteins, hoof meal, horn meal, poultry offal meal, dry greaves, fishmeal, dicalcium phosphate and gelatine. It directs that member states shall prohibit the use of processed animal proteins as food for farmed animals which are kept, fattened or bred for the production of food for human beings. Fishmeal may be given to non-ruminants and the prohibition does not apply to milk and milk products.

Meat by-products

Although animal foods based on mammalian meat are prohibited from being used as food for farmed animals in the UK and Europe, there are areas where this may be allowed. For this reason we have decided to include a few general comments on the nutritive value of such products. It may also be that, at some future date, current prohibitions on their use in Europe may be lifted.

The protein of meat by-products is of good quality (BV approximately 0.67 for adult man) and is particularly useful as a lysine supplement. Unfortunately it is a poor source of methionine and tryptophan. Various unidentified beneficial factors have been claimed to be present in meat meals, among them the enteric growth factor from the intestinal tract of swine, the *Ackerman* factor and a growth factor located in the ash.

Meat products are more valuable for simple-stomached than for ruminant animals since the latter have less need of a dietary supply of high-quality protein. The low methionine and tryptophan levels of the meals reduce their value since they cannot adequately make good the deficiencies of these amino acids in the high-cereal diets of pigs and poultry. This is especially so when high proportions of maize are given, maize being particularly low in tryptophan. Usually meat meal is given in conjunction with another animal or vegetable protein to complement its low content of methionine and tryptophan. Both meat meal and meat and bone meal are eaten readily by pigs and poultry and may be given at levels of up to 150 kg/t of the diet for laying hens and young pigs; for fattening pigs the level is usually kept below 100 kg/t. As well as being less beneficial for ruminants than for simple-stomached animals, these products are not readily accepted by ruminants and must be introduced into their diets gradually. Considerable care is required in storing meat products to prevent the development of rancidity and loss of vitamin potency.

Fishmeal

World production of fishmeal in 1999 was recorded as 6 247 000 tonnes, of which UK consumption was 180 000 tonnes. This compares with a consumption of 193 000 tonnes in 1996.

Fishmeal is produced by cooking fish, and pressing the cooked mass to remove most of the oil and water. The aqueous liquor is concentrated and added to the pressed mass and the whole dried. About 90 per cent of the raw material used in fishmeal production consists of oily species such as anchovies, capelin and menhaden, the remaining 10 per cent consisting of fish (plus some of the offal) of species such as haddock and cod.

Nutritionally, the drying process is very important since overdrying may significantly reduce the quality of the product. There are two main types of dryer: direct and indirect. In the former, hot air (at about 500 °C) is passed over the material as it is tumbled in a cylindrical drum. The temperature of the material should remain between 80 and 95 °C, but may be much higher if the process is not carefully controlled. In the indirect method the dryers are steam-jacketed cylinders or cylinders containing steam-heated discs, which again tumble the material during drying. The latter process is slower but more easily controlled. The dried product is ground so that less than 10 per cent passes a 1 mm screen and more than 90 per cent passes a 10 mm screen.

In well-produced meals the protein has a digestibility of between 0.93 and 0.95, but meals which are heated too strongly in processing may have values as low as 0.60. The quality of protein in fishmeal is high though variable as indicated by values of 0.36 and 0.82 quoted as the biological value for rats. Processing conditions, particularly the degree and length of time of heating, are probably the major determinants of protein quality, as shown by the figures for available lysine quoted in Table 24.4.

Values of 0.58 for $r = 0.02$, 0.44 for $r = 0.05$ and 0.38 for $r = 0.08$ (r = passage rate from the rumen: see p. 335) have been quoted for the rumen degradability of fishmeal protein and it is pre-eminent as a source of undegradable protein for ruminant animals.

The protein contents of various fishmeals vary over a range of about 500–750 g/kg but the composition of the protein is relatively constant. It is rich in the essential amino acids, particularly lysine, cystine, methionine and tryptophan, and is a valuable supplement to cereal-based diets, particularly where they contain much maize. The essential amino acid composition is compared with that of ideal protein (Table 13.7) in Box 24.3.

Fishmeals have a high mineral content (100–220 g/kg), which is of value nutritionally since it contains a high proportion of calcium and phosphorus and a number of desirable trace minerals including manganese, iron and iodine. They are a good source of B complex vitamins, particularly choline, B_{12} and riboflavin, and have enhanced nutritional value because of their content of growth factors known collectively as the animal protein factor (APF).

Table 24.4 Effect of various heat treatments on the available lysine contents of fishmeals

Treatment	Available lysine (g/kg CP)
Freeze dried	86
Oven dried at:	
105 °C for 6 h	83
170 °C for 6 h	69

Box 24.3

Ratio of the essential amino acids to lysine in fishmeal protein and ideal protein

Amino acid	Fishmeal	Ideal protein
Lysine	1.0	1.0
Methionine+cystine	0.45	0.50
Trytophan	0.14	0.14
Threonine	0.57	0.60
Leucine	1.01	1.00
Isoleucine	0.62	0.54
Valine	0.73	0.70
Histidine	0.29	0.33
Phenylalanine + tyrosine	1.04	0.96

Fishmeals contain high levels of polyunsaturated fatty acids and the ratio of omega-3 to the more abundant omega-6 acids is particularly desirable; they also contain significant amouts of eicosapentaenoic and docosohexenoic acids (Chapter 3). Increased levels of these substances are advatangeous to human beings consuming them.

The energy of fishmeals is present entirely in the form of fat and protein and is largely a reflection of the oil content. In the past, the energy value for animals has been underestimated, particularly in the case of ruminants, for which average values in the region of 14 MJ/kg of metabolisable energy are now accepted as realistic.

A wide variety of fishmeals are available depending upon the country of origin, the raw material and the process used. For convenience the meals currently available in the UK may be grouped into:

(a) South American fishmeals,
(b) herring-type fishmeals,
(c) UK-produced meal, which may contain a proportion of offal.

Typical figures for composition and nutritive value are given in Table 24.5.

The current trend in the marketing of fishmeals is towards specialised products tailored to suit particular species. Thus special low-temperature meals are produced for aquaculture and for early weaned pigs, and ruminant-grade products have strictly controlled levels of soluble nitrogen.

Fishmeals find their greatest use with simple-stomached animals. They are used mostly in diets for young animals, whose demand for protein and the indispensable amino acids is particularly high and for whom the growth-promoting effects of APF are valuable. Such diets may include up to 150 kg/t of fishmeal. With older animals, which need less protein, the

Table 24.5 Composition and nutritive value of some typical fishmeals

	Herring	South American	UK produced
Crude protein, g/kg	730	660	640
Oil, g/kg	70	60	65
Calcium, g/kg	20	45	80
Phosphorus, g/kg	15	30	50
ME, MJ/kg DM			
(ruminant)	17.8	14.6	14.6
(poultry)	14.9	13.9	13.4
DE, MJ/kg DM			
(pigs)	19.6	19.0	17.0

level of fishmeal in the diet is brought down to about 50 kg/t and it may be eliminated entirely from diets for those in the last stages of fattening. This is partly for economic reasons, since the protein needs of such animals are small, and partly to remove any possibility of a fishy taint in the finished carcass. This possibility must also be carefully considered with animals producing milk and eggs, which are vulnerable to taint development. Fully ruminant animals are able to obtain amino acids and B vitamins by microbial synthesis and the importance of fishmeal for such animals is as a source of undegradable protein. This is of particular importance for actively growing and pregnant animals. Rates of inclusion in the diet are usually about 50 kg/t. For lactating cows, the daily intake of fishmeal should be limited to not more than 1 kg. Above this the intake of oil could exceed 10 g/day, which would have detrimental effects on fermentation in the rumen.

In the UK, fishmeal is legally defined as '...the product obtained by processing whole or parts of fish from which part of the oil may have been removed and to which fish solubles may have been re-added'. Fish solubles are '...the stabilised product composed of press juice obtained during manufacture of fishmeal from which much of the fish oil and some of the water has been removed' (Feeding Stuffs (Amm) Regulations 1993). The nitrite content of fishmeals sold in the UK is strictly controlled by legislation. It may not be more than 60 mg/kg of food referred to a moisture content of 120 mg/kg and stated as sodium nitrite. This is equivalent to 14 mg nitrite nitrogen/kg DM.

Blood meal

In the UK, blood meal is defined as '...the product obtained by drying the blood of slaughtered warm-blooded animals. The product must be substantially free of foreign matter' (Feeding Stuffs (Amm) Regulations 1993).

It is manufactured by passing live steam through blood until the temperature reaches 100 °C. This ensures efficient sterilisation and causes the blood to clot. It is then drained, pressed to express occluded serum, dried by steam heating and ground.

Blood meal is a chocolate-coloured powder with a characteristic smell. It contains about 800 g/kg of protein, small amounts of ash and oil and about 100 g/kg of water. It is only important, nutritionally, as a source of protein. Blood meal is one of the richest sources of lysine and a rich source of arginine, methionine, cystine and leucine but is deficient in isoleucine and contains less glycine than either fish, meat, or meat and bonemeals. Owing to the poor balance of amino acids its biological value is low and in addition it has a low digestibility. It has the advantage, in certain situations, that its protein has a very low rumen degradability (about 0.20).

The meal is unpalatable and its use has reduced growth rates in poultry so that it is not recommended for young stock. For older birds, rates of inclusion are limited to about 10–20 kg/t of the diet. It should not be included in creep foods for pigs. Normal levels of inclusion for older animals are of the order of 50 kg/t of diet and it is usually used along with a high-quality protein source. At levels over 100 kg/t of the diet it tends to cause scouring and is best regarded as a food for boosting dietary lysine levels.

Hydrolysed feather meal

The Feeding Stuffs (Amm) Regulations 1993 define this material as the 'Product obtained by hydrolysing, drying and grinding poultry feathers'.

The meal is produced by steam cooking under pressures of 2.8–3.55 kg/cm^2 for 30–45 minutes, when temperatures of 140–150 °C are achieved.

Feather meal has a high protein content, typically about 850 g/kg, with individual samples varying from 610 to 930 g/kg. Histidine and lysine are joint first limiting amino acids with methionine being third limiting. Ileal digestibility is of the order of 0.5, with digestibility values for individual amino acids varying from 0.20 to 0.70. There is some evidence that protein digestibility may be improved if processing is carried out at higher pressures for a shorter time. Typically the meal has a digestible energy value for pigs of 12.6 MJ/kg DM and metabolisable energy values of 13.7, 12.5 and 13.6 MJ/kg DM for poultry, cattle and sheep, respectively. Variations in processing conditions may have a marked effect on nutritive value.

The meal is of low palatability and must be introduced into the diet gradually. Dietary rates of inclusion are generally low, being of the order of 25–30 kg/t of the total ration for adult ruminants, 25 kg/t for layers, broilers and turkeys, and 10 kg/t for calves, lambs, sows, and growing and finishing pigs. The meal is not used for weaner and creep-fed pigs or chicks.

There is a risk of contamination of the base material with *Salmonella* and it is important that strict control of processing conditions should be maintained in order to minimise the risk of this in the finished product.

24.5 Milk products

Whole milk

Whole milk from cows contains about 875 g/kg of water and 125 g/kg of dry matter, usually referred to as the total solids. Of this about 37.5 g/kg is fat. The remainder, termed the solids-not-fat (SNF), consists of protein (33 g/kg), lactose (47 g/kg) and ash (7.5 g/kg). Most of the fat consists of neutral triacylglycerols having a characteristically high proportion of fatty acids of low molecular weight and providing an excellent source of energy. It has about 2.25 times the energy value of the milk sugar or lactose. The crude protein fraction of milk is complex, about 5 per cent of the nitrogen being non-protein. Casein, the chief milk protein, contains about 78 per cent of the total nitrogen and is of excellent quality, but is marred by a slight deficiency of the sulphur-containing amino acids, cystine and methionine. Fortunately, β-lactoglobulin is rich in these acids so that the combined milk proteins have a biological value of about 0.85. The most economical use of the protein is in supplementing poor-quality proteins like those of the cereals, for which purpose it is better than either the meat or fish products. When milk products are used to replace fishmeal or meat and bone meal the diet must be supplemented with inorganic elements, particularly calcium and phosphorus, since the ash content of milk is low. Milk has a low magnesium content and is seriously deficient in iron. Normally milk is a good source of vitamin A but a poor one of vitamins D and E. It is a good source of thiamin and riboflavin and contains small amounts of vitamin B_{12}.

Whole milk is consumed by suckled calves, lambs, young dairy and bull calves, and animals being prepared for competition. Two milk by-products are widely used and are valuable foods for farm animals.

Skim milk

This is the residue after the cream has been separated from milk by centrifugal force. The fat content is very low, below 10 g/kg, and the gross energy is much reduced, about 1.5 MJ/kg as compared with 3.1 MJ/kg for whole milk. Removal of the fat in the cream also means that skim milk is a poor source of the fat-soluble vitamins, but it does result in a concentration of the SNF constituents. Skim milk finds its main use as a protein supplement in the diets of simple-stomached animals and is rarely used for ruminant animals; it is particularly effective in making good the amino acid deficiencies of the largely cereal diets of young pigs and poultry. It is usually given to pigs in the liquid state and is limited to a *per capita* consumption of 2.8–3.4 litres (3.0–3.6 kg) per day. When the price is suitable, it may be given *ad libitum*, and up to 23 litres (24 kg) per pig per day may be consumed along with about 1 kg of meal. Scouring may occur at these levels but can be avoided with reasonable care. Liquid skim milk must always be given in the same state, either fresh or sour, if digestive upsets are to be

Table 24.6 Effect of processing on the nutritive value of skim milk

	Digestibility of protein	Biological value of protein	Available lysine (g/kg CP)
Spray-dried	0.96	0.89	81
Roller-dried	0.92	0.82	59

avoided. It may be preserved by adding 1.5 litres of formalin to 1000 litres of skim milk. For feeding poultry, skim milk is normally used as a powder and may form up to 150 kg/t of the diet. It contains about 350 g/kg of protein, the quality of which varies according to the manufacturing process used: roller-dried skim milk is subjected to a higher drying temperature than the spray-dried product and has a lower digestibility and biological value (Table 24.6). For poultry, skim milk has the disadvantage of having a low cystine content.

Whey

When milk is treated with rennet in the process of cheese making, casein is precipitated and carries down with it most of the fat and about half the calcium and phosphorus. The remaining serum is known as whey, which, as a result of the partitioning of the milk constituents in the rennet coagulation, is a poorer source of energy (1.1 MJ/kg), fat-soluble vitamins, calcium and phosphorus. Quantitatively it is a poorer source of protein than milk but most of the protein is β-lactoglobulin and is of very good quality. Whey is usually given *ad libitum* in the liquid state, to pigs. Dried skim milk and whey find their main use as constituents of proprietary milk replacers for young calves. Unlike those of skim milk, whey proteins do not clot in the stomach and may give rise to digestive problems at high rates of inclusion.

24.6 Single-cell protein

Protein for feeding animals has been produced by microbial fermentation. Single-cell organisms such as yeasts and bacteria grow very quickly and can double their cell mass, even in large-scale industrial fermenters, in 3–4 hours. A range of nutrient substrates can be used including cereal grains, sugar beet, sugarcane and its by-products, hydrolysates from wood and plants, and waste products from food manufacture. Bacteria such as *Pseudomonas* spp. can be grown on more unconventional materials such as methanol, ethanol,

Table 24.7 Chemical composition of SCP grown on different substrates (g/kg DM) (After Schulz E and Oslage H J 1976 *Animal Feed Science and Technology* **1**: 9)

Substrates used	Microorganism	DM (g/kg)	Organic matter	Crude protein	Crude fat	Crude fibre	Ash
Gas oil	*Candida lipolytica*	916	914	678	25	44	86
Gas oil	*Candida lipolytica*	903	917	494	132	41	84
n-Paraffin	*Candida lipolytica*	932	934	644	92	47	66
n-Paraffin	*Candida lipolytica*	914	933	480	236	47	67
n-Alkanes	*Pichia guillerm*	971	941	501	122	76	59
Whey (lactic acid)	*Candida pseudotropicalis*	900	900	640	56	50	100
Methanol	*Candida boidinii*	938	939	388	77	107	61
Methanol	*Pseudomonas methylica*	967	903	819	79	5	97
Sulphite liquor	*Candida utilis*	917	925	553	79	13	75
Molasses	*Saccharomyces cerevisiae*	908	932	515	63	18	68
Extract of malt	*Saccharomyces carlsbergensis*	899	926	458	31	11	74

alkanes, alkanals and organic acids. The figures in Table 24.7 show that the material obtained by culturing different organisms varies considerably in composition.

The protein content of bacteria is higher than that of yeasts and contains higher concentrations of the sulphur-containing amino acids but a lower concentration of lysine. Single-cell protein (SCP) contains unusually high levels of nucleic acids, ranging from 50 to 120 g/kg DM in yeasts and from 80 to 160 g/kg DM in bacteria. Some of the purine and pyrimidine bases in these acids can be used for nucleic acid biosynthesis. Large amounts of uric acid or allantoin, the end-products of nucleic acid catabolism, are excreted in the urine of animals consuming SCP. The oil content of yeasts and bacteria varies from 25 to 236 g/kg DM and the oils themselves are rich in unsaturated fatty acids. Although SCP does contain a crude fibre fraction, which can be quite high in some yeasts, it is not composed of cellulose, hemicelluloses and lignin as in foods of plant origin; the fibre consists chiefly of glucans, mannans and chitin.

Studies with pigs have yielded digestibility coefficients for energy varying from 0.70 to 0.90 for yeasts grown on whey and *n*-paraffins, respectively, and a value of 0.80 was obtained for a bacterium grown on methanol. The inclusion of up to 150 kg/t of SCP in pig diets has given levels of animal production comparable with those obtained with diets containing soya bean meal or fishmeal. Similarly encouraging results have been obtained with calves, although it is usually recommended that the maximum level of inclusion of alkane yeasts in calf milk substitutes should be 80 kg/t.

In the case of poultry, dietary SCP concentrations of 20–50 kg/t have proved optimal for broilers and 100 kg/t has been suggested for diets for laying hens.

24.7 Synthetic amino acids

Meeting the requirements of non-ruminants for a balanced supply of essential amino acids requires the use of expensive protein sources, such as fishmeal, or high levels of less well-balanced protein sources such as soya bean meal (Chapter 13). Economics dictate that the latter option is usually taken, and, in order to meet the requirements for the limiting amino acid, an excess of total protein has to be supplied. This is deaminated and the nitrogen is excreted; both processes require energy, and such protein oversupply is wasteful in terms of both protein and energy metabolism. In addition, the excreted nitrogen may be a source of pollution in the environment.

An alternative to the use of high dietary levels of unbalanced proteins is to use them at lower levels but in combination with supplements of free amino acids. This reduces the total amount of protein given and, since the amino acid balance is more appropriate, the efficiency of energy and protein use is improved and less nitrogen is wasted. Amino acids can be produced industrially by chemical and microbial processes and are readily available.

Pig and poultry diets based on cereals and vegetable protein sources are now routinely supplemented with L-lysine hydrochloride (supplying 780 g lysine/kg), DL-methionine and L-threonine. A diet for a finishing pig, which has to contain 10 g lysine/kg, required a combination of 750 g barley and 250 g soya bean meal/kg, and this mix has a crude protein content of 185 g/kg (Appendix 2, Table 2.A.1.3). With the inclusion of 2 g of lysine hydrochloride, the same lysine content can be achieved with a mix of 808 g barley and 190 g soya bean meal, and the protein content is reduced to 165 g/kg. Such reductions in crude protein content have maintained a balanced supply of amino acids and resulted in improved rates of liveweight gain and food conversion efficiency. It is important that the supplementary acids are not used excessively to satisfy the animal's requirements, since this may bring about an undersupply of other essential amino acids.

24.8 Non-protein nitrogen compounds as protein sources

Non-protein nitrogen compounds are recognised as useful sources of nitrogen for ruminant animals, and simple nitrogenous compounds such as ammonium salts of organic acids can be utilised to a very limited extent by pigs and poultry. Commercially, non-protein nitrogen compounds are important for ruminants only. Their use depends upon the ability of the rumen microorganisms to use them in the synthesis of their own cellular tissues (Chapter 8), and they are thus able to satisfy the microbial portion of the animal's demand for nitrogen and, by way of the microbial protein, at least part of its nitrogen demand at tissue level. The compounds investigated include urea, ammonium salts of organic acids, inorganic ammonium salts and various amides as well as thiourea, hydrazine and biuret. Studies *in vitro* have shown that ammonium acetate, ammonium succinate, acetamide and diammonium phosphate are better substrates for microbial protein synthesis than urea, but from considerations of price, convenience, palatability and toxicity urea has been the most widely used and investigated non-protein nitrogen compound in foods for farm animals.

Urea

Urea is a white, crystalline, deliquescent solid with the formula:

$$
\begin{array}{c}
NH_2 \\
| \\
C{:}O \\
| \\
NH_2
\end{array}
$$

Pure urea has a nitrogen content of 466 g/kg, which is equivalent to a crude protein content of $466 \times 6.25 = 2913$ g/kg. 'Feed urea' contains an inert conditioner to keep it flowing freely and this reduces its nitrogen content to 464 g/kg, equivalent to 2900 g/kg crude protein. Urea is hydrolysed by the urease activity of the rumen microorganisms with the production of ammonia:

$$
\begin{array}{c}
NH_2 \\
| \\
C{:}O \ + \ H_2O \ \longrightarrow \ 2NH_3 \ + \ CO_2 \\
| \\
NH_2
\end{array}
$$

The ease and speed with which this reaction occurs when urea enters the rumen gives rise to two major problems owing to excessive absorption of ammonia from the rumen. Thus wastage of nitrogen may occur and there may be a danger of ammonia toxicity. This is characterised by a muscular twitching, ataxia, excessive salivation, tetany, bloat and respiratory defects (both rapid, shallow and slow, deep breathing have been reported). Dietary levels of urea vary in their effects and it is not possible to give accurate safety limits for any particular animal. Thus sheep receiving as little as 8.5 g per day have died whereas others have consumed up to 100 g per day without ill effects. Toxic symptoms appear when the ammonia level of peripheral blood exceeds 10 mg/kg and the lethal level is about 30 mg/kg. Such levels are usually associated with ruminal ammonia concentrations of about 800 mg/kg, the actual level depending upon pH. Ammonia, which is the actual toxic agent in urea poisoning, is most toxic at high ruminal pH owing to the increased permeability of the rumen wall to unionised ammonia compared with the ammonium ion, which predominates at low pH.

Urea should be given in such a way as to slow down its rate of breakdown and encourage ammonia utilisation for protein synthesis. It is most effective when given as a supplement to diets of low protein content, particularly if the protein is resistant to microbial breakdown. The diet should also contain a source of readily available energy so that microbial protein synthesis is enhanced and wastage reduced. At the same time, the entry of readily available carbohydrate into the rumen will bring about a rapid fall in rumen pH and so reduce the likelihood of toxicity. Many of the difficulties encountered with urea would be avoided if the amount and size of meals were restricted. To avoid the danger of toxicity, not more than one-third of the dietary nitrogen should be provided as urea and, where possible, this should be in the form of frequent and small intakes.

Urea, like other non-protein nitrogen sources, will not be used efficiently by the ruminant animal unless the diet does not contain sufficient degradable protein to satisfy the needs of its rumen microorganisms.

Although urea provides an acceptable protein source there is evidence that where it forms a major part of dietary nitrogen, deficiencies of the sulphur-containing amino acids may occur. In such cases supplementation of the diet with a sulphur source may be necessary. An allowance of 0.13 g of anhydrous sodium sulphate/per gram of urea is generally considered to be optimal. Urea does not provide energy, minerals or vitamins for the animal, and when it is used to replace conventional protein sources care must be taken to ensure that satisfactory dietary levels of these nutrients are maintained by adequate supplementation.

Urea is available in proprietary foods in several forms. It may be included in solid blocks which provide vitamin and mineral supplementation and contain a readily available source of energy, usually starch. Animals are allowed free access to the blocks, intake being restricted by the blocks having to be licked and by their high salt content. There is some danger of excessive urea intakes if the block should crumble or if there should be a readily available source of water that allows the animal to cope with the high salt intakes. Solutions of urea containing molasses as the energy source and carrying a variable amount of mineral and vitamin supplementation are also in use. Like the blocks they contain 50–60 g urea and about 250 g sugar per kilogram, and are supplied in special feeders in which the animal licks a ball floating in the solution.

Several proprietary concentrate foods contain urea at levels ranging from 10 to 30 g/kg. Protein supplements for balancing cereals usually contain about 100 g/kg urea but there are some highly concentrated products with as much as 500 g/kg. When urea is included in the concentrate diet thorough mixing is essential to prevent localised concentrations, which may have toxic effects. Urea is sometimes added to the roughage or concentrate portion of the ration as a solution containing about 350 g urea and 100 g molasses per kilogram.

Urea may be used for all classes of ruminants but is less effective in animals in which the rumen is not fully functional. Low-intensity, ranch-like conditions, in which animals are fed on diets containing poor-quality protein in low concentration, are very suitable for the use of block or liquid foods. Under such conditions, raising the nitrogen content of the diet may increase the digestibility and intake of the roughage owing to stimulation of microbial activity. Beef cattle and sheep consuming small amounts of concentrates are able to utilise urea efficiently, as are animals receiving large quantities of concentrate *ad libitum*, resulting in small and frequent intakes of food. Low-yielding dairy cows use urea-containing concentrates efficiently owing to their low concentrate intakes; cows of moderate and high yield receiving large meals of concentrates at milking do not. There is much evidence for reduced performance by such animals given diets containing urea.

When straws and other low-quality roughages are treated with ammonia (see p. 546) about 0.3–0.5 of the ammonia is retained by the roughage and may be utilised by the rumen microorganisms in the same way as ammonia derived from urea.

Biuret

Biuret is produced by heating urea. It is a colourless, crystalline compound with the formula:

$$NH_2 \cdot CO \cdot NH \cdot CO \cdot NH_2$$

It contains 408 g N/kg, equivalent to 2550 g CP/kg. Biuret is utilised by ruminants, but a considerable period of adaptation is required. Adaptation is speeded by inoculation with rumen liquor from an adapted rumen. The nitrogen of biuret is not as efficiently utilised as that of urea and it is very much more expensive. It has the considerable advantage that it is non-toxic even at levels very much higher than those likely to be found in foods.

Compound foods containing urea, biuret, urea phosphate or diureidoisobutane must by law carry a declaration of:

1. the name of the material,
2. the amount present,
3. the proportion of nitrogen, expressed as protein equivalent, provided by the non-protein nitrogenous content of the food,
4. directions for use specifying the animals for which the food is intended and the maximum level of non-protein nitrogen, which must not be exceeded in the daily ration.

Poultry waste

In the UK, poultry waste is legally defined as '…the product obtained by drying and grinding waste from slaughtered poultry. The product must be substantially free of feathers' (Feeding Stuffs Regulations 1993).

Despite aesthetic objections, dried poultry excreta have been successfully used for ruminants.

Poultry manures vary considerably in composition, depending upon their origins. That from caged layers has a lower fibre content than broiler litter, which has a base of the straw, wood chips or sawdust used as bedding. Broiler litters also vary in composition depending upon the number of batches put through between changes of bedding. Both types of waste have a high ash content, particularly that of layers, usually about 280 g/kg DM. Digestibility is low and, although metabolisable energy values of 6–9 have been quoted, 7.5 MJ/kg DM is probably a realistic figure. Protein (N × 6.25) contents are variable, between 250 and 350 g/kg DM, with a digestibility of about 0.65. Most of the nitrogen (600 g/kg at least) is present as non-protein compounds, mostly urates, which must first be converted into urea and then ammonia in order to be utilised by the animal. The conversion to urea is usually a slow process and wastage and the danger of toxicity are both less than with foods containing urea itself. Layer wastes are excellent sources of calcium (about 65 g/kg DM) but the ratio of calcium to phosphorus is rather wide at 3 : 1; broiler litters have less calcium with ratios closer to 1 : 1.

Dietary inclusion rates of up to 250 kg/t have been used for dairy cows and up to 400 kg/t for fattening cattle and have supported very acceptable levels of performance. Thus, dairy cows given 110 kg per tonne of diet to replace half the soya bean of a control diet yielded 20 kg milk, the same as the control, but gained only 0.58 kg/day compared with 0.95 kg/day for the controls. With fattening steers, concentrate diets containing wastes have supported gains of the order of 1 kg/day, but it has been estimated that for each inclusion of 100 kg excreta per tonne of diet, liveweight gains are reduced by about 40 g/kg.

One of the major anxieties constraining the use of poultry wastes in animal diets has been the fear of health hazards arising from the presence of pathogens such as *Salmonella* and the presence of pesticide and drug residues. The heat treatment involved in drying and the ensiling procedures used for storing the materials appear to offer satisfactory control of pathogens, and pesticides have not proved to be a problem. Drug residues may be a hazard but this can be overcome by having a withdrawal period of three weeks before slaughter.

The best method of treating poultry wastes for use as animal foods is by drying, but this is costly; ensiling, either alone, with forage, or with barley meal and malt, have all proved satisfactory.

The Feeding Stuffs Regulations 1991 state that 'no person shall sell, or have possession with a view to sale, for use as a compound feeding stuff, or use in a compound feeding stuff, any material which contains faeces, urine or separated digestive tract content resulting from the emptying or removal of the digestive tract, irrespective of any form of treatment or admixture'. Taken in conjunction with the requirements of the Processed Animal Protein Order (see above), this makes the use of poultry waste virtually impossible!

Poultry wastes must, by law, carry a declaration of the amount of protein equivalent of uric acid if 1 per cent or greater and of calcium if in excess of 2 per cent.

Summary

1. Oilseed cakes and meals are the residues remaining after oil is removed from oilseeds. They are rich in protein and are valuable foods for animals.

2. About 2.7 million tonnes were used by the UK food-compounding industry in 1999, of which soya bean meal contributed 38 per cent, rapeseed meal 23 per cent, and sunflower meal 18 per cent.

3. They usually have a high content of phosphorus, which tends to exacerbate the generally low calcium content.

4. The quality of oilseed protein is higher than that of the cereals, and the best may approach that of some animal proteins. In general, they have a low content of methionine and cystine, and a variable though usually low lysine content.

5. Oilseeds contain antinutritional factors, and some, such as castor oil seeds, are so toxic as to preclude their use as foods unless stringent conditions are attached to their use.

6. The use of leguminous seeds in the UK has been increasing in recent years, particularly that of peas. Protein quality is good, with methionine the main limiting amino acid.

7. Animal protein concentrates are used, primarily, in small amounts to make good deficiencies of certain indispensable amino acids, minerals and B vitamins in the diets of non-ruminant animals.

8. The use of various meat by-products as food for certain animals is currently prohibited in the UK and Europe.

9. About 180 000 tonnes of fishmeal was consumed by animals in the UK in 1999 (although its use is now restricted to non-ruminants). The quality of the protein is high and it is, in addition, an excellent source of calcium, phosphorus, choline, B_{12} and riboflavin. Fishmeals transmit high levels of polyunsaturated acids to animals consuming them, which may benefit human beings consuming their products

10. Milk products supply protein of high quality. They are used mainly in pig diets and for young animals in general.

11. The use of synthetic amino acids as dietary supplements has increased in the past few years and it is now routine to add L-lysine hydrochloride, DL-methionine and L-threonine to pig and poultry diets.

12. Non-protein materials such as urea are used to provide a supply of degradable nitrogen for the rumen bacteria and thus protein for the host ruminant. Care is needed to ensure an adequate supply of energy and sulphur for the bacteria.

Further reading

Aherne F X and Kenelly J J 1982 Oilseed meals for livestock feeding. In: Haresign W (ed.) *Recent Advances in Animal Nutrition*, London, Butterworth.

Altschul A M 1958 *Processed Plant Protein Foodstuffs*, New York, Academic Press.

Bell J M 1993 Factors affecting the nutritional value of canola meal – A review. *Canadian Journal of Animal Science* **73**: 679–97.

Briggs M H (ed.) 1967 *Urea as a Protein Supplement*, London, Pergamon Press.

Gillies M T 1978 *Animal Feeds from Waste Materials*, New Jersey, Noyes Data.

Liener I E 1969 *Toxic Constituents of Plant Foodstuffs*, New York, Academic Press.

Liener I E 1990 Naturally occurring toxic factors in animal feedstuffs. In: Wiseman J and Cole D J A (eds) *Feedstuff Evaluation*, London, Butterworth.

National Academy of Sciences 1976 *Urea and Other Non-protein Nitrogenous Compounds in Animal Nutrition*, Washington, NAS Publishing and Printing Office.

Takayuki S and Bjeldanes L F *Introduction to Food Toxicology*, New York, Academic Press.

25 Food additives

25.1   **Antibiotics**

25.2   **Probiotics**

25.3   **Oligosaccharides**

25.4   **Enzymes**

25.5   **Organic acids**

25.6   **Modifiers of rumen fermentation**

Food additives are materials that are administered to the animal to enhance the effectiveness of nutrients and exert their effects in the gut or on the gut wall cells. There are clear theoretical reasons why these additives should be effective but, given the dynamic nature of the gut physiology, it is often very difficult to demonstrate the effects in practice. For example, the effects of added organic acids within the gut lumen are difficult to quantify, even when re-entrant cannulae are employed.

25.1 Antibiotics

Antibiotics are chemical compounds which, when given in small amounts, halt the growth of bacteria. They are produced by other microorganisms, e.g. fungi, and are also synthesised in the laboratory. They are used at therapeutic levels, by injection or in food or water, to treat diseases caused by bacteria. In addition, sub-therapeutic levels of antibiotics are added to the food to enhance the rate of growth. The various groups of antibiotics act in different ways to reduce the numbers of specific bacteria in the gut (see Box 25.1), and thereby increase the efficiency of nutrient utilisation. This is brought about by:

- reduction or elimination of the activity of pathogenic bacteria which may cause subclinical infection, thus allowing the host to achieve production levels closer to the potential;
- elimination of those bacteria which produce toxins that reduce the growth of the host animal;
- stimulation of the growth of microorganisms that synthesise unidentified nutrients;

■ reduction of the growth of microorganisms that compete with the host animal (the fermentation of nutrients by bacteria is a wasteful process compared with direct absorption);
■ increased absorptive capacity of the small intestine through a decrease in the thickness of the intestinal wall.

These effects may be coupled with a reduced turnover of mucosal cells and reduced mucous secretion. The gut accounts for a large proportion of the energy and protein required to maintain an animal, and any reduction in the mass of the gut and cell turnover will release nutrients for other purposes such as growth.

Antibiotic growth promoters have been used mainly in pig and poultry foods, typically at levels of 20–40 mg/kg, which give improvements of 4–16 per cent in growth rate and 2–7 per cent in efficiency of food conversion. The response is greatest in young animals and those consuming diets containing vegetable protein rather than animal protein. Mortality in young pigs is also reduced. The effect is less in healthy herds and flocks. Young pre-ruminant calves also respond to growth-promoting antibiotics in the same manner as non-ruminants. Since ruminants depend primarily on bacteria in the rumen for their nutrient supply, the use of antibiotics in ruminant diets might be considered to be disadvantageous. However, certain antibiotics of the ionophore type (e.g. monensin sodium) have been found to be effective, particularly with low-roughage/high-concentrate diets, in controlling the proportion of undesirable bacteria in the rumen without disturbing the overall fermentation. There are small changes in the fermentation, however, such as reduced methane production and increased propionic acid proportions, which improve productivity. Monensin sodium at 20–30 mg/kg of food improves the efficiency of food conversion by increasing the rate of gain with the same food intake or by maintaining the rate of gain at a lower food intake. Amino acid degradation is decreased and the additional propionic acid spares amino acids for gluconeogenesis.

The widespread use of antibiotics coupled with the ability of resistant bacterial strains to evolve over a short period of time and transfer resistance to other strains has resulted in populations of bacteria that are resistant to many antibiotics. This has limited the effectiveness of antibiotic treatment against disease and, whereas at one time a bacterial infection could be treated with a range of antibiotics, now some bacteria are sensitive to a single antibiotic only. There is concern that if such bacteria gained resistance to this antibiotic then the disease would be untreatable. For this reason the use of antibiotics as growth promoters has been curtailed by legislation in recent years. In the EU in 2000, only four antibiotics were permitted for use as growth promoters: flavophospholipol, avilamycin, salinomycin sodium and monensin sodium.

Another material which enhances the growth of pigs, and is thought to act via the antibacterial effects described above, is copper (Chapter 6). When added to their food, as copper sulphate, at levels far in excess of the requirement for normal functions (e.g. 200 mg Cu/kg compared with 5 mg Cu/kg) pigs grow faster. Concern over the accumulation of copper in the

Box 25.1
Modes of action of antibiotics

Antibiotics halt the growth of bacteria by interfering with their cellular metabolism. There are four groups:

1. Antibiotics which interfere with the synthesis of the material forming the bacterial cell wall and cause the cell to burst. These are high molecular weight (>1200) compounds that act on Gram-positive bacteria. They are poorly absorbed by the host and thus are non-toxic, leave no detectable residues and have no withdrawal period (i.e. a period of time during which the compound must be removed from the food before the animal is slaughtered). Avoparcin and flavomycin are examples of this type of antibiotic.
2. Inhibitors of bacterial protein synthesis are also primarily active against Gram-positive bacteria and have a medium molecular weight (>500). Although they are absorbed to a greater extent than the higher molecular weight compounds, they do not have a withdrawal period. Examples of this type of antibiotic include tylosin and virginiamycin.
3. Inhibitors of bacterial DNA synthesis can have a broad spectrum of activity, have a low molecular weight (about 250) and require withdrawal periods. Nitrofurans and quinoxaline-N-oxides fall into this category of antibiotics.
4. Ionophore antibiotics interfere with the electrolyte balance (Na/K) of the bacterial cell by transporting potassium into the cell, which then has to use energy to pump it out. Eventually the ion pump fails to operate efficiently and potassium accumulates inside the cell. Water enters by osmosis and the cell ruptures. Monensin sodium is an example of this type of antibiotic.

environment and the potential toxic effect of applying manure of high copper content to pasture grazed by sheep, has resulted in legislation in the EU to control the amount of copper that may be added to pig foods. The food of young pigs up to 16 weeks of age may contain 175 mg Cu/kg but this must be reduced to 100 mg/kg thereafter, and from six months of age the maximum level is 35 mg/kg.

25.2 Probiotics

In contrast to the use of antibiotics as nutritional modifiers, which destroy bacteria, the inclusion of probiotics in foods is designed to encourage certain strains of bacteria in the gut at the expense of less desirable ones. A probiotic is defined as a live microbial food supplement that beneficially affects the host animal by improving the intestinal microbial balance. Although the digestive tract of all animals is sterile at birth, contact with the mother and the environment leads to the establishment of a varied microflora. The beneficial microorganisms produce enzymes which complement the digestive ability of the host and their presence provides a barrier against invading pathogens. Digestive upsets are common at times of stress (e.g. weaning), and feeding with desirable bacteria, such as *Lactobacilli*, in these situations is preferable to using antibiotics, which destroy the desirable bacteria as well as the harmful species.

It has been suggested that the desirable bacteria exert their effects in a number of ways:

■ Adhesion to the digestive tract wall to prevent colonisation by pathogenic microorganisms. Detrimental bacteria, such as *E. coli*, need to become attached to the gut wall to exert their harmful effects. Attachment is achieved by means of hair-like structures, called fimbriae, on the bacterial surface. The fimbriae are made up of proteins, called lectins, which recognise and selectively combine with specific oligosaccharide receptor sites on the gut wall. *Lactobacilli* successfully compete for these attachment sites, as shown in Fig. 25.1.

■ Neutralisation of enterotoxins produced by pathogenic bacteria which cause fluid loss. There is some evidence that live probiotic bacteria can neutralise these toxins but the active substance has not been identified.

■ Bactericidal activity: *Lactobacilli* ferment lactose to lactic acid, thereby reducing the pH to a level that harmful bacteria cannot tolerate. Hydrogen peroxide is also produced, which inhibits the growth of Gram-negative bacteria. It has also been reported that lactic acid producing bacteria of the *Streptococcus* and *Lactobacillus* species produce antibiotics.

■ Prevention of amine synthesis: coliform bacteria decarboxylate amino acids to produce amines, which irritate the gut, are toxic and are concurrent with the incidence of diarrhoea. If desirable bacteria prevent the coliforms proliferating then amine production will also be prevented.

■ Enhanced immune competence: oral inoculation of young pigs with *Lactobacilli* has elevated serum protein and white blood cell counts. This may aid the development of the immune system by stimulation of the production of antibodies and increased phagocytic activity.

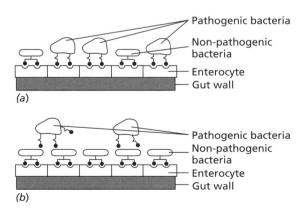

Fig. 25.1 (a) A mixed population of bacteria with substantial attachment of pathogenic bacteria. (b) Competitive exclusion of pathogens due to preferential attachment of non-pathogens. It should be noted that the recognition of receptor sites (carbohydrates) by the bacterial fimbriae (lectins) is very specific to different types of organism. (Adapted from Ewing W N and Cole D J A 1994 *The Living Gut*, Dungannon, Northern Ireland, Context.)

Other postulated effects include beneficial interaction with bile salts, increased digestive enzyme production and more efficient absorption of nutrients, and greater vitamin production.

In a review of the response of pigs of various ages to the administration of probiotics it was concluded that they were effective for young pigs, in which the digestive tract is still developing after weaning. However, they were less effective for growing and finishing pigs, which already have a balanced population of microorganisms. To be effective the desirable microorganism should not be harmful to the host animal, should be resistant to bile and acid, should colonise the gut efficiently, should inhibit pathogenic activity, and should be viable and stable under manufacturing and storage conditions.

In monogastric animals, strains of *Lactobacilli*, *Bacillus subtilis* and *Streptococci* have been used as probiotics. In ruminant animals, the application of yeast (*Saccharomyces cerevisiae*) in the form of live culture, or dead cells with culture extracts, has proved successful in beneficially modifying rumen fermentation.

Yeast cultures stimulate forage intake by increasing the rate of digestion of fibre in the rumen in the first 24 hours after its consumption. Overall digestibility is not affected. It is likely that this improvement in early digestion and intake is brought about by alterations in the numbers and species of microorganisms in the rumen. The precise means by which the effect is

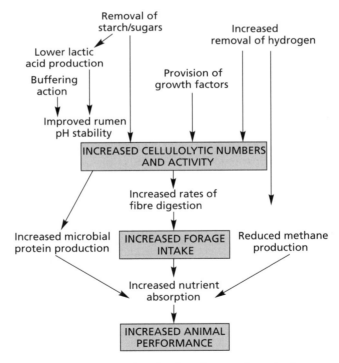

Fig. 25.2 Suggested modes of action of yeast cultures in the rumen. (Adapted from Offer N W 1991 Yeast culture: its role in maximising fibre digestion in the rumen. *Feed Compounder* January: 16–19.)

achieved have not yet been confirmed but there are a number of probable mechanisms (Fig. 25.2).

It is thought that metabolites of dead and live yeast cells (B vitamins, branched chain fatty acids, amino acids and peptides) stimulate the growth of the bacterial species *Megasphaera elsdenii*. This utilises the lactic acid produced from the rapid fermentation of starch and sugars associated with high-concentrate diets. Live yeasts ferment sugars derived from the degradation of starch, thus competing with the lactic acid producing bacteria, and thereby stabilise rumen pH and reduce the risk of acidosis. Live yeast cultures also scavenge oxygen in the rumen, helping to maintain anaerobic conditions and favouring the growth of cellulolytic bacteria. The increase in forage intake results in improved liveweight gain, milk yield and milk fat content. Improved fibre digestion has also been reported in horses when yeast cultures have been given.

25.3 Oligosaccharides

Oligosaccharides (2–20 monosaccharide units – see Chapter 2) have been claimed as beneficial nutritional modifiers for monogastric farm animals. They fall into the group of materials also known as prebiotics, which are defined as compounds, other than dietary nutrients, that modify the balance of the microfloral population by promoting the growth of beneficial bacteria and thereby provide a healthier intestinal environment. Oligosaccharides occur naturally in foods: soya bean meal, rapeseed meal and legumes contain α-galactooligosaccharides (GOS); cereals contain fructooligosaccharides (FOS); milk products have *trans*-galactooligosaccharides (TOS); yeast cell walls contain mannanoligosaccharides (MOS). They are also produced commercially.

It has been suggested that these compounds achieve their beneficial effects in the gut in two ways. First, although they are not easily digested by the host digestive enzymes, compounds such as FOS can be fermented by the favourable bacteria (e.g. *Bifidobacteria* and *Lactobacilli*), giving them a competitive advantage. This shifts the microbial population towards such microorganisms and away from the harmful species.

Second, the gut microbial population may be altered by the oligosaccharide interfering with the attachment of harmful bacteria to the gut wall. As a means of cell recognition all cell types have a unique configuration of carbohydrate-containing compounds (glycoproteins and glycolipids) on their surface. As described above in the section on probiotics, pathogenic bacterial cells have surface compounds called lectins which recognise these carbohydrates and by which they attach to the gut cells. Once attached the bacteria are able to multiply and produce their harmful effects. Species such as *Salmonella* and *E. coli* have a mannose-specific lectin which binds to mannose residues on the gut mucosal surface. By introducing mannose-containing compounds (MOS) into the diet the binding by pathogenic bacteria is disrupted and instead they bind to the oligosaccharide and are carried out

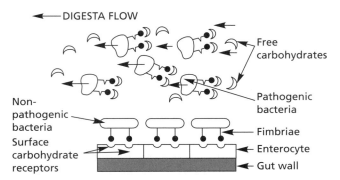

Fig. 25.3 The lectin–carbohydrate combination is specific to a particular organism. However, if the same carbohydrate (e.g. an oligosaccharide) is provided in the diet, harmful bacteria can be encouraged to attach to these and they do not adhere to the gut wall but are excreted without producing toxins. (From Ewing W N and Cole D J A 1994 *The Living Gut*, Dungannon, Northern Ireland, Context.)

of the gut with the passage of the digesta (Fig. 25.3). Yeasts have mannans in the cell wall structure and form the basis of some commercial products that are claimed to act in this way. Indeed the presence of such yeast fragments has been said to be the reason why yeast products are beneficial.

The efficacy of products containing oligosaccharides is currently the subject of active experimentation. There can be no guarantee that an oligosaccharide will favour the growth of beneficial species in a complex microflora such as that found in the pig's intestine. Experiments have shown that piglets given an oral challenge of *E. coli* responded to GOS with a reduced pH of ileal digesta and reduced population of coliforms. Supplements of FOS and TOS have reduced the numbers of aerobic bacteria in the gut of weaned piglets and there are reports of a reduced incidence of diarrhoea. Under farm conditions, improvements in gain and food conversion efficiency of the order of 4–6 per cent have been recorded. In other experiments reduced digesta pH has been reported but without a detectable change in the composition of the microflora, microbial metabolites or production responses. These conflicting results may have arisen because the diet already contained some oligosaccharides or because experimental conditions tend to be less stressful than those on farms.

25.4 Enzymes

The mammalian digestive enzymes were described in Chapter 8. Not all compounds in foods are broken down by these enzymes so some potential nutrients are unavailable. The chemical structure of plant cells (fibre), the limited time available for activity in certain parts of the gut and the presence of antinutritive compounds in some foods hinder the release of nutrients. Early experiments with supplemental enzymes to overcome these limitations yielded variable responses, mainly because they were impure materials. As a result of advances in biotechnology, more effective enzyme preparations can now be produced in large quantities and relatively inexpensively. Therefore, supplementation of the diet as a means of improving nutritive value is becoming commonplace.

The enzymes used as food additives act in a number of ways. First, they can improve the availability of plant storage polysaccharides (e.g. starch), oils and proteins, which are protected from digestive enzymes by the impermeable cell-wall structures. Thus cellulases can be used to break down cellulose, which is not degraded by endogenous mammalian enzymes. Even in ruminant animals, in which the rumen microflora possess cellulase, fibrolytic enzymes have been used but responses have been variable owing to the complex nature of the rumen fermentation system and other factors such as lignin encrustation of cellulose. Responses in pigs and poultry of the order of 5–10 per cent improvement in liveweight gain and 10 per cent improvement in food conversion efficiency have been recorded. Supplementation of a wheat by-product diet with cellulase increased the ileal digestibility of non-starch polysaccharides from 0.192 to 0.359 and crude protein from 0.65 to 0.71.

Second, enzymes are employed to destroy materials that interfere with the digestion, absorption and utilisation of nutrients. The major application in this area has been to break down carbohydrate fractions in cereal cell walls, the β-glucans and arabinoxylans, which are resistant to attack by digestive enzymes. These materials hinder the digestion and absorption of other nutrients by forming a viscous gum in the digestive tract. In pigs, it has been suggested that the increased viscosity brought about by β-glucans in the stomach contents may affect the natural sieving of particles. Thus large particles are suspended in the viscous digesta and pass through to the duodenum instead of falling to the bottom of the stomach, resulting in less efficient digestion. It is also possible that the increase in viscosity disturbs peristalsis and pancreatic secretion. In poultry, the viscous nature of the digesta results in poor performance and sticky droppings, which present a problem in the management of litter waste. Supplements of β-glucanase to barley-based diets improved the rate of gain and food conversion efficiency of broilers to values similar to those for wheat- or maize-based diets (Table 25.1). The β-glucanase also reduced litter problems and improved the metabolisable energy of the diet.

Supplementation of pig foods improves ileal and whole tract digestibility of nutrients. In breaking down the β-glucans, the enzymes reduce the viscosity of the digesta thus allowing better movement of endogenous enzymes through the mass and more efficient digestion and absorption of nutrients. It is still not clear whether viscosity *per se* is responsible for the effects or if it is an indicator of conditions pertaining in the gut lumen which cause the problems. The response differs between pigs and poultry because of differences in digestive physiology between the two species (see Box 25.2).

Phytase is in this category of enzyme. It releases the orthophosphate groups from phytic acid, which is only partially broken down by non-ruminants and is the major form of phosphorus in cereal grains and oilseeds (see p. 120). This results in a greater availability of phosphorus to the animal and the amount of

Table 25.1 Effect of β-glucanase addition to hulless barley (Scout) on food consumption, weight gain and food to gain ratio of broilers (Adapted from Rotter B A, Marquardt R R and Guenter W 1989 Enzymes in feed: they really can be made to work. *Alltech European Lecture Tour*, February–March 1989)

(a) Broilers 0–6 weeks

Diet type	Intake (g)	Gain (g)	Gain: food
Wheat	2910	1529	0.525
Barley	2941	1473	0.501
Barley + enzyme	2992	1638	0.547

(b) Broilers at 3 weeks

Diet type	Cumulative intake (g)	Bodyweight (g)	Gain: food
Maize	1002	756	0.713
Barley	987	593	0.569
Barley + enzyme	1001	746	0.705

inorganic phosphorus added to the diet can be reduced with beneficial effects on the environment through reduced phosphorus excretion.

In the third area of enzyme application, the aim is to supplement the enzyme complement of young animals, in which the rate of endogenous enzyme production may be limiting. Early weaned pigs have limited amylase, protease and lipase activity (see Chapter 8) and enhancement of the extent of digestion of nutrients would improve performance and reduce the incidence of the diarrhoea that results from undigested nutrients reaching the hind gut and being fermented by bacteria.

Finally, enzymes can increase the efficiency of utilisation of nutrients, in monogastric animals, by releasing them for uptake from the small intestine rather than allowing fermentation in the hind gut, which results in products of lower value to the animal (e.g. volatile fatty acids).

Box 25.2
Enzyme action in pigs and poultry

Enzymes have found major commercial application in pig and poultry diets and are extensively used for the latter. Differences in efficiency have been reported between the species, the effect generally being greater in poultry; this is possibly due to the differences in gut physiology between the two species. There is a greater reduction in the viscosity of digesta in the small intestine by enzymes in poultry, probably because of the lower water content than in pigs. The longer retention time in the stomach and acidic nature of stomach contents in the pig (around pH 3) is more destructive to enzymes than the milder conditions in the crop (pH 4–5). New commercial enzymes have been selected for their pH stability. Finally, pigs have a longer small intestine than poultry and the resultant increased retention time allows more time for the endogenous enzymes to act. Thus there is less opportunity to detect the effects of supplemental enzymes in pigs.

Table 25.2 Enzyme supplementation of diets

Diet constituent	Problem component	Enzyme
Barley and oats	β-glucan: glucose β-1;4 linked spine with β-1;3 side linkages; sequence of linkages varies between barley and oats and between cultivars	β-glucanase
Wheat (also rye and triticale)	Arabinoxylans: β-1;4 xylopyranose residues spine with terminal arabinofuranosyl substitutes	Pentosanase or xylanase
Fibrous materials	Cellulose	Cellulase
Soya bean and soya products	Limited digestion of protein in young animals	Proteinases
Grains and seeds	Low availability of phytin phosphorus in monogastrics	Phytase

In order to be effective when incorporated into the animal's diet, enzymes must survive storage at ambient temperature, the manufacturing process (heating and pelleting) and wide fluctuations in pH in the gut, be resistant to intestinal proteases and have specific activity on feed components in the upper digestive tract. The enzyme should be selected on the basis of its target substrate. Commercial enzymes are derived from a wide range of sources, particularly fungi such as *Trichoderma longibrachiatum*, *Aspergillus niger* and *Humicola insolens*. A summary of the application of enzymes is given in Table 25.2.

25.5 Organic acids

Efficient digestion of food in young pigs depends on the secretion of sufficient acid in the stomach. Acid additives have been used mainly in the diets of early weaned pigs. Here their use relates to the digestive physiology of the young pig and the replacement of sow's milk with solid food at weaning. The suckled piglet obtains nutrients on a little-and-often basis, sucking about once every hour. The milk clots in the stomach, and the limited secretion of hydrochloric acid is augmented by lactic acid produced from the fermentation of lactose by *Lactobacilli*. This ensures that the pH falls to a level that favours efficient protein hydrolysis and suppresses harmful bacteria. Thereafter there is a steady release of nutrients from the clot to the small intestine. The solid food presented at weaning is less digestible than milk, is consumed in larger, less frequent meals and has a higher buffering capacity. The limited hydrochloric acid production is not sufficient to ensure that the pH falls to a level which prevents the growth of pathogenic bacteria. In addition the nutrients may overload the immature digestive and absorptive capacity of the small intestine and their fermentation in the hind gut results in diarrhoea.

Box 25.3
Antimicrobial action of organic acids

Organic acids change from the undissociated to the dissociated form, depending on the pH of the environment. In the undissociated form they can diffuse rapidly through the semipermeable membrane into the cytoplasm of the microorganism. Once inside, at a pH of around 7, the acid becomes dissociated and acts by suppressing the cellular enzyme and transport systems. Acids with a high pKa (which are 50 per cent dissociated), longer chain length acids and unsaturated acids are the most effective.

In order to overcome these problems acids have been added to weaning diets to augment gastric HCl, giving a rapid fall in pH with beneficial effects on protein digestion and the gut microflora. Although inorganic acids have been used to supply hydrogen ions, and hence reduce pH, organic acids have additional desirable properties. In their salt form they are odourless and easy to handle; they lower the pH and acid binding capacity of the food; the anion has an antimicrobial effect on moulds and bacteria in the food (see Box 25.3), thereby improving food quality. The anion has been reported to act as a complexing agent for cations, such as Ca^{++} and Mg^{++}, thus improving their retention. It is also suggested that the presence of short chain fatty acids in the small intestine reduces the damage to the villi of the gut wall which is associated with weaning. Finally, organic acids, once absorbed by the pig, can be used as substrates in intermediary metabolism, principally as energy sources.

It has proved difficult to obtain experimental evidence for all of these purported effects of organic acids. In an extensive review Partenen and Mroz (1999) examined the evidence for their modes of action. Only a few studies have demonstrated a reduction in gastric pH, but the methodology of sampling is difficult given diurnal variations, differing proportions of food and secretions, and timing of sampling. A study is required with permanent fixed pH electrodes within the stomach. There is evidence of a positive effect on the ileal digestibility of amino acids and the retention of nutrients in growing pigs. The magnitude of the effect depends on the acid used and its concentration, the age of the pig and the composition of the diet. With regard to minerals, the absorption of calcium and phosphorus has been improved but variable results have been obtained for retention of these minerals. Addition of organic acids has had beneficial effects on microbial counts and species in some studies but there has been little work on the effects on gut morphology to test the theory of an improvement of the structure of the villi. Finally, in some studies, the addition of formic, fumaric and citric acids to the diet has increased the activity of enzymes associated with intermediary metabolism, indicating that the acids do contribute to nutrient metabolism.

Although the exact modes of action of the acids are not clear, their addition to pig diets has proved beneficial in terms of nutrient digestibility, growth and food conversion efficiency (Table 25.3). This is so, particularly

Table 25.3 Effect of graded levels of fumaric acid on the growth rate and food conversion efficiency of young pigs (10.0–18.7 kg) (Adapted from Easter R A 1988 Acidification of diets for pigs. In: Haresign W and Cole D J A (eds) *Recent Advances in Animal Nutrition – 1988*: London, Butterworth)

	Fumaric acid (%)				
	0	1	2	3	4
Diet pH	5.96	4.77	4.33	3.98	3.80
Liveweight gain, g/day	261	261	257	296	297
Food intake, g/day	501	484	445	493	493
Gain : food, kg gain/kg food	0.52	0.54	0.57	0.60	0.60

for newly weaned pigs, when coupled with good feeding management procedures.

Formic and propionic acids are more effective than fumaric or citric acids at the same rate of inclusion because the former have a lower molecular weight. Suggested levels of inclusion of acid (kg/tonne diet) are: formic 6 to 8; propionic 8 to 10; fumaric 12 to 15; citric 20 to 25, but recommendations vary.

25.6 Modifiers of rumen fermentation

Rumen fermentation processes are usually kept within narrow bounds and, as they form an integrated system, any changes made have more than one outcome. Thus modification of the processes to one desired target is difficult. Nevertheless, there are products which are used to modify the fermentation to achieve specific results (see also Chapter 8).

The use of *antibiotics* and *probiotics* was discussed above. Other rumen modifiers include *buffers*, *methane inhibitors* and *bloat-preventing substances*.

Buffers are included in ruminant diets to regulate rumen pH to levels that favour the activity of cellulolytic organisms (pH 6–7). Diets rich in readily fermentable carbohydrate create acidic conditions and increase lactic acid formation, both of which are detrimental to the cellulolytic bacteria. This reduces food intake and predisposes the animal to acidosis and laminitis. Secondary problems include reduced milk fat production, rumenitis, ketosis and liver abscesses. Chemicals such as sodium bicarbonate, sodium carbonate, calcium carbonate and magnesium oxide buffer the hydrogen ions and increase the dilution rate of the liquids in the rumen. This increases the efficiency of microbial protein synthesis and, as a result of the shorter retention time, allows starch and protein to escape to the intestines. The activity of cellulolytic bacteria and the proportion of acetate in the rumen volatile fatty acids are increased. In the USA it has become common to add up to 200 g $NaHCO_3$/day, and sometimes MgO in addition, to the rations of cows in early lactation. It has been suggested that, since bicarbonate buffers are expended once the CO_2 is released, it would be advantageous to use certain clays and bentonites, which contain insoluble and hence undegradable anions and would allow the buffering action to be carried further down the digestive tract.

Antibiotics such as virginiamycin and monensin have been used to select against bacteria that produce lactic acid. Certain probiotics encourage the uptake of lactic acid (see above).

The production of methane in the rumen is a wasteful process (up to 10 per cent of the gross energy) and contributes to the earth's so-called 'greenhouse gases'. Halogen compounds, particularly bromide and chloride homologues of methane, are toxic to the bacteria which produce methane. Suppression of methane production results in a reduction in the NAD/NADH ratio, which is unfavourable to anaerobic fermentation. However, the accompanying increase in propionate and butyrate production uses up the excess hydrogen, re-establishing the NAD/NADH ratio. Long chain polyunsaturated fatty acids also reduce methane production. Some studies have shown partial adaptation by the rumen microflora to these additives.

Under normal conditions rumen fermentation produces gases which, by the action of ruminal and reticular contractions, accumulate in the area known as the cardia, from which they are passed up the oesophagus in the process of eructation. Bloat occurs if the gas becomes trapped and the animal is not able to clear it. Rumen stasis due to acidosis is one of the causes of free gas bloat but on certain pastures, particularly those rich in legumes, and on high-concentrate (feedlot) diets the fermentation process in the rumen can result in rumen gases being trapped within fluid bubbles. Therefore, the gas cannot be eructated as normal and pressure builds up in the rumen. The condition is known as frothy bloat. In the case of legume bloat the frothing is due to chemicals associated with the plant. With high-concentrate diets the normal function of the rumen is disrupted by the rapid accumulation of fermentation acids and the release of bacterial mucopolysaccharides during cell lysis, which result in increased viscosity of the rumen fluid and the formation of the stable foam. In the case of pasture bloat there are management practices (e.g. strip grazing or use of grass/clover mixtures) to reduce its incidence but often these are insufficient. Antibiotics have been used to change the end-products of fermentation in order to produce fewer froth-forming materials. However, specific anti-foaming agents are more commonly used. These reduce the surface tension of the rumen fluid, preventing the formation of the stable foam. Vegetable oils, lecithin, animal fats, mineral oils, detergents (pluronics) and the synthetic polymer poloxalene fall into this category of agent. The additive alkylarylsulphonate prevents the release of pectin derivatives, which cause frothing. The main problem with such additives is the maintenance of an adequate concentration in the rumen, since liquids quickly pass through. Spraying the forage with oil has proved successful. In New Zealand, a product containing pluronic detergents and alcohol ethoxylate is used for pasture bloat but its effectiveness, and that of other pasture bloat remedies, for feedlot bloat is not known. For the latter type of bloat, the effectiveness of the ionophore antibiotics has been partially attributed to the lower absolute amount and daily variation of food intake. The addition of sodium chloride at 40 g/kg diet DM has been suggested as a preventive measure as this increases the rate of passage of fluids from the rumen, but it also depresses food intake.

Summary

1. Food additives are administered to the animal to improve the effectiveness of nutrients.

2. Antibiotics reduce the number of bacteria in the gut and increase the availability of nutrients to the animal. The widespread use of antibiotics has resulted in populations of bacteria that are resistant to many antibiotics and in several countries their use for growth promotion is restricted or forbidden.

3. Probiotics are live microbial supplements, which benefit the animal by improving the intestinal microbial balance. The beneficial microorganisms displace pathogenic bacteria and produce enzymes which complement the digestive ability of the host.

4. Oligosaccharides may be fermented by favourable bacteria, thereby shifting the population towards such microorganisms and away from harmful species. They also interfere with the attachment of harmful bacteria to the gut wall.

5. Exogenous enzymes are added to the animal's food to supplement its own digestive enzymes and to break down antinutritive fractions in foods.

6. The diets of young pigs may include organic acids, which reduce gut pH, with beneficial effects on protein digestion and control of the gut microflora.

7. Products which are used to modify rumen fermentation include buffers, which regulate rumen pH and favour the activity of cellulolytic bacteria, compounds to suppress methane production, and bloat-preventing compounds, which prevent the build-up of gas trapped in foam in the rumen fluid.

Further reading

Bedford M R and Schulze H 1998 Exogenous enzymes for pigs and poultry. *Nutrition Research Reviews* **11**: 91–114.

Ewing W N and Cole D J A 1994 *The Living Gut*, Dungannon, Northern Ireland, Context.

Jouany J P 1994 Methods of manipulating the microbial metabolism in the rumen. *Annales de Zootechnie*, **43**: 49–62.

Partenen K H and Mroz Z 1999 Organic acids for performance enhancement in pigs. *Nutrition Research Reviews* **12**: 117–45.

Appendix 1: Solutions to problems

10.1 Protein intake $= 2 \times 150 = 300$ g/day
Protein excretion $= 0.4 \times 175 = 70$ g/day
Digestibility $= 100(300 - 70)/300 = 77\%$

10.2 20 g of chromic oxide in the measured quantity of faeces collected (0.4 kg DM) would give a concentration of 50 g/kg, hence the actual and estimated quantities of faeces DM were in agreement.

10.3 Protein intake from soya bean meal $= 0.3 \times 450 = 135$ g/day
Hence protein intake from cereal $= 300 - 135 = 165$ g/day
Protein in faeces from soya bean meal $= 0.15 \times 135 = 20.2$ g/day
Hence protein in faeces from cereal $= 70 - 20.2 = 49.8$ g/day
Therefore digestibility of cereal protein $= 100(165 - 49.8)/165 = 69.8\%$

11.1 $ME = GE - \text{(energy in faeces, urine and methane)}$
$= (1.2 \times 19) - (6.0 + 1.56 + 1.80)$
$= 13.44$ MJ or $13.44/1.2 = 11.2$ MJ/kg

11.2 $HP = 8672 + 2214 - 112 - 111 = 10\ 660$ kJ (10.66 MJ)
$ER = 13.44 - 10.66 = 2.78$ MJ

11.3 Protein stored $= 10.4 \times 6.25 = 65$ g
Protein C stored $= 65 \times 0.512 = 33.28$ g
Fat C stored $= 182.5 - 33.28 = 149.22$ g
Fat stored $= 149.22/0.746 = 200$ g
Energy stored $= 65 \times 23.6 + 200 \times 39.3 = 9.39$ MJ

11.4 ME for maintenance $= 42/0.7 = 60$ MJ/day
ME for growth $= 8 \times 11 - 60 = 28$ MJ/day
$k_g = 12/28 = 0.43$

12.1 NE required (MJ/day)

Maintenance	2.4
Growth (0.2×15.6)	3.12
Total	5.52

Food and interface data

	Forage	Concentrate
ME (MJ kg/DM)	9	13
q_m	0.49	0.71
k_m	0.674	0.752
k_g	0.388	0.560
k_{mp}	0.476	0.630
NE (MJ/kg DM)	4.28	8.19

Net energy supply (MJ per head per day)
  From hay $= (0.5 \times 4.28) = 2.14$
  From concentrate $= (5.52 - 2.14) = 3.38$
Concentrate required $= (3.38/8.19) = 0.413$ kg per head per day

12.2 (a) $DE = 17.47 + (0.0079 \times 160) + (0.0158 \times 50) - (0.0331 \times 60) - (0.0140 \times 150)$
            $= 15.44$ MJ/kg DM

  (b) DE available (MJ/day)

For maintenance	7
For gain (15.44 − 7)	8.44
ME for gain (MJ/day) (8.44 × 0.96)	8.10
k values	
k_p	0.56
k_f	0.74
Mean $[(170 \times 0.56 + 200 \times 0.74)/370]$	0.657
Energy retained (MJ/day) (8.10 × 0.657)	5.32

Energy value of gain (MJ/kg) $[(170 \times 23.6 + 200 \times 39.3)/1000] = 11.87$
Gain (kg/day) (5.32/11.87) = 0.45

Appendix 2: **Notes on tables**

The data given in these tables have been compiled from a number of sources, a full list of which is given at the end of the appendix. Absence of figures does not imply a zero but merely that the information was not given in these sources.

Composition and nutritive value, Tables A2.1 and A2.2

The composition of a particular food is variable, and figures given in these tables should be regarded only as representative examples and not constant values. For more comprehensive data readers should consult the references.

Nutrient allowances, Tables A2.3–A2.13

The scientific feeding of farm animals is based on standards expressed in terms of either 'nutrient requirements' or 'nutrient allowances'. These terms are defined in Chapter 14. The figures in the tables are mainly nutrient requirements as they do not include safety margins. Only the figures for vitamins have safety margins. It has not been possible to include every class of farm animal in these tables; only a representative selection is given. For more detailed information readers should consult the relevant references.

Abbreviations used in the tables

ADF	Acid-detergent fibre
ADIN	Acid-detergent insoluble nitrogen
av.	Available
CF	Crude fibre
CP	Crude protein
D	Digestible organic matter in dry matter
DE	Digestible energy
dec.	Decorticated
DM	Dry matter
DMI	Dry matter intake
DUP	Digestible undegradable protein
EE	Ether extract

ERDP   Effective rumen-degradable protein
exp.   Expeller
extr.   Extracted
FME   Fermentable metabolisable energy
ME   Metabolisable energy
MP   Metabolisable protein
NDF   Neutral-detergent fibre
q_{m} Metabolisability of gross energy
W Liveweight

Table A2.1.1 Chemical composition of foods

Food	DM g/kg	Dry matter basis (g/kg)							
		CF	EE	Ash	CP	NDF	ADF	ADIN	Starch and sugar
Green crops									
Barley in flower	250	316	16	64	68				
Cabbage	150	160	47	107	160	244	136		320
Clover, red, early flowering	190	274	37	84	179				
Clover, white, early flowering	190	232	42	116	237	400	253		84
Grass, young (75–80 D)	200	130	55	105	156			1.3	
Grass, mature (60–65 D)	282	200	40	100	100			0.7	
Kale	140	179	36	136	157	243	197	2.3	284
Lucerne, early flowering	240	300	17	100	171				
Maize	190	289	26	63	89				
Rape	140	250	57	93	200				
Sugar beet tops	160	100	31	212	125				
Sugarcane	279	312	22	57	97				
Swede tops	120	125	42	183	192				
Silages									
Barley, whole crop	324	248	15	153	64	575	274	2.2	267
Grass, young	250	270	52	91	186			1.3	
Grass, mature	294	340	52	110	125			0.5	
Lucerne	250	296	84	100	168	495	406	1.7	16
Maize, whole crop	210	233	57	62	110	480	277	2.7	211
Potato	270	26	19	52	81				
Hays									
Clover, red	850	266	39	84	184			2.2	
Grass, poor quality	800	380	16	70	55	741	452	1.2	
Grass, good quality	900	298	18	82	110	650	364	0.5	
Lucerne, early flowering	850	302	13	95	225	493	375	2.1	
Dried herbage									
Grass	929	210	38	102	190	541	282	2.3	148
Lucerne	900	247	51	100	220	465	336	2.0	81
Straws									
Barley	860	394	21	53	38	811	509	1.0	32
Barley, ammoniated	871	450	15	46	70	778	542	2.0	20
Bean	860	501	9	53	52	778	542		19
Oat	860	394	22	57	34	749	523	0.6	20
Oat, ammoniated	843	431	18	66	75	735	522	1.0	16
Pea	860	410	19	77	105				
Wheat	860	417	15	71	34	809	502	0.8	
Wheat, ammoniated	869	434	13	56	68	773	544	1.5	13

Table A2.1.1 (cont.)

Food	DM g/kg	CF	EE	Ash	CP	NDF	ADF	ADIN	Starch and sugar
Roots and tubers									
Artichoke, Jerusalem	200	35	10	55	75				
Cassava	370	43	9	30	35	114	63		692
Fodder beet	183	56	3	81	63	136	72	0.9	660
Mangels	120	58	8	67	83				
Potatoes	210	38	5	43	90	73	44	1.6	638
Sugar beet pulp, dried	900	203	7	34	99	372	213	1.6	82
Sugar beet pulp, dried molassed	876	132	4	88	110	321	179	0.9	300
Sugar beet molasses	750	0	0	69	47				632
Sugarcane molasses	737	0	4	100	55				657
Swedes	120	100	17	58	108	140	125	0.2	587
Sweet potatoes	320	38	16	34	39			0.2	
Turnips	90	111	22	78	122			1.6	
Cereals and by-products									
Barley	860	53	17	26	108	201	64	0.4	599
Barley, brewer's grains	263	176	77	41	234	618	264	3.9	51
Barley, malt culms	900	156	22	80	271	463	163		171
Brewer's yeast, dried	900	2	11	102	443				
Grain distiller's grains	250				265	383		15.8	
Grain distiller's dark grains (maize)	890	89	55	52	351	303	193	14.3	108
Malt distiller's grains	248	199	86	34	211	673	694		23
Malt distiller's dark grains	907	121	67	60	275	420	175		65
Maize	860	24	42	13	98	117	28	1.3	717
Maize, flaked	900	17	49	10	110				
Maize gluten feed	900	39	38	28	262	383	114	1.4	210
Maize gluten meal	904	12	29	11	669	84	105	6.4	158
Millet	860	93	44	44	121				
Oats	860	105	49	33	109	310	149	0.4	482
Oats, naked	854	45	97	23	119	114	42	0.5	597
Oat husks	900	351	11	42	21				
Pot ale syrup	483	2	2	95	374	6			36
Rice, brown	907		23	9	111				
Rice, bran (extracted)	905		11	149	166	451	275	1.4	256
Rice, polished	860	17	5	9	77				
Rye	860	26	19	21	124	357			
Sorghum	860	21	43	27	108	107	57		745
Wheat	860	26	19	21	124	124	30	0.4	701
Wheat bran	880	114	45	67	170	475	137		259
Wheat feed	880	74	45	50	178	363	111	0.4	333
Wheat germ meal	889		82	48	279				

Table A2.1.1 (cont.)

Food	DM g/kg	Dry matter basis (g/kg)							
		CF	EE	Ash	CP	NDF	ADF	ADIN	Starch and sugar
Oilseed by-products									
Coconut meal	900	153	76	72	220			2.5	
Cottonseed meal, undecorticated	900	248	54	66	231			3.2	
Cottonseed meal, decorticated	900	87	89	74	457			2.0	
Groundnut meal, undecorticated	900	273	21	47	343				
Groundnut meal, decorticated	900	88	8	63	552	180	146	2.0	
Linseed meal	900	102	36	73	404	192	131	2.0	95
Palm kernel meal	900	167	10	44	227	693	470	3.0	51
Rapeseed meal	899	152	29	80	400	295	206	3.6	147
Soya bean meal	900	58	17	62	503	125	91	2.2	124
Soya bean meal, full fat	898	48	222	54	415	122	82		91
Sunflower meal, undecorticated	900	323	80	50	297			2.5	
Sunflower meal, decorticated	900	134	152	117	430			2.0	
Leguminous seeds									
Beans, field	860	80	15	36	275	168	123	0.5	412
Gram	860	57	13						
Peas	860	63	19	32	261	116	76	4.8	497
Animal by-products									
Blood meal	870		9	40	919			0	
Feather meal, hydrolysed	907	5	64	25	892				2
Fishmeal, UK-produced	915		69	238	699			0	
Fishmeal, herring	920		75	122	793			0	
Fishmeal, South American	900		60	197	733			0	
Meat meal	900	0	148	42	810			0	
Meat and bone meal	900	0	50	62	597			0	
Milk, cow's whole	128	0	305	55	266			0	
Milk, skim	100	0	70	80	350			0	
Milk, whey	66	0	30	106	106			0	

Table A2.1.2   Mineral contents of foods (DM basis)

Food	Calcium (g/kg)	Phosphorus (g/kg)	Magnesium (g/kg)	Sodium (g/kg)	Copper (mg/kg)	Manganese (mg/kg)	Zinc (mg/kg)	Cobalt (mg/kg)	Selenium (mg/kg)
Green crops									
Grass, close grazing	5.0	3.5	1.7	1.9	8.0	–	–	0.10	0.05
Grass, extensive grazing	4.8	2.8	1.7	1.7	7.0	16	5.0	0.08	0.04
Kale	21.0	3.2	2.5	2.0	4.5	38	–	0.10	0.05
Lucerne, late vegetative	21.9	3.3	2.7	2.1	11.0	41	–	0.17	–
Turnip tops	24.2	3.1	2.8	3.1	8.0	–	–	0.08	0.06
Silages									
Cereal, vegetative	4.0	2.7	1.0	1.8	6.0	80	25	0.07	0.06
Grass, early	8.0	4.0	3.0	3.0	11.0	90	25	–	0.10
Grass, mature	3.0	2.0	0.9	1.0	3.0	94	30	0.05	0.02
Hays									
Clover	15.3	2.5	4.3	1.9	11.0	73	17	0.16	–
Grass, poor quality	2.5	1.5	0.8	1.0	2.0	70	17	0.05	0.01
Grass, good quality	7.0	3.5	2.5	2.5	9.0	100	21	0.20	0.07
Lucerne, mature	11.3	1.8	2.7	0.8	14.0	44	24	0.09	–
Straws									
Barley	4.5	0.7	0.8	1.1	3.2	84	16	0.04	0.04
Oat	4.0	0.7	1.3	3.7	4.0	69	29	0.04	0.02
Roots and tubers									
Cassava, dried	2.0	1.0	–	0.2	–	20	–	–	–
Mangels	2.9	2.1	5.3	9.9	9.4	–	–	0.09	0.03
Potatoes	1.0	2.1	1.0	0.5	4.5	42	28	0.06	0.03
Sugar beet pulp, molassed, dried	5.7	0.8	2.4	2.5	11.0	51	32	0.10	0.02
Swedes	3.6	3.2	1.2	2.6	3.8	21	19	0.07	0.03
Turnips	5.0	3.6	1.4	2.2	2.7	35	36	0.04	0.03
Cereals and by-products									
Barley	0.5	4.0	1.3	0.2	4.8	18	19	0.04	0.02
Brewer's grains, dried	3.2	7.8	1.8	0.4	25.0	50	–	0.03	–
Brewer's yeast	1.3	15.1	2.5	0.8	35.3	6	42	–	–
Distiller's grains, malt	1.7	3.7	1.4	0.9	10.0	–	–	0.02	0.02
Maize	0.3	2.7	1.1	0.2	2.5	6	16	0.02	0.02
Maize gluten meal	1.6	5.0	0.6	1.0	30.0	8	190	0.08	–
Millet	0.6	3.1	1.8	0.4	24.4	32	16	0.04	–
Oats	0.8	3.7	1.3	0.2	3.6	42	41	0.04	0.03
Oat feed	1.5	2.9	1.0	0.2	3.9	–	–	0.04	0.03
Rice	0.7	3.2	1.5	0.6	3.0	20	17	0.05	–
Rye	0.7	3.7	1.4	0.3	8.0	66	36	–	–

Table A2.1.2 (cont.)

Food	Calcium (g/kg)	Phosphorus (g/kg)	Magnesium (g/kg)	Sodium (g/kg)	Copper (mg/kg)	Manganese (mg/kg)	Zinc (mg/kg)	Cobalt (mg/kg)	Selenium (mg/kg)
Sorghum	0.5	3.5	1.9	0.4	10.8	16	15	0.14	–
Wheat	0.5	3.5	1.2	0.1	5.0	42	50	0.05	0.02
Wheat bran	1.6	13.6	5.0	0.4	12.9	143	189	0.03	0.40
Wheat feed	1.1	8.0	3.3	0.4	17.5	–	–	0.03	0.04
Oilseeds and by-products									
Coconut meal	2.3	6.6	2.8	0.4	20.4	59	–	0.14	–
Cottonseed meal, dec.	1.9	12.4	5.0	0.6	16.0	25	79	0.05	–
Groundnut meal, dec.	2.9	6.8	1.7	0.8	17.0	29	22	0.12	–
Linseed meal	4.1	8.6	5.8	0.7	25.0	42	–	0.55	0.91
Soya bean meal	3.5	6.8	3.0	0.4	25.0	32	61	0.20	0.55
Leguminous seeds									
Beans	1.0	5.5	2.0	0.1	14.0	16	46	0.20	–
Peas	1.5	4.4	1.4	0.5	–	–	33	–	–
Animal by-products									
Fishmeal	79.0	44.0	3.6	4.5	9.0	21	119	0.14	2.00
Meat and bone meal	120	58.0	2.5	7.2	24.0	–	–	0.20	0.20
Whey, dried	9.2	8.2	1.4	7.0	50.0	6	3	0.13	–

Table A2.1.3 Amino acid composition of foods (g/kg) (fresh basis)

Food	DM (g/kg)	Nitrogen (g/kg)	Arginine	Cystine	Glycine	Histidine	Isoleucine	Leucine	Lysine	Methionine	Phenylalanine	Serine	Threonine	Tryptophan	Tyrosine	Valine
Green crops																
Dried grass	897	23.5	7.6	1.7	7.5	2.9	5.8	10.9	7.1	3.0	7.1	6.1	6.5	1.2	4.8	4.9
Dried lucerne	–	35.7	10.9	2.5	10.2	4.7	9.3	16.1	11.7	2.8	10.5	9.0	9.2	1.6	8.2	11.3
Cereals and by-products																
Barley	856	15.6	5.4	2.2	4.1	4.1	3.5	6.9	3.8	2.1	5.0	4.3	3.4	1.0	3.4	5.1
Brewer's yeast, dried	930	71.0	21.9	5.0	21.9	10.7	21.4	31.9	32.3	7.0	18.1	–	20.6	4.9	14.9	23.2
Distiller's dark grains	900	39.4	10.1	4.3	10.6	4.5	8.5	15.5	9.6	4.3	8.9	9.3	8.8	2.1	7.5	11.7
Distiller's solubles	–	42.9	3.8	2.4	12.9	4.0	8.0	13.0	6.8	3.4	7.7	6.4	6.0	3.6	8.5	12.8
Maize	852	13.5	4.3	1.9	3.3	2.6	3.0	11.1	2.5	2.3	4.5	4.3	3.2	0.4	3.9	4.3
Maize gluten meal	–	106.2	24.1	12.6	17.4	14.0	28.4	117.7	10.8	24.5	41.0	37.7	24.0	2.6	34.7	33.0
Oats	869	16.8	7.0	4.0	5.7	2.3	3.7	7.3	4.5	2.6	5.1	5.7	3.7	0.7	4.1	5.1
Rice, polished, broken (brewer's rice)	890	13.9	6.2	0.8	6.3	1.7	3.5	5.2	2.4	1.5	3.6	13.6	2.9	1.3	4.1	5.0
Sorghum	870	14.1	3.4	1.6	3.5	1.9	4.2	11.8	2.1	1.6	4.2	3.9	2.9	1.0	3.8	5.3
Wheat	858	16.2	5.2	2.3	4.1	2.5	3.5	7.1	3.1	2.1	4.8	4.8	3.1	1.2	3.3	4.5
Wheat feed	858	22.6	10.2	3.4	7.8	4.0	4.7	9.5	6.4	3.2	6.1	6.5	4.0	2.2	4.6	7.1

Table A2.1.3 (cont.)

Food	DM (g/kg)	Nitrogen (g/kg)	Arginine	Cystine	Glycine	Histidine	Isoleucine	Leucine	Lysine	Methionine	Phenylalanine	Serine	Threonine	Tryptophan	Tyrosine	Valine
Oilseed by-products																
Cottonseed meal	900	66.2	45.9	6.4	17.0	11.0	13.3	24.1	17.1	5.2	22.2	–	13.2	4.7	10.2	18.9
Groundnut meal	897	75.5	57.0	5.6	26.4	11.3	15.7	29.9	16.4	5.6	25.1	23.1	13.5	3.0	19.9	20.7
Lupinseed meal	–	60.8	42.7	6.2	13.6	8.5	16.5	26.8	17.0	3.0	13.1	17.4	12.1	1.8	18.5	14.6
Rapeseed meal	899	50.0	23.2	7.6	18.5	9.9	14.2	25.9	21.5	7.9	14.3	16.3	16.8	1.7	11.5	18.6
Soya bean meal	861	70.9	35.3	6.0	19.5	12.6	20.3	35.0	28.5	7.9	23.0	23.5	17.9	5.5	17.7	22.2
Sunflower seed meal	–	44.5	23.1	4.6	15.6	7.2	11.6	18.5	10.1	7.6	13.4	11.9	10.4	1.4	8.1	14.3
Leguminous seeds																
Beans (*Vicia faba*)	–	39.8	22.2	3.9	10.5	6.1	9.7	18.3	15.8	1.8	10.1	11.7	9.1	1.6	8.9	11.2
Peas (*Pisum sativum*)	–	31.4	17.1	3.0	8.7	5.3	8.2	14.4	15.2	2.5	8.9	9.5	8.0	0.9	7.2	9.2
Animal by-products																
Fishmeal	918	100.0	40.5	6.7	50.6	14.1	26.1	44.6	48.2	15.2	28.8	28.3	24.9	6.9	21.4	30.7
Meat and bonemeal	957	73.3	32.4	2.7	70.6	7.7	11.6	26.1	22.0	6.5	14.6	16.0	14.3	2.1	9.6	19.6
Whey, dried	930	19.2	3.4	3.0	3.0	1.8	8.2	11.9	9.7	1.9	3.3	3.2	8.9	1.9	2.5	6.8

Table A2.1.4   Vitamin potency of foods (fresh basis)

Food	Vitamin A potency[a] (iu/kg)	Vitamin E (iu/kg)	Thiamin (mg/kg)	Riboflavin (mg/kg)	Nicotinic acid (mg/kg)	Pantothenic acid (mg/kg)	Vitamin B_6 (mg/kg)	Vitamin B_{12} (mg/kg)	Choline (mg/kg)
Green crops									
Grass, dried	328	150	–	15.5	74	–	–	–	890
Lucerne, dried	267	200	–	16.6	43	–	–	0.003	1110
Cereals and by-products									
Barley	0.7	20	1.9	1.8	55	8	3.0	–	990
Brewer's yeast, dried	–	–	91.8	37.0	448	109	42.8	–	3984
Maize	5.0	22	3.5	1.0	24	4	7.0	–	620
Oats	0.6	20	6.0	1.1	12	–	1.0	–	946
Rice	–	12	–	0.4	15	–	–	–	780
Rye	0.2	17	3.6	1.6	19	8	2.6	–	419
Sorghum	0.7	12	4.0	1.1	41	12	3.2	–	450
Wheat	0.4	13	4.5	1.4	48	10	3.4	–	1090
Wheat, fine middlings	0.5	20	–	2.2	100	–	–	–	1110
Wheat, coarse middlings	0.4	57	–	2.4	95	–	–	–	1170
Oilseed by-products									
Coconut meal	–	16	–	3.3	27	–	–	–	1110
Cottonseed meal (dec. exp.)	0.3	39	6.4	5.1	38	10	5.3	–	2753
Groundnut meal (dec. extr.)	–	3	5.7	11.0	170	53	10.0	–	2396
Groundnut meal (dec. exp.)	0.3	3	7.1	5.2	166	47	10.0	–	1655
Linseed meal (extr.)	0.4	–	–	3.5	40	–	–	–	1660
Soya bean meal (extr.)	–	2	4.5	2.9	29	16	6.0	–	2794
Animal by-products									
Fishmeal	–	8	2.1	6.0	49	10	4.1	0.081	5180
Meat meal	–	1	0.2	5.5	57	5	3.0	0.068	2077
Milk, dried skim	0.3	1	–	21.0	12	–	–	0.055	1060

[a] For chicks. In the case of plant products, the values for pigs and ruminants are about half those quoted.

Table A2.2 Nutritive values of foods

Food	Ruminants ME (MJ/kg DM)	FME (MJ/kg DM)	ERDP (g/kg DM) 0.02[a]	0.05	0.08	DUP (g/kg DM) 0.02	0.05	0.08	Poultry ME (MJ/kg)	Pigs DE (MJ/kg)	Horses DE (MJ/kg DM)
Green crops											
Barley, in flower	10.0	9.4									
Cabbage	10.8	9.2								1.4	
Clover, red, early flowering	10.2	8.9									10.6
Clover, white, early flowering	9.0	7.5									10.5
Grass, young (75–80 D)	12.2	11.4	134	122	110	12	24	33			11.1
Grass, mature (60–65 D)	10.0	9.3	76	66	60	18	26	32			9.2
Kale	11.0	9.7	127	118	110	14	22	29			
Lucerne, early flowering	8.2	7.6									9.7
Maize	8.8	7.9									
Rape	9.5	7.5									
Sugar beet tops	9.9	8.8									
Sugarcane	8.9	8.1									
Swede tops	9.2	7.7									
Silages											
Barley, whole crop	8.7	7.8	45	42	40	4	7	10			
Grass, young	11.6	9.2	158	149	124	18	26	31			
Grass, mature	10.0	9.0	106	100	96	14	20	23			
Lucerne	8.5	7.7	150	144	139	7	24	27			10.5
Maize, whole crop	10.8	9.7	89	86	84	4	7	9			11.2
Potato	11.8	10.6									
Hays											
Clover, red	9.6	8.6									9.3
Grass, poor quality	7.0	6.3	36	30	29	10	14	19			7.7
Grass, good quality	9.5	8.6	75	63	55	29	40	47			9.1
Lucerne, early flowering	8.2	7.4	173	161	131	33	43	33			9.3
Dried herbage											
Grass	10.6	9.5	137	116	105	35	54	64	5.0	7.9	10.9
Lucerne	9.1	8.2	178	161	150	26	42	52	4.6	8.5	9.5

Table A2.2 (cont.)

Food	Ruminants		ERDP (g/kg DM)			DUP (g/kg DM)			Poultry	Pigs	Horses
	ME (MJ/kg DM)	FME (MJ/kg DM)	0.02[a]	0.05	0.08	0.02	0.05	0.08	ME (MJ/kg)	DE (MJ/kg)	DE (MJ/kg DM)
Straws											
Barley	6.5	5.8	25	23	21	6	8	10			6.8
Barley, ammoniated	7.5	7.0	50	48	46	7	8	10			
Bean	7.4	7.1									
Oat	7.0	6.7	23	20	18	6	9	11			6.8
Oat, ammoniated	7.8	7.2	30	23	21	35	41	43			
Pea	6.7	6.0									
Wheat	6.1	5.8	23	20	18	6	8	10			6.8
Wheat, ammoniated	7.4	6.9	48	47	46	9	11	11			
Roots and tubers											
Artichoke, Jerusalem	13.2	12.9									
Cassava	12.8	12.5							12.7[b]	13.4[b]	
Fodder beet	11.9	11.8	52	49	47	5	7	9		2.3	
Mangels	12.4	12.1									
Potatoes	12.5	12.3	74	71	68	6	8	11	12.1[b]	3.4	
Sugar beet pulp, dried	12.7	12.5	68	54	46	19	32	38			10.7
Sugar beet pulp, dried molassed	12.5	12.4	67	52	46	10	29	38		10.0	
Sugar beet molasses	12.9	12.9	37	37	37	10	10	10	7.9	10.3	14.2
Sugarcane molasses			44	44	44	11	11	11	7.9	10.3	14.6
Swedes	12.8	12.2									
Sweet potatoes	12.7	12.1									
Turnips	11.2	10.4	100	96	92	11	14	18			
Cereals and by-products											
Barley	12.8	12.2	96	90	85	8	14	18	11.4	12.9	15.4
Barley, brewer's grains	11.2	8.5	157	126	108	48	75	92	7.9		11.5
Barley, malt culms	11.2	10.4							8.4	8.0	11.5
Brewer's yeast, dried	11.7	11.3								14.0	

Table A2.2 (cont.)

Food	Ruminants ME (MJ/kg DM)	FME (MJ/kg DM)	ERDP (g/kg DM) 0.02[a]	0.05	0.08	DUP (g/kg DM) 0.02	0.05	0.08	Poultry ME (MJ/kg)	Pigs DE (MJ/kg)	Horses DE (MJ/kg DM)
Grain distiller's grains	14.7	10.9	185	155	108	35	61	104		3.3	
Grain distiller's dark grains	14.8	10.5	204	154	133	52	96	115		9.5	
Malt distiller's grains	10.2	7.2								2.8	
Malt distiller's dark grains	12.2	9.9								9.0	14.6
Malt distiller's dried solubles	12.4									14.0	
Maize	14.2	12.7	43	32	28	42	52	55	13.4	14.5	16.1
Maize, flaked	15.0	13.3								15.6	
Maize gluten feed	12.7	11.5	228	204	191	23	44	56	7.6	12.4	14.5
Maize gluten meal	17.5	16.4	348	233	182	253	355	404	12.3	13.8	
Millet	11.3	9.8							12.0		
Oats	12.0	10.3	86	85	84	18	19	20	10.0	11.4	13.4
Oats, naked	14.8	11.7	95	94	92	19	20	22	13.8		14.4
Oat husks	4.9	4.5							2.0	3.7	6.4
Pot ale syrup	15.4	15.3								6.2	
Rice, brown	15.0	14.8	101	91	85	50	59	65	15.0	15.3	15.9
Rice, bran (extracted)	11.0	7.9							6.0	8.3	12.3
Rice, polished	15.0										
Rye	14.0	13.3							12.1	13.9	16.1
Sorghum	13.4	11.9							12.8	14.2	14.9
Wheat	13.6	12.9	104	100	97	16	19	22	12.5	14.0	16.2
Wheat bran	10.1	8.5							8.5	9.8	13.8
Wheat feed	12.0	10.4	135	117	105	36	52	63	8.1	11.6	14.5
Wheat germ meal									11.1		
Oilseed by-products											
Coconut meal	12.7	10.0	134	92	75	63	101	117	6.9		
Cottonseed meal, undecorticated	8.5	6.6								10.1	
Cottonseed meal, decorticated	12.3	9.2	343	265	224	59	125	165	10.9		
Groundnut meal, undecorticated	9.2	8.5									12.6

Table A2.2 (cont.)

Food	Ruminants		ERDP (g/kg DM)			DUP (g/kg DM)			Poultry	Pigs	Horses
	ME (MJ/kg DM)	FME (MJ/kg DM)	0.02[a]	0.05	0.08	0.02	0.05	0.08	ME (MJ/kg)	DE (MJ/kg)	DE (MJ/kg DM)
Groundnut meal, decorticated	13.7	11.3	447	370	320	46	127	169	13.2	15.4	13.6
Linseed meal	11.9	8.9	323	283	259	61	98	120	8.7	13.1	15.8
Palm kernel meal	11.6	8.7	161	118	98	28	57	67	6.7	11.2	
Rapeseed meal	12.0	11.0	332	296	268	41	73	99	7.9	11.8	13.0
Soya bean meal	13.3	12.7	433	342	287	70	147	194	9.6	15.0	14.7
Soya bean meal, full fat	15.5	7.7							14.0	18.0	
Sunflower meal, undecorticated	9.5	6.7	258	232	211	21	45	63		9.0	
Sunflower meal, decorticated	10.4	10.0	374	335	305	27	54	74	8.3		11.7
Leguminous seeds											
Beans, field	13.4	13.0	228	209	195	39	57	69	10.0	12.7	
Gram	12.4	12.0									
Peas	13.8	13.3	211	191	177	12	30	42	11.2	13.6	14.4
Animal by-products											
Blood meal	12.8	12.5							13.0	15.2	
Feather meal, hydrolysed	12.5	10.3							12.8	12.4	
Fishmeal, UK-produced	14.6	12.2	447	350	308	182	350	391	12.9	15.5	13.4
Fishmeal, herring	17.8	15.2	508	397	349	206	397	444	15.2	18.0	
Fishmeal, South American	14.6	12.5	469	367	323	191	367	410	15.0	17.0	
Meat meal	16.3	11.1							15.7	16.8	
Meat meal, low fat										14.4	
Meat and bone meal	9.7	8.0							7.1	10.5	
Milk, cow's whole	20.2	9.5								3.0	
Milk, skim	15.3	12.9							10.9[b]	1.5	16.9
Milk, whey	14.5	13.5							12.0[b]	1.0	17.0

[a] Rumen outflow rate.
[b] Dried.

Table A2.3 Feeding standards for lactating and pregnant cattle
3.1 Daily requirements of cows producing milk of 38 g fat and 34 g protein/kg and weighing 550 kg. $q_{\mathrm{m}} = 0.55$

Milk yield (kg/day)	0	5	10	15	20	25	Days pregnant	
Liveweight change (kg/day)	0	+0.6	+0.4	+0.25	0	−0.2	230	260
DMI (kg)	9.7	11.8	14.0	16.1	17.2	17.2	10.7	10.7
ME (MJ)	57	110	127	146	161	180	69	80
MP (g)	261	582	791	1008	1211	1421	335	380
Ca (g)	22	39	55	72	87	100	33	39
P (g)	18	30	42	54	64	72	25	28
Mg (g)	10	13	17	21	24	28	13	13
Na (g)	4	7	10	13	16	19	6	6
Vit. A (iu)	←			55 000		→	← 55 000 →	
Vit. D (iu)	←			5500		→	← 5500 →	
Vit. E (iu)	←			330		→	← 330 →	

Table A2.3 (cont.)
3.2 Daily requirements of cows producing milk of 38 g fat and 34 g protein/kg and weighing 550 kg. $q_{\mathrm{m}} = 0.6$

Milk yield (kg/day)	0	5	10	15	20	25	Days pregnant	
Liveweight change (kg/day)	0	+ 0.6	+ 0.4	+ 0.25	0	−0.2	230	260
DMI (kg)	9.7	11.8	14.0	16.1	17.2	17.2	10.7	10.7
ME (MJ)	55	107	124	143	157	175	68	79
MP (g)	261	582	791	1008	1211	1421	335	380
Ca (g)	22	39	55	72	87	100	33	39
P (g)	18	30	42	54	64	72	25	28
Mg (g)	10	13	17	21	24	28	13	13
Na (g)	4	7	10	13	16	19	6	6
Vit. A (iu)	←			55 000		→	← 55 000 →	
Vit. D (iu)	←			5500		→	← 5500 →	
Vit. E (iu)	←			330		→	← 330 →	

Table A2.3 (cont.)
3.3 Daily requirements of cows producing milk of 38 g fat and 34 g protein/kg and weighing 550 kg. $q_m = 0.65$

Milk yield (kg/day)	0	5	10	15	20	25	30	Days pregnant	
Liveweight change (kg/day)	0	+ 0.6	+ 0.4	+ 0.25	0	−0.2	−0.4	230	260
DMI (kg)	9.7	11.8	14.0	16.1	17.2	17.2	17.2	10.7	10.7
ME (MJ)	54	105	121	139	153	171	190	66	78
MP (g)	261	582	791	1008	1211	1421	1631	335	380
Ca (g)	22	39	55	72	87	100	113	33	39
P (g)	18	30	42	54	64	72	79	25	28
Mg (g)	10	13	17	21	24	28	32	12	12
Na (g)	4	7	10	13	16	19	22	6	6
Vit. A (iu)	← 55 000 →							← 55 000 →	
Vit. D (iu)	← 5500 →							← 5500 →	
Vit. E (iu)	← 330 →							← 330 →	

Table A2.3 (cont.)
3.4 Daily requirements of cows producing milk of 38 g fat and 34 g protein/kg and weighing 650 kg. $q_m = 0.55$

Milk yield (kg/day)	0	5	10	15	20	25	Days pregnant	
Liveweight change (kg/day)	0	+0.6	+0.4	+0.25	0	−0.2	230	260
DMI (kg)	10.5	13	15	17	17.5	17. 5	12	12
ME (MJ)	64	117	134	153	168	187	80	93
MP (g)	296	616	826	1043	1246	1456	384	416
Ca (g)	25	42	58	75	89	102	37	43
P (g)	20	32	44	56	64	72	27	30
Mg (g)	12	15	19	23	26	30	14	14
Na (g)	5	8	11	14	17	20	8	8
Vit. A (iu)	← 65 000 →						← 65 000 →	
Vit. D (iu)	← 6500 →						← 6500 →	
Vit. E (iu)	← 385 →						← 385 →	

Table A2.3 (cont.)

3.5 Daily requirements of cows producing milk of 38 g fat and 34 g protein/kg and weighing 650 kg. $q_m = 0.6$

Milk yield (kg/day)	0	5	10	15	20	25	Days pregnant	
Liveweight change (kg/day)	0	+ 0.6	+ 0.4	+ 0.25	0	− 0.2	230	260
DMI (kg)	10.5	13	15	17	17.5	17.5	12	12
ME (MJ)	62	114	131	149	164	182	78	92
MP (g)	296	616	826	1043	1246	1456	384	416
Ca (g)	25	42	58	75	89	102	37	43
P (g)	20	32	44	56	64	72	27	30
Mg (g)	12	15	19	23	26	30	14	14
Na (g)	5	8	11	14	17	20	8	8
Vit. A (iu)			←	65 000		→	← 65 000 →	
Vit. D (iu)			←	6500		→	← 6500 →	
Vit. E (iu)			←	385		→	← 385 →	

Table A2.3 (cont.)

3.6 Daily requirements of cows producing milk of 38 g fat and 34 g protein/kg and weighing 650 kg. $q_m = 0.65$

Milk yield (kg/day)	0	5	10	15	20	25	30	Days pregnant	
Liveweight change (kg/day)	0	+ 0.6	+ 0.4	+ 0.25	0	− 0.2	− 0.4	230	260
DMI (kg)	10.5	13	15	17	17.5	17.5	18	12	12
ME (MJ)	61	111	127	146	159	178	196	77	91
MP (g)	296	616	826	1043	1246	1456	1666	384	416
Ca (g)	25	42	58	75	89	102	116	37	43
P (g)	20	32	44	56	64	72	81	27	30
Mg (g)	12	15	19	23	26	30	34	14	14
Na (g)	5	8	11	14	17	20	23	8	8
Vit. A (iu)			←	65 000			→	← 65 000 →	
Vit. D (iu)			←	6500			→	← 6500 →	
Vit. E (iu)			←	385			→	← 385 →	

Table A2.4 Feeding standards for growing cattle
4.1 Daily requirements of cattle of 200 kg liveweight

Category[a]	Component	Liveweight gain (kg/day)[b]					DM intake[b] (kg/day)
		0	0.5	0.75	1.0	1.25	
Heifers of	ME (MJ)	27/26	44/40	56/50	–/62	–/78	5.0/6.5
small-sized	MP (g)	122	224	270	313	377	
breeds							
Castrate males	ME (MJ)	27/26	40/37	49/45	–/54	–/65	5.0/6.5
of medium-sized	MP (g)	122	249	306	361	411	
breeds							
Bulls of	ME (MJ)	31/30	40/37	45/42	52/47	–/54	5.0/6.5
large-sized	MP (g)	122	274	343	408	469	
breeds							
All	Ca (g)	4	16	22	28	32	
breeds	P (g)	5	15	20	25	27	
	Mg (g)	3.5	4.8	5.4	6.0	6.6	
	Na (g)	1.5	2.3	2.7	3.1	3.5	
	Vit. A (iu)	←————————14 000————————→					
	Vit. D (iu)	←———————— 1200 ————————→					
	Vit. E (iu)	←———————— 115 ————————→					

[a] Categories: There is a range of nine categories of growing cattle (combinations of three breed sizes and three sexes). Of the three categories illustrated in this table, castrates of medium-sized breeds are the median category, and the other two are the extremes of the range.

[b] Where two values separated by a solidus (/) are given, the first is for diets with a metabolisability (q_m) of 0.55 and the second for diets with q_m of 0.65. A blank before the solidus indicates that the performance level is unlikely to be achieved with a diet of lower metabolisability.

Table A2.4 (cont.)
4.2 Daily requirements of cattle of 400 kg liveweight

Category[a]	Component	Liveweight gain (kg/day)[b]					DM intake[b] (kg/day)
		0	0.5	0.75	1.0	1.25	
Heifers of small-sized breeds	ME (MJ)	44/42	70/64	88/79	–/97	–/120	8.5/11.0
	MP (g)	206	297	338	377	413	
Castrate males of medium-sized breeds	ME (MJ)	44/42	64/59	79/71	95/84	–/100	8.5/11.0
	MP (g)	206	319	371	420	465	
Bulls of large-sized breeds	ME (MJ)	50/48	64/60	72/67	82/75	94/86	8.5/11.0
	MP (g)	206	342	404	463	517	
All breeds	Ca (g)	8	19	25	31	34	
	P (g)	9	19	24	31	31	
	Mg (g)	7.1	8.3	8.9	9.5	10.1	
	Na (g)	2.6	3.8	4.2	4.6	5.0	
	Vit. A (iu)	←		28 000		→	
	Vit. D (iu)	←		2400		→	
	Vit. E (iu)	←		195		→	

[a,b] See Table 4.1.

Table A2.5   Feeding standards for pregnant ewes
5.1 Daily requirements of ewes assuming zero weight change

Ewe weight (kg)	Component	Days from conception							
		Single lamb				Twin lambs			
		98	112	126	140	98	112	126	140
55	DMI (kg)	1.2	1.2	1.2	1.1	1.3	1.3	1.3	1.2
	ME (MJ)	7.9	8.8	9.9	11.3	8.9	10.2	12.0	14.4[a]
	MP (g)	72	76	82	89	77	84	93	104
	Ca (g)	3.1	3.8	4.5	5.1	3.9	5.0	6.2	7.2
	P (g)	2.7	2.9	3.0	3.0	3.2	3.5	3.8	3.8
	Mg (g)	1.0	1.0	1.3	1.3	1.0	1.0	1.5	1.5
	Na (g)	← 1.8 →				← 1.8 →			
	Vit. A (iu)	← 5500 →				← 5500 →			
	Vit. D (iu)	← 550 →				← 550 →			
	Vit. E (iu)	← 40 →				← 45 →			
75	DMI (kg)	1.5	1.5	1.5	1.4	1.6	1.6	1.6	1.5
	ME (MJ)	10.1	11.1	12.5	14.3	11.3	12.9	15.2	18.2[a]
	MP (g)	86	91	98	106	92	100	111	126
	Ca (g)	3.8	4.6	5.6	6.3	4.7	6.0	7.6	8.9
	P (g)	3.4	3.6	3.8	3.8	4.0	4.4	4.7	4.8
	Mg (g)	1.3	1.3	1.7	1.7	1.3	1.3	2.0	2.0
	Na (g)	← 2.5 →				← 2.5 →			
	Vit. A (iu)	← 7500 →				← 7500 →			
	Vit. D (iu)	← 750 →				← 750 →			
	Vit. E (iu)	← 55 →				← 60 →			

[a] These energy levels are not attainable owing to inadequate DMI.

Table A2.5 (cont.)

5.2 Daily requirements of 55 kg pregnant ewes losing 50 g liveweight/day and 75 kg ewes losing 75 g liveweight/day

Ewe weight (kg)	Component	Days from conception							
		Single lamb				Twin lambs			
		98	112	126	140	98	112	126	140
55	DMI (kg)	1.2	1.2	1.2	1.1	1.3	1.3	1.3	1.2
	ME (MJ)	6.0	6.8	7.9	9.3	6.9	8.2	10.0	12.3
	MP (g)	66	70	76	83	71	78	87	98
	Ca (g)	3.1	3.8	4.5	5.1	3.9	5.0	6.2	7.2
	P (g)	2.7	2.9	3.0	3.0	3.2	3.5	3.8	3.8
	Mg (g)	1.0	1.0	1.3	1.3	1.0	1.0	1.5	1.5
	Na (g)	← 1.8 →				← 1.8 →			
	Vit. A (iu)	← 5500 →				← 5500 →			
	Vit. D (iu)	← 550 →				← 550 →			
	Vit. E (iu)	← 40 →				← 45 →			
75	DMI (kg)	1.5	1.5	1.5	1.4	1.6	1.6	1.6	1.5
	ME (MJ)	7.1	8.1	9.4	11.2	8.2	9.9	12.2	15.1
	MP (g)	77	82	89	98	83	91	102	117
	Ca (g)	3.8	4.6	5.6	6.3	4.7	6.0	7.6	8.9
	P (g)	3.4	3.6	3.8	3.8	4.0	4.4	4.7	4.8
	Mg (g)	1.3	1.3	1.7	1.7	1.3	1.3	2.0	2.0
	Na (g)	← 2.5 →				← 2.5 →			
	Vit. A (iu)	← 7500 →				← 7500 →			
	Vit. D (iu)	← 750 →				← 750 →			
	Vit. E (iu)	← 55 →				← 60 →			

Table A2.6 Feeding standards for lactating ewes

6.1 Daily requirements of lactating ewes kept outdoors and assuming zero weight change. $q_m = 0.625$

Ewe weight (kg)	Component	Week of lactation					
		Single lamb			Twin lambs		
		1–4	5–8	9–12	1–4	5–8	9–12
55	DMI (kg)	1.5	1.7	1.6	1.6	1.8	1.7
	ME (MJ)	16.7	15.8	13.2	22.0[a]	20.0	16.0
	MP (g)	154	145	120	204	184	146
	Ca (g)	6.1	5.9	4.8	8.4	7.8	6.0
	P (g)	5.4	5.5	4.6	7.2	6.9	5.6
	Mg (g)	2.2	2.1	1.7	2.9	2.6	2.1
	Na (g)	2.1	2.0	1.9	2.4	2.3	2.0
	Vit. A (iu)	← 5500 →			← 5500 →		
	Vit. D (iu)	← 550 →			← 550 →		
	Vit. E (iu)	← 40 →			← 45 →		
75	DMI (kg)	1.9	2.2	2.0	2.0	2.3	2.1
	ME (MJ)	23.8[a]	22.3	17.7	30.9[a]	26.3	20.4
	MP (g)	223	209	164	289	246	191
	Ca (g)	9.0	8.8	6.7	12.0	10.5	7.9
	P (g)	8.0	8.1	6.4	10.2	9.4	7.4
	Mg (g)	3.3	3.1	2.5	4.2	3.6	2.9
	Na (g)	3.0	2.9	2.6	3.4	3.1	2.8
	Vit. A (iu)	← 7500 →			← 7500 →		
	Vit. D (iu)	← 750 →			← 750 →		
	Vit. E (iu)	← 55 →			← 60 →		

[a] These energy levels are not attainable owing to inadequate DMI.

Table A2.6 (cont.)

6.2 Daily requirements of 55 kg lactating ewes kept outdoors and losing 50 g liveweight/day and 75 kg ewes losing 75 g liveweight/day. $q_m = 0.625$

Ewe weight (kg)	Component	Week of lactation					
		Single lamb			Twin lambs		
		1–4	5–8	9–12	1–4	5–8	9–12
55	DMI (kg)	1.5	1.7	1.6	1.6	1.8	1.7
	ME (MJ)	14.9	14.0	11.5	20.2[a]	18.1	14.2
	MP (g)	148	139	114	198	179	140
	Ca (g)	6.1	5.9	4.8	8.4	7.8	6.0
	P (g)	5.4	5.5	4.6	7.2	6.9	5.6
	Mg (g)	2.2	2.1	1.7	2.9	2.6	2.1
	Na (g)	2.1	2.0	1.9	2.4	2.3	2.0
	Vit. A (iu)	← 5500 →			← 5500 →		
	Vit. D (iu)	← 550 →			← 550 →		
	Vit. E (iu)	← 40 →			← 45 →		
75	DMI (kg)	1.9	2.2	2.0	2.0	2.3	2.1
	ME (MJ)	21.1	19.6	15.0	28.1[a]	23.5	17.7
	MP (g)	214	200	155	280	237	182
	Ca (g)	9.0	8.8	6.7	12.0	10.5	7.9
	P (g)	8.0	8.1	6.4	10.2	9.4	7.4
	Mg (g)	3.3	3.1	2.5	4.2	3.6	2.9
	Na (g)	3.0	2.9	2.6	3.4	3.1	2.8
	Vit. A (iu)	← 7500 →			← 7500 →		
	Vit. D (iu)	← 750 →			← 750 →		
	Vit. E (iu)	← 55 →			← 60 →		

[a] These energy levels are not attainable owing to inadequate DMI.

Table A2.7 Feeding standards for growing lambs
7.1 Daily requirements of lambs of 20 kg liveweight

Category	Component	Liveweight gain (g/day)[a]				DM intake[a] (kg/day)
		0	50	100	150	
Females	ME (MJ)	3.4/3.2	4.5/4.2	5.8/5.3	–/6.5	0.46/0.56
	MP (g)	21[b]	45	58	71	
Castrate males	ME (MJ)	3.4/3.2	4.5/4.2	5.7/5.2	–/6.2	0.46/0.56
	MP (g)	21[b]	47	61	76	
Males	ME (MJ)	3.9/3.7	4.8/4.5	5.8/5.4	–/6.4	0.46/0.56
	MP (g)	21[b]	47	61	76	
All sexes	Ca (g)	0.7	1.6	2.5	3.4	
	P (g)	0.5	1.0	2.0	3.0	
	Mg (g)	0.38	0.50	0.61	0.72	
	Na (g)	0.57	0.63	0.69	0.75	
	Vit. A (iu)	←———— 660 ————→				
	Vit. D (iu)	←———— 120 ————→				
	Vit. E (iu)	←———— 21 ————→				

[a] Where two values separated by a solidus (/) are given, the first is for diets with a metabolisability (q_m) of 0.55 and the second for diets with q_m of 0.65. A blank before the solidus indicates that the performance level is unlikely to be achieved with a diet of lower metabolisability.

[b] This 'basal' requirement for maintenance does not include an allowance of protein for wool growth. If lambs were kept at a maintenance level they would continue to grow wool, and would need an additional 11 g/day of MP for this purpose.

Table A2.7 (cont.)
7.2 Daily requirements of
lambs of 35 kg liveweight

Category	Component	Liveweight gain (g/day)[a]				DM intake[a] (kg/day)
		0	50	100	150	
Females	ME (MJ)	5.2/5.0	7.0/6.5	9.1/8.2	–/10.2	0.77/0.92
	MP (g)	31[b]	54	65	77	
Castrate males	ME (MJ)	5.2/5.0	6.8/6.3	8.5/7.8	–/9.5	0.77/0.92
	MP (g)	31[b]	56	70	83	
Males	ME (MJ)	6.0/5.7	7.4/6.9	9.0/8.3	–/9.8	0.77/0.92
	MP (g)	31[b]	56	70	83	
All sexes	Ca (g)	0.9	1.7	2.5	3.4	
	P (g)	0.9	1.6	2.4	3.5	
	Mg (g)	0.62	0.74	0.85	1.00	
	Na (g)	0.99	1.05	1.10	1.15	
	Vit. A (iu)	←——— 1200 ———→				
	Vit. D (iu)	←——— 210 ———→				
	Vit. E (iu)	←——— 25 ———→				

[a,b] See Table 7.1.

Table A2.8 Dietary allowances (mg/kg DM) of trace elements for ruminants

		Cattle		Sheep	
Copper	Pre-ruminant calf		2	Pre-ruminant lamb	1
	Others		12	Growing lambs	3
				Others	6
Iron	Before weaning		30	All classes	30
	After weaning		40		
	>150 kg liveweight		30		
	Pregnant and lactating		40		
Iodine	Winter		0.5	Winter	0.5
	Summer		0.15	Summer	0.15
	Presence of goitrogens		2.00	Presence of goitrogens	2.00
Cobalt	All classes		0.11	All classes	0.11
Selenium	All classes		0.10	All classes	0.10
Zinc	All classes		40	All classes	40
Manganese	All classes		40	All classes	40

Table A2.9 Typical
dietary nutrient levels for
pigs (fresh basis)
9.1 Growing pigs

Component	Liveweight (kg)	
	20–50	50–90
Feed (kg/day)	1.22–2.0	2.2–2.7
Digestible energy (MJ/kg)	14.0	13.0
Crude protein (g/kg)	205	175
Ideal protein (g/kg)	165	145
Lysine (g/kg)	11.6	10.0
Methionine + cystine (g/kg)	5.8	5.0
Threonine (g/kg)	6.9	6.0
Tryptophan (g/kg)	1.7	1.4
Calcium (g/kg)	9.8	7.8
Phosphorus (g/kg)	7.0	5.9
Salt (g/kg)	3.2	3.0
Iron (mg/kg)[a]	62 (210)	57 (195)
Magnesium (mg/kg)	308	221
Zinc (mg/kg)[a]	56 (154)	47 (130)
Copper (mg/kg)	5.6	5.2
Manganese (mg/kg)	11.2	10
Iodine (mg/kg)[b]	0.15 (0.3)	0.14 (0.3)
Selenium (mg/kg)	0.15	0.14
Vitamin A (iu/kg)	8000	6000
Vitamin D (iu/kg)	1000	800
Vitamin E (iu/kg)	15	15
Vitamin K (mg/kg)	1	1
Essential fatty acids (g/kg)	10	7
Thiamin (mg/kg)	1.5	1.5
Riboflavin (mg/kg)	3.0	3.0
Nicotinic acid (mg/kg)	15	15
Pantothenic acid (mg/kg)	10	10
Pyridoxine (mg/kg)	2.5	2.5
Choline (mg/kg)	1000	1000
Biotin (mg/kg)	0.2	0.2
Vitamin B_{12} (mg/kg)	0.01	0.01

[a] Use higher level (in parentheses) when copper is used as a growth promoter.
[b] Use higher level (in parentheses) when diets contain rapeseed meal.

Table A2.9 (cont.)
9.2 Breeding pigs

Component	Sows		Boars
	Pregnancy	Lactation	
Feed (kg/day)	1.9–2.4	5.3–5.9[a]	2.2–2.6
Digestible energy (MJ/kg)	13.0	13.5	13.0
Crude protein (g/kg)	120–140	150–180	150
Lysine (g/kg)	4.8	8.0	6.5
Methionine + cystine (g/kg)	2.5	4.0	3.3
Threonine (g/kg)	2.9	4.8	3.9
Isoleucine (g/kg)	2.6	4.4	3.6
Leucine (g/kg)	4.8	9.0	6.5
Tryptophan (g/kg)	0.7	1.4	1.0
Calcium (g/kg)	8.5	8.8	8.5
Phosphorus (g/kg)	6.5	6.8	6.5
Salt (g/kg)	3.5	3.5	3.5
Iron (mg/kg)	60	60	60
Zinc (mg/kg)	50	50	100
Copper (mg/kg)	5	6	5
Manganese (mg/kg)	15	16	15
Iodine (mg/kg)	0.5[b]	0.5[b]	0.5[b]
Selenium (mg/kg)	0.15	0.15	0.15
Vitamin A (iu/kg)	6000	6000	6000
Vitamin D (iu/kg)	750	750	750
Vitamin E (iu/kg)	15	15	15
Vitamin K (mg/kg)	1	1	1
Essential fatty acids (g/kg)	7	7	7
Thiamin (mg/kg)	1.5	1.5	1.5
Riboflavin (mg/kg)	3.0	3.0	3.0
Nicotinic acid (mg/kg)	15	15	15
Pantothenic acid (mg/kg)	10	10	10
Pyridoxine (mg/kg)	1.5	1.5	1.5
Choline (mg/kg)	1500	1500	1500
Biotin (mg/kg)	0.30	0.30	0.30
Vitamin B_{12} (mg/kg)	0.015	0.015	0.015

[a] Or *ad libitum*.
[b] 0.9 in diets containing rapeseed meal.

Table A2.10 Typical dietary nutrient levels for poultry (fresh basis)
10.1 Chickens

	Growing chicks		Pullets 12–18 wks	Laying hens	Breeding hens	Broiler starter	Broiler finisher
	0–6 wks	6–12 wks					
ME (MJ/kg)	11.5	10.9	10.9	11.1	11.1	12.6	12.6
Crude protein (g/kg)	210	145	120	160	160	230	190
Amino acids (g/kg)							
Arginine	11	7.1	6.7	4.9	4.9	12.6	9.5
Glycine + serine	13.2	9.4	8.0	–	–	12.0	11.0
Histidine	5.1	3.3	2.4	1.6	1.6	5.0	5.0
Isoleucine	9	5.9	4.5	5.3	5.3	9.0	8.0
Leucine	14.7	9.9	8.4	6.6	6.6	16.0	13.0
Lysine	11	7.4	6.6	7.3	7.3	12.5	10.0
Methionine + cystine	9.2	6.2	4.5	5.5	4.6	9.2	8.0
Phenylalanine + tyrosine	15.8	10.8	8.0	7.0	7.0	15.8	14.0
Threonine	7.4	4.9	4.2	3.5	3.5	8.0	6.5
Tryptophan	2.0	1.4	1.2	1.4	1.4	2.3	1.9
Valine	10.4	6.6	5.3	5.3	5.3	10	9.0
Major minerals (g/kg)							
Calcium	12	10	8	35	33	12	10
Phosphorus (av.)	5	5	5	5	5	5	5
Magnesium	0.3[a]	0.3[a]	0.3[a]	0.3[a]	0.3[a]	0.3[a]	0.3[a]
Sodium	1.5	1.5	1.5	1.5	1.5	1.5	1.5
Potassium	3.0	–	–	–	–	3.0	3.0
Trace minerals (mg/kg)							
Copper	3.5[a]	3.5[a]	3.5[a]	3.5[a]	3.5[a]	3.5[a]	3.5[a]
Iodine	0.4[a]	0.4[a]	0.4[a]	0.4[a]	0.4[a]	0.4[a]	0.4[a]
Iron	80[a]	80[a]	80[a]	80[a]	80[a]	80[a]	45[a]
Manganese	100[a]	100[a]	100[a]	100[a]	100[a]	100[a]	100[a]
Zinc	50[a]	50[a]	50[a]	50[a]	50[a]	50[a]	50[a]
Selenium	0.15	–	–	–	–	0.15	0.15
Vitamins (iu/kg)							
A	2000[a]	2000[a]	2000[a]	6000[a]	6000[a]	2000[a]	2000[a]
D_3	600[a]	600[a]	600[a]	800[a]	800[a]	600[a]	600[a]
E	25[a]	25[a]	25[a]	25[a]	25[a]	25[a]	25[a]
Vitamins (mg/kg)							
K	1.3[a]	1.3[a]	1.3[a]	1.3[a]	1.3[a]	1.3[a]	1.3[a]
Thiamin	3	–	–	–	2	3	–
Riboflavin	4[a]	4[a]	4[a]	4[a]	4[a]	4[a]	4[a]
Nicotinic acid	28[a]	28[a]	28[a]	28[a]	28[a]	28[a]	28[a]
Pantothenic acid	10[a]	10[a]	10[a]	10[a]	10[a]	10[a]	10[a]
Choline	1300	–	–	–	1100	1300	1300
Vitamin B_{12}	–	–	–	–	0.01	–	–

[a]Added as supplement.

Table A2.10 (cont.)
10.2 Turkeys

	Growing chicks			Breeding turkeys
	0–6 wks	6–12 wks	12+ wks	
ME (MJ/kg)	12.6	11.9	11.9	11.3
Crude protein (g/kg)	300	260	180	160
Amino acids (g/kg)				
Arginine	16	13	8	5
Glycine + serine	9	–	7	–
Histidine	6	5	3.5	2
Isoleucine	11	10	5.5	5.5
Leucine	19	15	8	7
Lysine	17	13	8	7.5
Methionine + cystine	10	8	6	5.5
Phenylalanine + tyrosine	16	15	10	8
Threonine	10	9	5.5	4
Tryptophan	2.6	2.3	1.5	1.7
Valine	12	10	6	5
Major minerals (g/kg)				
Calcium	9	10	8	30
Phosphorus (av.)	4.5	5	4	5
Magnesium	0.36[a]	0.36[a]	0.36[a]	0.30[a]
Sodium	1.75	1.75	1.75	1.75
Potassium	–	–	–	–
Trace minerals (mg/kg)				
Copper	4.2[a]	4.2[a]	4.2[a]	3.5[a]
Iodine	0.48[a]	0.48[a]	0.48[a]	0.4[a]
Iron	96[a]	96[a]	96[a]	80[a]
Manganese	120[a]	120[a]	120[a]	100[a]
Zinc	60[a]	60[a]	60[a]	50[a]
Selenium	0.25	0.15	0.15	–
Vitamins (iu/kg)				
A	12 000[a]	12 000[a]	12 000[a]	10 000[a]
D$_3$	1800[a]	1800[a]	1800[a]	1500[a]
E	36[a]	36[a]	36[a]	30[a]
Vitamins (mg/kg)				
K	4.8[a]	4.8[a]	4.8[a]	4.0[a]
Thiamin	4.8[a]	4.8[a]	4.8[a]	4.0[a]
Riboflavin	12[a]	12[a]	12[a]	10[a]
Nicotinic acid	60[a]	60[a]	60[a]	50[a]
Pantothenic acid	19.2[a]	19.2[a]	19.2[a]	16[a]
Pyridoxine	6[a]	6[a]	6[a]	5[a]
Biotin	0.12[a]	0.12[a]	0.12[a]	0.1[a]
Folic acid	2.4[a]	2.4[a]	2.4[a]	2.0[a]
Vitamin B$_{12}$	0.024[a]	0.024[a]	0.024[a]	0.02[a]
Choline	1760	–	–	1350

[a]Added as supplement.

Table A2.11 Feeding standards for horses
11.1 Daily requirements of horses of 400 kg mature weight

	Weight (kg)	Gain (kg/day)	DE (MJ)	CP (g)	Lysine (g)	Ca (g)	P (g)	Mg (g)	K (g)	Vitamin A (iu)
Mature horses										
Maintenance	400		56	536	19	16	11	6.0	20.0	12 000
Stallions	400		70	670	23	20	15	7.7	25.5	18 000
Pregnant mares										
9 months	400		62	654	23	28	21	7.1	23.8	24 000
10 months			63	666	23	29	21	7.3	24.2	24 000
11 months			67	708	25	31	23	7.7	25.7	24 000
Lactating mares										
Foaling to 3 months	400		96	1141	40	45	29	8.7	36.8	24 000
3 months to weaning			82	839	29	29	18	6.9	26.4	24 000
Working horses										
Light work	400		70	670	23	20	15	7.7	25.5	18 000
Medium work			84	804	28	25	17	9.2	30.6	18 000
Intense work			112	1072	38	33	23	12.3	40.7	18 000
Growing horses										
6 months	180	0.55	54	643	27	25	14	3.4	10.7	8000
12 months	265	0.4	65	700	30	23	13	4.5	14.5	12 000
18 months	330	0.25	67	716	30	21	12	5.3	17.3	15 000
24 months	365	0.15	64	650	26	19	11	5.7	18.7	16 000

Table A2.11 (cont.)
11.2 Daily requirements of horses of 500 kg mature weight

	Weight (kg)	Gain (kg/day)	DE (MJ)	CP (g)	Lysine (g)	Ca (g)	P (g)	Mg (g)	K (g)	Vitamin A (iu)
Mature horses										
Maintenance	500		69	656	23	20	14	7.5	25.0	15 000
Stallions	500		86	820	29	25	18	9.4	31.2	22 000
Pregnant mares										
9 months	500		76	801	28	35	26	8.7	29.1	30 000
10 months			77	815	29	35	26	8.9	29.7	30 000
11 months			82	866	30	37	28	9.4	31.5	30 000
Lactating mares										
Foaling to 3 months	500		118	1427	50	56	36	10.9	46.0	30 000
3 months to weaning			102	1048	37	36	22	8.6	33.0	30 000
Working horses										
Light work	500		86	820	29	25	18	9.4	31.2	22 000
Medium work			103	984	34	30	21	11.3	37.4	22 000
Intense work			137	1312	46	40	29	15.1	49.9	22 000
Growing horses										
6 months	215	0.65	63	750	32	29	16	4.0	12.7	10 000
12 months	325	0.5	79	851	36	29	16	5.5	17.8	15 000
18 months	400	0.35	83	893	38	27	15	6.4	21.1	18 000
24 months	450	0.2	79	800	32	24	13	7.0	23.1	20 000

Table A2.11 (cont.)
11.3 Daily requirements of horses of 600 kg mature weight

	Weight (kg)	Gain (kg/day)	DE (MJ)	CP (g)	Lysine (g)	Ca (g)	P (g)	Mg (g)	K (g)	Vitamin A (iu)
Mature horses										
Maintenance	600		81	776	27	24	17	9.0	30.0	18 000
Stallions	600		102	970	34	30	21	11.2	36.9	27 000
Pregnant mares										
9 months	600		90	947	33	41	30	10.3	34.5	36 000
10 months			92	965	34	42	31	10.5	35.1	36 000
11 months			98	1024	36	44	33	11.2	37.2	36 000
Lactating mares										
Foaling to 3 months	600		141	1711	60	67	43	13.1	55.2	36 000
3 months to weaning			121	1258	44	43	27	10.4	39.6	36 000
Working horses										
Light work	600		102	970	34	30	21	11.2	36.9	27 000
Medium work			122	1164	41	36	25	13.4	44.2	27 000
Intense work			162	1552	54	47	34	17.8	59.0	27 000
Growing horses										
6 months	245	0.75	71	850	36	34	19	4.6	14.5	11 000
12 months	375	0.65	95	1023	43	36	20	6.4	20.7	17 000
18 months	475	0.45	101	1077	45	33	18	7.7	25.1	21 000
24 months	540	0.3	98	998	40	31	17	8.5	27.9	24 000

Table A2.12 Dietary allowances for horses of sodium, sulphur, trace minerals and vitamins

	Maintenance	Pregnant and lactating mares	Growing horses	Working horses
Sodium (g/kg DM)	1.0	1.0	1.0	3.0
Sulphur (g/kg DM)	1.5	1.5	1.5	1.5
Iron (mg/kg DM)	40	50	50	40
Manganese (mg/kg DM)	40	40	40	40
Copper (mg/kg DM)	10	10	10	10
Zinc (mg/kg DM)	40	40	40	40
Selenium (mg/kg DM)	0.1	0.1	0.1	0.1
Iodine (mg/kg DM)	0.1–0.6	0.1–0.6	0.1–0.6	0.1–0.6
Cobalt (mg/kg DM)	0.1	0.1	0.1	0.1
Vitamin A (iu/kg DM)	2000	3000	2000	2000
Vitamin D (iu/kg DM)	300	600	800	300
Vitamin E (iu/kg DM)	50	80	80	80
Thiamin (mg/kg DM)	3	3	3	5
Riboflavin (mg/kg DM)	2	2	2	2

Table A2.13 Water allowances for farm animals
13.1 Cattle and sheep

	kg water/kg DM intake		
	Environmental temperature (°C)		
	<16	16–20	>20
Cattle			
Calves, up to 6 weeks	7.0	8.0	9.0
Cattle, growing or adult, pregnant or non-pregnant	5.4	6.1	7.0
Sheep			
Lambs, up to 4 weeks	4.0	5.0	6.0
Sheep, growing or adult, non-pregnant	2.0	2.5	3.0
Ewes, mid pregnancy, twin bearing	3.3	4.1	4.9
Ewes, late pregnancy, twin bearing	4.4	5.5	6.6
Ewes, lactating, first month	4.0	5.0	6.0
Ewes, lactating, second/third month	3.0	3.7	4.5

Table A2.13 (cont.)
13.2 Lactating cows (600 kg liveweight)

Milk yield (kg/day)	Daily water intake (kg/head)		
	Environmental temperature (°C)		
	<16	16–20	>20
10	81	92	105
20	92	104	119
30	103	116	133
40	113	128	147

Table A2.13 (cont.)
13.3 Pigs

	Daily water intake (kg/head)
Growing pigs	1.5–2.0 at 15 kg liveweight, increasing to 6.0 at 90 kg liveweight
Non-pregnant sows	5.0
Pregnant sows	5.0–8.0
Lactating sows	15.0–20.0

Table A2.14 Values for metabolic liveweight ($W^{0.75}$) for weights at 10 kg intervals to 690 kg

Hundreds	Tens									
	0	10	20	30	40	50	60	70	80	90
0	0	5.6	9.5	12.8	15.9	18.8	21.6	24.2	26.8	29.2
100	31.6	34.0	36.3	38.5	40.7	42.9	45.0	47.1	49.1	51.2
200	53.2	55.2	57.1	59.1	61.0	62.9	64.8	66.6	68.4	70.3
300	72.1	73.9	75.7	77.4	79.2	80.9	82.6	84.4	86.1	87.8
400	89.4	91.1	92.8	94.4	96.1	97.7	99.3	100.9	102.6	104.2
500	105.7	107.3	108.9	110.5	112.0	113.6	115.1	116.7	118.2	119.7
600	121.2	122.7	124.2	125.7	127.2	128.7	130.2	131.7	133.2	134.6

References to appendix

ADAS 1986 Feeding by-products to pigs, Leaflet P3057, Alnwick, Northumberland, MAFF (Publications).

ADAS 1986 Feedingstuffs for pigs, Leaflet P3060, Alnwick, Northumberland, MAFF (Publications).

AFRC 1990 Technical Committee on Responses to Nutrients, Advisory Booklet, *Nutrient Requirements of Sows and Boars*, UK, HGM Publications.

AFRC 1990 Technical Committee on Responses to Nutrients, Report No. 6, *A Reappraisal of the Calcium and Phosphorus Requirements of Sheep and Cattle*. Nutrition Abstracts and Reviews (Series B) 61, 573–612.

AFRC 1993 Technical Committee on Responses to Nutrients, *Energy and Protein Requirements of Ruminants. An Advisory Manual*, Wallingford, UK, CAB International.

Agricultural Research Council, London 1975 *The Nutrient Requirements of Farm Livestock: No. 1, Poultry*, Farnham Royal, UK, Commonwealth Agricultural Bureaux.

Agricultural Research Council, London 1980 *The Nutrient Requirements of Ruminant Livestock*, Farnham Royal, UK, Commonwealth Agricultural Bureaux.

Agricultural Research Council, London 1981 *The Nutrient Requirements of Pigs*, Farnham Royal, UK, Commonwealth Agricultural Bureaux.

Black H, Edwards S, Kay M and Thomas S 1991 *Distillery By-Products as Feeds for Livestock*, Aberdeen, Scottish Agricultural College.

Bolton W and Blair R 1977 *Poultry Nutrition*, 4th edn, MAFF Bull. No. 174, London, HMSO.

Close W H 1994 Feeding new genotypes: establishing amino acid/energy requirements. In: Wiseman J, Cole D J A and Varley M A (eds), *Principles of Pig Science*, Nottingham University Press, pp 123–40.

MAFF 1990 *UK Tables of Nutritive Value and Chemical Composition of Feedingstuffs*, Aberdeen, Rowett Research Institute.

National Academy of Sciences/National Research Council 1989 *Nutrient Requirements of Poultry*, Washington.

National Academy of Sciences/National Research Council 1989 *Nutrient Requirements of Horses*, Washington.

Poultry Research Centre 1981 *Analytical Data of Poultry Feedstuffs, 1. General and Amino Acid Analyses, 1977–1980*, Occasional Publication No. 1, Roslin, UK, PRC.

Poultry Research Centre 1982 *Energy Values of Compound Poultry Feeds*, Occasional Publication No. 2, Roslin, UK, PRC.

Stranks M H, Cooke B C, Fairbairn C B, Fowler N G, Kirby P S, McCracken K J, Morgan C A, Palmer F G and Peers D 1988 Nutrient allowances for growing pigs, *Research and Development in Agriculture*, 5, 71–88.

Whittemore C T 1993 *The Science and Practice of Pig Production*. Harlow, UK, Longman.

Index

Aberdeen-Angus, 370, 383
abomasum, 164, 180, 185, 188
abortion
  from vitamin A deficiency, 80
  from manganese deficiency, 136
  from malnutrition, 405
  from ergot poisoning, 576
absorption, 163, 175–8, 187–91
acetic acid
  in intake regulation, 471
  in large intestine, 171
  in rumen, 183–5, 424
  in silage, 518–27
  metabolism, 199, 200, 213–14,
    229–34, 407
  utilization of
    for growth, 285–8
    for maintenance, 283–4
    for milk production, 413, 424
acetone, 22
acetyl coenzyme A, 91, 92, 148, 149,
  184, 207–35, 407
acid–base balance, 110, 115–18
acid-detergent fibre (ADF), 7, 247,
  255, 308
  in grass, 500, 542
  indigestible, 248
acid-detergent insoluble nitrogen
  (ADIN), 333, 340, 498
acid ether extract, 5
acidosis, 407, 628
Ackerman factor, 601
Acremonium loliae, 505
Acremonium coenophialum, 505
activators, 152, 159, 242
active transport, 175, 400
activity allowance, 357, 358, 426, 449
acyl-carrier protein, 98, 230
adenine, 66–8, 202, 226
adenosine, 66–8, 202, 225, 226
adenosine diphosphate (ADP),
  201–43

adenosine monophosphate (AMP), 67,
  67, 202, 204
adenosine triphosphate (ATP), 67,
  201–43, 270, 283, 361
Adolph E F, 467
adrenal corticosteroids, 52, 407
aflatoxin, 594, 595
Africa, 351
Agricultural and Food Research
  Council (UK) (AFRC), 305, 306,
  340, 349, 356–7, 365–72, 381,
  401–4, 422, 434
  energy system, 295, 300–5
  metabolizable protein system, 343–5,
    378, 430
  Technical Committee on Responses
    to Nutrients (TCORN), 333,
    343, 349, 357, 371, 381, 434,
    450, 477
Agricultural Research Council (UK)
  (ARC) see Agricultural and Food
  Research Council
Agropyron repens, 27
Agrostis spp. see bents
AIV silage, 526
alanine, 57, 64, 96, 218, 223–4, 518
albumins, 65, 331
aldaric acids, 20
aldehyde oxidase, 138
aldehydes, 14, 16, 42
aldonic acids, 20
alfalfa see lucerne
alimentary canal, 163–98
alkali disease, 140
alkaloids, 70–1, 556, 576, 599
alkalosis, 115
alkanes, 249
alkylarylsulphonate, 628
allantoin, 70, 267, 608
allometric equation, 368, 373, 385
alopecia, 509
aluminium, 109, 414

American net energy system, 302–5
amides, 70, 189, 314, 498
amines, 171, 189, 518, 524
    biogenic, 69
amino acids, 55–61, 64, 200, 237,
        313–43, 340
    absorption of, 176, 254
    antagonisms, 326
    as products of digestion, 168,
        169, 187
    as source of energy, 216–23
    assays, 324–6
    available, 324–5
    catabolism of, 217–24, 363
    chemical properties of, 59–60
    deamination of, 218, 518,
        523, 524
    deficiencies, 469
    dispensable, 318–24
    essential see indispensable
        amino acids
    formulae, 57, 58
    in cereals, 560–78
    in genetic engineering, 229
    in milk, 411
    in protein concentrates, 585–613
    in protein synthesis, 223–9
    in silage, 517–27
    indispensable see indispensable
        amino acids
    protection of, 194
    requirements of
        dog, 482
        horse, 377
        pig, 375–7, 455, 482
        man, 482
        poultry, 375–7, 398
    synthesis of, 91, 223–4, 229
    synthetic, 326, 375, 609
aminobutyric acid, 59, 518
aminoethylphosphoric acid, 331
aminopeptidases, 166, 169, 170
amino sugars, 19–20
ammonia
    in blood, 395
    in metabolism, 155, 171, 200,
        218–21
    in milk, 411
    in rumen, 187–90, 193, 254, 256,
        337, 547, 549, 610
    in silage, 478, 518–27
    treatment of grain, 581

    treatment of hay, 542
    treatment of straw, 257, 546–8
ammonium bispropionate, 541
amphipaths, 170
amygdalin, 22
amylase, 23, 166–9, 173, 174, 132
amylopectin, 25, 167, 561
amylose, 25, 561, 624
anabolism, 199, 404
anaemia
    copper deficiency, 129
    folic acid deficiency, 98–9
    haemolytic, 512, 597
    iron deficiency, 128, 177, 406
    pernicious, 103
    vitamin K deficiency, 89
    vitamin K excess, 105
androgens, 52, 384
anhydrous ammonia, 546
animal production level (APL), 297,
        299–302
animal products
    essentiality to man, 486
    future trends in output, 492
    non-food, 481
    objections to consumption, 435,
        487–8
animal protein concentrates, 600–7
animal protein factor (APF), 603
antibiotics, 178, 193, 488
    as feed additives, 488, 616–18, 628
antibodies, 172, 178
antioxidants, 42, 74, 87, 543, 588
apo-enzyme, 150
arabinans, 15, 25
arabinose, 15, 17, 22, 28, 29, 176
arabinoxylan, 623, 625
arachidonic acid, 36, 39–40
arginine, 219, 220, 498, 518, 595, 605
    as indispensable amino acid, 50,
        322, 375
    formula, 58
    requirement, 326
    synthesis, 224
Armsby H P, 294–5
arsenic, 109, 143, 414
artificially-dried forage, 507, 542–4
ascorbic acid see vitamin C
ash, 4, 252, 266, 280
    acid-insoluble, 248
asparagine, 57, 224, 498, 553
aspartic acid, 57, 70, 218–22

Aspergillus flavus, 594
associative effect of foods, 256, 288
Astragalus racemosus, 140
atherosclerosis, 50
atropine, 71
Australian Standing Committee on
    Agriculture, 300, 341, 357, 371,
    378, 478
availability of minerals, 113, 259, 380,
    435, 447, 451
Avena sativa see oats
avidin, 100
avilamycin, 617
Ayrshire, 125, 416–22, 435

bacteria
    in digestive tract, 171, 178, 180–96
    in silage, 517–19
    as silage additives, 525
    composition, 182
bagasse, 510, 544
Barber R S, 132
barley
    by-products, 564–70
    feed, 570
    grain, 268, 286, 287, 562–4, 579–81,
        624–5
    hay, 540
    straw, 544–5
    whole crop, 510
basal metabolism, 352, 363
beans, 597–9
    for silage, 508
Beckmann process, 546
bents (*Agrostis* spp.), 503
beri-beri, 73
berseem, 507
Beta vulgaris, 552
betaine, 69, 101, 555
bile, 168, 174, 176
    acids, 52, 620
bilirubin, 168
biliverdin, 168
biological value (BV), 320–3
    for wool growth, 378
    in factorial calculations, 456
    of food proteins, 585, 601–6
    of microbial protein, 329, 340
biotin, 90, 99–100, 123, 414
biuret, 190, 610, 612
Blaxter K L, 429
blind staggers, 140

bloat, 186, 508, 563, 610, 627–8
    preventing agents, 628
blocks (feed), 113, 611
blood, 415, 465
    dried, 604
    minerals in, 109, 118, 119, 125, 127
    sugar in, 235
blood meal, 604
body composition
    estimation of, 279–81
    condition scoring of, 392, 452
bomb calorimeter, 264, 279
bombesin, 62
bone, 84, 110, 117–21, 124, 366
bone flour, 118
boron, 109, 144, 414
botulism, 518
bovine spongiform encephalopathy
    (BSE), 178, 489, 600
bracken poisoning, 92
bran, 126, 555, 574, 575
Brassica campestris, 511, 552
Brassica napus, 511, 552, 592
Brassica oleracea, 511
brassicas, 135, 511–12, 552
Braude R, 132
brewer's grains, 267, 338, 555, 565–6
brewer's yeast, 567
Brody S, 353
bromine, 414
Brouwer E, 275
brown fat, 139, 361
Brunner's glands, 168
buffers, 180, 256, 335, 439, 515, 523,
    625, 627
butanoic acid *see* butyric acid
butylated hydroxyanisole, 42
butterfat, 38, 265, 412–57
butyric acid, 36, 171, 439
    in intake regulation, 471
    in milk, 412
    in rumen, 183–5
    in silage, 518–27
    utilization of, 199, 200, 212–3, 221,
        284, 287

cabbage, 89, 511
cadaverine, 69, 518
caeca, 164, 173, 195
caecum, 164, 171, 174, 195, 624–5
calcium, 117–19
    absorption of, 177, 178, 626

calcium – *continued*
  availability of, 260
  binding protein, 83
  deficiency symptoms, 84
  in animal body, 109
  in cereals, 561
  in eggs, 397
  in forages, 499, 504, 506
  in milk, 414, 419
  interrelations with
    manganese, 136
    phosphorus, 84, 118–20, 436
    vitamin D, 84
    zinc, 137
  requirements for
    egg production, 399
    lactation, 434–6, 447, 451, 456
    maintenance and growth, 379–81
    reproduction, 402, 403
  sources of, 118
calcium phosphate, 118
calf
  digestion in, 179, 194, 617
  mineral deficiencies in, 108–45
  vitamin deficiencies in, 73–105, 381
callose, 15, 27
calorie, 201, 468
calorimeter
  animal, 271–81, 354
  bomb, 264, 279
  gradient layer, 272
calorimetry, 271–81
Campylobacter spp, 488–9
cancer, 492
cannibalism, 123, 399
cannula, 165, 251, 253, 255, 474, 616
canola meal, *see* rapeseed meal
capric acid, 36, 412
caproic acid, 36, 42, 412
caprylic acid, 36, 412
carbamoyl phosphate, 219
carbohydrases,
carbohydrates, 2, 14–31
  absorption of, 176
  classification of, 14–16
  determination of, 4–9
  digestion of, 165–74, 182–7
  in eggs, 397
  in grasses, 497, 504
  in milk, 411, 415
  in silage, 516–27

  metabolism of, 199–214
    vitamins in, 75
  non-starch, 338
  non-structural, 7
  oxidation of, 205–9, 274
  synthesis of, 235–41
  utilisation of
    for maintenance, 284
    for production, 285–8
  water soluble, in grasses, 497
carbon balance, 278–9
carbonic anhydrase, 111, 137
carbon dioxide, 274
carboxyglutamic acid, 59, 89
carboxypeptidase, 111, 166, 169
carcass, 281, 319, 385, 367
carcinogens, 594
carmine, 246
carnaubic acid, 48
carnaubyl alcohol, 48
carnitine, 104
carotene, 76, 78, 81, 178, 400
  in grasses, 498, 543
  in hay, 538, 541
  in milk, 414
  (*see also* vitamin A)
Carpenter K J, 325–6
carrots, 23
caseins, 65, 265, 379, 411, 606
  hydrolysed, 317
cassava, 22, 557–8
castor bean, 583
cat, 1, 61, 167
catabolism, 199–223
catalase, 87,
cation–anion difference (CAD), 116
cattle
  feeding standards
    for growth, 365–86
    lactation, 421–41
    maintenance, 356–65
    reproduction, 390–5
    intake regulation, 475, 531
  milk fever, 116, 118
  (*see also* ruminants *and* individual breeds)
cellobiose, 15, 24, 181–3
cellulases, 182, 229, 250–1, 623, 625
cellulose, 15, 24, 27, 255, 257, 265, 546
  digestion
    in fowls, 174
    in horses, 196

in pigs, 171, 195
in ruminants, 179–96, 255
in grass, 497
Centrosema pubescens (centro), 507
cephalins, 46
cereal
consumption, 488
forage, 84, 510, 523
grains, 85, 93, 97, 99, 118, 257,
560–82
grain screenings, 578
hay, 539–40
processing, 257, 579–81
straws, 544–9
cerebrocortical necrosis, 92
cerebrosides, 45
ceruloplasmin, 111, 129
cetyl alcohol, 48
chaconine, 556
chaff, 544
Charolais, 372
chelates, 110, 114, 260
chemical score, 323, 327, 585
chemostatic theory of intake
regulation, 465–6, 471
chick *see* poultry
chitin, 19, 28, 608
chlorine, 109, 122, 414, 418
chloroform, 269
chlorophyll, 53, 78, 110, 556
choice feeding, 470–1
cholecalciferol, 51, 81–4 (*see also*
vitamin D)
cholecystokinin, 169, 466
cholesterol, 49–52, 65, 141, 168,
170, 378
in blood, 50, 490
choline, 46, 100–1, 414
chondroitin, 15, 29
chou moelier, 511
chromatography, 11
gas-liquid, 11, 37
chromic oxide, 249
chromium, 109, 141, 336
chromoproteins, 65
chylomicrons, 50, 176, 201
chymosin, 166, 168
chymotrypsin, 154, 166, 169, 586
chymotrypsinogen, 169
cimaterol, 384
citric acid, 148, 207–43, 626
citrulline, 219

Claviceps paspali, 505
climate, 358–62, 396, 407, 477, 478,
500, 501, 504
cloaca, 164, 174
clostridia, 69, 171, 506, 515–22, 524
clovers, 44, 89, 186, 250, 506–9, 540
clubbed down, 94
cobalamins *see* vitamin B$_{12}$
cobalt, 103, 109, 114, 132, 474, 499
cocaine, 71
cocksfoot, 38, 44, 502, 503, 540
cocoa bean meal, 596
coconut meal, 268, 589
cod liver oil, 38, 77, 80, 82, 440
codons, 226–7
coenzyme A, 90, 97, 123, 148, 149,
204–15
coenzymes, 150, 199–243
cold stress, 358–61, 477
coliform bacteria, 171, 519
collagen, 56, 64, 224
colon, 164, 171, 195
colostrum, 80, 82, 414
comparative nutrition, 481–3
comparative slaughter, 278–81,
319, 354
competitive inhibition, 157–9
Compositae, 27
condensed molasses solubles, 555
conjugated linoleic acid, 492
coniine, 71
cooking of foods, 196, 257, 556
copper, 109–11, 129, 138, 139, 260
as growth promoter, 132, 617
bolus, 114
deficiency symptoms, 129, 395, 406
in animal body, 109, 129
in egg production, 395, 397
in milk, 414
in pasture, 499
in wool production, 130, 379
sources of, 114, 131
toxicity of, 112, 131
copper oxide needles, 114
copper sulphate, 131, 193
coprophagy, 90, 195, 382
Cornell Net Carbohydrate and
Protein System, 7, 8, 11, 302,
335, 425, 478
coronary heart disease, 51, 490
corticosterone, 52
cortisol, 52, 407

cottonseed cake (or meal), 126, 321, 583–5, 588–9
coumestans, 509
cow *see* cattle *and* ruminant
crazy chick disease, 88
creatine, 205, 362, 410
creatine phosphokinase, 86
creatinine, 267, 362, 363
crop (of fowl), 164, 173, 466, 468
crop by-products, 487
crude fibre, 4, 5
    determination of, 5
    effect on digestibility, 255–6
    effect on milk composition, 439
crude protein, 4,
    as feeding standard, 313–15, 327
    components of, 4
    determination of, 5
    digestibility of, 247, 252, 255
cryptoxanthine, 78, 570
curled toe paralysis, 93
cutin, 48, 186
cyanocobalamin *see* vitamin B$_{12}$
cyanogenetic glycosides, 21–2, 135, 558, 590, 592
cysteine, 57, 59, 123, 130, 168, 218, 229, 324
cystine, 57, 59, 123, 323, 324, 328, 562
    association with methionine, 322
    in oilseed meals, 585–97
    in wool, 124, 378
cytidine, 225
cytochrome oxidase, 111, 129
cytochromes, 33, 65, 110, 111, 127, 138, 151, 203
cytosine, 66–8, 226

D value, 259, 510
Dactylis glomerata see cocksfoot
Dasytricha, 181
Daucus carota see carrots
day length, 477
deamination, 218, 518, 519, 523, 524
decanoic acid *see* capric acid
degradable protein *see* rumen degradable protein
7-dehydrocholesterol, 51, 82
dehydroretinol, 77
dementia, 547
deoxyadenosylcobalamin, 102
deoxyribonuclease, 166, 170

deoxyribonucleic acid (DNA), 20, 67, 68, 83, 99, 110, 170, 226, 228
deoxyribose, 20, 67, 68
deoxy sugars, 20
detoxification, 178
deuterium, 280
dextrins, 15, 26, 166, 167, 564
dhurrin, 22
diacylglycerols, 41, 169
diaminopimelic acid (DAPA), 331
dicoumarol, 89
dietary electrolyte balance, 115
digestibility, 245–62
    and intake, 468, 472–4, 478
    estimated from gas production, 251
    factors affecting, 255–8
    ileal, 255, 259, 315, 317, 324–5, 375, 623, 626
    in sacco, 251, 259, 332
    in vitro, 249–51, 259, 500
    measurement, 246–52, 258, 331
    of grass, 250, 499–505
    of hay, 538–42
    of proteins, 315
    of roots, 552
    of silage, 523–7
    of straw, 546–8
    standardised (SD), 316
digestibility coefficient, 245
digestible crude protein, 247, 315, 328
digestible energy, 266–8
    as feeding standard for pigs, 306, 453–4
digestible organic matter, 247, 254, 259, 308, 478
    apparently digested in the rumen (DOMADR), 337
digestion, 163–98
    in fowls, 173–4, 196
    in horses, 174–5, 196
    in pigs, 163–73
    in ruminants, 179–95
    rate of, 191, 192, 473
    studies of, 165
    work of, 283, 287
digestive tract, 163–5
    of ruminant, 179–80
    of horse, 196
diglycerides *see* diacylglycerols
dihydroxyacetone phosphate, 206, 215, 234
dihydroxycholecalciferol, 83, 116–17

dihydroxypyridine, 509
diiodotyrosine, 134
dilution rate, 191,
dilution techniques of estimating
    body composition, 280
dimethyl disulphide, 512
dipeptidases, 166, 170
disaccharides, 15, 22–4
dispensable amino acids, 224, 328
distiller's grains, 121, 567–9
    dark grains, 569–70
    supergrains, 569
distiller's solubles, 568–9
diureidoisobutane, 612
diverticulum, 173
DNA *see* deoxyribonucleic acid
docosahexaenoic acid, 39, 46
dodecanoic acid *see* lauric acid
dog, 1, 167
nutrient requirements, 482
Dolichos biflorus, 597
Dolichos lablab, 597
draff, 121, 127, 567–8 (*see also*
    distiller's grains)
dried grass, 542–4
dulcitol, 21
duodenal juice, 169
duodenum, 164–70, 173, 254, 468
Dutch
    net energy system, 301–4, 422
    protein system, 341

Echinochloa crusgalli see millet
EFA *see* essential fatty acids
egg, 396, 397
    hatchability, 400
    production, 117, 288, 396–400
    taint, 592
eicosanoic acid *see* arachidic acid
eicosanoids, 39–40
eicosapentaenoic acid, 39–40
Eijkmann C, 73
elastin, 64
Eleusine coracana see millet
Elsley F W H, 385
Embden and Meyerhof pathway,
    205–7, 212, 242
Emmans G C, 417
emulsification (during digestion),
    168
encephalomacia, 88
endergonic reactions, 202

endogenous urinary nitrogen
    see nitrogen
endopeptidases, 169
energy
    content of foods, 263–89
    content of liveweight gain, 367–73,
        425
    cost of eating, 270, 355–6
    digestible *see* digestible energy
    feed unit, 302
    intake, man, 484–7
    gross *see* gross energy
    metabolizable *see* metabolizable
        energy
    metabolism, 201–22
    models, 293, 304
    net *see* net energy
    productive values for poultry, 307
    requirement for
        egg production, 397
        growth, 370–3
        lactation, 421–8, 444, 449, 453–4
        maintenance, 352–62
        physical activity, 356
        pregnancy, 402–7
    systems, 292–312
Englyst H, 8
ensilage, 515–27
Enterobacteria, 519
Enterococcus faecalis, 517
enterokinase, 169
enterostatin, 62
enzootic ataxia, 129
enzymes, 146–62, 199–243
    acidity effects, 160
    allosteric effects, 159
        activator sites, 159
        inhibitory sites, 159
    classes, 147–9
    cofactors, 150
    concentration, 157
    digestive, 163–98
    feed additives, 257, 622–5
    genes for, 229
    inhibitors, 157–9, 257
    in feed evaluation, 251, 334
    in plants, 516, 521, 527, 537, 564
    microbial, 179–81, 537
    mode of action, 152–6, 624
    nature of, 149–52, 155
    nomenclature, 160
    silage additives, 524–7

enzymes – *continued*
  substrate concentration, 156
  temperature effects, 159
  specificity, 155–6
epimerases, 149
equine dysautonomia, 506
ergocalciferol, 51 (*see also* vitamin D)
ergosterol, 51, 82, 178, 498, 538
ergot, 71, 576
erucic acid, 38, 592
Erwinia herbicola, 519
erythrocuprein, 129
erythrose, 15
Escherichia coli, 89, 98, 229, 488, 519, 619
essential amino acids *see* indispensable amino acids
essential amino acid index (EAAI), 323
essential fatty acids (EFA), 39–40
  deficiency, 39
ethanol, 521, 523
ethanolamine, 46
ether extract, 5, 259, 440
ethylaminediaminetetraacetic acid (EDTA), 7, 114
euglobin, 411
exergonic reactions, 202
exogenous urinary nitrogen *see* nitrogen
exopeptidases, 169
exudative diathesis, 88, 139

facial excema, 505
facilitative transport (in absorption), 175
factorial method of estimating nutrient requirements, 350
  energy requirements, 421, 449, 453
  mineral requirements, 380, 434, 450, 456
  protein requirements, 364, 374, 430, 450, 455
faeces, 90, 172, 192, 246–55, 266, 267
  collection of, 246
falling disease, 130
fasting catabolism, 352
fasting metabolism, 352–4, 394
fats, 3, 33–44
  absorption of, 176
  and cardiovascular disease, 490
  brown, 33, 361

chemical properties of, 41–3
composition of, 37, 38
digestion of, 37, 166, 168–70, 190, 191
energy value, 32, 265
hydrogenation of, 43, 191
hydrolysis of, 41
melting point, 35–6
metabolism of, 214–17, 407
  vitamins in, 75
milk, 38, 412–16, 439–41
oxidation of, 41, 274
storage of, 278, 367–70
synthesis of, 229–35
taints in, 41, 47
utilization
  for maintenance, 284
  for production, 283, 285–7
fattening, 367–70, 383
fatty acids, 34–44, 148, 169, 200–17
  absorption of, 176
  chain elongation, 232
  cis and *trans* configuration, 35, 190
  desaturation, 232
  essential *see* essential fatty acids
  families of, 35
  in milk fat, 412–16, 439–40
  monounsaturated, 490
  nomenclature, 35–6
  oxidation of, 41, 214–17, 274
  polyunsaturated, 490
  saturated, 490
  synthesis of, 229–33
  see also volatile fatty acids
fatty liver and kidney syndrome (FLKS), 100
favism, 597
feathers, 64, 80, 95, 129, 142
  meal, 605
feedback inhibition, 159, 242
feeding standards
  for egg production, 396–400
  for growth, 365–80
  for lactation, 410–60
  for maintenance, 348–65, 379–83
  for pregnancy, 401–4
  for reproduction, 389–95
  for wool growth, 377–9
  safety factors in, 349, 427, 432
fermentable organic matter (FOM), 189, 337
ferric oxide (as marker), 246

ferritin, 104, 127, 128
fertilizers (nitrogenous), 126, 503, 552, 572
Festuca spp., 501, 503, 540
fibre, 255, 497, 620
    acid-detergent (ADF), 6, 247, 255, 308, 500, 542
    dietary, 6
    neutral-detergent fibre (NDF), 6, 255, 473
fimbriae, 619–22
Fingerling G, 306
fish meal, 601–4
    as source of minerals, 118, 123, 135, 138, 602
    as source of vitamins, 84, 89, 602
    protein, 321, 326, 374
Fisher C, 417
fistulated animals, 165, 249, 251, 253–5
flavophospholipol, 617
flame emission spectroscopy, 10
flavin adenine dinucleotide (FAD), 93, 202–3
flavin mononucleotide (FMN), 93, 150
flavoproteins, 65, 93
Floury-2 (maize), 571
flowmeal, 574
fluorine, 109, 112, 118, 140, 193, 414
fluoracetate dehydrogenase, 193
fluoro-2,4-dinitrobenzene (FDNB), 325
flushing, 392–3, 395
fodder beet, 551–4
    tops, 512
foetus, 400–6, 475
folacin, 414 (*see also* folic acid)
folic acid, 90, 98–9, 158, 186
food
    additives, 616–29
    digestibility of, 245–62
    digestion of, 163–98
    energy content of, 263–89
    energy value of, 292–312
    intake of, 192, 464–78, 531
    protein evaluation, 313–47
Forbes E B, 423
formaldehyde, 194, 522, 527
formalin, 379, 527, 607
formic acid, 184, 522, 527, 626
formononetin, 509
fowl *see* poultry

Fraps G S, 307
free radicals, 42, 79, 86–7, 105
French
    net energy system, 301, 307, 358, 373, 458
    protein system, 334, 341, 365, 377, 459
Fries J A, 423
Friesian, 416–22
fructans, 15, 18, 27, 182, 497, 507
fructooligosaccharides, 621
fructose, 15, 18, 21, 27, 170, 176, 182, 240, 497, 510, 518
fructose phosphate, 205, 206, 210, 240
fructosides, 21
fucose, 20
fumaric acid, 184, 207, 218–21, 626
fungi, 28, 181, 194, 509, 519, 538, 594
    cellulase of, 251
Funk C, 73

galactans, 15, 28
galactolipids, 44, 498
galactooligosaccharides, 621
galactosamine, 20, 29, 65
galactose, 15, 18, 20, 21, 23, 28, 29, 44, 170, 176, 240, 411
galactosides, 21
galacturonic acid, 29
gall bladder, 168
gallic acid, 42
Garrow J S, 482
gangliosides, 45
gastric juice, 168
gelatine, 76, 322
genetic engineering, 178, 193, 228–9, 384
genistein, 587
German
    energy systems, 294, 301–4
    protein system, 341
gizzard, 164, 173
gleptoferron, 128
gliadin, 574
Gliricidia sepium, 549
globulins, 64, 172
glucagon, 62
glucanase, 257
glucans, 15, 25–7, 257, 608
glucaric acid, 20
glucolipids, 33

gluconeogenesis, 285
  energy cost, 239
  from lactate, 236
  from amino acids, 237
  from glycerol, 238
  from propionate, 238
gluconic acid, 20
glucosaminans, 28
glucosamine, 15, 20, 28, 65
glucose, 15, 18–29, 265, 518
  as product of digestion, 167, 170, 265
  in grass, 497
  in intake regulation, 465
  in pregnancy toxaemia, 407
  metabolism of, 161, 182, 200–11,
      215, 221, 235–41, 265, 273, 441
  utilization of
    for foetal growth, 400
    for maintenance, 284
    for production, 285–8
glucose-1-phosphate, 19, 240, 241
glucose-6-phosphate, 19, 206, 240, 241
glucosides, 21, 174
glucosinolate, 592
glucostatic theory of intake
    regulation, 465
glucuronates, 267
glucuronic acid, 20, 29, 105
glutamic acid, 57, 59, 68, 98, 161, 168,
      498, 572, 574
  metabolism of, 218–24, 518
glutamic dehydrogenase, 223
glutamine, 57, 70, 220, 221, 498
glutathione peroxidase, 85–7, 111, 139
glutelin, 331, 571, 574
gluten, 572, 574
glycans, 14
glyceraldehyde, 15
glycerides, 34
glycerol, 34–48
  in milk fat, 412
  metabolism of, 200, 201, 215
  synthesis of, 234–6
glycerol-3-phosphate, 233
glycine, 57, 60, 64, 170, 177, 221, 518
  requirement by chick, 322, 375, 398
glycocholic acid, 52, 168
glycogen, 15, 26, 199, 200
  as energy source, 209
  digestion of, 167, 169
  reserves, 235
  synthesis of, 239

glycolipids, 15, 32–3, 44–5
glycolysis, 205–8
glycoproteins, 15, 18, 65
glycosides, 21–2
goat, requirements for lactation, 443–7
goitre, 134, 405, 509, 511, 592
goitrin, 135, 586, 592
goitrogens, 135, 509, 511
Gompertz growth equations, 402
gonadotrophic hormones, 393
gossypol, 586, 588
gramineae, 500–3, 560–81
grass, 496–506
  dried, 287, 542–3
    vitamins in, 543
  energy value, 300
  sickness, 506
  silage, 515–35
grazing
  animals, 350, 465, 476,
  time, 179, 476
  systems, 503
  energy cost of, 300
  pressure, 504
  zero, 504
grinding of foods, 185, 192, 253, 256,
      264, 269, 473, 542, 580
groats, 573
gross energy, 264–6,
  of milk, 422, 453
  of silage, 528
gross protein value (GPV), 318,
      326, 585
groundnut meal, 287, 326, 593–4
growth
  coefficient, 368–9
  compensatory, 385–6, 475
  control of, 383–6
  energy requirements for, 370–3
  energy utilization for, 285–8
  extra-uterine, 404
  feeding standards for, 365–86
  mineral and vitamin requirements
      for, 379–83
  modifying agents, 386
  protein requirements for, 374–77
guanidine, 317
guanine, 66–8, 225–7
guar, 583
Guernsey, 416–22
gulonolactone oxidase, 105
gums, 29

haem, 65, 151
haemoglobin, 65, 110, 127, 406,
    505, 512
haemolytic anaemia, 512, 597
haemosiderin, 127
hair, 129, 364, 430, 509
halibut liver oil, 77, 82
Hammond J, 331, 368, 384
hay, 247, 268, 269, 297, 338,
    536–42, 545
  chemical changes, 536–41
  losses, 541
  preservatives, 541–2
heat
  increment of foods, 270–89, 466
  increment of gestation, 403
  loss from body, 263–89, 351–62
  of combustion, 264
  of fermentation, 270, 283, 286
  production, 271–81, 351–62
  stress, 115, 361, 401, 477
heating of foods, 257, 510, 547,
    601, 625
hemicellulases, 182, 526
hemicelluloses, 15, 29, 255, 546
  digestion of, 171, 174, 182, 194
  in grass, 498, 499
  in silage, 517
hen *see* poultry
heptoses, 14–18, 209
herring meal, 38, 142, 603
heteroglycans, 15, 28–9
hexanoic acid *see* caproic acid
hexadecanoic acid, *see* palmitic acid
hexose–phosphate shunt *see* pentose
    phosphate pathway
hexoses, 14–18
hippuric acid, 267
histamine, 69, 518
histidine, 58, 60, 69, 99, 322, 326,
    456, 518
histones, 65
Hogg J, 132
Holcus lanatus (*see* Yorkshire fog)
holoenzyme, 150
holotrichs, 181
Holstein (Friesian), 370, 371, 391,
    416–22
homoarginine, 317
homoglycans, 15, 25–8
Hopkins F G, 73
hops, 555, 564, 567

Hordeum sativum see barley
hormones
  adrenal, 52, 407
  growth, 384
  gut, 169, 466
  metabolic, 124, 407, 414, 465–7
  sex, 40, 52, 384, 390–5, 405
  (*see also* individual hormones)
horse, 1, 92, 412, 543
  digestion in, 167, 171, 174–5, 196,
    579, 620
  energy systems, 307
  grass sickness, 506
  requirements
    energy, 357, 373, 458
    minerals, 459
    protein, 365, 377, 459
    vitamins, 77, 78, 81, 88, 89,
      103, 460
Huxley J S, 368
hyaluronic acid, 15, 29
hydrogen cyanide, 21–2, 590
hydrogen sulphide, 124
hydrolases, 148
hydrolysed feather meal, 605
hydroperoxide, 41
hydrogen peroxide, 548
hydroxybutyrate, 200, 212–13,
    413, 415
hydroxycholecalciferol, 83
hydroxycobalamin, 102
hydroxyl radical, 87
hydroxylysine, 56
hydroxyproline, 56
Hyparrhenia rufa
    see Jaragua grass
hyperketonaemia, 407
hypervitaminosis, 105
hypocalcaemia, 83, 118, 554
hypoglycaemia, 407
hypomagnesaemia, 125
hypomagnesaemic tetany, 125
hypothalamus, 465–7
hypoxanthine, 222

ideal protein, 327–8, 375, 384
ileum, 164, 253–5
ill thrift, 139
imidazoles, 547
immunoglobulins, 176, 400
indicators (marker substances),
    248–9, 331

indispensable amino acids, 60, 224, 341–2
  importance in protein evaluation, 318–28, 375–7
  requirements, 375–7, 398, 455
indole, 171
infertility, 80, 120, 394–5, 509
inhibitors, 159, 526–7
inoculants, 525
inosinic acid, 221–2
inositol, 46, 104, 414
in sacco techniques, 251–5,
insulin, 124, 393, 465
intake of food *see* food intake
international units, 79, 82, 85
intestine
  small, 166–70, 187, 468
  large, 171, 195
intrinsic factor, 103
inulin, 15, 27
invertase *see* sucrase
in vitro techniques, 249–51, 500
iodine, 109, 134, 511
  absorption, 178
  deficiency symptoms, 134
  in animal body, 134
  in eggs, 397
  in milk, 414
  requirements for
    egg production, 395, 399
    maintenance and growth, 380
    reproduction, 395, 406
iodothyronine deiodinase, 111, 139
ion pumping, 270
Ipomoea batatas see sweet potato
iron, 109, 111, 127
  absorption of, 128, 177, 178
  availability of, 128, 260
  deficiency symptoms, 127
  dextran, 128
  in animal body, 109, 127
  in eggs, 397
  in grass, 499
  in milk, 414
  in pregnancy, 406
  in proteins, 151
  requirement for egg production, 399
  sources of, 128
  toxicity, 129
isobutylidene diurea, 190
isocitric acid, 203
isoelectric point, 60

isoflavones, 509
isoleucine, 57, 60, 183, 322, 326, 397, 456
isomerases, 149
isoprene, 53
isothiocyanates, 592
isotopes, 259, 277, 280,
Isotrichidae, 181
Italian ryegrass, 250, 497, 503

jacobine, 71
James W P T, 482
Jaragua grass, 501, 502
Java beans, 21
Jersey, 125, 416–22, 476
joule, 201

kale, 89, 135, 511
Kellner O, 294, 295, 302, 303, 306, 423
keratin, 64, 378
kestose, 15, 24
keto acids, 524
ketoglutaric acid, 203, 207, 218–243
ketones, 42, 275, 407
ketoses, 176
ketosis, 275, 628
kilocalorie, 201
kilojoule, 201
Kjeldahl J, 5, 314
koilin, 173

lactase, 166, 170
lactalbumin, 411
lactation, 288, 336, 394, 410–60, 469, 475
  tetany, 125
lactic acid, 23, 174, 265, 620
  in metabolism, 147, 161, 171 173, 208, 210, 287
  in gut, 181, 183, 184, 579
  in silage, 517–27
lactic acid bacteria, 171, 181, 515–22, 525–6, 618–21
Lactobacillus brevis, 517
Lactobacillus plantarum, 517, 526
lactoglobulin, 411, 607
lactose, 15, 23
  digestion of, 170, 172, 625
  in milk, 411, 418–21, 449, 452, 606
  intolerance, 485
  synthesis of, 240, 241, 285

laminitis, 627
lanolin, 43
lard, 38
large intestine *see* intestine
lathyrism, 597
Lathyrus sativus, 597
laurel oil, 34
lauric acid, 34, 36, 38, 48
lean body mass, 280, 367
lecithinases, 166, 170
lecithins, 46, 101, 170, 628
lectins, 326, 619–22
legume, 473
  forages, 506–9
  hay, 540
  nutritional disorders, 508–9, 628
  seeds, 586, 597–600
  silage, 515
  straw, 544, 546
  trees and shrubs, 508
leptin, 62, 467
Leucaena leucocephala, 508, 509, 549
leucine, 57, 60, 169, 183, 218, 226,
  322, 326, 456, 595, 605
Leuconostoc mesenteroides, 517
leucotrienes, 39
levan, 15, 27
level of feeding, 258, 269, 299–301,
  432
ligases, 149
lignin, 29–30, 171, 544
  as indicator, 248
  effect on digestibility, 182, 186, 255
  in grass, 498, 499
limestone, 118
linamarin, 21–2, 558, 590
Lind J, 73
linoleic acid, 35–41
  conjugated, 492
  in cereals, 561
  in grasses, 498
  in milk fat, 412
  in rumen, 190
linolenic acid, 35–41
  in grasses, 498
  in milk fat, 412
  in rumen, 190
linseed, 40
  cyanogenetic glucosides in, 21–2,
  590
  meal, 126, 321, 590
  mucilage, 29, 590

lipase, 41, 52, 166–70, 174, 573, 624
lipids, 32–54
  classification, 32–3
  digestion of, 168–70
  in cereal grains, 560
  in eggs, 397
  in grasses, 498
  see also fats
lipoic acid, 104
lipoproteins, 65, 201, 441
  in blood, 50, 490
lipostatic theory of intake regulation,
  467
Listeriosis, 519
lithium, 109, 144
liver, 77
  fatty syndrome, 100
  importance in metabolism, 176, 188,
    199–202, 208, 210, 271
  necrosis, 594
  secretions, 168
Lolium multiflorum see Italian
  ryegrass
Lolium perenne see perennial ryegrass
losses of nutrients (silage), 521–2
lotaustralin, 22, 558
lucerne, 89, 250, 507, 524, 525
  dried, 542, 543
  hay, 185, 268, 284, 287, 538, 540
lupin seed meal, 395, 599
lupinin, 599
Lupinus spp., 599
lyases, 149
lysine, 56, 58, 60, 177, 218, 322–8, 518
  as indispensable amino acid, 60,
    322, 327–8, 377
  FDNB reactive, 325–6, 602
  in animal protein concentrates,
    600–6
  in cereal grains, 560, 562–78
  in grasses, 498
  in legume seeds, 598, 599
  in oilseed meals, 586–97
  requirement
    by poultry, 376, 398
    by pig, 377, 456
  synthesis, 229
lysophosphatidyl choline, 190
lysozyme, 167

Macroptilium atropurpureum
  see Siratro

magnesium, 109, 116, 124, 204,
  225, 349
  absorption of, 126
  availability of, 114, 259
  deficiency symtoms, 125
  in animal body, 125
  in eggs, 397
  in enzyme action, 151
  in grass, 499
  in milk, 414, 436, 447, 451
  losses from body, 259, 380, 447
  sources of, 126
  tetany, 125, 126
Maillard reactions, 63, 331, 340, 541
maintenance
  energy requirements for, 297–9,
    352–62, 396, 426, 444, 449, 453
  feeding standards for, 350–65
  mineral and vitamin requirements
    for, 379–83
  protein requirements for, 362–5
  utilization of energy for, 282–4, 296
maize, 284–7
  by-products, 571–2
  grain, 95, 268, 284–7, 321, 560–3,
    567, 570, 585, 624
  silage, 515, 523
  starch, 194, 439
  straw, 545
malic acid, 184, 210
malonyl coenzyme A, 230–1
malt culms, 555, 565–6
maltase, 166, 170, 173, 182
maltose, 15, 23, 26, 564
  digestion of, 167, 170, 182
mammary gland, 240, 403, 404,
  410–60
man, 77
  nutrient intakes, 484–7
nutrient requirements, 40, 482
manganese, 109, 136
  and reproduction, 136, 395
  deficiency symptoms, 136
  in animal body, 136
  in eggs, 397
  in enzyme action, 111, 230
  in grass, 499, 503
  requirements for egg production, 399
  toxicity, 137
mangel
  roots, 23, 552, 553
  tops, 512

Manihot esculenta see cassava
manioc see cassava
mannans, 15, 28, 608, 621
mannitol, 21, 518, 523
mannose, 15, 18, 21, 28, 176, 240
margarine, 43
markers, 192, 280, 336
mastitis, 418
matières azotés digestibles cheval
  (MADC), 459
mean retention time, 192
meat and bone meal, 118, 601
meat meal, 122, 128, 138, 326, 601
Medicago sativa see lucerne
Medicago truncatula, 509
megacalorie, 201
megajoule, 201
Melilotus albus, 89
melissic acid, 48
menadione, 88, 105
menaquinone, 89
menhaden oil, 38
Merino, 371
metabolic faecal nitrogen
  see nitrogen
metabolic liveweight, 353, 469, 475,
  584
metabolizability, 296–8
metabolizable energy, 266–89,
  295–305,
  fermentable (FME), 269, 337, 364,
    379
  of poultry diets, 267, 268, 309, 623
  prediction of, 308–9
  requirements for
    egg production, 396
    growth, 370–3
    maintenance, 356–62
    milk production, 421–58
  silage, 528
  systems of rationing, 293–309
  utilization, 281–9, 306
    for egg production, 288
    for growth, 285–8
    for heat production, 361
    for lactation, 288, 423–5
    for maintenance, 282–4
    for wool growth, 300
metabolizable protein, 343–5, 364–5,
  374, 430–452
metabolism, 199–243
  control of, 241–3

methaemoglobin, 505
methane, 180, 184–6, 193, 195,
    251, 265–70, 275–9, 352, 488,
    529, 566
  inhibitors, 193, 620, 627
methionine, 57, 99, 101, 124, 226,
    322–8
  as indispensable amino acid, 60
  association with cystine, 322, 328,
    376
  association with glycine, 375
  available, 324–5, 327
  in cereal grains, 560–3, 571, 572, 577
  in eggs, 397
  in protein concentrates, 585–613
  in wool, 378
  requirement by
    pig, 456
    poultry, 375, 398
methylcobalamin, 102
methylmalonyl coenzyme A, 103, 211
Michaels-Menten constant, 156
micelles, 170, 176, 190
microfibrils (of cellulose), 27
micronization, 257, 578
middlings, 575
milk
  by-products, 606–7
chemical composition of, 410–22, 443,
    448–9, 452,457
  consumption, 483–7
  digestion of, 172, 194
  fat-reduced, 491
  minerals in, 118, 120, 128, 130, 131,
    405–6, 414
  peptides in, 62
  protein, 321
  requirements for production of,
    410–60
  taints in, 552, 604
  utilization of energy for production
    of, 288, 299, 423–5
  vitamins in, 77, 82, 83, 96, 406, 414,
    437, 606
  yields of, 416–18, 428–30, 438, 442,
    447–8, 452–3, 457
milk fever, 116, 118, 438
millet, 562, 577
mimosine, 509
minerals, 109–45
  absorption of, 177
  and enzymes, 111

availability of, 112–13, 259, 380,
    435, 447, 451
  body compartments, 109
  deficiency symptoms, 97–145
  essential, 109
  in grass, 499, 498, 503
  in hay and straw, 545
  in legumes, 506, 549
  in milk, 414
  in tubers, 557
  requirements for
    egg production, 396–400
    lactation, 434–6, 447, 451,
      456, 459
    maintenance and growth, 379–83
  supplements, 113
Mitchell K G, 132
mitochondria, 33, 203, 230, 232
mobile nylon bag technique (MNBT),
    255
modified acid-detergent fibre
    (MADF), 7, 308
moisture, determination of, 4
molasses, 24, 127, 510, 525, 555–6,
    589, 611
Molinia caerulea see purple moor
    grass
molybdenum, 109, 130, 138
  in animal body, 138
  in enzymes, 111, 138
  in milk, 414
  in grass, 499, 503
  toxicity of, 130, 139, 395
monensin, 193, 270, 617, 627
monoacylglycerols, 41, 148, 169,
    170, 190
monoglycerides see monoacylglycerols
monoiodothyronine, 134
monopteroylglutamic acid, 98
monosaccharides, 16–22, 170
Morant S V, 417
morphine, 71
Morrison F B, 295
mouth (digestion in), 165–6
mucilages, 29, 590
mucin, 167
mucosal block theory, 128
mulberry heart disease, 87
mucopolysaccharides, 141
muscular dystrophy, 86
mycosterols, 49
mycotoxins, 505, 520, 538, 594

myelin, 47, 130
myopathy, 86, 139
myrosinase, 592
myricyl alcohol, 48
myricyl palmitate, 48
myristic acid, 36

naphthoquinones *see* vitamin K
National Research Council (USA), 305
near infrared reflectance spectroscopy
    (NIRR), 12, 308, 335
necrosis of the liver, 594
Neptunia amplexicaulis, 140
net energy, 266, 271
    determination of, 271–81
    requirements for
        growth, 370–3
        lactation, 421–8, 444, 449, 454
        maintenance, 352–8, 426
    systems for
        pigs and poultry, 306, 307
        ruminants, 294–305
    value of
        grass, 501, 504
        silage, 530
net protein retention (NPR), 318
net protein utilization (NPU), 321, 324
Neurath H, 63
neutral-detergent fibre (NDF), 6, 255,
    334, 336, 473
nickel, 43, 109, 142
nicotinamide, 90, 94–5
nicotinamide adenine dinucleotide
        (NAD), 90, 94, 147, 150,
        199–243, 628
nicotinamide adenine dinucleotide
        phosphate (NADP), 90, 95,
        199–243
nicotine, 71
nicotinic acid, 94, 383, 414
night blindness, 79
nitrates, 70, 189, 319, 331, 498, 505
nitrites, 70, 505, 604
nitrogen
    balance, 278–9, 319, 321, 364
    degradable *see* rumen degradable
        nitrogen
    endogenous urinary, 315–18, 320,
        363–4, 374, 378
        non-specific (NSE), 315–16
    exogenous urinary, 363–4
    isotope (^{15}N), 316–17
    faecal, 315, 319
    rumen, 337
    metabolic faecal, 252, 255, 317,
        320, 364
    undegradable *see* undegradable
        protein
    urinary, 274–6, 319, 362–4
    see also crude protein
nitrogen-free extractives, 5
non-glucogenic ratio (NGR), 440
non-protein nitrogenous (NPN)
        compounds, 65–72, 189, 190,
        609–12
Nordic (Scandinavian)
    energy systems, 302
    protein systems, 341
nucleic acids, 65–8, 170, 182, 339, 608
nucleosides, 66–8
nucleosidases, 166, 170
nucleotides, 66–8, 225

oat
    by-products, 468, 573–4
    dust, 573
    grain, 268, 287, 560–2, 572,
        580, 625
    hay, 540
    hulls, 573
    naked grain, 573
    straw, 544–5
oat hay poisoning, 70
octadecanoic acid *see* stearic acid
octanoic acid *see* caprylic acid
oesophageal groove, 179, 194
oesophagus, 164, 165, 167, 173
    249, 468
oestrogens, 52, 384, 509, 581
oil extraction, 583–6
oilseed cakes and meals, 583–97
oleic acid, 36, 38, 43
    in milk fat, 412, 413
    synthesis of, 232
oligo-1,6-glucosidase, 166, 170
oligosaccharides, 22–5, 167, 170,
    621–2
oligotrichs, 181
omasum, 164, 179
Onobrychis viciifolia see sainfoin
opaque-2 (maize), 571
Ophryoscolecidae, 181
organic acids (food additives), 625–7
ornithine, 219, 411

orotic acid, 104
Oryza sativa see rice
osteomalacia, 84, 118
ova, 389
oxalacetic acid, 148, 184, 207–38,
    407, 502
oxalates, 177, 260, 512, 596
oxalic acid, 512
oxidation
    in hay, 538
    in silage, 521
    of energy sources, 201–22, 274–6
    of fats, 41, 47
    of vitamins, 76, 78
oxidative decarboxylation, 91–2, 206
oxidative phosphorylation, 202–43
oxidoreductases, 147

palatability, 467, 474
palmitic acid, 36, 48, 216–17, 231–5,
    412, 413, 498
palmitoleic acid, 36, 232
palm kernel meal, 590
pancreas, 152, 168, 173
pancreatic juice, 169
pangamic acid, 104
Panicum miliaceum see millet
pantoic acid, 96
pantothenic acid, 90, 96–8, 405, 414
para-aminobenzoic acid, 157
parakeratosis, 137
parathyroid glands, 83, 117
parotid glands, 165
Paspalum dilatatum, 501, 505
Paspalum scorbiculatum see millet
passage rate (from rumen), 192, 256,
    336, 506, 509
passive transport (in absorption), 175
pasture, 44, 495–509 (*see also* grass)
peas, 321, 508, 599
peat scours, 130
pectic acid, 28, 182
pectic substances, 15, 28
pectin, 28, 182
Pediococcus pentosaceus, 517, 526
pelleting of foods, 185, 253, 256, 269,
    287, 288, 473, 542, 579, 580, 591
Pennisetum americanum see millet
pentanone, 42
pentosanase, 625
pentose phosphate pathway, 91,
    208–9

pentoses, 15, 17, 182, 518
pepsin, 166, 168, 250–1
pepsinogen, 167, 168
peptidases, 148, 169
peptide linkage, 62, 166, 168, 169, 227
peptides, 61–2, 166, 187, 338
    alimentation, 317
    chain formation, 226
    elongation, 227
    termination, 227
    opioid, 62
perennial ryegrass, 44, 250, 254, 500,
    501, 503, 505, 523, 537, 540, 543
perosis, 101, 136
peroxides, 86
Phaseolus spp., 597, 598
phenol, 42, 171
phenylalanine, 58, 168, 169, 218, 226
    as indispensable amino acid, 60,
        322, 456
    association with tyrosine, 322, 376
phenylethylamine, 69
Phleum pratense see timothy
phosphatases, 166, 170
phosphatidic acid, 46, 234
phosphatidylethanolamines, 46
phosphocreatine, 205
phosphoenolpyruvate, 203, 206,
    211–12
phosphogluconate oxidative pathway
    see pentose phosphate pathway
phosphoglycerate, 206, 502
phosphoglycerides, 46
phospholipids, 46, 52, 119, 190
    ether, 47
phosphoproteins, 65, 119
phosphoric acid
    in carbohydrates, 19
    in lipids, 45
    in metabolism, 201–43
    in nucleic acids, 65–68
    in proteins, 65
phosphorus, 109–21, 561
    absorption of, 177
    association with carbohydrates, 19
    availability of, 120, 260
    deficiency of 119, 289, 549
    in animal body, 119
    in eggs, 397
    in grass, 499
    in metabolism, 199–243
    in milk, 411, 414

phosphorus – *continued*
  interrelation with
    calcium, 84, 119
    magnesium, 121
    manganese, 136
    vitamin D, 84
    zinc, 137
  requirements for
    egg production, 399
    lactation, 434, 447, 451, 456, 459
    maintenance and growth, 380, 381
    reproduction, 394, 403
phosphorylation, 202
phosphoserine, 224
phosvitin, 65
phylloquinone, 88
phytase, 120, 120, 257, 260, 623, 625
phytates, 120, 177, 399, 561
Phytelaphas macrocarpa, 28
phytic acid, 114, 120, 260, 399, 596
phytosterols, 49–51, 178
pica, 120, 133
pig
  composition of, 370, 372, 384
  critical temperature of, 359
  digestion and absorption, 163–77,
      196, 616–27
  energy
    requirements of, 351, 372, 453–4
    systems for, 305–7
    utilization by, 284, 286
  feeding standards for
    growth, 370–83
    lactation, 452–7
    maintenance, 352–62
    reproduction, 389–95, 400–6
  intake regulation in, 465–71
  mineral requirements of, 379–81, 456
  protein evaluation for, 313–28
  protein requirements of, 375–7,
      455–6
  vitamin requirements of, 381–3, 457
piglet
  anaemia, 128
  digestion in, 172
pining, 132
pinocytosis, 176, 400
Pisum sativum see peas
Pithomyces chartarum, 505
pituitary hormones, 384, 393
placenta, 389, 395, 400–1, 405
pluronics, 628

pollution, 350, 488
poloxalene, 628
polyethylene glycol, 192
polyneuritis, 92
polynucleotides, 67
polypeptides, 62, 227
polysaccharides, 25–9, 167–72,
      182–7, 239
  non-starch (NSP), 7
polysomes, 226
polyunsaturated fatty acids, 35–40, 39,
      86, 87, 628
  ω-3, 51, 490
  ω-6, 51, 490
postabsorptive state, 352
pot ale, 568–9
potassium, 109, 121, 260, 380
  in animal body, 121, 280
  in eggs, 397
  in grass, 499
  in milk, 414
  in suint, 378
  interrelationship with magnesium,
      126
  losses from body, 380
potato, 196, 553, 556–70
poultry
  amino acid requirements of, 60,
      322, 375–7, 398
  digestibility measurements
      with, 246
  digestion in, 173, 196, 616–27
  energy
    requirements of, 353, 357, 373, 396
    systems for, 305–7
    utilization of, 280, 286, 288, 307,
      396
  feeding standards for, 373, 375–7,
      396–400
  mineral requirements of, 379–31,
      399
  protein
    evaluation of, 318–28
    requirements of, 375–7, 398
  vitamin requirements of, 103,
      381–3, 400
  waste, 190, 612, 613
pregnancy, 336, 400–7, 469, 475
  anabolism, 404
  toxaemia, 407
probiotics, 193, 618–21
procarboxypeptidase, 169

Processed Animal Protein Order, 600, 613
productive energy, 307
progesterone, 52, 393
prolamins, 331
proline, 56, 58, 64, 177, 183, 223, 226, 322
propionic acid, 171
  as food preservative, 541, 581
  efficiency of utilization
    for maintenance, 284
    for production, 287, 287
  in intake regulation, 471
  in rumen, 183–6, 193, 439–41, 627
  metabolism of, 103, 133, 210–12, 221
prostaglandins, 39–40, 51, 99
prosthetic groups, 65, 150
protamins, 65
protein concentrates, 583–613
protein efficiency ratio (PER), 318, 585
protein equivalent (PE), 329
proteins, 55–65
  absorption of, 176
  carrier, 176
  classification, 64–5
  degradable see rumen degradable protein
  denaturation of, 63
  digestibility of, 247, 254, 254–5
  digestion of, 166–74, 186–9, 192–5, 624
  equivalent, 329
  energy value of, 265, 372
  evaluation of, 313–45
  in animal body, 278, 367–73, 372, 383–6
  in cereals, 560–581
  in eggs, 397
  in grass, 497–8
  in milk, 410, 411, 418–21, 430, 445, 450, 452, 455
  in silage, 523–7, 530–1
  in wool, 378
  metabolizable, 343–5, 364
  microbial, 182, 187–8, 193, 329, 334–44, 364
  oxidation of, 274, 275
    vitamins in, 75
  properties of, 63
  requirements for

    egg production, 398
    growth, 374–77
    lactation, 430–4, 445, 450, 455
    maintenance, 362–5
    reproduction, 395, 402
    wool production, 377–9
  solubility, 339, 334
  storage of, 278, 367–70, 383
  structure of, 62–3
  synthesis of, 223–9
    energy cost, 228
  truly digestible, 339
  turnover, 224, 285, 362
  undegradable see undegradable protein
proteolysis, 517
proteolytic enzymes, 167–9
prothrombin, 89
protozoa, 181–2, 193–4, 339, 489
proventriculus, 164, 173
provitamins, 76, 78, 82
proximate analysis of foods, 4
pseudoglobulin, 411
Pseudomonas spp., 608, 609
Pteridium aquilinum, 92
pteroylmonoglutamic acid, 98
puberty, 366, 390–1
purines, 66–8, 99, 138, 170, 608
purple moor grass, 499, 500
putrescine, 69
pyridoxal, 90, 95–6
  phosphate, 96
pyridoxamine, 95–6
pyridoxine, 95–6
pyrimadines, 66–8, 170, 608
pyruvic acid, 92, 147, 161, 182, 183, 202–3, 205–8, 215, 518

quinine, 71
quinones, 42

rabbit, 195, 543
radioactive isotopes, 111, 121, 126, 259, 277, 280, 331
raffinose, 15, 24, 497
ragwort, 71
rancidity, 42, 589
rape, 135, 511
rate of passage, 192, 332, 335–6
rapeseed
  meal, 591–3
  oil, 38

rectum, 164
redgut, 509
rennin, 166, 168
reproduction, 389–409
repartitioning agents, 383–4
respiration chamber, 267, 272–7
respiratory exchange, 274–8
respiratory quotient, 274–6, 278, 352
respiration (in plants), 516
retention time, 174, 192
reticulum, 163, 179
retinaldehyde, 79
retinol see vitamin A
rhamnose, 20, 28, 29
rhodopsin, 79
riboflavin, 90, 92–4, 414, 498
ribonuclease, 166, 170
ribonucleic acid (RNA), 17, 67, 83, 99,
    110, 170, 224–8, 331
ribose, 15, 17, 65–8
ribosomes, 226–8
ribozymes, 149
ribulose, 15, 17, 209
ribulose bisphosphate carboxylase, 497
rice
    by-products, 194, 575
    grain, 562, 575–6
    straw, 256, 546, 548
ricinine, 71
ricinoleic acid, 38
rickets, 84, 118, 120, 399
roots, 551–6
rubidium, 109
rumen
    control of fermentation in, 193–5,
        620, 626–7
    digestion in, 164, 165, 179–95, 439
    dilution rate, 192
    dynamics, 191–2
    fill, 192, 472
    pH, 193, 256, 338, 580, 627
    protein breakdown in, 186–9,
        334–45
    synchrony, 194–5
    synthesis of vitamins in, 91, 92,
        103, 133, 191
rumen degradable protein (RDP), 187,
    190, 254, 329–45, 379, 395,
    430–4, 474, 530, 545
    effective (ERDP), 343–5, 365, 374,
        446, 450
rumenitis, 627

ruminants
    digestibility measurements with,
        246–50
    digestion in, 165, 179–95
    energy requirements of, 352–62,
        370–3, 403, 423–58
    energy systems for, 294–305
    intake regulation in, 471–8
    protein evaluation for, 328–45
    protein requirements of, 362–5, 374,
        378, 402, 430–52
    utilization of energy by, 281–9
rumination, 179, 355
rutin, 104
rye
    grain, 576
    straw, 546
ryegrass, 254, 268, 284–7,
    496–505, 540
    oil, 38
    staggers, 505
    see also Italian ryegrass and
        perennial ryegrass

saccharin, 467
Saccharum officinarum see sugar cane
sainfoin, 250, 507, 508
salinomycin, 193, 617
saliva, 165, 174, 179, 188, 220
Salmonella, 488, 613, 621
salt see sodium chloride
saponification, 41
saponine, 596
Scandinavian feed unit system,
    302, 304
scurvy, 73, 105
seaweed, 135
sebaceous glands, 378
Secale cereale see rye
secretin, 169
sedoheptulose, 15, 18
selenium, 109, 111, 114, 139–40,
    395, 499
    interrelationship with vitamin E,
        85–8, 139
    toxicity, 140
Selenomonas ruminantium, 180
sensory appraisal, 467
serine, 46, 57, 68, 99, 218, 223–4,
    226, 376
sesame seed meal, 595
Setaria italica see millet

shea nut meal, 596
sheep
  digestibility trials with, 246–8
  digestion in, 165, 179–95
  energy requirements of, 352–62,
    370–3, 400–4, 449
  energy utilization in, 281–9
  feeding standards for
    growth, 370–83
    lactation, 447–52
    maintenance, 350–65
    reproduction, 389–95, 400–4
  intake regulation in, 471–8, 531
  mineral requirements of, 379–81,
    394–5, 404–7, 450
  protein requirements of, 362–5, 374,
    378–9, 395, 450
  vitamin requirements of, 381–3,
    394–5, 404–7
  wool production, 377–9
Shorthorn, 419
silage, 189, 268, 284, 287, 338, 423,
  474, 507, 516–35
silica, 141, 256, 545, 546, 576
silicon, 109, 141, 414
silicosis, 141
silos, 516
Simmental, 372
single-cell protein, 607–8
Siratro, 507
skatole, 171
skim milk, 606
S-methylcysteine sulphoxide, 512
small intestine see intestine
soap, 41
sodium, 109, 122, 175, 260, 380, 474
  in animal body, 122, 402
  in eggs, 397
  in milk, 414, 436
  losses from body, 380
sodium bicarbonate, 256
sodium chloride, 122, 123
  requirements for, 399, 436
  toxicity, 123
sodium hydroxide treatment
  of grain, 579, 581
  of straw, 257, 546–8
solanidine, 556
solanine, 71, 556
Solanum tuberosum see potato
solids-not-fat (SNF), 410–29
sorbitol, 21

sorghum, 562, 577–8
soya bean, 583
  hay, 540
  meal, 100, 286, 287, 321, 326,
    586–8, 625
  oil, 441
specific gravity of animals, 281
spectroscopy
  atomic absorption, 10
  flame emission, 10
  inductively coupled plasma
    emission, 10
  near infrared reflectance (NIRR),
    12, 308, 335
  nuclear magnetic resonance
    (NMR), 12
spermaceti, 48
spermatozoa, 389–95, 509
sphagnum moss, 555
sphingomyelins, 47
sphingosine, 45, 47
spices, 467
splay-leg syndrome, 581
sporidesmin, 505
stachyose, 15, 25
standard reference weight (SRW),
  371, 478
stage of growth (of plants), 499–500,
  539
starch, 6, 15, 25, 265
  digestion of
    by fowls, 173
    by pigs, 167, 169, 196
    by ruminants, 181–3, 194
  in cereal grains, 560–80
  in legumes, 506
  in root crops, 556, 557
  utilization for
    maintenance, 284
    production, 287, 302
starch equivalemt, 295, 303
stearic acid, 36, 43, 190, 191
stereoisomers, 16
sterility, 395
steroids, 49–52, 384
sterols, 49–52, 83
Stickland reactions, 518
stiff lamb disease, 86
stocking rate, 504
stomach, 164–9, 174, 179, 466
  ulcers, 579
stravidin, 100

straw, 192, 544–9
  alkali-treated, 257, 546–8, 612
streptavidin, 100
streptococci, 23, 171, 181, 324, 619
strychnine, 71
suberin, 48
sublingual glands, 165
submandibular glands, 165
submaxillary glands, 165
succinic acid, 103, 183, 184, 207,
  211–12
succinyl coenzyme A, 91, 103, 184,
  211–12
sucrase, 23, 166, 170, 172
sucrose, 15, 22, 467, 497, 553
  digestion of, 166, 170, 182
sudoriferous glands, 378
sugar acids, 20
sugar alcohols, 21
sugar beet, 23, 69, 554
  molasses, 555
  pulp, 28, 170, 554
  tops, 512
sugarcane, 510
  molasses, 555–6
sugars, 6, 16–19, 611
  absorption of, 175, 176
  digestion of, 170–53, 302
  in grass, 497
  in roots, 551–5
  in silage, 516–27
suint, 378
sulphanilamide, 157
sulphite oxidase, 111, 138
sulphur, 55, 57, 109, 123, 377, 474,
  499, 503, 611
sunflower seed meal, 326, 594
superoxide dismutase, 87, 111
Synergistes jonesii, 509
surface area (of animals), 353
swayback, 130, 405
swede, 551–3
  roots, 552
  tops, 512
sweet clover disease, 89
sweet potato, 553, 558

taints, 41, 552, 592, 604
tallow, 38
tannins, 194, 256
  condensed, 508
taurocholic acid, 168

teart (pastures), 130, 139, 503
temperature, 159, 264, 336, 541
  542, 547, 600–6
  body, 358, 358, 466
  critical, 358–62, 477
terpenes, 53
tetany, 125, 610
tetradecanoic acid *see* myristic acid
tetrahydrofolic acid, 99
tetrahymena, 324
tetrahymenol, 324
tetraiodothyronine, 134, 139
tetrasaccharides, 15, 25
tetroses, 14, 15
theobromine, 596
thermal equivalent of oxygen, 274–5
thermostatic theory of intake
  regulation, 466
thiamin, 74, 76, 91–2, 123, 382, 414,
  575, 588
  pyrophosphate (TPP), 90–2
thiaminase, 92
thiamin diphosphate, 206
  see also thiamin pyrophosphate
thiocyanate, 511, 592
thioglucosides, 592
thiourea, 610
threonine, 57, 60, 218, 322, 397, 456
  synthesis, 229
thrombin, 59, 89
thromboxanes, 39, 51
thumps, 128
thymine, 66–8
thyroglobulin, 56
thyroid gland, 134, 511
thyroxine, 56, 111, 134, 139
timothy, 250, 501, 503, 540
tin, 109, 143, 414
titanium dioxide, 249
tocopherols, 42, 76, 85–7
  see also vitamin E
tocotrienols, 85
total digestible nutrients (TDN), 259,
  295, 302, 304
toxins, 71, 178, 616
trace elements, 109, 127–45
transamination, 60, 70, 218, 223
transferases, 125, 147
transferrin, 104
transgalactooligosaccharides, 621
transgenic animals, 229
transmethylation, 101

trefoil, 22
triacylglycerols, 34–8, 52, 148, 166–70, 190, 200, 201, 214, 229, 234–5, 411, 412, 441, 498
tricarboxylic acid cycle, 207–9, 210–12, 216, 218
Trifolium spp. *see* clovers
triglyceride *see* triacylglycerols
triiodothyronine, 56, 134, 139
trimethylamine, 69, 593
triolein, 36
trioses, 14, 15
tripalmitin, 221, 235, 274
trisaccharides, 15, 24
triticale, 576–7
Triticum aestivum see wheat
tritium, 280
true metabolizable energy, 267
true protein, 5, 315
trypsin, 152, 166, 169
  inhibitors, 257, 558, 586, 599
trypsinogen, 152, 169
tryptamine, 69
tryptophan, 58, 60, 94, 168, 169, 218, 322–8, 397–9, 456
  in cereal grains, 562
tubers, 553, 556–8
turacin, 129
turkeys, 99, 123, 375
turnip
  roots, 551–3
    tops, 512
tussock grass, 501
twin lamb disease, 407
tylosin, 618
tyramine, 69
tyrosine, 56, 58, 134, 169, 218, 322, 328

ubiquinone, 203
ultra-violet light, 51, 82
undegradable dietary protein (UDP), 339, 379, 395, 430–4, 446, 450, 545
  digestible (DUP), 344, 364, 374, 384, 395, 446, 450, 549
underfeeding, 349, 385, 391, 392, 404–7, 452, 475
unité fouragère cheval (UFC), 458
uracil, 66–8, 226–7
urea, 70
  in diet, 124, 329, 510, 548, 609–12

in milk, 410
in rumen, 188, 189, 548
in saliva, 188,
in tissues, 187, 195, 200
in urine, 188, 200, 267, 269, 363
metabolism of, 155, 187–9, 219, 220
treatment of foods, 547, 581
urease, 155, 189, 547, 610
uric acid, 70, 190, 220–1, 246, 608, 613
uridine diphosphate galactose, 239–41
uridine diphosphate glucose, 239
urine, 187, 246, 362
  energy losses in, 266–8, 275–8
  minerals in, 380
urolithiasis, 121
uronic acids, 21, 182
USA
  net energy system, 295, 302–5
  protein system, 341

valeric acid, 183, 440
valine, 57, 60, 91, 183, 187, 242, 322, 456
Van Es A J H, 423
vanadium, 109, 142
Van Soest P J, 6
vegetarians, 485
vernolic acid, 38
vetches, 508, 540
Vicia faba, 446, 597
vicianin, 22
villi, 164, 175
vinquish, 132
L-5-vinyl-2-oxazolidine-2-thione, 135, 592
  see also goitrin
virginiamycin, 618, 628
Virtanen A I, 526
vitamin A, 77–81, 561
  absorption of, 178
  chemical nature, 77
  hypervitaminosis, 105
  liver reserves, 77
  metabolism of, 79, 87
  requirements for
    egg production, 400
    growth, 382
    lactation, 437
    pregnancy, 405–6
  symptoms of deficiency, 79–81, 382, 395, 405

vitamin A$_2$, 77
vitamin B complex, 68, 76, 90–104
  requirements for
    egg production, 400
    growth, 382
  synthesis of
    in intestine, 171
    in rumen, 191
vitamin B$_1$ *see* thiamin
vitamin B$_2$ *see* riboflavin
vitamin B$_6$, 76, 95–6, 414
vitamin B$_{12}$, 76, 101–3, 110, 111, 132,
    134, 178, 186, 191, 414, 487
vitamin C, 76, 87, 104–5, 414
vitamin D, 74, 76, 81–4, 117, 561
  absorption of, 178
  and radiant energy, 82
  chemical nature, 51, 81
  hypervitaminosis, 105
  in roughages, 498, 543
  metabolism of, 83
  potency of different forms, 84
  requirements for
    egg production, 400
    lactation, 437
  symptoms of deficiency, 84
vitamin E, 42, 74, 76, 84–8, 543
  absorption of, 178
  chemical nature, 85
  in cereals, 85, 561, 575
  in grain, 85,
  in grasses, 498
  interrelationships with selenium,
    85–8, 139
  metabolism of, 85
  symptoms of deficiency, 86–8, 395
vitamin K, 76, 88–90
  absorption of, 178
  chemical nature, 88
  hypervitaminosis, 105
  metabolism of, 89
  symptoms of deficiency, 89–90
  synthesis of, 88, 191
vitamins, 4, 73–107
  absorption of, 52, 178
  and biochemistry, 74–5
  and energy utilization, 289
  in cereals, 561
  in dried grass, 543
  in milk, 414
  requirements for
    egg production, 400

lactation, 437, 457, 460
maintenance and growth, 381–3
supplementation, 74
synthesis of, 171, 191, 620
volatile fatty acids, 171, 195
  in intake regulation, 471
  in rumen, 181–5, 192
  metabolism of, 210–15
  utilisation for
    maintenance, 283, 284
    production, 285–9, 378, 485
voluntary food intake
    see food intake

wafering of foods, 256
water, 2
  in animal body, 280, 367
  in foods, 472, 536, 542, 552
  in salt poisoning, 399
  metabolic, 2
  requirements, 583
waxes, 48, 378
wheat
  grain, 27, 268, 560, 562, 574, 578,
    585, 624–5
  by-products, 268, 574–5
  hay, 540
  straw, 287, 546
whey, 606
white-fish meal *see* fishmeal
wilting, 521, 523–4
Wood P D P, 417
wool, 374, 377–9, 450
  loss of, 509
  minerals in, 124, 129, 140
  pigmentation, 129
  production of, 264, 378
  protein synthesis, 229

xanthine, 111, 222
xanthine oxidase, 138
xanthophylls, 78, 543
xerophthalmia, 80
xylans, 15, 17, 25, 29
xylanase, 625
xylose, 15, 17, 176, 182, 467
xylulose, 15, 17, 209

yeast, 171
  as feeds, 560–8
  as probiotic, 193, 620–2
  as source of minerals, 114, 136, 138

as source of vitamins, 84, 91–101
in cereal by-products, 564–7
in silage, 519
protein, 326
Yorkshire fog, 501, 503
ytterbium, 192

Zea mays see maize
zein, 571
zinc, 109, 111, 137
    absorption of, 114, 178
    deficiency symptoms, 137, 395, 399
    in animal body, 137
    in eggs, 397, 399
    in grass, 499
    in milk, 414
    requirements for
        egg production, 399
        reproduction, 395
    toxicity, 138
zoonoses, 488
zoosterols, 49
zwitter ions, 59
zymogens, 152, 169